# Marine Geochemistry

This book is dedicated with affection and gratitude
to Dr G. D. Nicholls, an innovative geochemist, and
a fine teacher who has the truly rare gift of being
able to inspire his students

# Marine Geochemistry

**Roy Chester**

*Department of Earth Sciences, University of Liverpool*

SECOND EDITION

**Blackwell
Science**

© 2000 by
Blackwell Science Ltd
Editorial Offices:
Osney Mead, Oxford OX2 0EL
25 John Street, London WC1N 2BL
23 Ainslie Place, Edinburgh EH3 6AJ
350 Main Street, Malden
  MA 02148-5018, USA
54 University Street, Carlton
  Victoria 3053, Australia
10, rue Casimir Delavigne
  75006 Paris, France

Other Editorial Offices:
Blackwell Wissenschafts-Verlag GmbH
Kurfürstendamm 57
10707 Berlin, Germany

Blackwell Science KK
MG Kodenmacho Building
7–10 Kodenmacho Nihombashi
Chuo-ku, Tokyo 104, Japan

First published 1990 by Unwin
Hyman Ltd
Reprinted 1993 by Chapman &
Hall Ltd
Second edition 2000

Set by Excel Typesetters Co., Hong
Kong
Printed and bound in Great Britain
by MPG Books Ltd, Bodmin,
Cornwall

The Blackwell Science logo is a
trade mark of Blackwell Science Ltd,
registered at the United Kingdom
Trade Marks Registry

A catalogue record for this title
is available from the British Library

ISBN 0-632-05432-8

Library of Congress
Cataloging-in-publication Data

Chester, R. (Roy), 1936–
    Marine geochemistry/Roy
    Chester. —[2nd ed.]
        p.    cm.
    Includes index.
    ISBN 0-632-05432-8
    1.  Chemical oceanography.   2.
    Marine sediments.
    3. Geochemistry.
    I.  Title.
    GC111.2.C47   1999
    551.46′01—dc21

DISTRIBUTORS
Marston Book Services Ltd
PO Box 269
Abingdon, Oxon OX14 4YN
(Orders: Tel: 01235 465500
        Fax: 01235 465555)

USA
Blackwell Science, Inc.
Commerce Place
350 Main Street
Malden, MA 02148-5018
(Orders: Tel: 800 759 6102
        781 388 8250
        Fax: 781 388 8255)

Canada
Login Brothers Book Company
324 Saulteaux Crescent
Winnipeg, Manitoba R3J 3T2
(Orders: Tel: 204 837 2987)

Australia
Blackwell Science Pty Ltd
54 University Street
Carlton, Victoria 3053
(Orders: Tel: 3 9347 0300
        Fax: 3 9347 5001)

For further information on
Blackwell Science, visit our website:
www.blackwell-science.com

# Contents

# Preface to the first edition

The past two or three decades have seen many important advances in our knowledge of the chemistry, physics, geology and biology of the oceans. It has also become apparent that in order to understand the manner in which the oceans work as a 'chemical system', it is necessary to use a framework which takes account of these interdisciplinary advances. *Marine Geochemistry* has been written in response to the need for a single state-of-the-art text that addresses the subject of treating the sea water, sediment and rock reservoirs as a unified system. In taking this approach, a process-orientated framework has been adopted in which the emphasis is placed on identifying key processes operating within the 'unified ocean'. In doing this, particular attention has been paid to making the text accessible to students from all disciplines in such a way that future advances can readily be understood.

I would like to express my thanks to those people who have helped with the writing of this volume. In particular, I wish to put on record my sincere appreciation of extremely helpful suggestions made by Professor John Edmond, FRS. In addition, I thank Dr S. Rowlatt for his comments on the sections covering the geochemistry of oceanic sediments, and Dr G. Wolff for his invaluable advice on the organic geochemistry of biota, water and sediments. It is a great pleasure to acknowledge the help of Dr K. J. T. Murphy, who gave so freely of his time at all stages in the preparation of the text. I also thank all those authors who have kindly allowed their diagrams and tables to be reproduced in the book. Many other people have influenced the way in which my thoughts have developed over the years, and to these friends and colleagues I owe a great debt of gratitude.

I would like to thank Unwin Hyman for their understanding during the preparation of the volume; Roger Jones for helping to develop the idea in the beginning, and Andy Oppenheimer, whose patience in handling the manuscript has known no bounds.

Finally, I would like to express my gratitude to my wife Alison, for all the devoted support she has given me during the writing of this book and at all other times.

R. Chester
*Liverpool*

# Preface to the second edition

The first edition of *Marine Geochemistry* was written in response to the need for a single state-of-the-art text that treats the oceans as a unified system. The original concept used an approach in which the emphasis was placed on identifying key processes operating within the 'unified ocean', and the format was designed to accommodate future advances in the subject. Since the first edition was written there have in fact been significant advances in several areas of marine geochemistry and the text of the present edition has been modified to accommodate them, while still retaining the original formula. Some of the modifications are essentially no more than 'fine tuning'. In contrast, others are more fundamental, and relate to advances which have provided fresh insights into marine processes. The areas in which our knowledge of oceanic processes have undergone fundamental conceptual changes include the following.

1 *Trace metal speciation.* When the first edition was written this topic was still in its relative infancy, but the field has expanded considerably over the past few years. In particular, speciation studies have allowed new theories in trace-metal–biota relationships to be established. For example, it has been suggested that feedback mechanisms between biological and chemical systems may be of the utmost importance in the 'high nitrogen, low productivity' (HNLP) regions of the oceans; with the biology strongly influencing the chemical speciation of the bioactive trace metals, and the speciation of the metals themselves influencing primary production, species composition and trophic structure.

2 *Carbon dioxide, and its role in world climate change.* New data have now become available on the oceanic carbon dioxide system, particularly on the role of new production in the drawdown of carbon dioxide from the atmosphere, and on the magnitudes of the carbon dioxide fluxes associated with the major oceanic source/sink regions.

3 *The transport of particulate material to the interior of the ocean.* Many of the Global Ocean Flux Study (GOFS) regional sediment-trap studies have been reported in the literature since the first edition was written, thus allowing a quantitative estimate of down-column fluxes to be made on an ocean-wide scale.

4 *Primary production and iron limitation.* A better understanding has now emerged of the status of different oceanic regions in marine primary productivity, and particularly on the potentially limiting role played by the micronutrient iron in areas where production is low in relation to the amount of the nutrients available; i.e. the HNLP regions.

5 *Colloids.* The importance of the role played by colloids in trace metal cycles has begun to be appreciated to a greater extent over the past few years, and will be a field for considerable further expansion.

6 *The preservation/destruction of organic matter in marine sediments.* In marine geochemistry, as in all scientific disciplines, certain topics are in vogue at particular times. If, on the basis of fundamental advances, the 1970s and 1980s could be regarded as having been the 'trace metal' decades in ocean research, the 1990s and beyond may well come to be viewed as the 'organic matter' decades. Already a number of new theories have emerged on the factors controlling the preservation of organic matter in marine sediments. Like all revolutionary theories these have been challenged, but they have opened up exciting avenues which will be further explored in the future.

The text of *Marine Geochemistry* has been modified to cover a number of these advances, and to accommodate them some material from the first edition has been omitted.

In addition to the acknowledgements made in the Preface to the first edition, I would like to express my personal thanks to my colleagues Dr H. Leach and Dr

R.G. Williams for their invaluable advice on various aspects of physical oceanography, and to Dr M. Ginger for his help in compiling Worksheet 9.1. I also would like to record my thanks to Val Hughes who undertook the mammoth task of typing the manuscript; a task she carried out with great forbearance and understanding.

I also would like to thank Ian Francis of Blackwell Science who encouraged me to write the second edition, and Jane Plowman and Jonathan Rowley for their understanding and patience during the writing and production of the book.

Finally, I again would like to express my gratitude to my wife Alison for her support and understanding during the writing of this second edition of the book.

R. Chester
Liverpool

# Acknowledgements

We are grateful to the following individuals and organizations who have kindly given permission for the reproduction of figures and tables.

Academic Press: Figs 3.5 (a, b (i, ii, v)), 4.4 (a, b), 5.1, 8.1, 8.2, 8.5, 8.6, 9.2 (b (i, iii)), 9.3, 9.6, 11.4, 13.5, 14.7, 15.6, 16.7, 17.1; Tables 5.1, 11.1 & 13.5; Worksheets 8.1 (a) & 14.2. American Geophysical Union: Figs 4.8, 13.2 (b), 15.4. American Journal of Science: Figs 14.9 (b), 15.9 (a, b). American Society of Limnology and Oceanography: Figs 6.3, 14.1. American Association for the Advancement of Science: Figs 3.1, 16.10. American Geophysical Union: Figs 4.3, 4.6, 6.6. American Journal of Science: Worksheet 14.4 (i, iia). American Society of Limnology and Oceanography: Fig. 11.7 (a), 11.8. Annual Reviews Inc: Fig. 6.5. CRC Press: Fig. 10.1. D.S. Cronan and S.A. Moorby: Fig. 15.7. Elsevier Science: Figs 3.5 (b (iii, iv)), 3.6, 4.1, 4.9, 5.2, 7.1 (a, b (i), c (i)), 7.3, 7.5, 9.2 (a), 9.4, 9.7, 9.9, 10.2, 10.3, 11.3, 11.5 (a, b, c, d, e), 11.7 (b), 11.9, 11.10, 11.11, 12.1, 12.3, 12.4, 12.7, 13.3 (a, b), 13.3 (a), 13.4, 14.3, 14.4, 14.5, 14.6, 14.8, 14.9 (a), 15.1, 15.2, 15.9 (c), 16.2, 16.3 (a, b), 16.4, 16.5, 16.6, 16.9, 16.12, 17.2; Tables 14.13 & 15.2; Worksheets 3.1 (i, ii), 7.3 (i–iv), 14.3 & 14.4 (ii b). G. Wolff: Worksheet 9.1. Geological Society of America: Figs 3.2, 13.1 (a), 15.5 (b). John Wiley & Sons Inc: Figs 4.7, 7.2, 7.4. Kluwer Academic Publishers: Figs 4.2 (a, b), 4.5, 8.4, 16.8. Macmillan Publishing Company: Fig. 7.1 (b (ii)). Munksgaard International Publishers Ltd: Worksheet 8.2 (i). Macmillan Magazines Ltd: Figs 9.1, 11.2, 16.11; Worksheet 7.2 (i, ii, iii). Oxford University Press: Fig. 13.2 (a); Worksheet 14.1. Plenum Publishing: Fig. 11.6; Tables 3.6 & 3.10. Prentice Hall Inc: Figs 7.1 (c ii), 9.2 (b ii), 13.1 (b), 15.5 (a). SCOPE/UNEP: Figs 6.1 (c), 6.2, 12.2; Tables 6.13 & 6.14. Springer-Verlag: Figs 4.10, 14.2. The Royal Society: Fig. 11.1. Unesco: Figs 3.3, 6.1 (a); Worksheet 7.1 (a, b). University of Rhode Island: Fig. 15.8. US National Academy of Science: Figs 12.5, 12.6; Table 12.2.

# Symbols and concentration units

## 1 General symbols

All symbols used in the present work are defined at the appropriate place in the text, which can be found by reference to the index at the end of the volume.

## 2 Units

The units defined below, and the symbols by which they are identified, are confined to a general list of those most commonly used in the present work; other units will be defined where necessary in the text itself. It must be noted that a number of traditional units have been retained as a matter of policy throughout the work because they are still widely used in the current as well as in the past literature; e.g. the litre has been used as a unit volume although IAPSO have recommended that for high-precision measurements of volume it be replaced by the cubic decimetre ($dm^3$). For a detailed treatment of the use of SI units in oceanography see the IAPSO recommendations published by Unesco (1985).

### Length

SI unit = metre
nm, nanometre = $10^{-9}$ m
μm, micrometre = $10^{-6}$ m
mm, millimetre = $10^{-3}$ m
cm, centimetre = $10^{-2}$ m
m, metre
km, kilometre = $10^3$ m

### Weight or mass

SI unit = kilogram
pg, picogram = $10^{-12}$ g
ng, nanogram = $10^{-9}$ g
μg, microgram = $10^{-6}$ g
mg, milligram = $10^{-3}$ g

g, gram
kg, kilogram = $10^3$ g
t, ton/tonne = $10^6$ g

### Volume

SI unit = cubic metre
$dm^3$, cubic decimetre = $10^{-3}$ $m^3$ = 1 litre
$m^3$, cubic metre
μl, microlitre = $10^{-6}$ l
ml, millilitre = $10^{-3}$ l
l, litre

### Time

SI unit = second
s = second
min = minute
h = hour
d = day
yr = year
Ma = million years = $10^6$ yr

### Concentration*

The SI unit for the amount of a substance is the mole. The most commonly used concentration for particu-

* *Note on the use of concentration units.* The concentration of dissolved elements is usually expressed in the text in the most widely used mole form. However, as it is still common practice for many authors to use the μg/ng g$^{-1}$ form for the expression of concentrations in particulates and sediments, the convention has been retained here. This does not present problems in the evaluation of elemental distribution patterns (or of the processes that control them) in either sea water or sediments. The approach adopted in the text, however, is to follow a global 'source–sink' journey, and in order to simplify and standardize assessments of the transport of elements from sea water to the sediment reservoir, mole concentrations have been converted to mass concentrations in water-column/sediment-surface flux calculations.

lates and sediments, however, is still mass per unit mass; e.g.

$\mu g\,g^{-1}$ = p.p.m. = parts per million
$ng\,g^{-1}$ = p.p.b. = parts per billion

A number of systems are currently in common use for expressing the concentration of solutes in sea water.

**1** The concentrations can be expressed in units of mass per unit volume or per unit mass of sea water: e.g. $g\,kg^{-1}$ or $mg\,kg^{-1}$ for major components, or $ng\,dm^{-3}$ or $ng\,kg^{-1}$ of sea water for trace elements; however, trace element concentrations are still often expressed in terms of $mass\,l^{-1}$. Examples of such concentrations are

$\mu g\,l^{-1} = 10^{-6}\,g\,l^{-1}$ (or $dm^{-3}$, or $kg^{-1}$)
$ng\,l^{-1} = 10^{-9}\,g\,l^{-1}$ (or $dm^{-3}$, or $kg^{-1}$)
$pg\,l^{-1} = 10^{-12}\,g\,l^{-1}$ (or $dm^{-3}$, or $kg^{-1}$)

**2** The most usual practice now is to use the mole as the unit of concentration for solutes in sea water. Examples of concentrations are

$\mu mol\,l^{-1} = \mu M = 10^{-6}\,mol\,l^{-1}$ (or $dm^{-3}$, or $kg^{-1}$)
$nmol\,l^{-1} = nM = 10^{-9}\,mol\,l^{-1}$ (or $dm^{-3}$, or $kg^{-1}$)
$pmol\,l^{-1} = pM = 10^{-12}\,mol\,l^{-1}$ (or $dm^{-3}$, or $kg^{-1}$)
$fmol\,l^{-1} = fM = 10^{-15}\,mol\,l^{-1}$ (or $dm^{-3}$, or $kg^{-1}$)

**3** Traditionally, the concentrations of the nutrients have often been expressed as $\mu g\text{-at}\,l^{-1}$, where

$\mu g\text{-at}\,l^{-1} = \mu g\text{-atoms}\,l^{-1} = (\mu g/\text{atomic weight})\,l^{-1}$

The atmospheric concentrations of particulate elements given in the text are expressed in the form

$\mu g\,m^{-3}$ of air = $10^{-6}\,g$ per cubic metre of air
$ng\,m^{-3}$ of air = $10^{-9}\,g$ per cubic metre of air

*Radioactivity*

SI unit = $Bq\,m^{-3}$ (Becquerels per cubic metre) or $Bq\,kg^{-1}$
d.p.m. = disintegrations per minute
Ci = curie; $1\,Ci = 3.7 \times 10^{10}\,Bq$

### 3  Some data that are useful for flux calculations

*Areas†*

Area of the oceans = $361\,110 \times 10^3\,km^2$

† Taken mainly from Baumgartner & Reichel (1975).

Area of the Atlantic Ocean (to c. 80°S) = $98\,013 \times 10^3\,km^2$
Area of the North Atlantic = $52\,264 \times 10^3\,km^2$
Area of the South Atlantic = $45\,749 \times 10^3\,km^2$
Area of the Indian Ocean (to c. 70°S) = $77\,700 \times 10^3\,km^2$
Area of the northern Indian Ocean = $12\,482 \times 10^3\,km^2$
Area of the southern Indian Ocean = $65\,218 \times 10^3\,km^2$
Area of the Pacific Ocean (to c. 80°S) = $176\,888 \times 10^3\,km^2$
Area of the North Pacific = $81\,390 \times 10^3\,km^2$
Area of the South Pacific = $95\,498 \times 10^3\,km^2$
Area of the continents = $148\,904 \times 10^3\,km^2$

*River transport*

River inflow into the North Atlantic Ocean = $11\,405$ $km^3\,yr^{-1}$
River inflow into the South Atlantic Ocean = $7946$ $km^3\,yr^{-1}$
River inflow into the northern Indian Ocean = $3247\,km^3\,yr^{-1}$
River inflow into the southern Indian Ocean = $2354\,km^3\,yr^{-1}$
River inflow into the North Pacific Ocean = $7678\,km^3\,yr^{-1}$
River inflow into the South Pacific Ocean = $4459\,km^3\,yr^{-1}$
Total river inflow to all oceans $\approx 37\,400\,km^3\,yr^{-1}$

*Atmospheric transport*

The area of the marine atmosphere is equal to the total area of the oceans; however, the atmospheric volume used for calculating atmospheric deposition fluxes depends on the scale height to which a component is dispersed, usually between about 3 and 5 km—see Section 6.2.1.1.

### Reference

Baumgartner, A. & Reichel, E. (1975) *The World Water Balance*. Amsterdam: Elsevier. Unesco 1985. Unesco Technical Paper on Marine Science, no. 32. Paris: Unesco.

# List of abbreviations and acronyms

| | |
|---|---|
| AABW | Antarctic Bottom Water |
| AAIW | Antarctic Intermediate Water |
| ACC | Antarctic Circumpolar Current |
| ACD | aragonite compensation depth |
| AEE | anomalously enriched elements |
| AIF | atmospheric interference factor |
| AIW | Atlantic Intermediate Water |
| ANTARES | Antarctic research in France |
| AOU | apparent oxygen utilization |
| | |
| BBL | benthic boundary layer |
| BIMS | bubble interfacial microlayer sampler |
| | |
| CFM | chlorofluoromethane |
| CLIMAPP | Climate Long-range Investigation Mapping and Prediction Study |
| COC | colloidal organic carbon |
| CPC | circumpolar current |
| CPI | carbon preference index |
| CPM | coarse particulate matter |
| CZCS | Coast Zone Colour Scanner |
| | |
| DBT | dibenzothiophene |
| DCAA | dissolved combined amino acids |
| DDT | dichlorodiphenyltrichloroethane |
| DFAA | dissolved free amino acids |
| DIC | dissolved inorganic carbon |
| DMDS | dimethyl disulphide |
| DMGe | dimethylgermanium |
| DMS | dimethyl sulphide |
| DMSP | dimethylsulphoniopropionate |
| DNA | deoxyribonucleic acid |
| DOC | dissolved organic carbon |
| DOM | dissolved organic matter/material |
| DON | dissolved organic nitrogen |
| DOP | dissolved organic phosphorus |
| DPASV | differential pulse anodic stripping voltammetry |

| | |
|---|---|
| DPCSV | differential pulse cathodic stripping voltammetry |
| DSC | deep-sea clay |
| DSDP | Deep Sea Drilling Project |
| DTAA | dissolved total amino acids |
| DTI | dissolved transport index |
| | |
| ECOMARGE | Ecosystèmes de Marges Continentales |
| EF | enrichment factor |
| ENSO | El Niño–Southern Oscillation |
| EPM | estuarine particulate matter |
| EPR | East Pacific Rise |
| ETNA | eastern tropical North Atlantic |
| | |
| FPM | fine particulate matter |
| | |
| GEOSECS | Geochemical Ocean Sections Study |
| GESAMP | Group of Experts on the Scientific Aspects of Marine Pollution |
| GOFS | Global Ocean Flux Study |
| GSC | Galapagos Spreading Centre |
| GWP | global warming potential |
| | |
| HAP | Hatteras Abyssal Plain |
| HC | high chlorophyll |
| HCB | hexachlorobenzene |
| HCH | hexachlorocyclohexane |
| HEBBLE | High Energy Benthic Boundary Layer Experiment |
| HG | Hanging Gardens (hydrothermal smoker) |
| HMW | high molecular weight |
| HNHP | high nitrogen, high productivity |
| HNLP | high nitrogen, low productivity |
| HNLC | high nitrate–low chlorophyll |
| HNLSLC | high-nitrate–low-silicate–low-chlorophyll |
| HTCO | high-temperature catalytic oxidation |

| | |
|---|---|
| IAPSO | International Association for the Physical Sciences of the Oceans |
| ICES | International Council for the Exploration of the Sea |
| IDP | interplanetary dust particles |
| IPPC | Intergovernmental Panel on Climate Control |
| ITCZ | Inter-Tropical Convergence Zone |
| JGOFS | Joint Global Ocean Flux Study |
| LAA | large amorphous aggregates |
| LMW | low molecular weight |
| LNLP | low nitrogen, low productivity |
| MANOP | Manganese Nodule Project |
| MAR | Mid-Atlantic Ridge |
| MIW | Mediterranean intermediate water |
| MMD | median mass diameter |
| MMGe | monomethylgermanium |
| MORT | mean ocean residence time |
| MSA | methane sulphonate |
| NABE | North Atlantic Bloom Experiment |
| NADW | North Atlantic Deep Water |
| NAO | North Atlantic Oscillation |
| NAP | Nares Abyssal Plain |
| NEE | non-enriched elements |
| NCP | net community production |
| NGS | National Geographic Smoker (hydrothermal smoker) |
| NODC | National Oceanographic Data Center |
| NSOS | non-sulphate oxidant suite |
| OBS | Ocean Bottom Seismometer (hydrothermal smoker) |
| ODZ | oxygen depleted zone |
| OMZ | oxygen minimum zone |
| OSOM | oxygen-sensitive organic matter |
| PAH | polycyclic aromatic hydrocarbons |
| PCB | polychlorinated biphenyl |
| PCP | pentachlorophenol |
| PF | polar front |
| POC | particulate organic carbon |
| POM | particulate organic matter |
| PTM | particulate trace metals |

| | |
|---|---|
| REF | relative oceanic enrichment factor |
| RNA | ribonucleic acid |
| RPM | river particulate material |
| RSM | river suspended material |
| SADS | SEREX Asian Dust Sampling |
| SCOC | sediment community oxygen consumption |
| SCOPE | Scientific Committee on Problems of the Environment |
| SCUMS | self-contained underway microlayer sampler |
| SEAREX | Sea–Air Exchange Program |
| SEEP | Shelf Edge Exchange Program |
| SFM | stagnant film model |
| SMOW | Standard Mean Ocean Water |
| SOM | shallow oxygen minimum |
| SPAN | SEREX South Pacific Aerosol Network |
| SS | Sargasso Sea |
| SW | South West (hydrothermal smoker) |
| TA | total alkalinity |
| TAG | Trans-Atlantic Geotransverse |
| TDL | theoretical dilution line |
| THAA | total hydrolysable amino acids |
| TSM | total suspended material |
| TTO | Transient Tracers in the Ocean |
| UNEP | United Nations Environment Programme |
| UNESCO | United Nations Educational, Scientific and Cultural Organization |
| VERTEX | Vertical Transport and Exchange |
| VOC | volatile organic carbon (organic carbon fraction of VOM) |
| VOM | vapour-phase organic material |
| VWM | volume weighted mean |
| WCO | wet chemical oxidation |
| WF | washout factor |
| WOCE | World Ocean Circulation Experiment |
| WSW | Weddell Sea water |
| WTNA | western tropical North Atlantic |

# 1 Introduction

The fundamental question underlying marine geochemistry is 'How do the oceans work as a chemical system?' At present, that question cannot be answered fully. The past three decades or so, however, have seen a number of 'quantum leaps' in our understanding of some aspects of marine geochemistry. Three principal factors have made these leaps possible:

1 advances in sampling and analytical techniques;

2 the development of theoretical concepts;

3 the setting up of large-scale international oceanographic programmes (e.g. DSDP, MANOP, HEBBLE, GEOSECS, TTO, VERTEX, JGOFS, SEAREX, WOCE), which have extended the marine geochemistry database to a global ocean scale.

## 1.1 Setting the background: a unified 'process-orientated' approach to marine geochemistry

Oceanography attracts scientists from a variety of disciplines, including chemistry, geology, physics, biology and meteorology. A knowledge of at least some aspects of marine geochemistry is an essential requirement for scientists from all these disciplines and for students who take courses in oceanography at any level. The present volume has been written, therefore, with the aim of bringing together the recent advances in marine geochemistry in a form that can be understood by all those scientists who use the oceans as a natural laboratory and not just by marine chemists themselves. One of the major problems involved in doing this, however, is to provide a coherent global ocean framework within which marine geochemistry can be described in a manner that cannot only relate readily to the other oceanographic disciplines but also can accommodate future advances in the subject. To develop such a framework, it is necessary to explore some of the basic concepts that underlie marine geochemistry.

Geochemical balance calculations show that a number of elements that could not have come from the weathering of igneous rocks are present at the Earth's surface. It is now generally accepted that these elements, which are termed the **excess volatiles**, have originated from the degassing of the Earth's interior. The excess volatiles, which include H and O (combined as $H_2O$), C, Cl, N, S, B, Br and F, are especially abundant in the atmosphere and the oceans. It is believed, therefore, that both the atmosphere and the oceans were generated by the degassing of the Earth's interior. In terms of global cycling, Mackenzie (1975) suggested that sedimentary rocks are the product of a long-term titration of primary igneous-rock minerals by acids associated with the excess volatiles, a process that can be expressed as

primary igneous-rock minerals + excess volatiles
  $\rightarrow$ sedimentary rocks + oceans + atmosphere

$$(1.1)$$

As this reaction proceeds, the seawater reservoir is continuously subjected to material fluxes, which are delivered along various pathways from external sources. The oceans therefore are a **flux-dominated** system. Sea water, however, is not a static reservoir in which the material has simply accumulated over geological time, otherwise it would have a very different composition from that which it has at present; for example, the material supplied over geological time far exceeds the amount now present in sea water. Further, the composition of sea water has not changed markedly over very long periods of time. Rather than acting as an **accumulator**, therefore, the flux-dominated seawater reservoir can be regarded as a **reactor**. It is the nature of the reactions that take place within the reservoir, i.e. the manner in which it responds to the material fluxes, which defines the composition of sea water via an input → internal reactivity → output cycle.

Traditionally, there have been two schools of thought on the overall nature of the processes that operate to control the composition of sea water.

**1** In the **equilibrium ocean** concept, a state of chemical equilibrium is presumed to exist between sea water and sediments via reactions that are reversible in nature. Thus, if the supply of dissolved elements to sea water were to increase, or decrease, the equilibrium reactions would change in the appropriate direction to accommodate the fluctuations.

**2** In the **steady-state ocean** concept, it is assumed that the input of material to the system is balanced by its output, i.e. the reactions involved proceed in one direction only. In this type of ocean, fluctuations in input magnitudes would simply result in changes in the rates of the removal reactions, and the concentrations of the reactants in sea water would be maintained.

At present, the generally held view supports the steady-state ocean concept. Whichever theory is accepted, however, it is apparent that the oceans must be treated as a **unified input–output type of system**, in which materials stored in the sea water, the sediment and the rock reservoirs interact, sometimes via recycling stages, to control the composition of sea water.

It is clear, therefore, that the first requirement necessary to address the question 'How do the oceans work as a chemical system?' is to treat the sea water, sediment and rock reservoirs as a unified system. It is also apparent that one of the keys to solving the question lies in understanding the nature of the chemical, physical and biological processes that control the composition of sea water, as this is the reservoir through which the material fluxes flow in the input → internal reactivity → output cycle. In order to provide a **unified ocean** framework within which to describe the recent advances in marine geochemistry in terms of this cycle, it is therefore necessary to understand the nature and magnitude of the fluxes that deliver material to the oceans (the input stage), the reactive processes associated with the throughput of the material through the seawater reservoir (the internal reactivity stage), and the nature and magnitude of the fluxes that take the material out of sea water into the sinks (the output stage).

The material that flows through the system includes inorganic and organic components in both dissolved and particulate forms, and a wide variety of these components will be described in the text. In order to avoid falling into the trap of not being able to see the wood for the trees in the morass of data, however, it is essential to recognize the importance of the processes that affect constituents in the source-to-sink cycle. Rather than taking an element-by-element 'periodic table' approach to marine geochemistry, the treatment adopted in the present volume will involve a **process-orientated** approach, in which the emphasis will be placed on identifying the key processes that operate within the cycle. The treatment will include both natural and anthropogenic materials, but it is not the intention to offer a specialized overview of marine pollution. This treatment does not in any way underrate the importance of marine pollution. Rather, it is directed towards the concept that it is necessary first to understand the natural processes that control the chemistry of the ocean system, because it is largely these same processes that affect the cycles of the anthropogenic constituents.

Since the oceans were first formed, sediments have stored material, and thus have recorded changes in environmental conditions. The emphasis in the present volume, however, is largely on the role that the sediments play in controlling the chemistry of the oceans. The diagenetic changes that have the most immediate effect on the composition of sea water take place in the upper few metres of the sediment column. For this reason attention will be focused on these surface deposits, and the role played by sediments in palaeooceanography will be touched upon only briefly.

In order to rationalize the process-orientated approach, special attention will be paid to a number of individual constituents, which can be used to elucidate certain key processes that play an important role in controlling the chemical composition of sea water. In selecting these process-orientated constituents it was necessary to recognize the flux-dominated nature of the seawater reservoir. The material fluxes that reach the oceans deliver both dissolved and particulate elements to sea water. It was pointed out above, however, that the amount of dissolved material in sea water is not simply the sum of the total amounts brought to the oceans over geological time. This was highlighted in the last century by Forchhammer (1865) when he wrote: 'Thus the quantity of the different elements in sea water is not proportional to the quantity of elements which river water pours into the sea, but is inversely proportional to *the facility with which the elements are made insoluble by general chemical or organo-chemical actions in the sea*' [my italics]. According to Goldberg (1963), this statement can be viewed as elegantly posing the theme of marine chemistry, and it is this 'facility with which the elements are made insoluble', and so are removed from the dissolved phase, which is central to our understanding of many of the factors that control the composition of sea water. This was highlighted more recently by Turekian (1977). In one of the most influential geochemical papers published in recent years, this author formally posed a question that had attracted the attention of marine geochemists for generations, and may be regarded as another expression of Forchhammer's statement, i.e. 'Why are the oceans so depleted in trace metals?' Turekian concluded that the answer lies in the role played by particles in the sequestration of reactive elements during every stage in the transport cycle from source to marine sink.

Ultimately, therefore, it is the transfer of dissolved constituents to the particulate phase, and the subsequent sinking of the particulate material, that is responsible for the removal of the dissolved constituents from sea water to the sediment sink. It must be stressed, however, that although dissolved → particulate transformations are the driving force behind the removal of most elements to the sediment sink, the transformations themselves involve a wide variety of biogeochemical processes. For example, Stumm & Morgan (1981) identified a number of chemical reactions and physicochemical processes that are important in setting the chemical composition of natural waters. These included acid–base reactions, oxidation–reduction reactions, complexation reactions between metals and ligands, adsorption processes at interfaces, the precipitation and dissolution of solid phases, gas–solution processes, and the distribution of solutes between aqueous and non-aqueous phases. The manner in which reactions and processes such as these, and those specifically associated with biota, interact to control the composition of sea water will be considered throughout the text. For the moment, however, they can be grouped simply under the general term **particulate ↔ dissolved** reactivity. The particulate material itself is delivered to the sediment surface mainly via the down-column sinking of large-sized organic aggregates as part of the oceanic **global carbon flux**. Thus, within the seawater reservoir, reactive elements undergo a continuous series of dissolved ↔ particulate transformations, which are coupled with the transport of biologically formed particle aggregates to the sea bed. Turekian (1977) aptly termed this overall process **the great particle conspiracy**. In the flux-dominated ocean system the manner in which this conspiracy operates to clean up sea water is intimately related to the oceanic throughput of externally transported, and internally generated, particulate matter. Further, it is apparent that several important aspects of the manner in which this **throughput cycle** operates to control the inorganic and organic compositions of both the seawater reservoir and the sediment sink can be assessed in terms of the oceanic fates of reactive trace elements and organic carbon.

Many of the most important thrusts in marine geochemistry over the past few years have used tracers to identify the processes that drive the system, and to establish the rates at which they operate. These tracers will be discussed at appropriate places in the text. The **tracer approach**, however, also has been adopted in a much broader sense in the present volume in that special attention will be paid to the trace elements and organic carbon in the source/input → internal reactivity → sink/output transport cycle. Both stable and radionuclide trace elements (e.g. the use of the 'time clock' Th isotopes as both transport and process indicators) are especially rewarding for the study of reactivity within the various stages of the cycle, and organic carbon is a vital constituent with respect to the oceanic biomass, the down-column transport of material to the sediment sink and sediment diagenesis.

To interpret the source/input → internal reactivity → sink/output transport cycle in a coherent and systematic manner, a three-stage approach will be adopted, which follows the cycle in terms of a **global journey**. In Part I, the movements of both dissolved and particulate components will be tracked along a variety of transport pathways from their original sources to the point at which they cross the interfaces at the land–sea, air–sea and rock–sea boundaries. In Part II, the processes that affect the components within the seawater reservoir will be described. In Part III, the components will be followed as they are transferred out of sea water into the main sediment sink, and the nature of the sediments themselves will be described. The treatment, however, is concerned mainly with the role played by the sediments as marine sinks for material that has flowed through the seawater reservoir. In this context, it is the processes that take place in the upper few metres of the sediments that have the most immediate effect on the composition of sea water. For this reason attention will be restricted mainly to the uppermost sediment sections, and no attempt will be made to evaluate the status of the whole sediment column in the history of the oceans.

The steps involved in the three-stage global journey are illustrated schematically in Fig. 1.1. This is not meant to be an all-embracing representation of reservoir interchange in the ocean system, but is simply intended to offer a general framework within which to describe the global journey. By directing the journey in this way, the intention therefore is to treat the seawater, sediment and rock phases as integral parts of a unified ocean system.

In addition to the advantages of treating the oceans as a single system, the treatment adopted here is important in order to assess the status of the marine environment in terms of planetary geochemistry. For example, according to Hedges (1992) there is a complex interplay of biological, geological and chemical processes by which materials and energy are

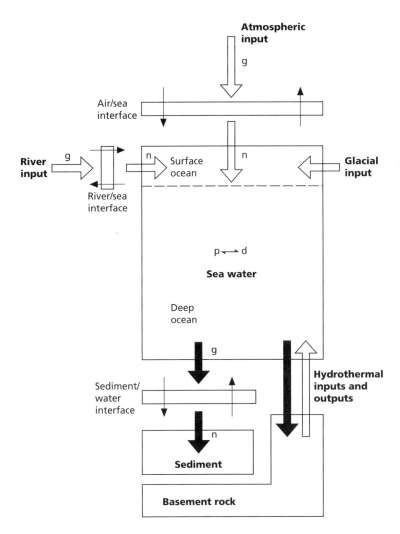

Fig. 1.1 A schematic representation of the source/input → seawater internal reactivity → sink/output global journey. The large open arrows indicate transport from material sources, and the large filled arrows indicate transport into material sinks; relative flux magnitudes are not shown. The small arrows indicate only that the strengths of the fluxes can be changed as they cross the various interfaces in the system; thus, g and n represent gross and net inputs or outputs, respectively. Material is brought to the oceans in both particulate and dissolved forms, but is transferred into the major sediment sink mainly as particulate matter. The removal of dissolved material to the sediment sink therefore usually requires its transformation to the particulate phase. This is shown by the p ⇌ d term. The intention here, however, is simply to indicate that internal particulate–dissolved reactivity occurs within the seawater reservoir, and it must be stressed that a wide variety of chemical reactions and physicochemical processes are involved in setting the composition of the water phase—see text. For convenience coastal zones are not shown.

exchanged and reused at the Earth's surface. These interreacting processes, which are termed **biogeo-chemical cycles**, are concentrated at interfaces and modified by feedback mechanisms. The cycles operate on time-scales of microseconds to eons, and occur in domains that range in size from a living cell to the entire ocean–atmosphere system, and interfaces in the oceans play a vital role in the biogeochemical cycles of some elements.

The volume has been written for scientists of all disciplines. To contain the text within a reasonable length, a basic knowledge of chemistry, physics, biology and geology has been assumed and the fundamental principles in these subjects, which are readily available in other textbooks, have not been reiterated here. As the volume is deliberately designed with a multidisciplinary readership in mind, however, an attempt has been made to treat the more advanced chemical and physical concepts in a generally descriptive manner, with appropriate references being given to direct the reader to the original sources. One of the major aims of marine geochemistry in recent years has been to model natural systems on the basis of theoretical concepts. To follow this approach it is necessary to have a more detailed understanding of the theory involved, and for this reason a series of Worksheets have been included in the text. Some of these Worksheets are used to describe a number of basic geochemical concepts; for example, those underlying redox reactions and the diffusion of solutes in interstitial waters. In others, however, the emphasis is placed on modelling a variety of geo-chemical systems using, where possible, actual examples from literature sources; for example, the topics covered include a sorptive equilibrium model for the removal of trace metals in estuaries, a stagnant film model for the exchange of gases across the air–sea interface, and a variety of models designed to describe solid-phase–dissolved-phase interactions in sediment interstitial waters.

Overall, therefore, the intention is to provide a unifying framework, which has been designed to bring a state-of-the-art assessment of marine geochemistry to the knowledge of a variety of ocean scientists in such a way that allows future advances to be understood within a meaningful context.

## References

Forchhammer, G. (1865) On the composition of sea water in the different parts of the ocean. *Philos. Trans. R. Soc. London*, **155**, 203–62.

Goldberg, E.D. (1963) The oceans as a chemical system. In *The Sea*, M.N. Hill (ed.), Vol. 2, 3–25. New York: Wiley Interscience.

Hedges, J.I. (1992) Global biogeochemical cycles: progress and problems. *Mar. Chem.*, **39**, 67–93.

Mackenzie, F.T. (1975) Sedimentary cycling and the evolution of the sea water. In *Chemical Oceanography*, J.P. Riley & G. Skirrow (eds), Vol. 1, 309–64. London: Academic Press.

Stumm, W. & Morgan, J.J. (1981) *Aquatic Chemistry*. New York: Wiley.

Turekian, K.K. (1977) The fate of metals in the oceans. *Geochim. Cosmochim. Acta*, **41**, 1139–44.

# Part I
# The Global Journey: Material Sources

# 2     The input of material to the ocean reservoir

The World Ocean may be regarded as a planetary dumping ground for material that originates in other geospheres, and to understand marine geochemistry it is necessary to evaluate the composition, flux rate and subsequent fate of the material that is delivered to the ocean reservoir.

## 2.1 The background

The major natural sources of the material that is injected into sea water are the continental crust, the oceanic crust and the atmosphere. Primary material is mobilized directly from the continental crust, mainly by low-temperature weathering processes and high-temperature volcanic activity. In addition, secondary (or pollutant) material is mobilized by a variety of anthropogenic 'weathering' processes, which often involve high temperatures. The various types of material released on the continents during both natural and anthropogenic processes include particulate, dissolved and gaseous phases, which are then moved around the surface of the planet by a number of transport pathways. The principal routes by which continentally mobilized material reaches the World Ocean are via fluvial (river), atmospheric and glacial transport. The relative importance of these pathways, however, varies considerably in both space and time. For example, atmospheric transport is strongest in low latitudes, where aeolian dust can be carried to the sea surface in the form of intermittent pulses. However, material is dispersed throughout the atmosphere over the whole ocean and is present, albeit sometimes at low concentrations, at all marine locations. Fluvial transport also delivers material to very large areas of the World Ocean, but glacial transport is much restricted in scope.

Water in the form of ice can act as a major mechanism for the physical mobilization of material on the Earth's surface. The magnitude of the transport of this material depends on the prevailing climatic regime. At present, the Earth is in an interglacial period and large-scale ice sheets are confined to the polar regions. Even under these conditions, however, glacial processes are a major contributor of material to the oceans. For example, Garrels & Mackenzie (1971) estimated that at present $\sim 20 \times 10^{14}\,\mathrm{g\,yr^{-1}}$ of crustal products are delivered to the World Ocean by glacial transport, of which $\sim 90\%$ is derived from Antarctica. Thus, ice transport is second only to fluvial run-off in the global supply of material to the marine environment. From the point of view of marine geochemistry, however, glacial transport is less important than either fluvial or atmospheric transport in the supply of material to the oceans on a global scale. There are two principal reasons for this.

1 Glacial transport is at present restricted largely to the polar regions, and so does not have the same global importance as either fluvial or atmospheric transport. For example, although ice-rafted material has been found in marine sediments from many areas, glacial marine sediments are confined largely to the polar regions around Antarctica, where they form a ring of sediment, and to areas in the Arctic Ocean (see Fig. 13.5a).

2 Water is the main agency involved in chemical weathering, and during glacial processes this water is locked in a solid form. As a result, there is a general absence of chemical weathering in the polar regions and therefore little release of elements into the soluble phase.

In general, therefore, glacial material does not make a significant global contribution to the dissolved pool of elements in sea water. It has been suggested, however, that one way in which glacially transported material can contribute to this dissolved pool is by the leaching of elements from fine-grained rock flour by sea water. In this context, Schutz & Turekian (1965) suggested that such a process might account for enrichments in Co, Ni and Ag in waters of intermediate depth south of 68°S. In general, however, it may be concluded that the effects resulting from the transport of material to the World Ocean by glacial processes are confined largely to the polar regions. The principal transport pathways that supply material derived from the continental crust to the oceans therefore are river run-off and atmospheric deposition.

Material also is supplied to the oceans from processes that affect the oceanic crust. These processes involve low-temperature weathering of the ocean basement rocks, mainly basalts, and high-temperature water–rock reactions associated with hydrothermal activity at spreading ridge centres. This hydrothermal activity, which can act as a source of some components and a sink for others, is now known to be of major importance in global geochemistry; for example, in terms of primary inputs it dominates the supply of dissolved manganese to the oceans. Although the extent to which this type of dissolved material is dispersed about the ocean is not yet clear, hydrothermal activity must still be regarded as a globally important mechanism for the supply of material to the seawater reservoir.

On a global scale, therefore, the main pathways by which material is brought to the oceans are:

**1** river run-off, which delivers material to the surface ocean at the land–sea boundaries;

**2** atmospheric deposition, which delivers material to all regions of the surface ocean;

**3** hydrothermal activity, which delivers material to deep and intermediate waters above the sea floor.

The manner in which these principal pathways operate is described individually in the next three chapters, and this is followed by an attempt to estimate the relative magnitudes of the material fluxes associated with them.

## References

Garrels, R.M. & Mackenzie, F.T. (1971) *Evolution of Sedimentary Rocks*. New York: Norton.

Schutz, D.F. & Turekian, K.K. (1965) The distribution of cobalt, nickel and silver in ocean water profiles around Pacific Antarctica. *J. Geophys. Res.*, 70, 5519–28.

# 3  The transport of material to the oceans: the river pathway

Much of the material mobilized during both natural crustal weathering and anthropogenic activities is dispersed by rivers, which transport the material towards the land–sea margins. In this sense, rivers may be regarded as the carriers of a wide variety of *chemical signals* to the World Ocean. The effect that these signals have on the chemistry of the ocean system may be assessed within the framework of three key questions (see e.g. Martin & Whitfield, 1983).

1 What is the quantity and chemical composition of the dissolved and particulate material carried by rivers?

2 What are the fates of these materials in the estuarine mixing zone?

3 What is the ultimate quantity and composition of the material that is exported from the estuarine zone and actually reaches the open ocean?

These questions will be addressed in this chapter, and in this way river-transported materials will be tracked on their journey from their source, across the estuarine (river–ocean) interface, through the coastal receiving zone and out into the open ocean.

## 3.1 Chemical signals transported by rivers

### 3.1.1 Introduction

River water contains a large range of inorganic and organic components in both dissolved and particulate forms. A note of caution, however, must be introduced before any attempt is made to assess the strengths of the chemical signals carried by rivers, especially with respect to trace elements. In attempting to describe the processes involved in river transport, and the strengths of the signals they generate, great care must be taken to assess the validity of the databases used and, where available, 'modern' (i.e. post *c*. 1975) trace-element data will be used in the present discussion of river-transported chemical signals.

### 3.1.2 The sources of dissolved and particulate material found in river waters

Water reaches the river environment either directly from the atmosphere or indirectly from surface run-off, underground water circulation and the discharge of waste solutions. The sources of the dissolved and particulate components that are found in the river water include rock weathering, the decomposition of organic material, wet and dry atmospheric deposition and, for some rivers, pollution. The source strengths are controlled by a number of complex, often interrelated, environmental factors that operate in an individual river basin; these factors include rock lithology, relief, climate, the extent of vegetative cover and the magnitude of pollutant inputs.

The various factors that are involved in setting the composition of river water are considered in the following sections, and to do this it is convenient to use a framework in which the **dissolved** and the **particulate** components are considered separately.

### 3.1.3 Major and trace elements: the dissolved river signal

#### 3.1.3.1 *Major elements*

The major element composition of rivers entering the principal oceans is given in Table 3.1, together with that of sea water. From the average river and seawater compositions given in this table it can be seen that there are a number of differences between these two types of surface water. The most important of these is that in river water there is a general dominance of **calcium** and **bicarbonate**, whereas in sea water

**Table 3.1** The major element composition of rivers draining into the oceans; units, mg l⁻¹ (data from Martin & Whitfield (1983) and Riley & Chester (1971)).

| Element | Atlantic | Indian | Arctic | Pacific | World average river water | Sea water |
|---|---|---|---|---|---|---|
| $Na^+$ | 4.2 | 8.5 | 8.8 | 5.2 | 5.3 | 10 733 |
| $K^+$ | 1.4 | 2.5 | 1.2 | 1.2 | 1.5 | 399 |
| $Ca^{2+}$ | 10.5 | 21.6 | 16.1 | 13.9 | 13.3 | 412 |
| $Mg^{2+}$ | 2.5 | 5.4 | 1.3 | 3.6 | 3.1 | 1 294 |
| $Cl^-$ | 5.7 | 6.8 | 11.8 | 5.1 | 6.0 | 19 344 |
| $SO_4^{2-}$ | 7.7 | 7.9 | 15.9 | 9.2 | 8.7 | 2 712 |
| $HCO_3^-$ | 37 | 94.9 | 63.5 | 55.4 | 51.7 | 142 |
| $SiO_2^{3-}$ | 9.9 | 14.7 | 5.1 | 11.7 | 10.7 | — |
| TDS* | 78.9 | 154.9 | 123.7 | 105.3 | 101.6 | — |

* TDS = total dissolved solids.

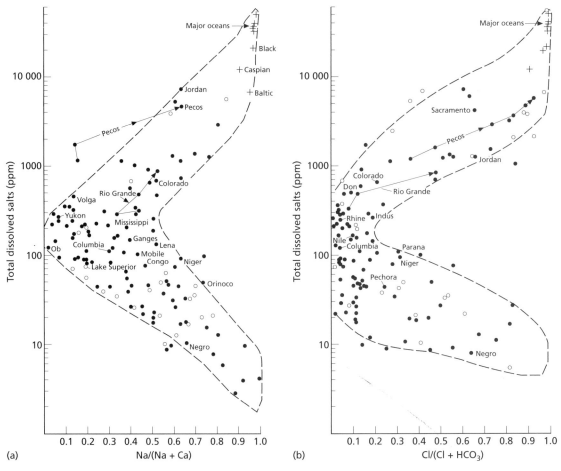

**Fig. 3.1** Processes controlling the composition of surface waters (from Gibbs, 1970). (a) Variations in the weight ratio Na/(Na + Ca) as a function of total dissolved salts. (b) Variations in the weight ratio Cl/(Cl + HCO₃) as a function of total dissolved salts. (c) Diagrammatic representation of the processes controlling end-member water compositions. See text for explanation.

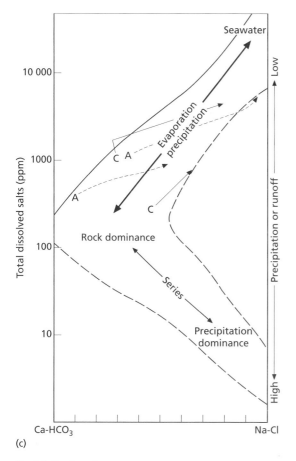

**Fig. 3.1** *Continued*

end-member surface waters. The cations that characterize the two principal water types are $Ca^{2+}$ for **fresh water** and $Na^+$ for **highly saline waters**, and Gibbs (1970) used variations in these two cations to establish compositional trends in world surface waters—see Fig. 3.1(a). He also demonstrated that the same general trends could be produced using variations in the principal anions in the two waters, i.e. $HCO_3^-$ for fresh water and $Cl^-$ for highly saline waters—see Fig. 3.1(b). By displaying the data in these two forms, Gibbs (1970) was able to produce a framework that could be used to characterize three end-member surface waters—see Fig. 3.1(c). These end-member waters were defined as follows:

1 A **precipitation- or rain-dominated** end-member, in which the total ionic content is relatively very low, and the $Na/(Na + Ca)$ and the $Cl/(Cl + HCO_3^-)$ ratios are both relatively high. Conditions that favour the formation of this end-member are low weathering intensity and low rates of evaporation.

2 A **rock-dominated** end-member, which is characterized by having an intermediate total ionic content and relatively low $Na/(Na + Ca)$ and $Cl/(Cl + HCO_3^-)$ ratios. This end-member is formed under conditions of high weathering intensity and low rates of evaporation.

3 An **evaporation–crystallization** end-member, which has a relatively very high total ionic content and also relatively high $Na/(Na + Ca)$ and $Cl/(Cl + HCO_3^-)$ ratios. Conditions that favour the formation of this end-member are high weathering intensity and high rates of evaporation.

Gibbs (1970) therefore was attempting to classify surface waters on the basis of the predominance of the principal external sources of the major ionic components, i.e. precipitation and rock weathering, and the operation of internal processes, such as evaporation and precipitation. The diagrams he produced, however, have received considerable criticism. For example, Feth (1971) suggested that the Pecos River, which was identified by Gibbs (1970) as belonging to the evaporation–crystallization end-member, has acquired its *major* increase in total dissolved salts from the inflow of groundwater brines; thus, it perhaps should be termed an evaporite end-member. Stallard & Edmond (1983) also demonstrated that the Amazon Basin rivers, which have relatively high total salt contents, have arisen primarily via the weathering of evaporites and carbonates.

**sodium** and **chloride** are the principal dissolved components contributing to the total ionic, i.e. salt, content. The major element composition of river water, however, is much more variable than that of sea water, and some idea of the extent of this variability can be seen from the data in Table 3.1. Maybeck (1981a) has ranked the global order of variability for the major dissolved constituents of river water as follows: $Cl^- > SO_4^{2-} > Ca^{2+} = Na^+ > Mg^{2+} > HCO_3^- > SiO_2 > K^+$. The major factors that control these variations are discussed below.

There are a number of types of water on the Earth's surface, which can be distinguished from each other on the basis of both their total ionic content (**salinity**) and the mutual proportions in which their various ions are present (**ionic ratios**). Gibbs (1970) used variations in both parameters to identify a number of

Stallard & Edmond (1981) also cast doubt on the existence of the rain-dominated end-member in the rivers of the Amazon Basin. By using chloride as a marine reference element, the authors were able to define the **cyclic salt** background for the Amazon surface waters, and drew the following conclusions from their data.

**1** Only $Na^+$, $Mg^{2+}$ and $SO_4^{2-}$, after $Cl^-$ the next three most abundant ions in sea water, exhibited significant cyclic contributions in any of the rivers in the basin. For a near-coastal river the cyclic source was dominant, and for one other river cyclic Na contributed ~50% to the surface water. For all other rivers, however, the cyclic Na and Mg was found to be minor compared with inputs from weathering.

**2** For the near-coastal river, ~15% of the Ca and K had a cyclic origin, but for all other rivers less than ~3% of the Ca and K had a marine origin. These estimates are considerably lower than those made by Gibbs (1970), who proposed that ~80% of the Na, K, Mg and Ca in the dilute lowland rivers of the Amazon Basin are cyclic.

These conclusions were confirmed in a later publication in which Stallard & Edmond (1983) concluded that it is rock weathering, and not precipitation input, which controls the major cation chemistry of the lowland rivers of the Amazon Basin, i.e. the rivers are *not* precipitation-dominated. Thus, the very existence of the rain-dominated river end-member was challenged.

It is apparent, therefore, that in the division of surface waters into the three end-members and the status of the rain-dominated and the evaporation–crystallization types must be seriously questioned. Nevertheless, there is no doubt that there are considerable variations in the total ionic content of river waters. This can be illustrated with respect to a number of individual river types (see Table 3.2), and in a general way the variations can be related to the Gibbs classification.

**1** Rivers with relatively small total ionic contents can be found:

(a) in catchments draining thoroughly leached areas of low relief where the rainfall is small, e.g. in some tropical regions of Africa and South America;

(b) in catchments that drain the crystalline shields, e.g. those of Canada, Africa and Brazil.

**Table 3.2** Average major element concentrations of rivers draining different catchment types; units, $mg\,l^{-1}$ (data from Maybeck, 1981a).

| Element | Rivers draining Canadian Shield | Mackenzie River; 'rock-dominated' end-member | Colorado River; 'evaporation–crystallization' end-member |
|---|---|---|---|
| $Na^+$ | 0.60 | 7.0 | 95 |
| $K^+$ | 0.40 | 1.1 | 5.0 |
| $Ca^{2+}$ | 3.3 | 33 | 83 |
| $Mg^{2+}$ | 0.7 | 10.4 | 24 |
| $Cl^-$ | 1.9 | 8.9 | 82 |
| $SO_4^{2-}$ | 1.9 | 36.1 | 270 |
| $HCO_3^-$ | 10.1 | 111 | 135 |
| TIC* | 18.9 | 207 | 694 |

* TIC = total ionic content.

For example, very 'pure' waters, with total ionic contents of ~$19\,mg\,l^{-1}$, are found on the Canadian Shield—see Table 3.2. It is waters such as these that will have their major ion composition most influenced by precipitation, even if they are not rain-dominated.

**2** As rock weathering becomes increasingly more important, the total ionic content of the river water increases. The Mackenzie River, which drains sedimentary and crystalline formations, is an example of a river having a rock-dominated water type. The average total ionic content of the Mackenzie River water is ~$200\,mg\,l^{-1}$, which is about an order of magnitude higher than that of the Canadian Shield rivers, and the concentration of $Ca^{2+}$ exceeds that of $Na^+$ by a factor of 4.7—see Table 3.2.

**3** Some river waters have relatively high total ionic contents and high Na/(Na + Ca) ratios. The Colorado River is an example of this type and has a total ionic content of ~$700\,mg\,l^{-1}$ and a $Na^+$ concentration slightly in excess of $Ca^{2+}$—see Table 3.2. It is probable, however, that the major ion composition of this river has been influenced more by the input of saline underground waters draining brine formations than by evaporation–crystallization processes.

Maybeck (1981a) took a global overview of the extent to which the three end-member waters are found on the Earth's surface. He concluded that the precipitation-dominated end-member (even if it exists at all) and the evaporation–crystallization (or evaporite) end-member together make up only around 2% only of the world's river waters, and that in fact ~98% of these surface waters are **rock-dominated** types.

Because the vast majority of the world's river waters belong to the rock-dominated category it is the extent to which the major rock-forming minerals are weathered, i.e. the influence of the **chemical composition** of the source rocks, that is the principal factor controlling the concentrations of the major ions in the waters. This can be illustrated with respect to variations in the major ion composition of rivers that drain a number of different rock types—see Table 3.3. From this table it can be seen, for example, that sedimentary rocks release greater quantities of $Ca^{2+}$, $Mg^{2+}$, $SO_4^{2-}$ and $HCO_3^-$ than do crystalline rocks. Maybeck (1981a) assessed the question of the chemical denudation rates of crustal rocks and concluded that:

1 chemical denudation products originate principally from sedimentary rocks, which contribute ~90% of the total products, ~66% being derived from carbonate deposits;

2 the relative rates at which the rocks are weathered follow the overall sequence evaporites $\gg$ carbonate rocks $\gg$ crystalline rocks, shales and sandstones.

These are general trends, however, and in practice the extent to which a crustal terrain is weathered depends on a complex of interrelated topographic and climatic factors.

**Table 3.3** Major ion composition of rivers draining different rock types; units, mg l$^{-1}$ (data from Maybeck, 1981a).

| Element | Plutonic and highly metamorphic rocks | Volcanic rocks | Sedimentary rocks |
|---------|---------------------------------------|----------------|-------------------|
| Na$^+$ | Lithological influence displaced by oceanic influence | | |
| K$^+$ | 1.0 | 1.5 | 1.0 |
| Ca$^{2+}$ | 4.0 | 8.0 | 30 |
| Mg$^{2+}$ | 1.0 | 3.0 | 8.0 |
| Cl$^-$ | Lithological influence displaced by oceanic influence | | |
| SO$_4^{2-}$ | 2.0 | 6.0 | 25 |
| HCO$_3^-$ | 15.0 | 45 | 100 |
| TIC* | 30 | 70 | 175 |

* TIC = total ionic content.

It is apparent, therefore, that river waters can be characterized on the basis of their major ionic constituents. The total concentrations, and mutual proportions, of these constituents are regulated by a variety of interrelated parameters. **Rock weathering**, however, is the principal control on the dissolved major element chemistry of the vast majority of the world's rivers, with regional variations being controlled by the lithological character of the individual catchment. The dissolved solid loads transported by rivers are correlated with mean annual run-off (see e.g. Walling & Webb, 1987), and although the *concentrations* of dissolved solids decrease with increasing run-off, as a result of a dilution effect, the *flux* of dissolved solids increases.

### 3.1.3.2 Trace elements

The sources that supply the major constituents to river waters (e.g. rock weathering, atmospheric deposition, pollution) also release trace elements into surface waters. Although relatively few reliable analyses have been reported for these trace constituents, a number of modern data sets have become available over the past few years and a summary of some of these is given in Table 3.4. The table includes the data set produced by Yeats & Bewers (1982). These authors made a compilation of the concentrations of dissolved trace elements in a number of major rivers and found, perhaps surprisingly, that there was a reasonable agreement between them. This led them to suggest that many of the observed differences in the composition of river water are probably a consequence of temporal variability, rather than being a result of major compositional variations. The factors that control the distributions of dissolved trace elements in large river systems, however, are considerably less well understood than those that govern the major constituents. The geology of a river catchment obviously will impose a fundamental constraint on the amounts of trace elements available for mobilization and transport. Despite this, some of the data available suggest that the relatively clear-cut rock–water chemistry relationships found for the major constituents do not apply universally to the trace elements. For example, in their study of the Mackenzie River system, Reeder *et al.* (1972) were unable to find any clear relationships between the dis-

**Table 3.4** Recent data on the concentrations of some dissolved elements in river water.

| Element | Data source* | | | | | | | | | | | | | | | | | | |
|---|---|---|---|---|---|---|---|---|---|---|---|---|---|---|---|---|---|---|---|
| | 1 | 2 | 3 | 4 | 5 | 6 | 7 | 8 | 9 | 10 | 11 | 12 | 13 | 14 | 15 | 16 | 17 | 18 | 19 |
| Fe ($\mu$g l⁻¹) | 55 | 20–75 | 5–60 | — | — | 130 | — | 4.9 | 30 | — | — | — | 30 | — | — | — | 125 | 40 | 40 |
| Mn ($\mu$g l⁻¹) | 6.3 | — | — | — | — | 8.3 | — | 10 | — | — | — | — | 19 | — | — | — | 12 | 8.2 | 8.2 |
| Al ($\mu$g l⁻¹) | 64 | — | — | — | — | 36 | — | — | — | — | — | — | 40 | — | — | — | 224 | 50 | 50 |
| Cd (ng l⁻¹) | 111 | 9–25 | 11–25 | 25–38 | — | — | — | 90 | — | — | — | 390 | — | — | — | — | 200 | — | 50† |
| Cu ($\mu$g l⁻¹) | 2.5 | 1.1–1.4 | 1.0–1.3 | 1.0–1.3 | 1.5 | 0.3 | — | 1.9 | — | — | 1.1 | 6.3 | 1.8 | — | — | — | 4.8 | 1.5 | 1.5 |
| Ni ($\mu$g l⁻¹) | 1.5 | 0.7–0.9 | 0.7–1.3 | 0.23 | 0.29 | — | — | 1.6 | — | 0.5 | — | — | — | — | — | — | 2.1 | 0.5 | 0.5 |
| Pb ($\mu$g l⁻¹) | — | 0.09–0.2 | 0.08–0.2 | — | — | — | — | — | — | — | — | — | — | — | — | — | — | 0.1 | 0.1 |
| Zn ($\mu$g l⁻¹) | 8.6 | 6–7 | 6–7 | — | — | — | — | 10 | — | — | — | 54 | — | — | — | 0.39 | 21 | 30 | 0.39 |
| Cr ($\mu$g l⁻¹) | — | — | — | — | — | — | 0.7 | — | — | — | — | — | — | — | — | — | — | 1.0 | 1.0 |
| Co ($\mu$g l⁻¹) | 0.15 | — | — | — | — | — | — | — | — | — | — | — | 0.06 | — | — | — | 0.14 | 0.2 | 0.2 |
| Ge (ng l⁻¹) | — | — | — | — | — | — | — | — | — | — | — | — | — | 8.1 | — | — | — | — | — |
| Sn (ng l⁻¹) | — | — | — | — | — | — | — | — | — | — | — | — | — | — | 1.4 | — | — | — | — |
| V ($\mu$g l⁻¹) | — | — | — | — | — | — | — | — | — | — | — | — | — | — | — | — | — | 1.0 | 1.0 |

* 1, St Lawrence river (Yeats & Bewers, 1982); 2, Gota river (Danielsson et al., 1983); 3, Nodre river (Danielsson et al., 1983); 4, Changjiang river (Edmond et al., 1985); 5, Amazon river (Boyle et al., 1982); 6, Zaire river (Maybeck, 1978); 7, St Lawrence estuary (Campbell & Yeats, 1984); 8, Mississippi river (Trefry & Presley, 1976); 9, Rhine river (Eisma, 1975); 10, Amazon river (Sclater et al., 1976); 11, Amazon river (Boyle et al., 1982); 12, Rhine river (Duinker & Kramer, 1977); 13, Amazon river (Gibbs, 1972, 1977); 14, Average inorganic Ge; remote, clean rivers (Froelich et al., 1985); 15, Geometric average 45 rivers (Byrd & Andreae, 1986); 16, Average South American and North American rivers (Shiller & Boyle, 1985); 17, Average of the data set given by Yeats & Bewers (1982); 18, Average global river water (Martin & Whitfield, 1983); 19, Average used in present work for flux calculations.
† An attempt to estimate a global average for Cd assuming a Cd concentration of 200 ng l⁻¹ for run-off from North America and Europe (~10% global run-off) and a concentration of 30 ng l⁻¹ for the remaining run-off.

tributions of Ni, Cu and Zn and rock lithology, thus giving an indication that factors in addition to relatively simple rock–water chemistry relationships control the concentrations of dissolved trace elements in river waters. These factors are related to the dissolved–particulate speciation of the elements and are influenced by the nature of the weathering solutions. The leaching mechanisms involved in the weathering of continental rocks are controlled mainly by carbonic acid and organic acids produced by biological activity, and to a lesser extent by mineral acids, the latter being enhanced in areas that receive acid rain. If organic acid leaching predominates, the solubility of many trace elements will be increased both as a result of the formation of complexes with organic ligands and by the stabilization of metal-containing colloids by organic material (GESAMP, 1987). For example, Windom & Smith (1985) gave data showing that the concentrations of dissolved Fe, Zn, Pb and Cd in river water increase as the concentration of dissolved organic carbon increases. Further evidence that factors other than simple rock–water relationships affect the dissolved concentrations of trace elements in rivers also has been provided by Shiller & Boyle (1985), who applied well tested modern analytical

techniques to the determination of dissolved Zn in a number of rivers draining both pristine and anthropogenically influenced catchments. The data reported by these authors, some of which are given in Table 3.4, showed that, in river systems which have suffered relatively little anthropogenic perturbation, dissolved Zn concentrations are typically only ~390 ng l⁻¹, whereas in systems influenced by anthropogenic inputs the concentrations are one to two orders of magnitude higher (see Table 3.4). One of the most significant findings to emerge from the study was that in the pristine rivers there was evidence of a degree of dependence between the dissolved Zn concentrations and the pH (acidity) of the waters, alkaline rivers being depleted in dissolved Zn relative to acidic systems. The authors concluded that this pH dependence did not reflect source rock composition, but was more likely to be chemical in nature, resulting from the adsorption of Zn on to, or its desorption from, suspended particulate matter. Fluvial dissolved Zn concentrations therefore were thought to be controlled by reversible adsorption–desorption reactions; however, as the concentration of dissolved organic carbon is related to pH, the increased trace element concentrations associated with decreasing pH could result from complexation with organic matter. A dependence of dissolved trace element concentrations on pH also has been suggested for Be in river waters by Measures & Edmond (1983), who showed that the mobility of the element was a strong function of pH, with acid streams (pH < 6) being strongly enriched with dissolved Be compared with alkaline carbonate rivers, in which the element had been flocculated or adsorbed on to particulate matter. Higher dissolved Fe concentrations are also found in rivers with relatively high pH values. The rock–water chemistry relationship, however, can also exert a control on the fluvial concentrations of trace elements. For example, Measures & Edmond (1983) reported data showing that, in addition to the pH dependence described above for Be, there was also a first-order separation between streams draining siliceous rocks ($^9$Be > 9 ng l$^{-1}$) and those draining uplifted Andean belts dominated by marine sediments ($^9$Be < 9 ng l$^{-1}$). The rock–water chemistry relationship was also demonstrated in a study reported by Froelich *et al.* (1985). These authors presented data on the concentrations of inorganic Ge in 56 rivers, which included seven of the largest in the world. Germanium concentrations averaged ~8 ng l$^{-1}$ in 'clean' rivers, compared with ~136 ng l$^{-1}$ in polluted systems, which again demonstrates the effect that anthropogenic inputs can have on fluvial trace element levels. The inorganic Ge exhibited a silicon-like behaviour pattern during continental weathering and the average naturally weathered fluvial flux carried a Ge:Si atom ratio signal of ~$0.7 \times 10^{-6}$, which is close to that in average continental granites. This rock–water chemistry relation, however, was perturbed in rivers that had suffered contaminant inputs, where the Gi:Si ratios could be up to 10 times higher than the natural background. The effect of anthropogenic inputs on the concentrations of trace elements in river waters has also been demonstrated for dissolved Sn by Byrd & Andreae (1986); these authors reported that the concentrations of dissolved Sn ranged from as low as ~0.2 ng l$^{-1}$ in pristine rivers to as high as ~500 ng l$^{-1}$ in very polluted systems, with an average concentration of 1.3 ng l$^{-1}$ in global river water.

It may be concluded, therefore, that the concentrations of many dissolved trace elements in river waters are influenced by several factors. These include:

1 geology of the river catchment;
2 chemical constraints within the aqueous system itself, chiefly particulate–dissolved equilibria, which can involve both inorganic and organic suspended solids (including biota) and are influenced by factors such as pH and the concentrations of complexing ligands;
3 anthropogenic inputs.

### 3.1.4 Major and trace elements: the particulate river signal

In the present context, river particulate material (RPM) refers to solids carried in suspension in the water phase, i.e. the suspended sediment load. River particulate material consists of a variety of components dispersed across a spectrum of particle sizes. These components include the following: primary aluminosilicate minerals, e.g. feldspars, amphiboles, pyroxenes, micas; secondary aluminosilicates, e.g. the clay minerals; quartz; carbonates; hydrous oxides of Al, Fe and Mn; and various organic components. In addition to the discrete oxides and organic solids, many of the individual suspended particle surfaces are coated with hydrous Mn and Fe oxides and/or organic substances.

The mineral composition of RPM represents that of fairly homogenized soil material from the river basin, and as a result each river tends to have an individual RPM mineral signature. This was demonstrated by Konta (1985), who gave data on the distributions of crystalline minerals in RPM from 12 major rivers. The results of the study may be summarized as follows.
1 Clay minerals, or sheet silicates, were the dominant crystalline components of the RPM, although the distributions of the individual minerals differed. Mica–illite minerals were the principal sheet silicates present and were found in all the RPM samples. Kaolinite was typically found in higher concentrations in RPM from tropical river systems where weathering intensity is relatively high, e.g. the Niger and the Orinoco. Chlorite was found in highest concentrations in kaolinite-poor RPM, and tended to be absent in RPM from rivers in tropical or subtropical areas of intense chemical weathering. Montmorillonite was found only in RPM from some tropical and subtropical rivers.

2 Significant quantities of quartz were present in RPM from all the rivers except one.

3 Other crystalline minerals found as components of RPM included acid plagioclase, potassium feldspar and amphiboles.

4 Calcite and/or dolomite were reported in RPM from seven of the rivers, but it was not known if these minerals were detrital or secondary in nature.

The crystalline components of RPM therefore are dominated by the clay minerals, and the distribution of these minerals reflects that in the basin soils, which itself is a function of source-rock composition and weathering intensity. As a result, the clays in RPM have a general latitudinal dependence; e.g. kaolinite has its highest concentrations in RPM from tropical regions. This imposition of latitudinal control on the distribution of clay minerals in soils is used in Section 15.1.1 to trace the dispersion of continentally derived solids throughout the oceans. The mineral composition of RPM, however, also is dependent on particle size. For example, the size distribution of RPM transported by the Amazon is illustrated in Fig. 3.2, and demonstates that whereas quartz and feldspar are found mainly in the >2 μm diameter fraction, mica and the clay minerals, kaolinite and montmorillonite are concentrated in the <2 μm fraction. This mineral size fractionation has important consequences for the 'environmental reactivity' of elements transported via RPM, and this topic is considered below when elemental partitioning among the various components of the solids is discussed. For a 'first look', however, the chemical composition of RPM will be described in terms of the total samples.

Martin & Maybeck (1979) have provided an estimate of the global chemical composition of the total samples of river particulate material. This estimate was based on the determination of 49 elements in RPM from a total of 20 of the world's major rivers, and as there are considerably fewer problems involved in the analysis of particulate than of dissolved trace elements, it provides a generally reliable compositional database for RPM. Highly polluted river systems were excluded from the study, and the findings therefore refer to RPM that has its composition controlled largely by natural processes. Under these conditions the RPM is derived mainly from the surface soil cover following the mechanical and chemical weathering of the surficial parent rocks. A partial data set for the chemical composition of RPM is given in Table 3.5, in order to indicate the variability between suspended solids in different rivers. On the basis of data such as these, Martin & Maybeck (1979) were able to identify a number of important trends in the chemistry of RPM.

1 For individual river systems the concentrations of the major elements in the RPM are not usually highly variable. For trace elements, the variations are much larger; however, they do not exceed one order of magnitude, and are much smaller than those found between different rivers.

2 Inter-river, i.e. geographical, variability in the composition of RPM differs considerably from one element to another. On the basis of the coefficient of variation ($s/x$) in the RPM, the elements were divided into a number of groups:

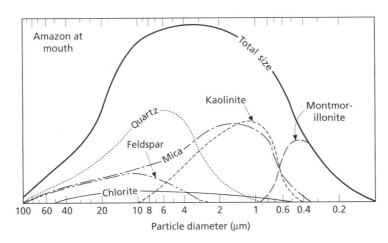

**Fig. 3.2** The size distribution of mineral phases transported by the Amazon River (from Gibbs, 1977).

**Table 3.5** The concentrations of some elements in RPM from selected individual rivers (data from Martin & Maybeck, 1979).

| River | Element ($\mu g\,g^{-1}$) | | | | | | | | | |
|---|---|---|---|---|---|---|---|---|---|---|
| | Al | Fe | Ca | Mn | Ni | Co | Cr | V | Cu | Zn |
| Amazon | 115 000 | 55 000 | 16 000 | 1030 | 105 | 41 | 193 | 232 | 266 | 426 |
| Colorado | 43 000 | 23 000 | 34 000 | 430 | 40 | 17 | 82 | — | — | — |
| Zaire | 117 000 | 71 000 | 8400 | 1400 | 74 | 25 | 175 | 163 | — | 400 |
| Ganges | 77 000 | 37 000 | 26 500 | 1000 | 80 | 14 | 71 | — | 30 | 163 |
| Garonne | 118 000 | 58 000 | 19 500 | 1700 | 33 | 39 | 255 | 150 | 51 | 874 |
| Mackenzie | 78 000 | 36 500 | 35 800 | 600 | 22 | 14 | 85 | — | 42 | 126 |
| Mekong | 112 000 | 56 000 | 5900 | 940 | 99 | 20 | 102 | 175 | 107 | 300 |
| Niger | 156 000 | 92 000 | 3300 | 650 | 120 | 40 | 150 | 180 | 60 | — |
| Nile | 98 000 | 108 000 | 40 000 | — | — | — | — | — | 39 | 93 |
| St Lawrence | 78 000 | 48 500 | 23 000 | 700 | — | — | 270 | — | 130 | 350 |

(a) those that are most consistent ($s/x < 0.2$), e.g. Si, Th and the rare earths Ce, Eu, La, Sm and Yb;
(b) those that exhibit slight variability ($s/x = 0.2$–$0.35$), e.g. Al, Fe, K, Hf, Sc, Ta and U;
(c) those that exhibit moderate variability ($s/x = 0.35$–$0.70$), e.g. Ag, Cr, Mg and Na;
(d) those that are highly variable ($s/x > 0.70$), e.g. Ca, Cs, Cu, Li, Mo, Ni, Pb, Sr and Zn.

The variability for some of the major elements in the RPM can be related to climatic–weathering intensity conditions in the river catchments. Many *tropical* rivers have large areas in their drainage basins in which the rate of mechanical erosion is generally low, and the RPM originates mainly from highly developed soil material that has undergone chemical weathering, i.e. **transport-limited** regimes. The RPM in this type of river is enriched in those elements that generally are relatively insoluble during chemical weathering, e.g. Al, Ti and Fe, and is depleted in the more soluble elements, which are leached into the weathering solutions, e.g. Na and Ca. In contrast, many *temperate* and *arctic* rivers have drainage basins in which mechanical erosion can greatly exceed chemical weathering, i.e. **weathering-limited** regimes. As a result, the parent material of the RPM in these rivers is either original rock debris or poorly weathered soils. Relative to the world average RPM, that found in temperate and arctic rivers tends to be depleted in Al, Ti and Fe, and enriched in Na and Ca, and its overall composition is closer to that of fresh rock than the RPM from tropical rivers.

It is apparent, therefore, that there is a considerable variability in the concentrations of some elements in RPM from different rivers, even when the material is derived from mainly natural sources. Nonetheless, the data for RPM given in Table 3.5 do offer an indication of chemical composition of the crust-derived solid material that is brought to the ocean margins by river transport. A more complete data set for the chemical composition of RPM is given in Table 3.6, together with the compositions of the continental rock and soil source materials and dissolved river constituents.

Up to this point we have described the elemental chemistry of RPM in terms of **total sample** composition. It was pointed out above, however, that RPM consists of a variety of individual components, which are present in a range of particle sizes. The elements in suspended particulates, and also in deposited sediments, are partitioned between these individual host components, some of which can bind them more strongly than others. In this respect, it is important to make a fundamental distinction between two genetically different element–host associations:

1 elements associated with the crystalline mineral matrix (the **detrital** or **residual** fraction), which are in an environmentally immobile form and largely have their concentrations fixed at the site of weathering;
2 elements associated with non-crystalline material (the **non-detrital** or **non-residual** fraction), which are in environmentally mobile forms and have their concentrations modified by dissolved ↔ particulate reactivity.

**Table 3.6** The average compositions of crustal rocks, soils, and dissolved and particulate river material (data from Martin & Whitfield, 1983; after Martin & Maybeck, 1979). Values given in parentheses are approximate.

| | Continents | | Rivers | | | Continents | | Rivers | |
|---|---|---|---|---|---|---|---|---|---|
| | Rock ($\mu g\,g^{-1}$) | Soils ($\mu g\,g^{-1}$) | Dissolved ($\mu g\,l^{-1}$) | Particulate ($\mu g\,g^{-1}$) | | Rock ($\mu g\,g^{-1}$) | Soils ($\mu g\,g^{-1}$) | Dissolved ($\mu g\,l^{-1}$) | Particulate ($\mu g\,g^{-1}$) |
| Ag | 0.07 | 0.05 | 0.3 | 0.07 | Mg | 16400 | 5000 | 3100 | 11800 |
| Al | 69300 | 71000 | 50 | 94000 | Mn | 720 | 1000 | 8.2 | 1050 |
| As | 7.9 | 6 | 1.7 | 5 | Mo | 1.7 | 1.2 | 0.5 | 3 |
| Au | 0.01 | 0.001 | 0.002 | 0.05 | Na | 14200 | 5000 | 5300 | 7100 |
| B | 65 | 10 | 18 | 70 | Nd | 37 | 35 | 0.04 | 35 |
| Ba | 445 | 500 | 60 | 600 | Ni | 49 | 50 | 0.5 | 90 |
| Br | 4 | 10 | 20 | 5 | P | 610 | 800 | 115 | 1150 |
| Ca | 45000 | 15000 | 13300 | 21500 | Pb | 16 | 35 | 0.1 | 100 |
| Cd | 0.2 | 0.35 | 0.02 | (1) | Pr | 9.6 | — | 0.007 | (8) |
| Ce | 86 | 50 | 0.08 | 95 | Rb | 112 | 150 | 1.5 | 100 |
| Co | 13 | 8 | 0.2 | 20 | Sb | 0.9 | 1 | 1 | 2.5 |
| Cr | 71 | 70 | 1 | 100 | Sc | 10.3 | 7 | 0.004 | 18 |
| Cs | 3.6 | 4 | 0.035 | 6 | Si | 275000 | 330000 | 5000 | 285000 |
| Cu | 32 | 30 | 1.5 | 100 | Sm | 7.1 | 4.5 | 0.008 | 7 |
| Er | 3.7 | 2 | 0.004 | (3) | Sr | 278 | 250 | 60 | 150 |
| Eu | 1.2 | 1 | 0.001 | 1.5 | Ta | 0.8 | 2 | <0.002 | 1.25 |
| Fe | 35900 | 40000 | 40 | 48000 | Tb | 1.05 | 0.7 | 0.001 | 1.0 |
| Ga | 16 | 20 | 0.09 | 25 | Th | 9.3 | 9 | 0.1 | 14 |
| Gd | 6.5 | 4 | 0.008 | (5) | Ti | 3800 | 5000 | 10 | 5600 |
| Hf | 5 | — | 0.01 | 6 | Tm | 0.5 | 0.6 | 0.001 | (0.4) |
| Ho | 1.6 | 0.6 | 0.001 | (1) | U | 3 | 2 | 0.24 | 3 |
| K | 24400 | 14000 | 1500 | 20000 | V | 97 | 90 | 1 | 170 |
| La | 41 | 40 | 0.05 | 45 | Y | 33 | 40 | — | 30 |
| Li | 42 | 25 | 12 | 25 | Yb | 3.5 | — | 0.004 | 3.5 |
| Lu | 0.45 | 0.4 | 0.001 | 0.5 | Zn | 127 | 90 | 30 | 250 |

A number of elemental associations are usually identified within the non-detrital fraction itself, e.g. an exchangeable fraction, a carbonate-associated fraction, a metal oxide-associated fraction, an organic-matter-associated fraction. A variety of techniques have been used to establish the partitioning of elements among these host components, and one of the most common involves the sequential leaching of the primary sample with a number of reagents that are designed progressively to take into solution elements associated with the individual hosts. Such sequential leaching techniques are open to a number of severe criticisms (see e.g. Chester, 1988), not the least of which is that the host fractions identified are operationally defined in terms of the technique used, and so are not necessarily analogues of natural binding fractions. However, when the various constraints on their use are recognized, sequential leaching techniques can provide important data on the manner in which elements are partitioned in materials such as suspended river particulates, sediments and aerosols.

A number of studies have been carried out on the partitioning of elements in RPM. Outstanding among these are the pioneer investigations reported by Gibbs (1973, 1977) on suspended particulates from the Amazon and Yukon rivers. These two rivers were selected because they are large, relatively unpolluted systems representative of the major rivers that have their catchments in tropical (Amazon) and subarctic (Yukon) land masses and drain a wide variety of rock types. In both studies, Gibbs gave data on the partitioning of a number of elements in river-transported phases. To do this, he distinguished between: elements in *solution* (i), and those associated with *particulates* in an ion-exchange phase (ii), a precipitated metallic oxide coating phase (iii), an organic matter phase (iv), and a crystalline (detrital or residual) phase (v). A summary of the partitioning data reported by Gibbs (1973) is given in Table 3.7. A number of general conclusions can be drawn from these data.

**1** The fraction of the total amounts of the metals transported in solution ranges from <1% for Fe to ~17% for Mn, but solution transport is not dominant for any of the metals.

**2** The partitioning signatures of Fe, Mn, Ni, Cu, Co and Cr are similar in RPM from both the tropical and the subarctic river systems.

**3** Cu and Cr are transported mainly in the crystalline particulate phase.

**4** Mn is transported principally in association with the precipitated metal oxide particulate phase.

**5** Most of the total amounts of Fe, Ni and Co are partitioned between the precipitated metal oxide and crystalline particulate phases.

**6** The non-crystalline carrier phases represent the particulate metal fraction which is most readily available to biota, and the most important of these carriers are the metal oxide coatings.

**7** There is a particle-size–concentration relationship for a number of elements in the RPM, with concentrations of Mn, Fe, Co, Ni and Cu all increasing dramatically with decreasing particle size.

It may be concluded, therefore, that in river systems that receive their supply of trace elements mainly from natural sources, crystalline, metal oxide and organic host fractions are the principal metal carriers. It is the non-crystalline phases that are the most readily environmentally available, and the proportions of particulate trace metals associated with these phases generally increase in river systems that receive inputs from pollutant sources. The various total element, solid-state partitioning, mineral and particle size data for RPM can be combined to develop the general concept that rivers transport two different types of suspended solids:

**1** a **trace-element-poor**, large-sized ($\geq 2\,\mu m$ diameter) fraction that consists mainly of crustal minerals such as quartz and the feldspars, a large proportion of the trace elements in this fraction being in crystalline solids and generally environmentally immobile;

**2** a **trace-element-rich**, small-sized ($\leq 2\,\mu m$ diameter), surface-active fraction that consists largely of clay minerals, organic matter and iron and manganese oxide surface coatings, a large proportion of the trace elements in this fraction being associated with non-crystalline carriers (e.g. oxide coatings and organic phases) and environmentally available.

It must be stressed that there is no sharp division between these two fractions, but the distinction between them is extremely important both (i) from the point of view of differential transport, i.e. the small-sized material can be carried for a longer period in suspension, and (ii) because it is the small-sized, non-crystalline, environmentally available trace elements that undergo the dissolved $\leftrightarrow$ particulate equilibria that play such an important role in controlling

**Table 3.7** The percentage partitioning of elements in RPM (data from Gibbs, 1973).

| River | Transport phase | Element | | | | | |
|---|---|---|---|---|---|---|---|
| | | Fe | Ni | Co | Cr | Cu | Mn |
| Amazon | Solution | 0.7 | 2.7 | 1.6 | 10.4 | 6.9 | 17.3 |
| | Ion exchange | 0.02 | 2.7 | 8.0 | 3.5 | 4.9 | 0.7 |
| | Metal oxide coating | 47.2 | 44.1 | 27.3 | 2.9 | 8.1 | 50 |
| | Organic matter | 6.5 | 12.7 | 19.3 | 7.6 | 5.8 | 4.7 |
| | Crystalline matrix | 45.5 | 37.7 | 43.9 | 75.6 | 74.3 | 27.2 |
| Yukon | Solution | 0.05 | 2.2 | 1.7 | 12.6 | 3.3 | 10.1 |
| | Ion exchange | 0.01 | 3.1 | 4.7 | 2.3 | 2.3 | 0.5 |
| | Metal oxide coating | 40.6 | 47.8 | 29.2 | 7.2 | 3.8 | 45.7 |
| | Organic matter | 11.0 | 16.0 | 12.9 | 13.2 | 3.3 | 6.6 |
| | Crystalline matrix | 48.2 | 31.0 | 51.4 | 64.5 | 87.3 | 37.1 |

the river → estuarine → coastal sea → open ocean transport–deposition cycles of many elements and in regulating the composition of sea water. The non-residual elements can undergo changes between the various solid-state carrier phases themselves. All of these particulate ↔ dissolved and solid-state changes are sensitive to variations in environmental parameters (e.g. the concentrations of particulate material and complexing ligands, redox potential, pH), so that during the global transportation cycle the mobile surface-associated elements can undergo considerable speciation migration.

### 3.1.5 Organic matter and nutrients

#### 3.1.5.1 Dissolved organic carbon (DOC)

A compilation of DOC concentration data for a number of rivers, both large and small, is given in Table 3.8, from which it can be seen that the concentrations range from as low as <1 mg l$^{-1}$ to ~50 mg l$^{-1}$. A number of authors have found a climate-related pattern in DOC concentrations in rivers. For

example, Thurman (1985) listed the following estimates for the average, and ranges (in parentheses), of DOC concentrations (mg l$^{-1}$) in a number of climatic zones:

| | |
|---|---|
| small rivers in Arctic and alpine environments | 2 (1–5) |
| taiga | 10 (8–25) |
| cool temperate | 3 (2–8) |
| warm temperate | 7 (3–15) |
| arid | 3 (2–10) |
| wet tropical | 6 (2–15) |
| rivers draining swamps and wetlands | 25 (5–60) |

Thus, the lowest values are found for rivers draining glacial and alpine environments and the highest for those draining swamp and wetland regions. These general trends are also reflected in the particulate organic carbon (POC) data given in Table 3.8.

The DOC in river waters originates mainly from three general sources:

1 phytosynthesis associated with fluvial production (autochthonous, mainly low molecular weight labile material);

**Table 3.8** The concentrations of DOC and POC in some rivers. Some of the data sets for both DOC and POC have been obtained from restricted river sections; further, they do not take account of seasonal fluctuations or particle size effects (POC). The data therefore should be regarded as offering no more than general indications of the DOC and POC concentrations to be found in rivers.

| Continent | River system | Classification* | DOC (mg l$^{-1}$) | POC (mg l$^{-1}$) | Continent | River system | Classification | DOC (µg g$^{-1}$) | POC (µg g$^{-1}$) |
|---|---|---|---|---|---|---|---|---|---|
| North America | North Dawes | — | 0.5† | — | Europe | Alpine streams | Rocky mountains | 1.5–5† | — |
| | Yukon | Taiga | 8.8‡ | 1.2‡ | | Ems–Dollart | — | 7–18† | — |
| | Mackenzie | Taiga | 4.5‡ | 3.2‡ | | Severn | — | 3.1–7.8† | — |
| | St Lawrence | Rocky Mountain | 3–5§ | 0.48‡ | | Rhone | Rocky mountains | 1.4¶ | 0.9¶ |
| | Mississippi | Temperate | 3.4–6.0† | 1.4‡ | Africa | Zaire | Tropical wet | 8.5** | 1.1** |
| | Missouri | Temperate | 1.9–9.0† | 20¶ | | Niger | — | 3.5** | 3.7** |
| | Sopchoppy | Temperate swamp | 6–52† | 1.6¶ | | Orange | — | 2.3** | 0.9** |
| | Satilla | Temperate swamp | 25–30† | — | | Gambia | — | 2.4** | 1.1** |
| South America | Amazon | Tropical wet | 3.5–6.5¶ | 1.1–8.2¶ | Asia | Yangsekiang | — | 5–23§ | — |
| | Orinoco | — | 2–5§ | <1–12‖ | | Brahmaputra | — | 1–6§ | 1–7†† |
| | Uruguay | — | 2–8§ | — | | Ganges | — | 1–9§ | 0.6–2.3†† |
| | Parana | — | <1–50‖ | <1–15‖ | | Indus | — | 2–22§ | 2.2†† |

* From Maybeck (1981b). † Mantoura & Woodward (1983). ‡ Telang *et al.* (1991). § Spitzy & Leenheer (1991). ¶ Maybeck (1981b). ‖ Depetris & Paolini (1991). ** Martins & Probst (1991). †† Subramanian & Ittekkot (1991).

2 the leaching of soils where the organic material is derived from plant and animal material via microbial activity (allochthonous, mainly high molecular weight refractory material);

3 anthropogenic inputs.

In non-polluted river systems the DOC pool thus contains organic matter synthesized and degraded in both the terrestrial (allochthonous) and aqueous (autochthonous) environments. According to Ertel *et al.* (1986) between ~40 and ~80% of fluvial DOC consists of combined humic substances, which generally are considered to be refractory material that can escape degradation in the fluvial–estuarine environment and so reach the ocean. Ertel *et al.* (1986) gave data on the humic and fulvic acid components of the DOC humic fraction in the Amazon River system. Both these components contain lignin (a phenolic polymer unique to vascular plants) and appear to be formed from the same allochthonous source material, but they differ in the extent to which they have suffered biodegradation in the soil, with fulvic acids being more oxidized than humic acids. In the Amazon system the fulvic acids do not undergo reactions with suspended particles and behave in a conservative manner. In contrast, humic acids can be adsorbed on to particle surfaces and also can undergo flocculation in the estuarine environment (see Sections 3.2.5 & 3.2.6). In view of this, Ertel *et al.* (1986) suggested that humic acids may not contribute carbon, or lignin, to the oceans, and that it is fulvic acids that represent the major proportion of the DOC input to sea water. The autochthonous DOC includes potentially very labile (metabolizable) biochemical material, such as proteins and carbohydrates, and less labile components, such as lipids and pigments; much of this labile material is likely to be degraded in the river or estuarine system and so will not escape into the open ocean (see also the refractory–labile classification of POC below).

### 3.1.5.2 *Particulate organic carbon (POC)*

Particulate organic carbon concentrations in a number of world rivers are given in Table 3.8, and lie in the range ~0.5 to ~20 mg l$^{-1}$, although higher values can be found in those rivers draining salt marshes and those that receive relatively large inputs of sewage and industrial wastes. Fluvial POC consists of living (i.e. bacteria and plankton) and non-living (detritus) fractions. In general, there is a decrease in fluvial POC with logarithmically increasing concentrations of total suspended material (TSM), which results from a reduction in primary productivity as a result of decreased light penetration arising from the presence of high suspended matter loads, and a dilution with mineral matter. It is apparent from Table 3.8 that rivers carry considerably more DOC than POC. The ratio between DOC and POC, however, also changes with increasing concentrations of TSM, varying from 10.8 at TSM concentrations <15 mg l$^{-1}$ to <1 at concentrations >500 mg l$^{-1}$ (Ittekkot & Laane, 1991). It is necessary to distinguish between **refractory** (non-metabolizable) and **labile** (metabolizable, or degradable) fractions of river-transported POC. This is an important distinction because, although the refractory fractions will survive microbial attack, the labile fractions are an important food source for organisms and can be lost within rivers, estuaries and the sea. Degens & Ittekkot (1985) used carbohydrates and proteins (amino acids; see Section 9.2.3.1), which are relatively labile, to identify the degradable fraction of fluvial POC, and concluded that between ~5% and ~30% of the POC carried by the world's rivers is labile in character. This estimate was later modified by Ittekkot & Laane (1991) to between 12% and 47%, with proteins (range, ~7–29%) exceeding carbohydrates (range, 5–17%). Ittekkot & Laane (1991) pointed out that labile POC decreased as the concentrations of TSM increased, with TSM in the range 1–150 mg l$^{-1}$ having an average labile content of ~35%, whereas for those with concentrations >150 mg l$^{-1}$ the labile fraction falls to ~15%. Taking into account the fact that the maximum transport of fluvial TSM occurs in the concentration range 500–1500 mg l$^{-1}$, Ittekkot & Laane (1991) estimated that globally ~17% of fluvially transported POC is labile in character. The labile fraction of POC, however, does not consist only of proteins and carbohydrates, and to take account of this Ittekkot (1988) doubled the labile fraction and obtained a global estimate of 35% for the labile fluvial POC fraction. The remainder of the fluvial POC, which is highly degraded, or refractory, in character, can escape to the oceans and so represents a significant source of organic carbon to marine sediments.

*3.1.5.3 The nutrients*

Nitrate, phosphates and silicate in river water are each derived from different sources.

*Nitrate.* Nitrate ($NO_3^-$), which originates mainly from soil leaching, terrestrial run-off (including that from fertilized soils) and waste inputs, is the most abundant stable inorganic species of nitrogen in well-oxygenated waters, but dissolved organic nitrogen may dominate in humid tropical and subarctic rivers. River water also contains particulate nitrogen, which is mainly biological in origin. The average concentration of total dissolved nitrogen in a wide range of unpolluted river systems has been estimated by Maybeck (1982) to be 375 µg l$^{-1}$, of which 115 µg l$^{-1}$ is present as dissolved inorganic species (very largely nitrate) and 260 µg l$^{-1}$ is in the form of dissolved organic species. Fewer data are available for the concentration of particulate nitrogen in river water, but Maybeck (1981a) estimated a value of ~560 µg l$^{-1}$ as a global river average. Biological reactions dominate the transfer of inorganic dissolved nitrogen to the particulate phase.

*Phosphates.* Phosphorus is present in river waters in dissolved and particulate forms. The dissolved phosphorus is mainly orthophosphate (principal species $HPO_4^{2-}$), together with dissolved organic phosphate and, in polluted systems, polyphosphate. According to Maybeck (1982) the global average river-water concentration of total dissolved phosphorus is 28 µg l$^{-1}$, and that of total particulate phosphorus is 530 µg l$^{-1}$, of which 320 µg l$^{-1}$ is in an inorganic form and 210 µm l$^{-1}$ is in an organic form. The sources of dissolved phosphate ($PO_4^{3-}$) in river waters include the weathering of crustal minerals (e.g. aluminium orthophosphate, apatite) and anthropogenic inputs (e.g. from the oxidation of urban and agricultural sewage and the breakdown of polyphosphates used in detergents). Dissolved phosphorus is removed during biological production, and is often considered to be the limiting nutrient in river systems. In addition, however, the concentrations of dissolved inorganic phosphorus in river waters are affected significantly by chemical processes involved in mineral–water equilibria, e.g. those involving adsorption on to phases such as clay minerals and ferric hydroxides.

*Silicate.* Dissolved reactive silicate is present in river waters almost exclusively as silicic acid ($H_4SiO_4$), and is derived mainly from the weathering of silicate and aluminosilicate minerals; i.e. unlike nitrogen and phosphorus, anthropogenic sources play a relatively minor role in the supply of dissolved silicon to rivers. Dissolved silicon is a major constituent of river water, making up ~10% of the total dissolved solids, and its global average concentration has been estimated to be 4.85 mg l$^{-1}$ (Maybeck, 1979). Silicon is also present in river water in a variety of particulate forms, which include inorganic minerals (e.g. quartz, aluminosilicates) and biological material (e.g. the opaline skeletons of diatoms).

### 3.1.6 Pollutant inputs

In addition to the mainly natural processes that have been discussed above, the input of pollutants, via industrial and domestic wastes, can exert a significant influence on the dissolved major and trace element compositions of some river waters, especially those of northern Europe and the USA. There are many examples of this in the literature, but a survey carried out on the Rhine will serve as an illustration. This river rises in the Swiss Alps, where it receives its water from a series of catchments that are relatively free from human influence, and subsequently flows through very heavily populated and industrialized areas before reaching the land–sea boundary. Zorbrist & Stumm (1981) presented compositional data which indicated that concentrations of dissolved constituents in the waters of the High Rhine were smaller than those in the waters of the Lower Rhine. Their findings are illustrated in Fig. 3.3, and show the extent to which pollution affects the dissolved species and how this varies from one component to another. At one extreme, the pollutant inputs have little or no influence on silicic acid and bicarbonate, but at the other extreme more than 90% of the sodium and chloride are anthropogenically derived. For the other constituents studied the effects of pollution lie between these two extremes.

Many other investigations have also revealed the influence that pollutant emissions have on river waters. For example, Van Bennekon & Salomons (1981) concluded that the global annual river discharge of nitrogen has been increased by a factor of about five, and that of phosphorus by a factor of

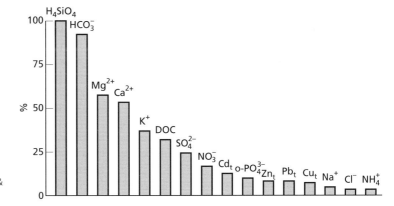

**Fig. 3.3** Components in 'pristine' Rhine river water as a percentage of the recent composition of the water (from Zorbrist & Stumm, 1981).

about four, as a result of anthropogenic inputs. The sulphate content of river waters has also been markedly raised as a result of pollution of the atmosphere, and in some rivers the original content has been doubled by inputs from this source (Maybeck, 1981a). The concentrations of some trace elements, especially those of the toxic heavy metals, can also suffer extreme perturbations from pollutant discharges in some rivers. Such discharges can increase the concentrations of metals such as Zn, Cd and Pb by factors as high as 500, and these may be even higher when acidic mine wastes drain into the rivers.

### 3.1.7 Relationships between the dissolved and particulate transport of elements in rivers

Material is transported by rivers in both dissolved and particulate forms, and in order to distinguish between the two phases Martin & Maybeck (1979) used the concept of a **dissolved transport index** (DTI), which expresses the amount of an element transported in solution as a percentage of its total (i.e. particulate + dissolved) transport. Dissolved transport indexes were calculated for a large range of elements found in river waters; however, it must be stressed that much of the dissolved trace element data used were suspect. Further, the DTIs reported by Martin & Maybeck (1979) are *global* averages and, because wide geographical variations can be found between individual rivers, the reliability of the values is probably no better than around ±20%. Despite these constraints, a number of general trends could be identified from the investigation. For example, the elements transported by rivers could be divided into

**Table 3.9** The dissolved transport index (DTI) in global river water (ranked in decreasing order) (data from Martin & Maybeck, 1979).

| | |
|---|---|
| 90–50% | Br, I, S, Cl, Ca, Na, Sr |
| 50–10% | Li, N, Sb, As, Mg, B, Mo, F, Cu, Zn, Ba, K |
| 10–1% | P, Ni, Si, Rb, U, Co, Mn, Cr, Th, Pb, V, Cs |
| 1–0.1% | Ga, Tm, Lu, Gd, Ti, Er, Nd, Ho, La, Sm, Tb, Yb, Fe, Eu, Ce, Pr, Al |

groups on the basis of their DTIs, and these are listed in Table 3.9. From this table it can be seen that elements exhibiting two extreme behaviour patterns can be distinguished.

**1** Elements having DTIs in the range ~90 to ~50% are carried mainly in solution by rivers; these include Na, Ca, Cl and S.

**2** Elements with DTIs <1% are transported almost exclusively in a particulate form; these include Al and Fe.

Other elements have dissolved–particulate transport behaviour patterns that lie between these two extremes.

### 3.1.8 Chemical signals carried by rivers: summary

**1** Rivers transport large quantitites of both dissolved and particulate components, and in this sense they may be regarded as carriers of a wide variety of chemical signals to the land–sea margins.

**2** The dissolved major element composition of river water is controlled mainly by the chemical composition of the source rock that is weathered in the catchment region. These relatively simple rock–water chemistry relationships, however, do not have the same degree of control on the dissolved trace element composition of river water, which appears to be strongly influenced by chemical constraints within the aqueous system itself; e.g. particulate–dissolved equilibria that involve both inorganic and biological particles, and which are influenced by factors such as pH and the concentrations of complexing ligands. Anthropogenic inputs also can influence the concentrations of trace metals in some river systems.

3 The chemical composition of RPM from different rivers systems shows considerable variation, some of which may result from climate-induced weathering intensity differences in catchment regions. Crystalline (residual), metal oxide and organic host fractions are the principal particulate trace-metal carrier phases in rivers that receive their trace elements mainly from natural sources.

4 There is a catchment-related pattern in the fluvial transport of POC and DOC by rivers, the highest concentrations being found in rivers draining swamp regions and the lowest in those flowing over glacial and alpine environments.

We have now tracked the transport of fluvial material to the river–ocean interface. Before reaching either the coastal receiving zone or the open regions of the sea itself, however, the river-transported material must pass through the estuarine environment. This environment can act as a filter, with the result that the fluvial signals can be severely modified before they are finally exported from the continents. The nature of these estuarine modifications is considered in the next section.

## 3.2 The modification of river-transported signals at the land–sea interface: estuaries

### 3.2.1 Introduction

Fairbridge (1980) defined an estuary as 'an inlet of the sea reaching into a river valley as far as the upper limit of the tidal rise'. This definition allows three distinct estuarine sections to be distinguished:

1 a marine (or lower) estuary, which is in connection with the sea;

2 a middle estuary, which is subject to strong sea-water–freshwater mixing;

3 an upper (or fluvial) estuary, which is characterized by freshwater inputs but which is subjected to daily tidal action.

Estuaries are therefore zones in which sea water is mixed with, and diluted by, fresh water. These two types of water have different compositions (see Table 3.1), and as a result estuaries are very complex environments in which the boundary conditions are extremely variable in both space and time. As a result, river-transported signals are subjected to a variety of physical, chemical and biological processes in the estuarine mixing zone. From this point of view, estuaries can be thought of as acting as **filters** of the river-transported chemical signals, which often can emerge from the mixing zone in a form that is considerably modified with respect to that which entered the system (see e.g. Schink, 1981). This concept of the estuarine filter is based on the fact that the mixing of the two very different end-member waters will result in the setting up of strong physicochemical gradients in an environment that is subjected to continuous variations in the supply of both matter and energy. It is these gradients that are the driving force behind the filter.

In the present section an attempt will be made to understand how estuaries work as chemical, physical and biological filters. Before doing this, however, it is necessary to draw attention to three important points.

1 The estuarine filter is selective in the manner in which it acts on different elements; for example, some dissolved species are simply diluted in an estuary and then carried out to sea, whereas others undergo reactions that lead to their addition to, or removal from, the dissolved phase.

2 The effects of the filter can vary widely from one estuary to another, so that it is difficult to identify global estuarine processes.

3 It is necessary to take into account the status of an estuary before any attempt is made to extrapolate its dynamics on to an ocean-wide scale.

Because of their environmental significance as material traps, estuaries require careful management, and as a result they have been the subject of considerable scientific interest over the past two or three decades. Much of this interest, however, has inevitably been directed towards relatively small urban estuarine systems, which often are highly perturbed by anthropogenic activities. Investigations of this type have provided invaluable insights into estuarine processes, but many of the estuaries themselves have little relevance on a natural global-ocean scale. Attempts have been made to study the chemical dynamics of major estuaries, such as those of the Amazon, the Zaire and the Changjiang (Yangtze) rivers, and it is the processes operating on the river signals in these large systems that will have a major influence on the chemistry of the oceans. In attempting to assess the importance of estuarine processes on fluvial signals it is necessary therefore to distinguish between such globally relevant estuaries and those that have a much more limited local effect. It must be stressed, however, that an approach which concentrates only on large river–estuarine systems has also important limitations. For example, Holland (1978) has pointed out that the combined run-off of the 20 largest rivers in the world accounts for only ~30% of the total global river run-off; further, these rivers drain the wettest areas of the globe. Thus, although processes operating in globally relevant estuaries may provide a better understanding of the effects that the estuarine filter has on the fluxes of material that reach the oceans, the fluxes themselves will be biased and may not represent truly global values.

### 3.2.2 The estuarine filter: the behaviour of elements in the estuarine mixing zone

The estuarine filter operates on dissolved and particulate material that flows through the system, and it can both modify and trap fluvially transported components within the mixing zone. The modification of the signals takes place via a number of chemical and biological processes that involve dissolved ↔ particulate speciation changes. The equilibria involved can go in either direction, i.e. particulate material can act either as a **source** of dissolved components, which are released into solution, or as a **sink** for dissolved components, which are removed from solution. As all the water in an estuary is eventually flushed out, usually on a time-scale of days or weeks, it is only the **sediment** that can act as an internal (i.e. estuarine) sink for elements that are brought into the system by river run-off. The sediments are not a static reservoir within the estuarine system, but are in fact subjected to a variety of physical, biological and chemical

processes (e.g. bioturbation, diagenesis), which can result in the recycling of deposited components back into the water compartment. These recycling processes include:

**1** the chemically driven diffusion of components from interstitial waters;

**2** the physically driven flushing out of interstitial waters into the overlying water column;

**3** the tidal resuspension of surface sediments, and sometimes their transfer from one part of an estuary to another.

The sediments themselves play a significant biogeochemical role in estuaries, and according to Bewers & Yeats (1980) they therefore can be regarded as acting as a third end-member in estuarine mixing processes, i.e. in addition to river water and sea water.

The physicochemical processes that control the estuarine filter therefore must be considered to operate in terms of a framework involving particulate–dissolved recycling associated with three estuarine end-members, i.e. river water, sea water (which may consist of more than one component end-member) and sediments. This is illustrated in a very simplified diagrammatic form in Fig. 3.4 in terms of the estuarine modification of the river signal. Within this framework the estuarine filter operates on the fluvial flux as it flows through the system. The filter does not, however, affect all fluvially transported chemical signals, and for this reason it is necessary as a first step to establish whether or not estuarine reactivity has actually taken place.

### 3.2.3 The identification of estuarine reactivity: the 'mixing graph' approach

One of the principal processes that modifies a river-transported chemical signal in an estuary is the physical mixing of fresh and saline waters of markedly different compositions along salinity (and other property) gradients. In the absence of any biogeochemical processes (reactivity) that lead to the addition or removal of a component, the physical mixing of the end-member waters would result in a linear relationship between the concentrations of a component and the proportions in which the two waters have undergone mixing (the salinity gradient), providing, that is, that the compositions of the end-members remain

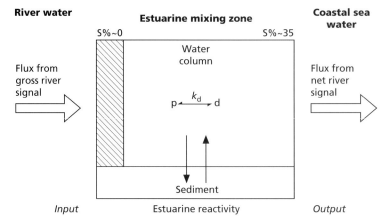

**River water**

**Estuarine mixing zone**

**Coastal sea water**

**Fig. 3.4** Simplified schematic representation of the modification of a river-transported signal in the estuarine environment. p $\xleftarrow{k_d}$ d indicates particulate–dissolved reactivity associated with physical, chemical and biological processes in the estuarine mixing zone. In natural waters the equilibrated partitioning of an element between dissolved and particulate phases can be described by a conditional partition coefficient $k_d$: $k_d = X/C$, where $X$ is the concentration of the exchangeable element on the particulate phase and $C$ is the concentration of the element in the dissolved phase—see Worksheet 3.1. ↑↓ Indicates two-way exchange of components between the water and sediment phases. ☐ Indicates the low-salinity zone of enhanced particulate–dissolved reactivity. For a discussion of gross and net river fluxes, see Section 6.1.

constant over a time approximating to the estuarine flushing time, and that there are no other sources or sinks for the components. This physical mixing relationship offers a useful baseline for assessing the effects that reactive biogeochemical processes have on the distribution of a component in an estuary. The technique used most commonly for this utilizes **mixing graphs** or **mixing diagrams**. In these diagrams, the concentrations of a component in a suite of samples (usually including the end-member waters) is plotted against a conservative index of mixing, i.e. a component whose concentrations in estuarine waters are controlled only by physical mixing. Salinity is the most widely used conservative index of mixing, although other parameters (e.g. chlorinity) also can be used for this purpose.

Mixing diagrams have been applied most commonly to dissolved components, and the theoretical relationships involved are illustrated in Fig. 3.5(a). If the distribution of a dissolved component is controlled only by physical mixing processes its concentrations in a suite of estuarine waters along a salinity gradient will tend to fall on a straight line, the **theoretical dilution line** (TDL), which joins the concentrations of the two end-members of the mixing series; under these conditions its behaviour is described as

being **conservative** or non-reactive. In contrast, when the component is involved in estuarine reactions that result in its addition to, or its loss from, the dissolved state its concentrations will deviate from the TDL, and its behaviour is termed **non-conservative** or reactive. For such a component the dissolved concentration data will lie above the TDL if it is added to solution, and below the TDL if it is removed from solution—see Fig. 3.5(a). Non-conservative behaviour therefore induces curvature in an estuarine mixing graph as concentrations deviate from the TDL. This curvature may be restricted to certain ranges of salinity, and thus can allow a geochemist to identify the specific estuarine zone in which the reactions take place.

The mixing diagram is a relatively simple concept for describing the behaviour of a dissolved component during estuarine mixing, but in practice it suffers from a number of fundamental problems. For example, although it is generally easy to identify conservative behaviour, deviations from the TDL are more difficult to interpret because they may result from factors other than estuarine reactivity; e.g. from the input of non-end-member waters.

There have been various attempts to rationalize the interpretation of estuarine mixing graphs. For

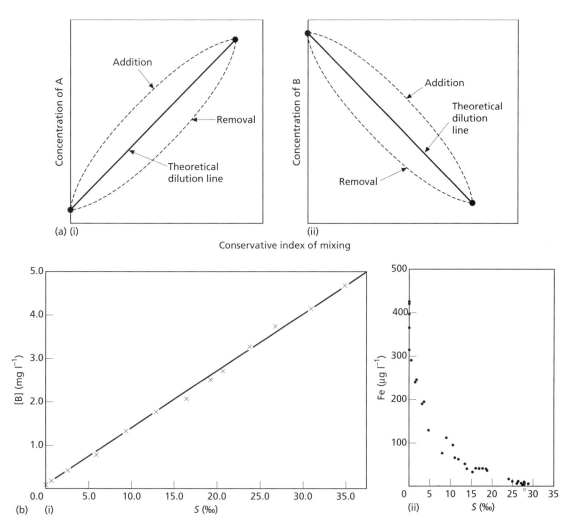

**Fig. 3.5** The behaviour of dissolved elements in estuaries.
(a) Estuarine mixing graphs for (i) a component with a concentration higher in sea water than in river water, and (ii) a component with a concentration higher in river water than in sea water (from Liss, 1976). (b) Estuarine mixing graphs and estuarine profiles for a series of representative dissolved elements. (i) Conservative behaviour of dissolved B in the Tamar estuary (from Liddicoat *et al.*, 1983). (ii) Non-conservative behaviour of dissolved Fe in the Beaulieu estuary showing removal at low salinity (from Holliday & Liss, 1976). (iii) Non-conservative behaviour of dissolved Ba in the Changjiang estuary showing addition to solution in the mid-salinity range (modified from Edmond *et al.*, 1985). (iv) Non-conservative behaviour of dissolved Cu in the Savannah estuary showing additions at both low and high salinities (from Windom *et al.*, 1983). Closed and open circles indicate data for different surveys. Broken curve indicates general trend of mixing curve only. (v) Non-conservative behaviour of dissolved Mn in the Tamar estuary showing both removal (low salinity) and addition (mid-range salinities) (modified from Knox *et al.*, 1981). $\Delta = Mn(II)$, $\bigcirc$ = total Mn; open symbols denote surface concentrations; closed symbols denote bottom concentrations.

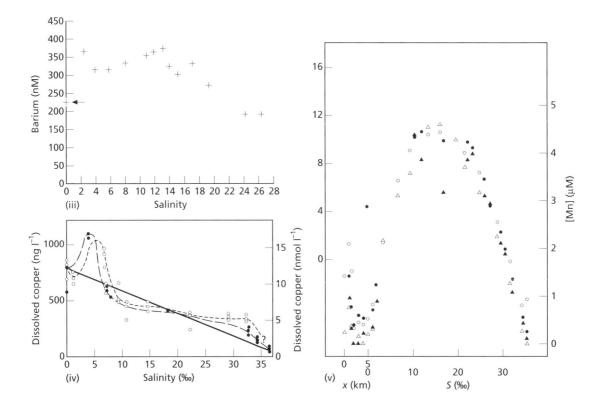

**Fig. 3.5** *Continued*

example, Boyle *et al.* (1974) derived a mathematical relationship for the variation of the flux of a dissolved component with salinity in an estuary. In doing this they developed a general model for mixing processes between river and ocean water in which definitive criteria were established for the identification of non-conservative behaviour among dissolved components. In this model it was assumed that the concentration of a dissolved component is a continuous, single-valued function of salinity, and the following relationship was developed for the variation of the flux of the dissolved component with salinity:

$$\frac{dQ_c}{dS} = -Q_w(S - S_r)\frac{d^2C}{dS^2} \tag{3.1}$$

where $Q_w$ is the flux of the river water, $Q_c$ is the flux of the dissolved component transported by the river water, $S_r$ is the salinity of the fluvial end-member, $S$ is the salinity at a given isohaline surface and $C$ is the concentration of the dissolved component at the same isohaline surface.

For conservative behaviour there is no gain or loss of the dissolved component during mixing, and hence

$$\frac{dQ_c}{dS} = 0 = \frac{d^2C}{dS^2} \tag{3.2}$$

and the plot of the concentration of the dissolved component against salinity will be a straight line.

For non-conservative behaviour, the second derivative $d^2C/dS^2$ will not be equal to zero, and the plot of the concentration of the dissolved component against salinity will be a curve described by eqn (3.1).

This explicit formulation of the mixing process, i.e. of $C(S)$, shows that over straight-line segments of the mixing curve simple two-end-member dilution processes are taking place, and that to establish non-conservative behaviour curvature must be demonstrated. By applying the model to various examples of the estuarine behaviour of dissolved components, the authors were able to demonstrate that previous examples of non-conservative 'curvature' behaviour could, in fact, be shown to fit to straight-line segments on the mixing curve. This was interpreted as resulting from the introduction of a third end-member component, i.e. in addition to the river-water and seawater end-members, into the mixing processes. This third end-member can be a tributary stream. More commonly, however, it is coastal or shelf water of intermediate salinity. For example, it was demonstrated that in the Mississippi estuary there is mixing between three end-members: Mississippi river water, shelf water of intermediate salinity and open Gulf water of high salinity. In situations such as this, the mixing between river water and sea water must therefore be interpreted in terms of two end-member sea waters, each of which can produce different straight-line relationships, which when combined can, under some circumstances, appear as curvature on the mixing diagram. The 'straight-line segment' approach to mixing diagrams outlined by Boyle *et al.* (1974) therefore offers a mathematical model for the interpretation of estuarine mixing processes.

Unless strict constraints are imposed, the mixing graph procedure can be both insensitive and imprecise. None the less, it has been used as an important first step in assessing the direction, if any, in which the estuarine filter has affected a river-transported signal, and in this respect a number of different estuarine behaviour patterns can be identified.

### 3.2.3.1 Conservative behaviour

Some elements appear to behave in a conservative manner in all estuarine situations. These include the major dissolved components that contribute to the salinity of sea water, e.g. $Na^+$, $K^+$, $Ca^{2+}$ and $SO_4^{2-}$. Boron, another major element, also seems to behave conservatively in many estuaries (see e.g. Fanning & Maynard, 1978). Liddicoat *et al.* (1983) demonstrated this for the Tamar estuary (UK), and concluded that the dissolved boron in the estuarine waters was derived almost entirely from sea water. The mixing graph for boron in the Tamar is illustrated in Fig. 3.5(b(i)), and provides an excellent

example of conservative behaviour during estuarine mixing.

### 3.2.3.2 Non-conservative behaviour

Mixing graphs that demonstrate that a component behaves in a non-conservative manner in estuaries can have a variety of forms, which can indicate the gain or loss of the dissolved component, and also can sometimes identify the zone in which the estuarine reactivity has taken place. A number of different types of estuarine mixing graphs are described below.

**1 Removal of dissolved components.** For these components estuarine concentrations fall below the TDL. An example of this type of mixing graph is illustrated for iron in Fig. 3.5(b(ii)), from which it is evident that the removal of dissolved iron has taken place largely in the low-salinity region.

**2 Addition of dissolved components.** For these components estuarine concentrations lie above the TDL. An example of this is provided by the behaviour of Ba in the Changjiang (Yangtze) estuary—see Fig. 3.5(b(iii)). A more complex mixing graph showing the addition of dissolved components has been reported by Windom *et al.* (1983) for the behaviour of copper in the Savannah estuary (USA). This graph is illustrated in Fig. 3.5(b(iv)) and indicates that there are additions to the fluvial Cu at both low (≤5‰) and high (≥20‰) salinities.

**3 Combined addition and removal.** This type of estuarine profile has been identified for dissolved manganese in the Tamar estuary (UK) by Knox *et al.* (1981). The estuarine profile is illustrated in Fig. 3.5(b(v)), and shows that dissolved Mn is removed from solution at low salinities, but is added to solution at mid-range salinities (see also Section 3.2.7.2).

It is apparent, therefore, that if sufficient constraints are applied mixing graphs can be used to identify: (i) whether or not a dissolved component has undergone estuarine reactivity; (ii) the direction, i.e. gain or loss, in which the dissolved component has been affected; and (iii) the general estuarine region (e.g. low-, mid-range, or high-salinity zones) over which the reactivity has been most effective. The mixing graph, however, suffers from one very fundamental limitation, that is, it does not provide information on *why* the component behaves in the way it does or on the *nature* of biogeochemical reactions that have caused the estuarine reactivity. In order to know how the estuarine filter operates it is therefore necessary to understand the nature of the reactive processes that occur in the mixing zone.

Biogeochemical reactivity in natural waters is controlled by a number of physicochemical parameters, which include pH, redox potential, salinity, and the concentrations of complexing ligands, nutrients, organic components and particulate matter. All these parameters undergo major variations in estuaries, as a result of which a variety of dissolved ↔ particulate transformations are generated in the mixing zone. These transformations are driven by physical, chemical and biological factors, and include: (i) sorption at the surfaces of suspended particles; (ii) precipitation; (iii) flocculation–aggregation; and (iv) uptake via biological processes. In general, the extent to which the transformations occur depends on the nature and concentrations of both the particulate and the dissolved components in the mixing zone.

### 3.2.4 Estuarine particulate matter (EPM)

#### 3.2.4.1 *The classification of EPM*

On the basis of the general scheme proposed by Salomons & Forstner (1984), the following classes of estuarine particulates can be identified.

**1 River-transported or fluvial particulates.** These solids, which are transferred across the river–estuarine boundary, include crustal weathering products (e.g. quartz, clay minerals), precipitated oxyhydroxides (principally those of iron and manganese), terrestrial organic components (e.g. plant remains, humic materials) and a variety of pollutants (e.g. fly ash, sewage).

**2 Atmospherically transported particulates.** The transfer of material across the atmosphere–estuarine boundary can be important in some estuaries. The atmospheric components involved in this transfer include crustal weathering products and pollutants such as fly ash.

**3 Ocean-transported particulates.** Particulate matter transferred across the ocean–estuarine boundary includes biogenous components of marine origin (e.g. skeletal debris, organic matter) and inorganic components (e.g. those originating in coastal sediments or formed in the marine water column).

**4 Estuarine-generated particulates.** This type of particulate material has an internal source and includes inorganic and organic flocculants and precipitates, and both living and non-living particulate organic matter. Of the various processes that lead to the formation of estuarine particulates,

    (a) flocculation,

    (b) precipitation and

    (c) the biological production of organic matter

are especially important.

Flocculation is a process that causes smaller particles (colloids or semi-colloids) to increase in size and form larger units, and has been described in detail by Potsma (1967) and Drever (1982). In estuaries, where the mixing of saline and fresh water leads to an increase in ionic strength, flocculation affects both organic and inorganic components; these include river-transported clay mineral suspensions, colloidal species of iron and dissolved organics (such as humic material). The aggregation of particles into larger sizes also can take place via biological mediation; e.g. through the production of faecal pellets by filter-feeding organisms (see Section 10.4).

Various types of precipitation processes occur in estuaries; of these, heterogeneous precipitation in the presence of particle clouds (e.g. in turbidity maxima) is especially important in the removal of dissolved Mn, and other metals, from solution.

The estuarine biomass is formed as a result of primary production, which is related to factors such as the supply of nutrients and the turbidity of the waters. According to the estimate made by Williams (1981), the global internal estuarine photosynthetic production is $\sim 5.2 \times 10^{14} \, g \, yr^{-1}$, which is $\sim 1.5\%$ of the total marine production if a figure of $3.6 \times 10^{16} \, g \, yr^{-1}$ is used for the latter (Section 9.1.3.3). Rivers transport $\sim 2 \times 10^{14} \, g \, yr^{-1}$ of carbon into estuaries, which is the same order of magnitude as that resulting from primary production. According to Reuther (1981), however, much of this imported carbon is refractory, i.e. resistant to microbial attack, and so does not enter into recycling within the estuarine system (see Section 3.1.5.2).

#### 3.2.4.2 *The distribution of EPM*

The physical processes that control the distribution of EPM involve water circulation patterns, gravitational settling and sediment deposition and resuspension.

All these processes, together with primary production, combine to set up the particle regime in an estuary.

Estuarine circulation patterns exert a fundamental influence on the processes that control the distribution of EPM. A number of characteristically different types of estuary can be distinguished on the basis of water circulation patterns (see Fig. 3.6), and these are described below in relation to the manner in which they constrain estuarine particle regimes, and therefore influence estuarine reactivity.

The most commonly occurring estuaries are of the positive type, i.e. loss from evaporation at the surface is less than the input of fresh water from rivers. The basic factor that determines the type of circulation in an estuary of this type is the part played by tidal currents (inflow of sea water) in relation to river flow (inflow of fresh water), and in a general way the type of circulation developed can be related to the dominance of one of these two water flow regimes.

In some estuaries the river flow is dominant. Under these conditions the less dense river water forms an upper layer over the more dense saline water, and the

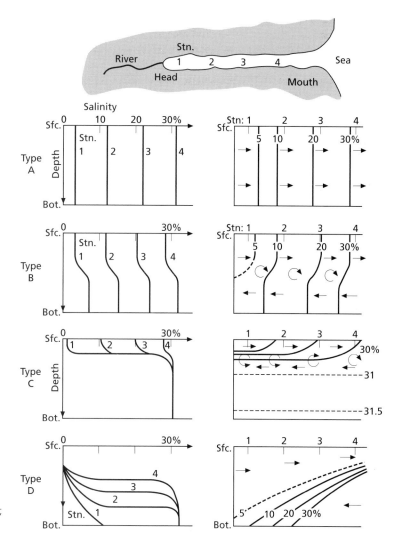

**Fig. 3.6** Salinity–depth profiles and longitudinal salinity sections in estuaries displaying different water circulation patterns (from Aston, 1978; after Pickard & Emery, 1982). Type A, a vertically mixed estuary; type B, a slightly stratified estuary; type C, a highly stratified estuary; type D, a salt-wedge estuary.

estuary becomes stratified. There are a number of types of stratified estuaries.

*Salt-wedge estuary.* When the circulation is almost completely dominated by river flow, a **salt-wedge** estuary can be formed. As the name implies, the salt water in this type of estuary extends into the river as a wedge under the freshwater outflow. A characteristic of the salt-wedge circulation pattern is that the outflowing water remains fresh up to the mouth of the river. There is also a steep density gradient between the fresh water and saline water, which prevents mixing between the two water layers until the river flow attains a critical velocity. This velocity is not attained in a salt-wedge estuary.

*Highly stratified estuary.* A second type of well developed vertical stratification, but one in which the saline layer is not confined to a wedge shape, can be formed in some estuaries. These estuaries are termed **highly stratified**, or are referred to as having a two-layer flow. Like the salt-wedge type, the river flow in a highly stratified estuary is large relative to the tidal flow and there is still a two-layer stratification. Now, however, the velocity of the seaward-flowing river water is sufficient to cause internal waves to break at the saline–freshwater interface. This results in the mixing of saline water from below into the upper freshwater layer, a process termed **entrainment**. This is a one-way process only and, in contrast to a salt-wedge type of circulation, it results in the salinity of the upper water layer increasing as it moves seawards.

Both Meade (1972), and later Potsma (1980), have pointed out that the distribution of suspended particulate matter in an estuary is largely controlled by the dynamics of the water circulation pattern. According to these authors, most of the suspended particulate material in both salt-wedge and highly stratified estuaries is of fluvial origin and is transported seawards in the upper water layer. The suspended matter does not thereafter enter the estuarine cycle proper.

*Partially stratified estuary.* As tidal forces become more important, an estuary will begin to oscillate. In a **partially stratified** (or partially mixed) estuary, there is still a two-layer structure developed, with fresh water at the surface overlying saline water at depth. In this situation, however, there is vertical mixing between the mainly inflowing bottom water and the mainly outflowing surface water. This mixing takes place through eddy diffusion, which, unlike entrainment, is a two-way process, i.e. salt water is mixed upwards and fresh water is mixed downwards, with the result that a layer of intermediate salinity is formed. In this type of estuary there are also longitudinal variations in salinity in both the upper and lower layers, and undiluted fresh water is found only at the head of the estuary.

In the partially stratified estuary, the lower-water landward flow is strong enough to move suspended material up the estuary to the head of the salt intrusion. This may include both fluvial material that has settled from the upper layer and marine material that has been transported landwards. A **turbidity maximum**, or particle cloud, may be developed at the head of the salt intrusion in the region where suspended solids can be transported from both up- and downstream directions. Such a turbidity maximum, which is a region of concentration of suspended material, is an important site for dissolved ↔ particulate interactions. In some estuaries very high concentrations (up to several hundred grams of sediment per litre) have been reported near the bed. This so-called **fluid mud**, which is built up of material sinking from the overlying turbidity maximum during neap tides, can undergo resuspension and restricted transport on spring tides.

*Vertically well-mixed estuary.* Under certain estuarine conditions tidal flow can become completely dominant. For example, in some estuaries, which usually have a small cross-section, a strong tidal flow can exceed the river flow and the velocity shear on the bottom may be of sufficient magnitude for turbulence to mix the water column completely. These estuaries are termed homogeneous or **fully mixed** estuaries and, because there is no appreciable river flow, the suspended particulates are concentrated in the nearshore region. There is some doubt, however, as to whether *completely* homogeneous estuaries actually exist in nature.

It is apparent, therefore, that the nature and strength of the circulation patterns in an estuary will exert a strong control on the suspended particle regime and can lead, in the case of the partially stratified type, to the development of zones of relatively high suspended solid concentrations (turbidity maxima), which are major sites for physical, chemical and biological reactions between particulate and dissolved species. It is in these zones, therefore, that the estuarine filter is especially active. Estuarine circulation also affects the residence time of the water in estuaries. This varies from a few days to a few months, and according to Duinker (1986) it increases with the increasing extent of vertical mixing; thus, residence times generally increase in the sequence salt-wedge < stratified < partially mixed < well-mixed estuarine types.

### 3.2.5 The concentration and nature of dissolved material in estuaries

For many dissolved components, the concentrations in river water exceed those in sea water (see Table 3.10), and as a result the initial **concentrations** of the dissolved components in estuaries are regulated by the extent to which the end-member waters have undergone mixing. The nature or **speciation** of the dissolved components is also affected by the water mixing. It is the speciation of an element between particulate, colloidal and dissolved (e.g. ion pairs, and both inorganic and organic complexes) forms, rather than its total concentration, that controls its environmental reactivity. Competitive complexing between the principal inorganic ligands (e.g. $Cl^-$, $SO_4^{2-}$, $CO_3^{2-}$, $OH^-$), the organic (e.g. humic material) ligands and particulate matter is the main factor that controls the nature of the inorganic species in natural waters. This topic is considered in more detail in Section 11.6.2, with respect to sea water. Elemental speciation, however, also is important in the estuarine mixing zone.

From the point of view of the behaviour of both dissolved and particulate elements during estuarine mixing, three of the most important physicochemical differences between the river-water and seawater end-members are those associated with the following parameters:

1 ionic strength (salinity), which varies from zero in the river end-member to $0.7\,M$ in the marine end-member;

2 ionic composition and the concentrations of complexing ligands—for example, in river water the most important ions are calcium and bicarbonate, whereas sea water is dominated by sodium and chloride;

3 pH—river waters can be either acidic or alkaline, with pH values that are usually in the range 5–8.

Thus, in some estuaries the pH of the river end-member may be similar to that of the seawater end-member (average pH ~8.2). River waters, however, usually are more acidic than sea water, with the result that under many estuarine conditions there is a pH gradient, which increases with increasing salinity. Changes in parameters such as those identified above will affect the speciation of elements as the two end-member waters undergo estuarine mixing.

pH and $P_e$ often are considered to be the **master variables** in natural waters (see Worksheet 14.1). $P_e$ is a convenient way of expressing the equilibrium redox potential, which is an important parameter in estuarine chemistry, because in estuaries elements can be exposed to reducing conditions in both the water column and sediment interstitial waters, and under some estuarine conditions sulphide complexing also must be taken into account in speciation modelling. Even with a knowledge of the master variables, however, there are many uncertainties involved in understanding the speciation of elements in any natural waters, and these are magnified when the mixed-water estuarine system is considered. For a detailed treatment of the fundamental concepts involved in chemical speciation in estuarine waters the reader is referred to the review presented by Dyrssen & Wedborg (1980), which covers the binding of trace metals with both inorganic and organic ligands and also includes complexing in anoxic waters.

At this stage, it is sufficient to understand that during the mixing of fresh and saline waters there is competition for dissolved trace elements between various complexing agents, such as those described above, and between these agents and the suspended particulate matter that is present in the resulting estuarine 'soup'.

### 3.2.6 Dissolved ↔ particulate interactions in estuaries

The concentrations of both complexing agents and particulate matter undergo large variations during

| Element | Rivers | | Ocean | |
|---|---|---|---|---|
| | Dissolved ($\mu$g l$^{-1}$) | Particulate ($\mu$g g$^{-1}$) | Water ($\mu$g l$^{-1}$) | Deep sea clays ($\mu$g g$^{-1}$) |
| Ag | 0.3 | 0.07 | 0.04 | 0.1 |
| Al | 50 | 94 000 | 0.5 | 95 000 |
| As | 1.7 | 5 | 1.5 | 13 |
| Au | 0.002 | 0.05 | 0.004 | 0.003 |
| B | 18 | 70 | 4440 | 220 |
| Ba | 60 | 600 | 20 | 1500 |
| Br | 20 | 5 | 67 000 | 100 |
| Ca | 13 300 | 21 500 | 412 000 | 10 000 |
| Cd | 0.02 | (1) | 0.01 | 0.23 |
| Ce | 0.08 | 95 | 0.001 | 100 |
| Co | 0.2 | 20 | 0.05 | 55 |
| Cr | 1 | 100 | 0.3 | 100 |
| Cs | 0.035 | 6 | 0.4 | 5 |
| Cu | 1.5 | 100 | 0.1 | 200 |
| Er | 0.004 | (3) | 0.0008 | 2.7 |
| Eu | 0.001 | 1.5 | 0.0001 | 1.5 |
| Fe | 40 | 48 000 | 2 | 60 000 |
| Ga | 0.09 | 25 | 0.03 | 20 |
| Gd | 0.008 | (5) | 0.0007 | 7.8 |
| Hf | 0.01 | 6 | 0.007 | 4.5 |
| Ho | 0.001 | (1) | 0.0002 | 1 |
| K | 1500 | 20 000 | 380 000 | 28 000 |
| La | 0.05 | 45 | 0.003 | 45 |
| Li | 12 | 25 | 180 | 45 |
| Lu | 0.001 | 0.5 | 0.0002 | 0.5 |
| Mg | 3100 | 11 800 | $1.29 \times 10^6$ | 18 000 |
| Mn | 8.2 | 1050 | 0.2 | 6000 |
| Mo | 0.5 | 3 | 10 | 8 |
| Na | 5300 | 7100 | $1.077 \times 10^7$ | 20 000 |
| Nd | 0.04 | 35 | 0.003 | 40 |
| Ni | 0.5 | 90 | 0.2 | 200 |
| P | 115 | 1150 | 60 | 1400 |
| Pb | 0.1 | 100 | 0.003 | 200 |
| Pr | 0.007 | (8) | 0.0006 | 9 |
| Rb | 1.5 | 100 | 120 | 110 |
| Sb | 1 | 2.5 | 0.24 | 0.8 |
| Sc | 0.004 | 18 | 0.0006 | 20 |
| Si | 5000 | 285 000 | 2000 | 283 000 |
| Sm | 0.008 | 7 | 0.0005 | 7.0 |
| Sr | 60 | 150 | 8000 | 250 |
| Ta | <0.002 | 1.25 | 0.002 | 1.0 |
| Tb | 0.001 | 1.0 | 0.0001 | 1.0 |
| Th | 0.1 | 14 | 0.01 | 10 |
| Ti | 10 | 5600 | 1 | 5700 |
| Tm | 0.001 | (0.4) | 0.0002 | 0.4 |
| U | 0.24 | 3 | 3.2 | 2.0 |
| V | 1 | 170 | 2.5 | 150 |
| Y | — | 30 | 0.0013 | 32 |
| Yb | 0.004 | 3.5 | 0.0008 | 3 |
| Zn | 30 | 250 | 0.1 | 120 |

**Table 3.10** The concentrations of dissolved elements in river and sea waters (data from Martin & Whitfield, 1983).

estuarine mixing. It is only particulate matter, however, that can be trapped in the mixing zone and this is coupled to dissolved material via particulate ↔ dissolved equilbria.

Processes that are involved in the addition or removal of dissolved components in the estuarine 'soup' include the following.

1 Flocculation, adsorption, precipitation and biological uptake, which result in the **removal** of components from the dissolved phase and their transfer to the particulate phase. It is these processes that can result in the retention of components in estuaries via particulate trapping in sediments.

2 Desorption from particle surfaces and the breakdown of organics, which result in the **addition** of components to the dissolved phase. This will result in the flushing out of the components if they remain in the dissolved phase.

3 Complexation and chelation reactions with inorganic and organic ligands, which **stabilize** components in the dissolved phase. These components will also be flushed out of an estuary with the water mass. The manner in which these different types of processes can be linked to (i) the kind and concentration of EPM, and (ii) the speciation of dissolved components, can be described in terms of a number of experimental approaches that have been used to model estuarine chemistry.

Laboratory studies designed to investigate chemical reactions in estuarine waters tend to be of two types:

1 those involving the mixing of end-member waters and the characterization of the reaction products formed;

2 those attempting to study individual reaction processes, such as adsorption–desorption, by experimentally varying the controlling parameters (e.g. pH, ionic strength).

A number of laboratory studies on the chemical reactions taking place in estuaries have involved the artificial mixing of both filtered and non-filtered samples of the fresh and saline end-member waters (see e.g. Coonley *et al.*, 1971; Duinker & Notling, 1977). This often is referred to as the **product approach,** and perhaps the most significant advances in our understanding of estuarine chemistry to emerge from this approach are those which have been reported by Sholkovitz and his co-workers.

The **Sholkovitz model** (Sholkovitz, 1978) was an

attempt to make a quantitative prediction of the reactivity of trace metals in the estuarine mixing zone. The model is based on the following assumptions.

1 A fraction of the dissolved trace metals in river water exist as colloids in association with colloidal forms of humic acids and hydrous iron oxides.

2 In the estuarine mixing zone the removal of the metals takes place via the flocculation of these colloids and/or their subsequent adsorption on to humic acids and hydrous iron oxide flocs.

3 The extent to which this removal takes place in the estuarine soup is determined by competition for the trace metals by seawater anions, humic acids and hydrous iron oxides, in the presence of seawater cations.

The Sholkovitz model was then tested in a variety of experiments using the product approach resulting from the mixing of river-water and seawater end-members. One of the most significant findings to emerge from the product-approach experiments was the importance of the role played by the rapidly flocculating humic fraction in the formation of metal humates. The importance of the flocculation of humics and hydrous iron oxides was demonstrated by Sholkovitz & Copland (1981), who carried out an investigation into the fates of dissolved Fe, Cu, Ni, Cd, Co and humic acids following the artificial mixing of river and sea water. Although they restricted their experimental work to the filtered water of a single organic-rich river, the River Luce (UK), they were able to demonstrate the effects that resulted from the flocculation of fluvial-transported colloids by the seawater cations encountered on estuarine mixing. In this process, $Ca^{2+}$ was found to be the main coagulating, i.e. colloid destabilizing, agent. On the basis of the River Luce data, the authors concluded that the extent of this flocculation removal (expressed as a percentage of the dissolved concentration) varied from *large* (e.g. Fe ~80%, humic acids ~60%, Cu ~40%), to *small* (e.g. Ni and Cd ~15%), to essentially *nothing* (e.g. Co and Mn ~3%). However, Sholkovitz & Copland (1981) pointed out that their findings, based on the organic-rich river water, should not be applied *a priori* to estuaries in general. For example. Hoyle *et al.* (1984) showed that dissolved rare-earth elements (REE) were flocculated in association with Fe–organic-matter colloids when water from the River Luce was mixed with sea water; however, REE removal did not occur when the river end-member used in the experiments was organic-poor in character. In addition to the cationic destabilization of fluvially transported colloidal humic substances, organic complexes can stabilize some dissolved elements in solution; for example, Waslenchuk & Windon (1978) concluded that the conservative transport of dissolved As in estuaries from the southeastern USA results from the stabilization of the element by organic complexes (see also Section 11.6.2).

In practice, estuarine dissolved ↔ particulate reactions depend on a number of variables. Several of these were considered by Salomons (1980) in a series of interlinked laboratory experiments in which the influence of pH, chlorinity and the concentration of suspended material on the adsorption of Zn and Cd was assessed under estuarine conditions. Three important conclusions regarding the nature of the competition for trace metals in the estuarine mixing zone can be drawn from this study.

1 The adsorption of both metals increased with increasing pH over the experimental range (pH 7.0–8.5).

2 The adsorption of Cd, and to a lesser extent that of Zn, decreased with increasing chlorinity, probably as a result of competition from the chloride ion for the complexation of the metals, thus keeping them in solution.

3 The adsorption of both elements increased with increasing turbidity, i.e. with an increase in the concentration of suspended matter. This apparently occurred to the extent that the suspended matter was able to compete effectively with chloride ions for metal complexation.

Extrapolated to real estuarine situations the latter two findings mean that the effectiveness of suspended particulate matter for the capture of dissolved Cd and Zn decreases as salinity increases, but that this effect can be overridden when the concentrations of the suspended particulates are relatively high, e.g. in the presence of a turbidity maximum, which can act as a zone of enhanced adsorption (Salomons & Forstner, 1984). A model designed to describe adsorption in the presence of a turbidity maximum in the estuarine environment is summarized in Worksheet 3.1.

The studies described above have focused largely

---

**Worksheet 3.1: A simple sorptive equilibrium model for the removal of trace metals in a low-salinity turbid region of an estuary**
(after Morris, 1986)

A variable fraction of the river influx of many dissolved trace metals is removed from solution in the low-salinity, high-turbidity, estuarine mixing zone. The equilibrated partitioning of a trace metal between the dissolved and particulate phases in natural waters can be described by the term $K_d = X/C$, where $K_d$ is a conditional partition coefficient, $X$ is the concentration (w/v) of exchangeable metal on the particulate phase, and $C$ is the dissolved metal concentration (w/v). To set up his sorptive equilibrium model, Morris (1986) made the following assumptions: (i) the diffusion of solutes from the estuary to the very low-salinity region (<0.2‰) is minor so that the river can be considered to be the only significant source of dissolved metal; and (ii) the salinity related changes in the conditional partition coefficient are negligible. The problems involved in describing the removal of the river influx of dissolved trace metals in the turbid low-salinity estuarine mixing zone therefore can be simplified to a consideration of the change in sorptive equilibrium, at constant $K_d$, induced by adding suspended particulates to the river water. Under these conditions, the mass balance for the conservation of a trace metal, for unit volume of river water, can be written:

*continued*

$$C_R + X_R P_R + X_S P_S = C + X(P_R + P_S) \tag{1}$$

where $C_R$ and $C$ are the equilibrated dissolved metal concentrations (w/v) in the river water and the turbidized water, respectively; $X_R$ and $X$ are the equilibrated exchangeable metal concentrations (w/w) on the particles in the river water and the turbidized water, respectively; $X_S$ is the exchangeable metal concentration (w/w) on the added particulate load; $P_R$ is the suspended particulated load (w/w) carried by the river and $P_S$ is the additional suspended particulate load (w/v). Substituting $X_R = K_d C_R$ and $X = K_d C$ into eqn (1), and introducing a term $\alpha = X_S / X$, yields:

$$C/C_R = (1 + K_d P_R)/[1 + K_d P_S (1 - \alpha)] \tag{2}$$

Equation (2) therefore predicts the change in dissolved metal concentration induced by adding particles to river water as a function of the particle load, the added particulate load, and $\alpha$ (the fraction of the exchangeable metal on the added particles relative to ultimate equilibration). For sorptive removal to occur, the added particles must have a lower concentration of exchangeable metal than the particles in the turbidized water, i.e. $\alpha < 1$.

Morris (1986) considered how the predicted concentration ratio ($C/C_R$) within the zone of removal varies with $K_d$, $P_R$ and $\alpha$. The relationship for $K_d$ is illustrated in Fig. (i), and shows how the predicted concentration ratio ($C/C_R$) varies with $K_d$ for additions of suspended particulate load ($P_S$) ranging from 0 to 1000 mg l$^{-1}$, with a representa-

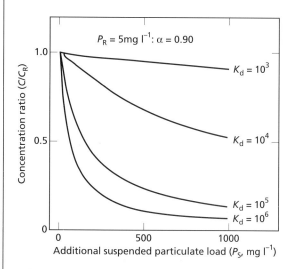

**Fig. (i)** The influence of the partition coefficient ($K_d$) on the sorptive uptake of fluvial dissolved constituents in the very-low-salinity, high-turbidity estuarine region.

*continued on p. 40*

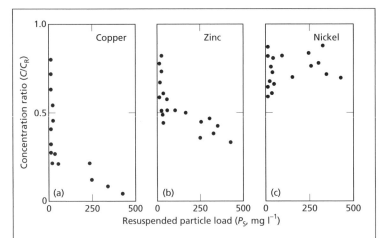

**Fig. (ii)** Variations in the concentration ratio ($C/C_R$) for dissolved (a) Cu, (b) Zn and (c) Ni with changes in suspended particulate load ($P_S$) in the very-low-salinity, high-turbidity zone in the Tamar estuary (UK) (from Morris, 1986).

tive depletion factor of $\alpha = 0.90$, and a river suspended particulate load ($P_R$) = 5 mg l⁻¹. The curves illustrate that at constant $P_R$ and $\alpha$, the degree of removal for any $P_S$ value increases (i.e. the ratio $C/C_R$ decreases) with increasing $K_d$, the most sensitive changes being in the $K_d$ range $1 \times 10^3$ to $1 \times 10^6$ mg l⁻¹. The extent of removal is also sensitive to changes in $P_R$ and $\alpha$. Morris (1986) then used data from the Tamar and plotted the concentration ratio ($C/C_R$) against the resuspended particulate load ($P_S$) for Cu, Zn and Ni. The results are illustrated in Fig. (ii), and show that for Cu and Zn the data correspond with the form of the relationship predicted by the model.

From the equilibrium sorption model it may be predicted that for elements with $K_d$ values higher than ~$1 \times 10^3$ there will be significant removal in the turbid waters. Data in the literature indicate that $K_d$ values in river water generally can exceed this value, and Morris (1986) therefore concluded that sorptive removal is potentially an important feature of the behaviour of trace metals in moderately to highly turbid estuaries.

on the removal of trace elements from solution on to particulate matter. In contrast, other laboratory experiments have shown that some elements can be added to the dissolved state via desorptive release from particulate matter during estuarine mixing (see e.g. Kharkar *et al.*, 1968; Van der Weijden *et al.*, 1977). This desorptive release, however, does not apparently affect all elements. For example, Li *et al.*

(1984) added radiotracer spikes to experiments in which river and sea water were mixed and concluded that Co, Mn, Cs, Cd, Zn and Ba will be desorbed from river suspended particulate material on estuarine mixing, whereas Fe, Sn, Bi, Ce and Hg will undergo removal by adsorption.

The findings deduced from laboratory studies can be used to interpret estuarine mixing diagrams. This can be illustrated with respect to the study reported by Windom *et al.* (1983), who carried out an investigation of the estuarine behaviour of copper in the Savannah and Ogeechee estuaries (USA). The field survey data showed that Cu behaved in a non-conservative manner, with additions to the dissolved fluvial load being found in both estuaries. The additions occurred at both low (<5‰) and high (>20‰) salinities, with a possible removal taking place at intermediate salinities (see Fig. 3.5b(iv)). The addition of Cu to the dissolved phase at the two ends of the estuarine salinity range implied that the element was being released from suspended particulate matter and/or bottom sediments. In order to distinguish between these two potential release mechanisms the authors carried out a series of experiments in which Ogeechee river water was mixed with mid-shelf sea water in various proportions before filtration and analysis of the samples. This series of experiments, in which only the water and its suspended particulate matter were used, thus eliminated any effects that would have arisen from the release of Cu from bottom sediments. The results revealed that, following an initial release of Cu at low salinities, the element behaved in an essentially conservative manner over the salinity range ~8 to ~34‰. The authors concluded, therefore, that the addition of Cu to solution at salinities <5‰ was a result of its release from suspended particulate matter, but that the addition at salinities >20‰ was a result of its mobilization from bottom or resuspended sediment.

It may be concluded that mixing graphs can be used to establish whether or not a dissolved component has suffered reactivity in the estuarine mixing zone, and laboratory experiments can be used to interpret the processes involved; although it must be remembered that such experiments never fully reproduce the complex estuarine situation. Despite this constraint, laboratory experiments have led to at least a first-order understanding of some of the processes, e.g. competition for dissolved components from complexing ligands and particulate material, which control particulate ↔ dissolved equilibria in the estuarine 'soup'. Perhaps the single most important finding to emerge from all this work, however, is that both the mixing graphs and the laboratory experiments have shown that some elements can, under different conditions, behave differently in the mixing zone. This reinforces the concept that for

many elements there is no such thing as a global estuarine behaviour pattern. For this reason it is necessary to look at the behaviour of components in a number of individual estuaries. With this in mind, an attempt will be made in the following section to summarize the behaviour patterns of some components in the estuarine mixing zone, and to relate these to both non-biological and biological controls; in doing this attention will be focused, where possible, on the large globally important estuarine systems.

### 3.2.7 The behaviour of individual components in the estuarine mixing zone

#### 3.2.7.1 *Total particulate material*

At the present day much of the particulate material that enters estuaries is trapped within the system. For example, only ~8% of the particulate matter entering the upper zone of the Scheldt estuary is exported to the North Sea (Duinker *et al.*, 1979); more than 90% of the RPM carried by the Mississippi is deposited in the delta (Trefry & Presley, 1976); more than 95% of the suspended solids in the Amazon settle out within the river mouth (Milliman *et al.*, 1975); over the entire St Lawrence system (i.e. upper estuary, lower estuary, gulf) ~90% of the particulate input is held back (Bewers & Yeats, 1977). Overall, therefore, it would appear that ~90% of RPM reaching the land–sea margins is retained in estuaries.

#### 3.2.7.2 *Iron and manganese*

Iron and manganese are extremely important elements in aquatic geochemical processes because their various oxide and hydroxide species act as scavengers for a variety of trace metals. The processes that control the estuarine chemistries of the two elements are different, however.

*Iron.* Iron is present in natural waters in a continuum of species ranging through true solution, colloidal suspension and particulate forms. Most determinations of dissolved iron given in the literature refer to species that have passed through a 0.45 µm filter, and this definition will be used here.

Dissolved iron in river water is present mainly as hydrous Fe(III) oxides, which are stabilized in colloidal dispersion by high-molecular-weight humic acids (see e.g. Boyle *et al.*, 1977; Sholkovitz *et al.*, 1978; Hunter, 1983). Humic material is affected by speciation changes on the mixing of fresh and saline waters, and the dominant mechanism for the removal of iron to the particulate state appears to be the flocculation of mixed iron-oxide–humic-matter colloids of fluvial origin, which undergo electrostatic and chemical destabilization during estuarine mixing, partly as a result of neutralization of colloid charges by marine cations. The coagulative removal of iron therefore takes place at low salinities, and a typical estuarine mixing graph for iron is given in Fig. 3.5(b(ii)).

Figueres *et al.* (1978) summarized the behaviour of iron in estuaries and found that in 15 out of the 16 systems considered, fluvial dissolved iron was removed during estuarine mixing; expressed in terms of the fluvial input, the removal ranged from ~70 to ~95%. This pattern has been confirmed with respect to the major estuaries, e.g. those of the Amazon, the Zaire and the Changjiang, which will have the greatest influence on oceanic chemistry. It may be concluded, therefore, that iron is one of the few elements for which a **global behaviour pattern**, i.e. removal at low salinities, has been demonstrated almost unambiguously for the estuarine environment.

*Manganese.* In contrast to iron, the distribution patterns for manganese in estuaries are variable and both gains and losses of dissolved manganese, and even conservative behaviour, have been reported in the literature. These variations can be related to the aqueous chemistry of the element.

Manganese is a redox-sensitive element that is biogeochemically reactive in the aqueous environment, and it readily undergoes transformations between the dissolved and particulate phases in response to physicochemical changes. Much of the dissolved manganese supplied by rivers is in the reduced Mn(II) state, and this can undergo oxidative conversion to particulate Mn(IV) as the physicochemical parameters change on the mixing of fresh and saline waters. The oxidation of Mn(II) is autocatalytic, the product being a solid manganese dioxide phase with a composition that depends on the reaction conditions, i.e. it is heterogeneous. The process is pH-dependent and also proceeds faster in the presence of a particulate phase that is able to adsorb the Mn(II), e.g. in the presence of a turbidity maximum. In estuarine waters, therefore, it is redox-driven processes involving dissolved

↔ particulate reactions that play the dominant role in the chemistry of manganese. These redox reactions also continue after the deposition of Mn(IV) particulates in the bottom sediments. Many estuarine sediments are reducing at depth so that the Mn(IV) solids can become reduced during diagenesis, thus releasing Mn(II) into the interstitial waters and setting up a concentration gradient with respect to the overlying water column, which usually contains smaller concentrations of dissolved manganese. Dissolved Mn migrates up this gradient and, if the upper sediments are reducing, it may escape into the overlying water column, either directly or during the tidal resuspension of bottom material. If the upper sediments are oxic, much of the dissolved Mn will be reprecipitated as Mn(IV) hydrous oxides. However, Mn in this form also can enter the water column during resuspension processes, and this has been demonstrated for the St Lawrence by Sundby *et al.* (1981). These authors showed that the manganese enrichment in the sediment surface layers was not uniform, but occurred mostly on discrete, small-sized (0.5–4 μm), Mn-enriched particles. These particles were also found in the water column, including the upper waters, and it was suggested that they can become caught up in the estuarine circulation pattern and, as a consequence of their small size, some may escape into the open ocean. This, therefore, provides a potential mechanism for the leaking of particulate manganese from the estuarine system.

Under certain estuarine conditions manganese can behave in a conservative manner. This was demonstrated for the Beaulieu estuary (UK) by Holliday & Liss (1976). Here, there was no evidence for the removal of dissolved manganese from solution, a situation that may be attributed to the low concentration of suspended solids and the rapid freshwater replacement time in the estuary (Mantoura & Morris, 1983). In many estuaries, however, dissolved manganese shows pronounced non-conservative behaviour patterns. These patterns can be of different types, but that reported by Knox *et al.* (1981) for the Tamar (UK) will serve to illustrate the behaviour of the element in an estuary where a turbidity maximum is present and where the bottom sediments are in a reduced condition. The principal features of the dissolved Mn profile in this estuary are (i) a decrease in the concentrations of dissolved manganese at low salinities; and (ii) a broad increase at higher, mid-estuarine, salinities (see Fig. 3.5b(v)). Essentially, this profile can be interpreted in terms of a fluvial input of Mn(II), which is strongly removed from solution by the oxidative precipitation of Mn(IV) at the turbidity maximum that is present in the low-salinity region of the Tamar estuary. This turbidity maximum removes a large quantity of the fluvially transported Mn(II) before it can enter the estuary proper, and in order to account for the mid-estuarine Mn(II) increase, the authors proposed that in this region interstitial waters from anoxic sediments had been swept into the overlying water column. Under these conditions, therefore, the redox-driven production of Mn(II) took place in the sediment reservoir. Different behaviour patterns of dissolved Mn, in which the production of Mn(II) takes place in the water column, have been reported for other estuaries, e.g. those of the Rhine and Scheldt (Duinker *et al.*, 1979).

It may be concluded, therefore, that profiles of dissolved manganese in estuarine surface waters are controlled by the fluvial input of dissolved Mn(II), the oxidative conversion of the dissolved Mn(II) to particulate Mn(IV), and the reduction and resolubilization of Mn(IV) in either the water column or the sediment compartments. However, owing to interestuarine variations in the parameters that control the redox-driven cycling of manganese, the element does not appear to exhibit a global estuarine mixing pattern. In many estuaries it appears that dissolved Mn is *removed* from solution, but this is not a universal feature, especially in large river–estuarine systems. For example, Edmond *et al.* (1985) reported that in the Changjiang estuary dissolved manganese appears to *increase* up to a salinity of $\sim 12\permil$, beyond which it behaves conservatively.

### 3.2.7.3 Barium

There is now a considerable body of evidence in the literature which suggests that there may be a common pattern in the behaviour of barium during estuarine mixing. This involves the desorption of the element from particulate matter at low salinities, often followed by conservative behaviour across the rest of the mixing zone (see e.g. Fig. 3.5b(iii)). Significantly, this estuarine production of barium has been reported for large river–estuarine systems, such as those of the Amazon (Boyle, 1976), the Mississippi (Hanor & Chan, 1977), the Zaire (Edmond *et al.*, 1978) and the

Changjiang (Edmond *et al.*, 1985); in the latter system the fluvial flux of dissolved barium was increased by 25% following the estuarine production of the element.

### 3.2.7.4 Boron

A number of authors have suggested that boron can be removed from solution during estuarine mixing (see e.g. Liss & Pointon, 1973). Fine-grained suspended material, especially illite-type clays, are thought to adsorb the boron from solutions, a process that is enhanced as salinity increases. Fanning & Maynard (1978), however, reported data showing that boron behaves conservatively in both the Zaire and Madalena river plumes (see also Fig. 3.5b(i)), thus casting doubt on the concept that the inorganic adsorption of the element on to suspended solids at the freshwater–salinewater interface is a globally significant geochemical process.

### 3.2.7.5 Aluminium

Several studies carried out on the behaviour of dissolved aluminium in small river–estuarine systems have indicated removal of the element at low salinities, either via flocculation, sorption on to resuspended sediment particles or precipitation with silica (see e.g. Hosokawa *et al.*, 1970; Hydes & Liss, 1977; Sholkovitz, 1978; Macklin & Aller, 1984). A consensus of the data in the literature therefore appears to indicate that the Al delivered to the oceans by fluvial sources may be largely removed in the estuarine and coastal zones before it can reach the open ocean (see e.g. Orinas & Bruland, 1985), the removal apparently varying between $\sim 20$ and $\gtrsim 50\%$ of the fluvial flux. It must be stressed, however, that few data exist on the behaviour of dissolved Al in the estuaries of large rivers. Van Bennekon & Jager (1978) carried out a preliminary study in the Zaire plume and reported that in this tropical system, which is characterized by low suspended solids, high dissolved organic matter loads and a short residence time of the water in the low-salinity region, there was a production of dissolved aluminium with maximum values at a salinity of $\sim 5\permil$. The authors concluded, however, that it is not possible to predict a global estuarine behaviour pattern for dissolved aluminium because variations in the concentrations and type of suspended solids, the amount and type of dissolved organics and the residence time of the water in different salinity regions must all be taken into account when assessing the fate of the element.

### 3.2.7.6 *The nutrients*

A variety of behaviour patterns have been reported for nitrogen, phosphorus and silicon in estuaries (see e.g. GESAMP, 1987). Nutrients are carried by rivers in both dissolved and particulate forms, and their estuarine reactivity can be brought about by chemical (e.g. adsorption or desorption) and biological (e.g. photosynthetic production) processes. In the absence of these processes, the nutrients can behave conservatively, with their distributions being controlled by the physical mixing of the river-water and seawater end-members.

In an attempt to elucidate the principals governing the fate of nutrients in estuaries, Kaul & Froelich (1984) modelled nutrient chemistry in a simple pristine system (the Ochlockonee River and Bay, USA). Using a combination of standard estuarine models, the authors were able to describe mathematically the long-term (14-month) profiles of silica, phosphate and nitrate in the estuary by deconvoluting them in terms of three component functions: i.e. linear physical mixing, removal by biological productivity and regeneration from biota. The results of the study may be summarized as follows.

**1** Around 30% of the dissolved fluvial silica flux was removed in the estuary by biological mechanisms. All of this silica, however, was regenerated within the estuary and released back into the water column. As a result, the flux of silica out of the estuary was virtually identical to the fluvial flux; i.e. little of the silica was trapped in estuarine sediments.

**2** Around 80% of the 'dissolved–reactive' phosphate, about one-third of which entered the system in a particulate-associated form, underwent biological removal in the estuary. Like silica, however, most of this was regenerated and ~100% of the fluvial phosphate entered the ocean.

**3** The authors were unable to construct a mass balance for nitrate because most of it enters and escapes the system in unmeasured forms; for example, as ammonia, or via denitrification to $N_2$ and $N_2O$.

Kaul & Froelich (1984) concluded, therefore, that estuaries are **geochemical transformers** that enhance the reactive phosphate flux to the oceans by releasing particulate phosphate, and that pass the fluvial silica flux virtually unaltered.

A number of studies have been carried out on the behaviour of the nutrients in globally important river systems and these can be illustrated with reference to the Amazon, the Changjiang and the Zaire estuarine systems.

*The Amazon estuary.* Data reported for the Amazon plume during a *summer* diatom bloom (Edmond *et al.*, 1981) showed that photosynthetic activity commenced when suspended particulates had decreased to ≤10 mg l⁻¹, which occurred at a salinity of ~7‰, thus making sufficient light available for primary production to be initiated. The diatom bloom occurred over the salinity range ~7–15‰, and here there was a complete depletion of dissolved nitrate and phosphate and a 25% depletion in dissolved silica. The underlying salt wedge became enriched in nutrients remineralized from sinking planktonic material, and mass balance calculations showed the following trends.

**1** Almost all the phosphorus was remineralized in the salt wedge, or at the sediment surface. Thus, phosphate was little affected by estuarine processes, because planktonic uptake and remineralization were in approximate balance.

**2** Only a minor part of the silica removed in production was remineralized, and ~20% of the total fluvially transported silica was transferred to the sediment sink as diatom tests. In a subsequent paper, however, DeMaster *et al.* (1983) estimated that only ~4% of the fluvial silica flux accumulates on the Amazon shelf as biogenic opal. It therefore would appear that between ~80% and ~95% of the fluvial silica escapes the Amazon plume.

**3** The remineralization of fixed nitrogen was only partial, with ~50% being transformed into forms other than nitrate or nitrite. Thus, only about half of the fixed nitrogen escaped the estuarine zone.

The finding that fluvially transported phosphate and silica essentially escaped from the Amazon plume under summer diatom bloom conditions was in agreement with the model predictions made by Kaul & Froelich (1984). Nutrient dynamics, however, can vary seasonally. In the *winter* months under conditions of very much reduced biological activity in the Amazon plume, phosphate concentrations increased at low salinies, probably as a result of desorption from particulate material (a chemical effect resulting in an additional fluvial source of dissolved phosphate), but both nitrate and silica were essentially conservative over the salinity range ~3 to ~20%.

*The Changjiang estuary.* In this estuary high concentrations of suspended particulates are maintained up to a salinity of $\geq$20‰, and as a result summer plankton blooms are found only on the inner shelf, at salinities greater than this. Edmond *et al.* (1985) have provided data on the seasonal behaviour of nutrients in the Changjiang estuary.

**1** In the *summer* months both silica and nitrate behaved conservatively out to salinities of ~20‰, but showed some depletion beyond this in the presence of plankton blooms, thus indicating biological removal. In *winter*, however, when there was no significant photosynthetic activity, both nutrients behaved in a conservative manner over the entire salinity range of the mixing zone. This was an important conclusion with regard to the estuarine behaviour of silica, because it showed that despite very high concentrations of suspended matter there was no significant removal by chemical processes associated with the particulates.

**2** In *summer* phosphate was almost completely depleted at salinities >18‰. In contrast, in the *winter* months, there was an increase in dissolved phosphate at low salinities (<5‰), indicating a desorption of phosphate from particulate material: a situation similar to that found in the Amazon plume.

*The Zaire estuary.* The nutrient relationships in this estuary are somewhat different from those in the Amazon and Changjiang systems, because photosynthetic activity in the Zaire does not occur at salinities less than ~30‰, and even at values above this the water is still relatively turbid and *in situ* production remains low (Cadee, 1978). Data on the nutrient dynamics in the Zaire estuary have been provided by Van Bennekon *et al.* (1978).

**1** Silica in the estuary is conservative out to salinities at which phytoplankton blooms occur, again indicating a lack of interaction with particulate matter, beyond which the concentrations decrease as a result of biological uptake.

**2** Phosphate profiles showed maximum concentrations at salinities of ~10‰, owing partly to desorption from suspended particulate matter, followed by uptake at higher salinities as a result of biological removal.

**3** Nitrate had a profile which was similar to that of phosphate, with highest concentrations at low salinities, but the addition to the fluvial source of nitrate mainly arose from the mixing of subsurface sea water, with higher nitrate concentrations, into the surface waters.

Nutrients are carried by fluvial transport in both particulate and dissolved forms. When sufficient dissolved nutrients are available, biological uptake is controlled by the availability of light. Turbidity has a major influence on light penetration and the distribution of suspended material appears to be the controlling variable on the onset of large-scale biological productivity in estuaries (see e.g. Milliman & Boyle, 1975). Once biological activity is initiated nutrients are removed from solution and the sinking of planktonic debris, containing nutrients in a particulate form, can lead to one of two general situations.

**1** The remineralization of the nutrients in the saline layer and their subsequent transport out to sea where, when the flow is debouched into open shelves, they can lead to an increase within subsurface waters and can sustain coastal productivity following upwelling.

**2** The trapping of the nutrients in the bottom sediment.

Although the picture is by no means clear yet, there is evidence that some nutrients may exhibit global estuarine patterns, and these can be summarized as follows.

**1** The fluvial supply of dissolved phosphate appears to be increased at low salinities via desorption from particulate matter. Phosphate is removed during photosynthetic production, but much of this is regenerated back into the water column from the sinking biota. As an overall result, estuarine processes can enhance the fluvial supply of dissolved phosphate to the oceans.

**2** Silica is removed during phytoplankton blooms but much of this is regenerated from sinking biota. In the absence of biological activity silica appears to behave conservatively and is not removed on to suspended particulates; the overall result is that the fluvial silica flux largely escapes estuaries and is delivered intact to the oceans. For example, DeMaster *et al.* (1983) estimated that ~95% of the silica supplied by the Amazon River is transported beyond the continental shelf to yield a net flux to the Atlantic Ocean of ~$4.7 \times 10^{13}$ g yr$^{-1}$.

3 The situation for nitrate is less clear, but it would seem that in some estuaries a considerable fraction of fluvial nitrate may be retained within the system.

In polluted estuaries enriched in nitrate and phosphate, however, the nutrient dynamics may be different from those in the large, relatively unpolluted estuaries.

### 3.2.7.7  Organic carbon

*Particulate organic carbon (POC).* In attempting to assess the fate of fluvial POC (and DOC) in the estuarine and coastal zones it is necessary to distinguish between a labile and a refractory fraction (see Section 9.2). The total fluvial flux of POC has been estimated to be $2.31 \times 10^{12}$ g yr$^{-1}$, and according to Ittekkot (1998) ~35% of this (i.e. $81 \times 10^{12}$ g yr$^{-1}$) is labile and may undergo oxidative destruction in estuaries and the sea. The remaining 65% (i.e. ~$150 \times 10^{12}$ g yr$^{-1}$) is refractory and can escape to the ocean environment to be accumulated in sediments.

*Dissolved organic carbon (DOC).* Data for the behaviour of DOC during estuarine mixing are somewhat contradictory. Evidence is available from both laboratory experiments and field surveys, which suggests that at least some fractions of fluvial DOC can be removed in estuaries and coastal waters by processes such as flocculation, adsorption on to particulates and precipitation (see e.g. Sieburth & Jensen, 1968; Menzel, 1974; Schultz & Calder, 1976; Sholkovitz, 1978; Hunter & Liss, 1979). These removal mechanisms would lead to the non-conservative behaviour of DOC at the land–sea margins. However, this was challenged by Mantoura & Woodward (1983). These authors reported data derived from a large-scale, two-and-a-half-year survey of the distribution, variability and chemical behaviour of DOC in the Severn estuary and Bristol Channel (UK), a system selected because of its rela-

tively long water residence times (100–200 days) and its particle-rich waters, both of which would favour the *in situ* detection of any of the DOC removal processes. The survey revealed, however, that even under these conditions DOC behaved in a conservative manner in the estuary. If such conservative nonreactivity is characteristic of the behaviour of DOC in *major*, i.e. globally significant, estuaries it will have extremely important implications for both the fluxes of DOC to the oceans (see e.g. Section 6.1.5) and the origins of the material in the deep-water oceanic DOC pool (see Section 9.2.2).

### 3.2.7.8  Trace metals

There is considerable scatter in the data for the behaviour of a number of trace metals in estuaries. Both conservative and non-conservative behaviour patterns have been reported for individual dissolved trace metals in different estuarine systems, and these patterns are obviously dependent on local conditions. For example, there is considerable evidence that trace metals such as Zn, Cu, Cd, Pb and Ni *can* exhibit non-conservative patterns during estuarine mixing, and Campbell & Yeats (1984) showed that ~50% of the dissolved Cr was removed at low salinities in the St Lawrence estuary; at salinities >5‰ the element behaved conservatively. Danielsson *et al.* (1983), however, gave data on the distributions of dissolved Fe, Pb, Cd, Cu, Ni and Zn in the Gota River estuary (Sweden), a relatively unpolluted salt-wedge system, and demonstrated that the processes which remove dissolved trace metals from solution did not operate very effectively in the system. The result of this was that, apart from iron, the metals behaved in an essentially conservative manner during estuarine mixing. Balls (1985) used salinity versus concentration plots to demonstrate the conservative behaviour of Cu and Cd in the Humber estuary (UK). It is apparent, therefore, that the non-conservative behaviour of some trace metals is not a universal feature of the estuarine environment. Further, even when non-conservative behaviour has been identified for trace metals it may involve either the loss of, or the production of, dissolved species. From the point of view of the influence that estuarine reactivity has on the trace metal chemistry of the World Ocean, it is therefore perhaps most sensible to concentrate on the behaviour of trace metals in the large estuarine systems. Although data on this topic are still relatively scarce, it is worthwhile summarizing the findings of a number of studies.

Boyle *et al.* (1982) concluded that Cu, Cd and Ni are usually unreactive during the mixing of river and sea water in the Amazon plume—a low organic and high particulate system. In contrast, Edmond *et al.* (1985) reported that in the Changjiang estuary Ni was desorbed from particulate matter at low salinities, then behaved conservatively at values above 8‰; however, Cu and Be behaved in a conservative manner over the entire mixing zone in the same estuary. Moore & Burton (1978) described the distribution of dissolved Cu in the Zaire estuary, and although trends in the data were not very clear there was a suggestion that Cu was desorbed at intermediate salinities. Both Boyle *et al.* (1982) and Edmond *et al.* (1985) have provided evidence that Cd undergoes desorption at low salinities in the Amazon and Changjiang estuaries, respectively. Measures & Edmond (1983) showed how dissolved Be can behave differently in different estuarine systems.

**1** In the Amazon system the element was removed dramatically in the early stages of mixing (up to a salinity of ~15‰) and then behaved in a conservative manner.

**2** In the Zaire system there was removal of Be at the onset of mixing followed by only minor non-conservative behaviour out to a salinity of ~28‰, after which there was a sharp drop in concentration (coincident with the onset of diatom growth), followed by conservative mixing out to open-ocean salinities.

**3** In contrast, in the Changjiang estuary Be exhibited approximately conservative behaviour; in this system, the high alkalinity of the river water had induced the flocculation or adsorption of the element from solution, with the result that dissolved concentrations were exceptionally low. Thus, it appears that, with the exception of iron, which is universally removed from solution (see above), there are few common trends in the conservative/non-conservative behaviour patterns of trace metals in estuaries. When non-conservative behaviour does occur, however, low-salinity **desorption**, i.e. gain to solution, followed by essentially conservative behaviour appears to be an important process in the behaviour of Ba, Ni and Cd in major estuarine systems.

The behaviour of elements in the estuarine mixing zone also can be dependent on speciation. For example, Froelich *et al.* (1985) found that although inorganic Ge follows silica and can behave in a non-conservative way in estuaries, monomethyl and dimethyl species of the element exhibited conservative behaviour patterns. Speciation differences can also affect the estuarine behaviour of other elements, such as Cu—see Section 11.6.2.

### 3.2.8 The estuarine modification of river-transported signals: summary

**1** Estuaries, which are located at the land–sea margins, exhibit large property gradients and can be regarded as acting as filters of river-transported chemical signals. This filter acts mainly via dissolved–particulate reactivity. This reactivity can be mediated by both physicochemical and biological processes, and is dependent on a range of interrelated factors, which include: the physical regime of an estuary, the residence time of the water, primary production, pH, differences in composition between the end-member waters, and the concentrations of suspended particulate material, organic and inorganic ligands and nutrients. The effects of the filter, however, can vary widely from one estuary to another, and it is difficult to identify global estuarine processes.

**2** The estuarine filter is also selective in the manner in which it acts on individual elements. Some dissolved elements are simply diluted on the mixing of river and sea water (conservative behaviour), whereas others undergo estuarine dissolved–particulate reactions, which lead to their addition to, or removal from, the dissolved phase (non-conservative behaviour).

**3** Particulate elements that are added to the dissolved phase (e.g. by desorption, biological degradation) increase the fluvial flux and can be carried out of the estuarine environment during tidal flushing. In contrast, elements that are added to the particulate phase (e.g. by adsorption, flocculation, biological uptake) can be trapped in the bottom sediments; however, processes such as nutrient regeneration, sediment resuspension and diagenetic release followed by the flushing out of interstitial waters can recycle these elements back into the water phase.

**4** The dissolved–particulate reactivity can take place over a number of estuarine regions. (a) Physicochemically mediated particulate–dissolved reactions are especially intense at the low-salinity region where the initial mixing of the fresh and saline end-member waters takes place. Here, processes such as flocculation (e.g. Fe) and particle adsorption–precipitation (e.g. Mn) are active in the removal of elements from solution, the latter being enhanced in the presence of a turbidity maximum. (b) Primary production, which involves the biologically mediated generation of particulate matter and the removal of nutrients from solution, is controlled largely by the availability of nutrients and the turbidity of the waters, and tends to reach a maximum at mid- and high-salinity ranges where the concentrations of particulate matter (turbidity) decrease and where nutrients are present in sufficient quantity to initiate production.

5  Around 90% of the particulate matter transported into estuaries via the fluvial flux is trapped in the estuarine environment under present-day conditions.

6  With the exception of iron, which is removed at low salinities, few dissolved elements exhibit a global estuarine behaviour pattern. For example, dissolved Cu has been shown to exhibit either conservative or non-conservative behaviour (including both addition to and removal from solution) in different estuaries. Evidence suggests, however, that desorption at low salinities, followed by conservative behaviour over the rest of the mixing zone, is an important process in the behaviour of Ba, Ni and Cd in **major** estuarine systems.

## References

Aston, S.R. (1978) Estuarine chemistry. In *Chemical Oceanography*, J.P. Riley & R. Chester (eds), Vol. 7, 361–440. London: Academic Press.

Balls, P.W. (1985) Copper, lead and cadmium in coastal waters of the western North Sea. *Mar. Chem.*, **15**, 363–78.

Bewers, J.M. & Yeats, P.A. (1977) Oceanic residence times of trace metals. *Nature*, **268**, 595–8.

Bewers, J.M. & Yeats, P.A. (1980) Behaviour of trace metals during estuarine mixing. In *River Inputs to Ocean Systems*, J.-M. Martin, J.D. Burton & D. Eisma (eds), 103–15. Paris: UNEP/Unesco.

Boyle, E.A. (1976) *The marine geochemistry of trace metals.* Thesis, MIT-WHOI, Cambridge, MA.

Boyle, E.A., Collier, R., Dengler, A.T., Edmond, J.M., Ng, A.C. & Stallard, R.F. (1974) On the chemical mass balance in estuaries. *Geochim. Cosmochim. Acta*, **38**, 1719–28.

Boyle, E.A., Edmond, J.M. & Sholkovitz, E.R. (1977) The mechanism of iron removal in estuaries. *Geochim. Cosmochim. Acta*, **41**, 1313–24.

Boyle, E.A., Huested, S.S. & Grant, B. (1982) The chemical mass balance of the Amazon plume—II. Copper, nickel and cadmium. *Deep-Sea Res.*, **29**, 1355–64.

Byrd, J.T. & Andreae, M.O. (1986) Geochemistry of tin in rivers and estuaries. *Geochim. Cosmochim. Acta*, **50**, 835–45.

Cadee, G.C. (1978). Primary production and chlorophyll in the Zaire river, estuary and plume. *Neth. J. Sea Res.*, **12**, 368–81.

Campbell, J.A. & Yeats, P.A. (1984) Dissolved chromium in the St. Lawrence estuary. *Estuarine Coastal Shelf Sci.*, **19**, 513–22.

Chester, R. (1988) The storage of metals in sediments. In *Workshop on Metals and Metalloids in the Hydrosphere*, Bochum 1987, Unesco International Hydrological Programme, 81–110.

Coonley, L.S., Baker, E.B. & Holland, H.D. (1971) Iron in the Mullica River and Great Bay, New Jersey. *Chem. Geol.*, **7**, 51–64.

Danielsson, L.G., Magnusson, B., Westerlund, S. & Zhang, K. (1983) Trace metals in the Gota River estuary. *Estuarine Coastal Shelf Sci.*, **17**, 73–85.

Degens, E.T. & Ittekkot, V. (1985) Particulate organic carbon: an overview. In *Transport of Carbon and Minerals in Major World Rivers, Part 3*, E.T. Degens, S. Kempe & R. Herrera (eds), 7–27. Hamburg: Mitt. Geol.-Palont. Inst. Univ. Hamburg, SCOPE/UNEP, Sonderband 58.

DeMaster, D.J., Knapp, G.B. & Nittrover, C.A. (1983) Biological uptake and accumulation of silica on the Amazon continental shelf. *Geochim. Cosmochim. Acta*, **47**, 1713–23.

Depetris, P.J. & Paolini, J.E. (1991) Biogeochemical aspects of South American rivers: the Parana and the Orinoco. In *Biogeochemistry of Major World Rivers*, E.T. Degens, S. Kempe & J.E. Richey (eds), 105–125. New York: Wiley.

Drever, J.L. (1982) *The Geochemistry of Natural Waters.* Englewood Cliffs, NJ: Prentice-Hall.

Duinker, J.C. (1986) Formation and transformation of element species in estuaries. In *The Importance of Chemical 'Speciation' in Environmental Processes*, M. Bernhard, F.T. Brinckman & P.J. Sadler (eds), 365–84. Berlin: Springer-Verlag.

Duinker, J.C. & Kramer, C.J.M. (1977) An experimental study on the speciation of dissolved zinc, cadmium, lead and copper in Rhine River and North Sea water, by differential pulsed anodic stripping voltammetry. *Mar. Chem.*, **5**, 207–28.

Duinker, J.C. & Notling, R.F. (1977) Dissolved and particulate trace metals in the Rhine Estuary and the Southern Bight. *Mar. Pollut. Bull.*, **8**, 65–71.

Duinker, J.C., Wollast, R. & Billen, G. (1979) Behaviour of manganese in the Rhine and Scheldt estuaries. II. Geochemical cycling. *Estuarine Coastal Shelf Sci.*, **9**, 727–38.

Dyrssen, D. & Wedborg, M. (1980) Major and minor elements, chemical speciation in estuarine waters. In *Chemistry and Biochemistry of Estuaries*, E. Olausson & I. Cato (eds), 71–120. New York: Wiley.

Edmond, J.M., Boyle, E.A., Drummond, D., Grant, B. & Mislick, T. (1978) Desorption of barium in the plume of the Zaire (Congo) River. *Neth. J. Sea Res.*, **12**, 324–8.

Edmond, J.M., Boyle, E.A., Grant, B. & Stallard, R.F. (1981) Chemical mass balance in the Amazon plume I: the nutrients. *Deep-Sea Res.*, **28**, 1339–74.

Edmond, J.M., Spivack, A., Grant, B.C., Ming-Hui, H., Zexiam, C., Sung, C. & Xiushau, Z. (1985) Chemical dynamics of the Changjiang estuary. *Cont. Shelf Res.*, **4**, 17–36.

Eisma, D. (1975) Dissolved iron in the Rhine estuary and the adjacent North Sea. *Neth. J. Sea Res.*, **9**, 222–30.

Ertel, J.R., Hedges, J.I., Devol, A.H., Richey, J.E. & Ribeiro, M. (1986) Dissolved humic substances of the Amazon River system. *Limnol. Oceanogr.*, **31**, 739–54.

Fairbridge, R.W. (1980) The estuary: its identification and geodynamic cycle. In *Chemistry and Biochemistry of Estuaries*, E. Olausson & I. Cato (eds), 1–36. New York: Wiley.

Fanning, K.A. & Maynard, V.I. (1978) Dissolved boron and nutrients in the mixing plumes of major tropical rivers. *Neth. J. Sea Res.*, **12**, 345–54.

Feth, J.H. (1971) Mechanisms controlling world water chemistry: evaporation–crystallization processes. *Science*, **172**, 870–2.

Figueres, G., Martin, J.M. & Maybeck, M. (1978) Iron behaviour in the Zaire estuary. *Neth. J. Sea Res.*, **12**, 329–37.

Froelich, P.N., Hambrick, G.A., Andreae, M.O. & Mortlock, R.A. (1985) The geochemistry of inorganic germanium in natural waters. *J. Geophys. Res.*, **90**, 1133–41.

GESAMP (1987) *Land/Sea Boundary Flux of Contaminants from Rivers*. Paris: Unesco.

Gibbs, R.J. (1970) Mechanisms controlling world river water chemistry. *Science*, **170**, 1088–90.

Gibbs, R.J. (1972) Water chemistry of the Amazon river. *Geochim. Cosmochim. Acta*, **36**, 1061–6.

Gibbs, R.J. (1973) Mechanisms of trace metal transport in rivers. *Science*, **180**, 71–3.

Gibbs, R.J. (1977) Transport phases of transition metals in the Amazon and Yukon rivers. *Bull. Geol. Soc. Am.*, **88**, 829–43.

Hanor, J.S. & Chan, L.A. (1977) Non-conservative behaviour of barium during mixing of Mississippi River and Gulf of Mexico waters. *Earth Planet. Sci. Lett.*, **37**, 242–50.

Holland, H.D. (1978) *The Chemistry of the Atmosphere and Oceans*. New York: Wiley Interscience.

Holliday, L.M. & Liss, P.S. (1976) The behaviour of dissolved iron, manganese and zinc in the Beaulieu Estuary. *Estuarine Coastal Mar. Sci.*, **4**, 349–53

Hosokawa, I.O., Ohshima, F. & Kondo, N. (1970) On the concentration of the dissolved chemical elements in the estuary of the Chikugogawa River. *J. Oceanogr. Soc. Jap.*, **26**, 1–5.

Hoyle, J., Elderfield, H., Gledhill, A. & Grieves, M. (1984) The behaviour of the rare earth elements during mixing of river and sea water. *Geochim. Cosmochim. Acta*, **48**, 148–9.

Hunter, K.A. (1983) On the estuarine mixing of dissolved substances in relation to colloid stability and surface properties. *Geochim. Cosmochim. Acta*, **47**, 467–73.

Hunter, K.A. & Liss, P.S. (1979) The surface charge of suspended particles in estuarine and coastal waters. *Nature*, **282**, 823–5.

Hydes, D.J. & Liss, P.S. (1977) The behaviour of dissolved Al in estuarine and coastal waters. *Estuarine Coastal Shelf Sci.*, **5**, 755–69.

Ittekkot, V. (1988) Global trends in the nature of organic matter in river suspensions. *Nature*, **332**, 436–8.

Ittekkot, V. & Laane, R.W.P.M. (1991) Fate of riverine particulate organic matter. In *Biogeochemistry of Major World Rivers*, E.T. Degens, S. Kempe & J.E. Richey (eds), 233–43. New York: Wiley.

Kaul, L.W. & Froelich, P.N. (1984) Modelling estuarine nutrient geochemistry in a simple system. *Geochim. Cosmochim. Acta*, **48**, 1417–33.

Kharkar, D.P., Turekian, K.K. & Bertine, K.K. (1968) Stream supply of dissolved Ag, Mo, Sb, Se, Cr, Co, Rb, and Cs to the oceans. *Geochim. Cosmochim. Acta*, **32**, 285–98.

Knox, S., Turner, D.R., Dickson, A.G., Liddicoat, M.I., Whitfield, M. & Butler, E.I. (1981) Statistical analysis of estuarine profiles: application to manganese and ammonium in the Tamar estuary. *Estuarine Coastal Shelf Sci.*, **13**, 357–71.

Konta, J. (1985) Mineralogy and chemical maturity of suspended matter in major rivers sampled under the SCOPE/UNEP Project. In *Transport of Carbon and Minerals in Major World Rivers, Part 3*, E.T. Degens, S. Kempe & R. Herrera (eds), 569–92. Hamberg: Mitt. Geol.-Palont. Inst. Univ. Hamburg, SCOPE/UNEP, Sonderband 58.

Li, Y.-H., Burkhard, L. & Teroaka, H. (1984) Desorption and coagulation of trace elements during estuarine mixing. *Geochim. Cosmochim. Acta*, **48**, 1879–84.

Liddicoat, M.I., Turner, D.R. & Whitfield, M. (1983) Conservative behaviour of boron in the Tamar Estuary. *Estuarine Coastal Shelf Sci.*, **17**, 467–72.

Liss, P.S. (1976) Conservative and non-conservative behaviour of dissolved constituents during estuarine mixing. In *Estuarine chemistry*, J.D. Burton & P.S. Liss (eds), 93–130. London: Academic Press.

Liss, P.S. & Pointon, M.J. (1973) Removal of dissolved boron and silicon during estuarine mixing of sea and river waters. *Geochim. Cosmochim. Acta*, **37**, 1493–8.

Macklin, J.E. & Aller, R.C. (1984) Dissolved Al in sediments and waters of the East China Sea: implications for authigenic mineral formation. *Geochim. Cosmochim. Acta*, **48**, 281–98.

Mantoura, R.F.C. & Morris, A.W. (1983) Measurement of chemical distributions and processes. In *Practical Procedures for Estuarine Studies*, A.W. Morris (ed.), 101–38. Plymouth: IMER.

Mantoura, R.F.C. & Woodward, E.M.S. (1983) Conservative behaviour of riverine dissolved organic carbon in the Severn Estuary: chemical and geochemical implications. *Geochim. Cosmochim. Acta*, **47**, 1293–309.

Martin, J.-M. & Maybeck, M. (1979) Elemental mass balance of material carried by major world rivers. *Mar. Chem.*, **7**, 173–206.

Martin, J.-M. & Whitfield, M. (1983) The significance of the river input of chemical elements to the ocean. In *Trace Metals in Sea Water*, C.S. Wong, E. Boyle, K.W. Bruland, J.D. Burton & E.D. Goldberg (eds), 265–96. New York: Plenum.

Martins, O. & Probst, J.-L. (1991) Biogeochemistry of major African rivers: Carbon and mineral transport. In *Biogeochemistry of Major World Rivers*, E.T. Degens, S. Kempe & J.E. Richey (eds), 127–55. New York: Wiley.

Maybeck, M. (1978) Note on dissolved elemental contents of the Zaire River. *Neth. J. Sea Res.*, **12**, 293–5.

Maybeck, M. (1979) Concentrations des eaux fluviales en elements majeurs et apports en solution aux oceans. *Rev. Geol. Dyn. Geogr. Phys.*, **21**, 215–46.

Maybeck, M. (1981a) Pathways of major elements from land to ocean through rivers. In *River Inputs to Ocean Systems*, J.-M. Martin, J.D. Burton & D. Eisma (eds), 18–30. Paris: UNEP/Unesco.

Maybeck, M. (1981b) River transport of organic carbon to the ocean. In *Flux of Carbon by Rivers to the Oceans*, 219–69. Washington, DC: U.S. Department of Energy.

Maybeck, M. (1982) Carbon, nitrogen and phosphorus transport by world rivers. *Am. J. Sci.*, **282**, 401–50.

Meade, R.H. (1972) Transport and deposition of sediments in estuaries. *Geol. Soc. Am. Mem.*, **133**, 91–120.

Measures, C.I. & Edmond, J.M. (1983) The geochemical cycle of $^9$Be: a reconnaissance. *Earth Planet. Sci. Lett.*, **66**, 101–10.

Menzel, D.W. (1974) Primary productivity, dissolved and particulate organic matter, and the sites of oxidation of organic matter. In *The Sea*, E.D. Goldberg (ed.), Vol. 5, 659–78. New York: Wiley Interscience.

Milliman, J.D. & Boyle, E.A. (1975) Biological uptake of dissolved silica in the Amazon River estuary. *Science*, **189**, 995–7.

Milliman, J.D., Summerhayes, C.P. & Barreto, H.T. (1975) Oceanography and suspended matter of the Amazon River, February–March 1973. *J. Sediment. Petrol.*, **45**, 189–206.

Moore, R.M. & Burton, J.D. (1978) Dissolved copper in the Zaire estuary. *Neth. J. Sea Res.*, **12**, 355–7.

Morris, A.W. (1986) Removal of trace metals in the very low salinity region of the Tamar Estuary, England. *Sci. Total Environ.*, **49**, 297–304.

Orians, K.J. & Bruland, K.W. (1985) Dissolved aluminium in the central North Pacific. *Nature*, **316**, 427–9.

Pickard, G.L. & Emery, W.J. (1982) *Descriptive Physical Oceanography*. Oxford: Pergamon Press.

Potsma, H. (1967) Sediment transport and sedimentation in the estuarine environment. In *Estuaries*, G.H. Lauff (ed.), 158–9. American Association for the Advancement of Science, Publication no. 83.

Potsma, H. (1980) Sediment transport and sedimentation. In *Chemistry and Biochemistry of Estuaries*, E. Olausson & I. Cato (eds), 153–86. New York: Wiley.

Reeder, S.W., Hitchon, B. & Levinson, A.A. (1972) Hydrogeochemistry of the surface waters of the Mackenzie River drainage basin, Canada—I. Factors controlling inorganic composition. *Geochim. Cosmochim. Acta*, **36**, 825–65.

Reuther, J.H. (1981) Chemical interactions involving the biosphere and fluxes of organic material in estuaries. In *River Inputs to Ocean Systems*, J.-M. Martin, J.D. Burton & D. Eisma (eds), 239–42. Paris: UNEP/Unesco.

Riley, J.P. & Chester, R. (1971) *Introduction to Marine Chemistry*. London: Academic Press.

Salomons, W. (1980) Adsorption processes and hydrodynamic conditions in estuaries. *Environ. Technol. Lett.*, **1**, 356–65.

Salomons, W. & Forstner, U. (1984) *Metals in the Hydrosphere*. Berlin: Springer-Verlag.

Schultz, D.J. & Calder, J.A. (1976) Organic $^{13}$C/$^{12}$C variations in estuarine sediments. *Geochim. Cosmochim. Acta*, **40**, 381–5.

Sclater, F.R., Boyle, E.A. & Edmond, J.M. (1976) On the marine geochemistry of nickel. *Earth Planet Sci. Lett.*, **31**, 119–28.

Shiller, A.M. & Boyle, E. (1985) Dissolved zinc in rivers. *Nature*, **317**, 49–51.

Shink, D. (1981) Behaviour of chemical species during estuarine mixing. In *River Inputs to Ocean Systems*, J.-M. Martin, J.D. Burton & D. Eisma (eds), 101–2. Paris: UNEP/Unesco.

Sholkovitz, E.R. (1978) The flocculation of dissolved Fe, Mn, Al, Cu, Ni, Co and Cd during estuarine mixing. *Earth Planet. Sci. Lett.*, **41**, 77–86.

Sholkovitz, E.R. & Copland, D. (1981) The coagulation, solubility and adsorption properties of Fe, Mn, Cu, Ni, Cd, Co and humic acids in river water. *Geochim. Cosmochim. Acta*, **45**, 181–9.

Sholkovitz, E.R., Boyle, E.A. & Price, N.B. (1978) The removal of dissolved humic acids and iron during estuarine mixing. *Earth Planet. Sci. Lett.*, **40**, 130–6.

Sieburth, J.M. & Jensen, A. (1968) Studies on algal substances in the sea. I. Gelbstoff (humic materials) in terrestrial and marine waters. *J. Exp. Mar. Biol. Ecol.*, **2**, 174–80.

Spitzy, A. & Leenheer, J. (1991) Dissolved organic carbon in rivers. In *Biogeochemistry of Major World Rivers*,

E.T. Degens, S. Kempe & J.E. Richey (eds), 213–32. New York: Wiley.

Stallard, R.F. & Edmond, J.M. (1981) Geochemistry of the Amazon 1. Precipitation chemistry and the marine contribution to the dissolved load at the time of peak discharge. *J. Geophys. Res.*, **86**, 9844–58.

Stallard, R.F. & Edmond, J.M. (1983) Geochemistry of the Amazon 2. The influence of geology and weathering environment on the dissolved load. *J. Geophys. Res.*, **88**, 9671–88.

Subramanian, V. & Ittekkot, V. (1991) Carbon transport by the Himalayan rivers. In *Biogeochemistry of Major World Rivers*, E.T. Degens, S. Kempe & J.E. Richey (eds), 157–68. New York: Wiley.

Sundby, B., Silverburg, N. & Chesselet, R. (1981) Pathways of manganese in an open estuarine system. *Geochim. Cosmochim. Acta*, **45**, 293–307.

Telang, S.A., Pocklington, R., Naidu, A.S., Romankevich, E.A., Gitelson, I.I. & Gladyshev, M.I. (1991) Carbon and mineral transport in major North American, Russian Arctic and Siberian rivers: the St. Lawrence, the Mackenzie, the Yukon, the Arctic Alaskan rivers, the Arctic Basin rivers in the Soviet Union, and the Yenisei. In *Biogeochemistry of Major World Rivers*, E.T. Degens, 3S. Kempe & J.E. Richey (eds), 75–104. New York: Wiley.

Thurman, E.M. (1985) *Organic Geochemistry of Natural Waters*. Boston, MA: Nijhoff and Junk.

Trefry, J.H. & Presely, B.J. (1976) Heavy metal transport from the Mississippi River to the Gulf of Mexico. In *Marine Pollutant Transport*, Windom, H.L. & Duce, R.A. (eds), 39–76. Lexington, MA: Lexington Books.

Van Bennekon, A.J. & Jager, J.E. (1978) Dissolved aluminium in the Zaire River plume. *Neth. J. Sea Res.*, **12**, 358–67.

Van Bennekon, A.J. & Salomons, W. (1981) Pathways of nutrients and organic matter from land to ocean through rivers. In *River Inputs to Ocean Systems*, J.-M. Martin, J.D. Burton & D. Eisma (eds), 33–51. Paris: UNEP/Unesco.

Van Bennekon, A.J., Berger, G.W., Helder, W. & De Vries, R.T.P. (1978) Nutrient distribution in the Zaire Estuary and river plume. *Neth. J. Sea Res.*, **12**, 296–323.

Van der Weijden, C.H., Arnoldus, M.J.H.L. & Meurs, C.J. (1977) Desorption of metals from suspended material in the Rhine estuary. *Neth. J. Sea Res.*, **11**, 130–45.

Walling, D.E. & Webb, B.W. (1987) Material transport by the world's rivers: evolving perspectives. In *Water for the Future: Hydrology in Perspective*, 313–29. Wallingford: International Association of Hydrological Sciences, Publication no. 164.

Waslenchuk, D.C. & Windom, H.L. (1978) Factors controlling the estuarine chemistry of arsenic. *Estuarine Coastal Shelf Sci.*, **7**, 455–62.

Williams, P.J. (1981) Primary productivity and heterotrophic activity in estuaries. In *River Inputs to Ocean Sytems*, J.-M. Martin, J.D. Burton & D. Eisma (eds), 243–9. Paris: UNEP/Unesco.

Windom, H.L. & Smith, R. (1985) Factors influencing the concentration and distribution of trace metals in the South Atlantic Bight. In *Oceanography of the Southeastern U.S. Continental Shelf*, L.P. Atkinson, D.W. Menzel & K.A. Bush (eds), 141–152. Washington, DC: AGU.

Windom, H.L., Wallace, G., Smith, R., Dudek, N., Maeda, M., Dulmage, R. & Storti, F. (1983) Behaviour of copper in southeastern United States estuaries. *Mar. Chem.*, **12**, 183–94.

Yeats, P.A. & Bewers, J.M. (1982) Discharge of metals from the St. Lawrence River. *Can. J. Earth Sci.*, **19**, 982–92.

Zorbrist, J. & Stumm, W. (1981) Chemical dynamics of the Rhine catchment area in Switzerland, extrapolation to the 'pristine' Rhine river input to the ocean. In *River Inputs to the Ocean Systems*, J.-M. Martin, J.D. Burton & D. Eisma (eds), 52–63. Paris: UNEP/Unesco.

# 4 The transport of material to the oceans: the atmospheric pathway

The trophosphere is a reservoir in which particles have a relatively short residence time, usually in the order of days for those with radii in the range ~0.1 to ~10 μm, and from which they are removed at about the same rate as they enter. The particles carried in the marine atmosphere have a different genetic history from those transported to the oceans via river run-off, one of the most important differences being that they do not undergo trapping, or modification, in the estuarine *filter* at the land–sea margins. At the present time, estuaries act as an effective trap for river-transported solids, holding back ~90% of river particulate material (RPM) (see Section 6.1.2). Potentially, therefore, the atmosphere is perhaps the most important pathway for the long-range transport of particulate material directly to open-ocean areas. This has become increasingly apparent over the past few decades. For example, Delany *et al.* (1967) concluded that the land-derived material in equatorial North Atlantic deep-sea sediments deposited to the east of and on the Mid-Atlantic Ridge has been derived wholly from wind transport. Such long-range transport, spanning several thousand kilometres, also has been found over the Pacific Ocean, and Blank *et al.* (1985) estimated that almost all of the non-biogenic material in deep-sea sediments in the central North Pacific is essentially aeolian in origin. In addition to supplying material to sediments the atmospheric components can have a pronounced effect on the biogeochemistry of the oceanic mixed layer. Atmospheric material, however, has to be transferred across an interface before it is introduced into the ocean system; this is the air–sea interface (or micro-layer), which is involved in the exchange of particulate and gaseous phases between sea water and the atmosphere.

## 4.1 Material transported via the atmosphere: the marine aerosol

### 4.1.1 Introduction

A suspension of solid and liquid material in a gaseous medium usually is referred to as an **aerosol**. Prospero *et al.* (1983) defined a number of aerosol types on the basis of their compositions and sources, and a summary of their classification is given in Table 4.1. It is important to stress at this stage that the components making up the world aerosol originate from two different types of processes: (i) the direct formation of particles (e.g. during crustal weathering, sea-salt generation, volcanic emissions); and (ii) the indirect formation of particles in the atmosphere itself by chemical reactions and by the condensation of gases and vapours. A generalized relationship between the processes responsible for the generation of aerosol particles and their size spectra is illustrated in Fig. 4.1. In this figure, the particles are divided into two broad groups, **fine** particles ($d < 2\,\mu m$) and **coarse** particles ($d > 2\,\mu m$), and there are three size maxima, two in the fine class and one in the coarse class. The maxima in the *fine* mode relate to two particle populations: (i) those in the Aitken nuclei range; and (ii) those in the accumulation range. **Aitken nuclei** originate predominantly from combustion processes, which are chiefly anthropogenic but also include volcanic emissions and biomass burning, and also from some non-combustion processes. Particles in the **accumulation** mode are thought to result primarily from the coagulation of Aitken nuclei into larger aggregates. In contrast, particles in the *coarse* mode have been formed by mechanical action, which can involve both low-temperature processes (e.g. the generation of crustal dusts and sea salts) and high-temperature processes (e.g. the formation of some industrial fly-ash).

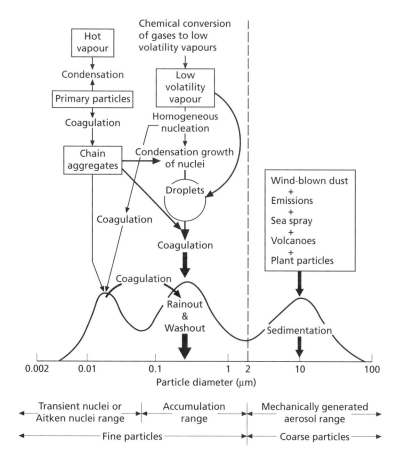

**Fig. 4.1** Schematic representation of the processes involved in the generation and removal of atmospheric particles (from Whitby, 1977).

**Table 4.1** The classification of aerosols on the basis of their composition or sources (based on Prospero *et al.*, 1983).

| Natural aerosols | Anthropogenic aerosols |
| --- | --- |
| Sea-spray residues | Direct anthropogenic particle |
| Windblown mineral dust | emissions |
| Volcanic effluvia | Products from the conversion of |
| Biogenic materials | anthropogenic gases |
| Smoke from the burning of land biota | |
| Natural gas-to-particle conversion products | |

The particle size of the marine aerosol has been described by Junge (1972), who identified five classes of size-related components in a North Atlantic aerosol. These are (i) particles with diameters >40 μm, (ii) sea-spray particles, (iii) mineral dust particles, (iv) tropospheric background particles and (v) particles with diameters <0.06 μm. The nature of the tropospheric background material has been the subject of much speculation, but it is generally thought that an aerosol of fairly uniform composition, dominated by sulphate, is present over ~85% of the troposphere. If the very small particles ($d < 0.06$ μm) are excluded, the marine aerosol therefore can be regarded as being composed of the tropospheric background material, upon which is superimposed continental dust (the mineral aerosol), sea spray (the sea-salt aerosol) and the large sized component (consisting of biological material, pollutants such as 'cokey balls', and giant sea-spray particles).

Close to the source, the composition of a specific aerosol component, such as mineral dust, will be related closely to that of the parent material. During their residence time in the atmosphere, however, both coarse and fine particles can undergo physical and chemical modifications, and the character of an aerosol will change with increasing distance from the source. For example, the concentration falls off, the particle size distribution is altered and the chemical composition is modified and tends to become more uniform. One result of aerosol transport is that the larger particles are removed more rapidly than small particles, so that aerosols over coastal regions will differ from those over distant open-ocean areas, which represent an integration of material from many sources (see e.g. Maring & Duce, 1990). Aerosols therefore may be regarded as the end-product of a complex series of processes and are not static components, but are best regarded in terms of a **dynamic aerosol continuum** (Prospero *et al.*, 1983). The physical and chemical composition of the marine aerosol therefore is extremely variable in both space and time, and the aerosol characteristics are governed by a combination of processes that are involved in the generation, conversion, transport and removal of particles. In the following sections the marine aerosol will therefore be discussed in terms of this general 'generation–conversion–transport–removal' framework.

### 4.1.2 The transport of aerosols within the troposphere

Most of the continentally derived material (both natural and anthropogenic) that contributes to the world aerosol is injected initially into the planetary boundary layer of the atmosphere. This is the layer in which the direct influence of the underlying surface is felt, and it has a height of $\sim$1000 to $\sim$1500 m over the land and $\sim$300 to $\sim$600 m over the sea (Hasse, 1983). The upper surface of the boundary layer is defined by an inversion, which inhibits the transfer of material to the upper atmosphere (Prospero, 1981). According to Prospero (1981), the primary transport path by which material generated close to the continents, i.e. within tens to hundreds of kilometres, reaches the sea surface may be via this marine boundary layer. Over longer distances, however, the major transport path is probably via the free troposphere above the marine boundary layer. Lead-210 (half-life 22 yr), which is supplied to the atmosphere from the radioactive decay of $^{222}$Rn (half-life 3.8 days) that escapes from soils, can be used as a tracer for the dispersal of continentally derived components

in the marine atmosphere (see e.g. Turekian *et al.*, 1989). It must also be remembered that, because most collections of the marine aerosol are made in the boundary layer, they are probably not representative of the concentrations at high levels in the troposphere.

The transport of aerosols within the troposphere takes place via the major wind systems, which operate on a global scale. For simplicity it is convenient to describe atmospheric circulation in terms of a meridional three-cell model. In this model the circulation in each hemisphere can be related to the presence of three cells of alternating belts of easterly and westerly zonal winds, the cells being separated by pressure zones. This is illustrated in Fig. 4.2(a) with respect to the Northern Hemisphere. The three pressure zones are:

1 the equatorial low-pressure belt or trough, originally termed the 'Doldrums' but now referred to as the Inter-Tropical Convergence Zone (ITCZ);

2 the subtropical high-pressure belt ($\sim$30°N): an area of dry sinking air characterized by calms and often called the 'Horse Latitudes';

3 the low-pressure belt at $\sim$60°N.

The main features of the circulation lying between these zones are illustrated in Fig. 4.2(b) and are described in very general terms below.

1 In the Hadley cell operating between the equatorial low-pressure and subtropical high-pressure belts, the prevailing zonal winds are northeasterly or southeasterly, and are termed the **trades**: the Northeast Trades in the Northern Hemisphere and the Southeast Trades in the Southern Hemisphere. These trades are directionally quite stable and have relatively constant velocities, in the range $\sim$16 to $\sim$32 km h$^{-1}$. The trades converge at the ITCZ. In some regions (e.g. India, Southeast Asia and China) the trade-wind pattern can be modified by the development of a **monsoon** system. Here there is seasonal reversal of wind direction in which transequatorial westerlies replace the regular trade-wind easterlies. These monsoons can result in the transfer of air across the Equator and so can initiate the inter-hemispheric exchange of aerosol components.

2 The circulation in the temperate or mid-latitude Ferrel cell, which lies between the subtropical high-pressure belt at $\sim$30°N and the low-pressure belt at $\sim$60°N, has a prevailing zonal wind that is westerly in direction. The **westerlies**, however, are not as direc-

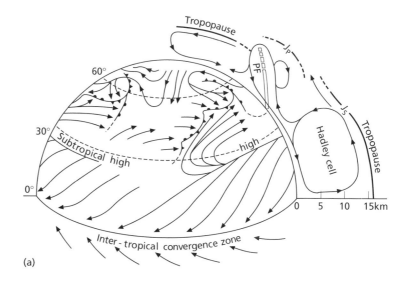

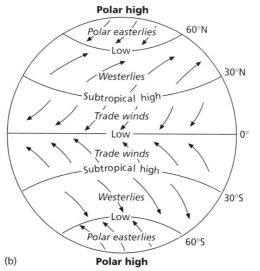

**Fig. 4.2** Circulation in the atmosphere. (a) Schematic representation of the general circulation in the atmosphere at the surface and a meridional cross-section (from Hasse, 1983; after Defant & Defant, 1958). $J_P$ and $J_S$ indicate the average positions of the Polar Front (PF) and the Subtropical Jet Streams. (b) Schematic representation of the surface wind distribution (from Iribarne & Cho, 1980).

tionally stable as the trades and exhibit considerable variability in both space and time. Further, they have a greater force than the trades, and are sometimes referred to as the 'upper westerlies' because their force increases with altitude, to reach a maximum at ~12 km. Cores of high-speed winds are embedded in the westerlies, the two main types being the Subtropical Jet Stream and the Polar Front Jet Stream, both of which are important for the transport of aerosols in the mid-latitudes.

3 A third or polar cell is found to the north of the low-pressure belt located at ~60°N. Here, the winds are the **polar easterlies**, although the atmospheric circulation system is extremely complex.

Aerosols are dispersed about the surface of the planet by these various zonal wind systems and by transfer between them. Such transfer, however, tends to be inhibited by the various pressure belts. Despite this, the inter-hemispheric transfer of aerosols can take place; for example, when the ITCZ is shifted across the Equator, or when it is absent during the development of some monsoon systems.

### 4.1.3 Sources of material to the marine atmosphere

The particulate and gaseous material supplied to the atmosphere can originate from either **natural** or **anthropogenic** sources, and when considering the elemental chemistry of both of these it is useful to distinguish between *low*-temperature and *high*-temperature generation processes (see Section 4.1.1 and Fig. 4.1). This is important because the forms in which the elements are present in aerosols, i.e. their speciation, can be strongly dependent on the temperature at which they were released from their parent hosts (see Section 4.2.1.3). Some of the more important sources of material to the world atmosphere are listed below.

#### 4.1.3.1 Natural sources

*The Earth's crust.* The Earth's surface can supply both particulate and gaseous components to the atmosphere. The generation of particulate crustal material involves a low-temperature mechanical mobilization of surface deposits by wind erosion, and these products are an important component of many marine aerosols.

*The oceans.* Like the crust, the surface of the ocean can supply material to the atmosphere in both particulate and gaseous forms. Particulate material is formed during the low-temperature generation of sea salts by mechanical action. Volatilization from the sea surface, in the form of a sea-to-air gaseous flux, also can be a source of some volatile components (see Section 8.2).

*Volcanic activity.* Volcanoes can release particulate material, e.g. ash, together with gaseous phases formed from high-temperature volatilization processes. These gaseous phases may undergo condensation reactions, which can result in the enrichment of the particulate material in some volatile elements.

*The biosphere.* The supply of material to the atmosphere from the biosphere occurs through the high-temperature burning of vegetation (phytomass) and by the emission of particulate and vapour phases from plant surfaces and soils. Phytomass burning includes natural forest and grassland fires, but most occurs from farming practices, especially the so-called slash and burn 'shifting' cultivation technique. Phytomass burning is probably more significant at low latitudes, and it has been estimated that >80% of the annual phytomass burn occurs in the 'developing' countries (see e.g. Suman, 1986). Material released during phytomass burning includes $CO_2$, $CH_4$ and particulates; a major component of the latter is charcoal (elemental carbon)—see e.g. Cachier *et al.* (1989) and Buat-Menard *et al.* (1989).

*Outer space.* Extraterrestrial sources provide a small, but interesting, supply of material to the atmosphere. This extraterrestrial material includes micrometeorites (cosmic spherules) and a number of cosmic-ray-produced radioactive or stable nuclides (see Section 15.4).

#### 4.1.3.2 Anthropogenic sources

There are a wide variety of anthropogenic processes that release both particulate and gaseous material into the atmosphere. These include fossil fuel burning, mining and the processing of ores, waste incineration, the production of chemicals, agricultural utilization, and numerous other industrial and social activities.

#### 4.1.3.3 Source strengths

Particle emission source strengths of some of the processes listed above are given in Table 4.2. The data are presented in terms of natural and anthropogenic emissions, and each of these is subdivided into **direct** and **particle conversion** generation processes. A number of points must be considered when any attempt is made to evaluate these data.

1 The data are average values, based on a number of wide-ranging estimates. This is illustrated by the inclusion of two data sets for the direct production of natural particles; however, for the purpose of making general comparisons, the data set provided by Prospero *et al.* (1983) will be used here.

2 The emission of natural particles considerably exceeds that from anthropogenic processes, especially for direct particle production.

3 For anthropogenic emissions, particle conversion

**Table 4.2** Estimates of the global emissons of particulate material to the atmosphere (units, $10^{12}$ g yr$^{-1}$).

| | | Global production | | |
|---|---|---|---|---|
| | Source | Prospero *et al.* (1983) | Nriagu (1979) | Lantzy & Mackenzie (1979) |
| Anthropogenic | Direct particle production | 30 | | |
| | Particles formed from gases: | | | |
| | converted sulphates | 200 | | |
| | others | 50 | | |
| | Total anthropogenic | 280 | | 200 |
| Natural | Direct particle production: | | | |
| | forest fires | 5 | 36 | |
| | volcanic emissions | 25 | 10 | |
| | vegetation | | 75 | |
| | crustal weathering (mineral dust) | 250 | 500 | |
| | sea salt | 500 | 1000 | |
| | Particles formed from gases: | | | |
| | converted sulphates | 335 | | |
| | others | 135 | | |
| | Total natural | 1250 | | |
| Total | | 1530 | | |

from gases exceeds that from direct production by an order of magnitude; however, direct production dominates the natural emission of particles. In total (i.e. anthropogenic + natural), direct particle production yields $\sim810 \times 10^{12}$ g yr$^{-1}$ and conversion from gases results in $\sim750 \times 10^{12}$ g yr$^{-1}$. It is apparent, therefore, that on a global scale the two processes have about the same magnitude. Sulphates are the principal particles derived from gas conversion, and $\sim535 \times 10^{12}$ g yr$^{-1}$ are produced in the atmosphere from a combination of natural and anthropogenic processes. This is similar in magnitude to sea-salt production ($\sim500 \times 10^{12}$ g yr$^{-1}$) and exceeds mineral dust generation ($\sim250 \times 10^{12}$ g yr$^{-1}$). The marine aerosol is therefore dominated by small-sized background **sulphate**, which is generated largely from gaseous precursors, and to a lesser extent nitrate (both of which are important oxidative end-members of the atmospheric sulphur and nitrogen cycles), together with directly generated **sea-salt** and **mineral dust**. The distributions of these components are considered in the following sections. Although sulphate, sea-salt and mineral dust can dominate the marine aerosol, however, they can be present in intimate mixtures rather than as individual components (see e.g. Andreae *et al.*, 1986).

### 4.1.4 The principal components of the marine aerosol

The largest global sources of marine aerosols are the sea surface (sea-salt), the Earth's crust (mineral dust) and anthropogenic processes (mainly sulphates). Sea-salts are supplied by all the surface ocean, but the mineral dust and sulphate sources are concentrated in specific belts. The principal sources of mineral dust and anthropogenic aerosols (essentially sulphates) are illustrated in Fig. 4.3, from which it can be seen that the continental sources of both the natural (dust) and anthropogenic (sulphate) aerosols are located principally in the Northern Hemisphere.

#### 4.1.4.1 *The mineral aerosol over the World Ocean*

Particulate material in the form of continental dust is mobilized into the atmosphere mainly by wind erosion. The process is strongly dependent on the nature of the surface cover in the source (or catchment) area, which itself is dependent on the prevailing geological, weathering and general climatic regimes. In regions having loose surface deposits, e.g. desert and arid land areas, there is a readily available reser-

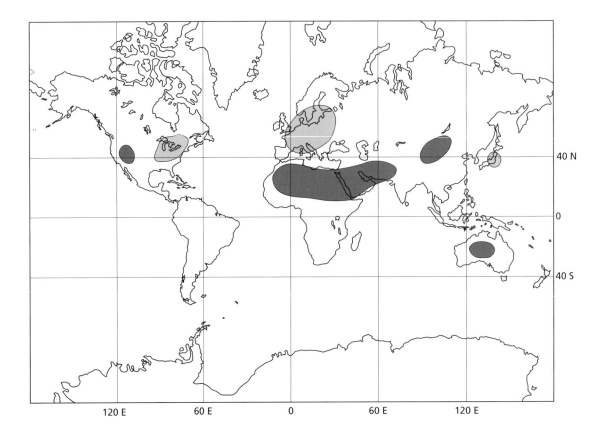

**Fig. 4.3** Major source regions of aerosol production (after Gilman & Garrett, 1994). Light grey areas represent regions of anthropogenic industrial emissions, and dark grey areas indicate regions of mineral aerosol production.

voir of particulate material that is susceptible to wind erosion and transport during dust storm events. In contrast to the conditions found in arid regions, surface covers of forest, grassland and snow or ice will considerably reduce the production rates of atmospheric dusts. Most mineral dust present in the atmosphere is produced from surface soils. In addition to the low-temperature mobilization of this surface material into the atmosphere, however, high-temperature volcanic activity can generate particles that form part of the mineral aerosol. This volcanic activity is a sporadic source, but at the time of large-scale eruptions very large quantities of material can be injected into the atmosphere and can affect world climate.

Junge (1979) estimated the magnitude of the inputs to the global tropospheric dust cycle, and a summary of the data is given in Table 4.3(a). Two important conclusions can be drawn from these data.
**1** The Northern Hemisphere has a higher tropospheric dust burden than the Southern Hemisphere, mainly as a result of the larger area of land mass in the northern latitudes.
**2** The Sahara Desert plays an extremely important role in the Northern Hemisphere dust cycle.
More recent estimates, however, have also indicated the importance of the Asian arid lands as sources of dust transported to the oceans in the Northern Hemisphere. A summary of some of these recent estimates is given in Table 4.3(b). It is apparent from this table that the dust flux to the western North Pacific ($\sim 300 \times 10^{12}\,g\,yr^{-1}$), much of which originates from Asian desert sources, has a similar magnitude to that for Saharan dust transported to the tropical North Atlantic. On the basis of the more recent data, the global dust source strength is $\sim 800 \times 10^{12}$ to $\sim 1100$

**Table 4.3** The global tropospheric dust cycle.

(a) After Junge (1979).

| Tropospheric region | Dust burden ($10^{12}$ g) | Source strength ($10^{12}$ g yr$^{-1}$) |
|---|---|---|
| Northern Hemisphere* | 3.0† | 150‡ |
| Southern Hemisphere* | 1.0† | 50‡ |
| Total troposphere* | 4.0† | 200‡ |
| Sahara plume | 1.2–4.0 | 60–200 |
| Total troposphere§ | 3.2–12.0 | 130–800 |

* These estimates disregard special production in deserts, particularly the Saharan plume in the North Atlantic.
† Estimated uncertainty factor about ±2.
‡ Source strength calculated from dust burden, assuming an aerosol residence time of 1 week. Estimated uncertainty factor about ±3.
§ After application of the above uncertainty factors.

(b) After Prospero (1981) and Prospero *et al.* (1989).

| Ocean region | Deposition rate $10^{-6}$ g cm$^2$ yr$^{-1}$ | Deposition rate $10^{12}$ g yr$^{-1}$ |
|---|---|---|
| North Atlantic, north of trades | 82 | 12 |
| North Atlantic trades | — | 100–400 |
| South Atlantic | 85 | 18–37 |
| Indian Ocean | 450 | 336 |
| North Pacific: | | |
|     western North Pacific | 5000 | 300 |
|     central and eastern | 11–62 | 30 |
| South Pacific | 5–64 | 18 |
| Entire Pacific | | 350 |
| All Oceans (minimum–maximum) | | 816–1135 |

**Table 4.4** Mineral aerosol loadings on an Arctic–Atlantic–Antarctic transect.

| Oceanic region | Mineral aerosol concentration (µg m$^{-3}$ of air) |
|---|---|
| Arctic | Variable; haze episodes |
| North Atlantic, westerlies | ~0.1–~2.5 |
| North Atlantic, Northeast Trades | <1.0–>10$^3$ |
| South Atlantic, Southeast Trades | ~0.1–~1.0 |
| South Atlantic, westerlies | <0.1 |
| Antarctic | <0.01 |

$\times 10^{12}$ g yr$^{-1}$; however, there is still considerable uncertainty in these estimates.

A large amount of data are now available for the concentrations of mineral aerosols in the marine atmosphere, and some of these are reviewed below in terms of the major oceans. It must be remembered, however, that the loadings cited include non-dust components.

*The Atlantic Ocean and surrounding waters.* In many ways the Atlantic Ocean and its surrounding waters provide a classic example of the factors that control the distributions of the mineral aerosol over marine areas. The reason for this is that the Atlantic is a relatively narrow elongated ocean, which, together with the polar waters on its northern and southern extremities, encompasses all the major wind systems and has a wide variety of particle catchment sources on its surrounding land masses. The latitudinal effects that these parameters exert on the transport of mineral aerosols can be illustrated in terms of an Arctic–North Atlantic–South Atlantic–Antarctic transect. Mineral aerosol dust loadings along this transect are listed in Table 4.4, from which it can be seen that there is a broad pattern in the data, with a general increase in the dust loadings towards lower latitudes in both hemispheres.

The **Arctic polar regions** are remote, but the atmosphere is less pristine than might be expected because it is subjected to seasonal injections of **Arctic haze**, which bring in a mixture of both mineral and pollutant components from distant sources such as Eurasia, the eastern USA and central Europe.

In the general region of the **North Atlantic westerlies** (~65 to ~40°N) mineral aerosol concentrations are probably <0.5 µg m$^{-3}$ of air, rising to ~2.5 µg m$^{-3}$ of air closer to the land masses. The westerlies transport aerosols generated in the USA–European pollution belt, but the forest–grass surface cover in the catchment inhibits the large-scale mobilization of crustal material.

The most striking feature in the distribution of the mineral aerosol over the Atlantic is the very high concentrations found in the **Northeast Trades** off the coast of West Africa. These winds transport crustal material originating in the Sahara Desert, and estimates of the Saharan dust burden carried over the Atlantic vary in the range $60 \times 12^{12}$ to $400 \times$

$12^{12}$ g yr$^{-1}$ (see Table 4.3). The offshore transport of this Saharan dust takes place mainly above the trade wind inversion in the 'Harmattan'. Mineral aerosol concentrations in the Atlantic Northeast Trades are among the highest found over the World Ocean and can reach values >$10^3$ μg m$^{-3}$ of air. However, the concentrations can be extremely variable. For example, at Sal Island, off the coast of West Africa, the mineral aerosol varied over the range 10–180 μg m$^{-3}$ of air during a 3-month period in 1974 (Savoie & Prospero, 1977). The variations result from outbreaks, or **pulses**, of Saharan dust-carrying air. The effect of these pulses was apparent in the data reported by Chester *et al.* (1984a), who found mineral aerosol concentrations in the Northeast Trades off West Africa to range between <1 μg m$^{-3}$ of air in quiet periods and ~700 μg m$^{-3}$ of air in dust storm outbreaks. Dust pulses therefore can be thought of as imposing intermittent concentration increases on to the background aerosol, and this is illustrated below with respect to the Pacific mineral aerosol (see Fig. 4.4b). Saharan dust can be transported all the way across the Atlantic, and even further west (see e.g. Delany *et al.*, 1967). It may be concluded, therefore, that there is a **dust envelope** over the North Atlantic in the area underlying the Northeast Trades, i.e. between ~30°N and ~5°N (see e.g. Chester *et al.*, 1979). The limits of the area over which the Saharan dust is transported vary seasonally, with a southerly shift in the dust envelope being evident in the winter months (Prospero, 1968). For example, over the Atlantic the zone of maximum transport shifts northwards from ~5°N during winter, when the dusts reach the northeastern coast of South America, to ~20°N in the summer, when they reach the Caribbean Sea: a seasonal migration which is associated with the latitudinal movement of large-scale atmospheric circulation.

Across the Equator in the Atlantic **Southeast Trades** the mineral aerosol concentrations fall off dramatically, and although few reliable data are available it would appear that in the region lying between the Equator and ~40°S the loadings are generally <~1 μg m$^{-3}$ of air (see e.g. Prospero, 1979; Chester *et al.*, 1984a). It is apparent, therefore, that the desert regions of southern Africa do not act as massive mineral aerosol reservoirs supplying material to the Southeast Trades (see e.g. Chester *et al.*, 1971).

Further south, the mineral aerosol concentrations continue to decrease, to reach values in the **South Atlantic westerlies** that are about an order of magnitude lower than those in the Atlantic Southeast Trades (see e.g. Chester *et al.*, 1984a).

In the pristine air over the snow-covered **Antarctic plateau** aerosol concentrations are extremely low, ranging from <0.004 μg m$^{-3}$ of air (winter) to ~0.01 μg m$^{-3}$ of air (summer); and in both seasons mineral components make up less than ~5% of the total aerosol populations.

*The Mediterranean Sea.* The Mediterranean Sea is confined to a narrow latitudinal band, and from the point of view of aerosol inputs it is of special interest because it lies between large source regions for both anthropogenic and mineral dust generation (see Fig. 4.3). As a result the Mediterranean Sea has contrasting aerosol-generation regimes on its opposite shores. On the northern shore it is bordered by nations having a variety of economies, ranging from industrial to agricultural, and air masses transported into the Mediterranean atmosphere from the north have often crossed part of the European 'pollution belt', which provides a continuous supply of anthropogenic 'background' aerosol. In contrast, it is bordered on its southern shore by the North African (Saharan) desert belt, which acts as a large reservoir for the supply of mineral dusts. These dusts are delivered in a seasonal pattern, mainly in the form of sporadic, i.e. non-continuous, dust 'pulses', which perturb the anthropogenic 'background' aerosol. Mineral dusts also can be supplied to the east of the region from the deserts of the Middle East. In addition to the European anthropogenic source and the Saharan and Middle Eastern desert sources, there are inputs of particulate material to the Mediterranean atmosphere from the Atlantic Ocean to the west, from the local sea surface and from volcanic activity in the region. Aerosols are transported into the Mediterranean atmosphere from all these sources, and the inputs vary seasonally. For example, there is maximum transport of African dust to central and eastern basins during the spring, and to the western and central basins during the summer: patterns that follow the distribution of Mediterranean cyclones. As a result the concentrations of mineral dust can exhibit wide concentration variations in the Mediterranean atmosphere. For example, Chester *et al.* (1984b)

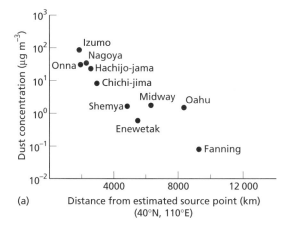

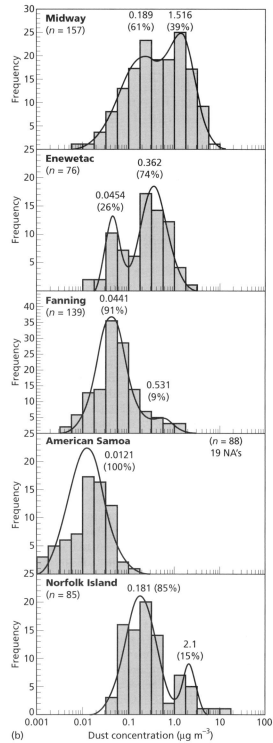

**Fig. 4.4** The distribution of mineral aerosol over the Pacific Ocean (from Prospero *et al.*, 1989). (a) The concentration of mineral aerosol over the North Pacific plotted as a function of distance from the estimated source point in Asia. (b) Frequency distributions of mineral aerosol concentrations over the Pacific Ocean. The mineral aerosols for the Midway and Enewetak Island sites in the North Pacific show a bimodal distribution, the lower mode representing the background aerosol and the upper mode representing dust pulses from Asian desert sources. In contrast, the aerosol frequency distributions for the Fanning Island site, which is close to the Equator, and the American Samoa site in the South Pacific, are essentially unimodal and represent background aerosol only. The locations of the island sites are shown in Fig. 4.5.

showed that incursions of Saharan dust could be identified in the lower troposphere over the Tyrrhenian Sea. The mineral dust concentrations in these pulses (average, $25\,\mu g\,m^{-3}$ of air) were an order of magnitude higher than those in the 'European background' air (average, $1.4\,\mu g\,m^{-3}$ of air). Pulses of this kind are intermittent, however, and make it difficult to identify an average mineral aerosol loading for the Mediterranean.

Evidence is now becoming available that long-term patterns in dust transport can be linked to large-scale variability in the atmosphere. In this context, the study reported by Moulin *et al.* (1997) provides a major landmark in our knowledge of the factors controlling the movement of atmospheric dust. These authors used daily satellite observations of airborne dusts to obtain an 11-yr regional-scale analysis of dust transport out of Africa to the Atlantic Ocean and the Mediterranean Sea. They were able to show that **seasonal** variability in dust transport to both marine regions could be explained by synoptic meteorology. The most important conclusion of the study, however, concerned the **interannual** variations in dust transport. Despite differences in the source regions and transport processes, these interannual variations were similar over both the Atlantic and the Mediterranean, which suggested that large-scale forces are constraining the dust export to both regions. Precipitation is one of the most efficient processes that can induce interannual variability in the transport of African dust, and Moulin *et al.* (1997) combined changes in precipitation patterns with those in atmospheric circulation to identify the constraints on dust transport out of Africa by linking them to the North Atlantic Oscillation (NAO). The NAO is a standing oscillation, or 'see-saw', between pressure differences in the Azores High and the Iceland Low pressure centres, and produces major changes in the meteorological regime. Winters with a high NAO index induce a northward shift of the North Atlantic westerlies, which provide much of the moisture to northern Africa and Europe, and results in drier conditions over southern Europe, the Mediterranean Sea and northern Africa. During years with a low NAO index, precipitation is likely to be greater over the Mediterranean Sea and parts of northern Africa, thus limiting both the uptake and transport of dust. Moulin *et al.* (1997) found that the NAO index covaries with desert dust transport for both the Mediterranean and

the Atlantic, and were thus able to establish that there is a large-scale climatic control on dust export from Africa, which is effected through changes in precipitation and atmospheric circulation over the regions of dust mobilization and transport. It is probable also that the NAO 'see-saw' will affect the transport and distribution of other aerosols, such as anthropogenic sulphates.

*The Indian Ocean.* The desert and arid regions surrounding the Northern Indian Ocean are a highly productive source of mineral aerosols. For example, in the Red Sea, aerosol loadings as high as $\sim 1200\,\mu g\,m^{-3}$ of air have been reported (Berry, 1990). Over the Arabian Sea an average loading of $16\,\mu g\,m^{-3}$ of air was reported by Chester *et al.* (1985); however, individual mineral aerosol concentrations ranged between $\sim 4$ and $\sim 250\,\mu g\,m^{-3}$ of air, clearly demonstrating the transport of dust 'pulses' from the desert regions of Iran–Makran and Rajasthan. The effect of this desert-derived material decreases with distance from the land sources. For example, Savoie *et al.* (1987) reported mineral dust loadings of $\sim 7.5\,\mu g\,m^{-3}$ of air at $\sim 10°N$ in the Indian Ocean, and Goldberg & Griffin (1970) estimated a loading of $\sim 7\,\mu g\,m^{-3}$ of air over the Bay of Bengal. To the south, Savoie *et al.* (1987) found an average of $0.27\,\mu g\,m^{-3}$ of air for the equatorial and Southern Indian Ocean, which is similar to the average of $0.16\,\mu g\,m^{-3}$ of air reported for the tropical and Southern Indian Ocean by Chester *et al.* (1991). Considerably further to the south, Gaudichet *et al.* (1989) found an average of $0.09–0.25\,\mu g\,m^{-3}$ of air at Amsterdam Island, indicating that the trend of decreasing mineral dust loadings continues into the southern regions of the Indian Ocean.

*The Pacific Ocean.* Most mineral aerosol data for the Pacific Ocean have been obtained from the SADS and SPAN networks set up as a part of the SEAREX Programme, and some of the sampling stations are shown in Fig. 4.5. The aerosol concentration data have been reviewed by Prospero *et al.* (1989). There are large spatial variations in dust concentrations in the Pacific and Prospero *et al.* (1989) distinguished four groupings on the basis of the annual mean mineral concentrations.

**1** The **North Pacific.** This has the highest dust concentrations, which are dominated by material sup-

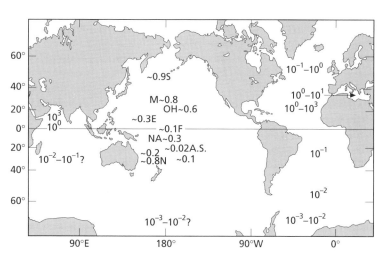

**Fig. 4.5** Tropospheric mineral aerosol concentrations over the World Ocean (units, µg m⁻³ of air) (from Chester, 1986). For most regions order-of-magnitude ranges are given, but mean values are listed for the Pacific Ocean where long-term collections have been made on island stations. The island stations indicated are: S, Shemya; M, Midway; Oh, Oahu; E, Enewetak; F, Fanning; NA, Nauru; A.S., American Samoa; N, Norfolk.

plied from the Asian deserts. Uematsu *et al.* (1983) identified two major trends in the dust data.

(a) A decreasing concentration gradient away from the Asian mainland, the highest concentrations being found in the high- and mid-latitude belt lying between ~50°N and ~20°N, with an average mineral aerosol loading for the three island stations in this belt (Shemya, Midway and Oahu) of 0.79 µg m⁻³ of air. The overall decrease in the concentrations of mineral aerosol with distance from source in the North Pacific is illustrated in Fig. 4.4(a).

(b) A well-defined seasonal variability in the transport of mineral dust over the North Pacific, with concentrations in the period February to June being an order of magnitude higher than those for the rest of the year. This seasonal pattern is similar to that of the frequency of dust storms in Asia, which exhibit a maximum in the spring.

2 The **equatorial Pacific**. The same spring-peak dust cycle was found in mineral aerosol loadings at Fanning Island close to the Equator, but the dust concentrations were considerably lower than those in the North Pacific group, with an average of 0.07 µg m⁻³. Fanning Island usually lies south of the ITCZ, and the appearance of Asian dust at the island suggests that small quantities of dust may be penetrating the convergence zone.

3 The **central South Pacific**. Mineral dust loadings at the stations in the belt lying between ~10°N and ~25°N, which includes American Samoa, were relatively small, and showed no evidence of an annual cycle. The loading at Samoa (average, 0.019 µg m⁻³) was the smallest recorded in the network, and is comparable to those found in the pristine Antarctic.

4 The **western South Pacific**. The highest concentrations of mineral dust found in the South Pacific were recorded at Norfolk Island, where there was a peak during the Australian summer, when dust loadings (average, 0.65 µg m⁻³) were about six times higher than during the winter. This station is the closest in the network to the Australian mainland and the mineral aerosol loadings probably reflect an input of dust from this continental source.

The dominant feature in the distribution of mineral dust over the Pacific Ocean is the seasonal input of dust 'pulses' from Asian sources to the North Pacific. The effect of these dust 'pulses' can be seen in concentration versus frequency diagrams for the mineral aerosol collected at a number of the Pacific network stations. A set of these diagrams is illustrated in Fig. 4.4(b). From this figure it can be seen that in the North Pacific the data for Midway and Enewetak exhibit biomodal distributions, the lower mode representing the background mineral aerosol and the upper mode representing perturbations from aerosols associated with the Asian dust storms; the percentages of the samples falling within each mode are shown in Fig. 4.4(b), from which it can be seen that at Enewetak ~75% of the mineral dust lies in the upper mode. In contrast, the data for Fanning Island, close to the Equator in the North Pacific, and American Samoa, in the South Pacific, are essentially unimodal and may be regarded as being background aerosol. At

Norfolk Island, which receives inputs from the Australian mainland, the aerosol belongs mainly to the background mode (85%), but there is evidence of dust events (15% of the total inputs) perturbing this background.

Mineral aerosol fluxes to the Pacific are given in Table 4.3(b), from which it is evident that the dust flux to the western North Pacific, mainly from Asian desert sources, is similar to that transported to the tropical North Atlantic from the Saharan Desert.

*The World Ocean: summary.* Following Chester (1986) the main features in the distribution of mineral aerosols over the World Ocean may be summarized as follows.

1 There is a mineral aerosol, or dust, veil present over all oceanic areas; however, the concentrations of material in the veil vary over several orders of magnitude in both space and time, ranging from $<\sim 10^{-3}\,\mu g$ $m^{-3}$ of air in the remote South Pacific to $>10^3\,\mu g\,m^{-3}$ of air in the Atlantic Northeast Trades off the coast of West Africa.

2 Desert and arid regions are the principal source reservoirs for mineral aerosols in the atmosphere, and because these regions tend to be concentrated into specific belts (see Fig. 4.3) they impose a general latitudinal control on the distribution of the mineral dust over marine areas. In the Atlantic the highest concentrations of the mineral aerosol are found at low latitudes (0–35°N), which reflects the influence of the North African deserts to the west, whereas in the Pacific the highest concentrations are located at mid-latitudes (20–50°N), reflecting the influence of the Asian deserts to the north.

3 A very important feature of the supply of material from the desert areas is that the mineral aerosol is often transported over the oceans in the form of pulses, which are related to dust-storm outbreaks on land. These pulses can bring relatively large quantities of mineral aerosol to the sea surface, and because of their intermittent nature they can impose short-term non-steady-state conditions on the upper water column. The pulses also can result in the transport of relatively large particles, e.g. with diameters of tens of micrometres, to open-ocean regions far from the land masses.

The general distribution of the mineral aerosol over the World Ocean is shown in Fig. 4.5.

### 4.1.4.2 *The sea-salt aerosol over the World Ocean*

The sea surface provides a vast reservoir for the generation of aerosols, and sea-salt is the largest component of the particulate matter that is cycled through the atmosphere. According to Berg & Winchester (1978), the bursting of bubbles produced by the trapping of air in surface water by breaking waves, or whitecaps, is the chief mechanism for the formation of particulate matter in the marine atmosphere, and the mechanism has been described in detail by Blandhard (1983). Sea spray produced by the direct shearing of droplets from wave crests, however, may play a role in the generation of sea-salts (see e.g. Lai & Shemdin, 1974; Koga, 1981).

Sea-salts have particle sizes ranging from $<0.2\,\mu m$ to $>200\,\mu m$ in diameter, with a distinct maximum in number distribution below a diameter of $2\,\mu m$. More than 90% of the sea-salt aerosol mass, however, is located in giant particles having a median mass diameter (MMD) between 2 and $20\,\mu m$, and McDonald *et al.* (1982) have shown that large and giant particles dominate the deposition of sea-salts, even though they are present in the air in only relatively low numbers.

The concentration of sea-salt in the marine atmosphere varies with both windspeed and altitude. In general, it appears that for winds between 5 and 35 m $s^{-1}$, the sea-salt concentration increases exponentially with windspeed, and at lower velocities the salt concentration falls off rapidly because few bubbles are produced by the sea under these conditions. The relationship between salt generation and windspeed is complex, however, and variations in salt concentration of a factor of more than two can occur at a given windspeed. Estimates of the global sea-salt production rate vary between $\sim 10^{15}$ and $\sim 10^{16}\,g\,yr^{-1}$ (Eriksson, 1959; Blanchard, 1963; Erikson & Duce, 1988).

Erickson *et al.* (1986) have produced global distribution maps of the concentration of atmospheric sea-salts at 15 m above the ocean surface. The maps for the summer and winter concentrations are illustrated in Fig. 4.6. The data generally support previous findings, and a number of overall conclusions can be drawn from the maps.

1 The high-latitude regions of both hemispheres usually have higher atmospheric sea-salt concentrations than do the low latitudes.

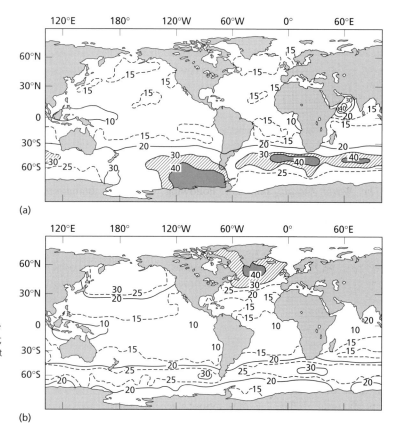

**Fig. 4.6** Sea-salt concentrations over the World Ocean (from Erickson *et al.*, 1986); the isopleths are lines of constant sea-salt concentration in $\mu g\,m^{-3}$ of air. (a) Sea-salt concentrations during the boreal summer (June–August). (b) Sea-salt concentrations during the boreal winter (December–February).

**2** The strong uniform surface winds in the Southern Hemisphere at high latitudes result in high, and relatively constant, atmospheric sea-salt concentrations in both winter and summer periods.

**3** The high-latitude winds in the Northern Hemisphere vary seasonally, and the atmospheric sea-salt concentrations exhibit a difference of a factor of three between winter and summer periods.

The overall distribution of the sea-salt aerosol differs from that of the mineral aerosol in that the highest sea-salt concentrations occur at high latitudes in areas of relatively strong winds, whereas the highest mineral aerosol concentrations are found in low-latitude areas off desert and arid regions (see Table 4.5).

### 4.1.4.3 *The sulphate aerosol over the World Ocean*

Sulphates, such as $(NH_4)_2SO_4$, are an important constituent of the tropospheric background aerosol. Sul-

**Table 4.5** Concentrations of sea-salt and mineral aerosols in the lower marine troposphere (data from Prospero, 1979).

| Oceanic region | Arithmetic mean concentration ($\mu g\,m^{-3}$ of air) | |
|---|---|---|
| | Mineral aerosol | Sea-salt aerosol |
| North Atlantic, 22°–64°N | 1.30 | 6.71 |
| North Atlantic, 0°–28°N | 36.6 | 11.2 |
| South Atlantic, 5°–35°S | 1.35 | 9.06 |
| South Atlantic, 5°–35°S including Cape of Good Hope | 1.22 | 11.3 |
| Pacific, 28°N–40°S | 0.58 | 8.44 |
| Mediterranean Sea | 4.57 | 6.98 |
| Indian Ocean, 15°S–7°N | 7.20 | 3.52 |

phates from both natural and pollutant sources can be transported into the marine atmosphere from the continents. In addition, sulphates can be generated from the sea surface, either directly during the formation of sea-salts or indirectly from gas-to-particle conversion reactions. The $SO_4^{2-}:Na^+$ ratio in sea water (0.25) has been used to estimate the sea-salt-associated sulphate in marine aerosols; thus, for ratios in excess of the seawater value, the non-sea-salt sulphate component is usually referred to by convention as **excess sulphate**.

Varhelyi & Gravenhorst (1983) compiled a comprehensive listing of the distribution of sulphates in the atmosphere over the World Ocean and a number of trends can be identified in the data.

1 The total sulphate concentrations range within about an order of magnitude ($\sim$0.9–9.4 µg sulphate $m^{-3}$ of air), with the highest values occurring near the continents. The highest values were found over the Mediterranean ($\sim$9.4 µg sulphate $m^{-3}$ of air), with concentrations of $\sim$1.5 to $\sim$2.4 µg sulphate $m^{-3}$ of air and $\sim$0.9–1.8 µg sulphate $m^{-3}$ of air being representative of the North Atlantic and the remainder of the World Ocean, respectively.

2 Concentrations of sea-salt sulphates are rather uniform, and although some variations are found, values generally lie in the range $\sim$0.6 to $\sim$1.2 µg sulphate $m^{-3}$ of air.

3 Excess sulphate concentrations decrease in the order: Mediterranean Sea ($\sim$8.8 µg sulphate $m^{-3}$ of air) $\gg$ North Atlantic ($\sim$0.0 to $\sim$1.2 µg sulphate $m^{-3}$ of air) > the rest of the World Ocean ($\sim$0.3 to $\sim$0.6 µg sulphate $m^{-3}$ of air).

It is apparent, therefore, that in the marine atmosphere there is a background of sea-salt-associated sulphate upon which is superimposed varying amounts of non-marine, or excess, sulphate. For example, Savoie *et al.* (1987) reported data on the distribution of excess sulphate in the Indian Ocean atmosphere, and showed that the concentrations varied widely from one region to another. The highest concentrations (2.7 µg $m^{-3}$ of air) were found over the northwestern Indian Ocean during dust outbreaks and the authors suggested that the excess sulphate, or its precursor, were transported together with the dust from the deserts of the Middle East. Between the dust outbreaks average concentrations of excess sulphate fell to 0.90 µg $m^{-3}$ of air. The lowest excess sulphate concentrations were reported for the Southern

Hemisphere air over the Indian Ocean, where they ranged from 0.3 µg $m^{-3}$ of air prior to the onset of upwelling, to 0.6 µg $m^{-3}$ of air after the onset of upwelling; the link between biota and the formation of excess sulphate is discussed below. In another study, Savoie *et al.* (1989) summarized data on the distribution of excess sulphate over the Pacific Ocean. They found that over the North Pacific excess sulphate had an average of $\sim$0.50 µg $m^{-3}$ of air, exhibiting a seasonal cycle that paralleled that of Asian dust, and estimated that $\sim$15–$\sim$25% of the excess sulphate in the northern mid-latitudes had an Asian source. Despite the transport from Asia to the northern regions, however, the highest average excess sulphate concentrations were found over the equatorial Pacific ($\sim$0.65 µg $m^{-3}$ of air), and even in the remote South Pacific the average concentration was 0.36 µg $m^{-3}$ of air.

Evidence for the origin of sulphates in the marine atmosphere has been provided from their particle size distributions (for a discussion of the element–particle-size relationships see Section 4.2.1.3). For example, Bonsang *et al.* (1980) reported that sulphates associated with large marine aerosol particles (diameter >2.5 µm) are characterized by having $Na:SO_4^{2-}$ ratios close to that of sea water, indicating that they have a marine origin from the generation of sea-salts. In contrast, $SO_4^{2-}:Na$ ratios in the aerosols increased sharply on smaller particles, to reach values 2–20 times greater than those for sea water on particles having diameters less than 0.6 µm. The excess sulphate in these marine aerosols therefore is present on submicrometre-sized particles, i.e. those in the accumulation mode (see Section 4.1), which are probably formed via gas-to-particle conversions. It would appear, therefore, that much of the excess sulphate in the marine aerosol has a gas precursor, which is most likely to be $SO_2$. In addition, substantial amounts of non-sea-salt sulphates in the marine aerosol can be derived directly from soil material (see e.g. Savoie *et al.*, 1987).

Over the continents particulate sulphates can be formed from precursor $SO_2$ that has an anthropogenic source and these can be transported to marine areas. In some regions, however, the excess sulphate is not associated with continentally transported material. For example, Bonsang *et al.* (1980) found that over some remote regions there was a good correlation between the concentrations of excess sulphate and $SO_2$, and concluded that both components have a common marine origin. The authors were also able to show that concentrations of $SO_2$ in the marine atmosphere increased as the primary production of the surface waters increased, implying that biological processes play a role in the origin of the $SO_2$. This raised a problem, however, because according to Nguyen & Bonsang (1979) $SO_2$ cannot emanate directly from sea water. To overcome this, Bonsang *et al.* (1980) suggested that dimethyl sulphide (DMS), formed during biological production, could be a major source of $SO_2$ to the marine atmosphere (see also Nguyen *et al.*, 1983), and subsequent studies confirmed that DMS is by far the most predominant volatile sulphur compound emitted to the atmosphere from the sea surface (see Andreae (1986) for a detailed discussion of this topic). In general, the marine biological cycle involved in the production of excess sulphate may be summarized as follows. Emissions of reduced sulphur gases are produced by biological activity in the form of DMS, which represents a major flux of sulphur to the marine atmosphere. The precursor of DMS, dimethylsulphoniopropionate (DMSP), is found in many species of marine plankton and it is now thought that most of the DMS in sea water results from the extracellular breakdown of DMSP rather than direct excretion from plankton cells. Once in the air, DMS can undergo a number of oxidation reactions, the most important probably being that with OH radicals to form methanesulphonate (MSA), sulphuric acid and $SO_2$, and so contributes to the acidity of aerosols and rain waters. The major oxidation product is $SO_2$, which can be oxidized further to form particulate sulphate. Savoie *et al.* (1989) concluded that marine organisms are responsible for the major fraction of excess sulphate over the entire Pacific Ocean, and are also important over other remote regions of the World Ocean. For example, the input to the oceans from the oxidation of reduced sulphur species originally emitted from the sea surface has been estimated to be $52\,Tg\,SO_4^{2-}\,yr^{-1}$ ($1\,Tg = 10^6$ metric tons $= 10^{12}\,g$), which amounts to ~70% of the total excess sulphate deposition ($73\,Tg\,SO_4^{2-}\,yr^{-1}$). Estimates of the marine DMS emissions to the atmosphere range from $16 \times 10^{12}$ to $40 \times 10^{12}\,g\,S\,yr^{-1}$, compared with a global anthropogenic sulphur emission of $93 \times 10^{12}\,g\,S\,yr^{-1}$ (see e.g. Malin *et al.*, 1993). The formation of particulate sulphate from an oceanic source means that the marine biomass may play a significant role in regulating climate, because the sulphate can form an important component of cloud condensation nuclei and so affect surface cloud cover and precipitation patterns, perhaps on a global scale.

It may be concluded, therefore, that over some continentally influenced marine regions sulphates formed via gas-to-particle conversion (e.g. from anthropogenically produced $SO_2$), together with soil-derived sulphates, can be transported into the atmosphere. Over most of the open-ocean surface, however, sea water itself is the only significant source of directly formed sulphate particles. Even so, this source cannot account for all the particulate sulphate in the marine aerosol and the excess sulphate, which is present largely on submicrometre-sized particles, is now thought to have formed mainly from the gas-to-particle conversion of $SO_2$, much of which has orginated from DMS emitted to the atmosphere following assimilatory biological reduction processes associated with the marine biomass.

### 4.1.4.4 Organic carbon in the marine aerosol

The atmosphere is a primary pathway for the transport of some types of organic substances to the oceans. The organic components in the atmosphere may be divided into particulate organic matter (POM), for which the organic carbon fraction is termed POC, and vapour-phase organic material (VOM), for which the organic carbon fraction is referred to as VOC. The principal sources for both POM and VOM in the atmosphere are vegetation, soils, the marine and freshwater biomass, biomass burning and a variety of anthropogenic activities; to these must be added the *in situ* production of some organic substances within the atmosphere itself. The sinks for all these organics can be related to the mechanisms that remove them from the atmosphere. For VOM, four removal mechanisms are important: conversion to POC, dry deposition, wet deposition and transformation to inorganic gaseous products. The removal of POM is dominated by wet and dry deposition. In addition, the oxidation of POM to inorganic gaseous products, e.g. $CO_2$, can result in its loss from the atmosphere.

The distributions and types of organic substances in the atmosphere have been described by Duce *et al.* (1983b) and Peltzer & Gagosian (1989).

*Particulate organic matter (POM)*. Particulate organic aerosols can be produced in two ways, either directly as particles or as a result of gas-to-particle conversions, and the resultant POM can be divided into non-viable and viable species.

The POC in the marine atmosphere is a complex mixture derived from a large variety of organic compounds, but most data for **non-viable POM** was confined originally to POC. From the study reported by Buat-Menard *et al.* (1989), concentrations of the carbonaceous aerosol in the marine atmosphere appear to vary over the range ~0.05 to ~1.20 μg C m$^{-3}$ of air, with values being significantly higher in the Northern (~0.4 to ~1.2 μg C m$^{-3}$ of air) than in the Southern Hemisphere (~0.05 to ~0.3 μg C m$^{-3}$ of air), suggesting that the Northern Hemisphere POC is dominated by continental sources. Buat-Menard *et al.* (1989) related the sources of POC in the marine atmosphere to particle size and reported that in the Northern Hemisphere marine-derived carbon (i.e. that on large particles, $r \geq 1.5$ μm) had a mean concentration of 0.07 μg C m$^{-3}$ of air, whereas the continentally derived POC (i.e. that on small particles, $r \leq 0.5$ μm) had a mean concentration of 0.45 μg m$^{-3}$ of air; for the Southern Hemisphere the concentrations were 0.07 and 0.06 μg C m$^{-3}$ of air for the marine-derived and continentally derived POC, respectively. Further, the isotopic signature of the atmospheric POC indicates a combustion-derived source for that in the Northern Hemisphere. In contrast, the atmospheric POC in the Southern Hemisphere is dominated by natural continental emissions. Clearly, therefore, two features are apparent in the distribution of POC in the marine atmosphere: (i) concentrations are higher in the Northern than in the Southern Hemisphere; and (ii) small-sized continentally derived material having an anthropogenic source dominates the atmospheric POC in the Northern Hemisphere, whereas marine-derived and natural continentally derived material contribute roughly the same amounts to the POC in the Southern Hemisphere. Buat-Menard *et al.* (1989) estimated that the global net ('wet' + 'dry') flux of carbon from the atmosphere to the ocean is ~26 Tg yr$^{-1}$ (i.e. $26 \times 10^{12}$ g yr$^{-1}$).

In addition to total particulate organic carbon, various authors have provided data on the distributions of individual organic compounds in the marine atmosphere. Much of the earlier work concentrated on dusts collected from the North Atlantic Northeast

Trades (see e.g. the data compiled by Simoneit, 1978). For example, Simoneit *et al.* (1977) showed that higher plant wax is present in dusts from these northeast trades and demonstrated the potential of aeolian transport to supply terrestrial lipids to the marine environment. Large-scale programmes were needed, however, to confirm the role of the atmosphere in the transport of organic material to the oceans. An outstanding example of such a programme was SEAREX, and considerable advances have been made to our understanding of the long-range transport of organic material to the sea surface as a result of data obtained from this programme. As part of SEAREX, a variety of organic source markers, selected to provide information on marine versus terrestrial sources, were determined in aerosols collected over the Pacific Ocean. The characteristic organic fractions of background aerosols from continental and coastal regions are dominated by a variety of compound classes of biogenic lipids, and a number of these source markers were determined in aerosols and rain samples from stations in the SEAREX network.

Peltzer & Gagosian (1989) provided SEAREX data on the following: *n*-alkanes, $C_{17}$–$C_{40}$ (terrestrial sources); fatty alcohols, $C_{21}$–$C_{36}$ (terrestrial source) and $C_{13}$–$C_{20}$ (marine source), and $C_{13}$–$C_{18}$ fatty acid salts (marine). Some of the principal findings arising from the study are summarized below.

**1** Organic carbon accounted for *c.* 10% of the total aerosol in the remote marine atmosphere. Less than ~1% of this organic material has been identified as individual compounds, yet it has provided valuable insights into the sources, transport paths, transformation mechanisms and processes controlling the fluxes of organic material to the sea surface.

**2** Over the Pacific, the concentrations of the terrestrially derived and the marine-derived compounds varied independently.

**3** **Terrestrially derived compounds.** The *n*-alkanes and the $C_{21}$–$C_{36}$ fatty alcohols were the most abundant of these compounds, which is consistent with their predominance in the epicuticular waxes of vascular plants. In contrast, the long-chain fatty acids, which are only minor constituents of the plant waxes, were present at lower concentrations in the aerosols. There were considerable variations in the concentrations of the terrestrially derived lipids both at individual sites and between sites. However, there was an overall trend for the concentrations to be highest in mid-latitudes and lowest in the tropics.

**4 Marine-derived compounds.** The $C_{13}$–$C_{18}$ fatty acid salts were the most abundant of these compounds at all the sites, with the $C_{13}$–$C_{20}$ fatty alcohols being typically the least abundant of all the lipid compounds.

**5** When the various lipid homologue distributions were related to long-range air mass trajectory analyses, a significant pattern of *regional* source marker relationships was found. For example, at Enewetak, a remote site in the North Pacific, there was a difference in the fatty alcohol distributions in aerosols collected in the dry high-dust and wet low-dust seasons. For the dry season the major homologue was $C_{28}$ and for the wet season it was $C_{30}$, a difference that is indicative of a change in climate of the source regions, with plants growing in tropical climates having higher molecular-weight homologues in their epicuticular waxes than those from temperate or subarctic climates.

**6** The SEAREX study also highlighted the importance of transformations that affect organic material in the marine atmosphere. For example, unsaturated fatty acids, which are characteristic of marine organisms, were absent from the atmospheric samples; however, their photochemical oxidation products were identified as a major class of organic compounds in both aerosols and rain water.

**7** The total lipid content of the background marine aerosol over the Pacific was 2–20 ng m$^{-3}$ of air.

**8** The major fluxes of the atmospherically transported organic material to the sea surface resulted from wet rather than dry deposition processes.

**9** There was evidence that some biogenic terrestrial material is protected from degradation in the marine environment to a greater extent than marine-derived material. This probably arises because the refractory terrestrial organics are protected by wax coatings and have already undergone significant degradation before reaching the sea surface.

The **viable POM** species in aerosols include fungi, bacteria, pollen, algae, yeasts, moulds, mycoplasma, viruses, phages, protozoa and nematodes, and a number of these have been identified in marine air. For example, Delany *et al.* (1967) noted the presence of various marine organisms, freshwater diatoms and fungi in aerosols collected at Barbados (North Atlantic) in the path of the Northeast Trades.

Folger (1970) reported the presence of phytoliths, freshwater diatoms, fungi, insect scales and plant tissue in the atmosphere over the North Atlantic. Various insects also have been found in marine air samples, and the atmosphere offers a pathway by which some plant and animal species can colonize remote islands.

*Vapour-phase organic matter (VOM).* Vapour-phase organics have varying degrees of chemical reactivity, and their atmospheric lifetimes therefore vary over a wide time-scale. In addition to methane, which can remain in the atmosphere for between 4 and 7 yr, a large range of vapour-phase organic compounds, which have considerably shorter lifetimes (~1–100 days), are present in marine air. For a detailed inventory of these non-methane vapour-phase organics see Duce *et al.* (1983b), Gagosian (1986) and Section 8.2.

In addition to natural organic material, the atmosphere is important in the transport of pollutant organics to the oceans; these include DDT residues and PCBs (see e.g. Goldberg, 1975). (DDT is a (complex) mixture containing mainly *p,p′*-dichlorodiphenyltrichloroethane and PCBs are polychlorinated biphenyls. They are used as insecticides.) Air–sea exchange is a critical link in this transport process, and differences in chemical properties of individual heavy synthetic organics can affect the mechanism by which they are transferred to the ocean surface and their reactivity in sea water—see Sections 6.2.3 and 6.4.3.2. For a review of this topic, and for comprehensive data on synthetic vapour-phase organics, the reader is referred to Duce *et al.* (1983b) and Atlas & Giam (1986).

## 4.2 The chemistry of the marine aerosol

### 4.2.1 Elemental composition

The concentrations of many particulate elements in the marine atmosphere vary over several orders of magnitude (see Table 4.6), and at any specific location the concentrations are dependent on a number of factors. These include: (i) the efficiency with which the host components containing the element are mobilized in the air; (ii) the time the components spend in the atmosphere and the distance they are transported through it, i.e. the extent to which the

**Table 4.6** The concentration ranges of some elements in the marine aerosol.

| Element | Estimated concentration range* (ng m$^{-3}$ of air) |
|---------|-----------------------------------------------------|
| Al | $1-10^4$ |
| Fe | $1-10^4$ |
| Mn | $0.1-10^2$ |
| Cu | $0.1-10^1$ |
| Zn | $0.1-10^2$ |
| Pb | $0.1-10^2$ |

* Approximate values only.

components are 'aged'; and (iii) the processes by which the components are removed from the air.

According to Berg & Winchester (1978) the composition of the marine aerosol resembles that expected from the mixing of finely divided materials from large-scale sources. The main sources of material to the atmosphere have been described in Section 4.1.3, and a source–control relationship offers a convenient framework within which to describe the chemistry of the marine aerosol.

One of the most common methods of relating an element in an aerosol to its source is by using a source indicator, or marker, that is derived predominantly from one specific source. In order to assess the enrichment of an element relative to a source it is common practice to define the excess, i.e. non-source fraction, in terms of an **enrichment factor** (EF), which is calculated with respect to an equation of the type:

$$EF_{source} = (E/I)_{air} / (E/I)_{source} \qquad (4.1)$$

in which $(E/I)_{air}$ is the ratio of concentrations of an element $E$ and the indicator element $I$ in the aerosol, and $(E/I)_{source}$ is the ratio of their concentrations in the source material.

The three most important sources for the supply of particulate matter to the marine atmosphere are: (i) the Earth's crust (the **crustal aerosol**), (ii) the ocean surface (the **sea-salt aerosol**), both of which involve natural low-temperature generation processes, and (iii) a variety of usually high-temperature anthropogenic processes (the **enriched aerosol**).

### 4.2.1.1 The crustal aerosol

Aluminium is the indicator element most commonly used for the crustal source and although there are problems involved in selecting a composition for the source (or precursor) material, that of the average crustal rock is frequently used for the calculation of $EF_{crust}$ values, according to the equation:

$$EF_{crust} = (E/Al)_{air} / (E/Al)_{crust} \qquad (4.2)$$

in which $(E/Al)_{air}$ is the ratio of the concentrations of an element $E$ and Al in the aerosol, and $(E/Al)_{crust}$ is the ratio of their concentrations in average crustal rocks.

Because of the various constraints involved in the calculation, $EF_{crust}$ values should be treated only as order-of-magnitude indicators of the crustal source. Thus, values close to unity are taken as an indication that an element has a mainly crustal origin, and those >10 are considered to indicate that a substantial portion of the element has a non-crustal origin.

Rahn (1976) and Rahn et al. (1979) have tabulated the $EF_{crust}$ values for some 70 elements in over 100 samples from the world aerosol, and a geometric mean for the whole population gives a first approximation of the most typical value for each element. These geometric means are plotted in Fig. 4.7, from which it can be seen that over half the elements have $EF_{crust}$ values that range between 1 and 10, indicating that they are present in the aerosol in roughly crustal proportions. These are termed the crustal or **non-enriched elements** (NEEs). The remaining elements have $EF_{crust}$ values in the range 10 to $\sim5 \times 10^3$, and their concentrations in the aerosol are not crust-controlled. These are referred to as the enriched or **anomalously enriched elements** (AEE).

The NEEs will almost always retain their character in all aerosols. It is important, however, to stress that the degree to which an AEE is actually enriched can vary considerably as the mutual proportions of the various components in an aerosol change. Data to illustrate this are given in Table 4.7, which lists the $EF_{crust}$ values for several elements from a variety of marine locations. These data show, for example, that $EF_{crust}$ values of the AEEs Cu, Pb and Zn can vary between <10 and >100. Thus, under certain conditions the $EF_{crust}$ values of the AEEs can indicate that they are present in crustal proportions. Some of the factors that control the source strengths, and so the $EF_{crust}$ values, of both the NEEs and the AEEs in the marine aerosol can be illustrated with respect to the distributions of Fe and Cu in the Atlantic atmosphere. The average Al, Cu and Fe concentrations, and

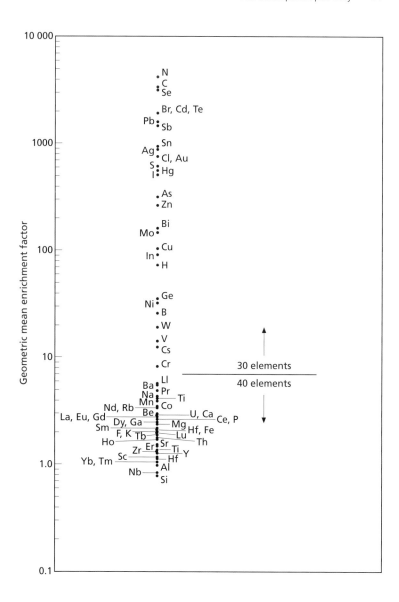

**Fig. 4.7** Geometric mean $EF_{crust}$ values for elements in the world aerosol (from Rahn *et al.*, 1979).

**Table 4.7** $EF_{crust}$ values for some anomalously enriched crustal elements in marine aerosols.

| | $EF_{crust}$ | | | |
|---|---|---|---|---|
| Element | North Atlantic, westerlies* | Bermuda (North Atlantic)† | Enewetak (North Pacific)‡ | North Atlantic, Northeast Trades§ |
| Cu | 120 | 9.6 | 3.2 | 1.2 |
| Zn | 110 | 26 | ~14 | 3.8 |
| Pb | 2200 | 170 | 40 | 9.1 |
| Cd | 730 | 570 | 130 | 9.4 |

* Data from Duce *et al.* (1975). † Data from Duce *et al.* (1976). ‡ Data from Duce *et al.* (1983a). § Data from Murphy (1985).

**Table 4.8** Average conentrations and $EF_{crust}$ values for Al, Fe and Cu in Atlantic aerosols (data from Murphy, 1985).

| | Element | | | | | |
|---|---|---|---|---|---|---|
| | | Fe | | | Cu | |
| Oceanic region | Al Concentration (ng m$^{-3}$ of air) | Concentration (ng m$^{-3}$ of air) | $EF_{crust}$ | | Concentration (ng m$^{-3}$ of air) | $EF_{crust}$ |
| North Atlantic | | | | | | |
| westerlies | 48 | 36 | 1.2 | | 1.0 | 30 |
| Straits of Gibralter | 827 | 626 | 1.1 | | 2.4 | 8.2 |
| Northeast Trades | 5925 | 3865 | 1.0 | | 4.5 | 1.2 |
| South Atlantic | | | | | | |
| Southeast Trades | 17 | 13 | 1.2 | | 0.30 | 44 |
| westerlies | 2.7 | 2.6 | 1.8 | | 0.29 | 225 |

$EF_{crust}$ values for Fe and Cu, from a number of aerosol populations sampled on a north–south Atlantic transport are listed in Table 4.8. Iron is a crustal element, and the Fe $EF_{crust}$ values are all <10 and exhibit little variation between the populations. In contrast, Cu is an AEE in the world aerosol, but Cu $EF_{crust}$ values in the Atlantic aerosol vary considerably. In most of the populations they are >10. However, the average Cu $EF_{crust}$ values fall in the aerosol collected over the Straits of Gibraltar and reach a minimum of around unity in the Northeast Trades population, which was sampled off the coast of West Africa, where pulses of Saharan dust-laden air are common—see Section 4.1.4.1. It is apparent, therefore, that although Cu is classified as an AEE in the world aerosol (average Cu $EF_{crust}$ value ~100), under some conditions crustal material can become the dominant source of the element and can completely mask the non-crustal components. It may be concluded, therefore, that although the atmospheric concentrations of the NEEs (e.g. Fe) vary considerably, their $EF_{crust}$ values remain relatively constant. In contrast, both the concentrations and $EF_{crust}$ values of the AEE vary, the latter being lowered by increases in the input of crustal material.

#### 4.2.1.2 The sea-salt aerosol

In order to assess the importance of the sea surface as a source for the marine aerosol, Na can be used as the marine indicator element and the precursor source composition is assumed to be that of bulk sea water. The $EF_{sea}$ is then calculated according to the equation

$$EF_{sea} = (E/Na)_{air}/(E/Na)_{sea\ water} \qquad (4.3)$$

in which $(E/Na)_{air}$ is the ratio of the concentrations of an element $E$ and Na in the aerosol, and $(E/Na)_{sea\ water}$ is the ratio of their concentrations in *bulk* sea water.

A selection of $EF_{sea}$ values for aerosols for a number of marine regions is given in Table 4.9. The regions are ranked on the basis of their distance from the major continental trace metal sources, and so represent increasingly more pristine marine environments. The elements in the table can be divided into three general groups.

**Group 1** contains K and Mg. For these two elements there is little intersite variation in their $EF_{sea}$ values, which are all around unity and indicate that the elements have a predominantly oceanic source at all locations.

**Group 2** contains Mo. This element has $EF_{sea}$ values that range up to ~100 close to the continents but fall to around unity in the remote South Atlantic westerlies. This indicates that in remote areas bulk sea water can be a significant source for this element (see e.g. Chester *et al.*, 1984a).

**Group 3** contains Al, Fe, Mn, Co, V, Cu, Pb and Zn. For these elements there is a progressive decrease in their $EF_{sea}$ values towards the more pristine oceanic environments. However, even at the remote sites the elements are still enriched, by factors ranging up to 600 000, relative to bulk sea water.

From their $EF_{sea}$ values, therefore, it would appear

**Table 4.9** $EF_{sea}$ values for a number of elements in the marine aerosol*.

| Element | $EF_{sea}$ North Atlantic | North Pacific, Hawaii | North Pacific, Enewetak |
|---|---|---|---|
| K | 1.3 | 1.1 | 1.3 |
| Mg | 0.9 | 1.0 | 1.1 |
| Al | $1 \times 10^6$ | $3 \times 10^5$ | $1 \times 10^5$ |
| Co | $8 \times 10^3$ | $<1 \times 10^3$ | $1 \times 10^3$ |
| Cu | $6 \times 10^4$ | $7 \times 10^3$ | $3 \times 10^3$ |
| Fe | $4 \times 10^7$ | $6 \times 10^5$ | $3 \times 10^4$ |
| Mn | $4 \times 10^5$ | $8 \times 10^3$ | $2 \times 10^4$ |
| Pb | $5 \times 10^5$ | $5 \times 10^5$ | $3 \times 10^4$ |
| V | $9 \times 10^3$ | $4 \times 10^2$ | $2 \times 10^2$ |
| Zn | $2 \times 10^6$ | $6 \times 10^5$ | $1 \times 10^5$ |
| Sc | $1 \times 10^5$ | — | $4 \times 10^4$ |

| | UK coastal aerosol | North Atlantic, westerlies | Atlantic, Northeast Trades | Atlantic, Southeast Trades | South Atlantic, westerlies |
|---|---|---|---|---|---|
| Mo | 116 | 20 | 13 | 9.5 | 1.4 |

* Data for Mo from Chester *et al.* (1984); data for all other elements from Weisel *et al.* (1984).

that the ocean is not a significant source for the elements in Group 3 above, even in areas remote from the influence of other sources. The calculation of the $EF_{sea}$ values, however, presents problems that are considerably more complex than those associated with $EF_{crust}$ values. These problems involve the selection of a composition for the oceanic precursor material, and the most serious of them arises in response to the nature of the processes involved in the generation of sea-salts. During the bursting of bubbles, which is the principal mechanism in sea-salt formation, part of the sea-surface microlayer can be skimmed off to be incorporated into the salt particles. This microlayer can contain concentrations of many trace metals, which are enhanced up to $10^3$–$10^7$ times relative to bulk sea water (see Section 4.3). As a result, considerable fractionation of these trace metals can take place during the formation of sea-salts, so that **bulk** sea water is not their immediate oceanic source. A number of attempts have been made to evaluate a more realistic source composition for material generated from the sea surface by relating the material directly to the microlayer itself. One of these involves the collection and analysis of bubble-produced sea-salts using the bubble interfacial microlayer sampler (BIMS) (see e.g. Fasching *et al.*, 1974; Piotrowicz *et*

*al.*, 1979). Weisel *et al.* (1984) carried out a series of BIMS experiments in the open-ocean western North Atlantic. During these experiments bubbles were artificially generated from water depths ranging from immediately below the surface to 1 m down, and the sea-salt particles formed were collected on filters and analysed subsequently for a series of elements. The net concentration of each element on the bubble-generated sea-salts was then used, together with its North Atlantic surface seawater concentration, to calculate a BIMS $EF_{sea}$ value. The geometric BIMS $EF_{sea}$ means for all bubble depths for the elements studied are given in Table 4.10, and on this basis, following Weisel *et al.* (1984), the elements can be divided into two broad groups.

**Group 1** contains K and Mg. These two elements have average BIMS $EF_{sea}$ values of unity, just as they did in the bulk aerosol (see above). That is, these elements, together with Na, retain their orginal bulk seawater ratios on the bubble-generated salts.

**Group 2** contains Al, Fe, Mn, Co, V, Cu, Zn, Pb and Sc. These elements have relatively high BIMS $EF_{sea}$ values, ranging from 10 for Sc to 20 000 for Zn, and apparently undergo fractionation at the sea surface.

Despite the very real difficulties inherent in assessing the fractionation of elements in the air–sea interface, the BIMS $EF_{sea}$ data provide at least a first approximation of the degree to which such fractionation has taken place. Further, by making the assumption that the BIMS $EF_{sea}$ values found for the North Atlantic apply globally, Weisel *et al.* (1984) introduced the important concept of a **relative oceanic enrichment factor** ($REF_{sea}$), which is the aerosol $EF_{sea}$ value (calculated with respect to bulk sea water) divided by the BIMS $EF_{sea}$ value. The $REF_{sea}$ values therefore directly related the enrichment of an element in a marine aerosol to the fractionation that occurs at the sea surface, and so should offer a more realistic assessment of the ocean as a source material. The $REF_{sea}$ values should approach unity if the sea surface is a significant source for an element, and this is evaluated with respect to a number of marine aerosol populations in Table 4.11. In this table, the BIMS $EF_{sea}$ values are expressed as a percentage of the $EF_{sea}$ value (i.e. that obtained using bulk sea water), and the $REF_{sea}$ values are also given. From these various data it can be seen that for Al, Co, Mn and Pb the BIMS $EF_{sea}$ values are <10% of the $EF_{sea}$ values, and for these elements the sea surface is a trivial source even at the remote site (Enewetak). For V, Cu and Zn, however, the BIMS $EF_{sea}$ values are >20% of the $EF_{sea}$ values for the Enewetak population, indicating that at some remote sites the sea surface becomes a non-trivial source for these elements; this was confirmed by Arimoto *et al.* (1987) for aerosols at the remote Samoa (South Pacific) site. It must be remembered, however, that particles deposited originally from the atmosphere to the sea surface can be recycled back into the air during sea-salt production (see e.g. Settle & Patterson, 1982), which makes it difficult to estimate the true *net* flux of elements from the ocean surface to the atmosphere. Another factor that can complicate the assessment of fluxes from the sea surface is the formation of particulate components from gaseous emissions derived from sea water, e.g. from biologically mediated reactions, because these conversion products will be associated with small-sized particles, and not the larger sized sea-salt particles (see e.g. Mosher *et al.*, 1987).

**Table 4.10** Bubble interfacial microlayer sampler (BIMS) $EF_{sea}$ values determined over the North Atlantic Ocean (data from Weisel *et al.*, 1984).

| Element | BIMS $EF_{sea}$ |
|---------|-----------------|
| K  | 1.0             |
| Mg | 1.0             |
| Al | $5 \times 10^3$ |
| Co | $6 \times 10^2$ |
| Cu | $8 \times 10^2$ |
| Fe | $1 \times 10^4$ |
| Mn | $1 \times 10^3$ |
| Pb | $4 \times 10^3$ |
| V  | $1 \times 10^2$ |
| Zn | $2 \times 10^4$ |

**Table 4.11** Sea-surface sources of elements in marine aerosol population*.

|  | North Atlantic | | North Pacific | | | |
|  |  |  | Hawaii | | Enewetak | |
| Element | BIMS $EF_{sea}$ as a percentage of aerosol $EF_{sea}$ | $REF_{sea}$ | BIMS $EF_{sea}$ as a percentage of aerosol $EF_{sea}$ | $REF_{sea}$ | BIMS $EF_{sea}$ as a percentage of aerosol $EF_{sea}$ | $REF_{sea}$ |
|---------|------|------|------|------|------|------|
| Al | 0.5  | 200  | 1.6  | 6    | 5    | 20   |
| Co | 0.75 | 133  | 6    | 17   | 6    | 17   |
| Mn | 0.25 | 400  | 12   | 8    | 5    | 20   |
| Cu | 1.3  | 75   | 11   | 8.75 | 27   | 3.75 |
| V  | 1.1  | 90   | 25   | 4    | 50   | 2    |
| Zn | 1    | 100  | 3    | 30   | 20   | 5    |
| Pb | 0.80 | 125  | 0.80 | 125  | 10   | 7.5  |

* Original data from Weisel *et al.* (1984).

### 4.2.1.3 The enriched aerosol

There are a number of processes that can supply the AEE to the atmosphere, and many of these are associated with some form of high-temperature generation (see Section 4.1.1). Unlike the low-temperature processes involved in the production of crustal and oceanic particulate material, however, the high-temperature processes do not, in general, have readily identifiable indicator elements that can be determined on a routine basis. Despite this, attempts have been made to identify the anthropogenic source, and element to element ratios can be used for this purpose. For example, Rahn (1981) used 'non-crustal Mn/non-crustal V' ratios as a tracer of large-scale sources of pollution in Arctic aerosols, and Rahn & Lowenthal (1984) used a seven 'element:element' ratio system to demonstrate that regional elemental tracers offer a way of determining the long-range sources of pollution in the atmosphere. Data for particulate anthropogenic trace metal emissions have also been used to establish metal sources, especially in coastal aerosols. For example, several authors have used metal/metal ratios from the data set provided by Pacyna et al. (1984) for the composition of anthropogenic emissions from Europe, to establish the sources of metals to the North Sea atmosphere (see e.g. Schneider, 1987; Yaaqub et al., 1991; Chester et al., 1994). In general, however, it is not the usual practice to calculate enrichment factors for high-temperature sources, and although the presence of elements associated with these sources is indicated by the fraction that is in excess of those accounted for by the $EF_{crust}$ and $REF_{sea}$ values, other approaches must be used for their actual identification. Two such approaches applicable to the marine aerosol will be discussed here; these involve the particle-size distribution and the solid-state speciation of elements in aerosols.

*The particle-size distribution of elements in the marine aerosol.* The particle size of an element in the atmosphere is controlled by the manner in which it is incorporated into its host component, which, in turn, is a function of its source. The overall particle-size–source relationships have been illustrated in Fig. 4.1, from which two general populations can be identified. The coarse particles are in the **mechanically generated** range and have diameters $> \sim 2\,\mu m$.

These include the crustal and sea-salt components, and during their formation these particles acquire their elements directly from the precursor material. In contrast, the fine particles are in the **Aitken nuclei** or **accumulation** range and have diameters $< \sim 2\,\mu m$. Those in the accumulation range are formed by processes such as condensation and gas-to-particle conversions, and elements associated with them can be acquired from the atmosphere itself. For example, during high-temperature processes (such as volcanic activity, fossil-fuel combustion, waste incineration and the processing of ore materials) some elements can be volatilized from the parent material in a vapour phase. They can be removed from this vapour phase via condensation and gas-to-particle processes during which the elements may be adsorbed on to the surfaces of the ambient aerosols. The processes are size-dependent because, in general, small particle condensation nuclei will have a large ratio of surface area to volume (see e.g. Rahn, 1976). As a result, many of the elements released during high-temperature processes become associated with small particles in the accumulation range, i.e. with diameters $< \sim 1{-}0.1\,\mu m$ — see Fig. 4.1.

Because of the different processes involved in the generation of their host particles, elements having a crustal or an oceanic source should be present in association with larger ($< \sim 1\,\mu m$ diameter) aerosols, whereas those having a high-temperature source should be found on smaller ($> \sim 1\,\mu m$ diameter) aerosols. There are also differences in the particle-size distributions of the mechanically generated crustal and sea-salt aerosols. The particle-size distribution of an element in an aerosol population should therefore offer an insight into its source. This has been demonstrated in a number of studies. For example, Fig. 4.8 illustrates the mass particle-size distributions of:

1 $^{210}Pb$, which can be used as an example of an element that has a mass predominantly associated with submicrometre-sized particles typical of many high-temperature-generated pollutant-derived elements;

2 Al, which is typical of crust-derived elements;

3 Na, which is typical of sea-salt-derived elements.

From data such as these, three general conclusions can be drawn regarding the particle size distribution of elements in the marine aerosol.

1 Sea-salt-associated elements have most of their total mass on particles with MMDs in the range

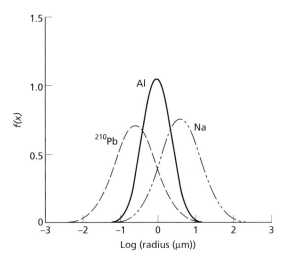

**Fig. 4.8** Mass size distributions of $^{210}$Pb, Al and Na in aerosols from the marine atmosphere (from Arimoto & Duce, 1986).

~3–7 µm; these elements include Na, Ca, Mg and K.

2 Crust-derived elements have most of their total mass on particles with MMDs in the range ~1–3 µm; these elements include the NEEs Al, Sc, Fe, Co, V, Cs, Ce, Rb, Eu, Hf and Th.

3 Elements associated with high-temperature sources (mainly anthropogenic) have most of their total mass on particles with MMDs <~0.5 µm; these elements include the AEEs Pb, Zn, Cu, Cd and Sb.

It may be concluded, therefore, that in an aerosol population the particle-size distributions of elements between the three size classes, together with EF data for the individual size classes, can be used to establish the predominance of their anthropogenic, crustal and sea-salt sources.

*The solid-state speciation of elements in the marine aerosol.* The partitioning of elements among the components of an aerosol, i.e. their solid-state speciation, can reveal information on their sources. This can be illustrated with respect to the study reported by Chester *et al.* (1989), in which a sequential leaching technique was used to establish the solid-state speciation of a series of trace metals in anthropogenic and crustal 'end-member' aerosols. The technique separated three trace-metal binding associations: (i) an exchangeable association; (ii) an oxide and carbonate association; and (iii) a refractory and organic

association. The results are illustrated in Fig. 4.9, and may be summarized as follows.

1 Aluminium and Fe are generally refractory in both aerosol 'end-members', i.e. greater than ~80% of the total Al (ΣAl) and greater than ~60% of the ΣFe are in refractory associations.

2 Manganese is speciated between all three binding associations in both 'end-members', with no single association being especially dominant.

3 Copper, Zn and Pb exhibit very different speciation signatures in two 'end-member' aerosols. In the crustal 'end-member' aerosol, ~65% of the ΣPb, ~75% of the ΣCu and ~90% of the ΣZn are in **refractory** associations. In contrast, in the anthropogenic 'end-member' ~50% of the ΣCu and ~90% of the ΣZn and the ΣPb are in **exchangeable** associations. The differences between the solid-state speciation signatures of Cu, Zn and Pb are entirely consistent with their different sources to the atmosphere for the two 'end-member' aerosols. Crustal dust is generated during low-temperature weathering–mobilization processes. In contrast, high-temperature processes are often involved in the emission of anthropogenic trace metals, such as Cu, Zn and Pb, to the atmosphere. During these high-temperature processes, the metals can be released from the parent material into the vapour phase from which subsequently they can be taken up by small-sized ambient particles in a weakly held exchangeable surface association (see particle-size discussion above).

In addition to the various chemical and particle-size indicators described above, which can be used to assess the general sources of elements to the atmosphere, the actual source regions supplying material to an individual air mass often can be determined using air mass back-trajectory techniques. These involve back-tracking the atmospheric path followed by an air mass in order to identify the potential aerosol source regions it has crossed during at least some part of its history, and they have been used in both short-range (see e.g. Chester *et al.*, 1994) and long-range (see e.g. Merrill, 1989) aerosol transport studies.

### 4.2.2 Elemental source strengths

There have been a number of attempts to estimate the global source strengths of elements to the atmosphere. There are large uncertainties involved in making these estimates and they should be treated at

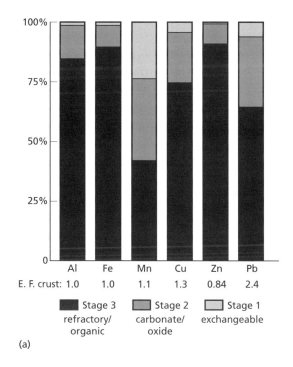

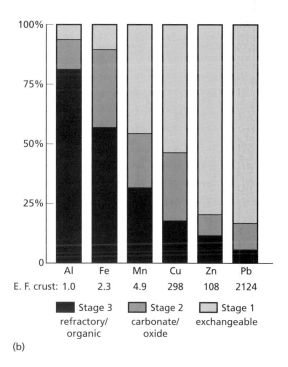

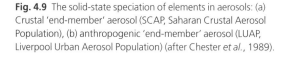

(a)

(b)

**Fig. 4.9** The solid-state speciation of elements in aerosols: (a) Crustal 'end-member' aerosol (SCAP, Saharan Crustal Aerosol Population), (b) anthropogenic 'end-member' aerosol (LUAP, Liverpool Urban Aerosol Population) (after Chester *et al.*, 1989).

best as no more than approximations. None the less, a number of overall trends in elemental atmospheric source strengths can be identified from the published estimates.

Lantzy & Mackenzie (1979) compiled data on the natural and anthropogenic global source of elements released into the atmosphere, and derived an **atmospheric interference factor** (AIF) calculated according to the equation

$$AIF = (E_a/E_n) \times 100 \qquad (4.4)$$

in which $E_a$ is the total anthropogenic emission of an element $E$, and $E_n$ is its total natural emission. Thus, an AIF of 100 indicates that the anthropogenic flux of an element is equal to its natural flux. The data for the global atmospheric source strengths, and the AIF values, for the various elements are listed in Table 4.12. The elements in this table fall into three main groups.

1 Al, Ti, Fe, Mn and Co have AIF values <100, and for these elements natural fluxes dominate their supply to the atmosphere.

2 Cr, V and Ni have AIF values in the range 100–500, i.e. their anthropogenic fluxes are in excess of their natural fluxes, but only by factors of less than five.

3 Sn, Cu, Cd, Zn, As, Se, Mo, Hg and Pb have AIF values in the range 500–35 000, and for these elements the atmospheric fluxes from anthropogenic sources greatly exceed those from natural sources.

It must be remembered, however, that the influx of anthropogenic components to the atmosphere is in the proportion ~90% to the Northern Hemisphere and only ~10% to the Southern Hemisphere (Robinson & Robbins, 1971).

A number of other authors have made estimates of the global source strengths of trace metals to the atmosphere. Those provided by Weisel *et al.* (1984) and Nriagu (1989), which are given in Table 4.13, are of interest because they offer a more detailed breakdown of some of the individual trace metal sources than was given by Lantzy & Mackenzie (1979), especially the ocean surface (sea-salt spray) and various biogenic sources. There are differences between the

**Table 4.12** Natural and anthropogenic sources of atmospheric emissions; units, $10^8$ g yr$^{-1}$ (from Lantzy & Mackenzie, 1979).

| Element | Continental dust flux | Volcanic dust flux | Volcanic gas flux | Industrial particulate emissions | Fossil fuel flux | Total emissions (industrial, fossil fuel) | Atmospheric* interference factor (%) |
|---|---|---|---|---|---|---|---|
| Al | 356 500 | 132 750 | 8.4 | 40 000 | 32 000 | 72 000 | 15 |
| Ti | 23 000 | 12 000 | — | 3600 | 1600 | 5200 | 15 |
| Sm | 32 | 9 | — | 7 | 5 | 12 | 29 |
| Fe | 190 000 | 87 750 | 3.7 | 75 000 | 32 000 | 107 000 | 39 |
| Mn | 4250 | 1800 | 2.1 | 3000 | 160 | 3160 | 52 |
| Co | 40 | 30 | 0.04 | 24 | 20 | 44 | 63 |
| Cr | 500 | 84 | 0.005 | 650 | 290 | 940 | 161 |
| V | 500 | 150 | 0.05 | 1000 | 1100 | 2100 | 323 |
| Ni | 200 | 83 | 0.0009 | 600 | 380 | 980 | 346 |
| Sn | 50 | 2.4 | 0.005 | 400 | 30 | 430 | 821 |
| Cu | 100 | 93 | 0.012 | 2200 | 430 | 2630 | 1363 |
| Cd | 2.5 | 0.4 | 0.001 | 40 | 15 | 55 | 1897 |
| Zn | 250 | 108 | 0.14 | 7000 | 1400 | 8400 | 2346 |
| As | 25 | 3 | 0.10 | 620 | 160 | 780 | 2786 |
| Se | 3 | 1 | 0.13 | 50 | 90 | 140 | 3390 |
| Sb | 9.5 | 0.3 | 0.013 | 200 | 180 | 380 | 3878 |
| Mo | 10 | 1.4 | 0.02 | 100 | 410 | 510 | 4474 |
| Ag | 0.5 | 0.1 | 0.0006 | 40 | 10 | 50 | 8333 |
| Hg | 0.3 | 0.1 | 0.001 | 50 | 60 | 110 | 27 500 |
| Pb | 50 | 8.7 | 0.012 | 16 000 | 4300 | 20 300 | 34 583 |

* Interference factor = (total emissions/continental + volcanic fluxes) × 100.

**Table 4.13** Global source strengths for atmospheric trace metals (units, $10^9$ g yr$^{-1}$) indicated by data from (1) Weisel *et al.* (1984) and (2) Nriagu (1989).

| Element | Natural | | | | | | | |
|---|---|---|---|---|---|---|---|---|
| | Crustal dusts | | Ocean | | Volcanoes | | Biogenic | |
| | 1 | 2 | 1* | 2 | 1 | 2 | 1† | 2‡ |
| Al | 20 000 | — | 200 | — | 700 | — | 40 | — |
| As | — | 2 | — | 1.7 | — | 3.8 | — | 4.0 |
| Cd | — | 0.20 | — | 0.06 | — | 0.82 | — | 0.35 |
| Co | 7 | 4.1 | 0.2 | 0.07 | 0.1 | 0.96 | 0.04 | 0.97 |
| Cr | — | 27 | — | 0.7 | — | 15 | — | 1.2 |
| Cu | 10 | 8.0 | 5 | 3.6 | 6 | 9.4 | 0.9 | 7.2 |
| Fe | 10 000 | — | 50 | — | 300 | — | 20 | — |
| Hg | — | 0.05 | — | 0.02 | — | 1.0 | — | 1.4 |
| K | 6000 | — | 30 000 | — | 200 | — | 100 | — |
| Mg | 3000 | — | 100 000 | — | 80 | — | 20 | — |
| Mn | 200 | 221 | 7 | 0.86 | 9 | 42 | 5 | 53 |
| Pb | 3 | 3.9 | 8 | 1.4 | 0.4 | 3.3 | 0.2 | 3.6 |
| Sb | — | 0.78 | — | 0.56 | — | 0.71 | — | 0.51 |
| Sc | 7 | — | 0.005 | — | 0.2 | — | 0.002 | — |
| Se | — | 0.18 | — | 0.55 | — | 0.95 | — | 7.7 |
| V | 30 | 33 | 10 | 3.1 | 0.7 | 5.6 | 0.2 | 3.0 |
| Zn | 80 | 19 | 100 | 0.44 | 10 | 9.6 | 10 | 16 |

*Continued*

data sets given by Weisel *et al.* (1984) and Nriagu (1989), particularly with respect to the importance of the biogenic source, but a number of common factors can be identified.

1  Natural sources make up >70% of the total atmospheric inputs of Al, Co, Sc, Se, K and Mg. Anthropogenic sources make up >70% of the total atmospheric inputs of Cd, Pb and possibly Zn. For Cd and Zn non-ferrous metal production dominates the anthropogenic emissions, whereas for Pb the emissions from car exhausts make up >75% of the anthropogenic input; however, this will decrease in importance as lead petrol additives continue to be phased out.

2  The oceanic source dominates the inputs of K and Mg, which are associated with the formation of seasalts, and on the basis of the BIMS data reported by Weisel *et al.* (1984) the sea surface also makes important contributions to the total atmospheric sources of V, Cu and Pb. It must be remembered, however, that elements released from the sea surface have been recy-

cled into the air, and do not take part in the *net* input of elements to the oceans from the atmosphere.

3  The crustal dust source accounts for >70% of the total natural inputs of Al, Fe, Sc, Co, Mn and V, between 30% and 70% of those of Cr, Cu, Pb, Sb and Zn, but <30% of those of As, Cd, Hg and Se.

4  For some trace metals that have only a relatively small crustal dust source, e.g. As, Cd, Hg and Se, volcanic and biogenic sources dominate the natural inputs to the atmosphere. According to Nriagu (1989) the importance of the biogenic source had been underestimated in many earlier assessments of trace metal emissions to the atmosphere. This biogenic source encompasses a number of individual sources, which include particulate and volatile emissions from both the continents and the oceans. For simplicity, the individual biogenic emissions are not listed in Table 4.13, but it is apparent that biogenic sources in general are important in the natural atmospheric cycles of As, Cd, Hg, Sb, Se and possibly Zn. Volcanic eruptions are sporadic, and this is taken into

**Table 4.13** *Continued*

Anthropogenic

| Fossil fuel | | Others | | Total natural | | Total anthropogenic | |
|---|---|---|---|---|---|---|---|
| 1 | 2§ | 1 | 2 | 1 | 2 | 1 | 2 |
| 2000 | — | 2000 | — | 20940 | — | 4000 | — |
| — | 2.2 | — | 16.8 | — | 12 | — | 19 |
| — | 0.79 | — | 6.9 | — | 1.4 | — | 7.7 |
| 0.9 | — | 2 | — | 7.3 | 6.1 | 2.9 | — |
| — | 12.4 | — | 18.1 | — | 44 | — | 30.5 |
| 2 | 7.1 | 50 | 26.6 | 22 | 28 | 35 | 52 |
| 2000 | — | 4000 | — | 10370 | — | 6000 | — |
| — | 2.3 | — | 1.3 | — | 2.5 | — | 3.6 |
| 200 | — | 200 | — | 36300 | — | 400 | — |
| 300 | — | 200 | — | 103100 | — | 500 | — |
| 8 | 12.4 | 400 | 26.5 | 221 | 317 | 408 | 39 |
| 4 | 12.7 | 400 | 319 | 11.5 | 12 | 404 | 332 |
| — | 1.3 | — | 2.25 | — | 2.6 | — | 3.5 |
| 0.8 | — | 0.4 | — | 7.2 | — | 1.2 | — |
| — | 2.3 | — | 1.5 | — | 9.4 | — | 3.8¶ |
| 20 | 84 | 2 | 2 | 41 | 45 | 22 | 86 |
| 80 | 16 | 200 | 115 | 200 | 45 | 280 | 132 |

* Based on BIMS $EF_{sea}$ data for all bubble depths. † Vegetation only. ‡ Including phytomass burning. § Coal, oil, wood (excluding gasoline). ¶ Particulate material only.

account in the data in Table 4.13, which are average values for the volcanic source. From the data in Table 4.13, it is apparent that this volcanic source plays a significant role in the natural atmospheric cycles of As, Cd, Cr, Cu, Hg, Sb and possibly Pb.

### 4.2.3 Geographical variations in the elemental composition of the marine aerosol

It was pointed out in Section 4.2.1 that the concentrations of many elements in the marine atmosphere vary over several orders of magnitude from one region to another. The reasons for these variations can now be assessed in the light of the factors discussed above. The overall composition of the marine aerosol is controlled by the extent to which components derived from the various sources are mixed together in the atmosphere. A considerable amount of data are now available on the concentrations of elements in marine aerosols from a wide variety of environments, and a compilation of some of the more recent data is given in Table 4.14. In this table the marine locations are arranged in an increasing order of remoteness from the primary, i.e. non-oceanic, sources. The principal trends in atmospheric concen-trations that emerge from this data set can be summarized as follows.

**1** The highest concentrations of the AEE are found over coastal seas that are relatively close to the continental anthropogenic sources.

**2** The highest concentrations of NEE are found over regions close to the continental arid land and desert sources.

**3** The concentrations of both the AEE and the NEE decrease with remoteness from the continental sources, and for those regions for which sufficient data are available the general rank order is: coastal seas > North Atlantic > North Pacific ≈ tropical Indian > South Pacific.

**4** The $EF_{crust}$ values are highest in aerosols from the coastal seas, but those for the Samoa aerosol are higher than those for the less remote Enewetak aerosol, which suggests that the residence times of the small-sized AEE-containing anthropogenic particles are longer than those of the larger NEE-containing mineral particles.

The various data in Table 4.14 offer an indication of the concentrations of elements in the marine atmosphere. These elements are available for deposition at the sea surface. Before entering the ocean

**Table 4.14** Concentration (1) and $EF_{crust}$ values (2) of trace metals in marine aerosols.

| Trace metal | Concentration units (m⁻³ of air) | Coastal regions: close to anthropogenic sources | | | | | | Coastal regions: close to crustal sources | | | | | |
| | | North Sea* | | Western Black Sea† | | Irish Sea coast‡ | | North Atlantic Northeast Trades§ | | Eastern Mediter-ranean¶ | | Northern Arabian Sea ‖ | |
| | | 1 | 2 | 1 | 2 | 1 | 2 | 1 | 2 | 1 | 2 | 1 | 2 |
|---|---|---|---|---|---|---|---|---|---|---|---|---|---|
| Al | ng | 294 | 1.0 | 540 | 1.0 | 286 | 1.0 | 5925 | 1.0 | 915 | 1.0 | 1227 | 1.0 |
| Fe | ng | 353 | 1.75 | 420 | 1.1 | 304 | 1.6 | 3685 | 1.0 | 570 | 0.9 | 790 | 1.0 |
| Mn | ng | 14.5 | 4.2 | 17 | 2.7 | 8.1 | 2.4 | 65 | 1.0 | 12 | 1.1 | 17 | 1.4 |
| Ni | ng | 3.8 | 14 | 4.9 | 10 | 3.4 | 13 | 6.6 | 1.2 | — | — | 2.0 | 2.3 |
| Co | pg | 250 | 2.8 | 250 | 1.5 | — | — | 2100 | 1.2 | — | — | 380 | 2.1 |
| Cr | ng | 4.7 | 13 | 9.0 | 14 | 2.0 | 5.8 | 10 | 1.4 | 1.6 | 1.4 | 3.0 | 2.4 |
| V | ng | — | — | 2.8 | 3.2 | 9.2 | 20 | 15 | 1.5 | — | — | 6.3 | 6.0 |
| Cu | ng | 6.3 | 30.5 | — | — | — | — | 4.5 | 1.1 | 4.9 | 8.0 | 2.6 | 7.2 |
| Zn | ng | 41 | 164 | 46 | 100 | 38 | 156 | 16 | 3.2 | — | — | 10 | 18 |
| Pb | ng | 34.5 | 781 | 60 | 741 | 55 | 1282 | 6.9 | 7.7 | 53 | 6.9 | 4.3 | 27 |
| Cd | pg | — | — | — | — | 640 | 932 | 120 | 8.3 | — | — | 45 | 18 |
| Se | ng | — | — | 0.73 | 2253 | — | — | — | — | — | — | — | — |
| Sb | pg | — | — | — | — | — | — | — | — | — | — | — | — |

*Continued*

system proper, however, they have to cross the air–sea interface, and the characteristics of this important transition zone are described in the next section.

## 4.3 Material transported via the atmosphere: the air–sea interface and the sea-surface microlayer

From the point of view of aerosol chemistry, the recycling of components is perhaps the most important process taking place at the air–sea interface. This interface, however, is also the site of a very specialized marine environment, the sea-surface **microlayer**, at which a number of other geochemically important processes occur.

Surface-active organic materials are found on the surfaces of all natural water bodies, including the oceans. These organic materials, which are usually of a biological origin but can also include anthropogenic substances (e.g. petroleum products), sometimes manifest themselves as visible slicks. Even in the absence of such slicks, however, the sea surface is covered by a thin organic film. This thin film, or microlayer, forms a distinct ecosystem and is an extremely important feature of the ocean reservoir.

The thickness of the microlayer has been variously reported to extend from that of a monomolecular layer to several hundred micrometres, but because it is notoriously difficult to sample it is usually defined operationally in terms of the device used to collect it. Some of these devices are more sophisticated than others; for example, a surface microlayer sampler, the self-contained underway microlayer sampler (SCUMS), which is designed to provide real-time information on interfacial chemical and biological components, has been described by Carlson et al. (1988). In practice, however, most collection techniques retrieve microlayer samples that are considerably diluted with underlying bulk sea water.

The microlayer is the site across which the atmosphere–ocean system interacts, i.e. where the sea 'breathes', and it has unique chemical, physical and biological properties, which are very different from those of the underlying sea water. Although there have been a number of major reviews of the microlayer (see e.g. MacIntyre, 1974; Liss, 1975; Duce & Hoffman, 1976; Hunter & Liss, 1981; Lion & Leckie, 1981a,b; Seiburth, 1983), our conceptual understanding of this marine phenomenon is still somewhat hazy and is continually evolving. There is, for example, contro-

**Table 4.14** *Continued*

Open ocean

| Tropical North Atlantic** | | South Atlantic westerlies§ | | Tropical Indian Ocean ‖ | | Tropical North Pacific (Enewetak)†† | | Tropical South Pacific (Samoa)‡‡ | |
|---|---|---|---|---|---|---|---|---|---|
| 1 | 2 | 1 | 2 | 1 | 2 | 1 | 2 | 1 | 2 |
| 160 | 1.0 | 2.7 | 1.0 | 11 | 1.0 | 21 | 1.0 | 0.72 | 1.0 |
| 100 | 0.9 | 2.6 | 1.4 | 8.8 | 1.1 | 17 | 1.2 | 0.21 | 0.4 |
| 2.2 | 1.2 | 0.11 | 3.5 | 0.16 | 1.2 | 0.29 | 1.2 | 0.005 | 0.6 |
| 0.64 | 4.4 | 0.02 | 8.1 | 0.043 | 4.4 | — | — | — | — |
| 80 | 1.6 | 10 | 12 | 21 | 6.1 | 8 | 1.3 | 0.37 | 1.7 |
| 0.43 | 2.2 | 0.17 | 52 | 0.066 | 8.3 | 0.09 | 3.5 | — | — |
| 0.54 | 2.1 | 0.03 | 6.8 | 0.023 | 1.4 | 0.08 | 2.3 | — | — |
| 0.79 | 7.4 | 0.29 | 161 | 0.077 | 10 | 0.045 | 3.2 | 0.013 | 27 |
| 4.4 | 32 | 1.8 | 784 | 0.10 | 13 | 0.17 | 9.5 | 0.07 | 114 |
| 9.9 | 407 | 0.97 | 2364 | 0.17 | 158 | 0.12 | 38 | 0.016 | 146 |
| — | — | — | — | 4.3 | 788 | 4 | 78 | — | — |
| 0.43 | 4405 | — | — | — | — | 0.13 | $10^3$ | 0.09 | $10^4$ |
| 110 | 286 | — | — | — | — | 4.0 | 79 | 0.20 | 116 |

* Chester *et al.* (1994). † Hacisalihoglu *et al.* (1992). ‡ Keyse (1996). § Murphy (1985). ¶ Saydam (1981). ‖ Chester *et al.* (1991).
** Buat-Menard & Chesselet (1979). †† Duce *et al.* (1983a). ‡‡ Arimoto *et al.* (1987).

versy over the organic composition of the microlayer. Lion & Leckie (1981a) have pointed out that most of the organics that have been reported to occur in the microlayer fall into two categories of surface-active material: type 1, e.g. fatty acids, alcohols and lipids; and type 2, which consists of proteinaceous substances. Carbohydrates, insoluble hydrocarbons and chlorinated hydrocarbons also are found in the microlayer. Originally, it was thought that in the absence of petroleum pollution, sea-surface films consisted matter of simple surfactants of type 1 (see e.g. Garrett, 1967), but this was thrown into doubt when it was found that type 2 materials were perhaps dominant (see e.g. Baier, 1970, 1972). More recently, however, it has been suggested that a third, humic-type, material makes up a large proportion of the organics in the microlayer. A large fraction of the dissolved organic matter (DOM) in sea water is uncharacterized (see Section 9.2.3.2) but is known to contain surface-active material, such as humic and fulvic acids (much of which originates from planktonic exudates). Hunter & Liss (1981) suggested that surfactants in the microlayer consist to a large extent of polymeric material arising mainly from that part of the uncharacterized DOM that is surface-active. Seiburth (1983) reviewed the data available on the organic composition of the microlayer and concluded that carbohydrates account for ~33% of the DOC in surface films, proteins ~13% and lipids ~3% (ratios that are similar to those in materials released from algal cultures), the remainder being thought to consist of condensed humic substances.

In the light of findings such as these, a different view of the microlayer began to emerge in the literature. By re-examining the original concepts, Seiburth (1983) proposed that substances such as carbohydrates, proteins, lipids and condensed humics, in both dissolved and colloidal forms, are advected through the mixed layer to adsorb on the 'solid' air–sea interface, where they form a microlayer (or surface film), which may be described in terms of a loose hydrated gel of intertangled macromolecules, both free and condensed, which is colonized by bacteria.

This then is the nature of the organic 'soup' that is present at the sea surface, and it is apparent from the various data given in the literature that relative to bulk sea water the microlayer is enhanced in a variety of substances. These include total suspended solids, particulate and dissolved organic carbon, organic and inorganic phosphorus, particulate and dissolved forms of nitrogen (excluding nitrate), bacteria and other micro-organisms, pollutants (such as DDT and PCBs) and trace metals. The enrichment of trace metals in the microlayer has been demonstrated by many workers (see e.g. Szekielda et al., 1972; Eisenreich et al., 1978; Pojasek & Zajicek, 1978; Pattenden et al., 1981; Hardy et al., 1985a,b). These enrichments are found mainly in the organic and particulate microlayer fractions, but not in the dissolved inorganic fractions. Lion & Leckie (1981a) summarized the various data given in the literature for the enhancement of trace metals in the microlayer (see Table 4.15) and drew a number of overall conclusions.

**1** The enhancement of trace metals in the microlayer is not a consistent phenomenon.

**2** The frequency and degree of enhancement increase with the presence of observable organic surface slicks.

**3** The relative amounts of both organically associated and particulate trace metals (PTMs) are higher in the microlayer than in bulk sea water, although the degree of enhancement varies widely.

It also may be concluded that the concentrations of elements such as Pb, Zn, Cu, Cd and Fe are higher in microlayer samples from polluted than from non-polluted coastal waters (see e.g. Hardy et al., 1985b). The question that then arises is 'What process, or processes, cause(s) these microlayer enrichments?' To address this question it is useful to refer to the model outlined by Lion & Leckie (1981b)—see Fig. 4.10. The principal features in this model can be described as follows. Trace metals are enhanced in the organic

**Table 4.15** Order-of-magnitude trace-metal enrichments in the sea-surface microlayer relative to bulk sea water (data from Lion & Leckie, 1981a).

| Element | Microlayer enrichment: no observable surface slick* | Microlayer enrichment: slick or foam samples* |
|---|---|---|
| Pb | $2-10^1$ | $10^3-10^7$ |
| Cu | $2-10^1$ | $10^2-10^5$ |
| Fe | $<1-10^1$ | $10^2-10^6$ |
| Ni | $1-2$ | $10^5$ |
| Zn | $1.5-3$ | $10^2-10^4$ |

* Approximate values only.

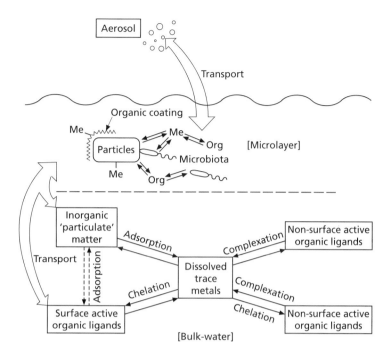

**Fig. 4.10** Schematic representation of the alternatives for the fates of trace metals at the air–sea interface (from Lion & Leckie, 1981b). The trace metals may be thought of as competing with each other, and with the major marine cations, for adsorption sites and for available complexing ligands. At equilibrium, the results of interactions of this type will reflect the differing aqueous chemistries (including speciation) of the various trace metals.

and particulate fractions of the microlayer (see above), e.g. by preferential transport via processes such as bubble flotation. Dissolved inorganic forms of trace metals, however, can also be transported into the microlayer, e.g. by turbulent mixing, where the enhancement of particulate matter and dissolved organic ligands provides the conditions for *in situ* adsorption and complexation. These various processes therefore combine to produce an enrichment of organically bound and particulate trace metals in the microlayer environment.

The question of the residence times of trace metals in the microlayer has been considered by a number of authors. For example, Hardy *et al.* (1985a) used laboratory derived data to set up a model designed to predict the behaviour of the metals in the microlayer under various conditions of biotic enhancement, windspeed and atmospheric PTM deposition rates. Mean predicted residence times for PTMs in the microlayer were in the range 1.5–15 h before they entered the water column.

The various components, including dissolved gases, that are concentrated in the microlayer are available for exchange across the air–sea interface and, in particular, the enhancement of trace metals in the microlayer can have an important effect on the

chemistry of the marine aerosol. For example, it was shown in Section 4.2.1.2 that during the formation of sea-salts by bubble bursting part of the microlayer can be skimmed off and transferred into the atmosphere. The microlayer is therefore the region in which aerosols are both deposited from the atmosphere into the sea and are injected from the sea into the atmosphere, i.e. it is a site of recycling.

On the basis of the brief discussion outlined above it therefore may be concluded that the microlayer is a complex and reactive environment that forms the interface separating the sea from the atmosphere.

## 4.4 The atmospheric pathway: summary

1 A particulate 'aerosol veil' is present over all the oceans, but the concentrations of material in the veil vary over several orders of magnitude, ranging from greater than $\sim 10^3\,\mu g\,m^{-3}$ of air off desert and arid regions to less than $\sim 10^{-3}\,\mu g\,m^{-3}$ of air in remote oceanic areas.

2 The principal sources of material to the marine aerosol veil are the Earth's crust (the crustal aerosol), the sea surface (the sea-salt aerosol) and a variety of mainly high-temperature anthropogenic processes (the enriched, or anthropogenic, aerosol).

3 There is a general tendency for the sources of both crustal and anthropogenic material to be concentrated in the Northern Hemisphere, where they occur in specific latitudinal belts.

4 The sources of anthropogenic material include episodes of 'high pollution' but, in general, anthropogenic components are supplied to the atmosphere on a more or less continuous basis and form part of the 'background' aerosol.

5 The global mineral dust source strength is ~100–~800 g yr⁻¹; however, unlike anthropogenic material, much of the dust from the major desert sources is delivered to the marine atmosphere in the form of intermittent, i.e. non-continuous, pulses.

6 The particulate elements in the marine aerosol have characteristic particle-size spectra constrained by their sources; sea-salt-associated elements have most of their total mass on particles with MMDs in the range ~3–~7 µm, crust-derived elements have most of their total mass on particles with MMDs in the range ~1–~3 µm, and anthropogenically derived elements have most of their total mass on particles with MMDs <0.5 µm.

7 The concentrations of many particulate elements in the marine aerosol vary over several orders of magnitude, the highest concentrations being found close to the continental (crustal and anthropogenic) sources and the lowest in remote pristine regions.

8 Atmospherically transported particulates have to cross the air–sea interface before entering the bulk ocean. This interface is the site of the sea-surface microlayer, which is an organic-rich, particle-rich and trace-metal-rich zone, and is one of the most reactive environments in the oceans.

## References

Andreae, M.O. (1986) The ocean as a source of atmospheric sulfur compounds. In *The Role of Air–Sea Exchange in Geochemical Cycling*, P. Buat-Menard (ed.), 331–62. Dordrecht: Reidel.

Andreae, M.O., Carlson, R.J., Bruynseels, F., Storm, H., van Grieken, R. & Maenhaut, W. (1986) Internal mixtures of sea salt, silicates, excess sulfate in marine aerosols. *Science*, 232, 1620–3.

Arimoto, R. & Duce, R.A. (1982) Dry deposition and the air–sea exchange of trace elements. *J. Geophys. Res.*, 91, 2787–92.

Arimoto, R., Duce, R.A., Ray, B.J., Hewitt, A.D. & Williams, J. (1987) Trace elements in the atmosphere of American Samoa: concentrations and deposition to the tropical South Pacific. *J. Geophys. Res.*, 92, 8465–79.

Atlas, S. & Giam, C.S. (1986) Sea–air exchange of high molecular weight synthetic organic compounds. In *The Role of Air–Sea Exchange in Geochemical Cycling*, P. Buat-Menard (ed.), 295–329. Dordrecht: Reidel.

Baier, R.E. (1970) Surface quality assessment in natural bodies of water. *Proceedings, 13th Conference on Great Lakes Research*, 114–27. International Association of Great Lakes Research.

Baier, R.E. (1972) Organic films on natural waters: their retrieval, identification, and modes of elimination. *J. Geophys. Res.*, 77, 5062–75.

Berg, W.W. & Winchester, J.W. (1978) Aerosol chemistry of the marine atmosphere. In *Chemical Oceanography*, J.P. Riley & R. Chester (eds), Vol. 7, 173–231. London: Academic Press.

Berry, S. (1990) *Solid state speciation and sea water solubility studies and trace metal chemistry of the Indian Ocean aerosol*. PhD Thesis, University of Liverpool.

Blanchard, D.C. (1963) Electrification of the atmosphere by particles from bubbles. *Prog. Oceanogr.*, 1, 71–202.

Blanchard, D.C. (1983) The production, distribution, and bacterial enrichment of the sea-salt aerosol. In *Air–Sea Exchange of Gases and Particles*, P.S. Liss & W.G.N. Slin (eds), 407–54. Dordrecht: Reidel.

Blank, M., Leinen, M. & Prospero, J.M. (1985) Major Asian aeolian inputs indicated by the mineralogy of aerosols and sediments in the western North Pacific. *Nature*, 314, 84–6.

Bonsang, B., Nguyen, B.C., Gaudry, A. & Lambert, G. (1980) Sulfate enrichment in marine aerosols owing to biogenic gaseous sulfur compounds. *J. Geophys. Res.*, 85, 7410–16.

Buat-Menard, P. & Chesselet, R. (1979) Variable influence of the atmospheric flux on the trace metal chemistry of oceanic suspended matter. *Earth Planet. Sci. Lett.*, 42, 399–411.

Buat-Menard, P., Cachier, H. & Chesselet, R. (1989) Sources of particulate carbon to the marine atmosphere. In *Chemical Oceanography*, J.P. Riley, R. Chester & R.A. Duce (eds), Vol. 10, 251–79. London: Academic Press.

Cachier, H., Bremond, M-P. & Buat-Menard, P. (1989) Carbonaceous aerosols from different tropical biomass burning sources. *Nature*, 340, 371–73.

Carlson, A.J., Cantey, J.L. & Cullen, J.J. (1988) Description of and results from a new surface microlayer sampling device. *Deep-Sea Res.*, 35, 1205–13.

Chester, R. (1986) The marine mineral aerosol. In *The Role of Air–Sea Exchange in Geochemical Cycling*, P. Buat-Menard (ed.), 443–76. Dordrecht: Reidel.

Chester, R., Elderfield, H. & Griffin, J.J. (1971) Dust transported in the northeast and southeast trade winds of the Atlantic Ocean. *Nature*, 233, 93–134.

Chester, R., Griffiths, A.G. & Hirst, J.M. (1979) The

influence of soil-sized atmospheric particulates on the elemental chemistry of the deep-sea sediments of the northeastern Atlantic. *Mar. Geol.*, **32**, 141–54.

Chester, R., Sharples, E.J. & Murphy, K.J.T. (1984a) The distribution of particulate Mo in the Atlantic aerosol. *Oceanol. Acta*, **7**, 441–50.

Chester, R., Sharples, E.J., Sanders, G.S. & Saydam, A.C. (1984b) Saharan dust incursion over the Tyrrhenian Sea. *Atmos. Environ.*, **18**, 929–35.

Chester, R., Sharples, E.J. & Sanders, G.S. (1985) The concentration of particulate aluminium and clay minerals in aerosols from the northern Arabian Sea. *J. Sed. Petrol.*, **55**, 37–41.

Chester, R., Murphy, K.J.T. & Lin, F.J. (1989) A three stage sequential leaching scheme for the characterization of the sources and environmental mobility of trace metals in the marine aerosol. *Environ. Technol. Lett.*, **10**, 887–900.

Chester, R., Berry, A.S. & Murphy, K.J.T. (1991) The distributions of particulate atmospheric trace metals and mineral aerosols over the Indian Ocean. *Mar. Chem.*, **34**, 261–90.

Chester, R., Murphy, K.J.T., Lin, F.J., Berry, A.S., Bradshaw, G.A. & Corcoran, P.A. (1993) Factors controlling the solubilities of trace metals from non-remote aerosols deposited to the sea surface by the 'dry' deposition mode. *Mar. Chem.*, **42**, 107–26.

Chester, R., Bradshaw, G.F. & Corcoran, P.A. (1994) Trace metal chemistry of the North Sea particulate aerosol: concentrations, sources and sea water fates. *Atmos. Environ.*, **28**, 2873–83.

Defant, A. & Defant, F. (1958) *Physikalische Dynamik der Atmosphäre*. Frankfurt: Akademie Verlagsgesellschaft.

Delany, A.C., Delany, A.C., Parkin, D.W., Griffin, J.J., G. Goldberg & Reinmann, B.E. (1967) Airborne dust collected at Barbados. *Geochim. Cosmochim. Acta*, **31**, 885–909.

Duce, R.A. & Hoffman, G.L. (1976) Atmospheric vanadium transport to the ocean. *Atmos. Environ.*, **10**, 989–96.

Duce, R.A., Hoffman, G.L. & Zoller, W. (1975) Atmospheric trace metals at remote Northern and Southern hemisphere sites: pollution or natural? *Science*, **187**, 59–61.

Duce, R.A., Hoffman, G.L., Ray, B.J., *et al.* (1976) Trace metals in the marine atmosphere: sources and fluxes. In *Marine Pollutant Transfer*, H.L. Windom & R.A. Duce (eds), 77–120. Lexington, MA: Lexington Books.

Duce, R.A., Arimoto, R., Ray, B.J., Unni, C.K. & Harder, P.J. (1983a) Atmospheric trace metals at Enewetak Atoll: I. Concentrations, sources and temporal variability. *J. Geophys. Res.*, **88**, 5321–42.

Duce, R.A., Mohnen, V.A., Zimmerman, P.R., *et al.* (1983b) Organic material in the global troposphere. *Rev. Geophys. Space Phys.*, **21**, 921–52.

Eisenreich, S.J., Elzerman, A.W. & Armstrong, D.E. (1978)

Enrichment of micronutrients, heavy metals and chlorinated hydrocarbons in wind-generated lake foam. *Environ. Sci. Technol.*, **12**, 413–22.

Erickson, D., Merrill, J.I. & Duce, R.A. (1986) Seasonal estimates of global atmospheric sea-salt distributions. *J. Geophys. Res.*, **91**, 1067–72.

Erickson, D.J. & Duce, R.A. (1988) On the global flux of atmospheric sea salt. *J. Geophys. Res.*, **93**, 14079–88.

Eriksson, E. (1959) The yearly circulation of chloride and sulphate in nature. Meteorological, geochemical, pedological implications. Part I. *Tellus*, **11**, 317–403.

Fasching, J.L., Courant, R.A., Duce, R.A. & Piotrowicz, S.R. (1974) A new surface microlayer sampler utilizing the bubble microtome. *J. Rech. Atmos.*, **8**, 649–52.

Folger, D.W. (1970) Wind transport of land-derived mineral, biogenic and industrial matter over the North Atlantic. *Deep-Sea Res.*, **17**, 337–52.

Gagosian, R.B. (1986) The air–sea exchange of particulate organic matter: the sources and long-range transport of lipids in aerosols. In *The Role of Air-Sea Exchange in Geochemical Cycling*, P. Buat-Menard (ed.), 409–42. Dordrecht: Reidel.

Gaudichet, A., Lefevre, R., Gaudry, A., Ardouin, B., Lambert, G. & Miller, J.M. (1989) Minerological composition of aerosols at Amsterdam Island. *Tellus*, **41B**, 344–52.

Garrett, W.D. (1967) The organic chemical composition of the ocean surface. *Deep-Sea Res.*, **14**, 221–7.

Gilman, C. & Garrett, C. (1994) Heat flux parameterisation for the Mediterranean Sea: the role of atmospheric aerosols and constraints from the water budget. *J. Geophys. Res.*, **99**, 5119–34.

Goldberg, E.D. (1975) Marine pollution. In *Chemical Oceanography*, J.P. Riley & G. Skirrow (eds), Vol. 3, 39–89. London: Academic Press.

Goldberg, E.D. & Griffin, J.J. (1970) The sediments of the northern Indian Ocean. *Deep-Sea Res.*, **17**, 513–37.

Hacisalihoglu, G., Eliyakut, F., Olmez, I., Balkas, T.I. & Tuncel, G. (1992) Chemical composition of particles in the Black Sea atmosphere. *Atmos. Environ.*, **26**, 3207–18.

Hardy, J.T., Apts, C.W., Crecelius, E.A. & Bloom, N.S. (1985a) Sea-surface microlayer metals enrichments in an urban and rural bay. *Estuarine Coastal Shelf Sci.*, **20**, 299–312.

Hardy, J.T., Apts, C.W., Crecelius, E.A. & Fellingham, G.W. (1985b) The sea-surface microlayer: fate and residence times of atmospheric metals. *Limnol. Oceanogr.*, **30**, 93–101.

Hasse, L. (1983) Introductory meteorology and fluid mechanics. In *Air–Sea Exchange of Gases and Particles*, P.S. Liss & W.G.N. Slin (eds), 1–51. Dordrecht: Reidel.

Hunter, K.A. & Liss, P.S. (1981) Organic sea surface films. In *Marine Organic Chemistry*, E.K. Duursma & R. Dawson (eds), 259–98. Amsterdam: Elsevier.

Irbarne, J.V. & Cho, H.-R. (1980) *Atmospheric Physics*. Dordrecht: Reidel.

Junge, C.E. (1972) Our knowledge of the physico-chemistry of aerosols in the undisturbed marine environment. *J. Geophys. Res.*, **77**, 5183–200.

Junge, C. (1979) The importance of mineral dust as an atmospheric constituent. In *Saharan Dust*, C. Morales (ed.), 243–66. New York: Wiley.

Keyse, S. (1996) *Factors controlling the trace metal chemistry of rainwaters*. PhD thesis, University of Liverpool.

Koga, M. (1981) Direct production of droplets from breaking wind waves — its observation by a multicolored overlapping exposure photography technique. *Tellus*, **33**, 552–63.

Lai, R.J. & Shemdin, O.H. (1974) Laboratory study of the generation of spray over water. *J. Geophys. Res.*, **79**, 3055–63.

Lantzy, R.J. & Mackenzie, F.T. (1979) Atmospheric trace metals: global cycles and assessment of man's impact. *Geochim. Cosmochim. Acta*, **43**, 511–25.

Lion, L.W. & Leckie, J.O. (1981a) The biogeochemistry of the air–sea interface. *Annu. Rev. Earth Planet. Sci.*, **9**, 449–86.

Lion, L.W. & Leckie, J.O. (1981b) Chemical speciation of trace metals at the air–sea interface: the application of an equilibrium model. *Environ. Geol.*, **3**, 293–314.

Liss, P.S. (1975) Chemistry of the sea surface microlayer. In *Chemical Oceanography*, J.P. Riley & G. Skirrow (eds), Vol. 3, 193–243. London: Academic Press.

McDonald, R.L., Unni, C.K. & Duce, R.A. (1982) Estimation of atmospheric sea salt dry deposition: wind speed and particle size dependence. *J. Geophys. Res.*, **87**, 1246–50.

MacIntyre, F. (1974) Chemical fractionation and sea-surface microlayer processes. In *The Sea*, E.D. Goldberg (ed.), Vol. 5, 245–99. New York: Interscience.

Malin, G.S., Turner, S., Liss, P., Holligan, P. & Harbour, D. (1993) Dimethylsulphide and dimethylsulphoniopropionate in the Northeast Atlantic during the summer coccolithophore bloom. *Deep-Sea Res.*, **40**, 1487–1508.

Maring, H.B. & Duce, R.A. (1990) The impact of atmospheric aerosols on trace metal chemistry in open surface sea water. 3. Lead. *J. Geophys. Res.*, **95**, 5341–47.

Merrill, J.T. (1989) Atmospheric long-range transport to the Pacific Ocean. In *Chemical Oceanography*, J.P. Riley, R. Chester & R.A. Duce (eds), Vol. 10, 15–20. London: Academic Press.

Mosher, B.W., Duce, R.A., Prospero, J.M. & Savoie, D.L. (1987) Atmospheric selenium: geographical distribution and ocean to atmosphere flux in the Pacific. *J. Geophys. Res.*, **92**, 13277–87.

Moulin, C., Lamber, C.E., Dulac, F. & Dayan, U. (1997) Control of atmospheric export of dust from North Africa by the North Atlantic Oscillation. *Nature*, **387**, 691–94.

Murphy, K.J.T. (1985) *The trace metal chemistry of the Atlantic aerosol*. PhD thesis, University of Liverpool.

Nguyen, B.C. & Bonsang, B. (1979) The ocean as a source and sink for natural and anthropogenic sulfur compounds. *XVIIth Assembl. Gen. Union Geodes. Int.*, Canberra, Australia, 1–15 December 1979.

Nguyen, B.C., Bonsang, B. & Gaudry, A. (1983) The role of the ocean in the global atmospheric sulfur cycle. *J. Geophys. Res.*, **88**, 10903–14.

Nriagu, J.O. (1979) Global inventory of natural and anthropogenic emissions of trace metals to the atmosphere. *Nature*, **279**, 409–11.

Nriagu, O.N. (1989) Natural versus anthropogenic emissions of trace metals to the atmosphere. In *Control and Fate of Atmospheric Trace Metals*, J.M. Pacyna & B. Ottar (eds), 3–13. Dordrecht: Kluwer Academic Publishers.

Pacyna, J.M., Semb, A. & Hanson, J.E. (1984) Emissions and long-range transport of trace elements in Europe. *Tellus*, **36B**, 163–178.

Pattenden, N.J., Cambray, R.S. & Playford, K. (1981) Trace and major elements in the sea-surface microlayer. *Geochim. Cosmochim. Acta*, **45**, 93–100.

Peltzer, E.T. & Gagosian, R.B. (1989) Oceanic geochemistry of aerosols over the Pacific Ocean. In *Chemical Oceanography*, J.P. Riley & R. Chester (eds), Vol. 10, 282–338. London, Academic Press.

Piotrowicz, S.R., Duce, R.A., Fasching, J.L. & Weisel, C.P. (1979) Bursting bubbles and their effect on the sea-to-air transport of Fe, Cu, and Zn. *Mar. Chem.*, **7**, 307–24.

Pojasek, R.B. & Zajicek, O.T. (1978) Surface microlayers and foams — source and metal transport in aquatic systems. *Water Res.*, **12**, 7–11.

Prospero, J.M. (1968) Atmospheric dust studies on Barbados. *Am. Meteorol. Soc. Bull.*, **49**, 645–52.

Prospero, J.M. (1979) Mineral and sea salt aerosol concentrations in various oceanic regions. *J. Geophys. Res.*, **84**, 725–31.

Prospero, J.M. (1981) Eolian transport to the World Ocean. In *The Sea*, C. Emiliani (ed.), Vol. 7, 801–74. New York: Interscience.

Prospero, J.M., Charlson, R.W., Mohnen, V., *et al.* (1983) The atmospheric aerosol system: an overview. *Rev. Geophys. Space Phys.*, **21**, 1607–29.

Prospero, J.M., Uematsu, M. & Savoie, D.L. (1989) Mineral aerosol transport to the Pacific Ocean. In *Chemical Oceanography*, J.P. Riley, R. Chester & R.A. Duce (eds), Vol. 10, 137–218. London: Academic Press.

Rahn, K.A. (1976) *The Chemical Composition of the Atmospheric Aerosol*. Kingston, RI: Technical Report, Graduate School of Oceanography, University of Rhode Island.

Rahn, K.A. (1981) The Mn/V ratio as a tracer of large-scale sources of pollution aerosol for the Arctic. *Atmos. Environ.*, **15**, 1547–64.

Rahn, K.A. & Lowenthal, D.H. (1984) Elemental tracers of distant regional pollution aerosols. *Science*, **223**, 132–39.

Rahn, K.A., Borys, R.D., Shaw, G.E., Schutz, L. & Jaenicke, R. (1979) Long-range impact of desert aerosol on atmospheric chemistry: two examples. In *Saharan Dust*, C. Morales (ed.), 243–66. New York: Wiley.

Robinson, E. & Robbins, R.D. (1971) *Emissions, Concentrations, and Fate of Particulate Atmospheric Pollutants.* Washington, DC: Publication 4067, American Petroleum Institute.

Savoie, D.L. & Prospero, J.M. (1977) Aerosol concentration statistics for the northern tropical Atlantic. *J. Geophys. Res.*, **82**, 5954–64.

Savoie, D.L., Prospero, J.M. & Nees, R.T. (1987) Nitrate, non-sea-salt sulphate, and mineral aerosol over the northwestern Indian Ocean. *J. Geophys. Res.*, **92**, 933–42.

Savoie, D.L., Prospero, J.M. & Saltzman, E.S. (1989) Nitrate, non-sea-salt sulphate and methanesulfonate over the Pacific Ocean. In *Chemical Oceanography*, J.P. Riley, R. Chester & R.A. Duce (eds), Vol. 10, 219–50. London: Academic Press.

Saydam, A.C. (1981) *The elemental chemistry of Eastern Mediterranean atmospheric particulates.* PhD thesis, University of Liverpool.

Schneider, B. (1987) Source characterization for atmospheric trace metals over Kiel Bight. *Atmos. Environ.*, **21**, 1275–83.

Seiburth, J. McN. (1983) Microbiological and organic-chemical processes in the surface and mixed layers. In *Air–Sea Exchange of Gases and Particles*, P.S. Liss & W.G.N. Slin (eds), 121–72. Dordrecht: Reidel.

Settle, D.M. & Patterson, C.C. (1982) Magnitudes and sources of precipitation and dry deposition fluxes of industrial and natural leads to the North Pacific at Enewetak. *J. Geophys. Res.*, **87**, 8857–69.

Simoneit, B.R.T. (1978) The organic chemistry of marine sediments. In *Chemical Oceanography*, J.P. Riley & R. Chester (eds), Vol. 7, 233–311. London: Academic Press.

Simoneit, B.R.T., Chester, R. & Eglinton, G. (1977) Biogenic lipids in particulates from the lower atmosphere over the eastern Atlantic. *Nature*, **267**, 682–85.

Suman, D.O. (1986) Charcoal production from agricultural burning in Central Panama and its deposition in the sediments of the Gulf of Panama. *Environ. Conservation*, **13**, 51–60.

Szekielda, K.M., Kupferman, S.L., Klemas, V. & Polis, D.F. (1972) Element enrichment in organic films and foams associated with aquatic frontal systems. *J. Geophys. Res.*, **77**, 5278–82.

Turekian, K.K., Graustein, W.C. & Cochrau, J.K. (1989) Lead-210 in the SEAREX program: an aerosol tracer across the Pacific. In *Chemical Oceanography*, J.P. Riley & R. Chester (eds), Vol. 10, 57–81. London: Academic Press.

Uematsu, M., Duce, R.A., Prospero, J.M., Chen, L., Merrill, J.I. & McDonald, R.L. (1983) Transport of mineral aerosol from Asia over the North Pacific Ocean. *J. Geophys. Res.*, **88**, 5343–52.

Varhelyi, G. & Gravenhorst, G. (1983) Production rate of airborne sea-salt sulfur deduced from chemical analysis of marine aerosols and precipitation. *J. Geophys. Res.*, **88**, 6737–51.

Weisel, C.P., Duce, R.A., Fasching, J.L. & Heaton, R.W. (1984) Estimates of the transport of trace metals from the ocean to the atmosphere. *J. Geophys. Res.*, **89**, 11607–18.

Whitby, K.T. (1977) The physical characteristics of sulphur aerosols. *Atmos. Environ.*, **12**, 135–59.

Yaaqub, R.R., Davies, T.D., Jickells, T.D. & Miller, J.M. (1991) Trace elements in daily collected aerosols at a site in southeast England. *Atmos. Environ.*, **25**, 985–96.

# 5 The transport of material to the oceans: the hydrothermal pathway

The upper part of the igneous oceanic basement (Layer 2 of the oceanic crust), which underlies the sediment column (Layer 1 of the oceanic crust), is composed predominantly of basaltic lavas and their intrusive equivalents. These basalts are by far the commonest rock type found on the sea bed, and the chemical compositions of a number of marine basalts are given in Table 5.1. These basaltic lavas can interact with sea water over a wide range of temperatures and timespans. In very general terms, the reactions can be classifed into three types:

1 those involving hydrothermal activity, either at high temperature at depth in the crust at the centres of sea-floor spreading or at intermediate temperatures on the ridge flanks;

2 those associated with the low-temperature weathering of basalt that has been exposed to sea water for relatively long periods of time, either at the sea floor or within the upper part of the basement;

3 those involving the extrusion of hot lava directly on to the sea bed.

## 5.1 Hydrothermal activity: high-temperature basalt–seawater reactions

The most dramatic manifestation of seawater–rock interaction has been identified only over the last few decades or so. This was the discovery that sea water convecting through newly generated oceanic crust at ridge-divergent plate boundaries during sea-floor spreading can play an important role in controlling the chemical mass balance of the oceans. Various lines of geological and geophysical evidence have now established that sea-floor spreading is the dominant process in the formation of the ocean basins. In this process, new oceanic crust (or lithosphere) is formed from molten rock (magma) at the spreading centres on the mid-ocean ridge system—see Fig. 5.1(a). The pre-existing, i.e. older, basalt sequences

are porous, and faults and large transform fractures are found at both fast- and slow-spreading ridges. In these regions cold sea water penetrates this highly permeable pre-existing crust around the centres, sometimes to a depth of several kilometres, where it undergoes heat-driven circulation and comes into contact with the zones of active magma intrusion. During this process the sea water undergoes drastic changes in composition to form high-temperature hydrothermal solutions, which emerge through the sea-floor venting system as hot springs that mix with the overlying sea water.

Elderfield (1976) made a distinction between geothermal and hydrothermal solutions. During the formation of **geothermal solutions** thermal transfer takes place between the circulating water and the heat source, but there is no chemical transfer between the two. These geothermal solutions can then extract, i.e. leach, chemical components from the rocks or sediments with which they come into contact as they circulate, with the result that they become mineralized in character. In contrast, **hydrothermal solutions** are involved in both thermal and chemical transfer from the heat source. During the chemical transfer the degassing of the mantle can release volatiles such as Hg, As, Ce and B, and gases such as non-radiogenic $^3H$. In addition to this chemical transfer from the heat source, hydrothermal solutions also can become mineralized by leaching metals from rocks and sediments as they circulate. This type of hydrothermal solution is generated at the East Pacific Rise (EPR), and the general features of the EPR hydrothermal model are illustrated in Fig. 5.1(b).

It is now known that large-scale hydrothermal activity is a ubiquitous concomitant to the production of new oceanic crust at the centres of the mid-ocean ridges (Edmond *et al.*, 1982), as was first predicted by Elder (1965). The process has far-reaching implications for Earth science, and such

**Table 5.1** Elemental composition of oceanic basalts*.

(a)  Major element composition (wt% oxides).

| Element | Oceanic island basalts | | Mid-Atlantic Ridge basalts | |
|---|---|---|---|---|
| | Average tholeiitic basalt | Average alkalic basalt | Tholeiitic basalt | High alumina basalt |
| $SiO_2$ | 49.36 | 46.46 | 50.47 | 48.13 |
| $TiO_2$ | 2.50 | 3.01 | 1.04 | 0.72 |
| $Al_2O_3$ | 13.94 | 14.64 | 15.93 | 17.07 |
| $Fe_2O_3$ | 3.03 | 3.37 | 0.95 | 1.17 |
| FeO | 8.53 | 9.11 | 7.88 | 8.65 |
| MnO | 0.16 | 0.14 | 0.13 | 0.13 |
| MgO | 8.44 | 8.19 | 8.75 | 10.29 |
| CaO | 10.30 | 10.33 | 11.38 | 11.26 |
| $Na_2O$ | 2.13 | 2.92 | 2.60 | 2.39 |
| $K_2O$ | 0.38 | 0.84 | 0.10 | 0.09 |
| $H_2O$ (total) | — | — | 0.59 | 0.29 |
| $P_2O_3$ | 0.26 | 0.37 | 0.11 | 0.10 |

(b)  Minor element composition ($\mu g\,g^{-1}$).

| Element | Oceanic alkali basalts | Oceanic tholeiitic basalts | Basalts; Mid-Atlantic Ridge | Basalts; Mid-Indian Ocean Ridge | Basalt; Tonga Island Arc |
|---|---|---|---|---|---|
| Fe | 82 600 | 76 800 | 63 616 | 68 282 | 54 008 |
| Mn | 1084 | 1239 | — | — | — |
| Cu | 36 | 77 | 87 | 90 | 51 |
| Ni | 51 | 97 | 123 | 242 | 25 |
| Co | 25 | 32 | 41 | 73 | 30 |
| Ga | 22 | 17 | 18 | 20 | 13 |
| Cr | 67 | 297 | 292 | 347 | 75 |
| V | 252 | 292 | 289 | 340 | 230 |
| Ba | 498 | 14 | 12 | — | 14 |
| Sr | 815 | 130 | 123 | 131 | 115 |

* From original data sources listed by Riley & Chester (1971), and Chester & Aston (1976). The central ocean areas are dominated by tholeiitic pillow basalts of layer 2A, which are generated at the spreading ridges and which are moved away during sea-floor spreading so that their ages increase with increasing distance from the ridge crests.

is its global extent that it has been estimated that the entire ocean circulates through the high-temperature ridge axial zone, and so undergoes reaction with fresh basalt, to emerge via the hydrothermal venting system every 8–10 Ma. The dramatic effect that the venting of these high-temperature hydrothermal solutions can have on the chemistry of sea water is illustrated in Fig. 5.2, which shows the increase in dissolved Mn around a vent system at the TAG site on the Mid-Atlantic Ridge. Because of the importance of basalt–seawater interactions associated with venting systems such as this, oceanic rock is now recognized as a major reservoir in global geochemistry.

Confirmation of the existence of the hydrothermal solutions has been provided by photographic and visual (submersible dives) sightings of the venting of hot springs on to the sea floor at various spreading centres. These sightings have also shown the presence of specialized biological communities around the hot springs. These include dense beds of clams, mussels and giant vestimentiferan tubeworms, for which the primary food source is chemosynthetic bacteria; this is an example of chemical, as opposed to photo-

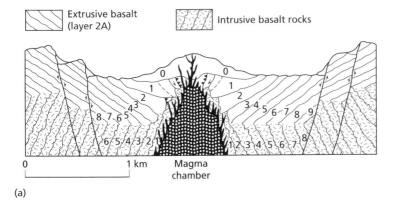

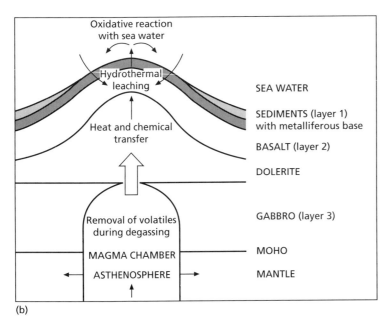

**Fig. 5.1** High-temperature hydrothermal activity at the spreading centres. (a) Preliminary structural model of the inner rift valley in the Famous area (Mid-Atlantic Ridge) showing the location of the magma chamber in relation to the generation of basalts (from Jones, 1978; after Moore *et al.*, 1974). Relative ages of the basalts are indicated by the numbers 0–9, with 0 being the youngest. (b) The generation of hydrothermal solutions: the East Pacific Rise model (from Elderfield, 1976). Sea water penetrates the porous rock at the spreading ridge centre and circulates through the underlying crust. During this process the sea water undergoes both thermal and chemical transfer from the magma heat source, and also leaches elements from the rocks and sediments with which it comes into contact before re-entering the seawater reservoir via a series of venting systems as mineralized solutions.

synthetic, production, and may have implications for the origin of life. For a detailed description of the biology of hydrothermal vent systems, and their status as environments for the origin of life, the reader is referred to the volumes edited by Childers (1988) and Holm (1992), respectively.

Following the formation of new crust, factors such as the deposition of sediments limit the access of sea water and so hinder the circulation in the basalt layer. Wolery & Sleep (1976) have estimated that on fast-spreading ridges hydrothermal activity stops after 6–14 Ma and that on slow-spreading ridges it ceases after 11–19 Ma. Thompson (1983) defines these

periods as the **active** phases of hydrothermal activity, during which new crust is continually exposed to circulating sea water and convection is vigorous and rapid. A **passive** phase continues on the flanks of the ridges, in the absence of the formation of new crust, and there is evidence that considerable convective circulation of sea water through Layer 2 continues for periods up to 50–70 Ma.

Hydrothermal activity at the spreading centres, during which the circulating sea water comes into contact with newly generated oceanic crust, leads to the production of high-temperature (~350°C), acidic, reducing solutions, which eventually are

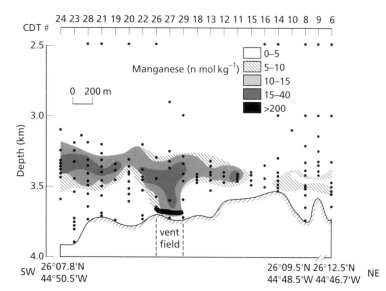

**Fig. 5.2** SW–NE transect for total reactive Mn at the TAG hydrothermal system on the Mid-Atlantic Ridge (from Klinkhammer *et al.*, 1986). Total reactive Mn is the Mn extracted from unacidified seawater samples by oxine after adjustment of the pH to 9, and represents mainly the dissolved fraction of the element.

mixed with ambient sea water that has a much lower initial temperature (~2°C). During the seawater–rock reaction the composition of the original sea water is considerably modified with respect to both major and minor constituents, gases (e.g. $^3$He and other rare gases, methane, hydrogen, carbon monoxide, carbon dioxide) and the isotopic ratios of various elements (e.g. O, Sr, Nd). The location of the zone in which the high-temperature hydrothermal solutions, rich in $H_2S$, mix with $O_2$-rich ambient sea water is critical in determining the composition of the hot-spring water that vents at the sea surface. To date, two general types of hydrothermal 'plumbing' systems have been identified with respect to the temperature at which fluids are debouched on to the sea floor.

*Low-temperature systems.* Hot springs found at the Galapagos Ridge, East Pacific (spreading rate ~3.5 cm yr$^{-1}$), are an example of a low-temperature hydrothermal system. Thompson (1983) described four active vents in the region, which were named Clambake, Dandelions, Garden of Eden and Oyster Beds after the benthic communities associated with them. Water flows on to the sea bed from these vents at a rate of 2–10 l s$^{-1}$, and has temperatures in the range 6–17°C. The chemistry of the Galapagos hydrothermal solutions has been described by Edmond *et al.* (1979). The authors found that there

was a negative magnesium–temperature relationship in the vented fluids, and as experimental laboratory data had shown that magnesium is completely removed from solution during basalt–seawater reactions (Bischoff & Dickson, 1975), they extrapolated the Mg–temperature relationship to zero concentration, which gave an intercept at ~350°C. Thus, by assuming that Galapagos vent solutions were a simple mixture of two end-members, i.e. a hot hydrothermal solution and cold sea water, they were able to demonstrate that the original hydrothermal fluid had a temperature of ~350°C and that it had mixed with sea water at depth within the venting system. During the mixing of hydrothermal solutions and sea water some elements form insoluble phases (e.g. sulphides, oxides), i.e. they behave non-conservatively, whereas others are unaffected by the mixing process, i.e. they are conservative. Edmond *et al.* (1979) extrapolated the concentration–temperature gradients for a number of conservative elements and so were able to estimate the composition of the 350°C hydrothermal end-member; the results are given in Table 5.2. These data were the first of their kind and led to the very important conclusion that, for some elements, fluxes into and out of the crestal ridge zone are comparable with, or greater than, those for river transport. It was also shown that the non-conservative elements (e.g. Cu, Ni, Cd, Se, V and Cr) were removed by the formation of insoluble phases at intermediate temper-

**Table 5.2** Chemical composition of the high-temperature (350°C) hydrothermal end-member solution*.

| Component | Galapagos vents | 21°N EPR vents | Sea water |
|---|---|---|---|
| Li ($\mu$mol kg$^{-1}$) | 1142–689 | 820 | 28 |
| K (mmol kg$^{-1}$) | 18.8 | 25.0 | 10.1 |
| Rb (mmol kg$^{-1}$) | 20.3–13.4 | 26.0 | 1.32 |
| Mg (mmol kg$^{-1}$) | 0 | 0 | 52.7 |
| Ca (mmol kg$^{-1}$) | 40.2–24.6 | 21.5 | 10.3 |
| Sr ($\mu$mol kg$^{-1}$) | 87 | 90 | 87 |
| Ba ($\mu$mol kg$^{-1}$) | 42.6–17.2 | 95–35 | 0.145 |
| Mn ($\mu$mol kg$^{-1}$) | 1140–360 | 610 | 0.002 |
| Fe ($\mu$mol kg$^{-1}$) | + | 1800 | – |
| Si ($\mu$mol kg$^{-1}$) | 21.9 | 21.5 | 0.160 |
| SO$_4^{2-}$ (mmol kg$^{-1}$) | 0 | 0 | 28.6 |
| H$_2$S (mmol kg$^{-1}$) | + | 6.5 | 0 |

* Data from Edmond *et al.* (1982). Data for the Galapagos vents is extrapolated to the 350°C end-member, the ranges resulting from different composition versus heat trends. Data for the EPR vents are based on direct observation of the high-temperature end-members.

+, non-conservative in subsurface mixing.

–, seawater concentration not known accurately.

atures during the mixing process, and the authors concluded that the elements were extracted not only from the hydrothermal solutions but also from the seawater end-member. For these elements, therefore, it was not possible to obtain concentration data for the 350°C end-member.

In the Galapagos system, therefore, reactions between sea water and hydrothermal solutions take place at depth within the venting system, leading to the precipitation of sulphide-forming metals below the sea surface. That is, only the last phases of hydrothermal fractionation occur on the sea bed in the Galapagos Rift Valley.

*High-temperature systems.* A variety of hydrothermal vents are located on the East Pacific Rise (EPR; spreading rate 6 cm yr$^{-1}$) at 21°N. These include: (i) inactive vents; (ii) warm springs of the Galapagos type, which discharge clear milky white plumes and are sometimes referred to as **white smokers**; and (iii) hot springs. The hot springs emit black plumes through tall (up to 10 m high) chimneys composed predominantly of the sulphides of Fe, Zn and Cu. These high-temperature systems are termed **black**

**smokers**, the colour arising from the precipitation of fine-grained sulphides as sea water is entrained in the plumes emitted from the chimneys, a process that took place at depth within the Galapagos white-smoker venting system. Black smokers, which grow in two stages (stage 1, the formation of a sulphate-dominated wall, followed by stage 2, the precipitation of sulphide minerals on the inner sides, and in pore spaces, of the walls), debouch high-temperature ($\sim$350°C), acidic, reducing, sulphide- and metal-rich hydrothermal solutions directly on to the sea floor, where they are rapidly mixed with ambient sea water.

The presence of high-temperature black smokers on the EPR allowed the pristine hydrothermal end-member to be sampled directly. The data obtained for this hydrothermal end-member have been reported by Edmond *et al.* (1982) and Von Damm *et al.* (1983). The data obtained for the EPR high-temperature fluids, which are summarized in Table 5.2, essentially confirm the validity of the extrapolated Galapagos hydrothermal end-member composition, and thus establish a general chemical uniformity in the solutions produced during the reactions of sea water with newly formed basaltic crust at high temperature at the centres of sea-floor spreading. It can be seen from the data in Table 5.2 that the high-temperature hydrothermal end-member has very special characteristics; thus, relative to sea water it is completely depleted in Mg and SO$_4^{2-}$, and is enriched in H$_2$S, Mn, Rb, Ca, Ba, K and Si. The Mg is removed by precipitation as sepiolite, which is formed in the altered basalt. During this process H$^+$ ions are generated, which make the solutions more acidic, so releasing elements from the basalts and keeping certain elements (e.g. Fe, Mn, Zn and Cu) in solution. Hydrothermal activity is also a sink for sulphate, e.g. by the precipitation of calcium sulphate, which subsequently is reduced by ferric iron in the basalts to give rise to a sulphide-dominated system (see e.g. Martin & Whitfield, 1983); for a detailed discussion of the fate of sulphate during hydrothermal mixing see McDuff & Edmond (1982).

A more detailed investigation of the hot springs in the 21°N EPR hydrothermal system was carried out by Von Damm *et al.* (1985a). Four hydrothermal fields were sampled, all of which vented high-temperature fluids, with the maximum temperatures being $\sim$350°C. The four hydrothermal smoker fields were designated National Geographic Smoker (NGS; venting solution exit temperature 273°C), Ocean Bottom Seismometer (OBS; venting solution exit temperature 350°C), South West (SW; venting solution exit temperature 355°C) and Hanging Garden (HG; venting solution exit temperature 351°C). The major ion data for the end-member venting solutions were consistent with the estimates made on the hot springs from the Galapagos Spreading Centre low-temperature venting system (see Table 5.3), and may be summarized as follows.

1 Li (up to ~50), K (~2.5), Rb (~25), Be (≳10³), Ca (up to ~2) and Ba (up to >10²) were enriched in the hydrothermal fluids; the figures in parentheses refer to the approximate enrichment over ambient sea-water concentrations.

2 Sr and Na exhibit both enrichments and depletions in the hydrothermal fluids relative to sea water.

3 Mg is depleted in the venting fluids relative to sea water, and is assumed to reach zero concentration in the hydrothermal end-member.

There are difficulties in the estimation of the hydrothermal end-member concentrations of many trace metals because of their precipitation as sulphides, which can be formed at depth in the system, in the chimney structures and in the hydrothermal plume itself; however, Mn, which rarely forms a sulphide, is an exception to this. The calculated high-temperature hydrothermal end-member concentrations of Mn, Fe, Co, Cu, Zn, Ag, Cd and Pb are given in Table 5.3. All these elements are enriched in the venting solutions relative to ambient sea water. One of the most striking features of the 21°N EPR data, however, is that there are considerable variations in the end-member concentrations of the trace metals at the various venting sites. Von Damm et al. (1985a) identified a number of factors that might contribute to variations of this type.

1 Differences in rock type; e.g. glasses react with sea water more rapidly than more crystalline forms.

2 Differences in temperature of the hydrothermal solutions at depth in the system.

3 Differences in the residence times, or flow rates, of the water in the fissure system: an increased residence time implies an increase in the time over which the water can react with the rock.

**Table 5.3** Chemical compositions of the individual high-temperature (~350°C) hydrothermal end-member venting solutions at 21°N on the EPR (data from Von Damm et al., 1985a).

| Component | Hydrothermal smoker field* | | | | Sea water |
| | NGS | OBS | SW | HG | |
|---|---|---|---|---|---|
| Li ($\mu$mol kg$^{-1}$) | 1033 | 891 | 899 | 1322 | 26 |
| Na (m kg$^{-1}$) | 510 | 432 | 439 | 433 | 464 |
| K (m kg$^{-1}$) | 25.8 | 23.2 | 23.2 | 23.9 | 9.79 |
| Rb ($\mu$mol kg$^{-1}$) | 31 | 28 | 27 | 33 | 1.3 |
| Be (nmol kg$^{-1}$) | 37 | 15 | 10 | 13 | 0.02 |
| Mg (m kg$^{-1}$) | 0 | 0 | 0 | 0 | 52.7 |
| Ca (m kg$^{-1}$) | 20.8 | 15.6 | 16.6 | 11.7 | 10.2 |
| Sr ($\mu$mol kg$^{-1}$) | 97 | 81 | 83 | 65 | 87 |
| Ba ($\mu$mol kg$^{-1}$) | >15 | >7 | >9 | >10 | 0.14 |
| Al ($\mu$mol kg$^{-1}$) | 4 | 5.2 | 4.7 | 4.5 | 0.005 |
| Mn ($\mu$mol kg$^{-1}$) | 1002 | 960 | 699 | 878 | <0.001 |
| Fe ($\mu$mol kg$^{-1}$) | 871 | 1664 | 750 | 2429 | <0.001 |
| Co (nmol kg$^{-1}$) | 22 | 213 | 66 | 227 | 0.03 |
| Cu ($\mu$mol kg$^{-1}$) | <0.02 | 35 | 9.7 | 44 | 0.007 |
| Zn ($\mu$mol kg$^{-1}$) | 40 | 106 | 89 | 106 | 0.01 |
| Ag (nmol kg$^{-1}$) | <1 | 38 | 26 | 37 | 0.02 |
| Cd (nmol kg$^{-1}$) | 17 | 155 | 144 | 180 | 1 |
| Pb (nmol kg$^{-1}$) | 183 | 308 | 194 | 359 | 0.01 |
| pH | 3.8 | 3.4 | 3.6 | 3.3 | 7.8 |
| Alk (meq) | −0.19 | −0.40 | −0.30 | −0.50 | 2.3 |
| NH$_4$ (m kg$^{-1}$) | <0.01 | <0.01 | <0.01 | <0.01 | <0.01 |

* See text for details of smoker fields.

4 Differences in the depth at which the basalt–seawater reaction takes place, which affect the path length of the hydrothermal cell.

5 Differences in the age of the hydrothermal systems; i.e. in older systems the water is flowing through rocks that have already undergone considerable leaching.

It was not possible to resolve the relative importance of these factors in the EPR system, but they serve to illustrate how the trace element composition of venting fluids can be influenced by local conditions.

From studies of low- and high-temperature venting fluids it has become apparent that the mass flux from hydrothermal systems occurs at high temperature. Originally, it was thought that high-temperature venting systems were more common on ridges formed at fast ($\sim$9–18 cm yr$^{-1}$) to intermediate ($\sim$5–9 cm yr$^{-1}$) spreading rates, but were fairly rare on those formed at slow ($\sim$1–5 cm yr$^{-1}$) rates; exceptions are those found at the Reykjanes Ridge and other 'hotspot'-influenced ridges. Rona *et al.* (1986), however, reported finding the first black smokers on a slow-spreading ridge in the TAG hydrothermal field on the Mid-Atlantic Ridge (MAR). Deposits of massive sulphides were found in association with the venting system in this region. Subsequently, data have become available which indicate that there are differences in the compositions of venting solutions associated with hydrothermal activity at slow and fast spreading ridges. For example, James *et al.* (1995) reported data for the chemistry of hydrothermal fluids from the Broken Spur at 29°N on the relatively slow-spreading MAR. A summary of their data is given in Table 5.4 and shows that relative to the 21°N EPR site the Broken Spur venting fluids are enriched in Li, but have lower concentrations of Rb, Mn and Sr. Further, James *et al.* (1995) concluded that a substantial removal of sea water B (>50%) had occurred in the low-temperature portion of the hydrothermal convection cell, and that the removal of sea water Sr

**Table 5.4** Chemical compositions of high-temperature hydrothermal venting solutions generated under different conditions.

| Component | High-temperature, sediment starved, slow-spreading ridge: Broken Spur solutions (MAR)* | High-termperature, sediment rich, fast-spreading ridge: Guaymas Basin solutions† | High-temperature, sediment-starved, fast-spreading ridge: 21°N EPR solutions‡ | Sea water |
|---|---|---|---|---|
| Li ($\mu$mol kg$^{-1}$) | 1006–1035 | 630–1076 | 891–1322 | 26 |
| Na (mmol kg$^{-1}$) | 419–422 | 475–513 | 432–510 | 464 |
| K (mmol kg$^{-1}$) | 18.1–19.6 | 37.1–49.2 | 23.2–25.8 | 9.79 |
| Rb ($\mu$mol kg$^{-1}$) | 13.0–13.6 | 57–86 | 27–33 | 1.3 |
| Be (nmol kg$^{-1}$) | — | 12–91 | 10–37 | 0.02 |
| Mg (mmol kg$^{-1}$) | — | 0 | 0 | 52.7 |
| Ca (mmol kg$^{-1}$) | 11.8–12.8 | 26.6–41.5 | 11.7–20.8 | 10.2 |
| Sr ($\mu$mol kg$^{-1}$) | 43–48 | 160–253 | 67–97 | 87 |
| Ba ($\mu$mol kg$^{-1}$) | >13–>21 | >7–>24 | >7–>15 | 0.14 |
| Al ($\mu$mol kg$^{-1}$) | — | 0.9–7.9 | 4.0–5.2 | 0.005 |
| Mn ($\mu$mol kg$^{-1}$) | 250–260 | 128–236 | 699–1002 | <0.001 |
| Fe ($\mu$mol kg$^{-1}$) | 1684–2156 | 17–180 | 750–2429 | <0.001 |
| Co (nmol kg$^{-1}$) | 130–422 | <5 | 22–227 | 0.03 |
| Cu ($\mu$mol kg$^{-1}$) | 28.3–68.6 | <0.02–1.1 | <0.02–44 | 0.007 |
| Zn ($\mu$mol kg$^{-1}$) | 40.8–88 | 0.1–40 | 40–106 | 0.01 |
| Ag (nmol kg$^{-1}$) | — | <1–230 | <1–38 | 0.02 |
| Cd (nmol kg$^{-1}$) | 75–145 | <10–46 | 17–180 | 1 |
| Pb (nmol kg$^{-1}$) | 221–376 | <20–652 | 183–359 | 0.01 |
| pH | — | 5.9 | 3.3–3.8 | 7.8 |
| Alk | — | 2.8–10.6 | −0.19 to −0.50 | 2.3 |
| NH$_4$ (mmol kg$^{-1}$) | — | 10.7–15.6 | <0.01 | <0.01 |

* James *et al.* (1995). † Von Damm *et al.* (1985b). ‡Von Damm *et al.* (1985a).

was around 10% greater at Broken Spur compared with vent sites in the Pacific where the spreading rates are faster.

The hydrothermal systems described above are associated with spreading ridge crests where the sedimentation rates are low and where the venting fluids debouch, via chimneys, directly into sea water: **sediment-starved** systems. In contrast, under some conditions the spreading axis can be buried under a blanket of sediment: **sediment-rich** systems. One example of this is found in the Red Sea where the axis is covered by marine evaporites. Another example of this kind of hydrothermal system, which has been described by Von Damm *et al.* (1985b), is located in the Guaymas Basin in the Gulf of California. North of the vent fields at 21°N, the EPR extends into the Gulf of California, and the spreading regime changes from a mature, open-ocean, sediment-starved type to an early opening, continental rifting type. An increase in sedimentation rate is associated with this change, and as a result the Guaymas Basin spreading axis is covered with a thick blanket of biogenous sediment that is rich in organic carbon derived from the highly productive overlying waters. The high-temperature (up to 315°C) Guaymas Basin venting solutions have a composition that is strikingly different from those characteristic of the sediment-starved, open-ocean systems, such as that found at 21°N on the EPR (see Table 5.4). In particular, whereas the venting solutions at 21°N on the EPR are rich in the ore-forming elements (Fe, Mn, Cu, Zn, Pb, Co and Cd), have an acidic pH (3.3–3.8) and have low concentrations of ammonium, those from the Guaymas Basin are depleted in the ore-forming elements, have a more alkaline pH (5.9) and contain ammonium as a major ion. Von Damm *et al.* (1985b) concluded that the Guaymas Basin venting solutions were formed as the result of a two-stage reaction process in which the hydrothermal solutions first reacted with the underlying basalt to produce a primary solution similar in composition to those found at the 21°N EPR sites, which then subsequently reacted with the biogenous sediments overlying the intrusion zone. During this second stage the pH of the primary hydrothermal solutions is raised by the dissolution of carbonate and by the thermocatalytic cracking of plankton carbon. As a result, sulphides are precipitated from the ascending solutions at depth in the sediment column,

and the Guaymas Basin is thus a site of the active formation of a **sediment-hosted** massive sulphide deposit. The authors concluded that sediment-covered hydrothermal systems probably are not quantitatively important for the formation of oxidized metal-rich sediments.

High-temperature reactions can also take place between basalt and sea water when hot lava is extruded directly on to the sea bed. Bonatti (1965) has classified these submarine volcanic eruptions into two general types:

1 a quiet type, in which the reaction between sea water and lava is essentially prevented by the instantaneous formation of a thin crust of glass;

2 a more violent or explosive type, in which the lava is shattered on contact with sea water and the large surface area then available allows considerable hydration of the glass.

In the explosive, **hyaloclastic**, eruption there is reaction between the basalt and sea water, with the result that new minerals such as palagonite, smectites and zeolites are formed and elements such as Ca, Na, K, Si, B, Mn, Zn and Cu are released into solution.

## 5.2 Hydrothermal activity: low-temperature basalt–seawater reactions

The upper 2–3 km of Layer 2 of the oceanic basement is a zone of chemical reaction between sea water and the crust as it moves away from the spreading centres. The extent to which an oceanic basalt undergoes low-temperature weathering reactions is a function of its age, i.e. of the time it has spent in contact with sea water, which is related to its distance from the generation centres. The low-temperature basalt–seawater reactions can take place both at the sea floor (as evidenced from dredged basalts) and at depth within Layer 2 of the oceanic crust (as evidenced from drilled basalts). Basalt–seawater reactions on the ocean floor take place in the presence of large volumes of water under oxidizing conditions. At depth within Layer 2, however, the rocks are in contact with much smaller volumes of water. In general, both field and laboratory experimental data indicate that the deeper basalts have undergone alteration, following reaction with sea water, in much the same way as surface basalts, although the reactions involved do not appear to have proceeded to the same extent.

Basalts dredged from the sea floor have invariably undergone some degree of reaction with sea water at the cold ambient bottom temperatures, and are often termed **weathered basalts**. During these low-temperature reactions, which have been reviewed by Honnerez (1981) and Thompson (1983), new mineral phases are formed and chemical transfer takes place between the rocks and sea water. Volcanic glass and bulk rock can be affected in different ways, but a number of general conclusions can be drawn regarding the low-temperature alteration of oceanic basalts (see e.g. Honnerez, 1981).

1  The oxidation and hydration of basalt is a ubiquitous phenomenon.

2  During basalt–seawater reactions there is often an uptake of K, Cs, Rb, B, Li and $^{18}O$ by the basalt (i.e. the rock acts as a *sink*, and the elements usually are incorporated into mineral phases formed during the alteration processes), and a loss of Ca, Mg and Si (i.e. the rock acts as a *source*).

3  Fe, Mn, Na, Cu, Ba and Sr exhibit no clear patterns, although the low-temperature weathering of basalt may provide a source for Fe and Mn to sea water (Elderfield, 1976).

4  Al, Ti, Y, Zr and the heavy rare earths show little or no change following basalt–seawater reactions.

5  Low-temperature alteration of the oceanic crust is a major sink for U, and according to Bloch (1980) may account for ~50% of the total amount of U supplied to the oceans at the present day.

### 5.3 The hydrothermal pathway: summary

1  The oceanic crust is a major reservoir in global geochemistry. To bring this reservoir on-line, a number of different types of reactions occur between sea water and the basalts of Layer 2.

2  These seawater reactions take place at a variety of temperatures and rock:water ratios, and can act as either source or sink terms in the marine budgets of some components.

3  The most dramatic hydrothermal activity is found at the spreading ridge centres, where cold sea water circulates through hot, newly formed, basaltic crust. In this process, the composition of the sea water undergoes extensive changes before emerging at the sea bed in the form of white smoker or black smoker hot springs.

In the present chapter attention has been confined to the effects of hydrothermal activity on sea water, and the magnitude of the fluxes involved are discussed in Section 6.3. The effects that hydrothermal processes have on marine geochemistry, however, are much wider than this. In particular, the hydrothermal activity of spreading centres results in the formation of a series of mineral precipitates and in the generation of a unique type of deep-sea sediment. The chemical dynamics involved in the formation of the hydrothermal precipitates are discussed in Section 15.3.6 in terms of a sequential precipitation model, and the generation of the hydrothermal sedimentary deposits is covered in Section 16.6.2. In this way, the full spectrum of hydrothermal activity in the oceans will be placed in a global marine context.

### References

Bischoff, J.L. & Dickson, F. (1975) Seawater–basalt interaction at 200°C and 500 bars: implications for origin of sea floor heavy metal deposits and regulation of seawater chemistry. *Earth Planet. Sci. Lett.*, **25**, 385–97.

Bloch, S. (1980) Some factors controlling the concentration of uranium in the World Ocean. *Geochim. Cosmochim. Acta*, **44**, 373–7.

Bonatti, E. (1965) Palagonite, hyaloclastites and alteration of volcanic glass in the ocean. *Bull. Volcanol.*, **28**, 257–69.

Childers, J.J. (1988) Hydrothermal vents. A case study of the biology and chemistry of a deep-sea hydrothermal vent of the Galapagos Rift. *Deep-Sea Res.*, **35**, 1677–1849.

Chester, R. & Aston, S.R. (1976) The geochemistry of deep-sea sediments. In *Chemical Oceanography*, J.P. Riley & R. Chester (eds), Vol. 6, 281–390. London: Academic Press.

Edmond, J.M., Measures, C.I., McDuff, R.E., *et al.* (1979) Crest hydrothermal activity and the balance of the major and minor elements in the ocean; the Galapagos data. *Earth Planet. Sci. Lett.*, **46**, 1–18.

Edmond, J.M., Von Damm, K.L., McDuff, R.E. & Measures, C.I. (1982) Chemistry of hot springs on the East Pacific Rise and their effluent dispersal. *Nature*, **297**, 187–91.

Elder, J.W. (1965) Physical processes in geothermal areas. In *Terrestrial Heat Flow*, W.H.K. Lee (ed.), 211–39. Washington, DC: American Geophysical Union Monograph no. 8.

Elderfield, H. (1976) Hydrogenous material in marine sediments: excluding manganese nodules. In *Chemical Oceanography*, J.P. Riley & R. Chester (eds), Vol. 5, 137–215. London: Academic Press.

Holm, N.G. (1992) *Marine Hydrothermal Systems and the Origin of Life*. Dordrecht: Kluwer Academic Publishers.

Honnerez, J. (1981) The aging of the oceanic crust at low temperature. In *The Sea*, C. Emillani (ed.), Vol. 7, 525–87. New York: Interscience.

James, R.H., Elderfield, H. & Palmer, M.R. (1995) The chemistry of hydrothermal fluids from the Broken Spur site, 29°N Mid-Atlantic Ridge. *Geochim. Cosmochim. Acta*, 59, 651–59.

Jones, E.J.W. (1978) Sea floor spreading and the evolution of the ocean basins. In *Chemical Oceanography*, Vol. 4, J.P. Riley & R. Chester (eds), 1–74. London: Academic Press.

Klinkhammer, G., Elderfield, H., Grieves, M., Rona, P. & Nelson, T. (1986) Manganese geochemistry near high temperature vents in the Mid-Atlantic Ridge rift valley. *Earth Planet. Sci. Lett.*, 80, 230–40.

McDuff, R.E. & Edmond, J.M. (1982) On the fate of sulphate during hydrothermal circulation at mid-ocean ridges. *Earth Planet. Sci. Lett.*, 57, 117–32.

Martin, J.-M. & Whitfield, M. (1983) The significance of the river input of chemical elements to the ocean. In *Trace Metals in Sea Water*, C.S. Wong, E. Boyle, K.W. Bruland, J.D. Burton & E.D. Goldberg (eds), 265–96. New York: Plenum.

Moore, J.G., Fleming, H.S. & Phillips, J.D. (1974) Preliminary model for extrusion and rifting at the axis of the Mid-Atlantic Ridge, 36°48′ north. *Geology*, 2, 437–40.

Riley, J.P. & Chester, R. (1971) *Introduction to Marine Chemistry*. London: Academic Press.

Rona, P.A., Klinkhammer, G., Nelson, T.A., Trefry, J.H. & Elderfield, H. (1986) Black smokers, massive sulphides and vent biota at the Mid-Atlantic Ridge. *Nature*, **321**, 33–7.

Thompson, G. (1983) Hydrothermal fluxes in the ocean. In *Chemical Oceanography*, J.P. Riley & R. Chester (eds), Vol. 8, 270–337. London: Academic Press.

Von Damm, K.L., Grant, B. & Edmond, J. (1983) Preliminary report on the chemistry of hydrothermal solutions at 21° north, East Pacific Rise. In *Hydrothermal Processes at Seafloor Spreading Centres*, P.A. Rona, K. Bostrom, L. Laubier & K.L. Smith (eds), 369–90. New York: Plenum.

Von Damm, K.L., Edmond, J.M., Grant, B., Measures, C.I., Walden, B. & Weiss, R.F. (1985a) Chemistry of submarine hydrothermal solutions at 21°N, East Pacific Rise. *Geochim. Cosmochim. Acta*, 49, 2197–220.

Von Damm, K.L., Edmond, J.M., Measures, C.I. & Grant, R. (1985b) Chemistry of submarine hydrothermal solutions at Guaymas Basin, Gulf of California. *Geochim. Cosmochim. Acta*, 49, 2221–37.

Wolery, T.J. & Sleep, N.H. (1976) Hydrothermal circulation and geochemical flux at mid-ocean ridges. *J. Geol.*, 84, 249–76.

# 6 The transport of material to the oceans: relative flux magnitudes

In the preceding chapters we have considered the three principal primary pathways by which material is transported to the World Ocean at the present day, i.e. river run-off, atmospheric deposition and hydrothermal exhalations. The magnitudes of the fluxes associated with each of these transport pathways are discussed below.

## 6.1 River fluxes to the oceans

### 6.1.1 Introduction
The flux of material transported by a river reflects a complex interaction between hydrological and chemical factors in the catchment system, and it must be stressed that there are considerable uncertainties associated with all fluvial flux estimates. To make global flux estimations, average concentrations of constituents in individual river systems are often used, e.g. in scaling-up procedures, but there are problems in assessing the extent to which the average values are meaningful. For example, concentrations of constituents in river water can be affected by a number of factors associated with variations in river flow. These include: long-term (e.g. annual) temporal variations at individual sampling stations, and spatial variations between individual stations; short-term fluctuations, e.g. those in response to storm events; fluctuations as a result of rare events, such as those brought about by severe drought and catastrophic flooding; and seasonal variations in biological production. Further, the scaling-up of a single river data set to a global ocean scale, even if the average concentrations used are representative of that river system, can bias flux estimates because hydrological, climatic and lithological controls on the composition of river water vary widely from one river catchment to another. One way of overcoming this difficulty is to use data from major river systems. This approach,

however, also has problems because the aggregate run-off from the 20 largest rivers in the world accounts for only ~30% of the global run-off. It therefore may be more reasonable to select rivers on the basis of particular catchment regimes rather than on the basis of size (see e.g. GESAMP, 1987). Despite these difficulties, however, it is still potentially rewarding to take a first look at the magnitude of fluvial fluxes to the oceans.

When the magnitude of the river-transported signal is considered it is necessary to distinguish between gross and net fluxes. In the present context, the following general definitions are adopted.

1 **Gross fluxes** are those transported by rivers to the marine boundary, which is taken here as the estuarine mixing zones.

2 **Net fluxes** are those that are transported out of the estuarine mixing zone in an offshore direction and so are discharged into coastal seas.

The simplest way of estimating gross fluvial fluxes, and the one adopted here, involves a scaling-up procedure, which can be represented by equations such as

$$RF_g = X_d Q + X_p M_p \qquad (6.1)$$

where $X_d$ is the average dissolved content of an element $X$ in river water, $Q$ is the annual river-water discharge to the oceans (usually assumed to be 37 400 km$^3$ yr$^{-1}$), $X_p$ is the average content of river particulate material (RPM) and $M_p$ is the river particulate discharge to the oceans (usually assumed to be $15.5 \times 10^{15}$ g yr$^{-1}$). For net fluxes, it is necessary to take account of the processes that occur in estuaries, and this is discussed below in relation to dissolved and particulate trace elements.

#### 6.1.2 The gross and net fluvial fluxes of total suspended material

##### 6.1.2.1 Gross flux

There have been various attempts to quantify the average global discharge of river suspended sediment to the land–ocean margins, and these have been reviewed by Walling & Webb (1987). Two of the most comprehensive assessments of the magnitudes of gross fluvial suspended load fluxes have been made by Holeman (1968) and Milliman & Meade (1983). A summary of the two data sets is given in Table 6.1,

and several features in the river sediment discharge pattern are illustrated in Fig. 6.1(a).

It can be seen from the data in Table 6.1 that there are a number of significant differences between the two RPM discharge estimates. From the point of view of global discharges to the World Ocean the most important of these can be summarized as follows.

**1** The Milliman & Meade (1983) estimate for the annual discharge of river sediment from Asia is less than half of that given by Holeman (1968). There is no doubt that these Asian rivers carry the highest sediment loads of any of the world's streams. Over much of the Asian continent, however, the sediment

**Table 6.1** The river discharge of suspended sediment to the oceans: gross flux estimates based on data sets from (1) Holeman (1968) and (2) Milliman & Meade (1983).

(a) Suspended sediment discharge from major rivers.

| River | Annual suspended sediment discharge ($\times 10^6$ t yr$^{-1}$) 1 | 2 |
|---|---|---|
| Hwang Ho (China) | 1890 | 1080 |
| Ganges (India) | 2180 | 1670 |
| Brahmaputra (Bangladesh) | | |
| Yangtze (China) | 502 | 478 |
| Indus (Pakistan) | 440 | 100 |
| Amazon (Brazil) | 364 | 900 |
| Mississippi (USA) | 349 | 210 |
| Irrawaddy (Burma) | 300 | 265 |
| Mekong (Thailand) | 170 | 160 |
| Red (North Vietnam) | 410 | — |
| Nile (Africa) | 111 | 0 |
| Zaire (Africa) | 64 | 43 |
| Niger (Nigeria) | 4 | 40 |
| St Lawrence (Canada) | 4 | 4 |

(b) Suspended sediment discharge in the oceans by rivers from the continents.

| Continental region | Drainage area ($10^6$ km$^2$) 1 | 2 | Sediment discharge ($10^6$ t yr$^{-1}$) 1 | 2 | Sediment yield (t km$^{-2}$ yr$^{-1}$) 1 | 2 |
|---|---|---|---|---|---|---|
| North and Central America | 20.48 | 17.50 | 1780 | 1462 | 87 | 84 |
| South America | 19.20 | 17.90 | 1090 | 1788 | 57 | 97 |
| Europe | 9.2 | 4.61 | 290 | 230 | 32 | 50 |
| Eurasian Arctic | — | 11.17 | — | 84 | — | 8 |
| Asia | 26.6 | 16.88 | 14480 | 6349 | 543 | 380 |
| Africa | 19.7 | 15.34 | 490 | 530 | 25 | 35 |
| Australia | 5.1 | 2.20 | 210 | 62 | 41 | 28 |
| Large Pacific Islands | — | 3.00 | — | 3000 | — | 1000 |

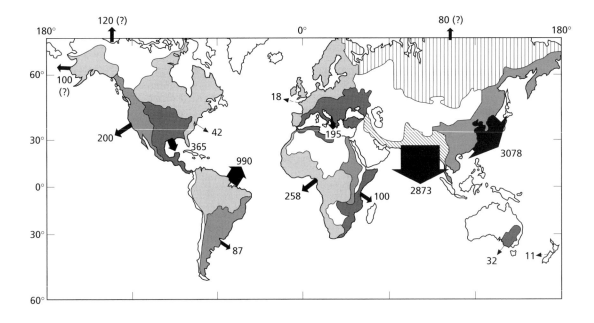

(a)

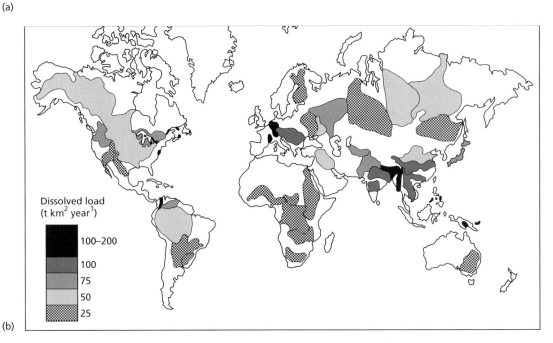

(b)

**Fig. 6.1** General trends in fluvial discharges of suspended and dissolved material to the ocean margins. (a) Discharge of suspended sediment (from Milliman, 1981); units in millions of metric tons (Mt) per year. (b) Discharge of total dissolved solutes (from Walling & Webb, 1987). (c) Relative fluvial suspended sediment and water discharges from the continents (from Degens & Ittekkot, 1985). NA, North America; SA, South America; AS, Asia; AF, Africa; AR, Arctic USSR; OC, Oceania; EU, Europe.

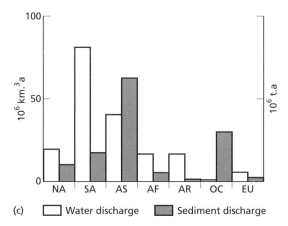

(c) ☐ Water discharge  ■ Sediment discharge

**Fig. 6.1** *Continued*

yield is in fact relatively small and the discrepancy between the two discharge estimates probably arose because Holeman (1968) did not take account of this in the scaling-up procedure used in calculating his average estimate of the Asian sediment discharge.

2 The estimate given by Milliman & Meade (1983) for sediment discharge from the Amazon, which is based on more recent measurements, is considerably higher than that listed by Holeman (1968) and, according to the later estimate, the Amazon now ranks third among the world's rivers for the discharge of sediment to the land–ocean boundary.

3 Holeman (1968) did not take into account the large islands of the western Pacific in making his global estimate of river sediment discharge.

Despite constraints imposed by uncertainties in the database, it is still possible to identify a number of general features in the geographical distributions of river suspended-sediment loads (i.e. excluding bedload, which probably makes up ~10% of the total sediment load) and their discharges to the land–ocean boundaries.

1 Most rivers have average suspended sediment loads in the range ~$10^2$–$10^3$ mg l$^{-1}$ (Milliman, 1981).

2 Mountain rivers have sediment yields that, on average, are around three times greater than those of plains rivers (Dedkov & Mozzherin, 1984).

3 The suspended loads transported to the ocean margins annually can vary over three orders of magnitude from one river to another (e.g. from the 1890 × $10^{12}$ g yr$^{-1}$ for the Hwang Ho, to the 4 × $10^{12}$ g yr$^{-1}$

for the St Lawrence—estimates from Holeman (1968)), and much of the suspended sediment delivered to the oceans is carried by a relatively small number of major rivers.

4 Overall, a large proportion of the total river suspended sediment discharged annually to the land–ocean boundary originates from the Asian continent. On the basis of the Holeman (1968) data set, the Asian river suspended-sediment input accounts for ~80% of the total discharge to the World Ocean, and still remains at ~50% in the revised estimates given by Milliman & Meade (1983), which include discharges from the Pacific islands.

5 There are a number of estimates in the literature for the total annual transport of river sediment to the ocean margins. In general, these appear to lie in the range ~13.5 × $10^{15}$ g yr$^{-1}$ (Milliman & Meade, 1983) to 18.3 × $10^{15}$ g yr$^{-1}$ (Holeman, 1968), with that of ~15.5 × $10^{15}$ g yr$^{-1}$ given by Martin & Maybeck (1979) lying between the two.

6 In a more recent estimate, however, Milliman & Syvitski (1992) suggested that sediment fluxes from small mountainous rivers have been greatly underestimated in previous estimates, perhaps by a factor of three, and that a conservative estimate of the combined river sediment discharge to the ocean margins might be as high as ~20 × $10^{15}$ g yr$^{-1}$.

7 Milliman (1981) has pointed out that whatever estimate is accepted for the present-day global river discharge of sediment to the oceans, it cannot be applied for forecasting future trends because discharges can vary dramatically with changes in climate and human land-use.

### 6.1.2.2 *Net flux*

Processes operating in the estuarine environment severely modify the gross RPM flux delivered to the land–sea margins, and according to Judson (1968) it is probable that at the present time ~90% of the particulate material transported by rivers is trapped in estuaries (see Section 3.2.7). If this estimate of ~90% for the estuarine removal of suspended particulates is applied to the global estimate of ~15.5 × $10^{15}$ g yr$^{-1}$ for the discharge of RPM (see Section 6.1.1) it would result in a *net* global river-transported flux of ~1.55 × $10^{15}$ g yr$^{-1}$. This is very close to the estimate of ~1.26 × $10^{15}$ g yr$^{-1}$ for the global flux of particulate material to the ocean made independently by Bewers

**Table 6.2** Gross fluvial fluxes of some dissolved and particulate major elements to the ocean margins (units, g yr$^{-1}$).

| Element | Dissolved flux* | Particulate flux‡ |
|---|---|---|
| B | $46.5 \times 10^9$ | $9.3 \times 10^{12}$ |
| Br | $743 \times 10^9$† | $77.5 \times 10^9$ |
| Ca | $481 \times 10^{12}$ | $345 \times 10^{12}$ |
| Cl | $245 \times 10^{12}$ | — |
| Cs | $1.3 \times 10^9$† | $93 \times 10^9$ |
| F | $3.1 \times 10^{12}$ | — |
| K | $70 \times 10^{12}$ | $310 \times 10^{12}$ |
| Li | $94 \times 10^9$ | $0.39 \times 10^{12}$ |
| Mg | $129 \times 10^{12}$ | $209 \times 10^{12}$ |
| Mo | $19 \times 10^9$† | $47 \times 10^9$ |
| Na | $131 \times 10^{12}$† | $110 \times 10^{12}$ |
| Rb | $32 \times 10^9$ | $1.55 \times 10^{12}$ |
| Sr | $2.3 \times 10^{12}$ | $2.3 \times 10^{12}$ |

* Based on data from Thompson (1983). † Based on data from Martin & Maybeck (1979). ‡ Calculated from data given by Martin & Maybeck (1979).

& Yeats (1977). A compilation of the gross particulate fluxes of some major elements is given in Table 6.2.

### 6.1.3 The gross and net fluvial fluxes of total dissolved material

According to Maybeck (1979) the total dissolved load transported to the oceans by rivers is $\sim 3.7 \times 10^{15}$ g yr$^{-1}$, which is around 25% of the suspended sediment load (using the estimate of $\sim 15.5 \times 10^{15}$ g yr$^{-1}$ given above), although it falls to around 20% if bedload is included in the particulate river load. Like the suspended sediment yields, there are patterns in the fluvial discharge of total solutes to the oceans. This was highlighted by Walling & Webb (1987), who produced a map showing the generalized global pattern of fluvial total solute yields—see Fig. 6.1(b). Global total solute discharge patterns reflect the combined influences of factors such as the magnitude of the river run-off, the catchment lithology and the climatic regime. Walling & Webb (1987) identified a number of general features in the global river discharge of total solutes that can be related to these run-off, geological and climatic controls.

1 The large dissolved solute discharge values for Asian rivers result from their high run-off, which promote enhanced transport.

2 The relatively high loads for European rivers reflect the predominance of sedimentary strata, including limestones, in their catchments.

3 Generally low dissolved loads are found for the rivers of Africa and Australia, which are a consequence of the presence of ancient basement rocks, with a low weathering susceptibility, in their catchments.

4 The extremely high dissolved solute loads for Burma (the Irrawaddy River) and Papua New Guinea (the Fly and Putari rivers) reflect a combination of high run-off, the availability of sedimentary rocks in the catchment regions and tropical temperatures, which enhance chemical weathering.

A comparison of fluvial water and sediment discharges from the continents is illustrated in Fig. 6.1(c).

To apply equations such as eqn (6.1) to dissolved components, it is necessary to have reliable data on their concentrations in river water. Data of this kind have been available for some time for a number of dissolved major components in river water, and the gross fluvial fluxes for several of these are given in Table 6.2, together with their gross particulate fluxes. Some dissolved elements exhibit unambiguous conservative behaviour in the estuarine mixing zone. For these elements, which include those major components contributing to the salinity of sea water, the gross river fluxes will not be markedly changed during estuarine mixing, and those listed in Table 6.2 can be assumed to represent the net fluxes.

### 6.1.4 The gross and net fluvial fluxes of dissolved and particulate trace elements

#### 6.1.4.1 Gross fluxes

Much of the concentration data in the literature for particulate elements in river water have been of an acceptable quality for some time. It was pointed out in Section 3.1.3, however, that it is only relatively recently that reliable data have become available on the concentrations of dissolved trace elements in river water, and even these data are still limited to a relatively small number of elements. In order to estimate order-of-magnitude gross fluxes of river-transported trace elements, an attempt therefore has been made to apply eqn (6.1) to a 'modern' database. To illustrate the concepts involved, attention has been limited to a

**Table 6.3** Gross fluvial fluxes of some dissolved and particulate trace elements to the ocean margins (units, $10^{12}$ g yr$^{-1}$): (1) scaled up from the average global dissolved river concentration given in Table 3.4, column 19—assuming a global river discharge of 37 400 km$^3$ yr$^{-1}$; (2) data from Yeats & Bewers (1982); (3) scaled up from the average composition and RPM given by Martin & Whitfield (1983)—assuming a total river particulate discharge load of $15.5 \times 10^{15}$ g yr$^{-1}$.

| Element | Dissolved gross flux | | Particulate gross flux | |
| | 1 | 2 | 2 | 3 |
| --- | --- | --- | --- | --- |
| Al | 1.9 | — | 1500 | 1460 |
| Fe | 1.5 | 0.16–2.2 | 730 | 745 |
| Mn | 0.31 | 0.25–0.63 | 17 | 16 |
| Ni | 0.019 | 0.011–0.057 | 1.5 | 1.4 |
| Co | 0.0075 | 0.0026–0.0065 | 0.32 | 0.31 |
| Cr | 0.037 | — | — | 1.55 |
| V | 0.037 | — | — | 2.6 |
| Cu | 0.056 | 0.061–0.088 | 1.6 | 1.55 |
| Pb | 0.0037 | — | — | 1.55 |
| Zn | 0.015 | — | 5.4 | 3.9 |
| Cd | 0.0015 | 0.0028–0.0043 | 0.026–0.03 | 0.016 |

**Table 6.4** Net fluvial fluxes of some particulate elements to the oceans (units, $10^{12}$ g yr$^{-1}$): (1) gross fluvial flux from Table 6.3, assuming an estuarine retention of 90% of RPM; (2) data from Bewers & Yeats (1977).

| Element | Particulate flux | |
| | 1 | 2 |
| --- | --- | --- |
| Al | 146 | — |
| Fe | 75 | 111 |
| Mn | 1.6 | 2.4 |
| Ni | 0.14 | 0.036 |
| Co | 0.03 | 0.017 |
| Cr | 0.155 | — |
| V | 0.26 | — |
| Cu | 0.155 | 0.036 |
| Pb | 0.155 | — |
| Zn | 0.39 | 0.018 |
| Cd | 0.0016 | 0.00008 |

number of process-orientated elements, which are used in Chapter 11, to describe the factors that control the distributions of trace elements in sea water. To make these gross flux estimates, data for the particulate phase were taken from Martin & Whitfield (1983), and for the dissolved phase the average global river-water concentration given in column 19 of Table 3.4 was used. The gross river fluxes obtained in this manner are listed in Table 6.3, together with those estimated by Yeats & Bewers (1982). These authors concluded that, in general, their flux estimates for individual rivers (expressed as ranges in Table 6.3) showed a remarkable degree of similarity, despite the fact that the river catchments involved: (i) varied from tropical to subarctic; (ii) included both polluted and unpolluted rivers; and (iii) had a wide range of suspended matter loads.

### 6.1.4.2 Net fluxes

The data given in Table 6.3 are for *gross* river-transported fluxes, i.e. those that enter the estuarine mixing zone. Processes operating in estuaries can modify both the dissolved and particulate phases in the river-water inputs, and as a result the strengths of the river-transported signals entering an estuary can be very different from those that are exported to the ocean (see Section 3.2).

It was shown above that ~90% of the particulate material transported by rivers is trapped in estuaries at the present time. It is apparent, therefore, that the gross river particulate fluxes given in Table 6.3 must be strongly reduced in order to derive *net* fluxes for the export of particulate components from estuaries. In order to offer a first look at these net fluxes, the gross particulate fluxes have been reduced by a global factor of 90%; the data are listed in Table 6.4, together with those derived by Bewers & Yeats (1977) from the St Lawrence experiment (see below).

Various attempts have been made over the past few years to estimate net river fluxes of dissolved trace elements from their gross fluxes by taking account of estuarine processes. For example, the **zero-salinity end-member** approach has been adopted in several investigations. This is an attempt to estimate the chemical composition of river water that has actually passed through the estuarine filter and reached coastal waters. Although there are a number of problems associated with this approach, it can still produce useful data. The zero-salinity end-member or, as it is sometimes called, the **effective river end-member**, can be identified from either estuarine or, under suitable conditions, coastal water data. A number of workers have used the estuarine mixing model outlined by Boyle *et al.* (1974) (see Section 3.2.3) to evaluate an effective river end-member by extrapolating linear element:salinity ratios back to a zero-salinity intercept. For example, Edmond *et al.* (1985) used this approach to evaluate the *net* river fluxes to the ocean from the Changjiang River and also quoted similar data for the Amazon; these data, together with those from some other rivers, are listed in Table 6.5. The application of the zero-salinity end-member approach for estimating the net fluvial input to coastal waters can be illustrated by two examples, selected to identify inputs from contrasting river catchments. These are (i) the eastern seaboard of the USA (Bruland & Franks, 1983); and (ii) the North Sea (Kremling, 1985); data for the two zero-salinity end-members are given in Table 6.5. It can be seen from this table that the concentrations of dissolved Ni, Cu and Cd in the North Sea zero-salinity end-member are higher than those for the USA eastern seaboard end-member. This probably reflects a combination of different estuarine processes in the two areas, together with a stronger pollutant influence from rivers draining into the North Sea.

The zero-salinity end-member approach yields estimates of the *net* flux of components from individual river-estuarine systems. These can then be extrapolated on to a global scale by some form of scaling-up procedure. Any attempt to extrapolate the effects of local estuarine removal processes to such a global scale inevitably will be fraught with very considerable

uncertainty. Despite this, Bewers & Yeats (1977) used data obtained from a mass-balance study carried out in the St Lawrence river to make such a net global flux estimate. To do this, the authors scaled up the St Lawrence data by taking account of global discharges of both river water and particulate material, thus correcting, for example, for the low suspended load in the St Lawrence system. The net fluvial World Ocean fluxes for the dissolved trace metals given by these authors are summarized in Table 6.5. It is of interest to compare the net global fluxes for dissolved trace metals estimated from the St Lawrence experiment with those scaled up from the zero-salinity end-member and effective river concentration estimates. The zero-salinity end-member concentration that was identified by Kremling (1985) for rivers discharging into the North Sea has apparently been affected by local river pollution. However, that derived by Bruland & Franks (1983) for rivers draining the eastern seaboard of the USA had a composition more similar to that of global river water. The data given by Bruland & Franks (1983) therefore can be scaled up to a global framework, and the net flux calculated in this way is given in Table 6.5. In addition, the Changjiang and Amazon fluxes have been scaled up to global estimates and these also are listed in Table 6.5. This table therefore offers a number of estimates of the *net* fluxes of dissolved components transported to the oceans via river run-off. Considering that these estimates have been derived by applying different

**Table 6.5** Net fluvial fluxes of some dissolved trace elements to the oceans.

| Element | Zero-salinity end-member | | Global net fluvial flux to oceans ($10^{12}$ g yr$^{-1}$) | | | |
|---|---|---|---|---|---|---|
| | North Sea ($\mu$g l$^{-1}$)* | USA eastern seaboard ($\mu$g l$^{-1}$)† | Estimated from zero-salinity end-member; USA eastern seaboard‡ | Estimated from St Lawrence estuary data§ | Estimated from Amazon data¶ | Estimated from Changjiang data¶ |
| Mn | 7.6 | 6.2 | 0.23 | 0.35 | — | — |
| Ni | 9.4 | 1.2 | 0.05 | 0.05 | 0.014 | 0.010 |
| Cu | 5.8 | 1.2 | 0.05 | 0.09 | 0.07 | 0.05 |
| Zn | — | 1.0 | 0.04 | 0.16 | — | — |
| Cd | 0.65 | 0.12 | 0.0046 | 0.0075 | 0.00045 | 0.0012 |
| Be | — | — | — | — | 0.00007 | 0.00006 |
| Ba | — | — | — | — | 3.1 | 1.9 |

* Data from Bruland & Franks (1983). † Data from Kremling (1985). ‡ Data scaled-up from column 2. § Data from Bewers & Yeats (1977). ¶ Data from Edmond *et al.* (1985).

approaches to data from a variety of river systems, there is a surprisingly good measure of agreement between them for some of the dissolved elements. By combining the data in Table 6.5 with those from a variety of other sources, it is possible to make a first-order estimate of the net fluvial fluxes of a number of dissolved and particulate elements to the World Ocean, and these are listed in Table 6.6. It should be pointed out that for some elements, e.g. Cd, the net dissolved fluxes exceed the gross fluxes, a situation that will arise when estuarine processes lead to the addition of dissolved components from the particulate phase.

**Table 6.6** Estimates of net global fluvial fluxes of some dissolved and particulate elements to the World Ocean (units, $10^{12}$ g yr$^{-1}$).

| Element | Dissolved flux* | Particulate flux† |
|---|---|---|
| Al | 0.95 | 146 |
| Fe | 0.30 | 75 |
| Mn | 0.30 | 1.6 |
| Ni | 0.03 | 0.14 |
| Co | 0.0075 | 0.03 |
| Cr | 0.0185 | 0.155 |
| V | 0.037 | 0.26 |
| Cu | 0.06 | 0.155 |
| Pb | 0.0037 | 0.155 |
| Zn | 0.025 | 0.39 |
| Cd | 0.0031 | 0.0016 |

\* Dissolved fluxes estimated as follows.

| | |
|---|---|
| Co, V, Pb | Gross fluxes only—see Table 6.3. |
| Zn, Mn | Based on the average of the net global flux estimates (Table 6.5) derived from the USA eastern seaboard zero-salinity end-member, the St Lawrence estuary data (Table 6.5) and the gross flux data in Table 6.3. |
| Cu, Cd, Ni | Based on the average of the net global flux estimates (Table 6.5) from the USA eastern seaboard zero-salinity end-member, the St Lawrence estuary data (Table 6.5), the Amazon River (Table 6.5) and the Changjiang River data (Table 6.5) and the gross flux data in Table 6.3. |
| Al | Gross flux from Table 6.3, reduced for an estuarine retention of 50% (see e.g. Maring & Duce, 1987). |
| Fe | Gross flux from Table 6.3, reduced for an estuarine retention of 80% (see e.g. Figueres *et al.*, 1978). |
| Cr | Gross flux from Table 6.3, reduced for an estuarine retention of 50% (see e.g. Campbell & Yeats, 1984). |

† All net particulate fluxes are from Table 6.4.

### 6.1.5 The gross and net fluvial fluxes of organic carbon and the nutrients

#### 6.1.5.1 Organic carbon

The database for the transport of organic carbon by rivers has been greatly expanded by the SCOPE/UNEP project *Transport of carbon and minerals in major world rivers*, and the results have been reviewed by Degens & Ittekkot (1985) and Degens *et al.* (1991a). Degens & Ittekkot (1985) estimated annual river fluxes of organic carbon to the ocean margins. The magnitudes of the carbon fluxes, expressed in relation to the major continental drainage regions, are illustrated in Fig. 6.2. For POC transport the river drainage areas are ranked Asia > North America > South America > Oceania > Africa > Europe, and for DOC transport the order is South America > Asia > Arctic > North America > Africa > Europe > Oceania. The SCOPE/UNEP data were subsequently revised to yield new estimates of organic carbon fluxes in which the continental rankings changed. The new data have been discussed by Degens *et al.* (1991b) and are listed in Table 6.7. The gross

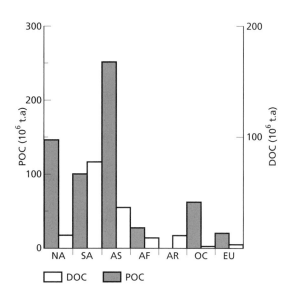

**Fig. 6.2** Annual fluvial inputs of particulate organic carbon (POC) and dissolved organic carbon (DOC) from the continents (from Degens & Ittekkot, 1985). NA, North America; SA, South America; AS, Asia; AF, Africa; AR, Arctic USSR; OC, Oceania; EU, Europe.

**Table 6.7** Fluvial organic carbon fluxes to the World Ocean (data from Degens *et al.*, 1991b).

| Continent | Total discharge ($km^3 yr^{-1}$) | Total suspended solids ($\times 10^{15} g yr^{-1}$) | DOC ($\times 10^{15} g yr^{-1}$) | POC ($\times 10^{15} g yr^{-1}$) | TOC‡ ($\times 10^{15} g yr^{-1}$) |
|---|---|---|---|---|---|
| South America | 11 039 | 1.93 | 0.044 | 0.024 | 0.067 |
| North America | 5840 | 1.83 | 0.034 | 0.0145 | 0.042 |
| Africa | 3409 | 0.211 | 0.025 | 0.008 | 0.033 |
| Asia | 12 205 | 11.2 | 0.094 | 0.13 | 0.17 |
| Europe* | 2826 | 0.42 | — | — | 0.024 |
| Total† | 35 319 | 15.6 | — | — | 0.336 |

* Selected rivers. † Excluding Australia. ‡ Approximate values of total organic carbon.

global TOC flux obtained from these data is $\sim 0.33 \times 10^{15} g yr^{-1}$. Dissolved organic carbon and POC values were not obtained for many of the individual river systems, but other estimates can be found in the literature. For example, Spitzy & Leenheer (1991) took the data available for a number of rivers and extrapolated them to estimate a global-scale DOC flux of $\sim 0.22 \times 10^{15} g yr^{-1}$, which may be compared to that of $0.5 \times 10^{15} g yr^{-1}$ given by Mantoura & Woodward (1983) and the range of $\sim 0.42 \times 10^{15}$ to $\sim 0.57 \times 10^{15} g yr^{-1}$ proposed on the basis of the original SCOPE/UNEP data. According to Mantoura & Woodward (1983), DOC can behave conservatively in the estuarine mixing zone, and on this basis the *net* DOC flux to the oceans should be similar to the gross flux, i.e. in the range $\sim 0.2 \times 10^{15} - \sim 0.6 \times 10^{15} g yr^{-1}$.

The SCOPE/UNEP estimate for the gross fluvial POC flux was in the range $\sim 0.11 \times 10^{15} - \sim 0.25 \times 10^{15} g yr^{-1}$, and the average proposed by Ittekkot (1988) was $\sim 0.23 \times 10^{15} g yr^{-1}$. Further, Ittekkot (1988) estimated that $\sim 65\%$ of the fluvial POC is highly refractory and so might escape the estuarine environment as a **net** POC flux, which is taken up by marine sediments, mainly in tropical and subtropical regions.

### 6.1.5.2 *The nutrients*

Natural gross fluvial nutrient fluxes to the land–sea margins have been estimated to be as follows: $\sim 14 \times 10^{12}$ to $\sim 15 \times 10^{12} g yr^{-1}$ for total dissolved nitrogen and $\sim 21 \times 10^{12} g yr^{-1}$ for particulate nitrogen (Van Bennekon & Salomons, 1981; Maybeck, 1982); $\sim 1 \times 10^{12} g yr^{-1}$ for total dissolved phosphorus and $\sim 20 \times 10^{12} g yr^{-1}$ for total particulate phosphorus

(Maybeck, 1982); and $181 \times 10^{12} g yr^{-1}$ for dissolved silicon (GESAMP, 1987). A number of authors have estimated the gross anthropogenic fluvial nutrient fluxes and these are in the ranges $\sim 7 \times 10^{12} - \sim 35 \times 10^{12} g yr^{-1}$ for dissolved nitrogen, and $\sim 0.6 \times 10^{12} - \sim 3.75 \times 10^{12} g yr^{-1}$ for dissolved phosphorus (GESAMP, 1987). It is apparent, therefore, that the anthropogenic fluxes of the dissolved nutrients are at least of the same order as, and may in fact exceed, those from natural sources. Net fluvial fluxes for the nutrients are difficult to estimate. However, the nutrients are extensively involved in biological, chemical and physical processes both in the estuarine and coastal sea zones, which can severely restrict their transport to the open ocean, for example, by trapping in sediments. It may be concluded, therefore, that the nutrients required to support primary production in the open ocean are supplied mainly by upwelling processes, or by vertical turbulent mixing (GESAMP, 1987).

## 6.2 Atmospheric fluxes to the oceans

### 6.2.1 Introduction

In order to estimate the *net* flux of a component from the atmosphere to the ocean it is necessary to know (i) its burden in the air; (ii) the rate at which it is deposited on to the sea surface; and (iii) the extent to which it is recycled back into the atmosphere.

### 6.2.1.1 *Atmospheric burden*

To determine the atmospheric burden of a component in a slice of the atmosphere (e.g. a $1 m^2$ column), data must be available on its concentration per unit of air (e.g. $\mu g\,m^{-3}$) and the height to which it is dispersed, i.e. the scale height. The global atmospheric concentrations of many elements vary over one to three orders of magnitude, and are often geographically dependent. Because of this, the assessment of the atmospheric burdens of elements has often been restricted to a local scale. As more reliable data have become available, however, it has been possible to make first-approximation estimates of the global atmospheric burdens of some elements (see e.g. Walsh *et al.*, 1979). In making these estimates it is assumed that the atmosphere is in a steady state, i.e. the rate of input of the component is equal to its rate of output, and the global burden ($C_T$) of the component is then computed from an equation of the type

$$C_T = C_O A S \qquad\qquad (6.2)$$

in which $C_O$ is the surface concentration of an element ($g\,m^{-3}$ of air), $A$ is the surface area of the atmosphere ($m^2$) and $S$ is the scale height, i.e. the height to which the atmospheric component is dispersed (usually assumed to be between ~3000 and ~5000 m).

### 6.2.1.2 *Rate of deposition*

Air to sea fluxes result from the removal of material that is present in the atmosphere. The removal of gaseous components is described in Section 8.2, and attention at this stage will be confined to aerosols.

The deposition of aerosols from the atmosphere is controlled by a combination of 'dry' (gravitational settling and turbulent diffusion) and 'wet' (precipitation scavenging) processes. In both depositional modes the total (i.e. particulate + dissolved) trace-metal concentrations reaching the sea surface from the atmosphere are dependent on the composition and atmospheric concentrations of the primary aerosol (Chapter 4). The principal constraint on the manner in which atmospheric trace metals enter the oceanic biogeochemical cycles, however, is the degree to which they are soluble in sea water, and with respect to the particulate/dissolved metal speciation the processes associated with the two deposition modes are geochemically different. In the 'dry' mode, aerosols are delivered directly to the sea surface, and trace metal solubility is constrained by *particle ↔ sea-water* reactivity. In contrast, in the 'wet' deposi-

tion mode, trace metal solubility is constrained initially by *particle ↔ rainwater* reactivity, and as some rain waters can have a pH as low as <3 this can result in a number of trace metals being highly soluble prior to the aerosols reaching the sea surface.

*'Dry deposition'.* During their residence time in the atmosphere aerosols can pass through a number of cloud systems that do not generate rain. Within these cloud systems the aerosol particles may be subjected to several 'wetting' and 'drying' cycles, during which they can be coated by solution films having pH values as low as ~2.0. The term 'dry' therefore is used in the present context to identify the aerosol deposition mode that does not involve an aqueous *deposition* phase. In the 'dry' deposition mode, therefore, although the aerosols may have passed through several 'wetting' and 'drying' cycles, they reach the sea surface by the direct air → sea surface route.

The 'dry' removal of particles from the atmosphere is a *continuous* process that is affected by a number of factors, which include windspeed and particle size. For example, McDonald *et al.* (1982) showed that the 'dry' deposition of sea-salts was dominated by large particles (MMD > 10 $\mu$m), which accounted for ~70% of the total salt deposition, although they made up only ~10% of the total mass. 'Dry' deposition therefore is especially important for the removal of large particles from the air.

*'Wet deposition'.* This involves the removal of both water-soluble gases and particulate material from the atmosphere by incorporation into precipitation scavenging in cloud droplets (in-cloud) or falling rain (below cloud) processes that are *random* in time. In the most general sense, 'wet' deposition rates depend on the concentrations of a component in rain and the total amount of rain that falls on to a surface. In this mode, therefore, aerosols reach the sea surface by the indirect air → rain → sea-surface route. To understand 'wet' deposition, it is therefore necessary to relate it to the elemental composition of rain water.

*Washout factors.* Washout factors, or scavenging ratios, are often used to determine the degree to which a component is removed from the air by rain. The washout factor (WF) is calculated from an equation of the type

$$WF = C_r/C_a \qquad (6.3)$$

in which $C_r$ is the concentration of a component in the rain and $C_a$ is its concentration in low-level air; sometimes an air density term is included in the equation. Values of $WF$ for the elements studied most commonly, lie in the range $\sim 10^2$–$\sim 10^3$.

### 6.2.1.3 Extent of recycling

The recycling of particulate components across the sea surface can occur during the generation of sea-salts. Data are now available on the degree to which some elements are fractionated at the ocean surface with respect to bulk sea water (see Section 4.2.1.2), and for these the extent of recycling across the air–sea interface can be estimated with some degree of certainty. For example, Arimoto *et al.* (1985) used a combination of aerosol, rain and seawater data to estimate that at Enewetak (North Pacific) the percentage of wet deposition associated with recycled sea-salts can be substantial; values for individual elements included 15% (Zn), 30% (V) and (48%) (Cu). Clearly, recycling must be taken into account when estimates are made of the *net* deposition fluxes of some elements to the sea surface from the atmosphere.

### 6.2.2 Atmospheric fluxes of trace elements

### 6.2.2.1 'Dry' deposition

Although the material reaching the sea surface by 'dry' deposition, i.e. the direct air → sea-surface route, will undergo changes brought about by seawater reactivity, it will initially have the same elemental composition as that of the material falling out from the parent aerosol: a particle size-dependent process, which changes with distance from the aerosol source.

### 6.2.2.2 'Wet' deposition: the elemental chemistry of marine rain water

In contrast to 'dry' deposition, aerosols involved in 'wet' deposition (precipitation scavenging), which reach the sea surface by the indirect air → rain → sea-surface route, can undergo considerable chemical changes before they reach the oceanic environment.

*Major ions in rain waters.* Data are now available on the major ion concentrations in rain waters from a number of marine locations. The principal major ions in world rain waters are dominated by the cations $H^+$, $NH_4^+$, $K^+$, $Ca^{2+}$ and $Mg^{2+}$, and the anions $SO_4^{2-}$, $NO_3^-$ and $Cl^-$. Scavenged sea-salt has a strong influence on the chemistry of precipitation over, and adjacent to, marine regions. Despite this, other sources impose fingerprints on the major ions in marine rain waters. For example, Church *et al.* (1982) have pointed out that even at remote marine locations alkali and alkaline earth cations in rain water can have a terrestrial dust source. The major anions in marine rain can also have contributions from non-sea-salt sources. For example, the sulphate in excess of that derived from sea-salt can arise from natural biogenic emissions, e.g. from compounds such as dimethyl sulphide (see Section 4.1.4.3), and from terrestrial anthropogenic sources. In marine-dominated rain, the proportions of the major ions are therefore influenced by variations in the inputs of sea-salt, crustal dust and anthropogenic components, the proportions of which differ with distance from the continental sources (see e.g. Galloway *et al.*, 1982).

A number of trends found in the major ion chemistry of marine rain waters can be illustrated with respect to samples taken at three contrasting marine environments. These are: (i) Lewes, Delaware, on the mid-Atlantic USA coast; (ii) Bermuda, in the North Atlantic; and (iii) Amsterdam Island, a remote site in the southern Indian Ocean. Data for the major ions in rain waters from these three sites are given in Table 6.8. The data for USA coast and Bermuda sites were reported by Church *et al.* (1982), who assessed the marine influence on precipitation arising from storms that leave the North American continent and transit over the western Atlantic. The findings of the study showed that sea-salt contribution (by weight) to the major ions rose from 54% at the coastal site to 80% at Bermuda. In contrast, the sulphate decreased from the coastal site to Bermuda; but even at Bermuda $\sim 50\%$ of the total sulphate was in excess of the sea-salt sodium. At the remote Amsterdam Island site the excess sulphate made up $\sim 20\%$ of the total sulphate.

The acidity of rain water is an important environmental parameter, and the free acidity can result from a number of proton donors, such as strong acids (e.g. $H_2SO_4$, $HNO_3$), weak organic acids (e.g. acetic acid, formic acid) or metal oxides (e.g. Al, Fe)—see e.g.

**Table 6.8** Major ions in marine-influenced rain water (volume weighted mean).

| | Amsterdam Island* | Bermuda† | Lewes, Delaware‡ |
|---|---|---|---|
| Percentage sea-salt (Na) | 97.7 | 80.3 | 54.1 |
| $\Sigma SO_4$ (% excess) | 29.2 (16.9) | 36.3 (51) | 62.5 (89) |
| $Na^+$ $\mu$eq $l^{-1}$ | 206.5 | 148 | 56.5 |
| $Mg^{2+}$ | 45.9 | 40 | 12.6 |
| $H^+$ (pH) | 8.8 (5.06) | 18.4 (4.74) | 53.4 (4.22) |
| $Ca^{2+}$ | 8.6 | 15.3 | 8.53 |
| $K^+$ | 4.4 | 4.03 | 1.91 |
| $NH_4^+$ | 1.8 | 4.54 | 18.9 |
| $Cl^-$ | 237.7 | 191 | 46.4 |
| $NO_3^-$ | 1.3 | 6.57 | 25.2 |

* Central Indian Ocean; data from Galloway & Gaudry (1984). † North Atlantic; data from Church *et al.* (1982). ‡ Eastern coast USA; data from Church *et al.* (1982).

Galloway *et al.* (1982). The pH of natural rain water is generally acidic ($\sim$5.0–$\sim$5.5) as a result of the equilibration of atmospheric $CO_2$ with precipitation (see e.g. Galloway *et al.*, 1982; Pszenny *et al.*, 1982). Rainwater pH can also be strongly influenced by both water-soluble and particulate aerosol components scavenged from the air. For example, precipitation on Bermuda, down wind of North American industrial areas, is acidified relative to equilibration with $CO_2$ by a factor of about eight (Church *et al.*, 1982). Sea-salt in the Bermuda precipitation should not neutralize more than $\sim$10% of the acidity, and Church *et al.* (1982) concluded that most of the acidity in excess of the natural equilibration results from the long-range transport of sulphur and nitrogen precursors in the marine troposphere, with sulphuric acid components dominating those of nitric acid. According to Galloway & Gaudry (1984) precipitation on Amsterdam Island has two components: one from a seawater source and the other contributing to the acidity of the precipitation. The component giving rise to the acidity is substantially smaller than the seawater component, and the main proton donors are $H_2SO_4$, low molecular weight organic acids (HCOOH and $CH_2COOH$) and $HNO_3$, with the maximum contribution to acidity being 30%, 25% and 15%, respectively. The authors reported that there is an interaction between the alkaline seawater component and the acid component, which results in an average loss of $\sim$10% of the original free acidity due to neutralization. The three principal sources for the acidic components in the rain water at Amsterdam Island were: (i) long-range transport from continental regions; and (iii) local island emissions. The authors concluded that the $SO_4^{2-}$ source is derived from the oxidation of reduced marine sulphur components, but that continental sources possibly influenced the $NO_3^-$; it was not possible to assign sources to the organic acids.

Seawater components can neutralize part of the acidity in precipitation. A more dramatic effect on the acidity of precipitation, however, can result from the scavenging of crustal dusts. For example, Loye-Pilot *et al.* (1986) demonstrated that the pH of western Mediterranean rain water is strongly influenced by the type of material scavenged from the air. For example, rain waters associated with air masses that had crossed western Europe, and had scavenged black particulate European 'background' material from the air, had pH values in the range 4.1–5.6. In contrast, the so-called 'red rains' associated with air masses that had crossed North African sources, and which had scavenged crust-dominated Saharan dust, had pH values as high as 6–7 as a result of the dissolution of calcium carbonate from the dusts.

*Trace metals in rain waters.* Trace metals in rain waters are derived from material scavenged from the air. The **total concentrations** (i.e. dissolved + particulate) of trace metals at any individual site will reflect the type of aerosol scavenged. Data are now available on the trace-metal compositions of rain waters from a number of coastal and marine regions, and a selection of these is given in Table 6.9. There are problems, however, in directly comparing the trace-metal concentrations from different individual sites. One reason for this is that the concentrations of the

**Table 6.9** Trace metal composition of marine-influenced rain waters (total trace metals; units, ng l⁻¹).

| Trace metal | North Sea | | | Mediterranean Sea | | North Atlantic | | | |
|---|---|---|---|---|---|---|---|---|---|
| | Northeast coast, Scotland* (VWM) | North coast, Germany† (VWM) | Open-sea‡ (VWM) | South coast, France§ (VWM) | Sardinia¶ (VWM) | Bermuda‖ (VWM) | Bantry Bay** (VWM) | North Pacific: Enewetak†† | South Pacific: Samoa‡‡ |
| Al | — | — | 21 | 144 | 883 | — | 3.62 | 2.1 | 16 |
| Fe | 88 | 18 | 31 | — | 519 | 4.8 | 8.06 | 1.0 | 0.42 |
| Mn | 3.8 | 4.2 | 3.6 | — | 8.0 | 0.27 | 0.13 | 0.012 | 0.020 |
| Cu | 2.3 | 1.7 | 0.98 | 2.8 | 2.9 | 0.66 | 0.86 | 0.013 | 0.021 |
| Zn | 13 | 25 | 7.6 | — | 16 | 1.15 | 8.05 | 0.052 | 1.6 |
| Pb | 4.0 | 6.4 | 3.5 | 3.7 | 1.6 | 0.77 | 0.51 | 0.035 | 0.014 |
| Cd | 0.68 | 0.48 | 0.08 | — | — | 0.06 | 0.04 | 0.0021 | — |
| Sb | — | 0.38 | 0.12 | — | — | — | — | — | — |
| Se | — | 0.52 | 0.34 | — | — | — | — | — | — |

* Balls (1989). † Stossel (1987). ‡ Chester *et al.* (1994). § Chester *et al.* (1997). ¶ Keyse (1996). ‖ Jickells *et al.* (1984).
** Lim *et al.* (1991). †† Arimoto *et al.* (1985). ‡‡ Arimoto *et al.* (1987).
VWM, volume weighted mean.

metals, and major ions, in the rain can change during the course of a rain event (see e.g. Lim *et al.*, 1991), and some trace metals can have higher concentrations in the early, relative to the later, precipitation. In addition, trace-metal concentrations can vary from one rain event to another at the same site. Because of this some authors quote their precipitation trace-metal data on a volume weighted mean (VWM) basis. This normalizes the precipitation concentration to the precipitation amount. Thus, according to Galloway & Gaudry (1984), it is as if all the precipitation at a single site collected over a given period was mixed in one container and the composition determined from that; most of the data sets given in Table 6.9 are expressed on a VWM basis.

The collection sites in Table 6.9 are ranked in terms of their remoteness from the major continental sources, and despite the problems inherent in comparing data sets from different sites, two overall trends in the distribution of trace metals in rain waters can be identified from the data in Table 6.9.

1 The concentrations of the NEEs, such as Al, are highest in the coastal rain waters where the atmosphere received inputs of crustal dust 'pulses', e.g. the Mediterranean Sea.

2 The concentrations of the AEEs, such as Pb and Zn, are highest in coastal regions, and decrease with the degree of remoteness of a site from the main continental regions.

These trends therefore reflect those found for the aerosols that are scavenged from the air by the precipitation (see Section 4.2.3).

### 6.2.2.3 Total atmospheric trace-element fluxes

Buat-Menard (1983) concluded that, in general, the net atmospheric fluxes of the AEEs (small particle size) to the sea surface primarily result from *wet* deposition over most marine areas, but that *dry* deposition is significant for sea-salt and mineral aerosols (large particle size). The data available in the literature tend to confirm these overall trends, although the situation is by no means absolutely clear. For example, at Enewetak, flux data (corrected for sea-surface recycling) showed that, although wet deposition exceeded dry deposition for Pb, V, Cd and Se, this was not the case for Cu and Zn. Further, at Enewetak, wet deposition was more important than dry deposition for Fe, although for Al dry deposition was an order of magnitude higher than wet removal.

Many of the early models used to estimate the atmospheric input of elements to the sea surface were inevitably somewhat crude, and total atmospheric deposition fluxes ($F$) were often calculated from an equation of the general type

$$F = CV \qquad (6.4)$$

in which $C$ is the mean atmospheric concentration of

an element and $V$ is the global deposition velocity. The simplest way of estimating the global deposition velocity of an element is by assuming that all deposition takes place by rain scavenging, which cleans the atmosphere around 40 times per year; see, for example, the model described by Bruland *et al.* (1974). In more complex models, attempts were made to take account of the actual deposition rates of elements to the sea surface, either by assuming a total deposition rate (see e.g. Buat-Menard & Chesselet, 1979) or by taking individual account of wet and dry deposition rates (see e.g. Duce *et al.*, 1976). Later, more advanced 'wet and dry' flux deposition models became available. For example, Arimoto *et al.* (1985) determined the gross and net atmospheric fluxes of a series of crustal and enriched elements to the sea surface at Enewetak. The findings are of particular interest because the models take account of sea-surface recycling, and thus offer an estimate of the true *net* deposition of the elements to the ocean surface from the atmosphere. Advanced models of this type were applied subsequently to the Samoa (South Pacific) aerosol (Arimoto *et al.*, 1987). The data sets provided by Arimoto *et al.* (1985, 1987) therefore represent the best available estimates of the net deposition of trace metals to the sea surface.

A summary of some of the data given in the literature for the fluxes of trace elements to the sea surface is given in Table 6.10. With the exception of the data sets for Enewetak and Samoa, most of the calculations do not take account of sea-surface recycling, with the result that the fluxes, especially those for the AEEs, will tend to be overestimated. Further, the fluxes have been obtained by different methods. Despite constraints such as these, however, a strong geographical trend can be identified from the data in Table 6.10, indicating that the strengths of the air-to-sea fluxes decrease by orders of magnitude as the degree of remoteness of a site from the major aerosol sources increases. For example, this trend is well developed for Pb, for which the atmospheric input from the North Atlantic westerlies is over 50 times greater than that from the South Pacific westerlies.

### 6.2.2.4 The fates of atmospherically transported trace elements in sea water

It is clear that the atmosphere provides an important pathway for the transport of trace metals to the oceans. The subsequent fate of the elements will depend on a number of factors. The initial constraint on the behaviour of atmospherically transported trace metals in the mixed layer will be imposed by the extent to which they are solubilized in sea water. This is important because the physical state (i.e. particulate or dissolved) of the metals affects both their subsequent involvement in the biogeochemical cycles and their residence times in sea water.

*'Dry' deposition.* Some of the early studies on the solubility of trace metals from aerosols indicated that the manner in which a metal is partitioned between crustal and non-crustal components exerts a major control on its fate in sea water (see e.g. Walsh & Duce, 1976; Hodge *et al.*, 1978). The general relationship between the solid state speciation of a trace metal in an aerosol (see Section 4.2.1.3) and its seawater solubility was confirmed by Chester *et al.* (1993), who showed that there is a well-developed relationship between the extent to which a trace metal is held in exchangeable associations in 'end-member' (crust-dominated and anthropogenic-dominated) aerosols and the extent to which it is soluble in sea water—see Fig. 6.3. On the basis of this 'percentage exchangeable versus percentage seawater soluble' relationship, Chester *et al.* (1993) divided the trace metals into a number of types; average percentage seawater solubility values for the crust-dominated and anthropogenic-dominated aerosols are given in Table 6.11.

Type 1 (e.g. Al, Fe) These metals are crust-controlled (i.e. they are NEEs and have $EF_{crust}$ values <10— see Section 4.2.1.1) in all marine aerosols. In both crust-dominated and anthropogenic-dominated aerosols, Al and Fe have less than ~10% of their total concentrations in exchangeable associations and also they are relatively insoluble in sea water (less than ~10% of their total concentrations). However, because they are present in relatively high concentrations in marine aerosols, even a solubility of a few per cent can release considerable quantities of the metals into sea water in a dissolved form. For example, the dissolution of Fe from aerosols can play an important role in primary productivity (see Section 9.1.3.2).

Type 2 (e.g. Mn) Manganese is usually crust-controlled (NEE) in marine aerosols, but unlike Al and Fe it is relatively soluble (~20%–~45%)

**Table 6.10** Atmospheric fluxes of trace metals to the sea surface; units, ng cm$^{-2}$ yr$^{-1}$.

| | New York Bight* | North Sea† | Western Mediterranean‡ | South Atlantic Bight§ | Bermuda¶ | North Atlantic; Northeast Trades‖ |
|---|---|---|---|---|---|---|
| Al | 6000 | 30 000 | 5000 | 2900 | 3900 | 97 000 |
| Sc | — | 5 | 1 | — | 0.6 | — |
| V | — | 480 | — | — | 5 | — |
| Cr | — | 210 | 49 | — | 9 | 111 |
| Mn | — | 920 | — | 60 | 45 | 570 |
| Fe | 5700 | 25 500 | 5100 | 5900 | 3000 | 48 000 |
| Co | — | 39 | 3.5 | — | 1.2 | 12 |
| Ni | — | 260 | — | 390 | 3 | 67 |
| Cu | — | 1300 | 96 | 220 | 30 | 48 |
| Zn | 1400 | 8950 | 1080 | 750 | 75 | 152 |
| As | — | 280 | 54 | 45 | 3 | — |
| Se | — | 22 | 48 | — | 3 | — |
| Ag | — | — | 3 | — | — | — |
| Cd | 30 | 43 | 13 | 9 | 4.5 | — |
| Sb | — | 58 | 48 | — | 1.0 | — |
| Au | — | — | 0.05 | — | — | — |
| Hg | — | — | 5 | 24 | — | — |
| Pb | 3900 | 2650 | 1050 | 660 | 100 | 32 |
| Th | — | 4 | 1.2 | — | — | — |

*Continued*

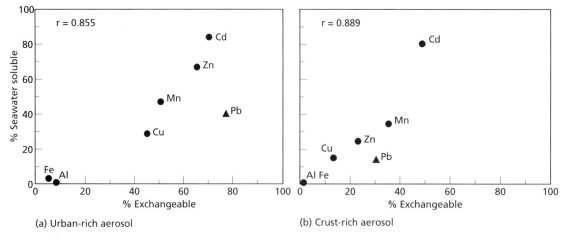

(a) Urban-rich aerosol   (b) Crust-rich aerosol

**Fig. 6.3** The relationship between the percentage of the total concentrations of trace metals that are soluble in sea water and the percentage in exchangeable associations in (a) anthropogenic-dominated and (b) crust-dominated aerosols (after Chester *et al.*, 1993).

**Table 6.10** *Continued*

| Tropical North Atlantic** | Tropical North Pacific: total net deposition†† | South Pacific: total net deposition‡‡ | North Atlantic; westerlies§§ | North Pacific; westerlies§§ | South Pacific; easterlies§§ |
|---|---|---|---|---|---|
| 5000 | 1200 | 132–1800 | — | — | — |
| 1.1 | 0.18 | 0.06 | — | — | — |
| 17 | 7.8 | — | — | — | — |
| 14 | — | — | — | — | — |
| 70 | 9.0 | 3.6 | — | — | — |
| 3200 | 560 | 47–337 | — | — | — |
| 2.7 | — | 0.25 | — | — | — |
| 20 | — | — | — | — | — |
| 25 | 8.9 | 4.4–7.9 | — | — | — |
| 130 | 67 | 5.8–2.4 | — | — | — |
| — | — | — | — | — | — |
| 14 | 4.2 | 0.8 | — | — | — |
| 0.9 | — | — | — | — | — |
| 5 | 0.35 | — | — | — | — |
| 3.5 | — | — | — | — | — |
| 0.1 | — | — | — | — | — |
| 2.1 | — | — | — | — | — |
| 310 | 7.0 | 1.4–2.8 | 170 | 50 | 3 |
| 0.9 | 0.61 | 0.036 | — | — | — |

\* Duce *et al.* (1976). † Cambray *et al.* (1975). ‡ Arnold *et al.* (1982). § Windom (1981). ¶ Duce *et al.* (1976). ‖ Chester *et al.* (1979).
\*\* Buat-Menard & Chesselet (1979). ††Arimoto *et al.* (1985). ‡‡Arimoto *et al.* (1987). §§Settle & Patterson (1982).

**Table 6.11** Seawater solubility of trace metals from atmospheric particulate aerosols; approximate percentage of total element soluble (after Chester *et al.*, 1996).

| | Coastal regions: anthropogenic-dominated aerosol | Coastal regions: crustal-dominated aerosol | Open-oceans: 'mixed-source' aerosols |
|---|---|---|---|
| Al | <10 | <10 | <10 |
| Fe | <10 | <10 | <10–50 |
| Mn | 45 | 20 | 20–50 |
| Ni | 50 | <20 | 20–50 |
| Co | 25 | <20 | <20–25 |
| Cr | 12 | <10 | 10 |
| V | 30 | <20 | <20–85 |
| Cu | 35 | <10 | <10–85 |
| Pb | 50 | <10 | <10–90 |
| Zn | 70 | <10 | <10–75 |
| Cd | 85 | 80 | 85 |

from both crust-dominated and anthropogenic-dominated aerosols. The reason why Mn is more soluble than the type 1 trace metals is that it has a higher percentage of its total concentration

($\sim$20–$\sim$$\leqslant$50%) in exchangeable associations in all marine aerosols.

**Type 3** (e.g. Cu, Zn and Pb) These metals have higher $EF_{crust}$ values and a higher proportion of their total concentrations in exchangeable associations in anthropogenic-dominated aerosols, in which they behave as AEEs, than in crust-dominated aerosols, in which they behave as NEEs. The higher solubilities of the type 3 metals from the anthropogenic-dominated ($\sim$35% of the total ($\Sigma$) Cu, $\sim$70% of the $\Sigma$Zn and $\sim$50% of the $\Sigma$Pb) than from crust-dominated aerosols (less than $\sim$10% of the $\Sigma$Cu, $\Sigma$Zn and $\Sigma$Pb) is thus entirely consistent with their solid-state speciation signatures in the two aerosol 'end-members'.

**Type 4 trace metals** (e.g. Cd) Like the type 3 metals, Cd has $EF_{crust}$ values that are higher in anthropogenic-dominated aerosols, in which Cd acts as an AEE, than in crust-dominated aerosols, in which it behaves as an NEE. However, Cd has a relatively high seawater solubility from both aerosol 'end-members' ($\sim$80% of the $\Sigma$Cd). The reason for this is that in both 'end-member' aerosols a relatively large fraction of the $\Sigma Cd$ ($\sim$55–$\sim$70%) is present in exchangeable associations—see Fig. 6.3.

It is unlikely that aerosols will retain their original identity as discrete particles in the upper water column following their deposition to the sea surface. For example, they may be coated with organics or enter the gut system of filter-feeding organisms, where the pH can be very much lower than that in sea water. Experimental studies therefore will reflect only the initial fates of atmospherically transported trace metals in the mixed layer. Despite difficulties such as these, a number of authors have attempted to estimate the solubilities of trace metals from aerosols (see e.g. Chester & Murphy, 1988; Duce *et al.*, 1991), and a compilation of the data available is listed in Table 6.11 in terms of three characteristically different oceanic environments, selected to illustrate how the solubility of a trace metal from an aerosol is dependent on its solid-state speciation, and therefore ultimately on its source (see Section 4.2.1.3). A number of overall trends can be identified from the data in Table 6.11.

1 In coastal regions with anthropogenic-dominated aerosols significant fractions of the AEEs are soluble in sea water.

2 In coastal regions with crust-dominated aerosols considerably smaller amounts of the AEEs, with the exception of Cd, will enter the dissolved phase on contact with sea water. In these areas, however, the atmospheric deposition fluxes of trace metals can be relatively very high, with the result that the soluble fractions of the metals may become significant, especially in those regions that receive dust 'pulses' (see Section 4.1.4.1); this is particularly relevant for Al and Fe, which are present in crustal dusts in relatively large concentrations.

3 Over open-ocean regions the solubilities of trace metals from aerosols will depend on the extent to which components from various sources (e.g. anthropogenic, crustal, marine) are mixed together in the total aerosol population.

It must be stressed, however, that the aerosol solubility data given in Table 6.11 are essentially crude and should be regarded as no more than first-order approximations.

*'Wet' deposition.* The trace-metal concentrations in rain waters reflect those in the aerosols that have been scavenged from the air so that the total precipitation fluxes, like the 'dry' fluxes, vary with the atmospheric concentrations of the metals. The dissolved/particulate speciation of trace metals following the 'dry' deposition of aerosols to the sea surface is a function of aerosol $\leftrightarrow$ seawater reactivity, and essentially is constrained by the solid-state speciation of the metals in the aerosols (see above). The dissolved/particulate speciation of the metals in 'wet' deposition, however, is a function of aerosol $\leftrightarrow$ rainwater reactivity, and so is constrained by both the solid-state speciation of the metals in the scavenged aerosols and the chemistry of the rain waters themselves. The aerosol $\leftrightarrow$ rainwater reactivity includes pH-dependent processes, such as adsorption/desorption and precipitation. As a result, although the overall dissolved–particulate speciation in rain waters depends on a number of complex interrelated factors, the solution pH exerts a fundamental control.

The average dissolved–particulate speciation of a series of trace metals in rain waters from several sites in the western Mediterranean is given in Table 6.12(a), and a number of general trends can be identified from these data (see e.g. Chester *et al.*, 1996).

1 Aluminium and Fe, which are NEEs in marine aerosols, are relatively insoluble (less than $\sim$20% of the $\Sigma Al$ and $\Sigma Fe$) in rain waters from all the sites.

2 Manganese, which is an NEE in marine aerosols, is generally relatively soluble ($\sim$60–70%) in rain waters from all the sites.

3 The solubilities of some trace metals (e.g. Cu, Zn, Pb), which can switch character between NEE and AEE behaviour in marine aerosols as a function of their source (see Section 4.2.1.1), vary considerably from one rainwater population to another. These variations can be related to the solution pH and the type of aerosol scavenged from the air. A number of authors have shown that the solubility of trace metals such as Cu, Zn and Pb in rain waters is a pH-dependent process (see e.g. Losno *et al.*, 1988; Chester *et al.*, 1990; Lim *et al.*, 1994; Guieu *et al.*, 1997). Further, the most striking feature of the pH–solubility plots for these trace metals in rain waters is the presence of the classical pH 'adsorption edge', which separates a 'low pH and high trace-metal solubility' region from a 'high pH and low trace-metal solubility' region. The pH of rain water can vary with the type of aerosol scavenged from the air (see Section 6.2.2.2); thus, the scavenging of

**Table 6.12** The dissolved–particulate speciation of trace metals in rain waters.

(a) The average dissolved–particulate speciation of trace metals in rain waters from a number of Mediterranean Sea sites; data given as percentage of the total trace metal concentration in the dissolved phase (data from Guieu et al., 1997).

|    | Cap Ferrat | Tour du Valat | Carpentras | Corsica |
|----|-----------|---------------|------------|---------|
| Al | 18 | 19 | — | 8 |
| Fe | — | 11 | — | 13 |
| Mn | 60 | 63 | — | 67 |
| Ni | 54 | 58 | — | — |
| Co | 61 | 50 | — | — |
| Cu | 82 | 71 | 66 | 49 |
| Zn | — | 68 | — | 76 |
| Pb | 65 | 52 | 80 | 48 |
| Cd | 92 | 75 | — | — |

(b) Total trace-metal concentrations and $EF_{crust}$ values in rain waters that have scavenged European 'background' and Saharan dust aerosols; concentration units: $\mu g\,l^{-1}$, % = percentage total concentration in dissolved phase (data from Chester et al., 1997).

| | Aerosol type scavenged | | | | | |
|----|---------------|------------|-----|------------|------------|-----|
| | European 'background'; sample 1, pH 3.95 | | | Saharan dust; sample 12, pH 6.55 | | |
| | Concentration | $EF_{crust}$ | % | Concentration | $EF_{crust}$ | % |
| Al | 140 | 1.0 | 35 | 1462 | 1.0 | 1.15 |
| Co | 0.078 | 1.9 | 67 | 0.58 | 1.3 | 0.50 |
| Ni | 1.88 | 15 | 69 | 2.26 | 1.7 | 23 |
| Cu | 5.79 | 59 | 86 | 4.0 | 3.9 | 30 |
| Pb | 26.2 | 1248 | 90 | 3.46 | 16 | 52 |

anthropogenic-dominated aerosols often leads to rain waters having a relatively low pH (less than ~5.0), and the scavenging of crust-dominated aerosols can result in rain waters having a relatively high pH (greater than ~6.5). This can be illustrated with respect to rain waters from the Mediterranean Sea region, where the scavenged aerosols are composed of a continuously supplied European anthropogenic-dominated 'background' material, upon which is superimposed sporadic intrusions of Saharan crustal dusts (see Section 4.1.4.1). The scavenging of these two types of 'end-member' aerosols by precipitation leads to different solution pH values—see Table 6.12(b). A range of pH values therefore can often be found for different rainwater events at an individual Mediterranean site. In addition, because scavenging is a function of particle size, pH variations can some-

times occur sequentially within the same rainwater event, as the scavenging processes associated with smaller sized AEEs are different from those associated with larger sized NEEs (see e.g. Chester et al., 1997). A typical 'solubility–pH' relationship for Pb in rain waters collected at Cap Ferrat, a western Mediterranean site on the southern coast of France, is illustrated in Fig. 6.4. From this figure it can be seen that the 'adsorption edge', which separates the 'low-pH–high-Pb solubility' region from the 'high-pH–low-Pb solubility' region lies between pH ~4.8 and ~5.8; i.e. on the more acidic side of the pH 'edge' ~80% of the $\Sigma$Pb is in a soluble form, whereas on the less acidic side only ~20% is soluble. Similar 'pH–solubility' diagrams have been reported for Zn, Cu and Cd (see e.g. Losno et al., 1988; Lim et al., 1994).

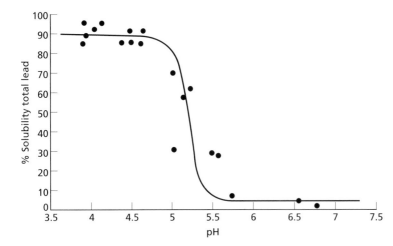

**Fig. 6.4** The pH–solubility relationship of Pb in western Mediterranean rain waters (after Chester et al., 1990).

It is apparent, therefore, that aerosol trace-metal solubility trends in both sea water and rain water can be related to the solid-state speciation of the metals, with the exchangeable associations (i) being the most soluble; and (ii) making up a higher fraction of the total metal in anthropogenic-dominated than in crust-dominated aerosols (see Section 4.2.1.3). The additional constraint that must be taken into account with respect to rain waters is the solution pH. This pH-driven solubility is also a function of the solid-state speciation of the metals in the scavenged aerosols. It must be remembered, however, that there is a coupling between the rainwater pH and the solid-state speciation in the scavenged aerosols because the type of aerosol scavenged has a strong influence on the pH. Thus, anthropogenic-dominated aerosols, which have relatively large fractions of the AEEs in exchangeable associations, give rise to rain waters with a low (more corrosive) pH and therefore a high AEE solubility. In contrast, crust-dominated aerosols, which have relatively small fractions of the AEEs in exchangeable associations, give rise to rain waters with a high (less corrosive) pH and therefore a lower solubility of any AEEs that are present.

Although the solution pH and the trace metal solid-state speciation in the scavenged aerosols may be regarded as the primary controls, a number of other factors must be taken into account when assessing trace-metal solubility in rain waters. For example, Guieu et al. (1997) highlighted the importance of particle concentrations on the solubilities of trace metals from aerosols in rain waters.

### 6.2.3 Atmospheric fluxes of organic matter

In terms of global cycles, the oceans provide a major sink for some atmospheric organic material, and although the data are still sparse several attempts have been made to estimate the global tropospheric burdens of organic matter. For example, Duce (1978) calculated that the non-methane global VOC burden is $\sim 50 \times 10^{12}$ g, and the total POC burden has been estimated to be $\sim 1 \times 10^{12}$–$5 \times 10^{12}$ g, of which $\sim 90\%$ is on particles with diameters $<1\,\mu$m (see e.g. Duce et al., 1983). In addition, the total amount of organic carbon cycling through the troposphere each year has been put at $\geq 800 \times 10^{12}$ g (see e.g. Duce, 1978; Robinson, 1978; Zimmerman et al., 1978). A number of first-order estimates have been made of the source–sink relationships of POM and VOM in the marine atmosphere. Thus, Williams (1975) calculated a wet deposition flux of carbon of $\sim 2.2 \times 10^{14}$ g yr$^{-1}$ to the ocean surface, which is somewhat smaller than that of $\sim 10 \times 10^{14}$ g yr$^{-1}$ estimated by Duce & Duursma (1977), although both are within the same order of magnitude. The dry deposition flux of organic material (carbon) to the oceans has been put at $\sim 6 \times 10^{12}$ g yr$^{-1}$ (Duce & Duursma, 1977). Thus, although it would appear that wet deposition is more important than dry fall-out in the fluxes of organic material (POC = POM $\times 0.7$) to the sea surface, all the estimates should be treated with great caution. Further complications to the assessment of the fluxes of organic material to the ocean arise because the sea surface itself can act as a source, as well as a sink, for both POM and VOM through processes such as gas exchange and bubble bursting (see Sections 4.1.4.2 and 8.2). For example, Duce (1978) calculated that $\sim 14 \times 10^{12}\,g\,yr^{-1}$ of organic carbon is produced by the ocean surface, with $\sim 90\%$ being found on particles $>1\,\mu m$ in diameter. Even when recycling is taken into account, however, it would appear that on the basis of the estimate given by Williams (1975) for the oceanic wet deposition flux of carbon ($\sim 2.2 \times 10^{14}\,g\,yr^{-1}$), the input of organic material to the ocean surface is greatly in excess of the output from the marine source. The overall result of this is that the oceans act as a major sink for atmospheric organic material (Duce *et al.*, 1983).

## 6.3 Hydrothermal fluxes to the oceans

Although the most dramatic hydrothermal activity is associated with high-temperature processes at the spreading ridge centres, basalt–seawater reactions take place over a variety of temperatures and rock : water ratios (see Chapter 5), and all reaction types must be taken into account when attempts are made to derive hydrothermal fluxes. Thompson (1983) distinguished four types of basalt–water reactions on the basis of their location, temperature and water : rock ratios.

1 Those involving **surface basement rocks**, which take place over long time periods away from the spreading centres at low temperature, high water : rock ratios.

2 Those involving **deeper basement rocks**, which take place within the basement over short time periods away from the spreading centres at low temperature, low water : rock ratios.

3 Those taking place at **the mid-ocean ridge flanks (off-axis activity)**. These are associated with hydrothermal circulation, but occur at medium temperature, medium water : rock ratios during the so-called passive circulation phase where the temperature of the water decreases away from the spreading centres.

4 Those taking place at **mid-ocean ridge axis (axial activity)**. These are associated with the dramatic venting of hydrothermal solutions at the spreading centres during the active phase of hydrothermal activity and take place at high temperature, low water : rock ratios.

Attention here will be focused on hydrothermal activity taking place at the crest and flank regions of the mid-ocean ridge system.

The first data available for the composition of hydrothermal venting solutions at the ridge crests was obtained from the low temperature, 'white smoker', Galapagos Spreading Centre (GSC), and Edmond *et al.* (1979) derived hydrothermal fluxes from these data. To do this the authors extrapolated the GSC venting solution composition to a 350°C high-temperature hydrothermal end-member and combined this with global heat-flux data estimated from hydrothermal $^3He$ : transported-heat ratios, to estimate global ocean hydrothermal fluxes for those elements that behaved in a conservative manner during the mixing of hydrothermal solutions and sea water in the 'white smoker' GSC system (see Section 5.1). Ridge crest (axial) hydrothermal fluxes obtained in this way are given in Table 6.13 (column A). Subsequently, data were obtained for 'black smoker' systems, which vent high-temperature fluids directly to sea water (see Section 5.1): for example, Von Damm *et al.* (1985) made a detailed study of high-temperature venting fluids in the 21°N EPR region (see Section 5.1). The data obtained allowed estimates to be made of the hydrothermal fluxes of the ore-forming metals (e.g. Mn, Fe, Co, Cu, Zn, Cd and Pb); this was not possible for the low-temperature GSC venting fluids because a large proportion of these ore-forming metals were precipitated at depth within the crust in this 'white smoker' system. Ridge-crest (axial) hydrothermal fluxes obtained from the directly measured high-temperature hydrothermal end-member are given in Table 6.13 (column B).

The axial hydrothermal flux estimates from both the GSC and the EPR systems were obtained using $^3He$:heat data to assess the extent of the hydrothermal input to the World Ocean. Elderfield & Schultz (1996), however, drew attention to the fact that there is a conflict between the different methods used to estimate axial chemical fluxes. The authors reviewed the various geophysical and geochemical methods of estimating hydrothermal fluxes, and concluded that the best estimate of the axial high-temperature hydrothermal water flux is 3 ($\pm 1.5$) $\times 10^{13}\,kg\,yr^{-1}$, which is better constrained than the $^3He$:heat data ($1 \times 10^{13}$ to $16 \times 10^{13}\,kg\,yr^{-1}$). This re-evaluated hydrothermal flux was then applied to the composition of 'black smoker' high-temperature venting solutions to yield the elemental hydrothermal fluxes given in Table 6.13 (column C). For most elements the fluxes listed by Elderfield & Schultz (1996), which were based on those presented originally by Kadko

**Table 6.13** Estimates of high-temperature axial hydrothermal fluxes; units, mol yr$^{-1}$.

| Element | A*: Galapagos Spreading Centre (GSC) | B†: East Pacific Rise 21°N | C‡: average axial high-temperature systems | D‡: fluvial fluxes |
|---|---|---|---|---|
| Li | $9.5–16 \times 10^{10}$ | $1.2–1.9 \times 10^{11}$ | $1.2–3.9 \times 10^{10}$ | $1.4 \times 10^{10}$ |
| K | $1.3 \times 10^{12}$ | $1.9–2.3 \times 10^{12}$ | $2.3–6.9 \times 10^{11}$ | $19 \times 10^{11}$ |
| Rb | $1.7–2.8 \times 10^{9}$ | $3.7–4.6 \times 10^{9}$ | $2.6–9.5 \times 10^{8}$ | $3.7 \times 10^{8}$ |
| Cs | — | — | $2.9–6.0 \times 10^{6}$ | $4.8 \times 10^{6}$ |
| Be | $1.6–5.3 \times 10^{6}$ | $1.4–5.3 \times 10^{6}$ | $3.0–12 \times 10^{5}$ | $370 \times 10^{5}$ |
| Mg | $-7.5 \times 10^{12}$ | $-7.7 \times 10^{12}$ | $-1.6 \times 10^{12}$ | $5.3 \times 10^{12}$ |
| Ca | $2.1–4.3 \times 10^{12}$ | $2.4–15 \times 10^{11}$ | $9–1300 \times 10^{9}$ | $12000 \times 10^{9}$ |
| Sr | $0$ | $-3.1–+1.4 \times 10^{9}$ | $0$ | $22 \times 10^{9}$ |
| Ba | $2.5–6.1 \times 10^{9}$ | $1.1–2.3 \times 10^{9}$ | $>2.4–13 \times 10^{8}$ | $100 \times 10^{8}$ |
| SO$_4$ | $-3.8 \times 10^{12}$ | $-4.0 \times 10^{12}$ | $-8.4 \times 10^{11}$ | $37 \times 10^{11}$ |
| Alk | — | — | $-7.2$ to $-9.9 \times 10^{10}$ | $3000 \times 10^{10}$ |
| Si | $3.1 \times 10^{12}$ | $2.2–2.8 \times 10^{12}$ | $4.3–6.6 \times 10^{11}$ | $64 \times 10^{11}$ |
| P | — | — | $-4.5 \times 10^{7}$ | $3300 \times 10^{7}$ |
| B | — | — | $1.1–4.5 \times 10^{9}$ | $54 \times 10^{9}$ |
| Al | — | $5.7–7.4 \times 10^{8}$ | $1.2–6.0 \times 10^{8}$ | $600 \times 10^{8}$ |
| Mn | $5.1–16 \times 10^{10}$ | $1.0–1.4 \times 10^{11}$ | $1.1–3.4 \times 10^{10}$ | $0.49 \times 10^{10}$ |
| Fe | — | $1.1–3.5 \times 10^{11}$ | $2.3–19 \times 10^{10}$ | $2.3 \times 10^{10}$ |
| Co | — | $3.1–32 \times 10^{6}$ | $6.6–68 \times 10^{5}$ | $1100 \times 10^{5}$ |
| Cu | — | $0–6.3 \times 10^{9}$ | $3.0–13 \times 10^{8}$ | $50 \times 10^{8}$ |
| Zn | — | $5.7–15 \times 10^{9}$ | $1.2–3.2 \times 10^{9}$ | $14 \times 10^{9}$ |
| Ag | — | $0–5.4 \times 10^{6}$ | $7.8–11 \times 10^{5}$ | $880 \times 10^{5}$ |
| Pb | — | $2.6–5.1 \times 10^{7}$ | $2.7–110 \times 10^{5}$ | $1500 \times 10^{5}$ |
| As | — | $0–6.5 \times 10^{7}$ | $0.9–140 \times 10^{5}$ | $7200 \times 10^{5}$ |
| Se | — | $0–1.0 \times 10^{7}$ | $3.0–220 \times 10^{4}$ | $790 \times 10^{5}$ |
| Cd | — | $2.3–26 \times 10^{6}$ | — | — |

* Data from Von Damm *et al.* (1985); extrapolated from low-temperature system. † Data from Von Damm *et al.* (1985); high-temperature system. ‡ Data from Elderfield & Schultz (1996).

*et al.* (1994), are at least one order of magnitude lower than those given by Von Damm *et al.* (1985), Mg, Al and Zn being exceptions to this.

Elderfield & Schultz (1996) also pointed out that there is considerable uncertainty in the relationship between axial high-temperature effluent with a deeper source and other hydrothermal activity, such as off-axis low-temperature effluent with a shallower source. For example, the alkalis Li, K, Rb and Cs have large high-temperature fluxes, but these are balanced by low-temperature removal as the crust is altered as it moves away from the ridge crests—see Table 6.14. Further, hydrothermal circulation on the ridge flanks is not well quantified. Kadko *et al.* (1994) estimated the flank fluxes for a series of elements, and the data are listed in Table 6.15. According to Elderfield & Schultz (1996) these flank fluxes should be regarded as no more than very approximate estimates, but it is

**Table 6.14** High-temperature axial hydrothermal vent input fluxes and low-temperature crustal alteration removal fluxes for the alkalis; units, mol yr$^{-1}$ (data from Elderfield & Schultz, 1996).

| Element | High-temperature input flux | Low-temperature removal flux |
|---|---|---|
| Li | $1.2–3.9 \times 10^{10}$ | $0.2–1.1 \times 10^{10}$ |
| K | $2.3–6.9 \times 10^{11}$ | $1.0–6.4 \times 10^{11}$ |
| Rb | $2.6–9.5 \times 10^{8}$ | $1.9–2.8 \times 10^{8}$ |
| Cs | $2.9–6.0 \times 10^{6}$ | $2.0–2.3 \times 10^{6}$ |

apparent that for some elements the hydrothermal source–sink relationship appears to switch character between the axial and off-axis environments. For example, the axial source fluxes of B and Ba are approximately balanced by the same order of magnitude ridge-flank sinks.

**Table 6.15** Ridge flank hydrothermal fluxes; units, mol yr$^{-1}$ (data from Elderfield & Schultz, 1996, after Kadke *et al.*, 1994).

| Element | Flank flux | Source (+) or sink (−) |
|---------|-----------|------------------------|
| B | $0.19–1.9 \times 10^{10}$ | − |
| U | $9.7 \times 10^{6}$ | − |
| Mg | $0.07–1.1 \times 10^{12}$ | − |
| Ca | $2.0–5.5 \times 10^{11}$ | + |
| Ba | $2.0 \times 10^{8}$ | − |
| Si | $1.3–1.8 \times 10^{12}$ | + |
| P | $3.2 \times 10^{9}$ | − |
| S | $8.0 \times 10^{12}$ | + |

**Table 6.16** Hydrothermal plume removal fluxes from sea water; units, mol yr$^{-1}$ (data from Elderfield & Schultz, 1996).

| Element | Plume removal flux | Fluvial input flux |
|---------|-------------------|---------------------|
| Cr | $4.8 \times 10^{7}$ | $6.3 \times 10^{8}$ |
| V | $4.3 \times 10^{8}$ | $5.9 \times 10^{8}$ |
| As | $1.8 \times 10^{8}$ | $8.8 \times 10^{8}$ |
| P | $1.1 \times 10^{10}$ | $3.3 \times 10^{10}$ |
| Mo | $1.9 \times 10^{6}$ | $2.0 \times 10^{8}$ |
| Be | $1.7 \times 10^{6}$ | $3.7 \times 10^{7}$ |
| Ce | $1.0 \times 10^{6}$ | $1.9 \times 10^{7}$ |
| Nd | $6.3 \times 10^{6}$ | $9.2 \times 10^{6}$ |
| Lu | $0.6 \times 10^{5}$ | $1.9 \times 10^{5}$ |
| U | $4.3 \times 10^{7}$ | $3.6 \times 10^{7}$ |

Three other factors must be taken into account when assessing the extent to which high-temperature axial hydrothermal fluxes affect the composition of sea water. Two of these factors are associated with the fact that hydrothermal exhalations form plumes above the venting sites as a result of the negative buoyancy of the hot hydrothermal fluids that rise above the sea floor. The third is related to the formation of sediments deposited around the venting centres.

### 6.3.1 Plume entrainment of sea water

As they rise the high-temperature fluids associated with axial hydrothermal venting systems entrain ambient sea water, with a consequent continuous increase in plume volume until neutral buoyancy is achieved, following which the plume disperses laterally. According to Elderfield & Schultz (1996) the water flux associated with this plume entrainment is $1.8 \times 10^{17}$ to $3.4 \times 10^{17}$ kg yr$^{-1}$, which, when compared with the axial hydrothermal flux ($\sim 3 \times 10^{13}$ kg yr$^{-1}$), yields an entrainment ratio, i.e. the ratio of entrained sea water to high-temperature fluid, of $\sim 10^{4}$. The global effect of this is that a very large volume of sea water passes through the plumes and is equivalent to an ocean water recycling time of $4 \times 10^{3}$ to $8 \times 10^{3}$ yr. There are, therefore, two seawater recycling stages associated with high-temperature axial hydrothermal activity. **Stage 1**, in which sea water recycles via circulation through the ridge system and changes composition as a result of high-temperature reactions involving the rocks it comes into contact with, to emerge at the sea bed as hydrothermal plumes. **Stage 2**, in which sea water circulates

through the hydrothermal plumes in the water column, and changes composition by reacting with the plumes. These two stages operate on different time-scales. Thus, although ocean water circulates through the mid-ocean ridge system in $\sim 10 \times 10^{6}$ yr, it is recycled through the hydrothermal plumes on a time-scale of thousands of years. This plume recycling has very important geochemical implications for seawater composition. The plumes are rich in particulates, e.g. from the oxidation and precipitation of $Fe^{2+}$ in the high-temperature venting solutions, which scavenge a wide range of dissolved trace metals, not only from the venting solutions themselves but also from the ambient sea water that is recycled through them. As a result of this, the hydrothermal plumes act as important chemical sinks for many of the dissolved trace metals in sea water, including oxyanions such as Mo, V and Cr for which hydrothermal activity is not a significant source—see Table 6.16. It may be concluded, therefore, that although hydrothermal activity at the spreading ridges acts as a source for many trace metals to the sea water (stage 1) when plume entrainment (stage 2) is included in the overall system, hydrothermal activity can act as a net water column (i.e. oceanic) **sink**, and not a net **source**, for some trace metals dissolved in sea water.

### 6.3.2 Plume dispersion

A major problem in estimating the global ocean importance of hydrothermal exhalations lies in assessing the extent to which the plumes, which may change composition as a result of seawater entrainment, are dispersed away from the venting centres. Edmond *et al.* (1982) combined data on the distributions of hydrothermally derived sediments and ridge-produced ³He to demonstrate that the dispersal of hydrothermal solutions is controlled by the global oceanic circulation at mid-water depths. Other workers have used Mn as a hydrothermal tracer. For example, Klinkhammer (1980) used dissolved Mn profiles to show the existence of hydrothermal vents on the EPR. He also reported anomalously high dissolved Mn concentrations in the bottom waters of the Guatemala Basin ~1000 km from the crest of the EPR. Hydrographic evidence suggested that this Mn anomaly is a hydrothermal signal, thus providing evidence of the dispersal of hydrothermal solutions over large distances. The extent to which the plume dispersal takes place may be controlled by the topography of the ridge region where the venting solutions are formed initially. In this context, Klinkhammer *et al.* (1985) used dissolved Mn as a tracer for hydrothermal activity on the Mid-Atlantic Ridge between 11°N and 26°N, and found that the plumes formed were confined to the rift valley and did not spill over into the adjacent deep ocean basins. Clearly, therefore, the extent to which hydrothermal plumes are dispersed from their point of origin can vary considerably depending on local conditions, and as a result of ridge geometry the solutions may be dispersed over much greater distances in the Pacific than in the Atlantic.

### 6.3.3  The formation of hydrothermal sediments

In addition to precipitation and deposition within the hydrothermal venting systems (e.g. as sulphides) some elements are incorporated within hydrothermal sediments surrounding the active ridge systems, and so are prevented from large-scale oceanic dispersal. These sediments, which are considered in Section 16.6.2, are especially rich in Fe, and may also retain a large fraction of other metals released from hydrothermal vents.

### 6.4  Relative magnitudes of the primary fluxes to the oceans

#### 6.4.1  Introduction

For many years it was thought that the oceans were a fluvially dominated system. This view has changed dramatically over the past few decades, however,

as the importance of atmospheric deposition and hydrothermal activity at the spreading centres has become increasingly recognized, and a quite different picture of the oceans has begun to emerge. In the three preceding sections an attempt has been made to estimate the fluxes of material delivered to the oceans from the major, global-scale, primary sources. These sources are fluvial, atmospheric and hydrothermal, although it must be remembered that hydrothermal activity also can act as a sink for some dissolved components as well as a source for others. The data sets for all these sources are still essentially crude, but it is now possible to make a number of general comparisons between them.

#### 6.4.2  Major elements and trace metals

##### 6.4.2.1  *Hydrothermal versus fluvial dissolved fluxes*

It was pointed out in Section 6.3 that the effects of axial high-temperature hydrothermal activity on the elemental chemistry of sea water are manifest in two principal ways: (i) by the plumes venting from the sea bed following the recycling of sea water through the hydrothermal axial system (stage 1); and (ii) by the plumes entraining sea water in the water column (stage 2).

Elderfield & Schultz (1996) derived a 'best estimate' of high-temperature axial hydrothermal plume fluxes (see Table 6.13, column C) and compared these with fluvial fluxes (see Table 6.13, column D); this comparison is illustrated in Fig. 6.5, in which the elements are ranked on the basis of the hydrothermal : river flux ratios in the order $10^2:1$, $1:1$, $1:10^2$. Those elements for which the concentrations in hydrothermal fluids exceed those from river water by a factor of $\gg 10^3$ have high-temperature hydrothermal fluxes that are approximately equal to, or exceed, those from fluvial inputs. These elements include Mn and Fe and the alkalis Li, K, Rb and Cs. Of these, Fe is taken up in vent chimney sulphide deposits and hydrothermal sediments, and the alkalis are removed in altered oceanic crust (see Section 6.3). Those elements that have hydrothermal fluxes between $1:1$ and $1:10^2$ of their fluvial fluxes include Si, K, B, Ba, Cu and Zn. In contrast, for Al and P the fluvial fluxes exceed the hydrothermal fluxes by two orders of magnitude. It is also evident that axial hydrothermal activity is an important sink for Mg in the oceans. Elderfield & Schultz (1996) also compared

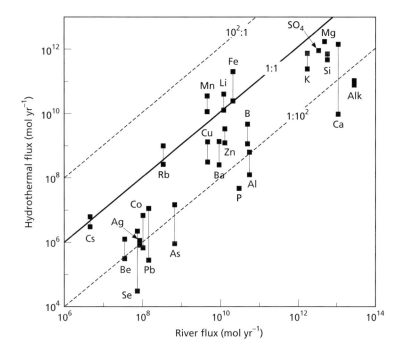

**Fig. 6.5** Comparison between hydrothermal and fluvial dissolved fluxes to the World Ocean (from Elderfield & Schultz, 1996).

the removal fluxes by plume entrainment in the water column with their fluvial fluxes—see Table 6.16. From this table it is evident that the plume removal fluxes of V, As, P, U and Nd are the same order of magnitude as their fluvial fluxes, indicating the importance of the hydrothermal plumes as a sink for these elements.

There are considerable uncertainties in the estimates of the elemental fluxes involved in hydrothermal activity. None the less, when axial hydrothermal activity is envisaged as a two-stage process, encompassing the recycling of sea water, firstly through the rocks at spreading ridge centres and secondly through the resulting venting plumes in the water column, then it must be considered to be an extremely important source or sink term in oceanic chemistry.

### 6.4.2.2 *Atmospheric versus fluvial fluxes*

A number of authors have attempted to make general comparisons between the strengths of the fluvial and atmospheric trace-metal fluxes to the oceans. These two fluxes are of special interest because they both deliver material directly to the surface ocean, where it can become involved in biogeochemical reactions in the particle-rich, biologically active, euphotic zone. In order to discuss fluvial–atmospheric flux compari-

sons it is convenient to divide the oceanic environment into coastal and open-ocean waters, because the source signals to the two regions vary considerably in strength.

*Coastal regions.* Because most of the particulate material transported by rivers is at present trapped in the estuarine environment (see Section 6.1.2.2), a number of the fluvial versus atmospheric flux comparisons for coastal regions have restricted the fluvial input to the dissolved phase. A summary of the data for some of these comparisons is given in Table 6.17(a) in the form of the ratios of the *total* atmospheric to the dissolved fluvial fluxes. From this table it is apparent that in the coastal environments studied the atmospheric input of trace metals is similar to, or greater than, that arising from river run-off. A significant portion of the AEEs in aerosols are soluble in sea water, however, and this should be taken into account when comparing the atmospheric to the fluvial dissolved inputs. Windom (1981) compared the dissolved fluvial and total atmospheric fluxes to the South Atlantic Bight, and in order to make the comparison more viable the seawater-soluble fractions of the atmospheric inputs have been calculated on the basis of the solubilities to be expected from polluted aerosols of the type thought to be deposited

(a) Ratio of total atmospheric flux to dissolved fluvial flux.

**Table 6.17** Relative fluvial and atmospheric fluxes to some coastal oceanic regions.

| Element | South Atlantic Bight* | New York Bight* | North Sea* | Western Mediterranean† |
|---|---|---|---|---|
| Fe | 5.8 | 6.4 | 1.7 | — |
| Mn | 0.6 | — | 0.8 | — |
| Cu | 1.9 | — | 1.9 | — |
| Ni | 1.7 | — | 1.3 | — |
| Pb | 9.5 | 20 | 6.8 | 6.2 |
| Zn | 2.3 | 3.1 | 1.9 | 0.8 |
| Cd | 2.7 | 3.1 | 1.1 | — |
| As | 2.1 | 1.0 | 1.7 | — |
| Hg | 22 | — | 2.1 | 0.8 |

(b) Estimated dissolved fluvial and soluble atmospheric fluxes to the South Atlantic Bight; units, $10^6$ g.

| Element | Fluvial flux* | Total atmospheric flux* | Soluble atmospheric flux‡ | Ratio: soluble atmospheric flux to fluvial flux |
|---|---|---|---|---|
| Fe | 950 | 5500 | 413 | 0.43 |
| Mn | 91 | 57 | 26 | 0.29 |
| Cu | 110 | 210 | 63 | 0.57 |
| Ni | 220 | 370 | 185 | 0.85 |
| Pb | 65 | 620 | 310 | 4.8 |
| Zn | 310 | 710 | 497 | 1.6 |
| Cd | 3 | 8 | 6.8 | 2.3 |

* Data from Windom (1981). † Data from Buat-Menard (1983). ‡ Solubility data from Table 6.11.

over the region. These data are listed in Table 6.17(b), and it can be seen that even after making this adjustment the dissolved atmospheric fluxes of Pb, Zn and Cd exceed their fluvial fluxes—see also Martin *et al.* (1989), who calculated similar budgets for Pb, Cu and Cd in the western Mediterranean Sea.

It may be concluded, therefore, that in coastal regions solubilization from atmospheric particulates can play a substantial role in governing the dissolved trace-metal burden of the surface sea waters. The overall influence that the atmospherically transported trace metals exert on the chemistry of the coastal waters will depend on physical processes, such as vertical and horizontal water mixing, and on biogeochemical cycling.

*Open ocean.* There have been a number of attempts to assess the relative magnitudes of the air to sea

fluxes of trace metals on a World Ocean scale, and two of these are summarized below.

1 In a major review, Duce *et al.* (1991) pointed out that both the 'dry' and the 'wet' depositional modes by which atmospheric trace metals reach the sea surface are particle-size dependent (see Section 4.2.1.3). On this basis, the authors divided atmospheric particulate trace metals into two groups. *Group 1* included the AEEs Pb, Cd, Zn, Cu, Ni, As, Hg and Sn, which are released into the atmosphere mainly from high-temperature processes and are found on particles with MMDs <1 μm. *Group 2* included the NEEs Al, Fe, Si and P. These are generated mainly from low-temperature crustal weathering and principal components of the mineral aerosol are found on particles with MMDs >1 μm. The global atmospheric fluxes for Pb and the mineral aerosol were then calculated by taking into account 'dry' deposition velocities and precipitation scavenging

ratios for each of a series of oceanic grids, and the concentration fields obtained are illustrated in Fig. 6.6(a,b). The atmospheric deposition fluxes of the other group 1 metals were normalized to that of Pb,

**Fig. 6.6** Global fluxes of (a) lead (units, μg m⁻² yr⁻¹) and (b) the mineral aerosol (mg m⁻² yr⁻¹) to the World Ocean (from Duce *et al.*, 1991).

and those of the group 2 metals to that of the mineral aerosol, and the data are summarized in Table 6.18. These data are for the total fluxes of the metals reaching the sea surface, and with respect to the entry of the metals into the biogeochemical cycles it is useful to establish the fractions of the total fluxes that are soluble in sea water. These were obtained using a series of metal solubility factors, and Duce *et al.*

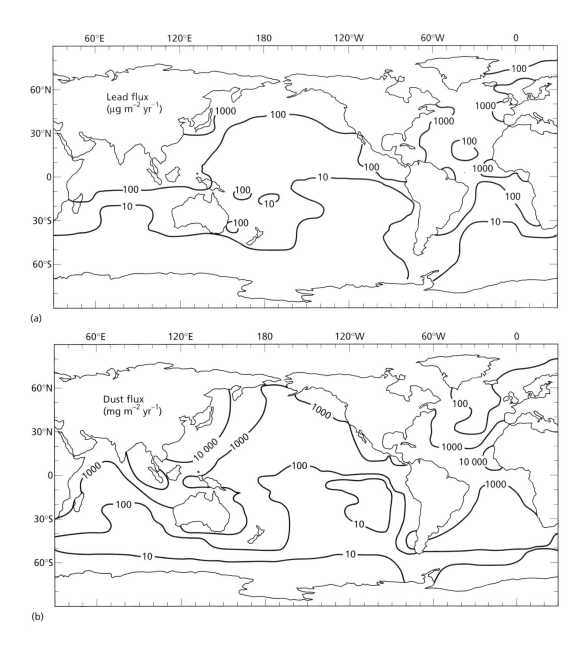

(a)

(b)

**Table 6.18** Atmospheric deposition fluxes to the ocean surface*.

| Element | Flux unit | North Atlantic | South Atlantic | North Pacific | South Pacific | North Indian | South Indian | Global total |
|---|---|---|---|---|---|---|---|---|
| Mineral aerosol | $10^{12}$ g yr$^{-1}$ | 220 | 24 | 480 | 39 | 100 | 44 | 910 |
| Al | $10^{12}$ g yr$^{-1}$ | 18 | 1.9 | 38 | 3.1 | 8.2 | 3.5 | 73 |
| Fe | $10^{12}$ g yr$^{-1}$ | 7.7 | 0.8 | 17 | 1.4 | 3.6 | 1.5 | 32 |
| Si | $10^{12}$ g yr$^{-1}$ | 68 | 7.4 | 150 | 12 | 31 | 14 | 280 |
| P | $10^{9}$ g yr$^{-1}$ | 230 | 25 | 500 | 41 | 110 | 46 | 950 |
| Ni | $10^{9}$ g yr$^{-1}$ | 14–19 | 1.1–1.4 | 4.5–5.9 | 0.6–0.7 | 1.2–1.6 | 0.6–0.8 | 22–29 |
| Cu | $10^{9}$ g yr$^{-1}$ | 10–39 | 0.8–2.9 | 3.2–12 | 0.4–1.5 | 0.9–3.3 | 0.4–1.6 | 16–60 |
| Pb | $10^{9}$ g yr$^{-1}$ | 57 | 4.2 | 18 | 2.2 | 4.8 | 2.4 | 89 |
| Zn | $10^{9}$ g yr$^{-1}$ | 29–170 | 2.1–13 | 9.1–54 | 1.1–6.6 | 2.4–14 | 1.2–7.2 | 45–270 |
| As | $10^{9}$ g yr$^{-1}$ | 5.1–4.6 | 0.4–0.34 | 1.6–1.4 | 0.2–0.18 | 0.4–0.4 | 0.2–0.2 | 7.9–7.1 |
| Cd | $10^{9}$ g yr$^{-1}$ | 1.5–2.9 | 0.11–0.21 | 0.47–0.90 | 0.06–0.11 | 0.12–0.24 | 0.06–0.12 | 2.3–4.5 |

* Data from Duce *et al.* (1991). Fluxes for Al, Fe, Si and P were normalized to that of the mineral aerosol. Fluxes for Ni, Cu, Zn, As and Cd were normalized to that of Pb; the first number in each pair was obtained using element to Pb ratios obtained from global atmospheric emissions, and the second was derived from element to Pb ratios measured in aerosol and rain samples.

(1991) then made a general comparison between particulate and dissolved atmospheric and fluvial fluxes to the oceans. The results are summarized in Table 6.19. From this table it can be seen that rivers are the most important sources of **particulate** trace metals to the oceans. It must be pointed out, however, that the fluvial data given in Table 6.19 are for gross fluxes; i.e. no correction has been made for estuarine retention, by which ~90% of the particulate fluvial input to the land–sea margins is stored. For the **dissolved** fluxes a different picture emerges. For P and Fe, and the mainly anthropogenic AEEs Cu, Ni and As, the dissolved atmospheric inputs appear to be similar for both fluvial and atmospheric sources, and the dissolved atmospheric inputs of Pb, Cd and Zn exceed those from fluvial sources. At present, ~75% of the atmospheric Pb arises from gasoline sources and the atmospheric input of the metal may decrease by a factor of about three within the next few decades as non-leaded gasoline continues to be phased in. Despite this, atmospheric Pb sources will still exceed fluvial sources.

2 Chester & Murphy (1990) made an estimate of the relative magnitudes of the fluvial and atmospheric sources of a series of dissolved trace metals to the World Ocean, the data being expressed as a flux to the sea surface in units of μg cm$^{-1}$ yr$^{-1}$. To make these estimates the following data were used.

(a) Net fluvial global-scale fluxes, calculated using data derived from a combination of the average

**Table 6.19** Atmospheric versus fluvial deposition of elements to the World Ocean; units, $10^{9}$ g yr$^{-1}$ (data from Duce *et al.*, 1991).

| Element | Atmospheric input | | Fluvial input | |
|---|---|---|---|---|
| | Dissolved | Particulate | Dissolved | Particulate |
| Fe | $3.2 \times 10^{3}$ | $29 \times 10^{3}$ | $1.1 \times 10^{3}$ | $110 \times 10^{3}$ |
| P | 310 | 640 | Total 300* | Total 300* |
| Ni | 8–11 | 14–17 | 11 | 1400 |
| Cu | 14–45 | 2–7 | 10 | 1500 |
| Pb | 80 | 10 | 2 | 1600 |
| Zn | 33–170 | 11–60 | 6 | 3900 |
| Cd | 1.9–3.3 | 0.4–0.7 | 0.3 | 15 |
| As | 2.3–5.0 | 1.3–2.9 | 10 | 80 |

* Total P input to marine sediments.

river-water concentrations given in Table 3.4 and the independently obtained estuarine effluxes listed in Table 6.6, were assumed to be spread in an even layer of 10.6 cm over the World Ocean surface (see e.g. Collier & Edmond, 1984). The net fluxes obtained are reproduced in Table 6.20.

(b) For the atmospheric inputs it was assumed that the deposition fluxes for the North Atlantic, North Pacific and South Pacific given in Table 6.10 offered a representative cross-section of those affecting all marine regions, and a weighted average deposition obtained from them was extrapolated to the whole ocean surface. This was

**Table 6.20** Net fluvial and atmospheric fluxes to the global sea surface; units, $\mu g\,cm^{-2}\,yr^{-1}$ (from Chester & Murphy, 1990).

| Element | Net global fluvial dissolved flux | | Net global total atmospheric input | | Estimated average seawater solubility from aerosols | Net global dissolved atmospheric flux |
|---|---|---|---|---|---|---|
| | Chester & Murphy (1990) | Collier & Edmond (1984) | Chester & Murphy (1990) | Collier & Edmond (1984) | | |
| Al | 0.27 | 0.13–0.67 | 1.85 | 0.27–5.4 | 5% | 0.088 |
| Fe | 0.085 | <0.73 | 1.0 | 1.12–3.35 | 7.5% | 0.075 |
| Mn | 0.085 | 0.093 | 0.021 | 0.016–0.044 | 35% | 0.0074 |
| Ni | 0.0082 | 0.0035 | <0.02 | 0.003 | 40% | <0.008 |
| Co | 0.0021 | — | 0.00068 | — | 22.5% | 0.00015 |
| Cr | 0.0052 | — | <0.014 | — | 10% | <0.0014 |
| V | 0.010 | — | <0.010 | — | 25% | <0.0028 |
| Cu | 0.018 | 0.019 | 0.010 | 0.00064–0.0095 | 30% | 0.0033 |
| Pb | 0.001 | — | 0.074 | — | 30% | 0.022 |
| Zn | 0.0079 | <0.0065 | 0.055 | 0.022–0.013 | 45% | 0.025 |
| Cd | 0.00088 | 0.00022 | <0.00096 | 0.0023 | 80% | <0.00077 |

then adjusted for the average seawater solubilities of the metals for mixed-source, open-ocean aerosols (Table 6.11), to yield the effective dissolved atmospheric flux. The atmospheric fluxes obtained in this way are reproduced in Table 6.20. Collier & Edmond (1984) also made a general comparison between fluvial and atmospheric trace metal inputs to the global ocean, and some of their data are included in Table 6.20.

A number of conclusions can be drawn from these first order comparisons, all of which, with the exception of that for the Cu fluxes, are in general agreement with the findings of Duce *et al.* (1991).

1 For Al, Mn and Cu the dissolved fluvial fluxes to the global ocean surface exceed the soluble atmospheric fluxes by an order of magnitude; for Cu, Duce *et al.* (1991) found that the two fluxes were similar.

2 For Fe and Ni the dissolved fluvial and soluble atmospheric inputs are the same order of magnitude.

3 For Zn and Pb the soluble atmospheric fluxes dominate the dissolved inputs of the two metals to the surface ocean.

The fluxes given in Table 6.20 are for the global ocean but the concentrations of trace metals in the atmosphere, and therefore their deposition fluxes, vary both spatially and temporally in the marine atmosphere. These variations can range over several orders of magnitude, and in an attempt to take account of them, Chester & Murphy (1990) calculated both

soluble atmospheric and dissolved fluvial fluxes on a regional basis. The data are listed in Table 6.21, from which it can be seen that soluble atmospheric deposition has the most influence on the surface waters of the North Atlantic and the least on those of the South Pacific. In terms of the open-ocean areas considered it is therefore the North Atlantic that receives the highest input of soluble trace metals from the atmosphere. The North Atlantic, however, also receives the highest inputs of fluvial dissolved metals, and the soluble atmospheric fluxes of Mn, Ni, Co, Cr, V and Cu are still an order of magnitude lower than the dissolved fluvial fluxes. For Al, Fe and Cd the two dissolved fluxes are the same order of magnitude, but for Zn and Pb the atmospheric fluxes are dominant.

Both river run-off and atmospheric transport deliver trace metals to the surface ocean, the difference between them being that river run-off enters the ocean reservoir at the land–sea margins, whereas transport via the atmosphere operates in both coastal and open-ocean regions. Atmospheric fingerprints imposed on the distributions of dissolved trace metals therefore should be more apparent in open-ocean than coastal waters. The extent to which these fingerprints are established, however, depends on the relative importance of other sources to surface waters. It will be shown in Chapter 11 that trace metals in the oceanic water column can be divided into 'scavenged-type' and 'nutrient-type' on the basis of their vertical

**Table 6.21** Atmospheric dissolved versus fluvial dissolved inputs to regions of the World Ocean; units, $\mu g\, cm^{-2}\, yr^{-1}$ (from Chester & Murphy, 1990).

| Element | North Atlantic | | North Pacific | | South Pacific | |
|---|---|---|---|---|---|---|
| | Fluvial dissolved flux* | Atmospheric dissolved flux | Fluvial dissolved flux | Atmospheric dissolved flux | Fluvial dissolved flux | Atmospheric dissolved flux |
| Al | 0.56 | 0.25 | 0.24 | 0.09 | 0.12 | 0.0066 |
| Fe | 0.18 | 0.24 | 0.076 | 0.062 | 0.038 | 0.0035 |
| Mn | 0.18 | 0.025 | 0.076 | 0.004 | 0.038 | 0.0013 |
| Ni | 0.017 | 0.008 | 0.0074 | — | 0.0037 | — |
| Co | 0.0044 | 0.00061 | 0.0019 | 0.00004 | 0.0009 | 0.000006 |
| Cr | 0.011 | 0.0014 | 0.0047 | — | 0.0023 | — |
| V | 0.022 | 0.0043 | 0.0094 | 0.0017 | 0.0047 | — |
| Cu | 0.037 | 0.0075 | 0.016 | 0.0021 | 0.0079 | 0.0013 |
| Pb | 0.0022 | 0.093 | 0.0009 | 0.0022 | 0.0005 | 0.0004 |
| Zn | 0.016 | 0.059 | 0.0071 | 0.029 | 0.0035 | 0.0026 |
| Cd | 0.0018 | 0.0016 | 0.0008 | 0.00023 | 0.0004 | — |

* No attempt has been made to adjust the North Atlantic fluvial flux for European and North American anthropogenic inputs.

water-column distributions, and this has a strong effect on the manner in which source fingerprints are recorded in open-ocean surface waters.

The '*scavenged-type*' trace metals have vertical water-column profiles that exhibit a surface enrichment and a subsurface depletion, indicative of a surface source and scavenging throughout the water column. It is these 'scavenged-type' trace metals, which include Al, Mn and Pb, that should retain the most identifiable open-ocean surface water fingerprints from atmospheric deposition, and such fingerprints have been reported in the literature. Perhaps the most distinctive of these have been found for Pb. It was shown above that the input of Pb to the oceans is dominated by atmospheric deposition. This is confirmed by the surface water distributions of dissolved Pb between the major oceans, which strongly reflects the magnitude of the atmospheric signal; i.e. the concentrations of dissolved Pb in the mixed layer of the North Atlantic are two to three times higher than those in the North Pacific. Further, the highest concentrations of dissolved Pb in the North Pacific are found in the central gyre, which can be explained only in terms of an atmospheric input (see Section 11.4). Atmospheric fingerprints in surface waters have also been found for Al and Mn. For example, Orians & Bruland (1985) showed that the concentra-

tions of dissolved Al in the North Atlantic were around eight times higher than those in the North Pacific, which is consistent with a stronger atmospheric signal to the North Atlantic. To confirm this, Kremling (1985) demonstrated that the input of Saharan dust 'pulses' can raise surface water dissolved Al concentrations in the region underlying the Northeast Trades. In addition, both Kremling (1985) and Statham & Burton (1986) provided evidence that the highest concentrations of dissolved Mn in the North Atlantic were found in the latitudes that receive large inputs of Saharan dusts.

The '*nutrient-type*' trace metals have vertical water-column profiles that exhibit a surface depletion and a subsurface enrichment, and are maintained by the involvement of the metals in the biological processes involved in the 'nutrient loop' (see Section 9.1), during which they are released at depth and recycled to surface waters. For this type of trace metal, therefore, the atmospheric fingerprints will be more difficult to identify in open-ocean surface waters because of their recycled supply from depth. Cadmium and Ni are examples of 'nutrient-type' trace metals. Bruland (1980) suggested that the main sources of these two trace metals to open-ocean surface waters were atmospheric deposition and upwelling from the vertical transport of deep water, and he estimated the magnitudes of the fluxes associated with these two supply mechanisms for Cd and Ni, together with those for Cu and Pb. The data are given in Table 6.22, from which it can be seen that the two fluxes vary in importance from one trace metal to another.

1  For the 'nutrient-type' trace metals Cd and Ni, the atmospheric flux is around one order of magnitude less than the upwelling flux. For these two metals, therefore, although atmospheric deposition is the principal *primary* surface water source, it is overwhelmed by Ni and Cd supplied from below by upwelling, with the result that the atmospheric fingerprint in surface waters will be masked.

2  Copper is taken up in surface waters by processes associated with biota but it also undergoes scavenging in deep waters, and so has a mixed 'nutrient–scavenging type' water column distribution. For Cu, the atmospheric and upwelling fluxes are approximately equal.

3  For the 'scavenged-type' trace metal Pb, deposition via the atmosphere is the only supply to the surface waters, and the atmospheric fingerprint should be relatively undisturbed—see above.

It may be concluded, therefore, that although atmospheric deposition may be the principal primary source of the dissolved 'nutrient-type' trace metals to open-ocean surface waters, the actual concentrations of the metals can be severely modified, and sometimes completely dominated, by secondary sources arising from the upwelling flux.

### 6.4.2.3  Comparison of all primary major element and trace metal fluxes to the oceans

The various elemental flux comparisons made above involved a variety of techniques, and it is worthwhile to attempt to synthesize the data in order to identify any broad trends. For this purpose a summary of the flux data is presented in Table 6.23.

1  In Table 6.23, column A, the gross total (i.e. dissolved and particulate) fluvial and total atmospheric inputs of the elements are listed. On this basis, it can be seen that fluvial inputs totally dominate the supply of the elements to the oceans. For the gross fluvial

**Table 6.22** Atmospheric and vertical mixing fluxes of some trace elements to the surface waters of the eastern North Pacific; units, $nmol\,cm^{-2}\,yr^{-1}$ (data from Bruland, 1980).

| Element | Atmospheric flux | Vertical mixing flux |
|---|---|---|
| Cd | 0.018 | 0.14 |
| Ni | 0.043 | 0.80 |
| Cu | 0.16 | 0.21 |
| Pb* | 0.18 | −0.01 |

* Pb is enriched in surface waters relative to deep water and thus has a negative vertical mixing term.

**Table 6.23** Elemental fluxes to the World Ocean from the major sources; units, $mol\,yr^{-1}$.

| | A | | B | | | C | | |
|---|---|---|---|---|---|---|---|---|
| Element | Fluvial gross flux: total (particulate + dissolved) | Atmospheric flux: total (particulate + soluble) | Fluvial net dissolved flux | Hydrothermal axial dissolved flux | Atmospheric soluble flux | Fluvial gross particulate flux | Fluvial net particulate flux | Atmospheric particulate flux |
| Al | $54 \times 10^{12}$ | $0.25 \times 10^{12}$ | $3.5–6.0 \times 10^{10}$ | $1.2–6.0 \times 10^8$ | $1.2 \times 10^{10}$ | $54 \times 10^{12}$ | $5.4 \times 10^{12}$ | $0.24 \times 10^{12}$ |
| Fe | $13.3 \times 10^{12}$ | $0.065 \times 10^{12}$ | $0.54–2.3 \times 10^{10}$ | $2.3–19 \times 10^{10}$ | $0.49 \times 10^{10}$ | $13 \times 10^{12}$ | $1.3 \times 10^{12}$ | $0.06 \times 10^{12}$ |
| Mn | $30 \times 10^{10}$ | $0.14 \times 10^{10}$ | $0.49–0.55 \times 10^{10}$ | $1.1–3.4 \times 10^{10}$ | $0.05 \times 10^{10}$ | $0.29 \times 10^{12}$ | $0.029 \times 10^{12}$ | $0.001 \times 10^{12}$ |
| Ni | $2.4 \times 10^{10}$ | $<0.12 \times 10^{10}$ | $0.05 \times 10^{10}$ | — | $<0.05 \times 10^{10}$ | $2 \times 10^{10}$ | $0.23 \times 10^{10}$ | $<0.075 \times 10^{10}$ |
| Co | $0.56 \times 10^{10}$ | $0.004 \times 10^{10}$ | $0.011–0.013 \times 10^{10}$ | $6.6–68 \times 10^5$ | $1 \times 10^7$ | $0.52 \times 10^{10}$ | $0.05 \times 10^{10}$ | $0.003 \times 10^{10}$ |
| Cr | $3.1 \times 10^{10}$ | $<0.10 \times 10^{10}$ | $0.036 \times 10^{10}$ | — | $<0.01 \times 10^{10}$ | $3 \times 10^{10}$ | $0.3 \times 10^{10}$ | $<0.09 \times 10^{10}$ |
| V | $5.2 \times 10^{10}$ | $<0.07 \times 10^{10}$ | $0.07 \times 10^{10}$ | — | $<0.02 \times 10^{10}$ | $5 \times 10^{10}$ | $0.5 \times 10^{10}$ | $<0.05 \times 10^{10}$ |
| Cu | $2.5 \times 10^{10}$ | $0.06 \times 10^{10}$ | $0.1–0.5 \times 10^{10}$ | $3–13 \times 10^8$ | $2 \times 10^8$ | $2.4 \times 10^{10}$ | $0.24 \times 10^{10}$ | $0.04 \times 10^{10}$ |
| Pb | $0.75 \times 10^{10}$ | $0.13 \times 10^{10}$ | $0.2–1.5 \times 10^8$ | $2.7–110 \times 10^5$ | $4 \times 10^8$ | $0.75 \times 10^{10}$ | $0.075 \times 10^{10}$ | $0.09 \times 10^{10}$ |
| Zn | $6.0 \times 10^{10}$ | $0.31 \times 10^{10}$ | $0.04–1.4 \times 10^{10}$ | $0.12–0.32 \times 10^{10}$ | $0.14 \times 10^{10}$ | $6 \times 10^{10}$ | $0.6 \times 10^{10}$ | $0.17 \times 10^{10}$ |
| Cd | $0.15 \times 10^9$ | $<0.03 \times 10^9$ | $3 \times 10^7$ | — | $<2 \times 10^7$ | $0.14 \times 10^9$ | $0.014 \times 10^9$ | $<0.7 \times 10^9$ |

inputs, however, estuarine trapping is ignored, and the situation changes when this is taken into account and also when dissolved hydrothermal inputs are considered.

2 In Table 6.23, column B, the dissolved hydrothermal, dissolved fluvial and soluble atmospheric fluxes of a series of elements are listed. From these data it is apparent that all three sources are important in the supply of the dissolved elements to the oceans. For dissolved Fe and Mn, however, hydrothermal inputs dominate, and for dissolved Pb the atmosphere is the major source.

3 In Table 6.23, column C, the gross and net fluvial and atmospheric particulate inputs of the elements are listed. On the basis of the gross inputs it can be seen that fluvial transport easily dominates the supply of the particulate elements to the oceans. Much of this particulate material, however, is trapped at the present day in estuaries, and a very different picture emerges when the net fluvial inputs are considered. Thus, on this basis, although the net particulate fluvial fluxes of most of the elements are still at least one order of magnitude in excess of the particulate atmospheric fluxes, for Pb and Zn the particulate atmospheric inputs are the same order of magnitude as the net river fluxes.

### 6.4.3 Nutrients and synthetic organic compounds

#### 6.4.3.1 Nitrogen nutrients

According to Duce *et al.* (1991) there are two classes of nitrogen species that can be utilized as nutrients.

1 **Oxidized nitrogen species**, including aerosol $NO_3^-$ and gas phase oxides of nitrogen (NO, $NO_2$, $HNO_3$ and related species). The total global source of oxidized nitrogen species is $\sim 50 \times 10^{12}\,g\,N\,yr^{-1}$, with fossil fuel combustion ($\sim 21 \times 10^{12}\,g\,N\,yr^{-1}$) and biomass burning ($\sim 4 \times 10^{12}$–$\sim 12 \times 10^{12}\,g\,N\,yr^{-1}$) sources being dominant. Over the ocean the sources of oxidized nitrogen species include NO production by lighting ($\sim 8 \times 10^{12}\,g\,N\,yr^{-1}$), and the downmixing of NO from the stratosphere ($\sim 0.5 \times 10^{12}\,g\,N\,yr^{-1}$). The only established oceanic source of oxidized nitrogen is from the photolysis of nitrite in surface waters, although this is relatively minor. Duce *et al.* (1991) concluded that over 80% of the sources of oxidized nitrogen species are found on land, particularly from anthropogenic sources in the Northern Hemisphere;

for example, $\sim 95\%$ of the nitrogen emitted from combustion sources to the global atmosphere originates in the Northern Hemisphere.

2 **Reduced nitrogen species**, including $NH_4^+$, gaseous $NH_3$ and related organic nitrogen species. The global budget for ammonia is only poorly understood; the most important sources include the breakdown of urea from domestic animals, soil emissions and biomass burning, which yield a total of $\sim 54 \times 10^{12}\,g\,N\,yr^{-1}$.

Using a variety of concentration data, deposition velocities and scavenging ratios, Duce *et al.* (1991) estimated the fluxes of various N species to the global ocean. The flux data are summarized in Table 6.24(a, b) on the basis of individual ocean basins, and in Table 6.24(c) the atmospheric nitrogen inputs are compared with those from fluvial sources. A number of trends can be identified from the data in the tables, and these are summarized below.

1 The atmospheric input of total N to the global ocean is $\sim 30.3 \times 10^{12}\,g\,N\,yr^{-1}$ (flux, $\sim 87\,mg\,N\,m^{-2}\,yr^{-1}$), which is made up of roughly equal amounts of oxidized ($\sim 13.4 \times 10^{12}\,g\,N\,yr^{-1}$) (flux, $\sim 38\,mg\,N\,m^{-2}\,yr^{-1}$) and reduced ($\sim 16.8 \times 10^{12}\,g\,N\,yr^{-1}$) (flux, $\sim 49\,mg\,N\,m^{-2}\,yr^{-1}$) nitrogen species. With the exception of those for the Indian Ocean, the oceanic fluxes of the oxidized and reduced nitrogen species are essentially equal.

2 The total nitrogen fluxes for the three northern oceans ($\sim 22 \times 10^{12}\,g\,N\,yr^{-1}$) (flux, $\sim 140\,mg\,N\,m^{-2}\,yr^{-1}$) are considerably higher than those for the southern oceans ($\sim 9 \times 10^{12}\,g\,N\,yr^{-1}$) (flux, $\sim 46\,mg\,N\,m^{-2}\,yr^{-1}$), and the largest total N fluxes are found for the North Atlantic and North Pacific. As a result of these large differences, the impact that atmospheric nitrogen will have on primary production could vary dramatically from one oceanic region to another.

3 The 'wet' removal of $NH_4^+$ accounts for the largest flux of reduced nitrogen to the oceans. This may play a significant role in the cycling of nitrogen species in the marine atmosphere, and could serve as an important source of nutrients to the oceans.

4 The atmospheric deposition of nitrogen species ($\sim 30 \times 10^{12}\,g\,N\,yr^{-1}$) is twice that of the *natural* fluvial input ($\sim 14 \times 10^{12}\,g\,N\,yr^{-1}$), but there is a substantial fluvial flux of *anthropogenic* nitrogen and the total (i.e. natural + anthropogenic) fluvial flux ($\sim 21 \times 10^{12}\,g\,N\,yr^{-1}$) is similar to that for the atmospheric input. The fluvial data are for gross fluxes, however,

**Table 6.24** Atmospheric input of nutrient nitrogen species to the World Ocean (data from Duce *et al.*, 1991).

(a)  Total deposition; units, $10^9$ g N yr$^{-1}$.

| Species | Deposition mode | North Atlantic | South Atlantic | North Pacific | South Pacific | North Indian | South Indian | Total |
|---|---|---|---|---|---|---|---|---|
| **Oxidized nitrogen** | | | | | | | | |
| $NO_3^-$ Wet | | 2730 | 330 | 2700 | 1370 | 490 | 820 | 8400 |
| $NO_3^-$ Dry | | 1110 | 190 | 760 | 410 | 140 | 220 | 2800 |
| $HNO_3$ Wet | | 380 | 50 | 380 | 210 | 74 | 120 | 1200 |
| $HNO_3$ Dry | | 320 | 77 | 300 | 170 | 43 | 94 | 1000 |
| $NO_x$ Wet | | 0 | 0 | 0 | 0 | 0 | 0 | 0 |
| $NO_x$ Dry | | 32 | 1 | 2 | 3 | 0 | 1 | 40 |
| Subtotal oxidized | | 4570 | 650 | 4140 | 2160 | 750 | 1260 | 13400 |
| **Reduced nitrogen** | | | | | | | | |
| $NH_4^+$ Wet | | 2740 | 610 | 4050 | 1900 | 1170 | 970 | 11400 |
| $NH_4^+$ Dry | | 530 | 160 | 630 | 290 | 170 | 170 | 1950 |
| $NH_3$ Wet | | 820 | 180 | 1220 | 570 | 350 | 290 | 3400 |
| Subtotal reduced | | 4090 | 950 | 5900 | 2760 | 1690 | 1430 | 16800 |
| **Total nitrogen deposition** | | 8700 | 1600 | 10000 | 4900 | 2400 | 2700 | 30300 |

(b)  Fluxes; units, mg N m$^{-2}$ yr$^{-1}$.

| Species | Deposition mode | North Atlantic | South Atlantic | North Pacific | South Pacific | North Indian | South Indian | Total |
|---|---|---|---|---|---|---|---|---|
| **Oxidized nitrogen** | | | | | | | | |
| $NO_3^-$ Wet | | 49 | 8 | 33 | 14 | 42 | 13 | 159 |
| $NO_3^-$ Dry | | 20 | 4 | 9 | 4 | 11 | 3 | 51 |
| $HNO_3$ Wet | | 7 | 1 | 5 | 2 | 6 | 2 | 23 |
| $HNO_3$ Dry | | 6 | 2 | 4 | 2 | 4 | 2 | 20 |
| $NO_x$ Wet | | 0 | 0 | 0 | 0 | 0 | 0 | 0 |
| $NO_x$ Dry | | 1 | 0 | 0 | 0 | 0 | 0 | 0 |
| Subtotal oxidized | | 83 | 15 | 51 | 22 | 63 | 20 | 254 |
| **Reduced nitrogen** | | | | | | | | |
| $NH_4^+$ Wet | | 49 | 14 | 50 | 20 | 99 | 16 | 248 |
| $NH_4^+$ Dry | | 9 | 4 | 8 | 3 | 15 | 3 | 42 |
| $NH_3$ Wet | | 15 | 4 | 15 | 6 | 30 | 5 | 75 |
| Subtotal reduced | | 73 | 22 | 73 | 29 | 144 | 24 | 365 |
| **Total nitrogen flux** | | 156 | 37 | 124 | 51 | 207 | 44 | 619 |

(c)  The atmospheric deposition of nitrogen species and gross fluvial dissolved nitrogen inputs; units, $10^{12}$ g N yr$^{-1}$.

| Atmospheric deposition | Fluvial fluxes | | Total |
|---|---|---|---|
| | Natural | Anthropogenic | |
| 30 | 14 | 7–35 | 21–49 |

and much of the fluvial input will be involved in bio-geochemical processes in the estuarine and coastal environments and so will not reach the open ocean.

Cornell *et al.* (1995) stressed the importance of the atmospheric input of dissolved organic nitrogen (DON) to the oceans. Dissolved organic nitrogen makes up half of the fluvial input of dissolved fixed nitrogen, but, as Cornell *et al.* (1995) pointed out, atmospheric fluxes usually include only the inorganic forms of nitrogen, such as nitrate and ammonia. By determining the DON in rain and snow the authors showed that it is a ubiquitous and significant component of precipitation, even in remote areas, and that it has a mainly anthropogenic origin. Cornell *et al.* (1995) concluded that on the basis of their data, which yielded an atmospheric flux of $\sim 30 \times 10^{12}$ to $\sim 85 \times 10^{12}\,g\,N\,yr^{-1}$, the inclusion of DON would increase the original estimates for the anthropogenic input of fixed nitrogen by a factor of 1.5. Because most of the DON in precipitation was thought to be anthropogenic in origin it would appear that humankind is perturbing the natural global nitrogen cycle to a much greater extent than at first thought, and that the perturbation is not confined to coastal regions. The potential effects of the large increase in fixed nitrogen inputs to the oceans from human activities will include increased marine primary production and, over a long time-scale, an increased burial of carbon.

### 6.4.3.2 Synthetic organic compounds

Many organic compounds have significant vapour pressures at ambient temperatures and therefore have major gas-phase components. By using published concentration data, and taking account of particle/gas partitioning to assess deposition processes, Duce *et al.* (1991) estimated the fluxes of polychlorinated biphenyls (PCBs), hexachlorobenzene (HCB), and the pesticides hexachlorocyclohexanes (HCHs), dichlorodiphenyltrichloroethanes (DDTs), chlordane and dieldrin to the World Ocean. The data are listed in Table 6.25. Table 6.25(a) gives the total deposition and mean fluxes of the organochlorines to different regions of the World Ocean; Table 6.25(b) provides estimates of the total deposition to the World Ocean in terms of the deposition mechanism ('wet' or 'dry'); and Table 6.25(c), compares the atmospheric and fluvial inputs of the organochlorines to the World

**Table 6.25** The atmospheric input of organochlorine compounds to the World Ocean (data from Duce *et al.*, 1991).

(a) Total deposition and mean fluxes*.

| Compound[†] | North Atlantic TD | North Atlantic MF | South Atlantic TD | South Atlantic MF | North Pacific TD | North Pacific MF | South Pacific TD | South Pacific MF | Indian TD | Indian MF | Global TD | Global MF |
|---|---|---|---|---|---|---|---|---|---|---|---|---|
| Σ HCH | 850 | 16 | 97 | 1.9 | 2600 | 30 | 470 | 4.3 | 700 | 10 | 4800 | 13 |
| HCB | 17 | 0.31 | 10 | 0.20 | 20 | 0.22 | 19 | 0.17 | 11 | 0.17 | 77 | 0.23 |
| Dieldrin | 17 | 0.30 | 2.0 | 0.04 | 8.9 | 0.10 | 9.5 | 0.09 | 6.0 | 0.09 | 43 | 0.11 |
| Σ DDT | 16 | 0.28 | 14 | 0.27 | 66 | 0.74 | 26 | 0.23 | 43 | 0.64 | 170 | 0.44 |
| Chlordane | 8.7 | 0.16 | 1.0 | 0.02 | 8.3 | 0.09 | 1.9 | 0.02 | 2.4 | 0.04 | 22 | 0.06 |
| Σ PCB | 100 | 1.8 | 14 | 0.27 | 36 | 0.40 | 29 | 0.26 | 52 | 0.77 | 240 | 0.64 |

* TD, total deposition (units, $10^6\,g\,yr^{-1}$); MF, mean flux (units, $\mu g\,m^{-2}\,yr^{-1}$). † See text.

(b) Depositional processes to the World Ocean expressed as a percentage of total deposition.

| Compound* | Particle Dry | Particle Wet | Gas Dry | Gas Wet |
|---|---|---|---|---|
| α-HCH | <0.1 | 0.1 | 38 | 62 |
| γ-HCH | <0.1 | <0.1 | 23 | 77 |
| HCB | 0.2 | 2.2 | 85 | 13 |
| Dieldrin | 0.3 | 13 | 54 | 33 |
| *pp'* DDT | 0.7 | 34 | 45 | 20 |
| Chlordane | 0.2 | 9.5 | 72 | 18 |
| Σ PCB | 0.6 | 23 | 65 | 11 |

* See text.

(c) Atmospheric and fluvial inputs to the World Ocean (units, $10^6\,g\,yr^{-1}$).

| Compound* | Atmospheric | Fluvial | Percentage atmospheric |
|---|---|---|---|
| Σ HCH | 4800 | 40–80 | 99 |
| HCB | 77 | 4 | 95 |
| Dieldrin | 43 | 4 | 91 |
| Σ DDT | 170 | 4 | 98 |
| Chlordane | 22 | 4 | 85 |
| Σ PCB | 240 | 40–80 | 80 |

* See text.

Ocean. The major trends in the data are summarized below.

**1** Because the major sources of the organochlorines are in the Northern Hemisphere, the dominant deposition of the compounds is to the North Atlantic and North Pacific.

**2** Because of different sources and transport regimes, different organochlorines predominate in individual ocean basins. For example, HCH and DDT compounds have their highest deposition rates in the North Pacific, which probably arises from sources in Asia. In contrast, PCBs and dieldrin have higher deposition rates in the North Atlantic than in the North Pacific as a result of sources in North America and Europe. Hexachlorobenzene and chlordane have deposition rates that are generally similar in the North Atlantic and North Pacific.

**3** The mechanisms of air–sea exchange differ for some organochlorines. Although the magnitudes of the direct gas exchange are uncertain, such a mechanism can account for between ~25% and ~85% of the total air–sea exchange of the organochlorines. The direct 'dry' deposition of particle-bound organic material accounts for <5% of the total particle deposition, and the primary differences in deposition mechanisms between the compounds occur in 'wet' deposition. Overall, particle scavenging by 'wet' deposition is most significant for the PCB and DDT compounds, and gas scavenging is the predominant mechanism for the removal of HCH compounds in rain.

**4** There is a general lack of data on the input of the organochlorines from large river systems, but on the basis of the limited data set given in Table 6.25(c), it is apparent that atmospheric input is the dominant transport route by which the organochlorines reach the World Ocean; however, it must be remembered that the estimates are still essentially very crude.

## 6.5 Relative magnitudes of the primary fluxes to the oceans: summary

**1** The major primary sources of both particulate and dissolved components to the oceans on a global scale are river run-off, atmospheric deposition and hydrothermal exhalations, with glacial sources being locally predominant in some polar regions.

**2** Of the major primary sources, river run-off and atmospheric deposition deliver their loads to the surface ocean, rivers to the ocean margins, and the atmosphere to the whole ocean surface. On a global basis, fluvial fluxes are generally greater than those resulting from atmospheric deposition, although there are some exceptions to this (e.g. atmospheric deposition dominates the input of Pb to almost the whole surface ocean). Retention in the estuarine and coastal zones, however, can mean that the net input of some fluvially transported components to the open ocean is less than that from atmospheric deposition. This atmospheric deposition retains its strongest surface water fingerprints for the 'scavenged-type' trace metals, such as Al, Mn and Pb.

**3** Hydrothermal activity can act as a source for some components and a sink for others. When it acts as a source, high-temperature axial hydrothermal activity delivers components to mid-depth and bottom waters via the formation of vent plumes following the circulation of sea water through the crust. Although the extent to which the components in these plumes are globally dispersed is not yet known with certainty, it is now recognized that for some elements (e.g. Fe, Mn) hydrothermal inputs can match, and sometimes exceed, fluvial inputs. In addition to being a source, the hydrothermal plumes can act as an oceanic sink for some dissolved elements as sea water is cycled through them in the water column.

We have now identified the principal primary sources that supply material to the World Ocean, and have described the pathways along which the material travels in order to enter the seawater reservoir. In the following sections, we will describe the physical and chemical nature of this reservoir, and discuss the mechanisms that keep it in motion.

## References

Arimoto, R., Duce, R.A., Ray, B.J. & Unni, C.K. (1985) Atmospheric trace elements at Enewetak Atoll: transport to the ocean by wet and dry deposition. *J. Geophys. Res.*, **90**, 2391–408.

Arimoto, R., Duce, R.A., Ray, B.J., Hewitt, A.D. & Williams, J. (1987) Trace elements in the atmosphere of American Samoa; concentrations and deposition to the tropical South Pacific. *J. Geophys. Res.*, **92**, 8465–79.

Arnold, M., Seghaier, A., Martin, D., Buat-Menard, P. & Chesselet, R. (1982) Geochimie de l'aerosol marin au-dessus la Mediteranee occidentale. *CIESMI J. Etud. Pollut. Mar. Mediterr.*, **VI**, 27–37.

Balls, P.W. (1989) Trace metal and major ion composition of precipitation at a North Sea coastal site. *Atmos. Environ.*, **23**, 2751–59.

Bewers, J.M. & Yeats, P.A. (1977) Oceanic residence times of trace metals. *Nature*, **258**, 595–8.

Boyle, E.A., Collier, R., Dengler, A.T., Edmond, J.M., Ng, A.C. & Stallard, R.F. (1974) On the chemical mass-balance in estuaries. *Geochim. Cosmochim. Acta*, **38**, 1719–28.

Bruland, K.W. (1980) Oceanographic distributions of cadmium, zinc, nickel and copper in the North Pacific. *Earth Planet. Sci. Lett.*, **77**, 176–98.

Bruland, K.W. & Franks, R.P. (1983) Mn, Ni, Zn and Cd in the western North Atlantic. In *Trace Metals in Sea Water*, C.S. Wong, E. Boyle, K.W. Bruland, J.D. Burton & E.D. Goldberg (eds), 395–414. New York: Plenum.

Bruland, K.W., Bertine, K., Koide, M. & Goldberg, E.D. (1974) History of metal pollution in southern California coastal zone. *Environ. Sci. Technol.*, **8**, 425–31.

Buat-Menard, P. (1983) Particle geochemistry in the atmosphere and oceans. In *Air–Sea Exchange of Gases and Particles*, P.S. Liss & W.G.N. Slinn (eds), 455–532. Dordrecht: Reidel.

Buat-Menard, P. & Chesselet, R. (1979) Variable influence of the atmospheric flux on the trace metal chemistry of oceanic suspended matter. *Earth Planet. Sci. Lett.*, **42**, 399–411.

Cambray, R.S., Jeffries, D.F. & Topping, G. (1975) *An Estimate of the Input of Atmospheric Trace Elements into the North Sea and the Clyde Sea (1972–73)*. Harwell: UK Atomic Energy Authority, Report AERE-R7733.

Campbell, J.A. & Yeats, J.M. (1984) Dissolved Cr in the St Lawrence estuary. *Estuar. Shelf Sci.*, **19**, 513–22.

Chester, R. & Murphy, K.J.T. (1990) Metals in the marine atmosphere. In *Heavy Metals in the Marine Environment*, R. Furness & P. Rainbow (eds), 27–49. Boca Raton, FL: CRC Press.

Chester, R., Griffiths, A.G. & Hirst, J.M. (1979) The influence of soil-sized atmospheric particulates on the elemental chemistry of the deep-sea sediments of the North Eastern Atlantic. *Mar. Geol.*, **32**, 141–54.

Chester, R., Nimmo, M., Murphy, K.J.T. & Nicolas, E. (1990) Atmospheric trace metals transported to the Western Mediterranean: data from a station on Cap Ferrat. *Water Poll. Res. Rep.*, **20**, 579–612.

Chester, R., Murphy, K.J.T., Lin, F.J., Berry, A.S., Bradshaw, G.A. & Corcoran, P.A. (1993) Factors controlling the solubilities of trace metals from non-remote aerosols deposited to the sea surface by the 'dry' deposition mode. *Mar. Chem.*, **42**, 107–26.

Chester, R., Bradshaw, G.F., Ottley, C.J., *et al.* (1994) The atmospheric distributions of trace metals, trace organics and nitrogen species over the North Sea. In *Understanding the North Sea System*, H. Charnock, K.R. Dyer, J.M. Huthnance, P.S. Liss, J.H. Simpson & P.B. Tett (eds). *Philos. Trans. R. Soc. London, Ser. A*, **343**, 545–56.

Chester, R., Nimmo, M. & Keyse, S. (1996) The influence of Saharan and Middle Eastern desert-derived dust on the trace metal compositions of Mediterranean aerosols and rainwaters: an overview. In *The Impact of Desert Dust across the Mediterranean*, S. Guerzoni & R. Chester (eds), 253–73. Dordrecht: Kluwer Academic Publishers.

Chester, R., Nimmo, M., Keyse, S. & Corcoran, P.A. (1997) Rainwater–aerosol trace metal relationships at Cap Ferrat: a coastal site in the Western Mediterranean. *Mar. Chem.*, **58**, 293–312.

Church, T.M., Galloway, J.N., Jickells, T.D. & Knap, A.H. (1982) The chemistry of precipitation at the mid-Atlantic coast on Bermuda. *J. Geophys. Res.*, **87**, 11013–18.

Collier, R.W. & Edmond, J.M. (1984) The trace element geochemistry of marine biogenic particulate matter. *Prog. Oceanogr.*, **13**, 113–99.

Cornell, S., Randell, A. & Jickells, T.D. (1995) Atmospheric inputs of dissolved organic nitrogen to the oceans. *Nature*, **376**, 243–45.

Dedkov, A.P. & Mozzherin, V.I. (1984) *Eroziya i stok Nanosov na Zemle*. Izdatatelstova Kazanskogo Universiteta.

Degens, E.T. & Ittekkot, V. (1985) Particulate organic carbon: an overview. In *Transport of Carbon and Minerals in Major World Rivers*, Part 3, E.T. Degens, S. Kempe & R. Herrera (eds), 7–27. Hamburg: Mitt. Geol.-Palont. Inst. Univ. Hamburg, SCOPE/UNEP, Sonderband 58.

Degens, E.T., Kempe, S. & Richey, J.E. (eds) (1991a) *SCOPE 42: Biogeochemistry of Major World Rivers*. Chichester: John Wiley & Sons.

Degens, E.T., Kempe, S. & Richey, J.E. (1991b) Summary: biogeochemistry of major world rivers. In *SCOPE 42: Biogeochemistry of Major World Rivers*, E.T. Degens, S. Kempe & J.E. Richey (eds), 323–47. Chichester: John Wiley & Sons.

Duce, R.A. (1978) Speculations on the budget of particulate and vapour phase non-methane organic carbon in the global troposphere. *Pure Appl. Geophys.*, **116**, 244–73.

Duce, R.A. & Duursma, E.K. (1977) Inputs of organic matter to the ocean. *Mar. Chem.*, **5**, 319–39.

Duce, R.A., Hoffman, G.L., Ray, B.J., *et al.* (1976) Trace metals in the marine atmosphere: sources and fluxes. In *Marine Pollutant Transfer*, H. Windom & R.A. Duce (eds), 77–119. Lexington: Heath.

Duce, R.A., Mohnen, V.A., Zimmerman, P.R., *et al.* (1983) Organic material in the global troposphere. *Rev. Geophys. Space Phys.*, **21**, 921–52.

Duce, R.A., Liss, P.S., Merrill, J.T., *et al.* (1991) The atmospheric input of trace species to the World Ocean. *Global Biogeochem. Cycles*, **5**, 193–259.

Edmond, J.M., Measures, C.I., McDuff, R.E., *et al.* (1979) Crest hydrothermal activity and the balance of the major and minor elements in the ocean; the Galapagos data. *Earth Planet. Sci. Lett.*, **46**, 1–18.

Edmond, J.M., Von Damm, K.L., McDuff, R.E. & Measures, C.I. (1982) Chemistry of hot springs on the East Pacific Rise and their effluent dispersal. *Nature*, **297**, 187–91.

Edmond, J.M., Spivack, A., Grant, B.C., *et al.* (1985) Chemical dynamics of the Changjiang estuary. *Cont. Shelf. Res.*, **4**, 17–30.

Elderfield, H. & Schultz, A. (1996) Mid-ocean ridge hydrothermal fluxes and the chemical composition of the ocean. *Annu. Rev. Earth Planet Sci.*, **24**, 191–224.

Figueres, G., Martin, J.M. & Maybeck, M. (1978) Iron behaviour in the Zaire estuary. *Neth. J. Sea Res.*, **12**, 327–37.

Galloway, J.N. & Gaudry, A. (1984) The composition of precipitation on Amsterdam Island, Indian ocean. *Atmos. Environ.*, **18**, 2649–56.

Galloway, J.N., Likens, G.E., Keene, W.C. & Miller, J.M. (1982) The composition of precipitation in remote areas of the world. *J. Geophys. Res.*, **87**, 8771–86.

GESAMP (1987) *Land/Sea Boundary Flux of Contaminants from Rivers*. Paris: Unesco.

Guieu, C., Chester, R., Nimmo, M., *et al.* (1997) Atmospheric input of dissolved and particulate metals to the North-Western Mediterranean Sea. *Deep Sea Res.*, **44**, 655–71.

Hodge, V., Johnson, S.R. & Goldberg, E.D. (1978) Influence of atmospherically transported aerosols on surface ocean water composition. *Geochem. J.*, **12**, 7–20.

Holeman, J.N. (1968) The sediment yield of major rivers of the world. *Water Resour. Res.*, **4**, 734–47.

Ittekkot, V. (1988) Global trends in the nature of organic matter in river suspensions. *Nature*, **332**, 436–8.

Jickells, T., Knapp, A.H. & Church, T.M. (1984) Trace metals in Bermuda rainwater. *J. Geophys. Res.*, **89**, 1423–8.

Judson, S. (1968) Erosion of the land (What's happening to our continents?). *Am. Sci.*, **5**, 514–16.

Kadko, D., Baker, E., Alt, J. & Baross, J. (1994) Global impact of submarine hydrothermal processes. Final Report. *Ridge/Vent Workshop*, 55 pp.

Keyse, S. (1996) *Factors controlling the solubility of trace metals in rainwater*. PhD Thesis, University of Liverpool.

Klinkhammer, G.P. (1980) Observations of the distribution of manganese over the East Pacific Rise. *Chem. Geol.*, **29**, 211–26.

Klinkhammer, G.P., Rona, P., Greaves, M. & Elderfield, H. (1985) Hydrothermal manganese plumes in the Mid-Atlantic Ridge rift valley. *Nature*, **314**, 727–31.

Kremling, K. (1985) The distribution of cadmium, copper, nickel, manganese and aluminium in surface waters of the open Atlantic and European shelf area. *Deep Sea Res.*, **32**, 531–55.

Lim, B., Jickells, T.D. & Davies, T.D. (1991) Sequential sampling of particles, major ions and total trace metals in wet deposition. *Atmos. Environ.*, **25**, 745–62.

Lim, B., Jickells, T.D., Colin, J.L. & Losno, R. (1994) Solubilities of Al, Pb, Cu and Zn in rain sampled in the marine environment over the North Atlantic Ocean and Mediterranean Sea. *Global Biogeochem. Cycles*, **8**, 349–62.

Losno, R., Bergametti, G. & Buat-Menard, P. (1988) Zinc partitioning in Mediterranean rainwater. *J. Geophys. Res. Lett.*, **15**, 1389–92.

Loye-Pilot, M.D., Martin, J.-M. & Morelli, J. (1986) Influence of Saharan dust on the rain acidity and atmospheric input to the Mediterranean. *Nature*, **321**, 427–28.

McDonald, R.L., Unni, C.K. & Duce, R.A. (1982) Estimation of atmospheric sea salt dry deposition; wind speed and particle size dependence. *J. Geophys. Res.*, **87**, 1246–50.

Mantoura, R.F.C. & Woodward, E.M.S. (1983) Conservative behaviour of riverine dissolved organic carbon in the Severn Estuary: chemical and geochemical implications. *Geochim. Cosmochim. Acta*, **47**, 1293–309.

Maring, H.B. & Duce, R.A. (1987) The impact of atmospheric aerosols on trace metal chemistry in open ocean surface waters. 1. Aluminium. *Earth Planet Sci. Lett.*, **84**, 381–92.

Martin, J.-M. & Maybeck, M. (1979) Elemental mass balance of material carried by major world rivers. *Mar. Chem.*, **7**, 173–206.

Martin, J.-M. & Whitfield, M. (1983) The significance of the river input of chemical elements to the ocean. In *Trace Metals in Sea Water*, C.S. Wong, E. Boyle, K.W. Bruland, J.D. Burton & E.D. Goldberg (eds), 265–96. New York: Plenum.

Martin, J.-M., Elbaz-Poulichet, F., Guieu, C., Loye-Pilot, M.D. & Han, G. (1989) River versus atmospheric input of material to the Mediterranean Sea: an overview. *Mar. Chem.*, **28**, 159–82.

Maybeck, M. (1979) Concentrations des eaux fluviales en elements majeurs et apports en solution aux oceans. *Rev. Geol. Dyn. Geogr. Phys.*, **21**, 215–46.

Maybeck, M. (1982) Carbon, nitrogen and phosphorus transport by world rivers. *Am. J. Sci.*, **282**, 401–50.

Milliman, J.D. (1981) Transfer of river-borne particulate material to the oceans. In *River Inputs to Ocean Systems*, J.-M. Martin, J.D. Burton & D. Eisma (eds), 5–12. Paris: UNEP/Unesco.

Milliman, J.D. & Meade, R.H. (1983) World-wide delivery of river sediment to the oceans. *J. Geol.*, **91**, 1–21.

Milliman, J.D. & Syvitski, J.P.M. (1992) Geomorphic/tectonic control of sediment discharge to the ocean: the importance of small mountain rivers. *J. Geol.*, **100**, 525–44.

Orians, K.J. & Bruland, K.W. (1985) Dissolved aluminium in the central North Pacific. *Nature*, **316**, 427–9.

Pszenny, A.A.P., MacIntyre, F. & Duce, R.A. (1982) Sea-salt and the acidity of marine rain on the windward coast of Samoa. *Geophys. Res. Lett.*, **9**, 751–54.

Robinson, E. (1978) Hydrocarbons in the atmosphere. *Pure Appl. Geophys.*, **116**, 327–84.

Settle, D.M. & Patterson, C.C. (1982) Magnitudes and sources of precipitation and dry deposition fluxes of industrial and natural lead to the North Pacific at Enewetak. *J. Geophys. Res.*, **87**, 8857–69.

Spitzy, A. & Leenheer, J. (1991) Dissolved organic carbon in rivers. In *SCOPE 42: Biogeochemistry of Major World Rivers*, E.T. Degens, S. Kempe & J.B. Richey (eds), 213–32. Chichester: John Wiley & Sons.

Statham, P.J. & Burton, J.D. (1986) Dissolved manganese in the North Atlantic Ocean, 0–35°N. *Earth Planet. Sci. Lett.*, **79**, 55–65.

Stossel, R.P. (1987) *Untersuchungen zur Nab und trokend-position von Schwermetallen auf der Insel der Pellworm.* Doctorgrades Dissertation, University of Hamburg.

Thompson, G. (1983) Hydrothermal fluxes in the ocean. In *Chemical Oceanography*, J.P. Riley & R. Chester (eds), Vol. 8, 270–337. London: Academic Press.

Van Bennekon, A.J. & Salomons, W. (1981) Pathways of nutrients and organic matter from land to ocean through rivers. In *River Inputs to Ocean Systems*, J.-M. Martin, J.D. Burton & D. Eisma (eds), 33–51. Paris: UNEP/Unesco.

Von Damm, K.L., Edmond, J.M., Grant, B., Measures, C.I., Walden, B. & Weiss, R.F. (1985) Chemistry of submarine hydrothermal solutions at 21°N, East Pacific Rise. *Geochim. Cosmochim. Acta*, **49**, 2197–220.

Walling, D.E. & Webb, D.W. (1987) Material transport by the world's rivers: evolving perspectives. In *Water for the Future: Hydrology in Perspective*, 313–29. Wallingford: International Association of Hydrological Sciences, Publication 164.

Walsh, P.R. & Duce, R.A. (1976) The solubilization of anthropogenic atmospheric vanadium in sea water. *Geophys. Res. Lett.*, **3**, 375–8.

Walsh, P.R., Duce, R.A. & Fasching, J.L. (1979) Considerations of the enrichment, sources and fluxes of arsenic in the troposphere. *J. Geophys. Res.*, **84**, 1719–26.

Williams, P.J. (1975) Biological and chemical aspects of dissolved organic material in sea water. In *Chemical Oceanography*, J.P. Riley & G. Skirrow (eds), Vol 2, 301–63. London: Academic Press.

Windom, H.L. (1981) Comparison of atmospheric and riverine transport of trace elements to the continental shelf environment. In *River Inputs to Ocean Systems*, J.-M. Martin, J.D. Burton & D. Eisma (eds), 360–9. Paris: UNEP/Unesco.

Yeats, P.A. & Bewers, J.M. (1982) Discharge of metals from the St. Lawrence River. *Can. J. Earth Sci.*, **19**, 982–92.

Zimmerman, P.R., Chatfield, R.B., Fishman, J., Crutzen, P.J. & Hanst, P.L. (1978) Estimates of the production of CO and $H_2$ from the oxidation of hydrocarbon emissions from vegetation. *Geophys. Res. Lett.*, **5**, 679–82.

# Part II
# The Global Journey: the Ocean Reservoir

# 7 Descriptive oceanography: water-column parameters

In the previous chapters, the various types of material that are brought to the oceans have been tracked along their transport pathways to the point where they cross the interfaces separating the sea from the other planetary reservoirs. Once they have entered the ocean reservoir these materials are subjected to a variety of physical, chemical and biological processes, which combine to control the composition of both the *sea water* and the *sediment* phases within it.

## 7.1 Introduction

In order to provide a framework within which to discuss the physical, chemical and biological oceanic processes, it is necessary first to describe the nature of the water itself, in terms of some of its fundamental properties, and then to define the circulation mechanisms that keep it in motion. This circulation controls the physical transport of water masses, and their associated dissolved and particulate constituents, from one part of the oceanic system to another in the form of **conservative** signals. Superimposed on these are **non-conservative** signals, which arise from the involvement of the constituents in the major biogeochemical cycles that operate within the World Ocean. Components are incorporated into bottom sediments mainly by the sinking of particulate material, and there is a continual movement of particulate components through the oceans. In order to understand how this throughput operates, a simple oceanic box model will be constructed. This will then be used, together with a variety of other approaches, as a framework within which to describe the oceanic distributions of a number of parameters that have proved especially rewarding for understanding the nature of the biogeochemical cycles in the sea; these parameters include dissolved oxygen and carbon dioxide, the nutrients, and both dissolved and particulate organic carbon. By means of this approach, the throughput of

materials in the oceans will be described in relation to organic matter, suspended particulates and dissolved trace metals as they are transported to the interface that separates the water column from the sediment sink.

## 7.2 Some fundamental properties of sea water

### 7.2.1 Introduction

The three fundamental properties of sea water that are of most interest to marine geochemists are **salinity** and **temperature**, which can be used to characterize water masses and which, together with pressure, fix the density of sea water, and **density** itself, which fixes the depth to which a water mass will settle and so drives thermohaline circulation. Each of these properties will be described individually below. It must be remembered, however, that they are interlinked, and the equation of state of sea water is a mathematical expression of the relationship between the temperature, pressure, salinity and density of sea water; it is used, for example, for the calculation of the density of sea water from the other parameters. Recently, a new International Equation of State of Seawater has been adopted, which is consistent with the definition of 'practical salinity' (see e.g. Millero *et al.*, 1980; Unesco, 1981); this equation of state is given in Worksheet 7.1.

### 7.2.2 Salinity

Salinity is a measure of the degree to which the water in the oceans is salty, and is a function of the weight of total solids dissolved in a quantity of sea water. The major ions in sea water, i.e. those that make a significant contribution to salinity, are listed in Table 7.1, and although the total salt content can vary, these

**Worksheet 7.1: The international equation of state of seawater, 1980** (from Unesco, 1981)

The equation of state of sea water is a mathematical expression that can be used to calculate the density of sea water from measurements of temperature, pressure and salinity, or from other parameters derived from these. The expression adopted by the Unesco/ICES/IAPSO Joint Panel on Oceanographic Tables and Standards as the International Equation of State of Seawater gives the density of sea water ($\rho$, $kg\,m^{-3}$) as a function of practical salinity ($S$), temperature ($t$, °C) and applied pressure ($p$, bar) in the form:

$$\rho(S,\ t,\ p) = \frac{\rho(S,\ t,\ 0)}{1 - p/K(S,\ t,\ p)}$$

where $\rho(S, t, 0)$ is the One Atmosphere International Equation of State, 1980, and $K(S, t, p)$ is the secant bulk modulus, the full definitions of which are given below.

*The One Atmosphere International Equation of State of Seawater, 1980*

*Definition*
The density ($\rho$, $kg\,m^{-3}$) of sea water at one standard atmosphere ($p = 0$) is to be computed from the practical salinity ($S$) and the temperature ($t$, °C) with the following equation:

$$\begin{aligned}
\rho(S,\ t,\ p) = \rho_w &+ (8.24493 \times 10^{-1} - 4.0899 \times 10^{-3}t \\
&+ 7.6438 \times 10^{-5}t^2 - 8.2467 \times 10^{-7}t^3 \\
&+ 5.3875 \times 10^{-9}t^4)S + (-5.72466 \times 10^{-3} \\
&+ 1.0227 \times 10^{-4}t - 1.6546 \times 10^{-6}t^2)S^{3/2} \\
&+ 4.8314 \times 10^{-4}S^2
\end{aligned}$$

where $\rho_w$, the density of the Standard Mean Ocean Water (SMOW) taken as pure water reference, is given by

$$\begin{aligned}
\rho_w = 999.842594 &+ 6.793952 \times 10^{-2}t - 9.095290 \times 10^{-3}t^2 \\
&+ 1.001685 \times 10^{-4}t^3 - 1.120083 \times 10^{-6}t^4 \\
&+ 6.536332 \times 10^{-9}t^5
\end{aligned}$$

The One Atmosphere International Equation of State of Seawater, 1980, is valid for practical salinity from 0 to 42 and temperature from −2 to 40°C.

*The High Pressure International Equation of State of Seawater, 1980*

*Definition*
The density ($\rho$, $kg\,m^{-3}$) of sea water at high pressure is to be computed from the practical salinity ($S$), the temperature ($t$, °C) and the applied pressure ($p$, bar) with the following equation:

*continued*

$$\rho(S,\ t,\ p) = \frac{\rho(S,\ t,\ 0)}{1 - p/K(S,\ t,\ p)}$$

where $\rho(S, t, 0)$ is the One Atmosphere International Equation of State, 1980, given above, and $K(S, t, p)$ is the secant bulk modulus given by

$$K(S,\ t,\ p) = K(S,\ t,\ 0) + Ap + Bp^2$$

where

$$\begin{aligned}K(S,\ t,\ 0) = K_w &+ (54.6746 - 0.603459t + 1.09987 \times 10^{-2}t^2 \\ &- 6.1670 \times 10^{-5}t^3)S + (7.944 \times 10^{-2} + 1.6483 \times 10^{-2}t \\ &- 5.3009 \times 10^{-4}t^2)S^{3/2}\end{aligned}$$

$$\begin{aligned}A = A_w &+ (2.2838 \times 10^{-3} - 1.0981 \times 10^{-5}t - 1.6078 \times 10^{-6}t^2)S \\ &+ 1.91075 \times 10^{-4}S^{3/2}\end{aligned}$$

$$B = B_w + (-9.9348 \times 10^{-7} + 2.0816 \times 10^{-8}t + 9.1697 \times 10^{-10}t^2)S$$

The pure water terms $K_w$, $A_w$ and $B_w$ of the secant bulk modulus are given by

$$\begin{aligned}K_w = 19\,652.21 &+ 148.4206t - 2.327\,105t^2 + 1.360\,477 \times 10^{-2}t^3 \\ &- 5.155\,288 \times 10^{-5}t^4\end{aligned}$$

$$\begin{aligned}A_w = 3.239\,908 &+ 1.437\,13 \times 10^{-3}t + 1.160\,92 \times 10^{-4}t^2 \\ &- 5.779\,05 \times 10^{-7}t^3\end{aligned}$$

$$B_w = 8.509\,35 \times 10^{-5} - 6.122\,93 \times 10^{-6}t + 5.2787 \times 10^{-8}t^2$$

The High Pressure International Equation of State of Seawater, 1980, is valid for practical salinity from 0 to 42, temperature from −2 to 40°C and applied pressure from 0 to 1000 bar.

constituents are always, or almost always, present in the same relative proportions. This **constancy of composition** of sea water has extremely important implications for oceanographers.

Historically, there have been a number of definitions of salinity. According to an International Commission set up in 1899 salinity was defined as 'the weight of inorganic salts in one kilogram of sea water, when all bromides and iodides are replaced by an equivalent quantity of chlorides, and all carbonates are replaced by an equivalent quantity of oxides'. In theory, perhaps the most simple direct method for the measurement of salinity is to evaporate the water to dryness and weigh the salt residue. In practice, however, gravimetric methods are tedious and time-consuming, and in order to design a routine method

for the determination of salinity the Commission made use of the concept of constancy of composition, which implies that it should be possible to use any of the major constituents as an index of salinity. It can be seen from Table 7.1 that chloride ions make up ~55% of the total dissolved salts in sea water, and reliable methods for the determination of chloride, using chemical titration with silver nitrate, were available at the time the Commission met. Chlorides, iodides and bromides have similar properties, and all of them react with silver nitrate to appear as chlorides in the titration.

An investigation was therefore made between **salinity** ($S$, ‰) and **chlorinity** ($Cl$, ‰), the latter being defined as 'the mass in grams of chlorine equivalent to the mass of halogens contained in one kilogram of sea

**Table 7.1** The major ions of sea water (data from Wilson, 1975).

| Ion | Concentration $g\,kg^{-1}$ at $S = 35‰$ |
|---|---|
| $Cl^-$ | 19.354 |
| $SO_4^{2-}$ | 2.712 |
| $Br^-$ | 0.0673 |
| $F^-$ | 0.0013 |
| $B$ | 0.0045 |
| $Na^+$ | 10.77 |
| $Mg^{2+}$ | 1.290 |
| $Ca^{2+}$ | 0.4121 |
| $K^+$ | 0.399 |
| $Sr^{2+}$ | 0.0079 |

water'. To examine the relationship, the salinities of nine sea waters were determined using an accurate gravimetric method. The chlorinities of the waters were then measured by titration, and the following relationship between salinity and chlorinity was established:

$$S(‰) = 1.805Cl(‰) + 0.030 \quad (7.1)$$

This is termed **chlorinity salinity**, and was used for many years as the working definition of salinity.

The position began to change, however, with the introduction of the conductimetric salinometer, and a new investigation was carried out into the interrelations between the *measured* parameters (e.g. chlorinity, conductivity ratio, refractive index) and the *derived* parameters (e.g. salinity, specific gravity) of sea water. The data were assessed by a Unesco Joint Panel, and on the basis of the relationship between chlorinity and the conductivity ratio a new equation for salinity was recommended:

$$S(‰) = 1.80655Cl(‰) \quad (7.2)$$

This is termed **conductivity salinity**, and the equation can be used to determine salinity from the chlorinity obtained by the titration method.

However, the increasing use, and greater convenience, of high-precision electrical conductivity measurements has meant that chlorinity titration is now no longer the preferred method for the determination of salinity. In the light of this, the Joint Panel proposed a method for the determination of salinity from conductivity ratios, using a polynomial that relates salinity to the conductivity ratio $R$ determined at 15°C at one standard atmosphere pressure:

$$S(‰) = -0.089\,96 + 28.2970R_{15} + 12.808\,32R_{15}^2$$
$$- 10.678\,69R_{15}^3 + 5.986\,24R_{15}^4 - 1.323\,11R_{15}^5$$

$$(7.3)$$

where $R_{15}$ is the conductivity ratio at 15°C, and is defined as the ratio of the conductivity of a seawater sample to that of one having an $S(‰)$ of 35‰ at 15°C under a pressure of one standard atmosphere. Tables are available for the conversion of $R_{15}$ values into salinity, and a second polynomial has been provided to permit the conversion of conductivity ratios measured at other temperatures. Salinities determined from these polynomials are termed **practical salinities**. Practical salinity is based on conductivity ratios and has no dimension, and terms such as ‰ have been abandoned to be replaced with $S = 2$ to $S = 42$; i.e. the range over which practical salinities are valid. However, it is still useful for many purposes to use the old ‰ values (see e.g. Duinker, 1986).

The major elements that contribute to salinity are described as being **conservative**, i.e. their concentration ratios remain constant, and their total concentrations can be changed only by physical processes. This is an operationally valid concept and it does appear that there are no significant variations in the ratios of sodium, potassium, sulphate, bromide and boron to chlorinity in sea water. Variations have been found, however, in the ratios of calcium, magnesium, strontium and fluoride to chlorinity (for a detailed review of the major elements in sea water—see Wilson (1975)). Despite the fact that the major elements are more or less conservative in sea water there are conditions under which their concentration ratios can vary considerably. Situations under which these atypical conditions can be found include those associated with estuaries and land-locked seas, anoxic basins, the freezing of sea water, the precipitation and dissolution of carbonate minerals, submarine volcanism, admixture with geological brines and evaporation in isolated basins.

Under *most* conditions, therefore, the major elements in sea water may be regarded as being present in almost constant proportions. However, the total salt content ($S$, ‰) can vary. This results from the operation of a number of processes. Those that *decrease* salinity include the influx of fresh water from precipitation (rainfall), land run-off and the melting of ice; and those that *increase* salinity include evaporation and the formation of ice (Bowden, 1975). A number of general trends can be identified in the distribution of salinity in the surface ocean.

**1** Salinity in surface ocean waters usually ranges between ~32‰ and ~37‰.

**2** Higher values are found in some semi-enclosed, mid-latitude seas where evaporation greatly exceeds precipitation and run-off. Examples of this are found in the Mediterranean Sea (range 37–39‰), and the Red Sea (range 40–41‰).

**3** In coastal waters run-off can result in a decrease in surface salinities.

The processes that cause the major variations in salinity occur at the surface of the ocean, and the most rapid variations usually are found within a few hundred metres or so of the air–sea interface. Regions in the water column over which rapid salinity changes take place are termed **haloclines** (see Fig. 7.1a). As there are no sources or sinks for salt in the deep layers of the ocean, the salinity derived at the surface can be changed only by the mixing of different water types with different initial salinities (see Section 7.3.4). In these deep-water layers, variations in salinity are smaller than those found at the surface. Different water masses, however, have individual 'salinity signatures', and these are extremely important in relation to deep-water circulation processes (see Section 7.3.3).

### 7.2.3 Temperature

Both horizontal and vertical temperature variations are found in the ocean. The main horizontal variations occur in the surface region, with temperatures ranging from ~28°C in the equatorial zones to as low as ~−2°C in the polar seas. There is also a vertical temperature stratification in the water column, with the warm surface layer being separated from the main body of the colder deep ocean by the **thermocline**. There are thus three main temperature zones in the vertical ocean.

**1 The surface ocean.** Here, the waters are heated by solar energy and the heat is mixed down to a depth of around 100–200 m, with the temperatures reflecting those of the latitude at which the water is found.

**2 The thermocline.** Below the surface, or mixed, layer, the temperatures decrease rapidly with depth

through a zone (the thermocline) that extends from the base of the surface layer down to as deep as ~1000 m in some places. A permanent **main thermocline** is found at low and mid-latitudes in all the major oceans, but it is absent at high latitudes. This main thermocline is usually less than 1000 m thick and its top is located at shallower depths in the equatorial than in the mid-latitude regions. Seasonal thermoclines are also found in parts of the ocean.

**3 The deep ocean.** Under the base of the thermocline the deep ocean extends to the bottom of the water column. The waters here are cold, most being at ~5°C, and although the temperature falls towards the bottom (to ~1°C) the rate of decrease is very much slower than that found in the thermocline. A generalized vertical temperature profile in the oceans, showing these various zones, is illustrated in Fig. 7.1(b).

If a sample of sea water is brought to the surface from depth without exchanging heat with its surroundings, the temperature will fall as a result of adiabatic cooling and so will be lower than the *in situ* temperature. The temperature that it would have at the surface under atmospheric pressure is termed the **potential temperature** ($\theta$). As $\theta$ is not a function of depth it is a more useful parameter than *in situ* temperature for characterizing water masses and vertical motion in the oceans. Potential temperature is now calculated using *in situ* temperature and the equation of state (see Section 7.2.1 and Worksheet 7.1).

### 7.2.4 Density

The density ($\rho$) of sea water is a function of temperature, salinity and pressure, and the equation of state is a mathematical expression used to calculate the density from measurements of these parameters. The density of sea water exceeds that of pure water owing to the presence of dissolved salts, and the densities of most surface sea waters lie in the range 1024–1028 kg m$^{-3}$. As the values always start with 1000, however, it is common practice to shorten them by introducing the quantity

$$\sigma_{s,t,p} = \rho_{s,t,p} - 1000 \qquad (7.4)$$

This is the *in situ* density, where subscripts $s,t,p$ indicate a function of salinity, temperature and pressure. Thus, the *in situ* density is the density of a sea water sample with the observed salinity and temperature,

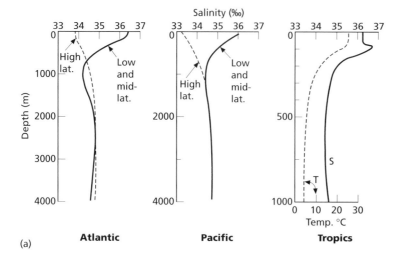

(a)

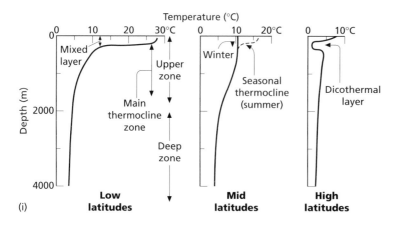

(i)

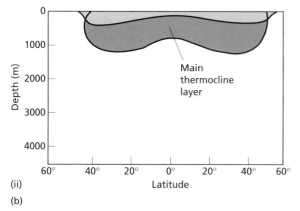

(ii)

(b)

**Fig. 7.1** The salinity, temperature and density structure of the oceans. (a) Depth distribution of salinity (from Pickard & Emery, 1982). (b) (i) Depth distribution of temperature (from Pickard & Emery, 1982); (ii) schematic representation of the thermocline (after Weihaupt, 1979). (c) (i) Depth distribution of density (from Pickard & Emery, 1982); (ii) schematic representation of the pycnocline (after Gross, 1977).

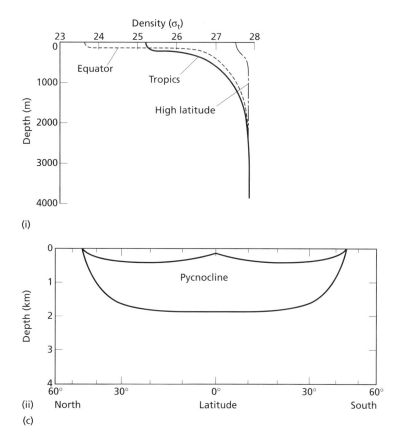

(i)

(ii)    North                          Latitude                         South

**Fig. 7.1** *Continued*    (c)

and the pressure found at the location of the sample. The effect of pressure on density, however, can often be ignored in descriptive oceanography, and for convenience atmospheric pressure is used, and the quantity

$$\sigma_t = \rho_{s,t,0} - 1000 \qquad (7.5)$$

is employed. This is termed 'sigma-tee', and seawater densities are most often quoted as $\sigma_t$ values. As these $\sigma_t$ values are a function of only temperature and salinity, they therefore can be plotted on to temperature–salinity diagrams, which then can be used in the identification of water masses. Alternatively, potential density can be used on potential temperature–salinity diagrams, potential density ($\rho_e$) being the quantity

$$\sigma_\theta = \rho_{s,\theta,0} - 1000 \qquad (7.6)$$

where $\theta$ is the potential temperature (see above). Lines of equal density in vertical or horizontal ocean sections are termed **isopycnals**, and isopycnal surfaces, or horizons, are often used in tracer circulation–distribution studies because they are the preferred surfaces along which lateral mixing takes place (see Section 7.4).

There is a vertical density stratification in the water column, with densities increasing with depth. The most dramatic change in density with depth is found in the upper water layer. In equatorial and mid-latitude regions there is a shallow layer of low-density water under the surface, below which the density increases rapidly. This zone of rapidly increasing density is termed the **pycnocline**, and it is an extremely important oceanographic feature because it acts as a barrier to the mixing of the low-density surface ocean and the high-density deep ocean. The pycnocline is formed in response to the combined rapid vertical changes in salinity (the halocline) and temperature (the thermocline). It was pointed out above, however, that the thermocline is absent in high-latitude waters, and the pycnocline is not developed here either—see Fig. 7.1(c). As a result, the water column at high latitudes is less stable than it is at lower latitudes, and in the absence of the mixing barrier the sinking of dense cold surface water can take place; it is this gravity-mediated sinking that is the driving force behind thermohaline circulation in the oceans (see Section 7.3.3).

## 7.3 Oceanic circulation

### 7.3.1 Introduction

The oceanic water-column distributions of many dissolved constituents (e.g. the trace elements) are controlled by a combination of a **transport process** signal, associated with oceanic circulation, and a **reactive process** signal, associated with the major biogeochemical cycles. In order to evaluate the overall factors that control the distributions of the dissolved constituents in sea water it is necessary therefore to be able to distinguish between the effects of the two signals, i.e. to extract reactive process information from the transport process, or *advective*, background. With this in mind, transport processes associated with oceanic circulation patterns will be described in the present section, the aim being simply to provide the basic information that is needed to interpret the reactive processes acting on dissolved constituents. This is especially relevant to the factors that control the distributions of the trace elements, a topic considered in Chapter 11. There is a wide spectrum of motions in the ocean and the processes that control them are complex. According to H. Leach (personal communication) the motions essentially can be subdivided into four types on the basis of the scales involved. These are:

**1** the **thermohaline circulation**, which operates on a global ocean-scale of $>10^4$ km;

**2** **gyres**, which operate on an ocean-basin scale of $10^3$ km;

**3** **mesoscale eddies**, which operate on scales of 10 to $10^2$ km;

**4** a variety of **small-scale motions** (e.g. surface waves, tidal currents, tidal mixing, Ekman layer transport), which operate on scales of $<10$ km to centimetres. Within the present context, however, it is only necessary to consider the most general features of oceanic circulation.

In the oceans, waters undergo both horizontal and vertical movements. The principal forces driving these movements are **wind** and **gravity**, and the resultant circulation patterns are modified by the effects of factors such as the rotation of the Earth, the topography of the sea bed and the positions of the land masses that form the ocean boundaries. Two basic types of circulation dominate the movement of water in the oceans. In the **surface ocean** the currents are wind-driven, and move in mainly horizontal flows at relatively high velocities. In the **deep ocean** circulation is driven largely by gravity (density changes) and the resultant thermohaline currents, which can have a vertical as well as a horizontal component, flow at relatively low velocities.

### 7.3.2 Surface water circulation

Circulation in the upper layer of the ocean is driven mainly by the response of the surface water to the movement of the wind in the major atmospheric circulation cells, and is constrained by the shape of the ocean basins. The currents that are generated across the air–sea interface decrease in strength downwards to fade away at relatively shallow depths, usually around 100 m. In terms of the major circulation patterns, the way in which the surface ocean responds to wind movement can be evaluated in terms of the two major wind systems, i.e. the trades and the westerlies. In average terms, the trades move diagonally from the east towards the Equator in low latitudes, and the westerlies move diagonally from the west away from the Equator at mid-latitudes. The principal result of these wind movements, combined with the constraint imposed by the shape of the ocean basins, is that in each hemisphere a series of large anticyclonic water circulation cells, termed **gyres**, are set up in subtropical and high-pressure regions. These gyres, which are the dominant feature of the surface water circulation system, are illustrated in Fig. 7.2; note that the water circulates clockwise in the Northern Hemisphere and anticlockwise in the Southern Hemisphere, with the result that there is **divergence** around the Equator.

**Boundary currents** are developed on the landward sides of the gyres as the land masses are encountered. These currents are more intense on the western sides of the ocean basins, and the western boundary currents (e.g. the Gulf Stream and the Kuroshio in the

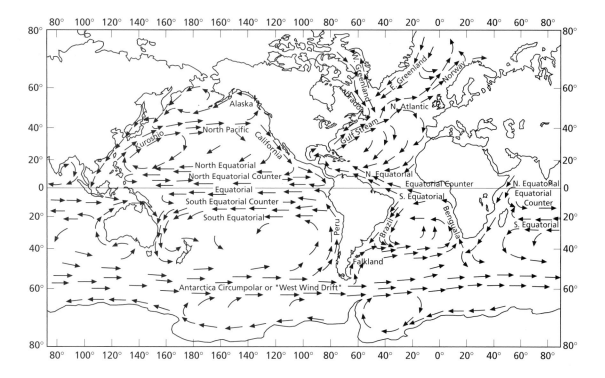

**Fig. 7.2** Surface water circulations in the World Ocean (from Stowe, 1979). Note the central gyre systems.

Northern Hemisphere; and the Brazil and Agulhas Currents in the Southern Hemisphere) are narrow and fast-flowing, and form a sharp break between coastal and open-ocean waters. In contrast, the return flows on the eastern sides of the gyres are less intense and more diffuse in nature.

The main northern and southern gyre flows in the Atlantic and the Pacific are separated by the **equatorial countercurrents.** These are particularly well developed in the Pacific; for example, the strong subsurface Cromwell Current has been traced for over 12 000 km. In the Atlantic, the countercurrents are less intense on the eastern than on the western side of the ocean, and there is an important link between the North and South Atlantic surface waters on the western side where part of the South Equatorial Current crosses the Equator to join the North Equatorial Current—see Fig. 7.2. At their southern edges the South Atlantic, South Pacific and Indian Ocean gyres open out into the Antarctic and are bounded by

the Circumpolar, or **West Wind Drift**, Current—see Fig. 7.2. A circumpolar current of this type is not, however, found at high latitudes in the Northern Hemisphere because of the presence of the land masses there.

The Indian Ocean is a special case in that its northern land boundary is located at relatively low latitudes (~20°N). As a result, surface circulation in the northern Indian Ocean is strongly influenced by conditions on the surrounding land masses and is affected by the monsoon systems. During the winter months, the North Equatorial Current is developed, as it is in the Atlantic and the Pacific, but during the summer in the southwest monsoon the current is replaced by a flow in the opposite direction. Surface circulation patterns in the southern Indian Ocean are, however, similar to those in the South Atlantic and South Pacific.

In addition to the generation of surface currents, wind flow also can be responsible for vertical movement in the oceans in the form of upwelling and downwelling (sometimes termed 'Ekmann pumping'). **Upwelling** is the process by which deep water is transferred to the surface, and is extremely important in controlling primary productivity because it brings nutrients from depth to the surface waters (see Section 9.1.3). This type of upwelling can occur under various conditions, but it is classically associated with the eastern boundary currents (i.e. on the western margins of the continents) in subtropical coastal regions where the wind drives the water offshore, and in order to conserve volume it is replaced by subsurface water from depths of around 200–400 m. **Coastal upwelling** is found (i) on the western margins of the continents off West Africa, Peru and Western Australia; (ii) off Arabia and the east coast of Asia under the influence of the monsoons; and (iii) around Antarctica. **Equatorial upwelling** occurs in non-coastal, or mid-ocean, areas where it can be initiated by, for example, diverging current systems, as happens in the equatorial Pacific. In other openocean areas where the thermocline acts as a mixing barrier, upwelling can take place owing to the erosion of the thermocline by turbulence as a result of an ocean–atmospheric coupling mechanism. This turbulent vertical transport through the thermocline occurs in pulses, and can have a major effect on the output of deep-water nutrients to the euphotic zone in oligotrophic open-ocean areas (see Section 9.1.1).

It may be concluded, therefore, that on a global scale wind-driven circulation, modified mainly by the Coriolis effect and constrained by the shape of the ocean basins, determines the surface current patterns in the oceans. In addition, tidal currents are present everywhere in the sea, but achieve a major importance only in coastal seas and estuaries.

### 7.3.3 Deep-water circulation

The movement of water in the deep ocean is mainly set up in response to gravity. The resultant **thermohaline circulation**, which results from the density difference between waters, is much slower than the surface circulation and involves vertical as well as horizontal motion. In the simplest sense, dense water will sink until it finds its own density level in the water column where it will then spread out horizontally along a density surface, which provides a preferred water transport horizon as well as setting up a density stratification in the oceans. Thus, if new dense water is produced continuously at the surface the displacement of subsurface water as the new water sinks will

lead to the vertical and horizontal thermohaline circulation that keeps the deep ocean in motion.

Mixing between the surface and deep ocean must take place across the pycnocline, which acts as a mixing barrier. The pycnocline is well developed in tropical and temperate waters, but is less well developed, or is absent entirely, at high latitudes where the density of the cold surface waters is greater. It is here, in these high-latitude regions where the mixing barrier does not operate efficiently, that the main sites are found for the sinking of the cold dense surface water, which is responsible for the ventilation of the deep sea.

A generalized deep-water circulation model was proposed by Stommel (1958), and this offers a convenient framework within which to outline the principal features in the transport of deep waters. Two primary sources of deep water are identified. Both are at high latitudes, one in the North Atlantic (the northern component) and one in the Antarctic (the southern component). The northern component deep-water source is in the Norwegian Sea and off the coast of southern Greenland. Here the waters, which have a low surface temperature and a high salinity, sink to form the **North Atlantic Deep Water** (NADW). The southern component originates in the Weddell Sea, where cold, dense, sub-ice water sinks to form **Antarctic Bottom Water** (AABW). The principal sources of deep water are located in the Atlantic and Antarctic, and it is possible to track the deep-water circulation path through the World Ocean from the Atlantic and Antarctic sources following the patterns suggested in Stommel's model, a characteristic feature of which is the existence of strong boundary currents on the western sides of the oceans (see Fig. 7.3).

Taking the pathway proposed in the model, the NADW spills out from the northern basins and begins its global oceanic 'grand tour'. At the start of the tour the NADW flows down the North Atlantic in the strong boundary currents at the western edge of the land masses, crosses the Equator into the South Atlantic and becomes underlain by the AABW from the southern sinking source. Deep-water leaves the Atlantic–Antarctic by moving eastwards below the tip of southern Africa to enter the Indian Ocean via the Circumpolar Current. Some of the deep-water circulates within the Indian Ocean, moving northwards along a western boundary current. Transport out of the ocean into the Pacific is again via the Antarctic

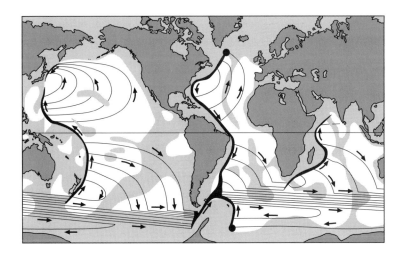

**Fig. 7.3** Deep-water circulation in the World Ocean (from Stommel, 1958). Note the strong western boundary currents.

Circumpolar Current. In the Pacific the water is moved northwards along a western boundary current through the South Pacific into the North Pacific. The overall global oceanic deep-water 'grand tour' is therefore down the Atlantic (~80 yr), through the Antarctic into the Indian Ocean and up through the South Pacific and then the North Pacific (~1000 yr). Thus, the deep-water of the North Pacific is the oldest in the World Ocean, and acts as a sink for other deep-waters. This pathway along which the deep ocean is ventilated has an important impact on the distributions of constituents such as nutrients, trace metals and particulate matter in the ocean system, and this is discussed in subsequent sections.

The water that sinks from the surface to the deep ocean must be replaced and in Stommel's model it is assumed that there is a uniform upwelling throughout the oceans, a process that must be distinguished from the localized (e.g. coastal) wind-driven upwelling described above. On the basis of radiotracer data, Broecker & Peng (1982) estimated that the upwelling of subsurface water occurs at a rate of ~3 m yr$^{-1}$, although somewhat higher values of ~5 to ~7 m yr$^{-1}$ have been proposed by Bolin $et$ $al.$ (1983). In order to reach the surface, deep-water has to cross the thermocline, which acts as a mixing barrier. Various studies, again using radiotracers, have demonstrated that the ventilation of the thermocline takes place, particularly in the equatorial regions of the oceans, from depths of at least 500 m (see e.g. Broecker & Peng, 1982). Large-scale upwelling of subsurface waters also occurs at high latitudes in the Antarctic, where the thermocline mixing barrier is absent.

Up to this point we have considered surface-water (wind driven) and deep-water (density driven) circulation as two separate systems. It is, however, artificial to divide the circulation into these components because nature can satisfy all the forcing mechanisms with a single integrated circulation (see e.g. Gordon, 1996). There have been a number of attempts to define this single circulation.

The principal feature within the surface ocean is the wind-driven gyre circulation, which retains water in the ocean basins. These gyres, however, are not closed and surface water is exchanged between the major oceans. This had led to the concept of a surface-water–deep-water circulation coupling, which acts as a global 'conveyor belt'. In this conveyor belt, the deep-water 'grand tour' transports colder water from the North Atlantic and Antarctic sinking centres to the North Pacific. In one scenario, the 'conveyor belt' was thought of as a **single** overturning global-scale cell that transports water around the planet from the North Atlantic sinking centre to the North Pacific (colder water) and back to the North Atlantic (warmer water). However, there has been considerable debate over the existence of this type of global circulation. In any global circulation model the return of the surface water to the sinking centres hinges on connecting passages between the major ocean basins. In this context, Macdonald & Wunsch (1996) presented a coherent global ocean circulation model, in which **two** nearly independent cells were envisaged, one connecting overturning in the Atlantic to other basins through the Southern Ocean, and the other connecting the Indian and Pacific basins through the Indonesian archipelago.

### 7.3.4 Mid-depth water circulation: water types and water masses

So far, a brief outline has been given of the circulation patterns at the upper and lower levels of the oceanic water column. It is now necessary to consider what happens in the mid-depth, or intermediate, water column, and for this the concept of a water mass will be introduced. Because there are no appreciable sinks for heat and salt in the interior of the ocean, the temperature and salinity of a water, both of which are conservative parameters, will have been fixed once it leaves the surface ocean and can be changed only by mixing with other waters having different properties. An exception to this is provided by heat transmitted through the oceanic basement, which can raise the temperature of bottom waters. Temperature and salinity therefore are very useful properties for characterizing sea waters of different origins, and a number of waters with different temperature–salinity signatures are found in vertical sections of the water column. The relationship between the temperature ($T$) and the salinity ($S$) of a water can be illustrated graphically on a *TS* diagram, or more usually a potential temperature–salinity diagram, which is one of the most widely used diagnostic tools in physical oceanography.

A **water type** is a body of unmixed water that is defined by a single temperature and a single salinity value, and therefore plots on a *TS* diagram as a single point. Thus, the temperature and salinity of a water type do not change. Changes do occur, however, on the mixing of water types and this leads to the identification of a **water mass**, which has a range of temperatures and salinities (and other properties) and is characterized on a *TS* diagram by a *TS* curve, instead of a single point. A well-defined water mass is one having a linear potential temperature–salinity relationship.

The structure of the water column can be elucidated in terms of water masses, and this can be illustrated with respect to the Atlantic—see Fig. 7.4. The NADW is formed at the northern source and sinks to the bottom to be overlain by the Atlantic Intermediate Water (AIW). The NADW flows northwards across the Equator where it is underlain by the AABW. Surface and near-surface waters in the South Atlantic are underlain by the Antarctic Intermediate Water (AAIW), which is formed by the sinking of surface water at the Antarctic convergence ($\sim$50°S), and extends into the North Atlantic as far as $\sim$10–20°N. Between $\sim$20 and $\sim$40°N there is an intrusion of high-salinity Mediterranean Water, which flows east to west.

## 7.4 Tracers

The water column is formed of a series of water types and water masses, which sink to a level dictated by their density, displacing other waters as they do so. As the formation of water types is a continuous process, the oceans are kept continually in motion as water circulates around the system. One of the principal thrusts in chemical oceanography over the past few years has been the use of various tracers to determine the rates at which the system operates. Tracers have also been used to elucidate the routes followed by components taking part in oceanic biogeochemical processes, and to study the rates at which the processes occur.

**Conservative tracers** have no oceanic sources, or

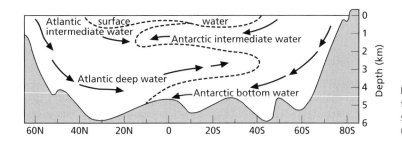

**Fig. 7.4** Cross-section of the Atlantic from Greenland to Antarctica, with schematic representation of water masses (from Stowe, 1979).

sinks, and their distributions depend only on the transport of water. In contrast, **non-conservative tracers** undergo a variety of reactions within the system; these include biogeochemical cycling, radioactive decay, isotopic fractionation and exchange with the atmosphere. Tracers used to date have included temperature, salinity, dissolved oxygen, the nutrients and, more recently, trace metals, stable and radioactive isotopes and the 'Freons'. For some oceanic experiments tracers are deliberately added to sea water; these include various dyes and substances such as $SF_6$ (sulphur hexafluoride).

The use of tracers such as these was considerably advanced by data obtained during GEOSECS (Geochemical Ocean Sections Study). This particularly was the case with respect to the penetration of anthropogenic tritium and $^{14}C$ from the atmosphere into the thermocline and deep water, thus allowing the surface waters of the World Ocean into which they initially entered to be labelled and traced during their oceanic passage. A detailed review of the findings that emerged from GEOSECS has been compiled by Campbell (1983). A follow-up programme, TTO (Transient Tracers in the Ocean), was started in 1980 and a series of papers on the initial results was published in 1985 in *J. Geophys. Res.*, **90**. Tracers also have been used extensively in the WOCE (World Ocean Circulation Experiment) programme. Tracers are proving to be invaluable tools for oceanographers, and an excellent overview of their use in ocean studies has been provided by Broecker & Peng (1982); this is one of the classic texts in chemical oceanography, and will serve for many years as the 'tracer Bible'. For this reason, tracers will not be treated in detail in the present volume, and only a very general summary of their applications will be given in order to relate their use to the wider aspects of marine geochemistry.

The tracers used in oceanography can be divided into three broad groups: those used to identify water masses, those used to study water transport, and those used to study geochemical processes.

### 7.4.1 Water-mass tracers

Parameters that have been used to define water masses include temperature, salinity, dissolved oxygen, phosphate, nitrate and silica. However,

dissolved oxygen, phosphate and nitrate are non-conservative tracers, i.e. their concentrations change during the residence of the water in the deep sea. Water masses originate from the mixing of water types and to unscramble the mixtures it is necessary therefore to use conservative tracers, i.e. those that are not changed by processes such as respiration or radioactive decay during their residence time in the deep sea. Tracers of this type include: (i) salinity (conductivity); (ii) temperature; (iii) $SiO_2$ and Ba; (iv) the isotopes $^2H$ and $^{18}O$; and (v) 'NO' and 'PO', which were introduced by Broecker (1974). For a detailed treatment of the use of these various tracers in the identification of water masses, the reader is referred to Broecker (1981), Broecker & Peng (1982) and Campbell (1983). Other types of tracers also are used for water-mass studies. These include the chlorofluoromethanes (CFMs). The two CFMs ($CCl_3F$ (Freon-11) and $CCl_2F_2$ (Freon-12)) measured most commonly have no known natural sources and have been increasing in concentration in the atmosphere since their commercial production started in the 1930s. The CFMs enter the ocean across the air–sea interface (see Section 8.2), and their concentrations in surface sea water are functions of their concentrations in the overlying atmosphere and their water solubilities, which are dependent on the temperature and salinity of the sea water. According to Fine & Molinari (1988), the CFMs have their oceanic sources at outcrops where water masses are ventilated by the atmosphere, and as a result relatively high CFM concentrations are indicative of waters that have had recent contact with the atmosphere at the source regions and have not suffered extensive dilution by mixing with other waters having lower CFM concentrations. Thus, as the CFMs are transported from the surface into the interior of the ocean their distributions can be used to trace mixing and circulation patterns. For example, Fine & Molinari (1988) used CFMs, together with other parameters, to investigate the extent of the Deep Western Boundary Current in the North Atlantic.

### 7.4.2 Water-transport tracers

The main emphasis on water transport in recent years has involved radioisotopes, which, for this purpose, can be classified into two types, depending on whether their distributions are influenced primarily by (i) transport in the water phase; or (ii) transport in the particulate phase. **Water tracers** include the following: the anthropogenically produced transient tracers $^{90}$Sr (half-life, 28.6 yr), $^{137}$Cs (half-life, 30.2 yr), $^{85}$Kr (half-life, 10.7 yr) and the Freons; the natural steady-state tracers $^{39}$Ar (half-life, 270 yr), $^{228}$Ra (half-life, 5.8 yr) and $^{32}$Si (half-life, 250 yr); and the mixed-origin tracers $^{14}$C (half-life, 5730 yr) and $^3$He (half-life, 12.4 yr). **Particulate-phase tracers** include: $^{210}$Pb (half-life, 22.3 yr), $^{228}$Th (half-life, 1.9 yr), $^{210}$Po (half-life, 0.38 yr) and $^{234}$Th (half-life, 0.07 yr). Together, these two types of tracers yield information on ventilation and mixing in the subsurface ocean and on the origin, movement and fate of particulate matter in the system. In addition to these classic tracers, a number of individual chemical elements have been used to trace water-mass movements. For example, Measures & Edmond (1988) used dissolved Al as a tracer for the outflow of Western Mediterranean Deep Water into the North Atlantic. Other tracers that have been used for the elucidation of circulation and mixing in the oceans include the rare-earth elements (REE) (see e.g. Elderfield, 1988) and Mn (see e.g. Burton & Statham, 1988).

### 7.4.3 Geochemical process tracers

Plotting down-column dissolved trace element concentrations against a conservative tracer, e.g. salinity or potential temperature, and interpreting the shape of the curve can provide evidence on geochemical processes; for example, a concave shape indicates deep-water removal, or scavenging, of the element. However, the natural series radioisotopes are the most widely used geochemical process tracers. These isotopic clocks, which operate on time-scales ranging from a few days (e.g. $^{234}$Th, half-life, 24 days), through a few tens of years (e.g. $^{210}$Pb, half-life, 22 yr), to several thousand years (e.g. $^{230}$Th, half-life, $7.52 \times 10^4$ yr), have been used for a wide variety of biogeochemical purposes. These include tracing the origin of atmospherically transported components (e.g. $^{210}$Pb), gas exchange across the air–sea interface (e.g. $^{222}$Rn), trace element scavenging in the water column (e.g. $^{230,228,234}$Th), trace element down-column fluxes (e.g. $^{234}$Th) and trace metal accumulation rates and bioturbation in sediments (e.g. $^{234}$Th, $^{210}$Pb). In addition, considerable use has been made in

the past of artificial radionuclides, and this was revived following the Chernobyl incident.

Tracers have provided information on a wide variety of oceanographic topics, and from the point of view of their importance to our understanding of the processes involved in marine geochemistry it is worthwhile highlighting a number of these.

**1 Deep-sea residence time.** According to Broecker (1981) the great triumph of natural radiocarbon measurements in the oceans has been the establishment of a 1000 yr time-scale for the residence of water in the deep sea.

**2 Thermocline ventilation time.** Carbon-14 and $^3$He data have been used to establish the ventilation time of the main oceanic thermocline. For example, the findings suggest that there is an upwelling flux in the equatorial regions that on a global scale is comparable to the flux of newly formed deep water.

**3 The rates of vertical mixing.** In the more simple box models used by marine chemists it is often the practice to separate a surface-ocean and a deep-ocean compartment. In order to apply such models to biogeochemical processes it is necessary to have at least an approximation of the rate at which the reservoirs mix. There are various problems involved in making such approximations, but Collier & Edmond (1984) used the $^{14}$C deep-water residence and a deep-water reservoir thickness of 3200 m to derive a mixing flux of 3.5 m yr$^{-1}$ for the mixing rate between the surface and deep ocean (see Section 7.5).

**4 The dispersion of hydrothermal solutions.** Helium isotopes have been used to predict the mid-depth circulation of hydrothermal fluids vented at sea-floor spreading centres (see Section 6.4).

**5 Oceanic biogeochemical processes.** Important advances have been made in our understanding of geochemical processes through the use of the radionuclide 'time clock' tracers. For example, Bacon & Anderson (1982) used Th isotopes to investigate particulate–dissolved exchange in the water column and demonstrated that reversible equilibria take place as the particulate matter falls to the sea bed. Other work of note includes that of Coale & Bruland (1985), who have shown that the removal rates of $^{234}$Th correlate closely with primary productivity, and that non-reversible equilibria are involved in dissolved–particulate interactions on these short time-scales.

**6 Palaeooceanography.** Oxygen isotopes in carbonates have been used for the identification of past climatic conditions (see e.g. Broecker, 1974).

**7 Atmospheric transport.** Lead-210 has been used as an atmospheric source indicator (see e.g. Schaule & Patterson, 1981).

**8 Gas exchange.** Radon-222, and its disequilibrium with the parent $^{226}$Rn, has been used as a gas tracer in the oceans to model air–sea exchange processes.

**9 Sedimentary processes.** Bacon & Rosholt (1984) and Thompson *et al.* (1984) have used thorium isotopes to elucidate trace element accumulation rates in deep-sea deposits. In addition, various isotopes have been used to investigate processes such as bioturbation, and to identify sources of individual components in sediments; the latter has involved elements such as the rare earths (see e.g. Bender *et al.*, 1971) and Sr isotopes (see e.g. Dasch *et al.*, 1971).

**10 The oceanic cycles of elements.** The extent to which the isotopes of an element undergo fractionation is often characteristic of specific geochemical processes, and isotopic tracers have proved to be extremely useful in elucidating the marine cycles of a number of elements. An example of this is given in Worksheet 7.2, with respect to the marine cycle of Sr.

---

**Worksheet 7.2: The use of isotopes in the study of the marine cycles of elements**

An example of how isotopic ratios can be used as 'process tracers' in marine geochemistry has been provided by Palmer & Elderfield (1985), who used Sr isotopes to evaluate long-term variations in the compositions of sea water by modelling recent data on $^{87}$Sr : $^{86}$Sr variations to establish the marine geochemical cycle of the element. A Sr isotope curve was produced from dated samples showing how the ratio has varied with time over the past 75 Ma—see Fig. (i).

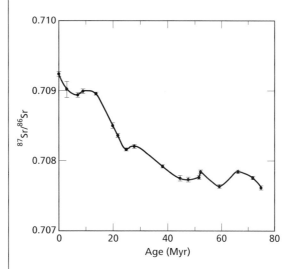

**Fig. (i)** The marine $^{87}$Sr : $^{86}$Sr ratio over the past 75 Ma (after Palmer & Elderfield, 1985).

*continued on p. 152*

The outstanding feature in the curve is the increase in $^{87}$Sr : $^{86}$Sr ratios over the past *c*. 25 Ma. Strontium isotopes can be fractionated during different geochemical processes, and the principal factors that were originally thought to control variations in the marine $^{87}$Sr : $^{86}$Sr ratio were: (i) the proportions of marine Sr derived from crustal sources (mainly granitic) and mantle sources (continental basalts and hydrothermal activity); (ii) the rate at which Sr is transported from the continents to the oceans; and (iii) the recycling of Sr from marine calcareous sediments. To compare the relative importance of processes such as these, a model was constructed for the marine geochemical cycle of Sr. This model incorporated the best currently available estimates of the magnitudes, and the Sr isotope ratios, associated with present-day fluxes, and is illustrated in Fig. (ii).

In the model the principal flux of dissolved Sr to sea water is via river run-off, with *c*. 75% originating from recycled Sr associated with the weathering of uplifted limestones, and the rest from silicate rocks. The other fluxes result from the diffusion of Sr from the pore waters of marine sediments (mainly carbonate-rich types), and from Sr isotope exchange between sea water and basalt during high-temperature hydrothermal activity at the ridge spreading centres. The authors then considered the effects that temporal variations in these fluxes would have on the marine $^{87}$Sr : $^{86}$Sr ratio. In this way, they were able to show that variations in the ratio resulting from changes in either the rates of hydrothermal activity, or in the recycling of Sr via diffusion from carbonate-rich sediments, were unable to account fully for the observed variations in the marine Sr isotopes. They concluded, therefore, that changes in the isotopic ratio of the fluvial flux must be a major control on the temporal variations in the $^{87}$Sr : $^{86}$Sr in sea water. Further, by modelling the various factors that can affect the Sr isotopic ratio in the fluvial flux, they concluded that the only processes that could adequately account for the temporal variations in the marine Sr isotopic ratios are changes in the $^{87}$Sr : $^{86}$Sr ratio derived from the weathering of silicate rocks. The changes in the Sr isotopic ratios derived from silicate rock weathering that are necessary to produce the observed marine $^{87}$Sr : $^{86}$Sr ratios are illustrated in Fig. (iii).

The principal feature of the curve illustrated in Fig. (iii) is the sharp rise in $^{87}$Sr : $^{86}$Sr ratios from *c*. 25 Ma BP to the present day. Palmer & Elderfield (1985) pointed out that this feature correlates with two important variables in geological history.

1 It is similar to the curve for the intensity of glacial activity over the past 75 Ma, thus providing evidence to confirm an earlier suggestion that increased glacial action had an effect on the marine $^{87}$Sr : $^{86}$Sr ratio by exposing, and increasing the weathering rates of, ancient shield regions in terrains that have relatively high $^{87}$Sr : $^{86}$Sr ratios.

*continued*

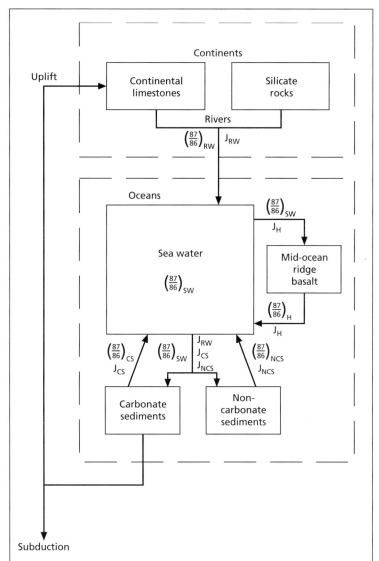

**Fig. (ii)**  A model for the marine geochemical cycle of Sr showing the magnitudes of the fluxes, $J$, involved together with their associated $^{87}Sr : ^{86}Sr$ isotopic ratios (87/86) (from Palmer & Elderfield, 1985). At steady state, $[87]_{SW} = ([87]_{RW}J_{RW} + [87]_H J_H + [87]_{CS}J_{CS} + [87]_{NCS}J_{NCS})/(J_{RW} + J_H + J_{CS} + J_{NCS})$ and $[86]_{SW} = ([86]_{RW}J_{RW} + [86]_H J_H + [86]_{CS}J_{CS} + [86]_{NCS}J_{NCS})/(J_{RW} + J_H + J_{CS} + J_{NCS})$, where SW = sea water, RW = river water, H = hydrothermal isotope exchange, CS = carbonate sediment pore waters and NCS = non-carbonate sediment pore waters. Magnitudes of present-day fluxes $(mol\,yr^{-1})$ are: $J_{RW} = 2.5 \times 10^{10}$, $J_{NCS} = 4.0 \times 10^8$, $J_{CS} = 3.0 \times 10^9$, $J_H = 1.2 \times 10^{10}$ or $0.38 \times 10^{10}$. $^{87}Sr$ ratios of present-day fluxes are: $(87/86)_{NCS} = 0.7064$, $(87/86)_{CS} = 0.7087$, $(87/86)_H = 0.7040$ and $(87/86)_{SW} = 0.70924$. Sensitivity of model to uncertainties in above values: resulting uncertainty in calculated $(87/86)_{SW}$ is $2 \times 10^{-4}$ for uncertainty in $J_{RW}$ and $1.4 \times 10^{-4}$ for uncertainty in $(87/86)_H$; all other variables produce negligible uncertainties in $(87/86)_{SW}$ or are considered in the original text.

*continued on p. 154*

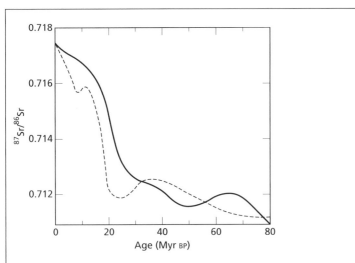

**Fig. (iii)** Variations in the $^{87}Sr:^{86}Sr$ ratio derived from the weathering of silicate rocks; the broken curve indicates qualitative variations in glacial intensity (from Palmer & Elderfield, 1985).

2 The Alpine and Himalayan orogenies occurred over the last 30 Ma and resulted in the uplift, exposure and accelerated erosion of rocks having relatively high $^{87}Sr:^{86}Sr$ ratios.

The authors concluded, therefore, that although hydrothermal fluxes and carbonate recycling are extremely important in determining the *absolute* value of the marine $^{87}Sr:^{86}Sr$ ratio at any one time, the principal control over the *variations* in the ratio during the past 75 Ma has been changes in the isotopic composition of the Sr derived from the weathering of silicate rocks.

Even from this very brief survey it is apparent that the utilization of tracers has provided invaluable insights into the processes involved in the mixing and transport of waters in the oceans, and especially into the rates at which the system operates. It is this mixing, and the subsequent transport of the waters, that controls the structure of the oceans. The transport of water during oceanic circulation also exerts a fundamental control on the distributions of both dissolved and particulate constituents in the system. This control is conservative, and compositional changes take place only as the result of the mixing of end-member waters that have different compositions acquired at their sources. Superimposed on this conservative circulation control is a second effect produced by the involvement of many constituents in the major biogeochemical processes that operate within the oceans, and which lead to non-conservative behaviour. It is possible therefore to make an important distinction between two genetically different types of signals in the oceans: (i) one resulting from the mixing and transport of water masses; and (ii) one resulting from internal biogeochemical oceanic reactivity.

In order to gain an understanding of how these two types of signals interact to control the oceanic cycles of elements it is useful to set up ocean models designed to trace the pathways by which materials move through the system. In the followng section a simple two-box model will therefore be outlined for use as a framework within which to describe the distributions of dissolved and particulate components in the oceans.

## 7.5 An ocean model

### 7.5.1 Introduction

Mixing in the water column involves a combination of advection and diffusion. **Advection** is a large-scale transport involving the net movement of water from one point in the ocean to another. This may occur either in a horizontal direction, e.g. along a major current system, or in a vertical direction, e.g. during upwelling or sinking. Superimposed on this net water transport are the effects of turbulent mixing, or turbulent exchange; this is a more or less random mixing caused by **diffusion**, in which an exchange of properties takes place between waters by eddy, and to a much lesser extent molecular, diffusion without any overall net transport of the water itself.

There are, however, problems involved in understanding the real nature of vertical and horizontal mixing in the oceans. Some of these problems have been identified by Broecker & Peng (1982), who distinguished between **isopycnal** and **diapycnal** mixing. Isopycnal surfaces, along which potential density remains constant, are the preferred surfaces along which lateral mixing takes place. Diapycnal mixing results when water is carried in a perpendicular direction across isopycnal surfaces. In the interior of the ocean isopycnal surfaces are almost horizontal, but at high latitudes they rise towards the surface, where they can outcrop. The authors suggested therefore that the concept of isopycnal and diapycnal mixing should replace that of horizontal and vertical mixing.

Although a wide variety of models are available, many marine chemists prefer to adopt a discontinuous, or **box model**, approach to gain a first-order, or perhaps even a zero-order, understanding of biogeochemical processes within the ocean system. Box models of varying complexity have been used for the interpretation of the distributions of various components in the World Ocean, using tracer-derived data for the rates at which the system operates. The general principles involved in these models are described below.

In the *steady-state* ocean a component is removed from the system at the same rate at which it is added, i.e. the input balances the output, and the concentrations at any point in the system do not change with time. To construct an oceanic model for a constituent it is necessary therefore to have data on (i) magnitude and rates of the primary input mechanisms; (ii) the rates of exchange between the oceanic reservoirs involved; and (iii) the magnitudes and rates of the primary output mechanisms. The *primary* inputs to the oceans are mainly via river run-off, atmospheric deposition and hydrothermal exhalations (see Chapters 3–6). Exchange between oceanic reservoirs occurs through the advective transport of water during thermohaline circulation (downwelling and upwelling), by turbulent exchange and by the sinking of particulate matter from the surface to the deep ocean. Output from the system for most elements is via the burial of particulate material in sediments, although reactions of various kinds between sea water and the basement rocks also can act as sinks for some elements (see Chapter 5). In the steady-state ocean, the amount of a component entering a reservoir must balance the amount leaving. For example, if the concentration of an element in the surface ocean is to remain constant, the gain from primary inputs and upwelling must be matched by the outputs from downwelling and particle sinking (Broecker, 1974). Thus, in the system as a whole, the 'through flux' of a constituent entering via the primary input mechanisms must equal the total permanently lost to the output mechanisms.

### 7.5.2 A simple two-box oceanic model

It was shown in Section 7.3 that different processes control the circulation in the surface ocean (wind-driven) and the deep ocean (gravity-driven). The simplest two-box model therefore divides the ocean into these two reservoirs, i.e. a **surface water reservoir** (~2% of the total volume of the ocean) and a **cold deep-water reservoir** (~80% of the total volume of the ocean), which are separated by the waters of the **thermocline** (~18% of the total volume of the ocean). Both reservoirs are assumed to be well mixed and interconnected, and they interact via vertical mixing, i.e. the downwelling of surface water and the upwelling of deep water and the winter turnover of the cooled upper layer, and via the sinking of coarse particulate matter (CPM; see Section 10.4) from the surface ocean.

This type of two-box model is a useful first approximation of the ocean system because the main barrier to mixing in the water column is the thermocline, which separates the two main reservoirs, the waters of which have different densities. Two other factors also are important in distinguishing between the two reservoirs:

1 two of the principal primary input mechanisms, i.e. river run-off and atmospheric deposition, supply material to the surface ocean;

2 the bulk of the phytoplankton and zooplankton biomass, which is involved in the oceanic cycles of many components, is found in the upper sunlit water layer of the surface reservoir (see Section 9.1.3.1).

A number of parameters are used to evaluate the rate of exchange of water between the surface and deep reservoirs. A $^{14}C$ residence time of ~900 yr has been proposed for the deep ocean. This means that at the rate at which deep-water circulates in the World Ocean it takes ~900 yr for water to go through the cycle of sinking at the poles, circulating through the deep ocean and returning to the surface. By rounding up the estimate to 1000 yr, Broecker & Peng (1982) calculated that the yearly volume of water exchange between the deep and surface ocean is equal to a layer ~300 cm thick with an area equal to that of the ocean, i.e. a mixing flux of 3 m yr$^{-1}$. Thus, the $^{14}C$ data indicate that upwelling brings an amount of water to the surface equal to an ocean-wide layer of

~300 cm. For comparative purposes, Broecker (1974) estimated that fluvial input to the ocean from continental run-off would yield a layer ~10 cm thick if it was spread over the entire ocean.

In order to illustrate the principles involved in incorporating data of this kind into oceanic box models, a generalized two-box model that has been applied to the marine geochemical cycles of trace metals is illustrated in Fig. 7.5.

## 7.6 Characterizing oceanic water-column sections

Various parameters that characterize the water in the oceans have now been described. From the viewpoint of marine geochemistry the question that must now be addressed is 'Which of these paremeters must be measured on a semi-routine basis during the occupation of water-column sections in order (i) to set the location in an oceanographic context, and (ii) to provide data that can be used to interpret biogeochemical processes?'

1 To set an individual seawater section into an oceanographic context, i.e. to describe the nature of the water, it is usual to prepare vertical profiles of temperature, salinity, dissolved oxygen and the nutrients; other parameters also may be measured during individual investigations. These various parameters, either singly or in combination, can be used for pur-

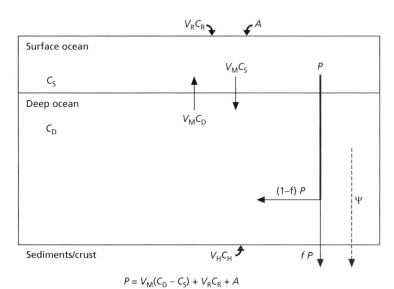

$$P = V_M(C_D - C_S) + V_R C_R + A$$

**Fig. 7.5** A generalized two-box ocean model applied to the marine geochemical cycles of trace elements (from Collier & Edmond, 1984): $P$, particulate matter flux out of the surface ocean; $V_R$, $C_R$, volume and concentration of an element in river water; $A$, atmospheric input; $C_S$, $C_D$, dissolved concentrations of an element in the surface and deep reservoirs; $f$, fraction of particulate matter preserved in sediments; $\psi$, additional particulate matter flux resulting from scavenging within the deep ocean; $V_H$, $C_H$, volume and concentration of an element in hydrothermal solutions. The particulate matter flux ($P$) is calculated by the mass balance of all other inputs and outputs to the surface ocean reservoir.

poses such as defining the depth of the mixed layer, describing the structure and location of the thermocline, and identifying the different water masses (and therefore water sources) that are cut by the section.

2 Data that are useful for the interpretation of biogeochemical processes include dissolved oxygen (to characterize redox-mediated reactions), the nutrients (to characterize the involvement of elements in biological cycles), total particulate matter (to characterize the involvement of elements in particle-scavenging reactions), chlorophyll (to characterize the standing crop and primary production), DOC and POC (to characterize the oceanic carbon cycle and global carbon flux), and a variety of the tracers listed in Section 7.4 (the choice depending on the nature of a specific investigation); however, it must be stressed that not all these parameters are measured on a purely routine basis. The use of a variety of water-column parameters in a marine geochemical context is illustrated in Worksheet 7.3. Further examples of

---

**Worksheet 7.3: Characterizing the water column for geochemical studies**

The distribution of 'reactive' dissolved constituents in sea water is controlled by a combination of oceanic circulation patterns and the effects of internal reactive processes. A large proportion of the chemical signal for these constituents, however, can be determined by long-distance advection and the mixing of water masses of different end-member compositions, and it is necessary therefore to extract chemical information from the advective background. One way of doing this for trace elements is to relate the distribution of the element to an 'analogue' species that has a well-understood distribution (see Section 11.1). This 'advective–chemical' approach, which combines data for a variety of water-column parameters, was used by Chan *et al.* (1977) in their study of the distribution of dissolved barium in the Atlantic Ocean. This study can be used to illustrate how a variety of the seawater parameters described in the text can be used both to characterize the water column and to interpret the biogeochemical controls on the distribution of trace elements in the oceans. In this context, the use of a number of these water-column parameters is described below in relation to a variety of geochemical applications.

*'Species analogue' interpretation of trace-element data*
Physical circulation is the dominant control on the distribution of dissolved Ba, and many other dissolved constituents, in the basins of the Atlantic, and in order to extract chemical information from the advective background Chan *et al.* (1977) related the distribution of the element to those of other species that have well-understood distributions. The results showed that the water-column distributions of barium mimic those of the refractory nutrient silicate. This relationship is illustrated in Fig. (i) for a station southeast of Iceland, and shows the close correspondence between silicate and Ba.

*continued on p. 158*

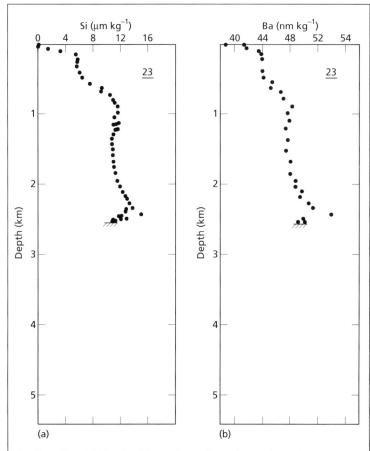

**Fig. (i)** Profiles of Si (a) and Ba (b) at station 23 (from Chan *et al.*, 1977).

Chan *et al.* (1977) were able to use this Ba–silicate correlation to extract chemical information from the advective background by demonstrating that the marine geochemistry of barium is dominated by its involvement in the oceanic biogeochemical cycle in which it takes part, in a deep-water regeneration cycle associated with a slowly dissolving refractory, non-labile, phase (mimicked by silicate), rather than a more rapidly dissolving, labile, tissue phase (mimicked by nitrate and phosphate) that is regenerated at higher levels in the water column. The Ba–silica relationship is not an exact one, however, and this can be seen in property–property diagrams for GEOSECS station 37; this is located in the western Atlantic at a site where the major water-mass cores are well developed. The silica–Ba plot for this station is illustrated in Fig. (ii). It is apparent that this plot consists of straight-line segments connecting end-member waters of different compositions; e.g. the Antarctic water masses are strongly enriched in silica relative to barium

*continued*

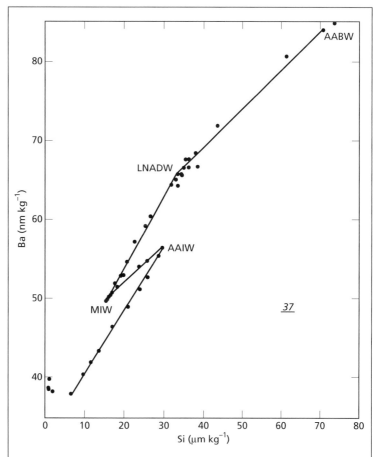

**Fig. (ii)** Plot of Ba against Si at station 37 (from Chan *et al.*, 1977).

compared with the North Atlantic deep-water. Property–property plots such as this therefore can prove extremely useful for elucidating the complex interplay between chemical and physical features in the distributions of dissolved constituents in the ocean.

In general, the oceanic distribution of Ba mimics that of silicate and further details on the physical features that affect Ba can be obtained by combining data for the element with that for silicate, which can be used to characterize water masses. Two such applications using this approach are described below.

*The application of property–property plots to distinguish between water masses*
Silicate is the 'analogue' for dissolved Ba, and as the silicate concentrations of water masses can differ, silicate–property plots can be used to

*continued on p. 160*

distinguish between individual water masses. This can be illustrated in terms of silicate–potential temperature plots for the waters of a series of high-latitude South Atlantic stations. In the region around the Weddell Sea the circulation is controlled by the interaction of the warmer, more silica-rich, Circumpolar Current (CPC) water with the colder, less silica-rich, Weddell Sea Water (WSW). The relationship can be displayed on a $\theta$–Si plot and is illustrated in Fig. (iii) for deep waters from a number of stations lying between the South Sandwich Trench and the Atlantic Indian Ridge.

Information can be extracted from the type of plot shown in Fig. (iii) in terms of identifying both (1) overall water-mass trends and (2) variations at individual stations.

1 The make-up of the deep-water column. The waters at these high-altitude stations have a range of properties intermediate between those of the CPC and WSW end-members. The property–property plot shows a broad silicate maximum centred around $\sim$–0.2°C, with values that are close to those for the CPC ($\sim$129 µmol kg$^{-1}$) at stations 86 and 87. Below the silicate maximum there is clear evidence of the presence of colder, less silica-rich Weddell Sea Water ($\theta$, –0.88; silicate,

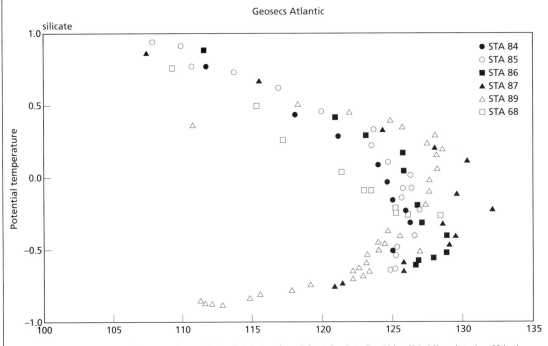

**Fig. (iii)** Si–$\theta$ plots for stations between the South Sandwich Trench and the Atlantic Indian Ridge (84–90) and station 68 in the Argentine Basin (after Chan *et al.*, 1977; not all original data points are plotted).

*continued*

111.5 µmol kg⁻¹), the most extreme values for the Weddell Sea Bottom Water being at station 89.

2 Effects at individual stations. On the basis of the data for the different stations, it would appear that the influence of the CPC proper on the deep-waters has its strongest effect at stations 86 and 87, but that it has a lesser effect on stations 84, 85 and 89.

*Characterizing the water column*

The hydrography of the Atlantic is dominated by the horizontal advection and vertical mixing of water mass cores formed at high latitudes. The formation processes in the Northern and Southern Hemispheres are different and therefore so are the physical and chemical properties of the waters produced in the two regions. In the North Atlantic the water masses are formed by the cooling and downwelling of surface waters that have tropical and subtropical origins, and the cores formed have *high salinities* and *low nutrient* concentrations. The waters are formed in a number of regions (see Section 7.3.3). The most extreme water types are produced in late winter in the Norwegian and Greenland Seas via a modification of high-salinity water transported northwards through the passage around Iceland. The densest water formed is stored within deep basins, but there is a southwards return flow of waters of less extreme properties, which are determined by the source depths at which they originated and vertical mixing both in the overflow regions themselves and during descent to the deep-ocean floor. The main flows of these water masses are through the Denmark Strait, the Southwestern Faroe Channel and across the Wyville–Thompson Ridge between Faroe and Orkney. These water masses, together with one formed in the Labrador Sea, follow a number of complex trajectories and are then propagated southwards as a composite termed the North Atlantic Deep Water (NADW). Mediterranean Deep Water, which outflows through the straits of Gibraltar, forms an additional core that spreads across the North Atlantic and forms the upper boundary of the deep-water that moves into the South Atlantic.

The great complexity of the water column, which is characteristic of the formation areas of the NADW, can be illustrated by the chemical and hydrographic data obtained for the station southeast of Iceland (station 23). These data can be used to characterize the water column in terms of water masses, and to illustrate this plots of salinity and silicate against potential temperature are shown in Fig. (iv); some features of this figure should be interpreted in conjunction with Fig. (i).

On the basis of these plots the characterization of the water column at station 23 can be summarized as follows. The upper part of the column is occupied by high-salinity waters of a central Atlantic origin.

*continued on p. 162*

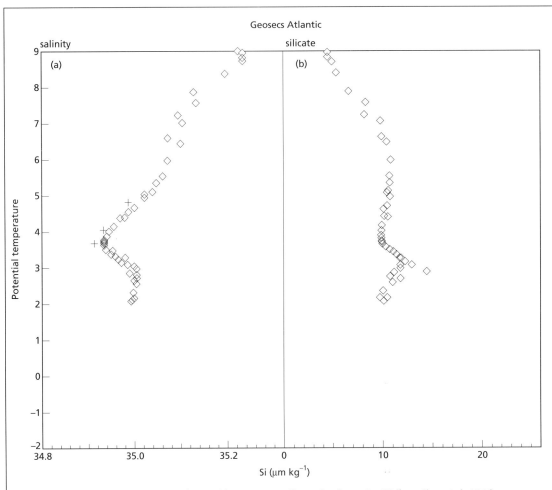

**Fig. (iv)** Potential temperature–salinity and potential temperature–silicate plots for station 23 (from Chan *et al.*, 1977).

At ~800 m there is an inflection in potential temperature and salinity and an increase in silicate. This, together with other properties not illustrated here (e.g. a pronounced oxygen minimum), indicates the presence of the core of the Mediterranean Intermediate Water (MIW). At depths below this there is a broad minimum in salinity and silicate (and an oxygen maximum), which shows the presence of the Labrador Sea Water. Below the salinity and silicate minima, i.e. at depths ≳1600 m, the silicate values increase to reach a well-defined maximum (15.2 μm kg⁻¹) at ~2400 m, then fall off rapidly towards the bottom (11.0 μm kg⁻¹) at 2516 m. The values for potential temperature and salinity also decrease in a similar manner. These features were interpreted as representing the residual core of the Antarctic Bottom Water (AABW; high silicate, intermediate salinity) overriding the Arctic

*continued*

Bottom Water (ABW; low silicate, high salinity) from the Faroe Channel; i.e. the AABW has penetrated to these high latitudes in the Northern Hemisphere. The θ–S properties of the ABW at station 23 indicate that the properties of the original Norwegian Sea Water, which has overflowed via the Faroe Channel, have been strongly modified by mixing with warmer, more saline Atlantic waters as it flowed down the slope.

the manner in which oceanographic-related and process-related water-column parameters are used in marine geochemistry will be described later in the text, when the factors controlling the elemental compositions of the seawater and the sediment reservoirs are described.

## 7.7 Water-column parameters: summary

**1** The oceanic water column can be divided into two layers: a thin, warm, less dense surface layer that caps a thick, cold, more dense deep-water layer. The two layers are separated by the thermocline and the pycnocline, which act as barriers to water mixing.

**2** Currents in the surface layer of the ocean are wind-driven, and the major features of the surface circulation are a series of large anticyclonic circulation cells, termed gyres, which have boundary currents on their landward sides.

**3** Circulation in the deep ocean is gravity-driven. Cold, relatively dense, surface water sinks at high latitudes where the pycnocline is less well developed, or is absent entirely. The main features of the bottom circulation are strong boundary currents on the western sides of the oceans. The deep-water undergoes a global oceanic 'grand tour', which takes it down the Atlantic, through the Antarctic into the Indian Ocean, up through the South Pacific and into the North Pacific. As a result, the deep-water of the North Pacific is the oldest in the World Ocean, and acts as a sink for other deep-waters.

**4** For many geochemical purposes, the ocean can be represented by a two-box model that distinguishes between

(a) a warm surface water reservoir (~2% of the total volume of the ocean) and

(b) a cold deep-water reservoir (~80% of the total volume of the ocean). The two layers are separated

by the waters of the thermocline (~18% of the total volume of the ocean).

**5** The structure of the oceans is controlled by the mixing and transport of water masses, and information on the processes involved, and on the rates at which they operate, can be gained by using a variety of oceanic tracers.

## References

Bacon, M.P. & Anderson, R.F. (1982) Distribution of thorium isotopes between dissolved and particulate forms in the deep sea. *J. Geophys. Res.*, **87**, 2045–56.

Bacon, M.P. & Rosholt, J.N. (1984) Accumulation rates of Th-230, Pa-231 and some transition metals on the Bermuda Rise. *Geochim. Cosmochim. Acta*, **48**, 651–66.

Bender, M., Broecker, W.S., Gornitz, V., *et al.* (1971) Geochemistry of three cores from the East Pacific Rise. *Earth Planet. Sci. Lett.*, **12**, 425–33.

Bolin, B., Bjorkstrom, A. & Holmen, K. (1983) The simultaneous use of tracers for ocean circulation studies. *Tellus*, **35B**, 206–36.

Bowden, K.F. (1975) Oceanic and estuarine mixing processes. In *Chemical Oceanography*, J.P. Riley & G. Skirrow (eds), Vol. 1, 1–41. London: Academic Press.

Broecker, W.S. (1974) *Chemical Oceanography*. New York: Harcourt Brace Jovanovich.

Broecker, W.S. (1981) Geochemical tracers and oceanic circulation. In *Evolution of Physical Oceanography*, B.A. Warren & C. Wunsch (eds), 434–60. Cambridge. MA: MIT Press.

Broecker, W.S. & Peng, T.-H. (1982) *Tracers in the Sea*. Palisades, NY: Lamont-Doherty Geological Observatory.

Burton, J.D. & Statham, P.J. (1988) Trace metals as tracers in the ocean. *Philos. Trans. R. Soc. London*, **325**, 127–45.

Campbell, J.A. (1983) The Geochemical Ocean Sections Study—GEOSECS. In *Chemical Oceanography*, J.P. Riley & R. Chester (eds), Vol. 8, 89–155. London: Academic Press.

Chan, L.H., Drummond, D., Edmond, J.M. & Grant, B. (1977) On the barium data from the GEOSECS Expedition. *Deep-Sea Res.*, **24**, 613–649.

Coale, K.H. & Bruland, K.W. (1985) $^{234}$Th : $^{238}$U disequilibria within the California Current. *Limnol. Oceanogr.*, **30**, 22–33.

Collier, R.W. & Edmond, J.M. (1984) The trace element geochemistry of marine biogenic particulate matter. *Prog. Oceanogr.*, **13**, 113–99.

Dasch, E.J., Dymond, J. & Heath, G.R. (1971) Isotopic analysis of metalliferous sediments from the East Pacific Rise. *Earth Planet. Sci. Lett.*, **13**, 175–80.

Duinker, J.C. (1986) Formation and transformation of element species in estuaries. In *The Importance of Chemical 'Speciation' in Environmental Processes*, M. Bernhard, F.E. Brinkman & P.J. Sadler (eds), 365–84. Berlin: Springer-Verlag.

Elderfield, H. (1988) The oceanic chemistry of the rare-earth elements. *Philos. Trans. R. Soc. London*, **35**, 105–26.

Fine, R.A. & Molinari, R.L. (1988) A continuous deep western boundary current between Abaco (26.5°N) and Barbados (13°N). *Deep Sea Res.*, **35**, 1441–50.

Gordon, A.L. (1996) Communication between oceans. *Nature*, **382**, 399–400.

Gross, M.G. (1977) *Oceanography—a View of the Earth*. Englewood Cliffs, NJ: Prentice Hall.

Macdonald, A.M. & Wunsch, C. (1996) An estimate of global ocean circulation and heat fluxes. *Nature*, **382**, 436–39.

Measures, C.I. & Edmonds, J.M. (1988) Aluminium as a tracer of the deep outflow from the Mediterranean. *J. Geophys. Res.*, **93**, 591–5.

Millero, F.J., Chen, C.-T., Bradshaw, A. & Schleicher, K. (1980) A new high pressure equation of state for sea water. *Deep Sea Res.*, **27**, 255–64.

Palmer, M.R. & Elderfield, H. (1985) Sr isotope composition of sea water over the past 75 Myr. *Nature*, **314**, 526–8.

Pickard, G.L. & Emery, W.J. (1982) *Descriptive Physical Oceanography*. Oxford: Pergamon Press.

Schaule, B.K. & Patterson, C.C. (1981) Lead concentrations in the northeast Pacific: evidence for global anthropogenic perturbations. *Earth Planet. Sci. Lett.*, **54**, 97–116.

Stommel, H. (1958) The abyssal circulation. *Deep Sea Res.*, **5**, 80–2.

Stowe, K.S. (1979) *Ocean Science*. New York: Wiley.

Thompson, J., Carpenter, M.S.N., Colley, S., Wilson, T.R.S., Elderfield, H. & Kennedy, H. (1984) Metal accumulation in northwest Atlantic pelagic sediments. *Geochim. Cosmochim. Acta*, **48**, 1935–48.

Unesco (1981) Unesco Technical Paper on Marine Science, no. 38. Paris: Unesco.

Weihaupt, J.G. (1979) *Exploration of the Oceans. An Introduction to Oceanography*. New York: Macmillan.

Wilson, T.R.S. (1975) Salinity and the major elements of sea water. In *Chemical Oceanography*, J.P. Riley & G. Skirrow (eds), Vol. 1, 365–413. London: Academic Press.

# 8 Dissolved gases in sea water

The atmosphere is the major source of gases to sea water. The atmosphere itself consists of a mixture of major, minor and trace gases, and the abundances of a number of these are given in Table 8.1. The ocean can act as either a source or a sink for atmospheric gases. The gases enter or leave the ocean via exchange across the air–sea interface, and are transported within the ocean reservoir by physical processes.

## 8.1 Introduction

During their residence in the sea some gases behave in a conservative manner. In contrast, other gases are reactive and take part in biological and chemical processes. Dissolved gases are important for a number of reasons, but from the point of view of marine geochemistry the most significant implications of their presence in sea water are related to the role they play in the oceanic biogeochemical cycles. In terms of global-scale processes within these cycles the two most important gases are oxygen and carbon dioxide, and attention here will therefore focus on the roles played by these two gases. Before looking at these roles in detail, however, it is necessary to review briefly the factors that control the ocean–atmosphere exchange of gases.

## 8.2 The exchange of gases across the air–sea interface

The solubility of a gas in sea water is an important factor in controlling its uptake by the oceans; thus, the more soluble gases partition in favour of the water phase, whereas the less soluble gases partition in favour of the atmosphere. The solubilities of gases in sea water are a function of temperature, salinity and pressure. A considerable amount of very accurate data are now available on the solubilities of various gases (e.g. oxygen, nitrogen, argon) in sea water, and equations have been derived to express the dependence of the solubility values on temperature and salinity; an example of this is given in Worksheet 8.1. All gases become more soluble, at least to some degree, in water as the temperature decreases, and this has important implications for the distributions of gases in surface waters, which exhibit considerable temperature variations from the equatorial regions to the poles.

Kester (1975) drew attention to the importance of the concept of partial pressure as a useful means of representing the composition of a gaseous mixture, e.g. the atmosphere. Thus, the total pressure exerted by a mixture of gases in a volume of the mixture is the sum of the partial pressures of the individual gases. This concept of partial pressure also can be applied to gas molecules dissolved in aqueous solution, and Henry's law describes the relationship between the partial pressure of a gas in solution ($P_G$) and its concentration ($C_G$); thus

$$P_G = K_G C_G \tag{8.1}$$

where $K_G$ is the Henry's law constant. In a very simplistic manner, therefore, the rate of transfer of a gas from the atmosphere is proportional to its partial pressures in the two reservoirs, i.e. the atmosphere and the sea, respectively. At equilibrium, when the partial pressure of a gas is the same in both the air and the water reservoirs, molecules enter and leave each phase at the same rate. When the partial pressure of the gas in one reservoir is higher than that in the other, however, there will be a *net* diffusive flow of gas into or out of the sea in response to the concentration gradient across the air–sea interface.

Following Liss (1983) and Liss & Merlivat (1986), a net gas flux ($F$) across the air–sea interface therefore must be driven by a concentration difference ($\Delta C$) between the air and the surface water, with the mag-

**Table 8.1** Non-variable gases in the atmosphere (after Richards, 1965).

| Gas | Concentration (%) | Gas | Concentration (p.p.m.) |
|-----|------------------|-----|------------------------|
| $N_2$ | 78.084 | Ne | 18.18 |
| $O_2$ | 20.946 | He | 5.24 |
| $CO_2$ | 0.033 | Kr | 1.14 |
| Ar | 0.394 | Xe | 0.087 |
| | | $H_2$ | 0.5 |
| | | $CH_4$ | 2.0 |
| | | $N_2O$ | 0.5 |

nitude and direction of the flux being proportional to the numerical value and sign of $\Delta C$; thus

$$F = K_{(T)w}\Delta C \qquad (8.2)$$

where $\Delta C$ is the concentration difference driving the flux ($F$) and the constant of proportionality $K_{(T)w}$, which links the flux and the concentration difference, has the dimensions of a velocity and may be termed the transfer coefficient, transfer velocity, or piston velocity.

The concentration difference ($\Delta C$) can be expressed as

$$\Delta C = C_a H^{-1} - C_w \qquad (8.3)$$

where $C_a$ and $C_w$ are the gas concentrations in the air and the water, respectively, and $H$ is the dimensionless Henry's law constant (expressed as the ratio of the concentration of gas in air to its concentration in non-ionized form in the water, at equilibrium).

The total transfer velocity can be expressed as

$$1/K_{(T)w} = 1/\alpha k_w + 1/Hk_a \qquad (8.4)$$

where $k_a$ and $k_w$ are the individual transfer velocities for chemically unreactive gases in the air and water phases, respectively, $\alpha$ is a factor that quantifies any enhancement of gas transfer in the water as a result of chemical reactions, and $H$ is as defined in eqn (8.3). According to Liss (1983), however, it is more convenient to think in terms of the reciprocal of the transfer velocity, which is a measure of the resistance to interfacial gas exchange. In this manner, eqn (8.4) can be expressed in terms of resistance to gas exchange when $1/K_{(T)w} = R_{(T)w}$, $1/\alpha k_w = r_w$ and $1/Hk_a = r_a$; thus

$$R_{(T)w} = r_w + r_a \qquad (8.5)$$

Liss (1983) pointed out that the phase with the resistance that controls the air–sea transfer of a gas can be identified from a knowledge of the magnitude of $r_w$ and $r_a$. On this basis, the author divided the principal gases into two groups.

1 Gases for which $r_a \gg r_w$. These gases generally have low $H$ values (high water solubilities), i.e. they partition dominantly into the water phase. Gases in this group include $SO_2$, $NH_3$ and HCl.

2 Gases for which $r_w \gg r_a$. Thus $r_w$ is the dominant resistance to transfer. These gases generally have high $H$ values (low water solubility) and include $O_2$, $CO_2$, CO, $CH_4$, $CH_3I$, MeI, $Me_2S$ and the inert gases. However, gases that are sufficiently rapidly reactive in sea water can have their transfer across the air–sea interface by physical mechanisms considerably enhanced by **chemical transfer processes**; this is particularly important for $CO_2$ (see Section 8.4.4). Liss & Merlivat (1986) pointed out that the majority of gases that are important in geochemical cycling fall into category 2.

The exchange or transfer of a gas across the air–sea interface is therefore dependent on:

1 the concentration difference across the air–sea interface, i.e. any flux must be driven by a **concentration gradient** and the magnitude of the diffusive flux is proportional to this gradient (the coefficient of molecular diffusion);

2 the **transfer coefficient** (piston velocity, transfer velocity).

In turn, these are dependent on physical factors such as wind velocity and temperature, and on the solubilities, diffusion rates and chemical reactivities (aqueous chemistry) of individual gases. Data are available on both the molecular diffusivities and solubilities of gases in sea water, and a summary of these is given in Table 8.2.

The main resistance to gas transfer is concentrated in a thin layer near the air–sea interface, and a number of approaches have been used to investigate gas exchange rates across this interface. These include:

1 theoretical models of the gas transfer processes;

2 laboratory experiments designed to quantify the parameters required for the flux equations, e.g. wind-tunnel experiments for investigating the dependence of transfer coefficients on windspeed;

3 field measurements of air–sea gas fluxes or transfer coefficients.

Each of these approaches is considered below.

**Worksheet 8.1:  The solubility of gases in sea water**

Weiss (1970) derived an equation to calculate the temperature and salinity dependence of gas solubilities in sea water from moist air. Thus:

$$\ln c^* = A_1 + A_2(100/T) + A_3\ln(T/100) + A_4(T/100)$$
$$+ S\text{‰}\left[B_1 + B_2(T/100) + B_3(T/100)^2\right] \qquad (1)$$

where $c^*$ is the solubility in sea water from water-saturated air at a total pressure of one atmosphere; the $A$s and $B$s are constants, the numerical values of which depend on the individual gas and the expression of the solubility; $T$ is the absolute temperature; and $S$ (‰) is the salinity. A data set derived from eqn (1) is given in Table (i) showing how the solubility of oxygen in sea water varies with temperature and salinity. There are various ways of expressing gas solubilities, and in order to illustrate the temperature and salinity variations the values listed are given in $\mu\text{mol}\,\text{kg}^{-1}$.

**Table (i)**  The solubility of oxygen in sea water; units, $\mu\text{mol}\,\text{kg}^{-1}$ (from Kester, 1975).

| $T$ (°C) | Salinity (‰) | | | | | | | | | | | | |
| --- | --- | --- | --- | --- | --- | --- | --- | --- | --- | --- | --- | --- | --- |
|  | 0 | 4 | 8 | 12 | 16 | 20 | 24 | 28 | 31 | 33 | 35 | 37 | 39 |
| −1 | 469.7 | 455.5 | 441.7 | 428.3 | 415.4 | 402.8 | 390.6 | 378.8 | 370.2 | 364.6 | 359.0 | 353.5 | 348.2 |
| 0 | 456.4 | 442.7 | 429.4 | 416.5 | 404.0 | 391.9 | 380.1 | 368.7 | 360.4 | 354.9 | 349.5 | 344.2 | 339.0 |
| 1 | 443.8 | 430.6 | 417.7 | 405.3 | 393.2 | 381.5 | 370.1 | 359.0 | 351.0 | 345.7 | 340.5 | 335.4 | 330.3 |
| 2 | 431.7 | 418.9 | 406.5 | 394.5 | 382.8 | 371.5 | 360.5 | 349.8 | 342.0 | 336.9 | 331.8 | 326.9 | 322.0 |
| 3 | 420.2 | 407.9 | 395.9 | 384.2 | 372.9 | 361.9 | 351.3 | 340.9 | 333.4 | 328.5 | 323.6 | 318.8 | 314.1 |
| 4 | 409.3 | 397.3 | 385.7 | 374.4 | 363.5 | 352.9 | 342.6 | 332.5 | 325.2 | 320.4 | 315.7 | 311.1 | 306.5 |
| 5 | 398.8 | 387.2 | 375.9 | 365.0 | 354.4 | 344.1 | 334.1 | 324.4 | 317.4 | 312.7 | 308.1 | 303.6 | 299.2 |
| 6 | 388.7 | 377.5 | 366.6 | 356.0 | 345.8 | 335.8 | 326.1 | 316.7 | 309.8 | 305.3 | 300.9 | 296.5 | 292.2 |
| 7 | 379.1 | 368.2 | 357.7 | 347.4 | 337.5 | 327.8 | 318.4 | 309.3 | 302.6 | 298.3 | 294.0 | 289.7 | 285.5 |
| 8 | 369.9 | 359.4 | 349.1 | 339.2 | 329.6 | 320.2 | 311.1 | 302.2 | 295.8 | 291.5 | 287.3 | 283.2 | 279.2 |
| 9 | 361.1 | 350.9 | 341.0 | 331.3 | 322.0 | 312.9 | 304.0 | 295.4 | 289.2 | 285.0 | 281.0 | 277.0 | 273.0 |
| 10 | 352.6 | 342.7 | 333.1 | 323.7 | 314.6 | 305.8 | 297.2 | 288.9 | 282.8 | 278.8 | 274.8 | 271.0 | 267.1 |
| 11 | 344.5 | 334.9 | 325.5 | 316.5 | 307.6 | 299.1 | 290.7 | 282.6 | 276.7 | 272.8 | 269.0 | 265.2 | 261.5 |
| 12 | 336.7 | 327.3 | 318.3 | 309.5 | 300.9 | 292.6 | 284.5 | 276.6 | 270.8 | 267.0 | 263.3 | 259.6 | 256.0 |
| 13 | 329.2 | 320.1 | 311.3 | 302.8 | 294.4 | 286.3 | 278.5 | 270.8 | 265.2 | 261.5 | 257.9 | 254.3 | 250.8 |
| 14 | 322.0 | 313.2 | 304.7 | 296.3 | 288.2 | 280.4 | 272.7 | 265.3 | 259.8 | 256.2 | 252.7 | 249.2 | 245.8 |
| 15 | 315.1 | 306.5 | 298.2 | 290.1 | 282.3 | 274.6 | 267.1 | 259.9 | 254.6 | 251.1 | 247.7 | 244.3 | 240.9 |
| 16 | 308.5 | 300.1 | 292.0 | 284.2 | 276.5 | 269.1 | 261.8 | 254.7 | 249.6 | 246.2 | 242.8 | 239.5 | 236.3 |
| 18 | 295.9 | 288.0 | 280.3 | 272.9 | 265.6 | 258.5 | 251.7 | 245.0 | 240.1 | 236.9 | 233.7 | 230.6 | 227.5 |
| 20 | 284.2 | 276.7 | 269.5 | 262.4 | 255.5 | 248.8 | 242.3 | 235.9 | 231.2 | 228.2 | 225.2 | 222.2 | 219.3 |
| 22 | 273.4 | 266.3 | 259.4 | 252.7 | 246.1 | 239.7 | 233.5 | 227.5 | 223.0 | 220.1 | 217.3 | 214.4 | 211.6 |
| 24 | 263.3 | 256.5 | 250.0 | 243.6 | 237.3 | 231.3 | 225.4 | 219.6 | 215.4 | 212.6 | 209.9 | 207.2 | 204.5 |
| 26 | 253.8 | 247.4 | 241.2 | 235.1 | 229.2 | 223.4 | 217.7 | 212.2 | 208.2 | 205.6 | 202.9 | 200.4 | 197.8 |
| 28 | 245.0 | 238.9 | 232.9 | 227.1 | 221.5 | 215.9 | 210.6 | 205.3 | 201.5 | 198.9 | 196.4 | 194.0 | 191.5 |
| 30 | 236.7 | 230.9 | 225.2 | 219.7 | 214.2 | 209.0 | 203.8 | 198.8 | 195.1 | 192.7 | 190.3 | 188.0 | 185.7 |
| 32 | 228.9 | 223.4 | 217.9 | 212.6 | 207.5 | 202.4 | 197.5 | 192.7 | 189.2 | 186.9 | 184.6 | 182.3 | 180.1 |

**Table 8.2** The solubilities and molecular diffusivities of some gases in sea water (data from Broecker & Peng, 1982).

| Gas | 0°C | | 24°C | |
| --- | --- | --- | --- | --- |
| | Solubility (cm³ l⁻¹) | Diffusion coefficient (×10⁻⁵ cm² s⁻¹) | Solubility (cm³ l⁻¹) | Diffusion coefficient (×10⁻⁵ cm² s⁻¹) |
| He | 7.8 | 2.0 | 7.4 | 4.0 |
| Ne | 10.1 | 1.4 | 8.6 | 2.8 |
| $N_2$ | 18.3 | 1.1 | 11.8 | 2.1 |
| $O_2$ | 38.7 | 1.2 | 23.7 | 2.3 |
| Ar | 42.1 | 0.8 | 26.0 | 1.5 |
| Kr | 85.6 | 0.7 | 46.2 | 1.4 |
| Xe | 192 | 0.7 | 99 | 1.4 |
| Rn | 406 | 0.7 | 186 | 1.4 |
| $CO_2$ | 1437 | 1.0 | 666 | 1.9 |
| $N_2O$ | 1071 | 1.0 | 476 | 2.0 |

*Theoretical models.* The ability to model the gas transfer process adequately is obviously important because it would permit transfer velocities to be determined from a knowledge of other parameters (e.g. windspeed), and also would allow measurements of the transfer velocity for one gas to be converted into equivalent values for another gas. Models of varying complexity have been used to describe the exchange of gases across the air–sea interface. These include the following.

1 The 'stagnant film model', in which the rate-controlling process is transport across the stagnant film layer by molecular diffusion.

2 The 'surface renewal model', which still retains the concept of a stagnant film, but where this film is periodically replaced by bulk sea water and the rate-determining step for gas transfer is the rate at which the film is replaced.

3 'Boundary-layer models', which are relatively complex models that make use of boundary-layer theories on the transfer of mass (or heat) at surfaces and apply them to gas exchange across an air–sea interface.

There are various problems involved in applying any of these models to field data (see e.g. Burton *et al.*, 1986). The classic stagnant film model (SFM) has been widely used in oceanography, however, and according to Broecker & Peng (1974) it does offer an adequate first-order approximation of the very complex processes that actually take place at the air–sea interface. For this reason the SFM is considered in more detail in Worksheet 8.2. In order to use the theoretical models it is necessary to have a knowl-

edge of, among other parameters, air–sea gas transfer coefficients (or transfer velocities) $(k_w)$. These can be measured under both laboratory and field conditions.

*Laboratory measurements.* Most laboratory experiments involve the use of wind tunnels to elucidate the dependence of gas transfer coefficients on wind velocity. Liss & Merlivat (1986) summarized the available wind tunnel data on the $k_w$–windspeed relationship and drew the following conclusions.

1 In 'smooth surface regimes' with windspeeds up to ~5 m s⁻¹ the water surface has only few waves and values of $k_w$ increase only gradually.

2 In 'rough surface regimes' with windspeeds between ~5 and ~12 m s⁻¹, waves are more common and there is a considerable increase in $k_w$ with increasing windspeed.

3 In 'breaking wave (bubble) regimes', which have windspeeds >10 m s⁻¹, the bubble bursting enhances gas transfer rates, which increase more strongly than in the other two regimes with increasing windspeed; the importance of bubbles in increasing air–sea gas fluxes has been discussed by Farmer *et al.* (1993).

In both the rough surface and breaking wave regimes, the dependence of $k_w$ on wind velocity varies for different gases; thus, less soluble gases (e.g. $O_2$) show the effect of bubble enhancement on $k_w$ at lower windspeeds than the more soluble gases (e.g. $CO_2$; see Table 8.2). However, there is considerable doubt as to the applicability of the laboratory experiments to real oceanic conditions, and according to Roether (1986) the transfer coefficients can be predicted only from field measurements.

**Worksheet 8.2:  The stagnant film model for gas exchange across the air–sea interface**

Following Broecker & Peng (1974, 1982), the concepts underlying the stagnant film model (SFM) can be summarized as follows. In the SFM it is assumed that the rate-limiting step to the transfer of a gas between the air and the water is molecular diffusion through a stagnant water film. The air above the film and the water below it are assumed to be well mixed, and the gas concentration at the top of the film is assumed to be in equilibrium with the air above it. If the partial pressure of the gas in the air yields a concentration of the gas at the top of the film that is different from that in the water below, then a concentration gradient is set up. Transfer of the gas through the film then results from molecular diffusion along the gradient from the high- to the low-concentration region.

The rate at which the diffusion occurs depends on a number of factors, which include the following.

**1** The rate at which the gas diffuses through sea water. A list of the molecular diffusivities of dissolved gases in sea water is given in Table 8.2, from which it can be seen that the diffusivities increase with temperature.

**2** The thickness of the film. As the thickness of the film increases, the time taken for a gas to diffuse through it also increases. The thickness of the film, and so the rate of gas exchange across it, is dependent on the degree of agitation of the sea surface by the wind; the stronger the wind, the thinner the film and the more rapid the exchange rate.

**3** The difference in the concentration of a gas between the air and the sea (disequilibrium magnitude). The larger the difference, the greater the concentration gradient and the faster the molecular diffusive transfer.

The SFM is illustrated diagrammatically in Fig. (i), with reference to radon exchange. The rate of transfer of the gas between the water and the air is controlled by the thickness of the stagnant boundary layer, through which the gas is transferred only by molecular diffusion. The overlying air and the underlying water are assumed to be well mixed. In the flux equation $D$ is the molecular diffusivity of the gas, $z$ is the thickness of the film, $C_s$ is the concentration of the gas at the bottom of the film, $\alpha$ is the solubility of the gas, and $p$ is the partial pressure of the gas in the air.

There are difficulties in applying the SFM to 'reactive' gases, such as $CO_2$, for which chemical, as well as physical, transport across the stagnant layer is important. For example, $CO_2$ can react via hydration to give, on ionization, $HCO_3^-$ and $CO_3^{2-}$ (see Section 8.4.4). If this hydration were rapid enough for carbonate equilibria to be reached at all points as it crossed the film, then concentration gradients would be set

*continued on p. 170*

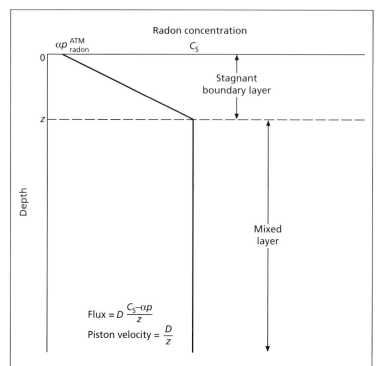

**Fig. (i)** The stagnant boundary layer model (from Broecker & Peng, 1974).

up for all components in the system, i.e. not simply for dissolved $CO_2$, and since the net rate of transport is the sum of all the individual carbon species along the individual gradients, the overall transfer rate would be enhanced. The effect of chemical enhancement on the exchange rate for $CO_2$ varies with the film thickness, although the average film thickness in the sea probably lies in the region where the chemical enhancement is negligible. Under some conditions, however, the air–sea exchange of $CO_2$ can be enhanced by the side effects of pH variations within the film, which are dependent on the equilibria in the carbonate system.

*Field measurements.* A number of field techniques are available for the measurement of $k_w$, and these have been summarized by Roether (1986). An example of how field data can be plugged into a theoretical model has been provided by Broecker & Peng (1974) who attempted to assess the applicability of the SFM to the exchange of gases across the ocean–atmosphere interface using radiocarbon and radon tracers to determine $k_w$ values. Their principal findings may be summarized as follows.

1 The mean stagnant film thickness for the World

Ocean was found to be between $40 \pm 30$ and $50 \pm 30$ μm, depending on the tracter technique used. The thickness appears to vary in inverse proportion to the square of the wind velocity, as predicted by Kanwisher (1963).

2 From a knowledge of the wind velocity, mixed-layer thickness and partial pressure difference between the air and the sea surface, the net transfer rate of most gases between the atmosphere and the sea can be predicted to within an accuracy of $\sim \pm 30\%$.

**3** Using the average film-layer thickness it is possible to compute the mean residence times of various gases in the atmosphere with respect to transfer to the mixed layer, and in the mixed layer with respect to transfer to the atmosphere. The average residence time for transfer from the atmosphere to the mixed sea-surface layer is $\sim$300 yr for gases with solubilities in the 'normal' range (see Table 8.2); for $CO_2$ it is $\sim$8 yr. The average residence time for transfer from the mixed layer of the sea to the atmosphere for most gases is $\sim$1 month. For $CO_2$ the situation is more complex. The mixed layer of the ocean probably achieves chemical equilibrium with $CO_2$ in the atmosphere in 1.5 yr. This leads to the important conclusion that, because the replacement time of the mixed layer by mixing with underlying water is around a decade, the rate-limiting step in the removal of anthropogenic (fossil fuel) $CO_2$ from the atmosphere is vertical mixing within the sea, rather than gas transfer of $CO_2$ across the air–sea interface. Except for gases with very high solubilities (e.g. $SO_2$, $CO_2$) the sea is not an effective sink for anthropogenic gases, as their entry into the system would require many hundreds of years.

Some of the constraints affecting the air–sea exchange of trace gases have been discussed above, but the overall process is extremely complex and for a detailed review of the subject the reader is referred to Erickson (1993). This author considered a number of the constraints involved, including, in addition to windspeed, the influence of thermal stability at the air–sea interface on transfer velocities, and proposed a stability dependent theory for air–sea gas exchange.

Once a gas is dissolved in sea water, i.e. it has diffused across the interface into the liquid phase, the processes that determine its oceanic distribution depend on the nature of the gas itself. The distribution of those gases that generally are regarded as being **non-reactive** in sea water, e.g. nitrogen and the inert gases, is controlled by physical processes and by the effects of temperature and salinity on their solubility (Kester, 1975). It is usual to express variations in the non-reactive gases in terms of their percentage saturation. To do this, the observed concentration of the gas in the sea is given as the percentage of its solubility in pure water of the same temperature and salinity. Thus

$$\text{saturation } (\%) = 100 \times G/G' \tag{8.6}$$

where $G$ is the observed gas content in sea water, and $G'$ is the solubility of the gas in pure water having the observed temperature and salinity (Richards, 1965; Riley & Chester, 1971).

Factors in addition to those described above affect the distributions of the **reactive** gases, such as oxygen and carbon dioxide, in the oceans. Although oxygen is in fact relatively chemically inert in sea water, it is involved in biological processes and these strongly affect its distribution in the water column. Carbon dioxide also takes part in biological processes, and the competing processes of photosynthesis (utilization of $CO_2$, liberation of $O_2$) and respiration (utilization of $O_2$, liberation of $CO_2$) are the cause of many of the *in situ* changes in the concentrations of the two gases in the sea. In addition, $CO_2$ is extremely reactive in sea water. Oxygen and carbon dioxide therefore behave in a non-conservative manner in the oceans.

All gases present in the atmosphere are found to some extent in solution in sea water. In the present context, attention will be confined mainly to those gases that play a major role in the marine biogeochemical cycles. These are chiefly oxygen and carbon dioxide; the importance of $SO_2$ (and other sulphur species) in the marine aerosol sulphate cycle has been discussed in Section 4.1.4.3, and the role of $H_2S$ in redox-mediated reactions is covered in Section 11.5.6. Before describing the factors that affect the distributions of oxygen and carbon dioxide in the oceans, however, it is worthwhile briefly reviewing the work carried out on some of the other gases that are dissolved in sea water.

The oceans can act as a source or a sink for atmospheric gases, and air–sea exchange can play a significant role in the global geochemical cycles of some gases. A number of attempts have been made to evaluate the importance of this air–sea exchange in the geochemical cycles of various gases. For example, Liss (1983) used the then currently available data to draw up a compilation of the air–sea fluxes of a number of gases. Some of these data are reproduced in Table 8.3, in which the magnitudes of the fluxes indicate whether the sea surface acts as a source (sea $\rightarrow$ air) or a sink (air $\rightarrow$ sea). The magnitudes of the fluxes of the gases are also included in the table, so that the importance of the air–sea exchange flux can be evaluated within their global cycles. For example, it can be seen that for methane ($CH_4$) the sea surface acts

**Table 8.3** Net global fluxes of some gases across the air–sea interface.

| Gas | Global air–sea direction* | Flux magnitude† | Data source |
|---|---|---|---|
| $CH_4$ Anthropogenic | + | $10^{12}$–$10^{13}$ | ‡ |
| $CO_2$ | – | $6 \times 10^{15}$ | ‡ |
| $N_2O$ | + | $6 \times 10^{12}$ | ‡ |
| $CCl_4$ | – | $0$–$10^{10}$ | ‡ |
| $CCl_3F$ | – | $0$–$5 \times 10^{9}$ | ‡ |
| $CH_3I$ | + | $3$–$13 \times 10^{11}$ | ‡ |
| CO | + | $100 \pm 90 \times 10^{12}$ | § |
| $H_2$ | + | $4 \pm 2 \times 10^{12}$ | § |
| Hg | + | $\sim 2 \times 10^{9}$ | ¶ |

* + indicates sea → air flux direction; – indicates air → sea flux direction.
† Units in grams (of the compound, where applicable) per year.
‡ Data from Liss (1983).
§ Data from Conrad & Seiler (1986).
¶ Data from Fitzgerald (1986).

as a source; however, the sea → air flux ($\sim 10^{12}$–$10^{13}$ g yr⁻¹) is only a few per cent of the total amount of $CH_4$ from natural terrestrial sources ($\sim 10^{15}$ g yr⁻¹). Methyl iodide ($CH_3I$) also has a flux in the sea → air direction, and recent data indicate that it is in the range $\sim 3 \times 10^{11}$–$13 \times 10^{11}$ g yr⁻¹. It has been estimated that a sea to air flux of some form of volatile iodine amounting to $\sim 5 \times 10^{11}$ g yr⁻¹ is required to balance the global geochemical cycle of iodine, and it is apparent that the $CH_3I$ out-of-sea flux can account for between $\sim 50$ and 250% of this. Liss (1986) concludes, therefore, that the emissions from the sea surface are important, if not dominant, in the global geochemical cycling of iodine. The sea surface also acts as a source for atmospheric CO and $H_2$ (see Table 8.3), although the marine inputs amount to only $\sim 3\%$ of the total CO and $\sim 5\%$ of the total $H_2$ emitted to the atmosphere. In contrast, for some gases the sea acts as a sink. Carbon tetrachloride ($CCl_4$) and trichlorofluoromethane (Freon-11) are examples of such gases, and both are of special interest because anthropogenic sources dominate their inputs to the atmosphere; in fact, Freon-11 has no natural sources. Data reported by Liss & Slater (1974) for the North and South Atlantic indicated that there was a sea →

air flux of $\sim 1.4 \times 10^{10}$ g yr⁻¹ for $CCl_4$ ($\sim 2\%$ of the total anthropogenic production rate at the time), with a mean undersaturation of sea water with respect to the air of $\sim 10\%$. A decade later, however, data from the same region (Hunter-Smith *et al.*, 1983) indicated that the air and sea water were essentially at equilibrium, thus implying that there was no net $CCl_4$ flux at this time. A similar situation has also been found for Freon-11. Thus, although Liss & Slater (1974) estimated that there was an air–sea Freon-11 flux of $\sim 5.4 \times 10^{9}$ g yr⁻¹, more recent data have indicated that there is a saturation equilibrium for the gas between the surface ocean and the atmosphere. These time changes in the air–sea flux relations for the two gases probably result from decreases in their release rates to the atmosphere.

Nitrous oxide ($N_2O$) is an important gas because of its contribution to the 'greenhouse effect' (see Section 8.4.2), and its deleterious effect on the ozone layer. There is a flux of $N_2O$ from the sea to the atmosphere, and thus the oceans represent a potential source of the gas to the atmosphere. Recent studies, however, have indicated that there is a strong spatial heterogeneity in the oceanic $N_2O$ source term. For example, Law & Owens (1990) reported that upwelling in the northwest Indian Ocean represents one of the most significant marine sources of $N_2O$, contributing between 5 and 18% of the total marine flux from a surface area of only 0.5% of the World Ocean. Exchange across the air–sea interface also is important in the global cycle of mercury. For example, Fitzgerald (1986) concluded that although there are many deficiencies in the database the contribution of mercury from sea-surface sources ($\sim 2 \times 10^{9}$ g yr⁻¹) could account for $\sim 30$–40% of the total emissions of the element to the atmosphere (see also, Fitzgerald, 1989).

## 8.3 Dissolved oxygen in sea water

The vertical and horizontal distributions of oxygen in the oceans reflect a balance between: (i) input across the air–sea interface from the atmosphere; (ii) involvement in biological processes; and (iii) physical transport. The various factors that control the distribution of dissolved oxygen in the sea lead to a number of pronounced features in its vertical profiles. These are illustrated in Fig. 8.1(a), and can be summarized as follows.

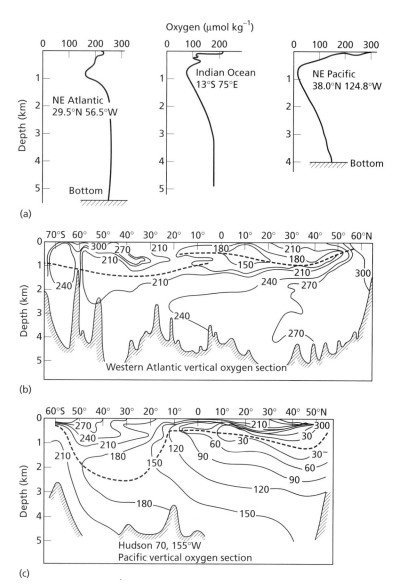

**Fig. 8.1** The vertical distribution of oxygen in the oceans (modified from Kester (1975), which lists the original data sources). (a) Vertical profiles for the Atlantic, Indian and Pacific Oceans. (b) Vertical section—western Atlantic. (c) Vertical section—central Pacific. The broken curves in (b) and (c) indicate the depth of the oxygen minimum surface.

**1** Oxygen in the surface, or mixed, layer is derived from exchange with the atmosphere so that its concentration is determined largely by its solubility in sea water, and on a global basis the concentrations of dissolved oxygen in sea water are greater in cold high-latitude waters than in those from the warmer subtropical regions. This atmospheric supply is supplemented by oxygen released during photosynthesis, a process that may be represented as

$$CO_2 + H_2O = CH_2O + O_2 \qquad (8.7)$$

where $CH_2O$ indicates carbohydrate material (Richards, 1965). The rate of exchange with the atmosphere, however, is much faster than the rate at which the internal oceanic processes take place. As a result, photosynthesis usually does not lead to an excess of oxygen in the surface layer; nonetheless, the surface ocean usually does have a slight oxygen supersaturation (~5%), which may result from the trapping of air bubbles (Broecker, 1974). Under some conditions, however, the exchange of photosynthetically produced oxygen with the atmosphere can be blocked (e.g. by a density cap formed by summer warming of the surface layer), with the result that photosynthesis can produce an oxygen saturation. For example, a summer subsurface **shallow oxygen maximum** (SOM) has been reported in a number of nutrient-poor oligotrophic marine regions (see e.g. Shulenberger & Reid, 1981).

2 Below the zone in which photosynthesis takes place there is a decrease in dissolved oxygen owing to its consumption as a result of respiration and the decay of organic matter. The change in concentration at these depths may be either gentle or sharp.

3 **Oxygen minima** are a characteristic feature in many marine areas. An *in situ* consumption of oxygen is necessary to sustain the layers, and this may arise from the oxidative decomposition of sinking detritus that has accumulated at a particular depth. Oxygen minimum zones (OMZs), or oxygen depleted zones (ODZs), can be large features. Intense zones are especially well developed close to regions of upwelling and high primary productivity north and south of the Equator in the eastern Pacific (~100–

900 m depth), where they advance westwards off the coasts of North and South America, and in the northern Arabian Sea (~100–1000 m depth) — see Fig. 8.2. Although the oceanographic features leading to the formation of the zones are not fully understood, a number of theories have been advanced to explain their existence. These include:

(a) slow advection of water into the zones, allowing long periods for the decomposition of organic matter to consume the oxygen;

(b) large local oxygen consumption rates as a result of enhanced production in surface waters;

(c) low oxygen concentrations in waters advected into the zones.

Olsen *et al.* (1993) considered these various theories, and concluded that in the northern Arabian Sea the near-zero oxygen concentration is maintained by moderate consumption applied to waters that initially have low oxygen concentrations and are advected through the zone at moderate speed. Packard *et al.* (1988) suggested that the OMZ in the Alboran Sea (western Mediterranean) is the result of a chain of processes, which start with the nutrient enrichment of Atlantic water, flowing into the Mediterranean, by mixing in the Straits of Gibraltar. The enriched water supplements upwelled water along the Spanish coast where plankton blooms are developed. The blooms are transported to the convergence zone in the centre of the Alboran gyre, which acts as a plankton and POM trap. Dead

**Fig. 8.2** The distribution of oxygen-deficient water (<20 µg atom l⁻¹) in intermediate and deep waters from the World Ocean (from Deuser, 1975).

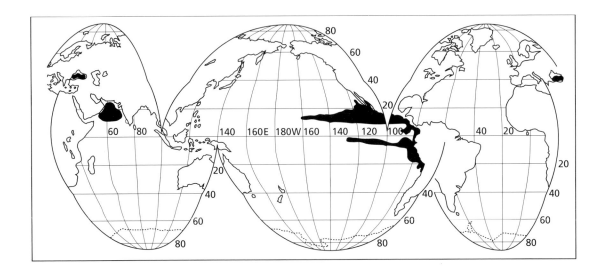

plankton and faecal material rain down into the Leventine Intermediate Water, where they are metabolized by bacteria, a process which consumes oxygen and maintains the most intense OMZ in the Mediterranean Sea.

4 All *in situ* processes in the deep ocean lead to a decrease in dissolved oxygen, so that a depth depletion of the gas might be expected. However, relatively little organic matter reaches the deep ocean, and as a result there is little consumption of oxygen. In fact, dissolved oxygen concentrations usually show a gradual increase from the base of the minimum layers to the bottom of the water column. This is a consequence of the deep-water thermohaline circulation system in which oxygen-rich high-latitude surface waters sink and are transported along the global 'grand tour' (see Section 7.3.3). Once the oxygen has been transported to deep water it is removed from contact with the atmosphere, and it is also taken out of the euphotic zone in which photosynthetic reactions take place. None the less, there is *some* oxygen utilization by animals and bacteria in deep water, and this oxygen cannot be replaced by exchange with the atmosphere or by plant activity. During the 'grand tour' therefore dissolved oxygen becomes depleted, with the overall result that concentrations decrease from the waters of the deep Atlantic to those of the deep Pacific; this is the reverse of the situation for the nutrients (see Section 9.1.2.2). In order to evaluate the amount of oxygen that is biologically consumed in the deep ocean the concept of **oxygen utilization** is often used. This is a measure of the oxygen that has been utilized, rather than the amount that remains in the waters, and is based on the premise that in surface water the oxygen is present at almost equilibrium, or saturation, values with the overlying atmosphere. In contrast to surface waters, deep waters are highly undersaturated with respect to dissolved oxygen, and a measure of the amount of oxygen that has been utilized can therefore be obtainied from the difference between the saturation and the observed oxygen contents. It is usual, however, to use the term **apparent oxygen utilization** (AOU) because surface water can be up to ~5% supersaturated with dissolved oxygen (see above). The AOU is thus a measure of the change in dissolved oxygen that has taken place after the waters have left the surface. Values of AOU are lowest in the Atlantic (less utilization) and highest in the Pacific (more utilization), as the deep water 'grand

tour' progresses along its path and utilization and water-mass mixing take place. It must be stressed, however, that most oxygen utilization occurs in surface waters.

There are some conditions in the oceans, and in other bodies of water, where the subsurface circulation, and so the supply of dissolved oxygen, is restricted. Here, the oxidative decomposition of organic matter utilizes, and can sometimes exhaust, the dissolved oxygen so that **anoxic** conditions are set up. Such anoxic conditions actually prevail in only a very small proportion of the World Ocean, examples being found in fjords, certain basins (e.g. the Gotland Basin, the Black Sea) and in deep-sea trenches (e.g. the Cariaco Trench).

According to Kester (1975), ocean-scale sections of the vertical and horizontal distributions of dissolved oxygen reflect a number of major features in water circulation patterns. This is evident in the oxygen distribution diagrams given in Fig. 8.1(b,c), and these can be used to illustrate in a general way how dissolved oxygen can be utilized for the characterization of water masses. In this context, Kester (1975) identified a number of features in these diagrams.

1 Antarctic Intermediate Water can be seen as an intrusion of oxygen-rich water in both the Atlantic and Pacific Oceans, extending from the surface at ~50°S to ~800 m at ~20°S.

2 North Atlantic Deep Water is evident as oxygen-rich deep water extending from ~0 to ~2000 m at ~60°N to an oxygen maximum at ~3000 m in the south equatorial Atlantic.

3 The northward flow of North Atlantic Deep Water and Antarctic Bottom Water into the deep Pacific gives rise to a progressive decrease in the oxygen content from south to north.

4 Waters at intermediate depths are more deficient in oxygen in the North Pacific than in the North Atlantic.

The two common isotopes of oxygen are $^{16}O$ and the stable trace species $^{18}O$. These isotopes are fractionated during biochemical reactions so that the organic matter has an $^{18}O : ^{16}O$ ratio that is lower than that in sea water. The fractionation is also affected by temperature changes. For example, the $^{18}O : ^{16}O$ ratios of carbonate shells reflect the temperature of the water in which they grew, and this has been used in evaluating past climatic changes. In addition, $^{18}O : ^{16}O$ ratios can be used as water-mass tracers; this is because surface waters can be characterized on the basis of their $^{18}O$ contents, which tend to be enriched at low latitudes.

## 8.4 Dissolved carbon dioxide in sea water: the dissolved $CO_2$ cycle

### 8.4.1 Introduction

Carbon dioxide is transferred into the oceanic biosphere via both the photosynthesis of marine plants and the formation of carbonate shells by plants and animals, and therefore is involved in the formation of both soft organic tissues and hard skeletal meterial. As carbon dioxide is removed from the waters by these biological reactions it is being continuously added from the atmosphere, and at greater depths in the water column it is regenerated by the oxidative destruction of organic matter and so increases in concentration below the surface. Thus, typical water-column profiles of carbon dioxide show a depletion in the surface layers and an overall increase towards the base of the euphotic zone. In this respect the behaviour of carbon dioxide and oxygen may be thought of as being a mirror image of each other. Carbon dioxide, however, also differs from oxygen in that it is extremely chemically *reactive* in sea water, and this reactivity within the $CO_2$ system has a number of profound effects on the chemistry of the oceans. Perhaps the most important of these is the control it maintains on the pH of sea water. Another important feature of the reactivity of the gas in sea water is that the oceans act as a sink for excess atmospheric $CO_2$ and therefore as a regulator of planetary $CO_2$.

Carbon dioxide is taken into sea water via exchange with the atmosphere. Photosynthesis, primary production and the organic carbon cycle are covered in Section 9.3, and at this stage attention will be confined mainly to the marine $CO_2$ cycle.

### 8.4.2 $CO_2$ and world climate

The biogeochemical cycles of C, N, P, S and O play significant roles in controlling the global environment, and the oceans, which cover three-quarters of the Earth's surface, are important in these cycles. This especially is true for carbon, for which the oceans are a major reservoir, and one of the most important

recent thrusts in oceanography has been to elucidate the role played by the **oceanic carbon flux** in the global carbon cycle. This role has received increasing attention because of the effects that anthropogenic intervention can have on climate cycles by influencing the atmospheric levels of carbon dioxide and other 'greenhouse' gases.

**Climate** describes the long-term behaviour of the **weather**, and in the past this had included a number of climatic extremes, as evidenced in ice cores and deep-sea sediments. The driving force for both weather and climate is solar energy. Climate is dependent upon the radiative balance of the atmosphere, and is governed by interactions between the atmosphere and a number of terrestrial reservoirs, which include the oceans, the ice-caps, the land surfaces and various ecosystems. Heat is redistributed around the planet by the global wind system and oceanic currents. Because of their capacity for heat storage the oceans provide the main natural control on the heat-retaining properties of the atmosphere. As a result, the oceans play a dominant role in determining climate, and water movements (both horizontal and vertical) in the oceans act like a large flywheel that drives global climate. In this way, long-term changes in the interactions between the atmosphere and the oceans may account for much of the climatic variation over the past few thousand years as the 'flywheel' has speeded up or slowed down.

The oceans operate on different time-scales to the atmosphere. For example, the ocean eddies correspond to cyclonic features in the atmosphere, but whereas the latter exist on scales of days/weeks, eddies can last for many months. Other oceanic features that are involved in the transport of heat persist for considerably longer. Thus, ocean gyres take around a decade to circumnavigate an ocean basin. Further, the global journey associated with the deep circulation conveyor belt operates over a time-scale of millennia (see Section 7.3.3), which means that the deep ocean can sequester heat absorbed at the surface for periods of 1000 yr or more.

Climate also can be affected by the additional warming of the ocean caused when the activities of humankind enhance the natural 'greenhouse' effect. The mean temperature of the Earth is constrained by the balance between energy from the sun coming in, visible radiation (sunlight) which warms it, and infra-red radiation going out, which cools it down.

This is the **radiative forcing** on climate. There are a number of factors which can change the Earth's radiation balance, and these are termed the **climate forcing agents**. Apart from changes in solar radiation itself, the most important climate forcing agents are associated with the '**greenhouse effect**'.

The atmosphere is relatively transparent to the shortwave solar radiation involved in warming the Earth, but many atmospheric trace gases absorb some of the longwave (infra-red) radiation, emitted from the surface, which cools the Earth. As a result, the atmosphere acts like a blanket, preventing much of the infra-red radiation from escaping into space and so makes the Earth warmer. This is analogous to a greenhouse in which the glass allows sunlight in, but prevents some of the infra-red radiation leaving. The 'greenhouse effect' is a natural phenomenon that prevents the Earth from freezing as the trace gases in the atmosphere absorb emitted heat, and the mean temperature of the Earth is at present ~32°C warmer than it would be if the natural 'greenhouse' gases were not present. However, as the result of anthropogenic intervention the concentrations of a number of the 'greenhouse' gases have increased markedly over the past 100 yr, and are expected to produce significant global warming in the next century. In fact, there is evidence that greenhouse warming is already taking place; e.g. the global mean air temperature has increased by ~0.3–0.6°C over the last century.

The principal natural 'greenhouse' gas is water vapour, but this is not influenced significantly by emissions related to human activity, and it is the gases which have strong anthropogenic sources that give rise to concern. Of these, carbon dioxide ($CO_2$), methane ($CH_4$), nitrous oxide ($N_2O$) and the halocarbons, especially the chlorofluorocarbons (CFCs), and in the lower atmosphere ozone ($O_3$), are the most important. The effectiveness of a 'greenhouse' gas in influencing the Earth's radiative budget depends on its concentration in the atmosphere and its ability to absorb outgoing terrestrial radiation (Watson *et al.*, 1990).

Carbon dioxide is the least effective 'greenhouse' gas per kilogram emitted to the atmosphere and the **global warming potential** (GWP) index defines the time-integrated warming effect of a unit mass of a 'greenhouse' gas relative to that of $CO_2$ (GWP = 1). The GWPs of the 'greenhouse' gases, extrapolated over a 100-yr period, increase in the order (GWP): carbon dioxide (1) < methane (21) < nitrous oxide (290) < CFCs (3500–7300). However, the *contribution* of a 'greenhouse' gas to global warming is a function of the product of the GWP and the amount of a gas emitted, and 1990 emissions for the 'greenhouse' gases have been estimated to be: carbon dioxide (26 000 Tg) >> methane (300 Tg) >> nitrous oxide (6 Tg) >> CFCs (7 Tg); 1 Tg = $10^6$ metric tons = $10^{12}$ g. Carbon dioxide therefore has the greatest potential contribution to global warming. Thus, over the period 1980–1990, the contribution from each of the anthropogenic 'greenhouse gases' to changes in the radiative forcing of the climate has been estimated to be: carbon dioxide (55%), CFCs (24%), methane (15%), and nitrous oxide (6%) (Houghton *et al.*, 1990). At present, the increase in the concentration of carbon dioxide is therefore the most important single agent in the radiative forcing of the climate.

### 8.4.3 $CO_2$ — the oceanic role

Carbon in a variety of forms (e.g. $CO_2$, carbonates, organic matter) is exchanged naturally between the large global reservoirs of the atmosphere, the oceans (including biota) and the terrestrial biosphere. There have been a number of recent attempts to quantify the magnitude of both the global carbon reservoirs and the exchanges between them, and Fig. 8.3 summarizes the various data. Figure 8.3(a) presents a detailed assessment of the global carbon cycle and the fluxes involved for (i) a reconstructed pre-industrial scenario and (ii) the contemporary scenario, and Fig. 8.3(b) summarizes the data for the contemporary scenario. The largest natural exchanges occur between the atmosphere and the land biota, and the atmosphere and the surface ocean. The exchange of $CO_2$ between the atmosphere and the surface ocean is extremely important, and on time-scales of decades, or more, the $CO_2$ concentration of the atmosphere is controlled mainly by exchange with the oceans, which are the largest of the carbon reservoirs (see Fig. 8.3); the total, i.e. surface and deep ocean, contains ~39 000 Gt C; 1 Gt = $10^9$ metric tons = $10^{12}$ kg = $10^{15}$ g. It is apparent, therefore, that there is around 20 times more $CO_2$ dissolved in sea water than occurs on land in plants, animals and soil, and the release of just 2% of the $CO_2$ stored in the oceans would double the level of atmospheric $CO_2$. Further,

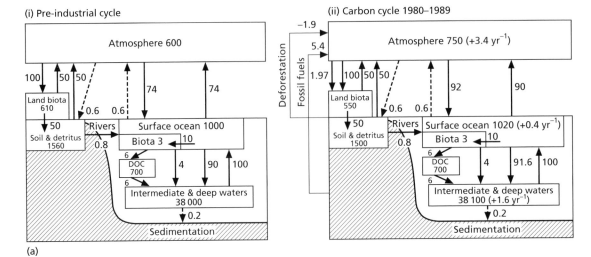

(i) Pre-industrial cycle

(ii) Carbon cycle 1980–1989

(a)

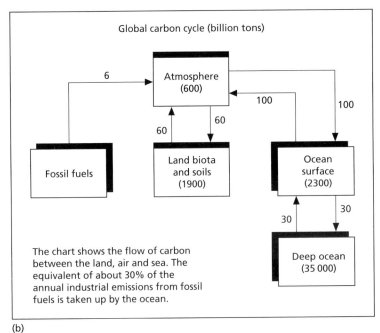

Global carbon cycle (billion tons)

The chart shows the flow of carbon between the land, air and sea. The equivalent of about 30% of the annual industrial emissions from fossil fuels is taken up by the ocean.

(b)

**Fig. 8.3** Global carbon cycle. (a) Reservoirs and fluxes for (i) a reconstructed pre-industrial scenario, and (ii) the contemporary scenario. Units: reservoirs (Gt C), fluxes (Gt C yr⁻¹). (From Sarmiento & Siegenthaler, 1993.) (b) Reservoirs and fluxes for the contemporary scenario. Units: reservoirs (billion tons C), fluxes (billion tons C yr⁻¹). (From Takahasi, 1989.)

around 15 times as much $CO_2$ is taken up, and released, by natural marine processes than the total produced by the burning of fossil fuel, deforestation and other human activities.

The oceanic role in the global carbon budget is discussed in Section 8.4.6.2. At this stage, however, it is useful to put the situation in a general context, and the carbon budget produced by the Intergovernmental Panel on Climate Exchange (IPCC) will serve for

this purpose, although subsequent modifications to this budget have been suggested (see Section 8.4.6.2). The IPCC budget is given in Table 8.4, and a number of points can be made from the data. Each year anthropogenic processes add ~5.3 Gt C as $CO_2$ to the atmosphere from the burning of fossil fuels, and ~1.6 Gt through land use changes, mainly via the loss of tropical forests. This anthropogenic $CO_2$ is partitioned between the atmosphere, the ocean and the

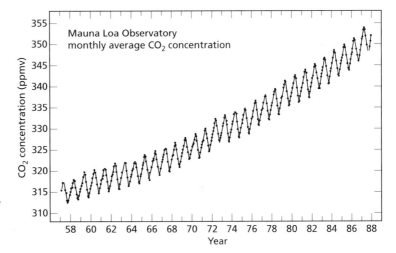

**Fig. 8.4** Monthly average atmospheric $CO_2$ concentrations over the period 1958–1988, as measured at Mauna Loa, Hawaii (after data from C.D. Keeling). Dots are monthly averages; smooth curve is a fit. Seasonal variations result mainly from the drawdown and production of $CO_2$ by terrestrial biota.

**Table 8.4** The Intergovernmental Panel on Climate Exchange (IPCC) 1980–9 global $CO_2$ budget; units, $Gt C yr^{-1}$ (data from Watson et al., 1990).

| | | |
|---|---|---|
| $CO_2$ source | Fossil fuels | $5.3 \pm 0.5$ |
| | Deforestation | $1.6 \pm 1.0$ |
| | Total | $7.0 \pm 1.2$ |
| $CO_2$ sinks | Atmosphere | $3.4 \pm 0.2$ |
| | Oceans | $2.0 \pm 0.8$ |
| | Total | $5.4 \pm 0.8$ |
| Imbalance | Inferred terrestrial sink | $1.6 \pm 1.4$ |

terrestrial biosphere. The oceans take up ~33% of the anthropogenic $CO_2$ released annually to the atmosphere. This $CO_2$ does not therefore affect the Earth's radiation balance, with the result that the oceanic uptake of anthropogenic $CO_2$ can mitigate against global warming. Anthropogenic processes mean that, allowing for losses in the oceans, an excess of ~3.4 Gt C, in the form of $CO_2$, remains in the atmosphere each year, and $CO_2$ levels began to increase in the 19th century with the onset of the industrial revolution and increasing deforestation (see Fig. 8.4). As a result, the concentration of $CO_2$ in the atmosphere has increased from a pre-industrial level of 280 p.p.m.v. (parts per million by volume) to the 1990 level of 353 p.p.m.v.; i.e. an increase of ~0.5% $yr^{-1}$. There is a general proportionality between the rising atmospheric $CO_2$ concentrations and industrial $CO_2$ emissions. This proportionality,

however, was perturbed in the 1890s by a disproportionally high rate in the rise of atmospheric $CO_2$, followed by a pronounced slowing down of the growth rate. Keeling & Peng (1995) suggested that these interannual extremes in the rate of rise of atmospheric $CO_2$ resulted mainly from variations in global air temperatures, which altered both the terrestrial biospheric and the oceanic carbon sinks, and possibly also by precipitation. Another consequence of the global carbon budget is that there is a net imbalance of ~1.6 Gt C between the anthropogenic $CO_2$ emission sources (~7 Gt C), and the atmospheric retention (~3.4 Gt C) and oceanic sink (~2.0 Gt C) terms; this is the so-called 'missing $CO_2$', and is discussed in Section 8.4.6.2.

The role played by the oceans in the global $CO_2$ cycle involves a complex interplay between a number of carbon reservoirs; these include the atmosphere, the surface ocean, the deep ocean, marine biota and deep-sea carbonate sediments. Traditionally, the oceanic role has been viewed in terms of a series of 'pumps' that (i) initially draw $CO_2$ down into the system and then (ii) transport it within the system itself. However, there is some confusion within the literature on the definitions of these 'pumps'. In an attempt to simplify the situation, we will consider the factors that constrain the introduction of $CO_2$ into surface seawater, those that transport either $CO_2$ itself, or carbon, out of the surface layer, and the manner in which deep-sea carbonates affect the system. In this approach, reference will be made to the tranditional 'pump' terminology, but the emphasis will be on the processes involved.

1 The introduction of $CO_2$ into the oceans. This takes place via the physico-chemical dissolution of the gas in sea water. The net flux of $CO_2$ into, or out of, the oceans across the air–sea interface is driven by the difference ($\Delta pCO_2$) between the partial pressure of $CO_2$ in the atmosphere and the equilibrium partial pressure of $CO_2$ in surface seawater. Two factors in particular can mediate to set the $pCO_2$ in surface seawater.

(a) The temperature effect. In general, $CO_2$ is less soluble in warm low-latitude waters and more soluble in cold high-latitude waters.

(b) The biological effect. The transfer of $CO_2$ into the oceans is affected by phytoplankton, especially under bloom conditions, when the photosynthetic growth of the organisms utilizes $CO_2$ (see Section 9.2.2.2), resulting in the **drawdown** of the gas into sea water.

The transfer of $CO_2$ into the oceans from the atmosphere is a function of the solubility of the gas in sea water, and is often referred to as the '**solubility pump**'.

The transfer occurs across the air–sea interface into the thin surface-water oceanic reservoir (i.e. the mixed layer, or seasonal boundary layer). This caps the deep-water reservoir, which forms the bulk of the ocean system and comprises by far the largest oceanic carbon reservoir. Although still small relative to the deep ocean, the capacity of the surface ocean to take up $CO_2$ is enhanced by the transfer of carbon out of the mixed layer to deep waters, and this takes place via two other 'pumps' that influence the way $pCO_2$ is set in surface waters.

2 Deep-water transfer by water mixing; the '**physical pump**'. This transfer of $CO_2$ to depth takes place through downward mixing and down-welling associated with the formation of deep-water masses and the initiation of the thermohaline oceanic circulation system (see Section 7.3.3). The 'physical pump' is strongly coupled to the 'solubility pump' because $CO_2$ is more soluble in cold high-latitude waters where the deep-water transfer takes place. The 'physical pump' can also help to set the $pCO_2$ in surface seawater via the upwelling of deep-water.

3 Deep-water transfer by biological intervention; the '**biological pump**'. The 'biological pump' transports fixed carbon from surface waters via the vertical gravitational settling of the biogenic debris produced in the euphotic zone (see Chapters 10 and 12). Most of the organic debris is destroyed in the upper water layers ($< \sim 1000\,m$) and is recycled there. The 'biological pump' transport to deep waters therefore refers to the less than $\sim 5\%$–$10\%$ of the new surface production that escapes recycling in the 'nutrient loop' and sinks into deep water via the *global carbon flux* (see Section 12.1): a process that functions at all latitudes. Several pathways are involved in this 'biological pump'. These include: the sinking of plant and animal debris containing both organic carbon (the '*biological organic carbon pump*') and calcium carbonate (the '*biological carbonate pump*'), the sinking of faecal pellets, and the downward advection and diffusion of dissolved organic and inorganic carbon. The net effect of the 'biological pump' is to reduce the $pCO_2$ in surface waters, causing an enhancement in the drawdown of $CO_2$ from the atmosphere initiated by phytoplankton growth. The 'biological pump' as defined here, i.e. that fraction of production that escapes recycling, is a one-way system, and the material transported to deep water can be returned to the surface only by water-mass transport associated with the thermohaline system.

4 Within the deep-water reservoir, $CO_2$ can be utilized in the dissolution of sedimentary carbonates, with the result that when the deep-waters are returned to the surface they can absorb more $CO_2$ from the atmosphere (see below).

The division of the oceans into surface and deep-water reservoirs is very important from the point of view of the global $CO_2$ cycle, not only because the deep-water reservoir is much larger than the surface reservoir, but also because the $CO_2$ in the two reservoirs has very different equilibration times with the atmosphere. The main reason for this is that although deep-water formation can occur over time-scales of months or seasons, once it is formed the deep-water is out of contact with the atmosphere for as long as 1000 yr.

It may be concluded, therefore, that a combination of physico-chemical conditions (the 'solubility pump'), water circulation patterns (the 'physical pump'), and biological intervention (the 'biological pump') govern (i) $CO_2$ transfer rates across the air–sea interface; (ii) $CO_2$ solubility in sea water; and (iii) the bulk transport of carbon within the oceans. The various stages in the oceanic $CO_2$ cycle operate on different time-scales. Because $CO_2$ is withdrawn from surface waters by photosynthesis and returned usually within days, or weeks, when organic compounds are broken down by plant, animal and microbial respiration, short-term spatial and seasonal changes in biological activity have an important influence on $pCO_2$ in the surface ocean. The 'biological pump' is also relatively rapid, removing carbon from the surface waters at sinking rates of $\sim 150\,m\,day^{-1}$ (see Section 12.1). In contrast it can take 100–1000 yr before the thermohaline circulation returns downwelled surface water back to the ocean surface.

Various aspects of the global oceanic $CO_2$ cycle are discussed below.

### 8.4.4 Parameters in the seawater $CO_2$ system

The chemical cycle of $CO_2$ in the oceans is governed by a series of equilibria, which can be expressed as follows (see e.g. Skirrow, 1975; Unesco, 1987).

**1** The $CO_2$ in the atmosphere equilibrates with sea water via exchange across the air–sea interface; thus

$$CO_2(gas) \leftrightharpoons CO_2 \qquad (8.8)$$
$$\text{(dissolved)}$$

**2** The dissolved $CO_2$ then becomes hydrated; thus

$$CO_2 + H_2O \leftrightharpoons H_2CO_3 \qquad (8.9)$$
$$\text{(carbonic acid)}$$

**3** The carbonic acid undergoes very rapid dissociation; thus

$$H_2CO_3 \leftrightharpoons H^+ + HCO_3^- \qquad (8.10)$$
$$\text{(bicarbonate ion)}$$

and

$$HCO_3^- \leftrightharpoons H^+ + CO_3^{2-} \qquad (8.11)$$
$$\text{(carbonate ion)}$$

These carbon-dioxide–seawater equilibria are temperature- and pressure-dependent, and the relative proportions of the species are set by the pH of the system; transitions between the species take place in a direction that tends to maintain a constant pH, i.e. the carbonate in sea water is a buffering system. The major parameters in the seawater–carbon-dioxide systems are therefore $CO_2$, $H_2CO_3$, $HCO_3^-$ and $CO_3^{2-}$, and the total carbon dioxide content, or total inorganic carbon ($\Sigma CO_2$), is given by the sum of the concentrations ($C$) of all the species:

$$\Sigma CO_2 = c_{CO_2} + c_{H_2CO_3} + c_{HCO_3} + c_{CO_3} \qquad (8.12)$$

Only two of the parameters in the oceanic carbonate system can be measured directly; these are $\Sigma CO_2$ and the equilibrium partial pressure of carbon dioxide ($pCO_2$). In this manner, therefore, $pCO_2$ is a measure of the dissolved carbon dioxide, and $\Sigma CO_2$ is a term that refers to the chemistry of the carbonate system in the oceans.

Carbon dioxide enters sea water as the atmosphere and the ocean attempt to achieve equilibrium, and in doing so it changes the chemistry of the system. Two of the keys to understanding these changes are linked to the pH and the alkalinity of the oceans.

*pH*

Traditionally, pH has been defined operationally as

$$pH = -\log_{10}\alpha_H \qquad (8.13)$$

where $\alpha_H$ is the activity of the hydrogen ion. Thus, pH is the negative log of the hydrogen ion concentration, and the pH of a solution is a measure of its acidity in terms of some operational scale (see also Worksheet 14.1). In terms of modern electrochemical theory, however, the situation is much more complex than this. Precise potentiometric techniques are available for the measurement of pH, but there is disagreement over the scales that should be used for saline waters. The theoretical concepts involved in the measurement of pH have been described by Stumm & Morgan (1981), and the various problems inherent in the selection of pH scales for marine and estuarine waters have been reviewed by Culberson (1981), Dickson (1984) and Millero (1986). For the present purposes, however, it is sufficient to understand that, by taking the appropriate precautions, it is possible to make highly reproducible pH measurements in sea water that are normally adequate for most purposes in marine chemistry.

In sea water the equilibrium between the components of the carbon dioxide system is controlled by the pH. At the pH range normally found for sea water, >99% of the dissolved $CO_2$ is present in the form of carbonate ($CO_3^{2-}$) and bicarbonate ($HCO_3^-$) ions. Thus, according to Skirrow (1975) for many purposes eqn (8.12) can be rewritten so that

$$\Sigma CO_2 = c_{HCO_3} + c_{CO_3} \qquad (8.14)$$

The pH range in normal open-ocean sea water is ~7.5–8.4, and Sillen (1963) proposed that on a geological time-scale this pH is controlled by chemical equilibria between the water and the common minerals of marine sediments (see Section 15.1.3). On time-scales of hundreds, or thousands, of years, however, it is changes in the equilibria between dissolved carbon dioxide, bicarbonate ion, carbonate ion and hydrogen ion that provide the principal pH-regulating system in the ocean (Skirrow, 1975).

*Alkalinity*

From the pH range given above it can be seen that sea water is slightly alkaline in character, a property that arises from the dissolution of basic minerals in sea water. The **total alkalinity** (TA) is the buffering capacity of natural waters and is equal to the charges of all the weak ions in solution (Stumm & Morgan, 1981). Total alkalinity is an important physicochemical property of sea water and plays a critical role in several chemical and biological processes. The reasons for this have been summarized by Burke & Atkinson (1988) as follows.

1 $HCO_3^-$ and $CO_3^{2-}$ ions are the major anions of weak acids in sea water, so that changes in total positive charge resulting from alterations in the ratios of cations can be accompanied by shifts in TA ($HCO_3^-$ → $CO_3^{2-}$).

2 The precipitation of calcium carbonate decreases the TA, and as a result TA is a measure of calcification and other biogeochemical processes involving species of $CO_2$.

3 Net photosynthesis and respiration of biological communities change the concentration of dissolved inorganic carbon (DIC); the DIC of a water sample therefore can be calculated from its pH and TA.

Overall, alkalinity is a major factor in the oceanic carbon dioxide system; for example, changes in alkalinity affect the regulation of the ocean–atmosphere $CO_2$ equilibrium and the dissolution–preservation patterns of biogenic carbonates. Alkalinity therefore is one of the parameters used in assessing the status of the oceans in the global carbon cycle.

The alkalinity of sea water can be expressed in terms of the amount of strong acid necessary to bring its reaction to some standard specified end-point in a given volume of solution. Alkalinity therefore has been determined using a titrimetric technique, and

historically total alkalinity (TA) has been defined as the number of equivalents of strong acid required to neutralize 1 l of sea water to an end-point corresponding to the formation of carbonic acid from carbonate. Rakestraw (1949) attempted to redefine the historical concept of alkalinity on a more rigorous basis by relating it to the Lowry–Brønsted definition, in which an acid is a proton donor and a base is a proton acceptor. Thus, alkalinity can be defined as the excess of bases (proton acceptors) over acids (proton donors) in sea water. The only anions of weak acids that are present at significant concentrations in sea water are the bicarbonate, the carbonate and the borate ions, and originally it was considered that it is these which contribute to the alkalinity. For the determination of alkalinity by titration with a strong acid (e.g. HCl) an end-point was therefore selected at which the bicarbonate, carbonate and borate ions are completely combined with protons to form $H_2CO_3$ and $H_3BO_3$, and the expression for the total alkalinity of sea water (in units of equivalents per litre, eq. $1^{-1}$) was therefore commonly given as follows (see e.g. Dickson, 1981):

$$TA = [HCO_3] + 2[CO_3] + [B(OH)_4] + [OH] - [H] \tag{8.15}$$

which defines the equivalence point for an alkalinity determination at which

$$[H] = [HCO_3] + 2[CO_3] + [B(OH)_4] + [OH] \tag{8.16}$$

For example, the alkalinity of sea water was defined by Riley & Chester (1971) as

$$\text{Alkalinity(eq } 1^{-1}) = c_{HCO_3^-} + 2c_{CO_3^{2-}} + c_{B(OH)_4}$$
$$+ (c_{HO} - c_H) \tag{8.17}$$

where $c_{HCO_3^-}$, $2c_{CO_3^{2-}}$ and $c_{B(OH)_4}$ are the equilibrium concentrations of the ions (the concentrations of the doubly charged ions are multiplied by 2), and the term $(c_{HO} - c_H)$ is included to take account of the fact that the concentrations of the various anions determined by titration are greater than their equilibrium concentrations by $(c_{HO} - c_H)$ because of the reactions of the type

$$HCO_3^- + H_2O = H_2CO_3 + OH^- \tag{8.18}$$

The contributions made to the alkalinity of sea water by the carbonate species are termed the **carbon-**

ate alkalinity (i.e. $c_{HCO_3^-} + 2c_{CO_3^{2-}}$), and as bicarbonate and carbonate are the dominant anions of weak acids in sea water, carbonate alkalinity $\cong$ total alkalinity; that is, it is largely bicarbonate and carbonate ions which give sea water its alkalinity. Various workers, however, have drawn attention to the fact that sea water also contains a variety of other bases (such as $NH_3$, $Si(OH)_3$, $O^-$, $SO_4^{2-}$), which will react with hydrogen ions in an alkalinity titration. In view of this, a number of attempts have been made both to expand eqn (8.16) to take account of these other acid–base systems for the definition of total alkalinity, and to design instrumentation for its precise determination; for a detailed description of the principles underlying this modern approach to alkalinity, the reader is referred to Edmond (1970), Hansson & Jagner (1973), Dickson (1981) and Brewer *et al.* (1986).

Much of the more recent data using the modern approach to alkalinity in the oceans has been generated by the GEOSECS and, subsequently, the TTO programmes. In the GEOSECS programme around 6000 alkalinity measurements were made in waters from the major oceans, and the data have been described by Takahasi *et al.* (1980, 1981) and reviewed by Campbell (1983). Alkalinity in surface waters is well correlated with salinity (at a salinity of 35‰ the alkalinity is ~2300 eq. l⁻¹), and variations in

alkalinity are the result mainly of differences in salinity (see e.g. Brewer *et al.*, 1986). As a result, different water masses often have characteristic alkalinities, and clear linear alkalinity–salinity trends are found in all the major oceans. As Campbell (1983) has pointed out, however, processes such as the precipitation and dissolution of calcium carbonate, and the removal and regeneration of nitrate, affect alkalinity and contribute to its non-conservative behaviour.

The geographical distribution of alkalinity can be illustrated with respect to the GEOSECS data reported by Takahasi *et al.* (1980) and Campbell (1983) for the surface waters of the Atlantic Ocean lying between ~40°S and ~60°S (Fig. 8.5). The alkalinity can be expressed as a linear function of salinity, and three clear trends are evident in the data.

1 A warm water trend, which is defined by waters having a temperature >10°C and probably represents the North and South Atlantic Central Waters.

2 The Antarctic trend, which is defined by water having a temperature of 2°C and probably represents Weddell Sea and Circumpolar Water.

3 The sub-Antarctic transition, which forms a mixing line between the two major trends.

The alkalinity differences between the two major trends reflect differences in calcium concentrations resulting from the extraction of calcium, as $CaCO_3$, in the warm surface water and the dissolution of

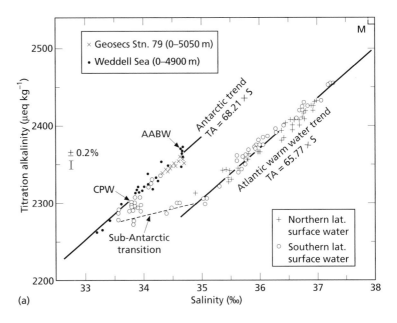

**Fig. 8.5** The alkalinity–salinity relationship for the Atlantic Ocean (from Campbell 1983, after Takahasi *et al.* 1980). Data are from GEOSECS stations (40°N–60°S) for water depths less than 50 m.

CaCO₃ in deep-water, which outcrops in the Antarctic. The GEOSECS data also provided information on the vertical distribution of alkalinity in the oceans. In general, there is a gradual increase in alkalinity with depth below the mixed layer (see Fig. 8.6(b)); there also is an increase in alkalinity in the deep-waters of the Pacific, at the end of the 'grand tour'.

From the point of view of marine geochemistry, the two most important features in the distribution of alkalinity in the oceans are: (i) an increase in cold high-latitude relative to warm low-latitude surface water; and (ii) an increase in deep relative to surface waters. These variations are important because they can be related to the production and dissolution of

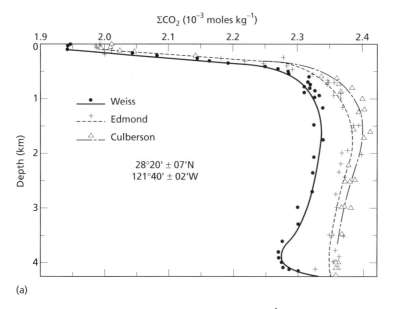

(a)

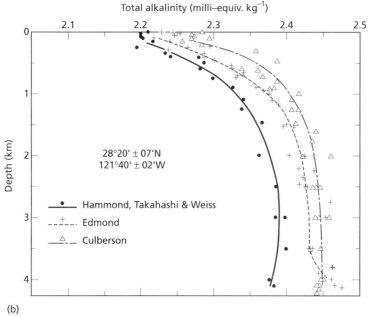

(b)

**Fig. 8.6** Vertical profiles of carbonate system parameters at GEOSECS Pacific Station 28°20′N, 121°W (modified from Skirrow, 1975; data sources given in original publication): (a) $\Sigma CO_2$; (b) total alkalinity; (c) $pCO_2$.

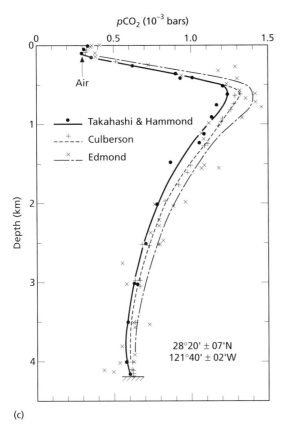

(c)

**Fig. 8.6** *Continued*

picture of how the system varies on a global basis. Variations in the oceanic $CO_2$ system arise in response to a number of factors, which can be divided into two general types:

1 External factors, such as temperature and pressure changes, water-mass circulation and water mixing.

2 Internal factors, which involve *material exchange* within the oceanic carbon dioxide system itself. The principal effects of material exchange are those associated with the gain, or loss, of either carbon dioxide or calcium carbonate. The loss of $CO_2$ during photosynthesis, or evasion to the atmosphere, leads to an increase in pH and a decrease in $pCO_2$ and $\Sigma CO_2$ in the surface layer. In contrast, the production of $CO_2$ by respiration, and the oxidative destruction of organic matter, lead to an increase in $pCO_2$ and $\Sigma CO_2$. This also lowers the pH of the deeper waters, making them more corrosive to calcium carbonate. In turn, as the carbonates dissolve they raise the $\Sigma CO_2$ concentration and the alkalinity of the deep waters.

### 8.4.5 The uptake of $CO_2$ by the oceans

According to Takahasi (1989) three principal factors govern the ocean's capacity to hold $CO_2$. These are:

1 the unique chemical properties of $CO_2$ in sea water, i.e. the 'solubility pump' and associated reactions in the carbonate system;

2 the 'biological pump' that transports $CO_2$ from the surface to the deep ocean;

3 the rate and pattern of ocean water circulation (the 'physical pump').

Source–sink terms in the oceanic $CO_2$ cycle are affected by the complex interplay between these parameters and, in turn, are intimately related to the atmospheric $CO_2$ cycle. A direct biological response to raised levels of atmospheric $CO_2$ is not to be expected because phytoplankton are not carbon-limited. Marine production is sensitive to nutrient availability and hydrographic conditions, however, which means that temperature-mediated changes in these properties could induce considerable indirect biological responses to changes of $CO_2$ in the atmosphere. Further, Riebesell *et al.* (1993) proposed that the supply of $CO_2$ for phytosynthesis can in fact restrict the rate of new primary production in diatoms.

As a gas, $CO_2$ is chemically stable, but dissolved in

calcium carbonate in the oceans. Thus, in warm low-latitude waters the relatively low alkalinity values probably result from the rapid growth of carbonate-secreting organisms and the associated uptake of nutrients during primary production, both of which lead to a lowering of the alkalinity. In deep-waters, the increases in alkalinity reflect the presence of excess calcium, which is released as a result of the increasing extent of carbonate dissolution with depth in the water column and in the underlying sediment (see Section 15.2.4.1).

Any two of the four properties, i.e. $pCO_2$, $\Sigma CO_2$, pH and alkalinity, together with measurements of temperature and pressure, can be used to compute the remaining properties in the oceanic carbon dioxide system, and it is now possible to build up a general sea water it becomes very reactive and is involved in a complex series of reactions with other seawater constituents (see Section 8.4.4). The time taken for such reactions to reach equilibrium, their effects on pH and alkalinity, and water mixing rates near to the surface film provide the first-order controls on $CO_2$ exchanges across the air–sea boundary layer.

Carbon dioxide reacts with sea water to form carbonic acid ($H_2CO_3$), bicarbonate ions ($HCO_3^-$) and carbonate ions ($CO_3^{2-}$), and it is via the reactions with **carbonate ions** that the water absorbs $CO_2$ (see below). Vertical profiles of the carbonate system in the sea show a number of distinctive features, the most striking being the differences between surface and deep waters; profiles of $\Sigma CO_2$, total alkalinity and $\rho CO_2$ are given in Fig. 8.6(a–c). From this figure it is evident that $\Sigma CO_2$ and $\rho CO_2$ increase rapidly to ~1000m. Some of this vertical gradient can be attributed to ocean circulation patterns and water-mass movement, but the gradient is influenced most strongly by the operation of the 'biological pump'. It is the $<\sim 5\%$ to ~10% of surface production which escapes recycling that is transported to deep waters, but most of the organic debris transported down the water column by the 'biological pump' is destroyed in the upper water layers ($<\sim 1000$m), leading to the recycling of nutrients and to the release of $CO_2$ into the water. As the strength of the 'biological pump' increases, the 'drawdown' of $CO_2$ from the atmosphere increases and the $CO_2$ content of the atmosphere decreases. Thus, the operation of both the organic carbon and the carbonate components of the 'biological pump' leads to a surface depletion and a subsurface enrichment of $\Sigma CO_2$, and both $\Sigma CO_2$ and $\rho CO_2$ reach a maximum corresponding approximately to the oxygen minimum (see Section 8.3). Below this maximum, the values of $\rho CO_2$ and $\Sigma CO_2$ fall and then either fall further towards deeper water or become more or less constant with depth. In these deep waters, which are out of contact with the atmosphere, $\Sigma CO_2$ is changed by the mixing of different water masses, by the dissolution of carbonate shells, and by the decomposition of the relatively small amount of organic matter that reaches these depths. Overall, therefore, the shapes of the deep-water profiles of the parameters of the carbonate system depend on: (i) the flux of organic matter from the upper layers of the water column and its rate of oxidation; (ii) the rate of dissolution of calcium car-

bonate; and (iii) the characteristics of, and circulation patterns in, the various water masses cut by the vertical profile. However, although the downward increase in $\Sigma CO_2$ is a universal oceanic feature, the depth–concentration profiles of $\Sigma CO_2$ vary from one ocean to another. In particular, the deep-waters of the North Pacific are among those having the highest concentrations of $\Sigma CO_2$, and the deep-waters of the North Atlantic are among those having the lowest concentrations. This difference can be attributed to the pattern of oceanic circulation during which the North Pacific deep-waters are replaced at a slower rate than those of the North Atlantic, resulting in the accumulation of a large quantity of 'biological pump' products in the Pacific deep-waters (see Section 9.1.2.2).

The depth distribution of $\Sigma CO_2$ in the oceans also has another important consequence in that the ocean system comprises a large body of deep water highly supersaturated with $CO_2$, which is capped with a thin layer of warm and less dense water. The two layers are separated by the thermocline (see Section 7.2.3), which acts as a barrier and prevents the rapid transfer of $CO_2$ from the deep-ocean $CO_2$ reservoir to the atmosphere. According to Takahasi (1989), the capacity of the oceans to take up $CO_2$ therefore is not governed simply by the degree of $CO_2$ saturation in the deep-waters, which are the main oceanic $CO_2$ reservoir, but is also dependent on the water circulation patterns and the effect of the 'biological pump'.

According to Sundquist *et al.* (1979) the capacity of sea water to dissolve atmospheric $CO_2$ is enhanced by the capacity of the ocean system to buffer the associated changes in seawater chemistry. This buffering capacity depends on the conditions and reactions that control the partitioning of carbon within the oceans. The buffer factor, or Revelle factor, may be written

$$R = \frac{(d\rho CO_2 / \rho CO_2)TA, T, S}{(dT_{CO_2} / T_{CO_2})TA, T, S} \tag{8.19}$$

where $T_{CO_2}$ is the total concentration of carbon dioxide in all its forms, $\rho CO_2$ is the partial pressure of carbon dioxide gas, $TA$ is the total alkalinity, $T$ is the temperature and $S$ the salinity. This homogeneous buffering reaction determines how much additional $CO_2$ can be dissolved in surface sea water in response to a $\rho CO_2$ increase in the atmosphere. The buffer factor varies with temperature and has a value of ~10. Thus, a change of ~10% in $pCO_2$ produces a change of only ~1% in $CO_2$. However, $R$ changes as the $CO_2$ in the oceans rises. For example, as the $CO_2$ content of the atmosphere increases, the concentration of dissolved $CO_2$ will also increase to re-establish equilibrium. Because $CO_2$ is in equilibrium with bicarbonate, carbonate and hydrogen ions, the solution of the gas will be accompanied by a decrease in pH, and the value of $R$ increases; i.e. the resistance of the aqueous system to change increases, and the ocean absorbs proportionally less $CO_2$ and the atmospheric fraction increases. The system is complex, however, and is sensitive to the alkalinity: $\Sigma CO_2$ ratio, and therefore to pH.

The buffering capacity of the water results mainly from the presence of carbonate ion, which can combine with $CO_2$ to form bicarbonate and, as Brewer (1983) has pointed out, the principal effect of injecting additional $CO_2$ into the surface ocean therefore is to consume carbonate ion; thus

$$CO_2 + CO_3^{2-} + H_2O = 2HCO_3^- \qquad (8.20)$$

Although the reaction does not proceed completely to the right because of the buffer factor, the result is that much of the additional, or excess, $CO_2$ entering the oceans is converted to bicarbonate. The reaction increases the total carbon in the water, but it does not affect its charge balance or alkalinity. One of the major factors that can change the alkalinity of sea water, however, is the dissolution of calcium carbonate and this can occur via deep-sea carbonate sediments according to the reaction:

$$CO_2 + CaCO_3 + H_2O \rightleftharpoons 2HCO_3^- + Ca^{2+} \qquad (8.21)$$

Once $CO_2$ enters the surface ocean it can be transferred to the deep ocean by the operation of the 'physical' and 'biological' pumps. In the deep ocean the oxidation of downward sinking biogenic debris increases the $CO_2$ of the deep-waters and lowers the pH, thus making them more corrosive towards calcium carbonate. If carbonate dissolution occurs then both the alkalinity and the $\Sigma CO_2$ increase, but the **net** effect of the alkalinity increase would be to enhance the ocean's capacity for $CO_2$ uptake by maintaining the buffer factor and providing carbonate ions, which can be consumed by additional $CO_2$. This will have little immediate effect on atmospheric $CO_2$ levels because it is sequestered in the deep ocean,

but once deep-waters are returned to the surface after the dissolution of carbonate sediments they can draw down $CO_2$ from the atmosphere. On long time-scales, therefore, any increase in the drawdown of $CO_2$ into the ocean surface from fossil fuel $CO_2$ will induce the dissolution of carbonate sediments. The overall effect of this may be that the calcium carbonate stored in marine sediments will ultimately, on long time-scales (i.e. several hundreds of years), neutralize the $CO_2$ generated by fossil fuel combustion by acting as a vast buffer system.

### 8.4.6 Oceanic $CO_2$ budgets

The chemical driving force for ocean–atmosphere $CO_2$ exchange is the difference between the partial pressure of $CO_2$ ($pCO_2$) in the surface ocean waters and the overlying air. This difference is termed $\Delta pCO_2$ (see also Section 8.2 and eqn (8.2)) and is a measure of whether the sea water is undersaturated, or supersaturated, with respect to $CO_2$. When the $pCO_2$ in sea water exceeds that in the overlying atmosphere, $CO_2$ should escape into the air (fugicity), and when the $pCO_2$ in sea water is less than that in the atmosphere, $CO_2$ drawdown should occur. The $\Delta pCO_2$ values are not zero, and it is evident that the atmosphere is out of equilibrium with the surface ocean. One reason for the $pCO_2$ differences between the two reservoirs arises because the gas transfer rates of $CO_2$ to, and from, the atmosphere are much slower than the rates of temperature changes and biological processes, both of which are linked intimately with the fate of $CO_2$ in sea water.

#### 8.4.6.1 $CO_2$ sources and sinks in the surface ocean

Several new data sets have become available recently on the source–sink characteristics of $CO_2$ in surface ocean waters. In particular, those derived from the Joint Global Ocean Flux Study (JGOFS) have contributed to our understanding of the dynamics of the ocean's role in the global $CO_2$ cycle. The JGOFS involves the measurement of a wide variety of basic oceanographic parameters in order to investigate the movement of carbon in the oceans between the air–sea interface, i.e. **the upper boundary condition**, and the sediment surface, i.e. **the lower boundary condition**. The JGOFS data for a variety of contrasting oceanic environments has now been reported, and

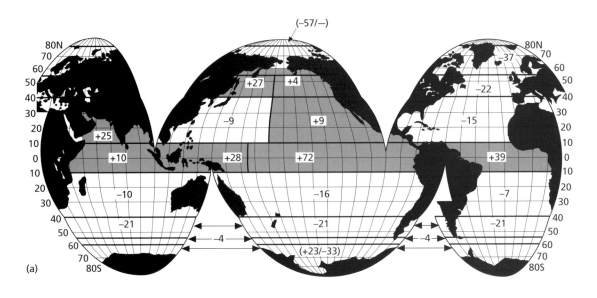

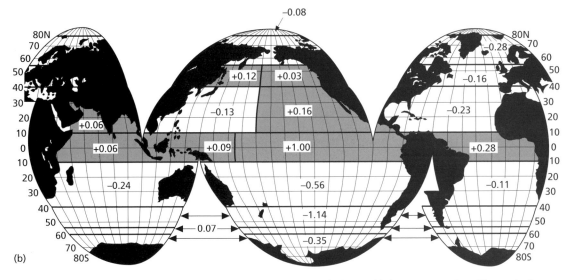

**Fig. 8.7** The $CO_2$ ocean–atmosphere system. (a) The mean annual differences of the partial pressure of carbon dioxide ($pCO_2$) between the surface ocean and the overlying air ($\Delta pCO_2$). Units are in microatmospheres. Positive values (+) indicate that the ocean is a source of $CO_2$ to the atmosphere; the major oceanic source regions are identified as hatched areas. Negative values (–) indicate that the ocean is a sink for atmospheric $CO_2$. (b) The mean annual net $CO_2$ transfer fluxes across the sea surface. Units are in Gt C yr$^{-1}$. Positive values (+) indicate oceanic source fluxes; the major source regions are identified as hatched areas, with the equatorial Pacific being the most intense $CO_2$ source. Negative values (–) indicate oceanic sink fluxes, the sub-Antarctic belt (~40°S–50°S) being the most intense $CO_2$ sink. (From Takahasi, 1989.)

some of the more important findings are included in the present discussion.

A number of features in the $CO_2$ ocean–atmosphere system are illustrated in Fig. 8.7; Fig. 8.7(a) gives the mean annual differences of $pCO_2$ between the surface ocean and the overlying air ($\Delta pCO_2$), and Fig. 8.7(b) shows the mean annual net $CO_2$ transfer fluxes across the sea surface. In general, the features in Fig. 8.7 mirror the complex interrelationships between the 'solubility' and the 'biological' 'physical' pumps.

In Fig. 8.7(a), $\rho CO_2$ differences between the sea surface and the atmosphere ($\Delta \rho CO_2$) are shown for normal, i.e. non-El Niño, years (see below). In the figure, positive $\rho CO_2$ differences between surface sea water and the air indicate regions where the ocean surface is a **source** of atmospheric $CO_2$, and negative values indicate regions where the ocean is a **sink** for atmospheric $CO_2$. Several regional trends in surface seawater $CO_2$ source–sink relationships can be identified from Fig. 8.7(a). It must be pointed out, however, that although oceanic $CO_2$ sinks and sources have been measured over many areas of the Northern Hemisphere and equatorial regions, they are poorly known over the Southern Hemisphere oceans. Further, Fig. 8.7(a) has been compiled on a mean annual basis, and therefore takes no account of the seasonal changes which occur in some oceanic regions where the $CO_2$ source–sink character can switch.

Mean surface water temperatures vary from $\sim -2.0°C$ in the polar sea to $\sim 28°C$ in the equatorial oceans, and the two main global-scale features in the $\rho CO_2$ map mimic these temperature differences.

1 In general, warm equatorial waters, especially those in the Pacific, are supersaturated with $CO_2$ and hence are strong **sources** of $CO_2$ to the atmosphere.

2 In contrast, colder high-latitude waters, such as those of the northern North Atlantic and the Southern Hemisphere oceans, are $CO_2$ **sinks**.

Overall, these patterns of the uptake of $CO_2$ in temperate and subpolar regions (particularly in the Northern Atlantic), and the release of $CO_2$ in equatorial regions (particularly the Pacific), closely match the global patterns of oceanic circulation; i.e. with surface water cooling and sinking in the uptake regions, and deep-waters upwelling and warming in the release regions. The lateral movements of upper and near-bottom waters in opposite directions across most ocean basins complete the physical 'ocean conveyor belt' for the transport of carbon around the globe. The individual $CO_2$ source and sink regions are discussed below.

*The equatorial Pacific.* The $CO_2$ source character of the equatorial Pacific waters is the result of a combination of (i) an intense upwelling in the eastern equa-

torial Pacific that brings deep $CO_2$-rich, high-nutrient waters to the surface; and (ii) surface warming: a combined effect which overcomes $\rho CO_2$ lowering caused by photosynthesis. In 'normal' years, the equatorial Pacific is the major oceanic source of $CO_2$, supplying $\sim 1$ Gt of carbon as $CO_2$ to the atmosphere annually, and is the atmosphere's largest natural source of $CO_2$ (Tans *et al.*, 1990). Its role as a $CO_2$ source, however, is strongly affected during the climatic changes associated with **El Niño** events.

The El Niño–Southern Oscillation (ENSO) phenomenon arises from an ocean–atmosphere coupling. Normally, a relatively small area of warm surface water exists in the western tropical Pacific (the western Pacific warm pool), and gives rise to the western trade winds. These trade winds, which prevail over the tropical Pacific, drive warm surface waters westwards and expose colder water at the surface to the east. However, when the area of the warm surface water increases, during the Southern Oscillation, there is a change in the prevailing westerly trade winds across the Pacific. This is characterized first by a build up in the strength of the trades, which is followed by a decline. As the winds decline, the water that they originally propelled westwards falls back eastward. This weakens equatorial divergence and diminishes equatorial upwelling, and the high $\rho CO_2$ waters normally present in the equatorial Pacific are capped by a blanket of warmer low-$\rho CO_2$ water moved eastwards from the western equatorial Pacific during an El Niño event. During these events the $CO_2$ in the waters can be reduced almost to that of the atmospheric value, and hence the $CO_2$ source can diminish.

The 1991–1992 El Niño event took place during the equatorial Pacific JGOFS (for a review of the JGOFS of the equatorial Pacific see Murray, 1995). During the event, the flux of $CO_2$ to the atmosphere was reduced by a factor of about three from its average of $\sim 1$ Gt C yr$^{-1}$ to a level of $\sim 0.3$ Gt C yr$^{-1}$. The change was controlled largely by ocean physics rather than biology. Upwelling continued during the El Niño event, and although macronutrients were always available in excess of biological requirements, water with a lower $CO_2$ content was upwelled as a consequence of the depression of the nutricline by Kelvin waves.

The data obtained from the equatorial Pacific JGOFS allowed a number of speculations to be made on how long-term changes in equatorial oceanography could influence controls on the ocean–atmosphere cycling of carbon. For example, since around 1975, the sea-surface temperature in the western Pacific has increased and the western Pacific warm pool has shifted eastwards, which may result in more persistent El Niño conditions. In turn, this would produce a significant long-term reduction in the flux of $CO_2$ from the equatorial sea-surface to the atmosphere.

*The North Atlantic.* The global ocean $CO_2$ sink is dominated by the North Atlantic, despite its small area in comparison with the North Pacific. The Atlantic north of 40°N and the Norwegian–Greenland Seas are strong $CO_2$ sinks, and the low $pCO_2$ in these regions is attributed mainly to the rapid cooling of the warm Gulf Stream water and its northward extension, which flows into the Arctic Ocean, and to photosynthesis that occurs in the water during the spring and summer months, e.g. during the spring bloom—see below. In the winter months $pCO_2$ in the surface waters in the northern Atlantic increases mainly as a result of the convective mixing of deep-waters rich in $CO_2$ and nutrients, and the area becomes a $CO_2$ source; however, this weak winter $CO_2$ source is neither strong enough, nor widespread enough, to offset the strong $CO_2$ sink conditions existing during most of the year.

Data for the JGOFS North Atlantic Bloom Experiment (NABE) have been reported in Ducklow & Harris (1993), and wider aspects of the role of the North Atlantic in the global carbon cycle have been described in Eglinton *et al.* (1995).

*The NABE.* The spring bloom in the North Atlantic is one of the most conspicuous seasonal events in the World Ocean. In the eastern North Atlantic, deep convection in winter supplies the upper ocean with nitrate, which supports new primary production following restratification of the water column in April–May. The bloom follows the classic sequence of an open-ocean spring bloom event, with rapid growth of the phytoplankton following the development of a thermocline and an increasingly favourable light regime. The bloom is driven by an excess of production over consumption, and because only a fraction of the phytoplankton are consumed by zooplankton there is an accumulation of biogenic material in the surface waters. The Coastal Zone Colour Scanner (CZCS) indicates that the bloom appears as a sudden explosion of ocean colour (from phytoplankton cells) filling the North Atlantic basin north of ~40° latitude in April and May each year. The NABE was carried out in a region that has a strong mesoscale eddy structure and horizontal advection, and one of the most important findings to emerge from the study was that the structure of the chlorophyll fields, indicating production, was coincident at the same spatial scales as the physical field, thus identifying intimate casual connections between the mesoscale circulation and the biological dynamics of the bloom.

During the bloom, the $CO_2$ in the surface ocean was highly variable on short spatial scales, showing a fine-scale patchiness that was well correlated with plankton chlorophyll (biological activity) and surface temperature. The drawdown of $CO_2$ corresponding to the spring bloom was rapid and dramatic and the surface waters remained substantially undersaturated with respect to the atmosphere through the summer, implying a large $CO_2$ flux into the ocean through this period. Later in the summer, phytoplankton abundance and productivity declined as surface nutrients were used up, and there was increased grazing by zooplankton. This resulted in an important switch from 'new' to 'regenerated' production (see Section 9.1.1), with a decreased use of nitrate from deeper water and a greater use of recycled nitrogen in the form of ammonium from zooplankton activity. Because carbon also is increasingly recycled under these latter conditions, regenerated production is less effective than new production in causing $CO_2$ drawdown from the atmosphere, and hence there is less net export of carbon from the upper ocean. The NABE therefore went some way to answering the question 'what controls the variability of $CO_2$ in the surface ocean?' by establishing that in temperate seas the variability is strongly tied to the bloom dynamics.

Overall, the results of the NABE study have shown that $CO_2$ exchanges in spring and summer are dominated by the patterns of biological activity in the upper ocean, which are, in turn, controlled by circulation features in the 10–100 km range.

*The high-latitude Pacific Ocean.* In contrast to the North Atlantic high-latitude areas, those of the North Pacific are a strong $CO_2$ source on an annual basis.

During the winter months this source is as strong as that in the equatorial Pacific. This is due partly to the fact that the North Pacific deep-waters, which have upwelled to the surface, have the highest concentrations of $CO_2$ in the World Ocean. During the summer months, the surface water $CO_2$ in this area is drawn down, mainly because of intense photosynthetic activity, but on an annual basis the winter source condition predominates.

*The Southern Ocean.* Data for the Southern Ocean JGOFS, which was carried out in the Bellingshausen Sea during October–December 1992, have been reported in Turner *et al.* (1995).

Biogeochemical carbon fluxes in the Southern Ocean are controlled by a number of factors, some of which are unique to the region. Two of these are of particular importance.

1 Seasonal differences in euphotic zone productivity are considerable. In most parts of the World Ocean, primary productivity is limited by the availability of dissolved inorganic nutrients (see Section 9.1.3), and in seasonally stratified waters the annual burst of phytoplankton growth results in a depletion of the nutrient pool. Although nutrient depletion does take place in some local environments that show a shallowing of the mixed layer (e.g. from icemelt-induced stability), it does not occur over most of the Southern Ocean and primary production is generally lower than that which might be expected from the high concentrations of inorganic nutrients in the surface waters (i.e. it is a HNLP region—see Section 9.1.3.2). The availability of iron might be a cause of this, although this is still somewhat speculative (see Section 9.1.3.1).

2 The Southern Ocean is strongly influenced by sea-ice, which affects the light environment of the waters, which in turn is a major control on productivity in the Southern Ocean.

A seasonally persistent $CO_2$ sink forms a belt that almost encircles the sub-Antarctic between 40° and 50° latitude. Here, south-flowing warm subtropical waters cool rapidly as they flow towards the higher southern latitudes, which quickly decreases their $\rho CO_2$. The south-flowing subtropical waters mix with north-flowing sub-Antarctic waters, which also have low $\rho CO_2$ values from the photosynthetic utilization of $CO_2$. Seasonal changes in $\rho CO_2$ have been observed for the Weddell Sea, near Antarctica. Here,

during the winter, the surface water is chilled and becomes more dense than the deep underlying water, thus setting off the deep-water convective circulation (see Section 7.3.3). This causes nutrient-rich deep-waters, which are also rich in $CO_2$, to rise to the surface. As a result, the $CO_2$ trapped in deep-water is released to the atmosphere in winter. In contrast, in spring and summer more sunlight becomes available, stimulating rapid phytoplankton growth in the surface ocean waters, which utilizes the $CO_2$ and nutrients that were brought up to the surface during the previous winter and reduces the $\rho CO_2$ in the surface water to below the atmospheric level. This occurs despite the substantial warming of the water, which by itself causes the $\rho CO_2$ in the water to increase. Thus, the proliferation of phytoplankton in the summer is the main driving force responsible for the ocean becoming a $CO_2$ sink in these high-latitude regions.

*The central ocean gyres.* Biological productivity in the great central ocean gyres is very low and, as a result of this, seasonal changes in the $\rho CO_2$ in these regions are driven mainly by temperature changes.

### 8.4.6.2 $CO_2$ ocean fluxes

Before attempting to assess the rates of $CO_2$ exchange in the individual oceanic source and sink regions, it is necessary to draw attention to three factors that affect the oceanic $CO_2$ cycle. These are the 'missing' $CO_2$, the inter-hemispheric distribution of atmospheric $CO_2$, and seasonal variations in atmospheric concentrations of $CO_2$.

1 **The 'missing' $CO_2$.** The predicted increase in atmospheric $CO_2$ from anthropogenic inputs is larger than that actually observed. For example, in the IPCC $CO_2$ budget discussed below (column 2, Table 8.5) it is apparent that the accumulation of $CO_2$ in the atmosphere ($3.4\,Gt\,C\,yr^{-1}$) accounts for only ~50% of the anthropogenic $CO_2$ released from fossil fuel burning and land-use practices ($\sim7.0\,Gt\,C\,yr^{-1}$). When account is taken of the uptake of $CO_2$ into the oceans there is a net imbalance of $\sim1.6\,Gt\,CO_2$. This has led to the concept of the existence of a 'missing' $CO_2$ sink, and leads to the question 'where has the missing $CO_2$ gone to?' The unaccounted for $CO_2$ may result, in part, from the so-called 'fertilization effect', in which increased $CO_2$ levels increase continental

**Table 8.5** Estimates of the present-day global $CO_2$ budget; units, Gt C yr$^{-1}$.

|  | Takahasi (1989) | IPCC* | Tans et al. (1990)† | Revised IPCC‡ | Revised Tans et al.§ |
|---|---|---|---|---|---|
| **$CO_2$ source:** |  |  |  |  |  |
| fossil fuels |  | 5.3 ± 0.5 | 5.3 | 5.4 | 5.3 |
| deforestation |  | 1.6 ± 1.0 | 0.0–3.2 | 1.6 | 0.0–3.2 |
| total | ~5.3 | 7.0 ± 1.2 | 5.3–8.5 | 7.0 | 5.3–8.5 |
| **$CO_2$ sinks:** |  |  |  |  |  |
| atmosphere | ~3.0 | 3.4 ± 0.2 | 3.0 | 3.4 | 3.0 |
| oceans | ~1.6 | 2.0 ± 0.8 | 0.3–0.8 | 1.7–2.8 | 1.1–2.4 |
| total | ~4.6 | 5.4 ± 0.8 | 3.3–3.8 | 5.1–6.2 | 4.1–5.4 |
| Imbalance (inferred terrestrial sink) | ~0.7 | 1.6 ± 1.4 | 2.0–4.7 | 0.8–1.9 | 1.2–3.1 |

* Watson et al. (1990).
† Given in Sarmiento & Sundquist (1992).
‡ The IPCC budget revised by Sarmiento & Sundquist (1992) for an oceanic sink of 1.7–2.8 Gt C yr$^{-1}$ (see text).
§ The Tans et al. budget revised by Sarmiento & Sundquist (1992) to take account of the net geochemical flux of carbon to the oceans from rivers and rain, the 'skin temperature' of oceanic surface waters and carbon monoxide oxidation to $CO_2$—see text.

and oceanic primary production. Alternatively, there may be an as yet unidentified $CO_2$ sink associated with the terrestrial or oceanic biosphere. Another key question associated with the '$CO_2$–global warming' problem is 'what would happen to the "missing" $CO_2$ if the global climate became warmer?' If the proportion of missing $CO_2$ decreases in response to global warming, industrial $CO_2$ would accumulate in the atmosphere at a faster rate; this is a **positive feedback** condition that could lead to a faster rate of global warming. In contrast, if the global warming causes the carbon reservoirs to absorb more $CO_2$, e.g. by the more rapid growth of plants, then $CO_2$-induced global warming would cause a reduction in the $CO_2$ accumulation rate in the atmosphere, thus slowing down the global warming rate; this is a **negative feedback** condition. It therefore is important to identify the 'missing' $CO_2$ sink, and a considerable amount of research has been directed towards this in recent years, especially with respect to the role played by the oceans in the global $CO_2$ cycle. The problem of the 'missing $CO_2$ sink' is also linked to inter-hemispheric variations in the distribution of atmospheric $CO_2$ discussed next.

**2 The inter-hemispheric distribution of atmospheric $CO_2$.** Around 95% of the industrial emissions of $CO_2$ to the atmosphere are released into the Northern Hemisphere. During the 1980s, the $CO_2$ content at high northern latitudes was, on average, greater than that at southern high latitudes by ~3 μatm (Broecker & Peng, 1992). This difference, however, is only about half that predicted on the basis of model calculations (~5.7–7.3 μatm), given that nearly all of the industrial $CO_2$ is released into the Northern Hemisphere and that there is a limited rate of atmospheric transport of anthropogenic $CO_2$ from the Northern to the Southern Hemisphere. To account for this discrepancy (i.e. the 'missing $CO_2$') it has been suggested that an as yet unknown **$CO_2$ sink** must be located in northern latitudes. It is generally thought that this is most likely to be a terrestrial and not an oceanic sink.

**3 Seasonal variations.** There is a seasonal cycle in atmospheric $CO_2$ concentrations, which is considerably more pronounced in the Northern than in the Southern Hemisphere. The seasonality is dominated by the uptake and release of $CO_2$ by land plants, but also occurs in the oceans. In general, the equatorial Pacific has only a relatively minor seasonal variation during non-El Niño years. In contrast, other oceanic regions can exhibit much stronger seasonal variations. For example, according to Takahasi et al. (1992) the sub-Arctic western North Pacific is a strong $CO_2$ source during the northern winter, but switches to become a sink in the summer. Similar seasonal changes also have been found in parts of the Southern Ocean.

The atmosphere integrates the signals from all $CO_2$ sources and sinks, and modelling the oceanic $CO_2$ budget is difficult because account has to be taken of water circulation patterns (the 'physical pump'), $CO_2$ chemistry (the 'solubility pump'), and biological processes (the 'biological pump'). Despite these difficulties, there have been various attempts to model the oceanic $CO_2$ source–sink system using a variety of techniques. The models include those using the following approaches.

**1 Ocean uptake, or 'gas' models,** in which the net flux of $CO_2$ into, or out of, the oceans is given by the product of the gas transfer coefficient and $\Delta\rho CO_2$, i.e. the $CO_2$ partial pressure difference between the sea surface and the atmosphere (see e.g. Takahasi, 1989).

**2 Synoptic models,** which use observed air-to-sea $CO_2$ differences to constrain the interpretation of $CO_2$ variations in the atmosphere, and combine these with inverse atmospheric transport models to separate the effects of local sources and sinks from long-range transport and so yield the synoptic $CO_2$ fluxes (see e.g. Tans et al., 1990).

**3 Tracer-calibrated models,** e.g. one-dimensional (see e.g. Broecker & Peng, 1982) and three-dimensional (see e.g. Sarmiento et al., 1992) ocean models validated using bomb [14]C data.

**4 $\delta^{13}C/^{12}C$ models,** which involve the partitioning of oceanic and terrestrial $CO_2$ sources on the basis of atmospheric $CO_2$ $\delta^{13}C/^{12}C$ measurements. The underlying rationale here is that the atmospheric $^{13}C/^{12}C$ ratio has been decreasing as a result of the release of isotopically light $CO_2$ from fossil fuel emission to the atmosphere. Plant photosynthesis on land and in the ocean also discriminates against $^{13}C$. In contrast, the dissolution of $CO_2$ in the ocean by gas exchange has little effect on the ratio because even if the reason for the oceanic uptake is the biological removal of carbon in the surface ocean, the large amount of carbon in the ocean reservoir strongly dilutes the isotopic effect of photosynthesis (see e.g. Ciais et al., 1995). As a result, present-day $^{13}C/^{12}C$ disequilibrium can be used to estimate the net air–sea $CO_2$ flux.

**5 Circulation–heat-budget models** (see e.g. Broecker & Peng, 1992; Keeling & Peng, 1995).

There have been a number of estimates of the recent global $CO_2$ budget (i.e. including anthropogenic emissions), which highlight the oceanic role in the global cycle. Several of these estimates have involved the different models described above, and a representative sample is discussed below with some of the relevant data being presented in Table 8.5.

*The Takahasi (1989) budget.* Takahasi estimated the mean annual net $CO_2$ transfer across the ocean surface by multiplying the sea/air $CO_2$ partial pressure difference ($\Delta\rho CO_2$; the chemical 'driving force') by the gas transfer coefficient, or transfer velocity, for $CO_2$ (the rate at which the driving force operates). Data for the budget are given in column 2, Table 8.5, and the flux values obtained, expressed in gigatonnes of carbon per year, are illustrated in Fig. 8.7(b). The principal trends in the data are summarized below; however, it must be stressed that the values are annual mean averages that do not take account of seasonal variations.

**1** The total oceanic release of $CO_2$ is ~1.7 Gt C yr$^{-1}$, and the total oceanic uptake is ~3.3 Gt C yr$^{-1}$, giving rise to a mean annual flux of ~1.6 Gt C yr$^{-1}$ from the atmosphere into the oceans (i.e. the difference between the sink and source terms). As ~5.3 Gt of carbon is being added to the atmosphere at present, and ~3 Gt of this is found in the atmosphere and 1.6 Gt is taken up by the oceans (~30% of the anthropogenic carbon), this leaves ~0.7 Gt unaccounted for; i.e. the 'missing' $CO_2$ according to this budget.

**2** The equatorial Pacific is the largest $CO_2$ **source** region, releasing ~1.0 Gt C yr$^{-1}$, which accounts for ~60% of the total oceanic $CO_2$ source (~1.7 Gt C yr$^{-1}$ in this estimate), which highlights the importance of diminishing the source during El Niño events (see above).

**3** The sub-Antarctic belt is the largest $CO_2$ sink (~1.14 Gt C yr$^{-1}$), which accounts for ~35% of the total oceanic release of $CO_2$.

*The Intergovernmental Panel on Climate Change (IPCC) budget (Watson et al., 1990).* The IPCC $CO_2$ global budget for the decade 1980–1989 is given in column 3, Table 8.5. This budget was based on models of ocean $CO_2$ uptake, and it can be seen that estimated emissions exceed the sum of the atmospheric accumulation plus the estimated oceanic uptake, and yield a net imbalance of 1.6 Gt C yr$^{-1}$; i.e. the 'missing' $CO_2$ according to this budget.

*The Tans et al. (1990) budget.* Tans et al. (1990) estimated the synoptic air-to-sea $CO_2$ flux by combining an atmospheric transport model with observed air-to-sea $CO_2$ differences, and their budget is given in column 4, Table 8.5. The authors concluded that the limited rate of atmospheric transport of anthropogenic $CO_2$ from the Northern Hemisphere restricts oceanic $CO_2$ uptake south of the Equator, and that the air-to-sea observations set limits on oceanic $CO_2$ absorption in the Northern Hemisphere. Carbon dioxide uptake constrained in this way is ~0.3 to ~0.8 Gt C yr$^{-1}$, which is considerably lower than the values in both the Takahasi and the IPCC models. The $CO_2$ imbalance was significantly larger (2.0–4.7 Gt C yr$^{-1}$) than that in both the Takahasi (0.7 Gt C yr$^{-1}$) and the IPCC (1.6 Gt C yr$^{-1}$) estimates, thus implying the existence of a much larger northern terrestrial $CO_2$ sink.

*The Sarmiento & Sundquist (1992) budget.* Despite the different approaches used, the above models predict that the oceans are a net sink for $CO_2$. Further, although the predicted size of the sink varies, all the estimates require that the global $CO_2$ budget needs to be balanced by a large unknown terrestrial sink. However, the size of the unknown terrestrial sink has been questioned. For example, Sarmiento & Sundquist (1992) concluded that the size of the oceanic sink in the Tans *et al.* (1990) estimate was incompatible with many oceanic models, and suggested that a more reasonable estimate for the sink was between a lower limit of 1.7 Gt C yr$^{-1}$ and an upper limit of 2.8 Gt C yr$^{-1}$. They further suggested that the Tans *et al.* (1990) synoptic model must be adjusted by taking account of the following: (i) the net 'geochemical' flux of carbon to the oceans from rivers and rain; (ii) the 'skin temperatures' of oceanic surface water, which usually are colder than the bulk temperatures normally used in determining air–sea $pCO_2$ differences and which lead to an increase of 3–4% in $CO_2$ solubility; and (iii) carbon monoxide oxidation to $CO_2$. The Tans *et al.* (1990) global $CO_2$ budget, modified for these effects, is given in column 6, Table 8.5, from which it is apparent that the total oceanic uptake is now increased to 1.1–2.4 Gt C yr$^{-1}$, and overlaps the 1.7–2.8 Gt C yr$^{-1}$ range suggested by Sarmiento & Sundquist (1992). These modifications also have been incorporated into the revised IPCC budget (column 5, Table 8.5); i.e. the IPCC and Tans *et al.* (1990) oceanic $CO_2$ uptake budgets have been reconciled. However, although the size of the inferred

terrestrial $CO_2$ sink in the two revised budgets has been reduced to ~0.8–3.1 Gt C yr$^{-1}$, it is still significant.

*The Ciais et al. (1995) budget.* These authors used an isotopic $\delta^{13}C$ technique to determine the net partitioning of $CO_2$ between the ocean and terrestrial ecosystems as a function of latitude and time. The authors concluded that:
1 there is a major sink of ~3.5 Gt C yr$^{-1}$ in temperate and boreal forests, and a major ocean sink of ~2.7 Gt C yr$^{-1}$ from 15° to 90°N;
2 the equatorial oceans account for an annual release of ~0.9 Gt C;
3 seasonality at northern mid-latitudes results in a rapid ocean uptake in spring, which was attributed to a bloom of phytoplankton activity.

*Enting & Mainsbridge (1991).* These authors estimated the surface distribution of atmospheric $CO_2$ in space and time, and concluded that there is an oceanic sink in the Southern Hemisphere of ~1.3 Gt C yr$^{-1}$, and a northern sink of ~1.9 Gt C yr$^{-1}$.

*Hesshaimer et al. (1994).* These authors reviewed the use of radiocarbon to validate oceanic carbon cycling models, and suggested the data implied that the oceans take up 25% less anthropogenic $CO_2$ than was thought previously.

It has been suggested that changes in the atmospheric levels of $CO_2$ can be correlated with large-scale changes in ocean and atmospheric circulation associated with El Niño events, which usually result in an increase in atmospheric $CO_2$ levels. Sarmiento (1993), however, proposed that El Niño-stimulated $CO_2$ rises in the atmosphere can be stalled, and cited the effects of the Mount Pinatubo volcanic eruption, which resulted in a fall in the atmospheric $CO_2$ level, as an example. Sarmiento (1993) suggested that perhaps a widespread iron fertilization of the oceans had occurred in regions already having excess surface nutrients, e.g. the Southern Ocean, as a result of the dispersion of aerosols from the Mount Pinatubo eruption. Another suggestion was that cooling induced by the eruption could affect the capacity of the oceans to act as a $CO_2$ sink; however, Ciais *et al.* (1995) did not consider that the drop in temperature associated with the eruption was sufficient to account for all of the increased $CO_2$ uptake.

On the basis of the Takahasi (1989) budget it is apparent that in the Northern Hemisphere, the Pacific, Atlantic and Indian Oceans north of 15°N have a net uptake of ~0.6 Gt of carbon as $CO_2$ annually, and that the equatorial belt (15°N–15°S) of the three oceans releases ~1.3 Gt of carbon as $CO_2$ during a normal, i.e. non-El Niño, year. Because the equatorial release exceeds uptake in the Northern Hemisphere, balancing the World Ocean $CO_2$ budget hinges on the sink intensity of the Southern Hemisphere oceans. In this context, the Southern Ocean is potentially very important. The reasons for this are that it covers a large area (15–20%) of the World Ocean, and the high winds and rough seas mean that the presence of even a small gradient in the partial pressure of $CO_2$ between the air and the surface ocean would drive an appreciable $CO_2$ flux. In this context, Robertson & Watson (1995) found a summer sink for atmospheric $CO_2$ between 88°W and 80°E in the Southern Ocean. Although no winter data are available, the size of the sink in summer suggests, if representative of the Southern Ocean as a whole, a drawdown of $CO_2$ of ~0.35–0.50 Gt C over a period of 4 months. The overall status of the Southern Ocean in the oceanic $CO_2$ budget is, however, still unclear.

Around 99% of the $CO_2$ supplied to the atmosphere at present is injected into the Northern Hemisphere, and there is only a limited transport of this anthropogenic $CO_2$ between the Northern and Southern Hemispheres (see above). Tans *et al.* (1990) concluded that this limited rate for the atmospheric transport of $CO_2$ from the Northern Hemisphere restricts the uptake of $CO_2$ in the Southern Hemisphere and requires the existence of a large unknown $CO_2$ sink in the Northern Hemisphere. Using model calculations, however, Sarmiento & Sundquist (1992) suggested that a large fraction of the 'geochemical' $CO_2$ efflux must effectively be leaving the ocean in the Southern Hemisphere. To overcome the problem of a restricted transport of atmospheric $CO_2$ constrained by the inter-hemispheric $CO_2$ gradient, the authors postulated that a non-atmospheric mechanism may be involved in the transfer of this 'geochemical' $CO_2$ efflux (~0.4–0.7 Gt C yr$^{-1}$) from the Northern to the Southern Hemisphere.

One such non-atmospheric mechanism may be a north to south **inter-hemispheric oceanic pump**, as proposed by Broecker & Peng (1992). These authors pointed out that although anthropogenic $CO_2$ emissions have given rise to a north to south inter-hemispheric $CO_2$ gradient in the atmosphere it has been suggested that prior to the Industrial Revolution natural $CO_2$ sources set up a reverse south to north gradient, which drove ~1 Gt of carbon annually through the *atmosphere* from the Southern to the Northern Hemisphere. At steady state, this flux must have been balanced by a counter flow of carbon from the north to the south through the *ocean*. The North Atlantic is unique in that its surface water is exchanged with other oceans by thermohaline and not wind-driven transport, and Broecker & Peng (1992) suggested that the North Atlantic 'conveyor belt' circulation was the most likely method for the inter-hemispheric transport of $CO_2$ because it carried water from the Northern Hemisphere, which has been in contact with the Northern Hemisphere atmosphere before sinking into the deep sea, into the Southern Hemisphere; thus, it can act as an inter-hemispheric 'pump'. Broecker & Peng (1992) estimated the magnitude of this natural flux, and concluded that before the Industrial Revolution, deep-water formed in the North Atlantic carried *c.* 0.6 Gt of carbon annually to the Southern Hemisphere. This estimate is close to the 'geochemical' fluxes (~0.4–0.7 Gt C yr$^{-1}$) estimated by Sarmiento & Sundquist (1992) and the existence of the inter-hemispheric carbon 'pump' raises the question of the need for a large terrestrial carbon sink in the Northern Hemisphere, such as that suggested by Tans *et al.* (1990), to balance the present-day global carbon budget.

### 8.4.7 The uptake of $CO_2$ by the oceans: synthesis

The three major reservoirs through which carbon dioxide exchanges are the atmosphere, the terrestrial biosphere and the oceans. The oceans are the largest of the rapidly exchanging carbon reservoirs and contain about 60 times as much carbon as the atmosphere, most of which is dissolved $CO_2$ in the form of bicarbonate. Not all of the excess, i.e. anthropogenic, $CO_2$ injected into the atmosphere enters the oceans, but because they do act as a sink for the gas much research has been directed towards predicting the environmental impact that excess $CO_2$ will have both on the ocean system itself and on the global climate in general.

There are three stages in the uptake of excess $CO_2$ by the oceans.

1 The thin, well mixed, surface layer will establish equilibrium with $CO_2$ in the atmosphere relatively quickly (a few years), so that on time-scales of decades the concentration of $CO_2$ in the atmosphere is controlled mainly by exchange with the oceans. The capacity of the surface ocean to take up excess $CO_2$ is enhanced by the buffer mechanism that converts the gas to bicarbonate.

2 The capacity of the surface ocean to take up additional $CO_2$ is increased by the transport of $CO_2$ to deep water, which effectively reduces surface water $pCO_2$. This occurs by two main processes.

(a) By the sinking of organic debris via the 'biological pump'. The mechanism operates on short time-scales at all latitudes, and it has been estimated that it removes ~4 Gt C yr$^{-1}$ from the surface ocean.

(b) By downward water mixing associated with the oceanic thermohaline circulation: the 'physical pump'. This process operates at high latitudes on long time-scales associated with deep-water transit, and in most oceanic regions only the top thousand or so metres of the ocean have yet acquired significant amounts of excess $CO_2$.

Because most anthropogenic $CO_2$ is emitted at low and mid-latitudes in the Northern Hemisphere, and because there is limited north to south atmospheric transport, it might be supposed that on short time-scales much of this $CO_2$ leaves the surface seawater layer via the 'biological pump'. Although it is extremely important in the natural carbon cycle, Watson *et al.* (1990) questioned the efficiency of the 'biological pump' to sequester excess $CO_2$ because marine biota do not respond **directly** to increases in $CO_2$. It was pointed out in Section 8.4.5, however, that the biota are sensitive to other factors that could induce **indirect** responses to changes in the concentration of atmospheric $CO_2$.

3 Both the 'physical pump' and the 'biological pump' mechanisms trap $CO_2$ in the deep waters, and this can be returned to the surface ocean only by the long-timescale thermohaline circulation. As the $CO_2$ content of the deep-water rises, however, water masses now supersaturated with respect to carbonate

minerals will become undersaturated and carbonates that are in contact with them will begin to dissolve. Thus, the capacity of the ocean to absorb $CO_2$ will be further enhanced on a long time-scale by the utilization of $CO_2$ during the dissolution of sedimentary carbonate, with the result that when the waters are returned to the surface they can absorb further $CO_2$; for a detailed discussion of the effect of deep-sea calcite and organic carbon preservation on atmospheric $CO_2$ concentrations, see Archer & Maier-Reimer (1994).

With respect to the question 'how will the oceans cope with excess $CO_2$?', the conclusions drawn by Broecker & Peng (1982) provide a useful summary of the possible consequences of an oceanic invasion of $CO_2$: a summary that can be related to the various stages involved in the uptake of excess $CO_2$.

1 The times required for the ocean system to equilibrate with the carbon dioxide in the atmosphere are about a few years for the surface ocean, several tens of years for the main thermocline, and hundreds of years for the deep sea. Thus, the thermocline acts as a barrier for the oceanic equilibrium with atmospheric $CO_2$, and equilibrium between the atmosphere and the total ocean will take many centuries.

2 The dissolution of calcite from deep-sea sediments, which raises the alkalinity, affects the bottom waters and therefore only equilibrates with the atmosphere when the waters are brought to the surface in a time period that requires several thousands of years. Further, the excess calcium released from the carbonates in a dissolved form will only be removed from sea water in many tens of thousands of years.

It is apparent, therefore, that the various reservoirs, or boxes, in the oceans equilibrate with the atmosphere on very different time-scales, ranging up to tens of thousands of years. Broecker & Peng (1982) concluded, therefore, that as a consequence of this, carbon dioxide added to the atmosphere by human activity over the next two centuries will have an effect on ocean chemistry even on geological time-scales.

## 8.5 Dissolved gases in sea water: summary

1 The oceans can act as either a source or a sink for atmospheric gases, which leave or enter the system via exchange across the air–sea interface. Attention here has been focused on oxygen and carbon dioxide,

both of which play important roles in the biogeochemistry of sea water.

2 Dissolved oxygen is released during the photosynthetic production of biological soft-tissue material in the euphotic zone and is consumed during the oxidative destruction of the organic matter at greater depths in the water column. Oxygen minima, which occur at depths ranging between $\sim 100$ and $\sim 1000$ m, are found in various parts of the ocean; these are important features that can mediate redox reactions in the water column. There is relatively little consumption of dissolved oxygen in deep waters. Apparent oxygen utilization values, which are a measure of the amount of the gas that has been utilized after the waters have left the surface, follow the deep water 'global grand tour' and are lowest in Atlantic subsurface waters (less utilization) and highest in the Pacific (more utilization). Dissolved oxygen can be used to characterize water masses.

3 Carbon dioxide is consumed during photosynthesis, during which carbon enters both the soft tissue and the hard skeletal phases of organisms, and is released during the oxidative destruction of organic matter. In addition, carbon dioxide is extremely reactive in sea water and the complex equilibria involved have a number of profound effects on the chemistry of the oceans. The manner in which the oceanic carbon dioxide system operates can be summarized as follows. Carbon dioxide is pulled down into surface sea water from the atmosphere in response to a combination of biological and physical processes, but because the charge balance is not affected this does not result in a change in the alkalinity of sea water. The air–sea fluxes of $CO_2$ vary with latitude, large positive net fluxes (net evasions from the ocean) being found in tropical warm water latitudes where $pCO_2$ is high, and net negative fluxes (net invasion into the ocean) occurring in colder waters where the $pCO_2$ values are low. Once it is drawn down into the surface ocean, $CO_2$ enters into a biological cycle and also takes part in a complex inorganic carbonate chemistry. In the biological cycle carbon dioxide is utilized to fix carbon into organic tissue by the photosynthetic activity of phytoplankton. This removal of $CO_2$ raises the pH, and the uptake of the nutrients, nitrate and phosphate that occurs at the same time raises both the pH and the alkalinity of the waters, so that the biological cycle also affects the inorganic carbonate chemistry. As the organisms die, a small frac-

tion ($\sim 5\%$) of the organic material escapes to carry the fixed $CO_2$ to deep waters (see Section 9.3). However, the vast majority of the organic tissue undergoes oxidative regeneration, with the release of $CO_2$ and the nutrients into waters of intermediate depths. The release of the gas lowers the pH of the waters and makes them more corrosive to calcium carbonate material. The dissolution of the carbonates alters the charge balance of the system and changes the alkalinity. As the waters of the World Ocean move along the 'grand tour' they become progressively depleted in oxygen and enriched in $CO_2$ and the nutrients. As a result, the dissolution horizon below which calcium carbonate dissolves (see Section 15.2.4.1) shoals to shallower depths in the North Pacific relative to the North Atlantic. The calcium carbonate dissolved from sediments further increases the $CO_2$ content of the bottom waters and also raises the alkalinity of the system. The oceans are a major sink for excess $CO_2$ in the atmosphere. This excess $CO_2$ enters the seawater carbonate system and passes through the surface ocean, the main thermocline and the deep sea, and equilibrates with the various reservoirs on different time-scales, which apparently range from tens to tens of thousands of years.

# References

Archer, D. & Maier-Reimer, E. (1994) Effect of deep-sea sedimentary calcite preservation on atmospheric $CO_2$ concentration. *Nature*, 367, 260–63.

Brewer, P.G. (1983) Carbon dioxide and the oceans. In *Changing climate*, 186–215. Report of the Carbon Dioxide Assessment Committee. Washington, DC: National Academy Press.

Broecker, W.S. (1974) *Chemical Oceanography*. New York: Harcourt Brace Jovanovich.

Broecker, W.S. & Peng, T.-H. (1974) Gas exchange rates between air and sea. *Tellus*, 26, 21–35.

Broecker, W.S. & Peng, T.-H. (1982) *Tracers in the Sea*. Palisades: Lamont-Doherty Geological Observatory.

Broecker, W.S. & Peng, T.-H. (1992) Interhemispheric transport of carbon dioxide by ocean circulation. *Nature*, 356, 587–9.

Burke, C.M. & Atkinson, M.J. (1988) Measurement of total alkalinity in hypersaline waters. *Mar. Chem.*, 25, 49–55.

Burton, J.D., Brewer, P.G. & Chesselet, R. (eds) (1986) *Dynamic Processes in the Chemistry of the Upper Ocean*. New York: Plenum.

Campbell, J.A. (1983) The Geochemical Ocean Sections Study—GEOSECS. In *Chemical Oceanography*, J.P. Riley

& R. Chester (eds), Vol. 8, 89–155. London: Academic Press.

Ciais, P., Tans, P.P., White, J.W.C., *et al.* (1995) Partitioning of ocean and land uptake of $CO_2$ as inferred by $\delta^{13}$ measurements from the NOAA/CMDL Global Air Sampling Network. *J. Geophys. Res.*, **99**, 5051–70.

Conrad, R. & Seiler, W. (1986) Exchange of CO and $H_2$ between ocean and atmosphere. In *The Role of Air–Sea Exchange in Geochemical Cycling*, P. Buat-Menard (ed.), 269–82. Dordrecht: Reidel.

Culberson, C.H. (1981) Direct potentiometry. In *Marine Electrochemistry*, M. Whitfield & D. Jagner (eds), 187–261. Chichester: Wiley.

Deuser, W.G. (1975) Reducing environments. In *Chemical Oceanography*, J.P. Riley & G. Skirrow (eds), Vol. 3, 1–37. London: Academic Press.

Dickson, A.G. (1981) An exact definition of total alkalinity and a procedure for the estimation of alkalinity and total inorganic carbon from titration data. *Deep Sea Res.*, **28**, 609–23.

Dickson, A.G. (1984) pH scales and proton-transfer reactions in saline media such as sea water. *Geochim. Cosmochim. Acta*, **48**, 2299–308.

Ducklow, H.W. & Harris, R.P. (eds) (1993) Topical studies in oceanography. JGOFS: the North Atlantic Bloom Experiment. *Deep Sea Res.*, **40**.

Edmond, J.M. (1970) High precision determination of titration alkalinity and the total carbon dioxide content of sea water by potentiometric titration. *Deep Sea Res.*, **17**, 737–50.

Eglinton, G., Elderfield, H., Whitfield, M. & Williams, P.J. Le B. (eds) (1995) The role of the North Atlantic in the global carbon cycle. *Philos. Trans. R. Soc. London*, **348**.

Enting, I.G. & Mainsbridge, J.V. (1991) Latitudinal distribution of sources and sinks of $CO_2$: results of an inversion study. *Tellus*, **43**, 156–70.

Erickson, D.J. (1993) A stability dependent theory for air–sea gas exchange. *J. Geophys. Res.*, **98**, 8471–88.

Farmer, D.M., McNeil, C.L. & Johnson, B.D. (1993) Evidence for the importance of bubbles in increasing air–sea flux. *Nature*, **361**, 620–23.

Fitzgerald, W.F. (1986) Cycling of mercury between the atmosphere and oceans. In *The Role of Air–Sea Exchange in Geochemical Cycling*, P. Buat-Menard (ed.), 363–408. Dordrecht: Reidel.

Fitzgerald, W.F. (1989) Atmospheric and oceanic cycling of mercury. In *Chemical Oceanography*, J.P. Riley, R. Chester & R.A. Duce (eds), Vol. 10, 151–86. London: Academic Press.

Hansson, I. & Jagner, D. (1973) Evaluation of the accuracy of Gran plots by means of computer calcuations. Application to the potentiometric titration of the total alkalinity and carbonate content in sea water. *Anal. Chim. Acta*, **65**, 363–72.

Hesshaimer, V., Helmann, M. & Levin, I. (1994) Radiocarbon evidence for a smaller oceanic carbon dioxide sink than previously believed. *Nature*, **370**, 201–05.

Houghton, J.T., Jenkins, G.J. & Ephraums, J.J. (eds) (1990) *Climate Change: the IPCC Scientific Assessment.* Cambridge: Cambridge University Press.

Hunter-Smith, R.J., Balls, P.W. & Liss, P.S. (1983) Henry's law constants and the air–sea exchange of various low molecular weight halocarbon gases. *Tellus*, **35**, 170–6.

Kanwisher, J. (1963) On the exchange of gases between the atmosphere and the sea. *Deep Sea Res.*, **10**, 195–207.

Keeling, R.F. & Peng, T.-H. (1995) Transport of heat, $CO_2$ and $O_2$ by the Atlantic's thermohaline circulation. In *The Role of the North Atlantic in the Global Carbon Cycle*, G. Eglinton, H. Elderfield, M. Whitfield & P.J. Le B. Williams (eds), *Philos. Trans. R. Soc. London*, **348**, 133–42.

Kester, D.R. (1975) Dissolved gases other than $CO_2$. In *Chemical Oceanography*, J.P. Riley & G. Skirrow (eds), Vol. 1, 497–589. London: Academic Press.

Law, C.S. & Owens, N.J.P. (1990) Significant flux of atmospheric nitrous oxide from the northwest Indian Ocean. *Nature*, **346**, 826–28.

Liss, P.S. (1983) Gas transfer: experiments and geochemical implications. In *Air–Sea Exchange of Gases and Particles*, P.S. Liss & W.G. Slinn (eds), 241–98. Dordrecht: Reidel.

Liss, P.S. (1986) The air–sea exchange of low molecular weight halocarbon gases. In *The Role of Air–Sea Exchange in Geochemical Cycling*, P. Buat-Menard (ed.), 283–94. Dordrecht: Reidel.

Liss, P.S. & Merlivat, L. (1986) Air–sea exchange rates: introduction and synthesis. In *The Role of Air–Sea Exchange in Geochemical Cycling*, P. Buat-Menard (ed.), 113–27. Dordrecht: Reidel.

Liss, P.S. & Slater, P.G. (1974) Flux of gases across the air–sea interface. *Nature*, **247**, 181–4.

Millero, F.J. (1986) The pH of estuarine waters. *Limnol. Oceanogr.*, **31**, 839–47.

Murray, J.W. (ed.) (1995) Topical studies in oceanography. A U.S. JGOFS process study in the equatorial Pacific. *Deep Sea Res.*, **42**.

Olsen, D.B., Hitchcock, G.L., Fine, R.A. & Warren, B.A. (1993) Maintenance of the low-oxygen layer in the central Arabian Sea. *Deep Sea Res.*, **40**, 673–85.

Packard, T.T., Minas, H.J., Coste, B., *et al.* (1988) Formation of the Alboran oxygen minimum zone. *Deep Sea Res.*, **35**, 1111-8.

Rakestraw, N.W. (1949) The conception of alkalinity of excess base in sea water. *J. Mar. Res.*, **8**, 14–20.

Richards, F.A. (1965) Dissolved gases other than carbon dioxide. In *Chemical Oceanography*, 1st edn, J.P. Riley & G. Skirrow (eds), 197–225. London: Academic Press.

Riebesell, U., Wolf-Gladrow, D.A. & Smetacek, V. (1993) Carbon dioxide limitation of marine phytoplankton growth rates. *Nature*, **361**, 249–51.

Riley, J.P. & Chester, R. (1971) *Introduction to Marine Chemistry*. London: Academic Press.

Robertson, J.E. & Watson, A.J. (1995) A summer-time sink for atmospheric carbon dioxide in the Southern Ocean between 88°W and 80°E. In *Topical Studies in Oceanography. Southern Ocean JGOFS: the U.K. 'Sterna' Study in the Bellingshausen Sea*, D. Turner, N. Owens & J. Priddle (eds). *Deep Sea Res.*, **42**, 1081–91.

Roether, W. (1986) Field measurements of gas exchange. In *Dynamic Processes in the Chemistry of the Upper Ocean*, J.D. Burton, P.G. Brewer & R. Chesselet (eds), 117–28. New York: Plenum.

Sarmiento, J.L. (1993) Atmospheric $CO_2$ stalled. *Nature*, **365**, 697–8.

Sarmiento, J.L. & Sundquist, E.T. (1992) Revised budget for the oceanic uptake of anthropogenic carbon dioxide. *Nature*, **356**, 589–93.

Sarmiento, J.L., Orr, J.C. & Siegenthaler, U. (1992) A perturbation simulation of $CO_2$ uptake in an ocean general circulation model. *J. Geophys. Res.*, **97**, 3621–45.

Shulenberger, E. & Reid, J.L. (1981) The Pacific shallow oxygen maximum, deep chlorophyll maximum, and primary productivity, reconsidered. *Deep Sea Res.*, **28**, 901–19.

Sillen, L.G. (1963) How has sea water got its present composition? *Sven. Kem. Tidskr.*, **75**, 161–77.

Skirrow, G. (1975) The dissolved gases—carbon dioxide. In *Chemical Oceanography*, J.P. Riley & G. Skirrow (eds), Vol. 2, 1–192. London: Academic Press.

Stumm, W. & Morgan, J.J. (1981) *Aquatic Chemistry*. New York: Wiley.

Sundquist, E.T., Plummer, L.N. & Wigley, T.M.L. (1979) Carbon dioxide in the ocean surface: the homogeneous buffer factor. *Science*, **204**, 1203–5.

Takahasi, T. (1989) The carbon dioxide puzzle. *Oceanus*, **32**, 22–9.

Takahasi, T., Broecker, W.S., Werner, S.R. & Bainbridge, A.E. (1980) Carbonate chemistry of the surface waters of the World Ocean. In *Isotope Marine Chemistry*, E.D. Goldberg, Y. Horibe & K. Saruhashi (eds), 291–326. Tokyo: Uchida Rokahuho.

Takahasi, T., Broecker, W.S. & Bainbridge, A.E. (1981) The alkalinity and total carbon dioxide concentration in the world oceans. In *Carbon Cycle Modelling*, Scope 16, B. Bolin (ed.), 159–99. New York: Wiley.

Takahasi, T., Tans, P.E. & Fung, I.Y. (1992) Balancing the budget. *Oceanus*, **35**, 18–28.

Tans, P.P., Fung, I.Y. & Takahasi, T. (1990) Observational constraints on the global atmospheric $CO_2$ budget. *Science*, **247**, 1431–38.

Turner, D., Owens, N. & Priddle, J. (eds) (1995) Topical Studies in Oceanography. Southern Ocean JGOFS: the U.K. 'Sterna' Study in the Bellingshausen Sea. *Deep Sea Res.*, **42**.

Unesco (1987) Unesco Technical Paper on Marine Science, no. 51. Paris: Unesco.

Watson, R.T., Rodhe, H., Oeschger, H. & Siegenthaler, U. (1990) Greenhouse gases and aerosols. In *Climate Change: the IPCC Scientific Assessment*, J.T. Houghton, G.J. Jenking & J.J. Ephraums (eds), 1–40. Cambridge: Cambridge University Press.

Weiss, R.F. (1970) The solubility of nitrogen, oxygen and argon in water and sea water. *Deep Sea Res.*, **17**, 721–35.

# 9 Nutrients, organic carbon and the carbon cycle in sea water

The manner in which the carbon cycle operates in the ocean is of prime importance to marine geochemistry because the down-column transport of particulate organic carbon, i.e. the *global carbon flux*, drives the processes that control the removal of material from the water column and its incorporation into the sediment sink.

## 9.1 The nutrients and primary production in sea water

### 9.1.1 Introduction

Parsons (1975) defines a nutrient element as one that is involved functionally in the processes of living organisms. He points out that in oceanography, however, the term has been applied almost exclusively to **nitrate, phosphate** and **silicate**, and attention here initially will be focused on these traditional nutrients.

*Nitrogen nutrients.* Nitrogen is present in sea water as: (i) molecular nitrogen; (ii) fixed inorganic salts, such as nitrate nitrogen ($NO_3$-N), nitrite nitrogen ($NO_2$-N) and ammonia ($NH_3$-N); (iii) a range of organic nitrogen compounds associated with organisms, e.g. amino acids and urea; and (iv) particulate nitrogen.

Nitrogen is brought to the oceans from fluvial and atmospheric sources, by diffusion from sediments and by *in situ* nitrogen fixation. Nitrogen **fixation** is the process by which organisms can assimilate, or fix, molecular nitrogen. Nitrogen fixation in the oceans can be carried out by only a few phytoplankton, mainly the cyanobacteria, and most oceanic phytoplankton are not nitrogen fixers. Because of this they utilize 'pre-fixed' nitrogen in the form of dissolved species, with a preference for nitrate, nitrite and ammonia. The input of fixed nitrogen to the oceans by *in situ* fixation and by fluvial and atmospheric fluxes is an important control on marine productivity on long time-scales, and hence on ocean–atmosphere $CO_2$ exchange, which affects global climate. The utilization of the fixed nitrogen by phytoplankton takes place in the euphotic zone and some of the nitrogenous nutrients are released in a soluble form within this zone. The remainder are transported out by sinking particulates, and a large fraction of these is released back into the solution at depth in the water column by remineralization of the organic material. This occurs mainly via bacterial mediation, the final inorganic end-product being nitrate. The oxidation of ammonia to nitrite and then nitrate is termed **nitrification**, and is mediated by nitrifying bacteria. The reverse process, which is termed **denitrification**, is mediated by denitrifying bacteria, mainly in anoxic sediments, and is the dominant mechanism for the removal of fixed nitrogen from the biosphere.

*Phosphorus nutrients.* There are a variety of forms of phosphorus in sea water. These include dissolved inorganic phosphorus (predominantly orthophosphate ions, $HPO_4^{2-}$), organic phosphorus and particulate phosphorus; however, phytoplankton normally satisfy their phosphorus requirements by direct assimilation of orthophosphate. Phosphate, like nitrate, is also released back into the water column during the oxidative destruction of organic tissues. Most of the regeneration of phosphorus probably takes place via bacterial decomposition, which leads to the formation of orthophosphates, although chemical decomposition also may occur.

*Inorganic* nitrate and phosphate are not the only forms of nitrogen and phosphorus that can be used by organisms for their nutritional needs, and Jackson & Williams (1985) have discussed the importance of dissolved *organic* nitrogen (DON) and phosphorus

(DOP) in the nutrient economy of the ocean. The data provided by these authors showed that DON and DOP concentrations increase as those of nitrate and phosphate decrease, indicating a change in the dominant chemical form of nitrogen from nitrate to DON and of phosphorus from phosphate to DOP. The authors concluded that both DON and DOP constitute significant fractions of the oceanic nutrient pools, especially in the euphotic zone, with the labile fractions of DON and DOP being important sources of nitrogen and phosphorus for phytoplankton in oligotrophic oceanic areas.

*Silicon nutrients.* Organisms such as diatoms and radiolarians require silicate for shell formation. Silicon is supplied to the oceans in both dissolved and particulate forms via river run-off, atmospheric deposition and glacial weathering (especially from the Antarctic), and dissolved silicon in sea water is probably present as orthosilicic acid, $Si(OH)_4$. The particulate forms of the element include a wide variety of silicate and aluminosilicate minerals, together with diatom and radiolarian shells, which contain silica in the form of opal (see Section 15.2.2). Silica from the skeletal or hard parts of the organisms is released back into the water column during down-column solution; this process does not appear to involve bacterial action, but it is probably aided by passage through the gut of other organisms in the foodchain and expulsion as a faecal material.

When considering the macronutrients* in terms of their oceanic chemistries, it is useful to distinguish silicate from nitrate and phosphate. The reason for this is that both the latter two nutrients are involved in nutrition and are incorporated into soft tissues, whereas silicate is used only in the building of hard skeletal parts. None the less, all three nutrients are initially removed from solution by organisms, mainly phytoplankton, during primary production in the euphotic zone.

---

* Nitrate, phosphate and silicate have traditionally been referred to as 'micronutrients'. More recently, the term 'micronutrient' has been applied to essential trace metals, such as iron. To avoid confusion, therefore, in the present text nitrate, phosphate and silicate will be termed 'macronutrients' and the trace metals will be referred to as micronutrients'; although on the basis of their concentrations the terms 'micronutrients' and 'nanonutrients', respectively, would perhaps be more appropriate.

One of the most important concepts to emerge in the nutrient field in recent years has been the distinction between new and regenerated production (see e.g. Dugdale & Goering, 1967; Eppley & Peterson, 1979). This concept is related to the way in which nutrients are supplied to the euphotic zone, and the processes involved can be illustrated with respect to nitrogen. The supply of nitrogenous nutrients required during primary production can be related to two different types of sources:

1 A new supply to the euphotic layer from river run-off, atmospheric deposition, upwelling of deep-water and nitrogen fixation, mainly in the form of nitrate.

2 A regenerated supply from the short-term recycling processes within the euphotic layer itself, mainly in the form of ammonia, with lesser amounts of urea and amino acids, arising from the excretory activities of animals and the metabolism of heterotrophic micro-organisms, i.e. this supply is derived from phytoplankton via the food web.

The system is illustrated diagrammatically in Fig. 9.1. Eppley & Peterson (1979) pointed out that in an ideal closed system, with steady-state standing stocks and fluxes, the cycling of nutrients through an enclosed food web could continue indefinitely. In the real ocean system, however, there is a loss of organic material from the euphotic zone to deep water, e.g. by the sinking of faecal pellets and 'marine snow' (see Section 10.4), and this is compensated by the input of new nutrients into the euphotic zone. Primary production associated with the newly available nitrogen (e.g. nitrate) is termed **new production**, and that resulting from the nitrogen recycled in the euphotic zone (e.g. ammonia) is referred to as **regenerated production**. Thus, the organic matter sinking out of the euphotic zone (export production) represents the new production (Martin *et al.*, 1987), i.e. the new production is that part of the primary production which is available for export, and it is this that drives the downward flux of organic matter (the global carbon flux) to deep waters (see Sections 9.3 & 10.4). New production, as a percentage of the total primary production, ranges from less than ~10% in the oligotrophic waters of the subtropical gyres to greater than ~20% in coastal upwelling regions (see e.g. Bruland, 1980; see Table 9.3(a)).

Regenerated production arises from nutrients that

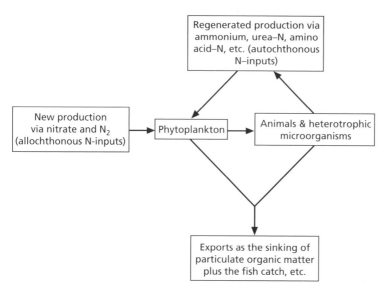

**Fig. 9.1** The production system of the surface ocean, illustrating the concepts of new and regenerated production (from Eppley & Peterson, 1979).

undergo short-term recycling in the euphotic zone. In contrast, new production requires the input of new nutrients into the euphotic zone; e.g. from (i) atmospheric deposition, (ii) a deep-water flux via eddy diffusion and vertical advection and, for nitrogen, (iii) fixation (see e.g. Jenkins & Goldman, 1985; Duce, 1986), and (iv) possibly the vertical migration of diatom mats (see e.g. Villareal *et al.*, 1993).

### 9.1.2 The distributions of nutrients in the oceans

Over the past three decades or so, the database on the distributions of nutrients in the oceans has been extended considerably by information obtained in a variety of ways. Data on the spatial distributions of the nutrients have been derived from sources such as: (i) large-scale expeditions (e.g. GEOSECS, NORPAX and TTO); (ii) attempts to produce global ocean nutrient maps from other parameters, such as temperature and $\sigma_t$, see eqn (7.5), (e.g. the study reported by Kamykowski & Zentara (1986) using the NODC data set); and (iii) remote sensing (see e.g. Traganza *et al.*, 1983). Data from sources such as these, together with older archival material, can be used to build up a general picture of the distribution of nutrients in the oceans.

Redfield (1934, 1958) showed that the concentrations of the major nutrients, such as nitrate and phosphate, in sea water change in relation to fixed concentration ratios (stoichiometry) in organisms (see Section 9.2.3.1 for a discussion of the Redfield ratios), and therefore implied that it is organisms which control the concentrations and distributions of the nutrients in sea water. As a result of this, linear relationships exist between the concentrations of dissolved nutrients. For example, nitrate and phosphate exhibit such a linear relationship in sea water (see Fig. 9.2a); however, there are two general exceptions to this.

1 In anoxic regions in which nitrate is used in the destruction of organic matter (see Section 14.3.2), phosphate can increase with a corresponding decrease in nitrate.

2 Nitrate concentrations can approach, and sometimes reach, zero concentrations in nutrient-starved regions that have low concentrations of phosphate (see e.g. Jackson, 1988).

Three features are apparent in the overall distributions of the nutrients in various seawater environments:

1 nutrient concentrations in surface waters are highest in coastal areas and regions of upwelling;

2 deep-water concentrations are considerably higher than in those of open-ocean surface waters;

3 deep-water concentrations are higher in the Pacific than in the Atlantic.

### 9.1.2.1 The water-column profiles of the nutrients

Nutrients are taken up in the euphotic zone and released back into sea water following the remineralization of sinking detritus, and the concentrations thus build up in deeper water from which they can be brought to the surface again via physical processes, such as upwelling and seasonal increases in depth of the wind-mixed upper layer (stratification): i.e. the '**biological pump**'. The overall effect of these processes is to generate vertical nutrient distributions that exhibit a characteristic '**surface-water depletion, deep-water enrichment**' profile. Such vertical water-column profiles are of particular interest in marine geochemistry because when they are exhibited by other constituents they indicate that these also have been affected by processes associated with living organisms, i.e. they have entered the oceanic biomass cycle. Thus, nitrate and phosphate can be used as analogues for elements that are taken into organic tissue phases, and silicate can be used as an analogue for elements incorporated into the skeletal parts of organisms. In this way it is possible to distinguish between organic tissue and skeletal trace-metal-particulate carrier phases; this type of approach is discussed in detail in Section 11.6.3.

Nitrate, phosphate and silicate, for which precise analytical techniques are available, are convenient indicators of the involvement of nitrogen, phosphorus and silicon, respectively, in the oceanic nutrient cycles.

*Nitrate and phosphate.* Vertical water-column profiles of nitrate and phosphate in the major oceans are illustrated in Fig. 9.2 (b(i) and (ii), respectively). There are considerable similarities between the profiles of the two nutrients and both profiles can be subdivided into a number of layers. The general features can be identified as follows:

1 A surface layer in which nitrate and phosphate are heavily depleted by biological uptake.

2 A layer in which the concentrations increase rapidly with depth as a result of their regeneration from the sinking biomass. Sometimes a layer of maximum concentration is found for both nutrients between ~500 and ~1500 m; this is particularly well developed in Atlantic Ocean profiles.

3 A thick layer in which the concentrations vary little with depth.

Within this overall pattern, however, the individual profiles for the two nutrients differ in detail between the various oceans. This can be illustrated with respect to nitrate. The inter-oceanic variations in the depth profiles of this nutrient have been described by Sharp (1983), and can be summarized with reference to Fig. 9.2(b(i)). The North Atlantic profile exhibits a pronounced nitrate maximum, which is found near the oxygen minimum layer, below which the concentration is generally uniform. A nitrate maximum associated with an oxygen minimum is also found in the South Atlantic, but here it is broader and deeper than in the North Atlantic, and a secondary maximum is located in the deep-water. The South Pacific also has a broader, deeper, nitrate maximum than the North Pacific. The Antarctic has a very shallow nitrate maximum and a generally constant deep-water value (N content $32 \mu g \, atom \, l^{-1}$), which is characteristic of the waters of this region. In addition to variations in the detailed shapes of the nitrate profiles, concentrations also differ between the oceans. For example, the concentration of nitrate at the maximum in the North Pacific (N content $\sim 45 \mu g \, atom \, l^{-1}$) is higher by a factor of two than in the North Atlantic (N content $\sim 22 \mu g \, atom \, l^{-1}$) (see below).

*Silicate.* Although silicate is incorporated into the shells of organisms, and not their soft tissues, it still has a nutrient-type distribution profile in the water column; some typical profiles are illustrated in Fig. 9.2(b(iii)). The silicate profile, however, does differ from those of nitrate and phosphate in that the return of silicate to the water column by shell dissolution can occur at different depths to that of the soft tissue regeneration of the other two nutrients—see Fig. 9.2. In general, the concentration of silicate increases down the water column, as a result of the dissolution of shell material, to around 1000 m, but does not usually exhibit a very distinct maximum at this depth. Below this, silicate tends to remain fairly constant to the sea bottom. Silicate is depleted in the surface layers of all the oceans, but the degree of depletion with respect to deep-water concentrations differs from ocean to ocean, with, for example, the concentration of dissolved silicate being very much higher in the North Pacific than in the North Atlantic deep-waters.

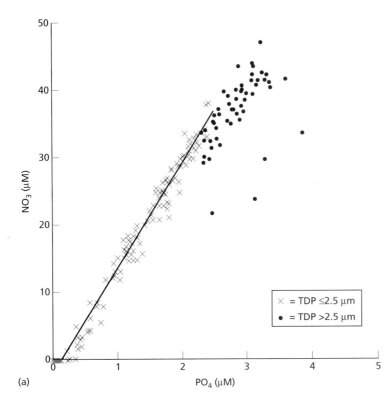

**Fig. 9.2** The distributions of nutrients in the oceans. (a) Nitrate versus phosphate relationships (from Jackson & Williams, 1985). The data are taken from water-column profiles in the Pacific Ocean. The least-squares fit to a linear relationship for all data points is $[NO_3 (\mu M)] = 13.2[PO_4 (\mu M)] + 0.23$, $r^2 = 0.935$, $n = 186$. For samples with total dissolved phosphorus (TDP) $\leqslant 2.5\,\mu M$, the relationship is $[NO_3 (\mu M)] = 15.5[PO_4 (\mu M)] - 2.27$, $r^2 = 0.978$, $n = 134$ (solid line). (b) Typical vertical profiles of nutrients in the oceans: (i) nitrate (from Sharp, 1983; plots constructed from GEOSECS data); (ii) phosphate (from Sverdrup *et al.*, 1942); (iii) silicate (from Armstrong, 1965).

### 9.1.2.2 The horizontal distribution and circulation of nutrients in the oceans

In Section 7.1, attention was drawn to the fact that the chemical signals of non-conservative constituents in the oceans are controlled by circulation patterns that transport the constituents from one part of the sea to another and mix water masses, upon which are superimposed the effects of internal oceanic reactions arising from the involvement of the constituents in biogeochemical cycles. The nutrients provide a classic example of how this **twofold signal control** operates. The shapes of the vertical oceanic profiles of nitrate, phosphate and silicate described above were interpreted in terms of the involvement of the nutrients with biota. During this, they are removed from solution into particulate phases of the biomass, including both tissue and skeletal parts of organisms, in the upper water layer and are subsequently regenerated back into solution at depth, i.e. a non-conservative signal. The horizontal distributions of the nutrients illustrate the control by circulation patterns.

Data have become available on the large-scale horizontal distributions of nutrients in sea water, both for general oceanic sections (see e.g. Sharp, 1983 (nitrate); Tsuchiya, 1985 (phosphate)) and for specific isopycnal surfaces (see e.g. Kawase & Sarmiento, 1985; Takahasi *et al.*, 1985). Nitrate can be used to illustrate the factors that control the horizontal distribution of a nutrient in the World Ocean. Sharp (1983) used GEOSECS data to compile latitudinal profiles through a multiple ocean basin section. The section is illustrated in Fig. 9.3, and the principal features in the nitrate distribution are summarized below.

1 Deep-waters of the Indian and Pacific Oceans are derived partially from the North Atlantic and Antarctic sinking regions (see Section 7.3.3), and North Atlantic Deep Water (NADW) can be seen in the section at a depth lying between 1.5 and 4 km in the western Atlantic (isopleths for N at 16–35 $\mu g$ atom kg$^{-1}$).

2 Antarctic waters show little variation with depth and are the source of the Antarctic Bottom Water

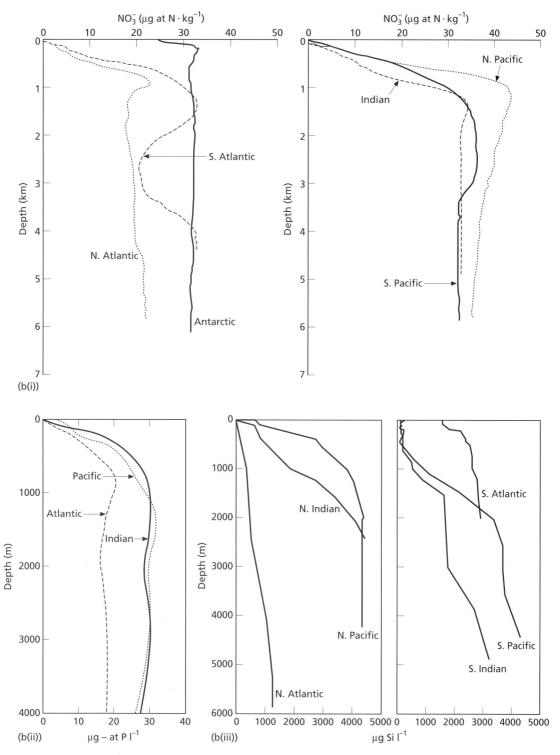

**Fig. 9.2** *Continued*

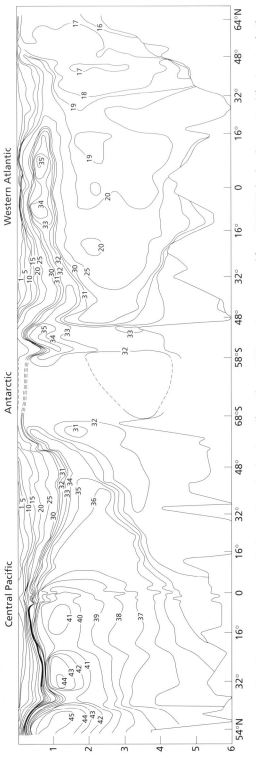

**Fig. 9.3** Central Pacific–Antarctic–western Atlantic oceanic section for nitrate (from Sharp, 1983; section constructed from GEOSECS data); units, µg-atom N kg⁻¹, water depths in kilometres.

(AABW), which has isopleths for N of 32–33 μg atom kg$^{-1}$.

3 In the western Atlantic NADW is overlain by Antarctic Intermediate Water (AAIW), which appears as a distinct intrusion extending as far as ~30°N. Although the AAIW is less distinct in the Pacific Ocean it is apparent that the source of the subsurface waters is largely in the Antarctic.

4 Equatorial upwelling can be seen in both the Atlantic and Pacific, but it is more pronounced in the Pacific.

5 The highest nitrate concentrations are found in the deep waters of the Pacific Ocean.

The type of nutrient distribution described above can be related to deep-water circulation patterns, and Broecker (1974) has suggested that the following sequence governs the nutrient build-up in the deep waters of the Pacific (see Section 9.1.2.1). The waters of the high-latitude sink (see Section 7.3.3) are depleted in nutrients by biological activity, but as they move along the global deep-water circulation path some biogenic debris containing nutrients falls from the surface to deep water as particulate material, and this is supplemented by mixing with other nutrient-carrying deep waters from the Indian and Pacific Oceans. For every unit of deep-water formed at the sinking sources, one unit must return to the surface by upwelling, a process that occurs over the entire ocean but tends to be more active in certain regions (see Section 7.3.2). This upwelled water, which is rich in nutrients, is then transported by surface currents. Instead of following the surface circulation patterns, however, the nutrients themselves are utilized by organisms, extracted from the water, and carried downwards via the organic particulate matter. A large fraction of these nutrients is regenerated, but some nutrients are carried to the deep sea in association with the particulate matter that escapes destruction (see Section 9.3). Thus, as Broecker & Peng (1982) have pointed out, the nutrients can be transported by deep currents but not by surface currents, the net result being that there is a steady push of nutrients towards the deep Pacific, which lies at the end of the transport path on the deep-water global 'grand tour' (see Section 7.3.3). The distributions of the nutrients in the oceans therefore are controlled by an interaction, or balance, between oceanic circulation and biological activity.

### 9.1.3 Nutrients and primary production in the oceans

The nutrients are intimately involved in the generation of biota, and it is convenient to describe primary production at this stage.

#### 9.1.3.1 *Primary production: plankton*

**Plankton** are microscopic free-floating, or weakly swimming, organisms that are passively distributed by ocean currents. **Eukaryotes** are unicellular, or multicellular, organisms having nuclear membranes and DNA that is arranged in the form of chromosomes. **Prokaryotes** are unicellular organisms that do not possess nuclear membranes and do not have their DNA in the form of chromosomes. Plankton are subdivided into phytoplankton (the 'grass' of the sea) and zooplankton (the 'grazers' of the sea). The **phytoplankton** are autotrophs, i.e. they manufacture their organic constituents from inorganic compounds and do not rely on other organisms as an energy source. Important members of the phytoplankton include diatoms, cocolithophores, silicoflagellates and dinoflagellates, all of which are eukaryotes. The only major group of autotrophic prokaryotes are the cyanobacteria. These cyanobacteria are unique in the oceans because they can fix dissolved gaseous nitrogen ($N_2$), whereas other phytoplankton can only utilize 'pre-fixed' forms of nitrogen (see Section 9.1). **Zooplankton** are heterotrophs, i.e. they obtain energy supplies and organic substrates by feeding directly, or indirectly, on autotrophs. The zooplankton include copepods, pteropods, radiolarians and foraminifera; the latter two are protists (i.e. unicellular eukaryotes).

Marine autotrophs (e.g. the algal biomass) are primary producers. The photo-autotrophic biomass (phytoplankton) is the most important primary source of organic carbon in the oceans, and primary production is the initial stage in the marine food-chain, which subsequently involves a number of trophic levels. For example, herbivorous zooplankton (the 'grazers') consume the phytoplankton (the 'grass') during secondary production. These herbivorous zooplankton are, in turn, fed upon by carnivorous zooplankton and fish species (tertiary production).

A number of size-based classifications have been

proposed for the planktonic community. The size categories identified by Lalli & Parsons (1993) include the following: femtoplankton (0.02–0.2 μm), picoplankton (0.2–2.0 μm), nanoplankton (2.0–20 μm) and microplankton (20–200 μm). Viruses are part of the femtoplankton, and bacteria are placed in the picoplankton. It was the identification of the **nanoplankton**, however, which changed our view of the oceanic plankton community. Because of their relatively small size, the nanoplankton were often missed by the early collection techniques, and it is now thought that small phytoplankton and zooplankton are responsible for a large fraction of oceanic primary production and its consumption in secondary production.

The process of **photosynthesis**, in which organic compounds are synthesized from inorganic constituents present in sea water during the growth of marine plants (phytoplankton), is usually termed **primary production**, although this is not a biochemically rigorous definition. In the process, which includes the absorption of solar energy and the assimilation of carbon dioxide, the chlorophyll-containing plants synthesize complex organic molecules from inorganic starting materials, the principal products being oxygen and food sustances such as carbohydrates (see eqn 9.1). It is now the convention to distinguish between **new** primary production, which arises from nutrients transported into the euphotic zone (e.g. from upwelling and atmospheric deposition), and **regenerated** production, which utilizes nutrients that are recycled within the upper layers of the ocean (see Section 9.1.1).

Photosynthesis requires light in the wavelength range 370–720 nm, and even in clear tropical waters only ~1% of the visible light energy penetrates to a depth of ~100 m. Photosynthesis is therefore generally restricted to the upper 100 m or so of the open-ocean water column, although it may, in fact, be inhibited at the surface layer itself because of the effects of high light intensity, and as a result the zone of *maximum* production is often found a few metres below the surface. Gross primary production decreases with depth, but the loss of carbon through respiration remains constant because phytoplankton respire during the hours of sunlight and darkness. The depth at which gross primary production and respiration balance is termed the **compensation depth**, and at this position net production is zero.

Net primary production is positive in the water column above the compensation depth, and this is the region referred to as the **euphotic zone**. In the aphotic zone, which lies below the compensation depth, respiration exceeds gross production and there is a net loss of organic material.

### 9.1.3.2 Primary production: controls

In its simplest form a limiting factor for a specific activity may be defined as the critical minimum required to sustain that activity and, in addition to **light availability**, the factors that limit primary production include **nutrient availability** and **zooplankton grazing**. Primary production in the oceans is thus controlled by a complex combination of physical (e.g. light, temperature), biological (e.g. growth rate) and chemical (e.g. availability of nutrients) variables. According to Neinhuis (1981), however, the availability of nutrients is probably the limiting factor in most marine systems. For example, Platt *et al.* (1992) examined data obtained over several years and suggested that the parameters in the 'photosynthesis–light curve' in the western North Atlantic provided direct evidence for the nutrient control of photosynthesis in the open ocean.

*Nutrient availability.* The availability of nutrients can limit phytoplankton growth and the limiting nutrients were traditionally considered to be the macronutrients nitrate and phosphate, although silicate can limit diatom growth. In addition, there is a growing realization of the importance of the micronutrients, especially iron, in nutrient limitation. It must be stressed, however, that the concept of a limiting nutrient strictly refers only to new production, because productivity can be maintained in the presence of low nutrient concentrations by recycling. Most phytoplankton have an absolute requirement for one or more nutrients, but the concept of a 'limiting nutrient' is complex. Nutrient availability can regulate rate processes, such as photosynthesis, or the final yield of a plant crop. In practice, the two have often been blurred with the result that nutrient limitation is sometimes applied to growth rates of phytoplankton and sometimes to the limitation of the standing crop, although the limitation of the standing crop is not necessarily accompanied by a severe limitation of growth rates (see e.g. Cullen *et al.*, 1992).

**1** *Nitrate limitation.* Geographical variations in marine primary production are often interpreted in terms of nutrient limitation. Further, the relative availability of nutrients can be used to classify the marine environment in terms of primary production regions. For example, nitrate is usually considered to be the most important limiting nutrient for primary production in the marine environment, and the trophic status of a marine ecosystem can be assessed by the rate of injection of new nitrogen, which maintains production levels. On this basis, three oceanic regions can be identified (see e.g. Dugdale & Wilkerson, 1992).

(a) **Eutrophic** regions, where vertical advection and irradiance led to highly productive regimes (high nitrogen, high productivity; HNHP regions).
(b) **Oligotrophic** regions, which have low nutrient supplies and are poorly productive (low nitrogen, low productivity; LNLP regions).
(c) Regions with **high surface nitrogen levels, but low productivity** (high nitrogen, low productivity; HNLP regions); for example, ~29% of the open-ocean surface waters have sufficient light and nutrient concentrations, yet the standing crops of phytoplankton remain low (Martin *et al.*, 1994).

It also is important to relate the distribution of phytoplankton species, especially with respect to cell size, to this trophic-status classification of oceanic waters. According to Dugdale & Wilkerson (1992) the upward gradient in the concentration of chlorophyll between oligotrophic and eutrophic conditions is associated with a gradient in phytoplankton cell size. New production is dominated by organisms with larger cells, such as chain-forming diatoms, which predominate in eutrophic regimes. In contrast, open-ocean productivity is dominated by regenerated production involving nanoplankton and picoplankton. This cell-size difference may well play an important role in sustaining the HNLP regions—see below.

The existence of the HNLP regions is a paradox that has given rise to considerable speculation—see e.g. the volume edited by Chisholm & Morel (1991a) for a discussion of the factors that control phytoplankton production in nutrient-rich open-ocean areas. In the HNLP regions, chiefly the equatorial Pacific, the subarctic Pacific and the Southern Ocean, phytoplankton do not exhaust nitrate, or phosphate, in the surface waters. The result of this is what Chisholm & Morel (1991b) described as a 'slippage'

in the global coupling between the supply of nutrients to surface waters and organic synthesis, the outcome being a less efficient transfer of carbon from surface to deep-waters via the **biological pump**. In this respect, the HNLP regions are potentially important because increased production in them could remove significant amounts of carbon dioxide from the atmosphere (see Section 8.4.6).

In addition to the HNHP, LNLP and HNLP categories, various authors have proposed more detailed biogeochemical classifications of trophic areas in the oceans. For example, Campbell & Aarup (1992) used mean surface chlorophyll images, based on 5 yr of Coastal Zone Colour Scanner (CZCS) data, to subdivide the North Atlantic into three zones (mid-latitude, subtropical and subpolar), each having distinct seasonal surface chlorophyll patterns and 'new' production characteritics. In another study, Sathyendranath *et al.* (1995) classified waters in the North Atlantic basin into four primary biogeochemical domains, each having unique physical characteristics likely to dominate their phytoplankton ecology. These domains were: polar, west-wind, trades and coastal, each of which could be subdivided into smaller provinces.

Nitrate has usually been considered to be the most important limiting nutrient for primary production in the marine environment, with other nutrients being less significant, but conditions under which phosphate and silicate, and also iron, can act to limit production have been described.

**2** *Phosphate limitation.* It has been suggested that in certain oceanic regions phosphate can act as the limiting nutrient for phytoplankton growth. The Mediterranean Sea is, in general, depleted in both nitrate and phosphate compared with the adjacent Atlantic Ocean. This results from an inflow of nutrient-poor surface waters, and the balancing outflow of nutrient-rich intermediate and deep waters. The western basin of the Mediterranean may be nitrate-limited (see e.g. Owens *et al.*, 1989), but Krom *et al.* (1991) suggested that in parts of the eastern Mediterranean Sea the phytoplankton production in surface waters may be strongly phosphate-limited, probably as a result of the removal of phosphate by adsorption on to iron-rich dust particles, following the input of aerosols from the Saharan region.

**3** *Silicate limitation.* Nutrient limitation is especially interesting with respect to the HNLP regions identified above. At first sight, these HNLP regions offer an apparent paradox: nitrogen levels are high enough to support well-developed growth, but production is low. However, HNLP regions are often characterized by low silicate, as well as by high nitrate concentrations, in the mixed layer. Because of this, Dugdale *et al.* (1995) suggested that HNLP regions would better be described as high-nitrate–low-silicate–low-chlorophyll (HNLSLC) regimes; chlorophyll, a photosynthetic pigment, is often used as a measure of primary production. The low silicate levels in these regions may be the key to understanding the paradox because, although nitrogen is readily available, many of the HNLSCL regions are diatom-dominated and therefore rely on the availability of sufficient silicate for primary production. Because of this, silicate may limit diatom growth in such regions.

In this context, Dugdale *et al.* (1995) have described the operation of a 'silicate pump' (see Fig. 9.4), which operates in diatom-dominated communities to enhance the loss of silicate from the euphotic zone to deep-water. Unlike N and P, which are found in a number of inorganic and organic forms in sea water, silicon occurs predominantly as orthosilicic acid ($Si(OH)_4$), and it is regenerated by the dissolution of opaline $SiO_2$ and not by organic degradation.

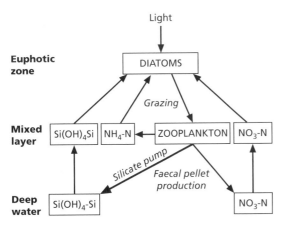

**Fig. 9.4** The operation of the 'silicate pump'. Idealized flow diagram illustrating the flow of nitrogen and silicon ('silicate pump') in a diatom-dominated grazed system (after Dugdale *et al.*, 1995). In this type of system the 'silicate pump' for the export of silicate to depth occurs via zooplankton faecal pellets following diatom phytoplankton grazing.

During the operation of the 'silicate pump', once silicate has been incorporated into the diatoms it is mainly exported to depth through the sinking of phytoplankton particulate debris, or silicate-rich faecal pellets, and so leaves the recycled pool. The operation of the 'silicate pump' therefore results in low-silicate–high-nitrate conditions in the mixed layer because nitrate is more readily recycled in the 'grazing loop'. Diatoms have an absolute requirement for silicate, and the continual export of silicon at a higher rate than nitrogen therefore means that the 'silicate pump' drives the system to silicate limitation. Dugdale *et al.* (1995) described a coupled silicon–nitrogen model to describe phytoplankton metabolism, and to evaluate the differential flow of these nutrients into, and out of, the euphotic zone in a diatom-dominated HNLSLC community. In these silicate-controlled systems the export production, i.e. that lost to deep water, of silicon and nitrogen is not equivalent. The reason for this is that the export production of silicon is controlled by the input of silicate, but that of nitrogen is controlled by the grazing rate and regeneration recycling.

The 'silicate pump' invokes the differential export of silicon to deep-water as the controlling factor on new production in HNLSLC regions, in which it is the silica dynamics that control new production: a concept that challenges the more traditional view that nitrate concentrations drive both new and export production. Dugdale *et al.* (1995) also suggest that the 'silicate pump' model provides an alternative to the 'iron hypothesis' (see below) for the control of productivity in some HNLC regions.

4 *Iron limitation.* Iron, a constituent of many oxidizing metallo-enzymes, respiratory pigments and proteins, is essential for life. Phytoplankton require iron for the synthesis of chlorophyll and nitrate reductase, and the idea that iron can act as a limiting nutrient on primary production is not a new one. For example, in his review article on the topic Martin (1992) quotes references to the importance of iron in limiting phytoplankton production as far back as the 1930s. However, there are a number of problems with respect to evaluating the importance of iron as a limiting nutrient. One of the most important of these is the low concentrations of iron in open-ocean waters. For example, using phytoplankton C:Fe ratios of $10^3$–$10^4$ C to 1 Fe as being indicative of plankton requirements, and relating these to the

nitrate and iron concentrations in open-ocean waters, Martin (1992) pointed out that on the basis of the iron requirement only ~10% of the available nitrate would be utilized by phytoplankton. That is, at the low concentrations present iron would run out before nitrate was depleted, and most open-ocean waters would be infertile because upwelled deep-waters bringing nutrients to the mixed layer are very poor in iron. To overcome this, it has been suggested that the iron requirement must be supplied from other sources, and for the open-ocean environments it has been proposed that the long-range transport of atmospheric dust, from which the iron is leached into sea water, can provide such a source (see e.g. Moore *et al.*, 1984; Duce, 1986).

Martin & Gordon (1988) provided data on the distribution of dissolved and particulate iron in northeast Pacific waters. The water-column profiles of dissolved iron at several stations showed a number of common features:

(a) very low dissolved iron levels in surface waters ($<0.1 \, nmol \, l^{-1}$);

(b) increasing levels with depth;

(c) maxima (~$1.0$–$1.3 \, nmol \, l^{-1}$) in association with the oxygen minima.

These features, together with the similarity between dissolved iron and nitrate profiles, suggest that iron is actively removed by phytoplankton from surface water, with the coincidence of the iron maxima and the oxygen minimum resulting from the release of iron following the destruction of organic matter from sinking particles during the 'nutrient loop' (see Section 9.1.1). The authors then attempted to evaluate the iron data in terms of phytoplankton requirements in three contrasting oceanic environments.

(a) Highly productive upwelling environments off the California coast. In these environments there will be an increased demand for iron, resulting from production following the upwelling of nutrient-rich waters. It was suggested that this demand could be met by iron supplied from a number of sources, which included upwelling of deep-water, release via diagenetic processes from shelf and slope sediments, and atmospheric transport; riverine input was not important in this region.

(b) Low productivity open-ocean environments. The authors concluded that these waters required a source of iron in addition to the upwelling of deep-

water, and this was thought to be from atmospheric dust inputs.

(c) Environments where the atmospheric input of iron may be inadequate, e.g. the northeast Pacific Subarctic, and the Southern Ocean.

In both these latter environments nitrate values are high, but production is low (i.e. the HNLP regions identified above). For these regions, in which there is an insufficient supply of atmospheric iron to the surface waters, it was suggested that iron itself was the limiting nutrient; i.e. it is iron that turns on the 'biological pump'. Barber & Chavez (1991) assessed the factors controlling primary productivity in the equatorial Pacific and related them to the iron input pattern.

(a) In the low-nutrient western Pacific the productivity rate is a linear function of nutrient concentration and the aeolian supply of iron is adequate to permit the productivity rate to be set by the subsurface nutrient concentration.

(b) In the nutrient-rich eastern Pacific productivity is limited and independent of nutrient concentration as a result of the inadequate supply of atmospheric iron.

(c) Around the Galapagos Islands primary production is not limited because the input of sedimentary iron is sufficient to support nutrient-regulated productivity rates.

Several authors have carried out laboratory 'bottle' simulations and have shown that phytoplankton growth rates increase in response to the addition of iron to the system. Further, there are experimental data which indicate that the addition of iron can result in significant shifts from picoplankton to diatom-dominated assemblages (see e.g. Chavez *et al.*, 1991). The concept that iron concentrations in open-ocean waters have to be supplemented with atmospheric iron for primary production has been questioned by Sunda *et al.* (1991). On the basis of laboratory experiments, these authors suggested that the open-ocean phytoplankton Fe requirement is much smaller than first thought, with a C:Fe ratio of 500 000:1. Martin (1992), however, summarized experiments carried out by his group on equatorial Pacific phytoplankton, which he suggested demonstrated that the production rates reported by Sunda *et al.* (1991) were achieved only after the addition of atmospherically leached iron. Martin (1992) further suggested that there may be a 'threshold' iron concentration, below which phytoplankton growth is severely limited but above which additional iron has little effect; the 'threshold' value varies from one oceanic region to another, but is probably <0.5 nmol l$^{-1}$. Hudson & Morell (1990) studied the kinetics of iron uptake on cell surfaces and proposed that there is a very sharp cut-off between organisms that can grow maximally at oceanic iron concentrations (e.g. those studied by Sunda *et al.* (1991), which have C:Fe ratios of 500 000:1) and those that can grow only after additional inputs of iron into the system (e.g. when an atmospheric dust event has injected leachable iron into the system). Martin (1992), however, suggested that the amount of iron supplied to surface open-ocean waters via upwelling is too low even for phytoplankton with C:Fe ratios of the order 500 000:1, and concluded that maximal phytoplankton growth cannot occur in the open-ocean without the addition of iron from atmospheric dust.

Coale *et al.* (1996a) proposed that in the equatorial Pacific biological production is controlled not simply by the input of atmospheric iron but also by iron from upwelled waters. Despite this, it is now generally accepted that the *major* source of iron to remote open-ocean regions is via the deposition of atmospheric dust particles. Finden *et al.* (1984), however, have shown that iron oxyhydroxide particles and iron colloids are not directly available to phytoplankton. The bioavailable forms of iron are thought to be dissolved Fe(II) (which is rapidly converted into Fe(III), and Fe(III) itself, which is the thermodynamically stable form in oxygenated sea water; it also has been suggested recently that as much as ~99% of the dissolved iron in the oceanic environment is organically complexed (see e.g. van den Berg, 1995). What is apparent is that the iron derived from aerosols must undergo solubilization before it becomes available to support phytoplankton growth.

Iron delivered to the sea surface by 'wet' deposition (see Section 6.2.2) may undergo reductive dissolution to Fe(II) as a result of the oxidation of $SO_2$ (Zhuang *et al.*, 1992). Duce & Tindale (1991) have suggested that this iron would be immediately available as a nutrient to phytoplankton. According to Johnson *et al.* (1994), however, Fe$^{II}$ is oxidized to Fe$^{III}$ with a half-life of ~2 min in surface ocean water, and adsorption on to particles is likely to render the iron biologically unavailable in a relatively short time.

Some process therefore is required to solubilize the iron from aerosols in a form that is *available* to phytoplankton. Siderophores are compounds with a high affinity for ferric iron that are secreted by microorganisms in response to low iron environments, and the production of marine siderophores increases iron solubility and its bioavailability to some organisms. Reid *et al.* (1993) identified a siderophore from a marine bacterium that may solubilize iron, but siderophores have not been reported in sea water and only a few marine organisms have the capability to produce them. Johnson *et al.* (1994) addressed the problem of the bioavailability of iron to plankton and concluded that a model of iron cycling in which there is: (i) photoreductive dissolution of colloidal iron, followed by (ii) its subsequent oxidation, and (iii) a biological uptake of the dissolved Fe(III), may have a significant impact on iron availability to phytoplankton in the open-ocean. Zhuang & Duce (1993) considered the fate of atmospheric iron after it enters sea water, and concluded that although adsorption on to particulate matter is the predominant mechanism controlling dissolved iron concentrations in open-ocean waters, the addition of mineral aerosol particles from episodic events would result in a net addition of dissolved iron. It may be concluded, therefore, that although the mechanism(s) governing the solubility of iron from aerosols is still unclear, the dissolved iron in open-ocean waters has a predominantly atmospheric source. The strength of this source has been estimated by Duce & Tindale (1991), who provided data on the atmospheric input of mineral aerosols and dissolved iron to a number of regions in the World Ocean. These data are summarized in Table 9.1, together with a comparison of atmospheric versus fluvial iron inputs, and the magnitude of the total iron (i.e. particulate + dissolved flux) is illustrated in Fig. 9.5. Duce & Tindale (1991) concluded that the atmosphere is a major transport path for both mineral matter and dissolved iron to the open-ocean, and is probably the dominant source of nutrient iron to the photic zone, where it exceeds the fluvial input by a factor of about three (see Table 9.1b).

Laboratory simulations on phytoplankton growth controls have been criticized as not being representative of a community response of the kind that would occur on an oceanic-scale. The 'iron hypothesis' can be in fact tested on such an oceanic-scale, however, by

**Table 9.1** The atmospheric input of mineral matter, total iron and dissolved iron to the World Ocean and a comparison with fluvial inputs (data from Duce & Tindale, 1991).

(a) Atmospheric input; units, $10^{12}$ g yr$^{-1}$.

| Oceanic region | Mineral dust | Total iron | Dissolved iron |
|---|---|---|---|
| North Pacific | 480 | 16.5 | 1.7 |
| South Pacific | 39 | 1.4 | 0.14 |
| North Atlantic | 220 | 7.7 | 0.77 |
| South Atlantic | 24 | 0.84 | 0.08 |
| North Indian | 100 | 3.5 | 0.35 |
| South Indian | 44 | 1.5 | 0.15 |
| Global total | 900 | 32 | 3.2 |

(b) Atmospheric versus fluvial inputs; units, $10^{12}$ g yr$^{-1}$.

| Source | Dissolved iron | Particulate iron |
|---|---|---|
| Atmosphere | ~3 | ~32 |
| Rivers | ~1 | ~110 |

**Fig. 9.5** Estimated atmospheric flux of total (particulate + dissolved) iron to the World Ocean (from Duce & Tindale, 1991).

both regional comparisons and intervention experiments. Three recent studies designed to test the hypothesis can be used to illustrate this.

(a) Data for a large-scale 'iron enrichment' intervention experiment (IronEx I) were reported by Martin *et al.* (1994). It involved a mesoscale experiment in which an area of 64 km$^2$ in a patch of sea, defined by $SF_6$ (sulphur hexafluoride) tracer, in the open equatorial Pacific was seeded with iron. The findings showed an unequivocal biological response to the added iron. This was evidenced by the fact that compared with the patch before the iron-seeding, and to conditions outside the patch, there was a doubling of the plant biomass, a threefold increase in chlorophyll and a fourfold doubling in plant production. Further, a large physiological response occurred within the first 24 h (the time required to seed the patch with iron). However, although the results showed this clear and unambiguous physiological response to the addition of iron, the patch of ocean was subducted beneath a layer of less dense sea water 4 days after being seeded with iron. This resulted in a much smaller than predicted biological and geochemical response. A number of hypotheses were proposed to explain this. (i) Iron was rapidly lost from the seeded patch; (ii) the subduction of the patch to

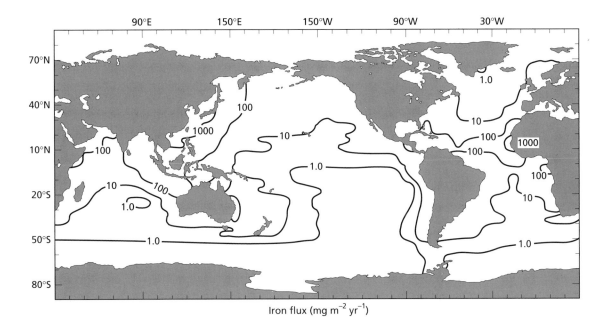

Iron flux (mg m$^{-2}$ yr$^{-1}$)

lower light levels minimized the photodissolution of iron colloids and decreased rates of bioavailable iron production; (iii) zooplankton quickly cropped the increase in phytoplankton biomass; and (iv) another nutrient, e.g. zinc or silicate, became limiting, and so prevented further growth. The next experiment (IronEx II) was initiated to test these hypotheses.

(b) The data for IronEx II have been reported by Coale *et al.* (1996b). Patches of surface water, again in the equatorial Pacific, were seeded with iron in the concentrations expected from natural events and a massive phytoplankton bloom was triggered, which consumed large quantities of carbon dioxide and nitrate that the plants could not fully utilize under natural conditions. In addition, the bloom was not checked by either zooplankton grazing or secondary nutrient limitation. The data obtained from IronEX II therefore provided unequivocal evidence supporting the hypothesis that the high nitrogen, low chlorophyll (HNLC) condition of these Pacific waters results from the iron-limitation of phytoplankton growth.

Iron-stimulated phytoplankton growth can affect the oceanic–atmospheric cycles of some climate-influencing components (see Section 8.4.2), and this is illustrated below with respect to carbon dioxide and dimethyl sulphide.

The behaviour of carbon dioxide during IronEx II has been described by Cooper *et al.* (1996). The intervention experiment stimulated a large biologically induced uptake of surface water $CO_2$, and the fugicity (escaping tendency) of $CO_2$ in the centre of the iron-seeded patch fell from a background value of ~510 μatm to ~410 μatm. This corresponded to a transient 60% decrease in the natural sea-to-air $CO_2$ flux, which is driven by the difference in partial pressure across the air–sea interface (see Section 8.2). However, modelling studies have shown that on a long-term basis the atmospheric partial pressure of $CO_2$ is insensitive to biological activity in the equatorial Pacific. In contrast, it has been estimated that in the Southern Ocean increased macronutrient utilization would result in decreases in the atmospheric $CO_2$ by between 6% and 21%, so that iron-induced enhanced nutrient utilization would alter the size of the present-day Southern Ocean $CO_2$ sink (see Section 8.4.6.1). This potentially important role for iron in South-

ern Ocean carbon cycling also has been highlighted by Martin (1990), who suggested that biological production in the region, stimulated by iron fertilization during the Last Glacial Maximum, was responsible for low atmospheric levels of $CO_2$ during this period. This is supported by the data provided by Kumar *et al.* (1995), who showed that there was an increased export of carbon to sub-Antarctic sediments when there was a higher influx of iron during the Last Glacial Maximum.

Increases in the seawater concentrations of dimethyl sulphide (DMS—see Section 4.1.4.3) by a factor of about 3.5 during IronEx II have been reported by Turner *et al.* (1996). Because oxidation of DMS is involved in the production of atmospheric sulphate particles, which can exert a climate cooling effect, Turner *et al.* (1996) concluded that the increases in DMS during iron fertilization provide direct support for an important link in the iron–DMS–climate hypothesis. Further, at least in the equatorial Pacific, changes in DMS production may have a more significant effect on climate than iron-induced transient changes in the uptake of $CO_2$.

(c) This study involved a comparison of natural levels of productivity in oceanic regions with differing iron concentrations. De Baar *et al.* (1995) reported data on the importance of iron for plankton blooms in two contrasting regions of the Southern Ocean: (i) the southerly branch of the Antarctic Circumpolar Current (ACC), where the upwelling of deep-waters supplies sufficient iron in the surface waters to sustain a moderate primary production but does not permit blooms to develop; and (ii) the fast-flowing, iron-rich jet of the polar front (PF), where spring blooms are produced and phytoplankton biomass is an order of magnitude higher than in the southern ACC waters. The authors reported that the iron-rich PF waters were sharply delineated from the adjacent iron-poor waters, and concluded that iron availability was the critical factor in allowing bloom production to occur.

*Grazing.* Grazing is also a mechanism by which the phytoplankton standing crop can be limited (see e.g. Walsh, 1976; Frost, 1991). In the 'grazing hypothesis' it is assumed that grazing by the heterotrophic community keeps the autotrophic standing crop low and so can lead to the presence of unused nutrients in the euphotic zone. In terms of this concept, therefore, HNLC environments may simply represent nothing other than the repression of the phytoplankton standing crop under conditions of high heterotrophic activity.

Minas & Minas (1992) pointed out that there is a great range in the rate of increase in phytoplankton standing stock, when expressed in terms of chlorophyll, in tropical upwelling systems. The authors used a model that compared net community production (NCP) rates with observed chlorophyll increase rates to assess the relative importance of the 'grazing limitation' and the 'iron limitation' hypotheses in relation to phytoplankton dynamics in a number of marine upwelling systems. They demonstrated that in a 'chlorophyll versus time' framework the whole spectrum of plankton growth rates can be observed in these systems, from very fast to extremely slow. On the basis of plankton growth rates, the authors were able to classify tropical oceanic upwelling regions into two 'end-member' types.

1 High chlorophyll (**HC**) regions, which have a fast-growing chlorophyll standing stock. These regions of high speed planktonic development include the upwelling systems off the northwest and southwest African coasts, and the Peru coastal upwelling during the weak upwelling season.

2 High nitrogen, low chlorophyll (**HNLC**) regions, which have a very low, or stationary, chlorophyll standing stock. These regions of slow speed planktonic development include the Pacific equatorial upwelling system with extreme HNLC conditions, and the Peru coastal upwelling area during the main upwelling season.

Between these two 'end-members' are a series of waters having intermediate plankton development and chlorophyll characteristics.

Minas & Minas (1992) concluded that grazing appears to be the principal factor for lowering the rate of increase in standing stock, especially in moderate and strong HNLC waters, and suggested that phytoplankton populations must present high specific growth rates on a daily rhythm in order to overcome the grazing effect (e.g. in HC regions). The situation is complicated, however, by the fact that phytoplankton blooms can occur away from upwelling sources, and also within some of the source regions themselves, e.g. the Costa Rica dome, which apparently can

switch from HNLC to 'normal' conditions. This raises the question of whether these plankton outbursts are a response to atmospheric iron inputs? One of the principal arguments advanced by the authors in defence of the 'grazing hypothesis' is that in a 'chlorophyll versus time' framework (i.e. the fast–slow phytoplankton development classification) the whole spectrum of growth velocities, from fast to intermediate to slow, can be found, whereas iron limitation should not permit such a variety of growth rates to occur. For Antarctic waters, however, Minas & Minas (1992) suggested that both the 'grazing limitation' and the 'iron limitation' hypotheses may be operative.

*The HNLP regions: synthesis.* It is apparent that a number of hypotheses have been proposed to explain the existence of the HNLP regions. According to Frost (1991), however, it is unlikely that any single factor is responsible for the setting up of all these regions. For example, the author pointed out that surface waters of the equatorial Pacific upwelling zone and the open-ocean subarctic Pacific may retain nutrients because of the grazing control of phytoplankton stocks, coupled with the preferential utilization of $NH_4$ by the phytoplankton. He stressed, however, that this grazing control may be possible because only certain phytoplankton, particularly fast-growing, inefficiently grazed species, are limited by a trace nutrient such as iron; i.e. grazing control may be exerted on a phytoplankton assemblage structured by iron limitation. Cullen (1991) also assessed the importance of:

1 the grazing hypothesis, in which it is considered that specific growth rates of phytoplankton are maximal and environmental stability allows the development of a balanced food web which maintains a low standing crop of phytoplankton;

2 the iron hypothesis, in which it is proposed that the standing crop of phytoplankton is constrained by the availability of iron, and with more iron available the standing crop of phytoplankton would increase and nitrate would be depleted despite the grazing effect.

The author highlighted a key point when he stressed, like Frost (1991), that iron limitation exerts a selective pressure on the phytoplankton population in different ways and may in fact regulate phytoplankton species composition. With respect to the equatorial Pacific, Cullen (1991) drew attention to the fact that although organisms with small cells dominate the planktonic system, those with large cells probably determine the degree of nutrient utilization. Further, because organisms with larger cells are poorer competitors for nutrients they are more likely to be limited by iron than organisms with smaller cells. As a result of this, iron might regulate productivity by increasing the growth rates of the larger diatoms, thereby changing the foodweb structure. Thus, in the equatorial Pacific grazing may control the populations of the dominant small (pico- and nanoplankton) cells, but the supply of iron may regulate nutrient uptake by limiting the specific growth rates of the larger cells. Miller *et al.* (1991) concluded that iron limitation also operates in the subarctic Pacific, where it gives rise to a plankton population dominated by small cells, which are themselves not limited by iron availability but which are constrained by light illumination and grazing; the efficient recycling of nitrogen (as $NH_3$) by the grazers leaves the system persistently rich in major nutrients. In contrast, the situation may differ in the cold physically perturbed Antarctic waters; for example, Mitchell *et al.* (1991) used a model to predict that in the region of the Antarctic Circumpolar Current (ACC), where mixed layers are in excess of 50 m deep, phytoplankton would not utilize >10% of the available macronutrients as a result of light limitation.

There are a number of processes and feedback loops associated with the IronEx large-scale intervention experiments that can give rise to problems (see e.g. Fuhrman & Capone, 1991), but there can be little doubt that such experiments have provided powerful evidence in support of the iron-limitation hypothesis. It is also probable that iron limitation itself, by constraining the growth of larger cells, promotes small-cell grazing control in HNLP regions. It may be concluded, therefore, that although the HNLP regions still remain a paradox, they apparently exist in response to a complex environmental web involving a variety of interlinked processes in which iron limitation does not act in isolation. For example, Chavez *et al.* (1991) suggested that iron limitation alone cannot explain the low levels of phytoplankton biomass in the equatorial Pacific, and that in this ecosystem growth rates balance loss rates (from grazing and sinking) and keep the populations at their observed levels. However, growth rates of the larger cell organisms can increase with the addition of iron, and if grazing and sinking balance growth, then only a small increase in specific growth rate is required for the biomass to accumulate. The authors conclude, therefore, that in the equatorial Pacific, biomass regulation of the larger size fraction of the phytoplankton occurs via a combination of (i) grazing, (ii) iron limitation, and (iii) sinking, and that understanding the factors which control the larger phytoplankton is therefore fundamental to unravelling the HNLP paradigm.

### 9.1.3.3 *Primary production: global patterns*

The geographical distribution of primary production in the oceans is illustrated in Fig. 9.6. This figure was compiled from the data given by Koblentz-Mishk *et al.* (1970) and the values for some regions have been amended subsequently; for example, Chavez *et al.* (1996) reported that the mean productivity for the equatorial Pacific was 900 mg C m$^{-2}$ day$^{-1}$, almost twice the value quoted by Chavez & Barber (1987) and four times the value (200 mg C m$^{-2}$ day$^{-1}$) given by Koblentz-Mishke *et al.* (1970) for the same region of the equatorial Pacific. More recent data sources, however, e.g. those based on chlorophyll distributions, have confirmed the overall trends in the Koblentz-Mishke *et al.* (1970) primary production map.

The trophic status of the HNHP, LNLP and HNLP marine ecosystems identified in Section 9.1.3.2 was assessed on the basis of the rate of injection of new nitrogen, which maintains production levels. Thus, **new production**, which is defined by the introduction of new nitrate to the euphotic zone, and **export production**, which is defined as the fraction of new production exported as particulate carbon, are major variables that characterize the efficiency of carbon and nutrient cycling. In this context, it is important to remember that new production and export production are equal only if steady-state conditions are operative and if all the export is in the form of POC, which may not be the case in some regions, such as the HNLP equatorial Pacific, where a large fraction of new production may be transported out of the area as DOC (see e.g. Bacon *et al.*, 1996); however, Murray *et al.* (1996) concluded that a significant export of DOC need not be invoked to balance new and export production in the region. The *f*-ratio was introduced by Eppley & Peterson (1979) to express the propor-

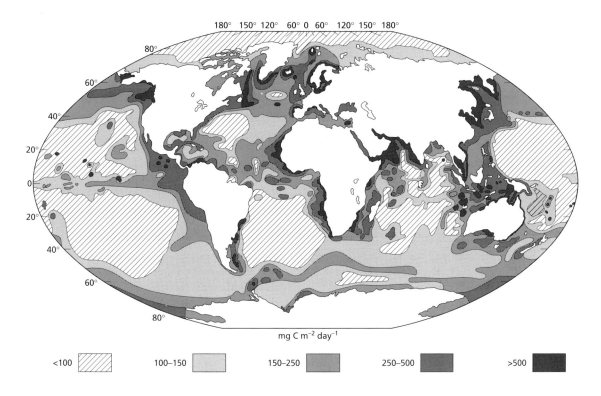

mg C m$^{-2}$ day$^{-1}$

| <100 | 100–150 | 150–250 | 250–500 | >500 |

**Fig. 9.6** Primary production in the World Ocean (from Degens & Mopper, 1976; after Koblentz-Mishke *et al.*, 1970).

tion of new to total production; the *f*-ratio is highest in eutrophic upwelling zones (values as high as ~0.8) and lowest in oligotrophic zones (values as low as ~0.1). Another variable in the system is the specific nitrate uptake rate (VNO$_3$) (see e.g. Dugdale & Wilkerson, 1992). A summary of some of the more important data in the oceanic 'primary production system' is presented in Table 9.2. In Table 9.2(a), estimates of the total global primary production and 'new' production are given in relation to oceanic provinces, and in Table 9.2(b) a variety of parameters are listed for the major trophic status regions in the oceans. Although it must be stressed that some of the parameters are relative, rather than absolute values, several conclusions can be drawn from Table 9.2, and by combining the data with the distribution patterns in Fig. 9.6, a number of major trends can be identified in the global oceanic distribution of primary production.

**1** Primary production is generally greater in coastal than in open-ocean waters, i.e. the so-called land-mass effect; for a detailed discussion of this phenomenon see Strickland (1965).

**2** Specific regions of high productivity (**HNHP eutrophic zones**) are found, especially in areas of upwelling. It was shown in Section 9.1.2 that in the water column the concentrations of nutrient elements are greater at depth than at the surface, where they are depleted by biological activity. During upwelling these nutrient-rich subsurface waters are brought to the surface, and so lead to an increase in primary productivity. High primary production is therefore found in many regions of upwelling: these include the shelves off the western margins of West Africa, Namibia, Peru, the western USA, India and southeast Asia. From the data in Table 9.2(b) it can be seen that the HNHP eutrophic coastal upwelling regions have relatively **high** values of (i) primary production rates, (ii) nitrate concentrations, (iii) VNO$_3$ and (iv) *f*-ratios. For detailed descriptions of the processes involved in upwelling, the reader is referred to Summerhayes *et al.* (1992).

**3** Regions of low fertility (**LNLP oligotrophic zones**) are found where water mixing is minimal, for example, where a deep permanent thermocline acts as

**Table 9.2**  Oceanic primary production.

(a)  Estimate of total primary production and new production (data from Martin *et al.* (1987) and Karl *et al.* (1996)).

| Oceanic province | Percentage of ocean | Area ($10^{12}$ m²) | Mean production (mg C m$^{-2}$ day$^{-1}$) | Global total production (Gt yr$^{-1}$)* | Percentage global total production | New production (mg C m$^{-2}$ day$^{-1}$) | Global new production† (Gt yr$^{-1}$) | Percentage global new production |
|---|---|---|---|---|---|---|---|---|
| Open-ocean | 90 | 326 | 356–463 | 42–55 | 82–86 | 29–49 | 3.5–5.9 | 70–80 |
| Coastal zone | 9.9 | 36 | 685 | 9 | 14–18 | 115 | 1.5 | 20–30 |
| Upwelling areas | 0.1 | 0.36 | 1150 | 0.15 | 0.2–0.3 | 238 | 0.03 | 0.4–0.6 |
| Total | | 362 | 2191–2298 | 51–64 | | | 5.0–7.5 | |

* 1 Gt = $10^{15}$ g. † New production = C flux at 100 m.

(b)  Typical values of a number of parameters in oceanic primary production.

| Oceanic region* | Primary production† (mg C m$^{-2}$ day$^{-1}$) | $NO_3$‡ (µM) | $VNO_3$‡ (day$^{-1}$) | f‡ (%) |
|---|---|---|---|---|
| HNHP | >~500 | ~15–~30 | 0.2–0.8 | ~60–~80 |
| LNLP | <~100 | <~0.05 | <0.03 | <~20 |
| HNLP | ~100–~250 | ~15–~30 | <0.05 | ~30–~50 |

* HNHP, high nitrogen–high productivity; LNLP, low nitrogen–low productivity; HNLP, high nitrogen–low productivity.
† Data after Koblentz-Mishk (1970); some of the data may be low in absolute values, but they are consistent in relative terms—see text.
‡ Data after Dugdale & Wilkerson (1992).

a barrier between nutrient-rich subsurface waters and the surface layer. Oligotrophic regions occur in the central ocean gyres, which lie between *c.* 10° and *c.* 40° north and south of the Equator (see Section 7.3.2). It is apparent from the data in Table 9.2(b) that the LNLP oligotrophic regions have relatively **low** values of (i) primary production rates, (ii) nitrate concentrations, (iii) $VNO_3$ and (iv) *f*-ratios.

4  In open-ocean areas, upwelling caused by divergence leads to intermediate values of primary production in a number of regions, one of the most important being the 'strip' in the equatorial Pacific. Another area of divergence giving rise to intermediate primary production rates lies around Antarctica. It has now become apparent, however, that although the Southern Ocean around Antarctica and the equatorial Pacific are biologically active systems, productivity in part of these regions is, in fact, considerably less than would be expected. This is because there are sufficient nutrients present in their surface waters to promote much higher productivity; i.e. some of the

nutrients are not utilized for primary production. These are the **high nitrogen, low productivity (HNLP)** regions described in Section 9.1.3.2, and from the data in Table 9.2(b) it is apparent that these HNLP regions have relatively high values of nitrate concentrations, but only intermediate values of primary production rates, $VNO_3$ and *f*-ratios. In these HNLP regions it is apparent, therefore, that primary production is not nitrate limited, and it is here that iron limitation may be the control on primary production (see Section 9.1.3.2).

It is apparent from this brief description of the major trends in primary productivity that the overall pattern is one of a 'ring' of high-fertility HNHP zones around the edges of the ocean basins, surrounding generally low-fertility LNHP zones in their centres. Where open-ocean divergence occurs, e.g. in the equatorial Pacific, HNLP zones can be developed in which the major nutrients are not fully utilized for primary production. Production here therefore is limited by some other factor, e.g. the availability of

**Table 9.3** Sources, reservoirs and sinks for organic carbon in the oceans (approximate values).

|  | Data | Williams (1975) | Hedges (1992) |
|---|---|---|---|
| Sources ($10^{15}$ g C yr$^{-1}$) | Atmospheric input | 0.22 | 0.1 |
|  | Fluvial input: POC | Total fluvial | 0.2 |
|  | Fluvial input: DOC | OC ≈ 0.18 | 0.2 |
|  | Net primary productivity | 36 | 50 |
|  | Input to DOC: |  |  |
|  | (1) phytoplankton excretions (10% productivity) | 3.6 |  |
|  | (2) resistant phytoplankton material (5% productivity) | 1.8 |  |
| Reservoirs ($10^{15}$ g C) | POC (non-living) | 30 | 32 |
|  | POC (living biomass) | 0.5 | 2 |
|  | DOC | 1000 | 665 |
| Sinks ($10^{15}$ g C) | Nearshore sediments | 0.003 | 0.126 |
|  | Pelagic sediments | 0.09 | 0.005 |
|  | All marine sediments | 0.093 | 0.126 |

iron. Within this overall pattern, primary production can be affected by **plankton blooms**. These blooms develop when the numbers of a particular phytoplankton species suddenly increase under favourable conditions. Blooms can occur in both coastal and open-ocean regions and often are triggered on a seasonal basis. Diatoms are the dominant phytoplankton during the spring blooms that occur in temperate and subpolar waters. A prime example of an extensive spring bloom is that found in the North Atlantic following restratification of the water column in April and May of each year (see Section 8.4.6.1), although blooms can occur in the absence of stratification (see e.g. Townsend *et al.*, 1992).

The earlier estimates of the total magnitude of primary production were ~25 × 10$^{15}$–36 × 10$^{15}$ g C yr$^{-1}$. More recent estimates, however, give a range of ~40 × 10$^{15}$–160 × 10$^{15}$ g C yr$^{-1}$ (Williams & Druffel, 1987), and Hedges & Keil (1995) quote a value of ~50 × 10$^{15}$ g C yr$^{-1}$. In general, ~80% of primary production occurs in open-ocean waters, and ~20% in coastal regions—see Table 9.2(a).

#### 9.1.4 The nutrients in the oceans: summary

**1** Nitrate and phosphate are involved in nutritional processes and are incorporated into the soft-tissue phases of plankton, whereas silicate is involved only in building the hard skeletal parts of organisms.
**2** New nutrients are supplied to the euphotic zone by processes such as upwelling and erosion of the thermocline from below, and atmospheric deposition;

these nutrients take part in new production and are transported out of surface waters via sinking particulates. Regenerated nutrients are recycled within the euphotic zone and take part in regenerated production.
**3** The vertical water-column profiles of the nutrients exhibit a characteristic surface-depletion–depth-enrichment effect as a result of their uptake in surface waters during primary production and their release at depth following the remineralization of sinking detritus.
**4** The horizontal distribution of the nutrients is controlled by water circulation patterns, upon which are superimposed the effects of internal bigeochemical reactivity. As a result, there is a steady push of nutrients towards the deep Pacific, which lies at the end of the deep-water global transport path, leading to their build-up in these waters.

### 9.2 Organic matter in the sea

#### 9.2.1 Introduction

For analytical convenience, the organic matter in aquatic systems is usually divided into two principal fractions.
**1** The fraction passing through a 0.45 μm membrane filter includes material in true solution, together with colloidal components, and is termed **dissolved organic matter** (DOM); its carbon content is classed as **dissolved organic carbon** (DOC). It has been estimated that as much as 50% of the DOM in sea water

may be present as high molecular weight (MW > 1000) colloids (see e.g. Carlson *et al.*, 1985), and the importance of the colloidal DOM fraction has been discussed by Lee & Wakeham (1992).

**2** The filter-retained material is referred to as **particulate organic matter** (POM), and its carbon content is termed **particulate organic carbon** (POC).

Some authors also distinguish a third fraction, termed **volatile organic carbon** (VOC).

Estimates of the amount of organic carbon in the POC and DOC reservoirs have varied over the years, and a summary of some of the most recent findings are given in Table 9.3. From the data in this table it is apparent that the DOC pool, which contains between $\sim 665 \times 10^{15}$ and $\sim 1000 \times 10^{15}\,g\,C$ is the principal reservoir for organic carbon in the oceans. This sea-water DOC provides a major dynamic pool in the global carbon cycle. The POC pool is dominated by the detrital (non-living) fraction, which holds $\sim 30 \times 10^{15}\,g\,C$. Little organic carbon is stored in the living plankton biomass ($\sim 0.5 \times 10^{15}$–$2.0 \times 10^{15}\,g\,C$), and on a whole ocean basis this reservoir is trivial compared with the seawater DOC and the non-living POC (Olsen *et al.*, 1985); however, in the euphotic zone, where most primary production occurs, plankton and bacteria become more important.

### 9.2.2 The sources of oceanic organic matter

The organic matter in sea water can be divided into two genetic classes. That having an external source is termed **allochthonous**, and that originating in the ocean itself, i.e. from internal sources, is referred to as **autochthonous**.

#### 9.2.2.1 *The sources of allochthonous organic matter*

This type of organic matter is brought to the oceans chiefly by river run-off, with smaller inputs from atmospheric transport; these inputs have been discussed in Sections 3.1.5 and 4.1.1.4, respectively.

#### 9.2.2.2 *The sources of autochthonous organic matter in the oceans*

The principal mechanism by which organic carbon is initially produced in the oceans is *in situ* photosynthesis by phytoplankton, which may be represented crudely by an equation of the general type

$$6CO_2 + 6H_2O = (C_6H_{12}O_6) + 6O_2 \qquad (9.1)$$
$$\text{(uptake)} \qquad \text{(release)}$$

where $C_6H_{12}O_6$ is carbohydrate. The resulting biotic carbon may form part of the POC pool, or it may enter the DOC pool through processes such as the exudation of organics by phytoplankton, excretion by zooplankton and post-decay reactions. The $CO_2$ involved in photosynthesis may be present as free dissolved $CO_2$, or as bound bicarbonate or carbonate ions (see Section 8.4.5), and because the concentration of $\Sigma CO_2$ is high enough it does not limit photosynthesis.

Until a few years ago, little was known of the importance of photochemical reactions in the sea, apart from those involved in photosynthesis. It is now recognized, however, that such reactions can play a role in the production, transformation and destruction of organic matter in the sea; e.g. by affecting its oxidation state. Photochemical reactions also may be significant in the breakdown of refractory high molecular weight DOM to biologically labile low molecular weight compounds in the euphotic zone, thus providing a pathway by which some refractory DOM may be recycled (Kieber *et al.*, 1989). For a discussion of marine photochemistry, the reader is referred to the reviews by Zafiriou (1983, 1986) and Lee & Wakeham (1992).

*Autochthonous POC: the living POC fraction.* This is composed largely of plankton, and these biota have been described in Section 9.1.3.1.

*Autochthonous POC: the non-living fraction (detritus).* This consists mainly of dead organisms, faecal material, organic aggregates of various types, and other complex organic particles. In open-ocean waters much of this non-living POC is derived from phytoplankton. Cawet (1981) has identified four major processes that are involved in the formation of non-living POC: (i) the direct formation of detritus (e.g. organic fragments, faecal pellets); (ii) the agglomeration of bacteria; (iii) the aggregation of organic molecules by bubbling in the surface layers; and (iv) the flocculation or adsorption of DOC on to mineral particles. Fellows *et al.* (1981) have also suggested that bacteria, and other micro-organisms, can repackage soluble nutrients into POC. This DOC → bacteria → POC route can result in increases in POC at depth in the water column and so enhance its supply to the deep-water pool.

*Autochthonous DOC.* Dissolved organic carbon is the principal reservoir for organic carbon in the oceans (see Section 9.2.1), but there is disagreement in the literature over the relative importance of the internal (autochthonous) and external (allochthonous) sources of DOC in sea water. For example, Mantoura & Woodward (1983) showed that the DOC in the Severn Estuary (UK) behaved in a conservative manner (see Section 3.2.7.7), and pointed out that the consequence of this conservative delivery of fluvial DOC is that, on a global-scale, river inputs will make a significant contribution, up to as much as *c.* 50%, to the oceanic DOC pool. Other lines of evidence, however, put the terrestrial contribution to this oceanic DOC pool at a much lower figure. Thus, Meyers-Schulte & Hedges (1986) showed that there is an absence of lignin, which is derived only from terrestrial higher plants, in open-ocean humic substances from the equatorial Pacific, and on the basis of this estimated that less than *c.* 10% of oceanic DOC is terrestrial in origin. Further, Williams & Gordon (1970) and Eadie *et al.* (1978) gave data on the $\delta^{13}$C signature of oceanic DOC, which suggested that it originates mainly from marine-derived organic carbon. Williams & Druffel (1987) also concluded that various radiocarbon and lignin data precluded a terrestrially derived DOC component exceeding *c.* 10% of the total oceanic DOC pool.

### 9.2.3 The distribution and composition of oceanic organic matter

#### 9.2.3.1 The distribution and composition of POM in the oceans

The POM in the oceans consists of living organisms and dead material (**detritus**), and can originate from both marine and terrestrial sources.

With respect to autochthonous POC, some aspects of the major element composition of organic tissue are relatively well understood. One reason for this is that the C:N:P ratios of both plant and animal tissue are fairly constant. These are often termed the **Redfield ratios**, after Redfield (1934, 1958). These ratios indicate that the major plant nutrients, phosphate and nitrate, change concentrations in sea water

in a fixed stoichiometry that is the same as the N and P stoichiometry of plankton. Thus, the nutrients are removed from sea water during photosynthesis in the proportions required by the biomass. The elemental composition of phytoplankton and zooplankton is very similar, and C:N:P proportions of 106:16:1 are commonly used for their organic tissue material. These are the Redfield ratios and represent the preformed composition of the organic material that undergoes oxidative destruction in the euphotic zone according to the relationship C:N:P:$O_2$ = 106:16:1:138, which corresponds to the consumption of 138 moles of oxygen to produce 106 atoms of carbon, 16 atoms of nitrogen and 1 atom of phosphorus. More recently, however, it has been suggested that the accepted Redfield ratios need revision. For example, Takahasi *et al.* (1985) used a variety of chemical data from isopycnal surfaces to estimate the composition of the organic matter oxidized within the thermocline of the Atlantic and Indian Oceans, and concluded that the composition is better represented by C:N:P ratios of 122(±18):16:1. Watson & Whitfield (1985) modelled the composition of particles in the global ocean and also concluded that the original Redfield ratios should be changed, suggesting that the material leaving the euphotic zone has a composition corresponding to $C_{org}$:N:P:Si:$C_{inorg}$ = 126:15.7:1:23.5:23.0. There also is evidence that the C:N ratio in the organic material exported from the upper waters in coastal regions may deviate from the 6.6 'Redfield' ratio by the elevated consumption of carbon, and Sambrotto *et al.* (1993) suggested that this also may occur in open-ocean waters.

The soft, i.e. non-skeletal, parts of organisms are composed mainly of proteins, carbohydrates, lipids and, in higher plants, lignins. The oceanic living biomass is dominated by plankton, and the structures of some of the major metabolites (i.e. substances that take part in the processes of metabolism) found in these organisms are illustrated in Worksheet 9.1.

Terrestrial, i.e. allochthonous, sources also provide a variety of particulate, and dissolved, organic matter to the oceans. This organic matter covers a wide variety of compounds; including hydrocarbons, fatty acids, carbohydrates, fatty alcohols and sterols, natural polymers such as lignin, cutin and chitin, and possibly amino acids. Some of these compounds are exclusive to terrestrial biota. For example, lignins are complex phenolic polymers that are unique to the

**Worksheet 9.1: Some organic compounds found in plankton**

A variety of **metabolites**, i.e. substances that take part in the processes of metabolism, are associated with marine organisms. The most important of these are described below, and a number of organic structures are illustrated.

*Proteins*

Proteins, a major structural component of living tissue, are complex nitrogenous organic compounds, which are polymers of **α-amino acids** assembled in chains. Globular proteins in the form of **enzymes** catalyse biochemical reactions. In addition to acting as enzymes, proteins serve as antibodies, transporters, receptors and metabolic regulators. The most abundant amino acids found in the proteins of plankton are glycine, alanine, glutamic acid and aspartic acid.

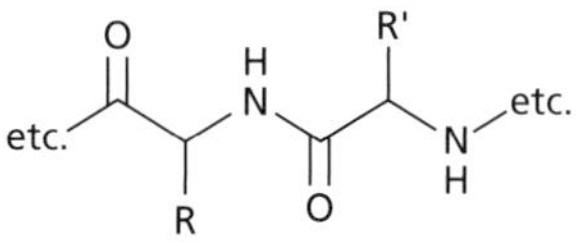

Biopolymers of amino acids e.g.

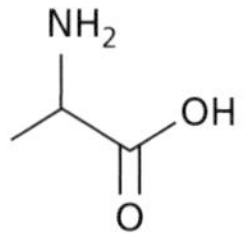

**Alanine** 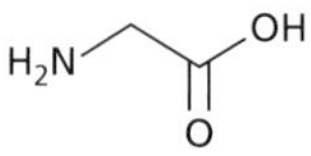          **Glycine**

*Carbohydrates*

Carbohydrates include sugars, starches and cellulose. Carbohydrates form the supporting tissues of phytoplankton and also act as energy sources; for example, catabolism of carbohydrates can provide most of the energy requirement for the cell. Carbohydrates can be classified into **monosaccharides**, which are subdivided into pentoses (e.g. ribose) and hexoses (e.g. glucose), **disaccharides** (e.g. sucrose), which are sugars, and **polysaccharides**, which include starches, cellulose and the mucopolysaccharide chitin.

**Glucose**

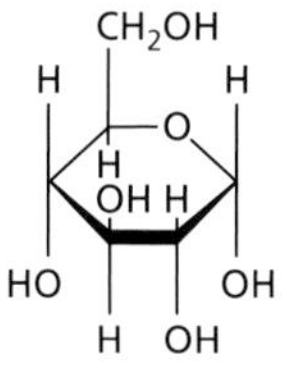

monosaccharide

*continued*

**Cellulose**

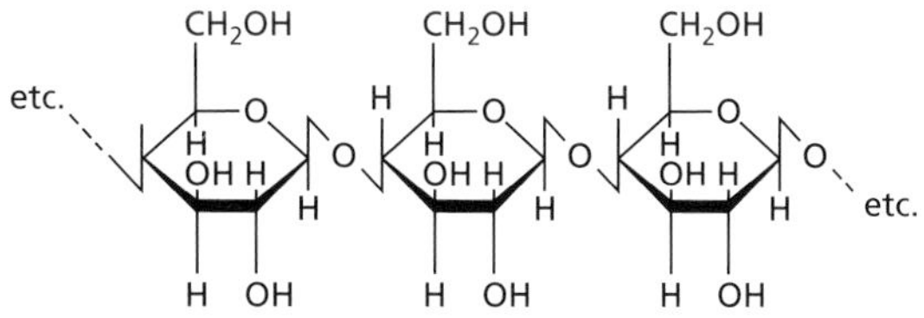

polysaccharide

**Chitin**

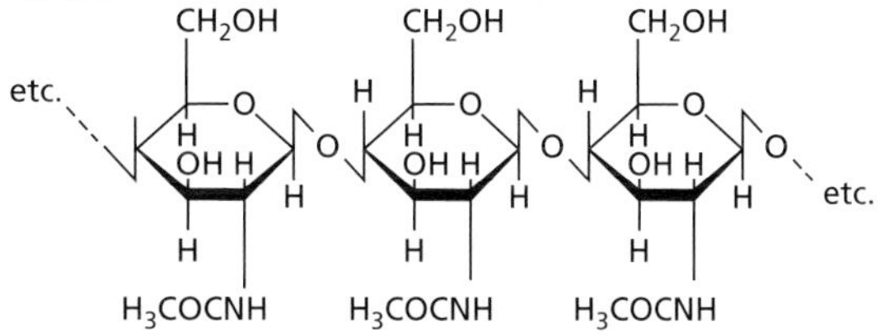

mucopolysaccharide

*Lipids*

Lipid is a general term that applies to all substances produced by organisms that are insoluble in water but can be extracted by some kind of organic solvent. In this broad definition the term lipid covers a wide variety of compound classes, including the photosynthetic pigments. More usually, lipids refer to fats and waxes, phospholipids and glycolipids, sterols and some hydrocarbons (e.g. *n*-alkanes).

Fats, which are utilized mainly as energy stores in the energy budget of organisms, are triglycerides formed by the combination of glycerol and fatty acids.

Wax esters, which have a function as protective coatings, are esters of fatty acids and long-chain alcohols. They also can act as food reserves in many species of marine organisms and are abundant in zooplankton.

Phospholipids are glycerides (fatty acid esters of glycerol) with both phosphoric acid and fatty acid units, and are the most abundant lipids in phytoplankton. They can serve as membrane constituents, buoyancy controls and thermal or mechanical insulators.

Glycolipids are combinations of lipids and sugars and are found in cell membranes and walls.

Sterols are the hormonal regulators of growth, respiration and reproduction in most marine organisms, and also may act as membrane rigidifiers. In addition, sterols can be present as lipoproteins (e.g. cholesterol).

*continued on p. 224*

**Wax ester**

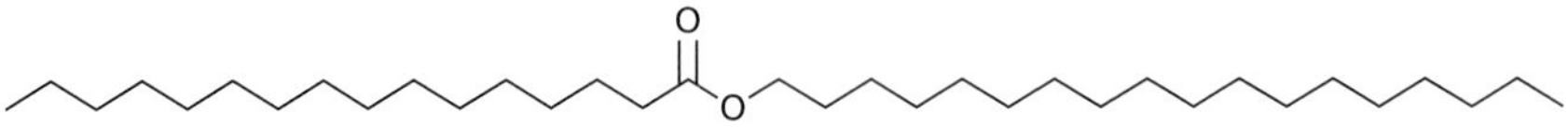

**Phospholipid**

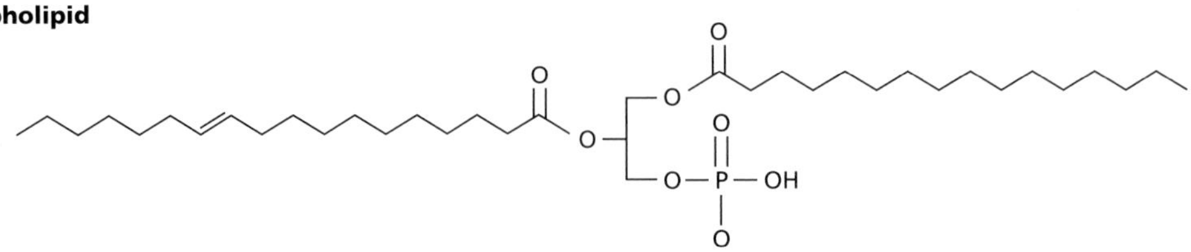

1-palmitoyl-2-palmitoleoyl-3-phosphatidic acid

**Sterols**

e.g. Cholesterol

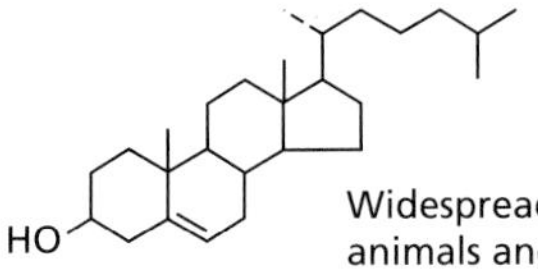

HO

Widespread distribution in
animals and plants

e.g. Dinosterol

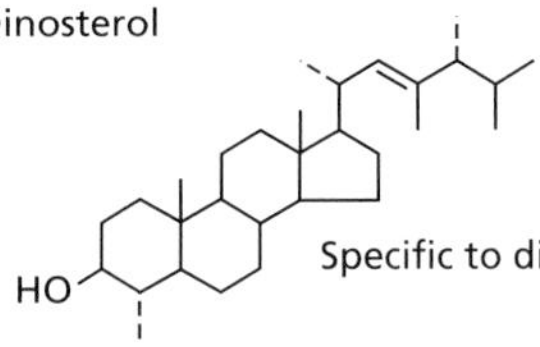

HO

Specific to dinoflagellate algae

*Pigments*

Phytosynthetic coloured pigments such as chlorophyll (green), and the carotenoids (e.g. carotene and fucoxantin; orange/red), are used by plants to absorb and transfer light energy during photosynthesis. Most phytoplankton have **chlorophyll** as their primary photosynthetic pigment, and four major types of chlorophylls have been identified (Chl *a*, Chl *b*, Chl *c* and Chl *d*), although for chlorophylls *c* and *d* there are a number of compounds with distinct structures. Chl *a* is the primary pigment in algae, higher plants and cyanobacteria, with other forms acting as accessory pigments to pass on light energy captured by the *a* form. Photosynthetic bacteria utilize a distinct set of pigments known as *bacteriochlorophylls* (*a* to *e*), with the *a* form being the primary pigment. The distribution of chlorophyll is not homogeneous throughout the oceanic water column, but is characterized by a **global-scale maximum** often found near the bottom of the euphotic zone in the vicinity of the thermocline. The concentrations of chlorophylls in surface waters are generally higher in euphotic than in oligotrophic regions, and chlorophyll *a* has been used as an indicator of primary production.

*continued*

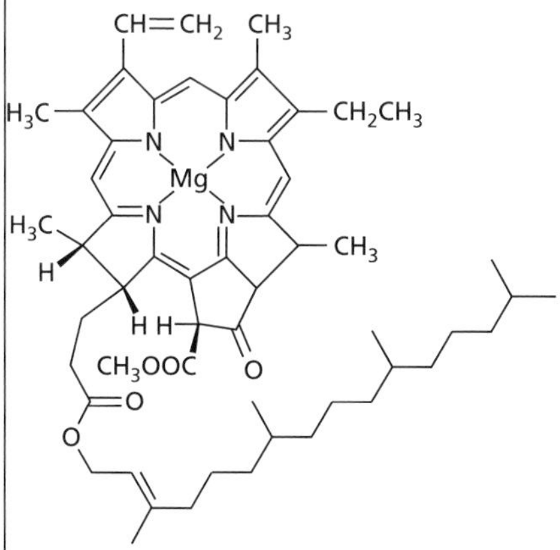

Chlorophyll *a*

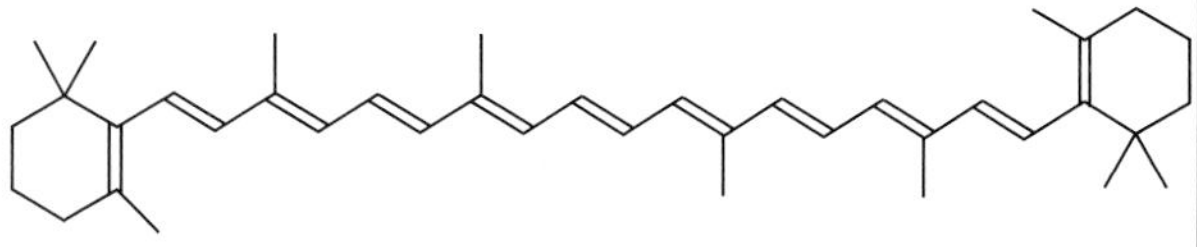

β-Carotene

### Nucleic acids

Ribonucleic acid (RNA) and deoxyribonucleic acid (DNA) are involved in the storage and transmission of genetic information and are made up from nucleotide units composed of a phosphate, a pentose sugar and a nitrogen-containing organic base. The DNA encodes the amino acid sequences of all the proteins synthesized by a cell. The RNA molecules are vital intermediates in the production of protein from this genetic code.

### Other metabolites

Various other organic compounds are also found in plankton. These include vitamins, co-enzymes and hydrocarbons. According to Gagosian & Lee (1981) the principal hydrocarbons in algae are alkenes, together with lesser amounts of *n*-alkanes, branched alkanes and cyclic alkanes. The *n*-alkanes can be used to distinguish marine from terrestrial organic matter inputs (see Section 9.2.3.2).

tissues of vascular land plants. Other compounds can be derived from both terrestrial and marine sources, but sometimes can be distinguished from each other on the basis of their detailed chemistries. This 'biomarker' approach can be illustrated with reference to a number of compounds. For example: (i) terrestrial *n*-alkanes derived from plant waxes have their principal homologues in the series $C_{23}$–$C_{35}$, whereas those derived from marine plankton are dominated by homologues in the series $C_{15}$–$C_{21}$; (ii)

terrestrial fatty acids derived from plant waxes have homologues in the series $C_{14}$–$C_{36}$, whereas those from marine plankton sources are generally in the range $C_{12}$–$C_{24}$ (see Peltzer & Gagosian, 1989).

Only a small fraction of deep-water POM, probably only a few per cent, is composed of living material, the remainder being *refractory* in character. The refractory deep-water POM consists of two fractions: (i) a small-sized fraction, which makes up the bulk of the POM, and (ii) a larger sized fraction, con-

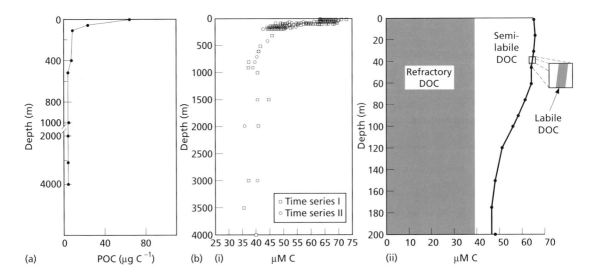

**Fig. 9.7** Vertical profiles of dissolved and particulate organic carbon in the oceanic water column. (a) POC profile at 36°20′N: 67°50′W in the Atlantic Ocean (after Williams, 1971). (b) DOC profiles at the JGOFS Equatorial Pacific Time Series site at 0°, 140°W in the central Pacific (from Carlson & Ducklow, 1995). (i) DOC profile from 0 to 4000 m. (ii) Partitioning of the bulk DOC pool in the upper 200 m into refractory, semi-labile and labile pools.

sisting mainly of faecal pellets and organic debris (marine snow), which is utilized in the foodchain. The refractory POM is considered in more detail in Section 9.3, where the oceanic organic carbon cycle is described.

The POC (living + non-living components) constitutes an average of less than ~5% of the TOC in sea water. In most surface waters (0 to ~300 m) the living fraction is the most important source of POC. In these surface waters POC concentrations vary both geographically and seasonally following the patterns of primary production, and appear to lie in the range ~0.02–0.1 mg Cl$^{-1}$ (~1.8 to ~8.3 μM C)—see Cawet (1981) to ~0.3 mg l$^{-1}$. In deeper waters the POC concentrations fall, and range from ~0.004 to ~0.03 mg Cl$^{-1}$ (~0.33 to ~2.5 μM)—see Cawet (1981), and have an average of ~0.01 mg l$^{-1}$. Thus, there is an overall decrease in the concentrations of POC with depth in the water column—see Fig. 9.7(a). Further, in deep-waters the POC, much of which is non-living and refractory, is considerably less vari-

able in concentration than it is in the surface layer. Nonetheless, variation in the concentrations of POC with depth in the water column can occur when the profile section cuts more than one water mass.

### 9.2.3.2 The distribution and composition of DOM in the oceans

Dissolved organic carbon makes up, on average, ~95% of the TOC in the oceans. Originally, it was thought that the distribution of DOC in sea water could be described in terms of three principal features:

1 DOC concentrations could vary in surface waters (less than ~300 m), but were relatively invariant throughout the deep ocean;

2 on average, surface water DOC concentrations (~1 to ~1.5 mg Cl$^{-1}$; ~80 to ~120 μM C) were about twice those in deep-water (~0.5 mg Cl$^{-1}$; ~40 μM C)—see Peltzer & Hayward (1996);

3 there was no evidence of any significant, or regular, differences in the concentrations of DOC between the oceans or within their climatic zones (see e.g. Williams, 1975).

There were always difficulties in determining the concentrations of DOC in sea water, and much of the original work was carried out using wet chemical oxidation (WCO) techniques to oxidize DOM to $CO_2$. Although WCO techniques have been shown to produce essentially complete conversion of both simple organic molecules and biopolymers to $CO_2$, conversion efficiencies for natural DOM mixtures were unknown. Subsequently, there was what appeared to be a 'quantum breakthrough' in DOC analysis with the introduction of a high-temperature catalytic oxidation (HTCO) technique, originally for the determination of DON (Suzuki et al., 1985) and subsequently for DOC (Sugimura & Suzuki, 1988). The DOC concentrations found initially by the HTCO technique were significantly higher than those determined previously using the WCO approach. This suggested that the WCO techniques were measuring only part of the DOC in sea water, and were thus underestimating the size of the DOC oceanic pool, the 'missing' fraction probably being labile in character. The HTOC technique also cast doubt on the shapes of the DOC concentration–depth profiles obtained previously, which showed a generally uniform distribution and did not suggest a clear relationship between DOC and apparent oxygen utilization (AOU)—see Section 8.3. Sugimura & Suzuki (1988) suggested that, because the decomposition of organic matter consumes dissolved oxygen in sea water, the DOC concentrations should be inversely related to the AOU, and the HTOC data showed that this indeed was the case.

The HTOC data appeared to have profound consequences for chemical oceanography and a number of meetings and intercalibration exercises were arranged to test the new technique and to assess its implications. Suzuki (1993), however, threw the whole question of the validity of the HTOC technique in doubt. He had been concerned that he could not reproduce the strong relationship between DOC and AOU originally found in the equatorial Pacific waters, and when he re-evaluated the original data he concluded that the accuracy of the elevated DOC concentrations reported by Sugimura & Suzuki (1988) for the Pacific were inadequate to support the arguments on DOC inventories in the oceans. The retraction of the Suzuki et al. (1985) and Sugimura & Suzuki (1988) papers was a great disappointment for many chemical oceanographers, and it appeared that the seawater DOM pool still remained one of the largest and least understood reservoirs of carbon at the Earth's surface.

It is of interest at this stage to consider two other parameters that have been used to characterize the DOM in sea water, i.e. molecular weight and age.

**1** The molecular weight of oceanic DOM. In the absence of chemical data, molecular weight is often used to characterize the DOM in sea water into low molecular weight (LMW) and high molecular weight (HMW) material, some of the latter probably being colloidal in nature. There is no general agreement on a 'cut-off' between LMW and HMW material, but values of 500 and 1000 daltons (Da) have been used for this purpose by some workers.

**2** The age of oceanic DOM. A considerable body of research has been directed towards establishing the 'age' of the DOC in sea water. A number of workers carried out $^{14}C$ age-determination on the WCO-determined DOC in ocean waters, with one of the most widely quoted values being that of ∼3500 yr for the DOC in deep-water given by Williams et al. (1969). More recent values, however, suggest that the DOC is older than this. For example, Williams & Druffel (1987) carried out a study in the waters of the north central Pacific from the surface to a depth of 5720 m. They gave mean apparent 'ages' for the WCO-determined surface water DOC of 1300 yr BP and 6000 yr BP for the DOC found from 900 m to 5720 m. On the basis of this ∼6000 yr 'age' for the deep-water DOC, a total DOC deep-water pool of $0.6 \times 10^{18}$ g C, and an assumed steady-state flux within a mixture of uniform $^{14}C$ composition, Williams & Druffel (1987) estimated a global ocean turnover rate of marine DOC to be ∼$0.1 \times 10^{15}$ g C yr$^{-1}$. In a later publication using a HTOC technique, Druffel et al. (1989) estimated an apparent 'age' of ∼3000 yr for deep-water DOC. Mopper et al. (1991) suggested that although the 'old' DOC is resistant to biological degradation it is readily attacked by oxidative photochemical reactions during its passage through the sunlit layer of the oceans, and that photochemical degradation may be the rate-limiting step for the removal of a large fraction of the oceanic DOC. Bushaw et al. (1996) also proposed that the larger molecules in the DOC pool may photochemically release nitrogen-rich compounds, including ammonium, which are biologically available. Mopper et al. (1991) estimated that the oceanic residence time of the biologically refractory, photochemically reactive DOC is 500–2100 yr. This is less than the average age of ∼3000–6000 yr, and may be explained by the injection of 'old' carbon from sediments. These 'ages' for the deep-water DOC are larger than the mixing times of the oceans (~1600 yr), so it would appear that the bulk of the refractory DOC is recycled several times before being deposited in sediments. The estimates are for an average 'age' of unfractionated deep-water DOM, however, and as the molecular components of DOM are diverse, it is unlikely that they are all of the same 'age'. It may be, therefore, that some of the DOC is in fact very old and that the rest undergoes a relatively rapid oceanic turnover time (see also, Mantoura & Woodward, 1983); i.e. the apparent old 'age' for deep-water oceanic DOC is the result of mixing different organic carbon fractions. For example, by applying a mass balance approach to $\Delta^{14}C$ data, Druffel et al. (1992) suggested that a fraction of the DOC pool might be composed of young material. A similar conclusion was reached by Santschi et al. (1995), who provided direct evidence for the existence of a contemporary $^{14}C$ age (i.e. younger than a few decades) for a high molecular weight (HMW) fraction of colloidal organic carbon (COC) in the upper water column of the Gulf of Mexico and the Middle Atlantic Bight. This HMW COC, which constituted between ~3% and ~5% of the total DOC, was derived from living POM and underwent rapid (days) cycling. In contrast, a significant fraction of the low molecular weight (LMW) DOM was composed of older (~380–4500 yr BP) more refractory COC, which cycled on much longer time-scales. It was suggested also that the COC was composed of the mixing of three end-members: fluvial (terrestrial), near-shore and off-shore (oceanic). The relatively short turnover times (~1–30 days) found for some of the COC support the hypothesis that the $^{14}C$ ages for the colloidal fractions of DOC result from the mixing of several HMW and LMW end-members having slow and fast turnover rates, as proposed by Mantoura & Woodward (1983).

Where do we now stand with respect to our understanding of the distribution and composition of DOC in the oceans in the light of the HTOC controversy?

Sharp et al. (1995) reported that when proper precautions were taken, agreement to within ~70% could be obtained for DOC concentrations determined by HTOC and WCO techniques. This is important because if the absolute concentrations of DOC obtained by the HTOC technique can be verified, the speed and improved precision of the HTOC technique will allow small but real differences in bulk DOC to be determined. For example, in a recent study, Carlson & Ducklow (1995) provided DOC data for the equatorial Pacific as part of the JGOFS EqPac study (see Section 8.4.6.1). They used a HTOC technique and reported that no elevated DOC values were found over those reported previously for WCO determinations. The data obtained by Carlson & Ducklow (1995) therefore allowed them to identify small-scale variations in the DOC depth profile. The concentration of DOC was high in surface waters (~65 μM C), then decreased exponentially below 60 m to an average of ~50 μM C at 150 m. Below 500 m to 4000 m the DOC profile remained fairly constant at ~39 μM C—see Fig. 9.7(b(i)). Thus, the original WCO DOC profiles were confirmed. The authors also were able to partition the bulk DOM in the waters of the central equatorial Pacific into the three individual pools. In the surface waters the bulk DOM consisted of all types of organic matter, ranging from extremely labile to extremely refractory. Below the euphotic zone (average depth 120 m) biologically available DOM is utilized, generating a residue of more refractory DOM. At water depths greater than 1000 m the relatively invariant concentrations of DOC indicated that it was composed of biologically resistant refractory components. The authors estimated that in the upper 200 m of the central equatorial Pacific ~70% of the bulk DOC was in the **refractory** pool, with most of the remaining DOC being assigned to a **semi-labile** pool, with a turnover of months to years, and a small percentage being in a **highly labile** pool, with a turnover of hours to days—see Fig. 9.7(b(ii)).

Only a small fraction, perhaps ~10–20%, of the DOM in sea water has been characterized. This is the labile fraction and consists mainly of compounds such as lipids, carbohydrates, amino acids, urea and pigments, all of which typically are associated with the biochemistry of living organisms (see Worksheet 9.1), and the biomass itself is probably the principal source of these compounds. In general, there are three pathways by which the autochthonous POC produced in oceans by biota can contribute to the oceanic DOC pool: (i) exudation by phytoplankton; (ii) excretion by zooplankton; and (iii) postdeath organism decay processes. During their lifetime phytoplankton release some of their phytosynthetically fixed carbon to the surrounding waters by metabolic processes. Following the death of both phytoplankton and zooplankton, decomposition occurs via the action of autolytic enzymes present in the tissues and by bacteria that have colonized the material. During these processes the DOC released into the water can include both biologically labile and refractory fractions. Estimates of the amount of phytosynthetically fixed carbon released as extracellular products vary, but Williams (1975) concluded that ~10% is probably representative of offshore plankton communities, which would yield an addition of $\sim 3 \times 10^{15}$–$5 \times 10^{15}\,\mathrm{g\,yr^{-1}}$ for total oceanic primary production estimates of $\sim 30 \times 10^{15}$–$50 \times 10^{15}\,\mathrm{g\,C\,yr^{-1}}$ (see Section 9.1.3.3 and Table 9.2).

The characterized fraction of the DOM also contains various organic pollutants, such as petroleum residues, DDT and PCBs. A number of the characterized DOM compounds are described below.

*Lipids.* These include *n*-alkanes ($C_{16}$–$C_{32}$), pristane, fatty acid esters, aliphatic hydrocarbons, triglyceride, sterols and free fatty acids, principally palmatic, oleic, myristic and stearic acids (see e.g. Ehrardt *et al.*, 1980; Kennicutt & Jeffrey, 1981; Kattner *et al.*, 1983; Parrish & Wangersky, 1988). Hydrocarbons are present in all marine organisms, but usually account for only ~1% of the total lipids. The non-volatile natural hydrocarbons ($>C_{14}$) in the marine environment include saturated aliphatic hydrocarbons (e.g. *n*-alkanes, regular branched isoprenoids, branched alkanes), unsaturated aliphatic hydrocarbons (e.g. *n*-alkenes), saturated alicyclic hydrocarbons (e.g. cyclanes), unsaturated alicyclic hydrocarbons (e.g. cyclenes) and aromatic hydrocarbons (e.g. retene, perylene). The pollutants include *n*-alkanes, isoalkanes and aromatic compounds from oil pollution, polycyclic aromatic hydrocarbons (PAHs), polychlorinated biphenyls (PCBs), and the biocides DDT and pentachlorophenol (PCP). The *n*-alkanes are the dominant constituents of natural hydrocarbons in the marine environment (Saliot, 1981). The *n*-alkanes in sea water originate from: (i) natural internal sources, i.e. the oceanic biomass; (ii) natural external terrestrial sources (mainly associated with higher plant metabolism); and (iii) anthropogenic sources (e.g. oil pollution). These sources sometimes can be distinguished by characterizing the *n*-alkanes on the basis of their carbon numbers and their **carbon preference index** (CPI). In this context

$$\mathrm{CPI} = \frac{\text{sum of odd carbon } n\text{-alkane concentrations}}{\text{sum of even carbon } n\text{-alkane concentrations}}$$

Thus, a CPI of 1 indicates no carbon preference (see e.g. Kennicutt & Jeffrey, 1981). *n*-Alkanes in the range $C_{17}$–$C_{22}$ are indicative of a phytoplankton source (see e.g. Blomer, 1970), whereas the heavier *n*-alkanes in the range $C_{28}$–$C_{32}$ originate from terrestrial plants (see e.g. Eglinton & Hamilton, 1963). The *n*-alkanes of plants and organisms generally show an odd carbon preference. In contrast, the *n*-alkanes of oils usually have a CPI of around 1.

*Carbohydrates.* The dissolved carbohydrates reported to be present in sea water during an algal bloom include laminaribiose, laminaritriose, glycosylglycerols, sucrose and raffinose (Sugugawa *et al.*, 1985). These low molecular weight carbohydrates also are found in dinoflagellate cells, and are thought to have been derived from phytoplankton.

*Amino acids.* These include dissolved free amino acids (DFAA), dissolved combined amino acids (DCAA) and dissolved total amino acids (DTAA). Lee & Bada (1977) reported that in the equatorial Pacific and the Sargasso Sea the concentrations of DCAA were much higher than those of DFAA, and showed a maximum in the euphotic zone where the concentrations of phytoplankton and zooplankton were higher. Further, the concentrations of the DCAA were higher at all water depths in the productive equatorial Pacific than in the oligotrophic Sargasso Sea. Lee & Bada (1977) also found that DFAA concentrations were small and relatively invariant in the water column. Liebeziet *et al.* (1980), however, reported that there was a significant variation in the concentrations of DFAA with depth in the Sargasso Sea. These authors provided data on the distributions of individual DFAA (threonine, serine, glutamic acid, aspartic acid, glycine, alanine) and found that concentrations were enriched at the upper boundaries of the seasonal pycnocline. These enhancements were attributed to increased auto- and heterotrophic activity, leading to the production of dissolved organic compounds in the sharp density layer, where there is a concentration of bacteria and zooplankton that take advantage of the energy-rich compounds concentrated at the discontinuities in the water column.

*Halogenated organics.* A number of organisms are known to produce halogen-containing organic compounds, with bromine rather than chlorine being the dominant halogen present. This topic has been reviewed by Fenical (1981).

*Dissolved organosulphur compounds.* These compounds include:
1 those having natural sources (e.g. dimethyl sulphide (DMS—see Section 4.1.4.3), dimethyl disulphide (DMDS), carbon disulphide, methylmercaptan (MeSH), dibenzothiophene (DBT), sulphur-containing amino acids;
2 those derived from anthropogenic sources (e.g. DBT from oil spills and diphenylsulphone).

The description of dissolved organic compounds given above has been confined mainly to the bulk water column. It must be stressed, however, that a considerable concentration of both DOC and POC is found in the sea-surface microlayer. For example, the total DOC is enhanced in the microlayer, relative to bulk sea water, by a factor of about 1.5 to about 3, and the individual compounds contributing to this enrichment include surfactant lipids (including hydrocarbons), carbohydrates and amino acids. The composition of the dissolved and particulate material in the microlayer is discussed in Section 4.3.

The bulk of the DOM in sea water has not yet been characterized, although it is known to be inert, i.e. largely refractory to biological degradation and chemical oxidation. Traditionally, the uncharacterized DOM had been given the name **Gelbstoff** because of the yellow colour it imparts to sea water. Gelbstoff is not a single component, but is a complex mixture of macromolecules of humic and lignin-type material, and was thought to be analogous to terrestrial soil residues. Recent studies, however, have suggested that terrestrial material does not make a significant contribution to marine DOM; for example, fluvial DOM contains substantial quantities of lignin, derived from terrestrial vascular plants, and marine DOM contains only a small fraction of lignin, indicating that if terrestrial lignin is present it has been altered chemically or biologically to the extent that its structure has been destroyed (Lee &

Henrichs, 1993). It therefore is probable that the uncharacterized DOM in sea water is composed of large, 'old', complex humic-like molecules, which probably have a mainly autochthonous source. Originally, it was believed that the autochthonous DOM was simply the residue remaining after the decomposition of phytoplankton debris. However, there now seems to be a general consensus that the 'humic' substances in sea water are derived from reactions involving simple organic compounds produced by living organisms; these 'humic' building blocks include amino acids, sugars, lipids, etc. According to Lee & Henrichs (1993) both small (<500 Da) and larger molecules are released from phytoplankton, and larger molecules dominate the material excreted. For example, Chrost & Faust (1983) reported that of the organic carbon released by algae, 19% was <500 Da, 30% between 10 000 and 30 000 Da, and 15% >> 300 000 Da. The low molecular weight fractions are dominated by amino acids, glycollate and sugar alcohols. The composition of the larger molecules is unknown, but it has been suggested that they are composed mainly of extracellular polysaccharide. It is thought generally that the low molecular weight fraction is utilized more rapidly by bacteria; e.g. the very labile fraction, which has turnover times of hours or days, and is involved in fueling the 'microbial loop' in surface waters. Amon & Benner (1994), however, have suggested that the bulk of oceanic DOM is made up of small (low molecular weight) molecules that cycle slowly and are relatively unavailable to micro-organisms. There is also evidence that some high molecular weight material excreted by phytoplankton can be turned over rapidly, although a small fraction is refractory. Moreover, Lee & Henrichs (1993) have pointed out that there are difficulties with the concept that polysaccharides constitute a large proportion of the DOM pool in sea water. It is apparent, therefore, that a major fraction of sea water DOM still remains to be characterized. In this context, Jickells *et al.* (1991) suggested that colloids of HMW terrestrial and marine biopolymers (e.g. polysaccharides, lignin, protein) and altered heteropolycondensates (humic material) could account for the uncharacterized DOM. According to Hedges (1978) condensation products of sugars and amino acids may also contribute to the uncharacterized DOM.

*To summarize.* Although the jury are still out with respect to the origin, absolute concentrations, and therefore the overall distribution of DOC in the oceans, a number of general conclusions can be drawn on the basis of our present knowledge.

**1** The depth profile of DOC in the water column generally shows higher values in the surface layer, indicating net production of DOC in the euphotic zone, decreasing to a generally invariant lower concentrations in deep water.

**2** Within the upper water column the vertical profiles of DOC can vary seasonally. For example, Carlson *et al.* (1994) showed that in the oligotrophic Sargasso Sea, DOC depth profiles in the upper 250 m are strongly influenced by a variety of factors, including convective mixing and vernal restratification. Thus, increased primary production in the early spring led to the accumulation of DOC, which was partly consumed in the summer and autumn, and in winter DOC was homogeneously distributed in this section of the water column—see Fig. 9.8. In contrast, in the equatorial Pacific there was no seasonal variability in the DOC in the upper 200 m of the water column (Carlson & Ducklow, 1995).

**3** As DOM is cycled through the heterotrophic bacteria, the residual material becomes more refractive and eventually biologically resistant and, in a very broad sense, the DOM in sea water can be divided into at least three pools.

(a) A **highly labile** DOM pool, containing biologically available organic matter that turns over on time-scales of hours to days;

(b) a **semi-labile** pool, containing organic matter that turns over on a seasonal time-scale;

(c) a **refractory** pool, which is 'old' and turns over on a relatively long time-scale of several thousands of years.

The refractory DOM is representative of deep-water DOM.

**4** The percentage of refractory material decreases in concentration from surface to deep waters. In surface waters the accumulation of labile DOC depends on the coupling of production and consumption (grazing rate) processes. When they are *tightly coupled* it is unlikely that highly labile DOC will accumulate in the euphotic zone, and most of the labile DOC will be in the semi-labile class. The central equatorial Pacific is an example of such a system. Here, ~70% of the bulk DOC in the upper ~200 m consists principally of refractory components, with the remainder being semi-labile and only a small percentage being highly labile. In contrast, during the North Atlantic Bloom Experiment (the NABE, see Section 8.4.6.1) Kirchman *et al.* (1991) reported that highly labile DOC,

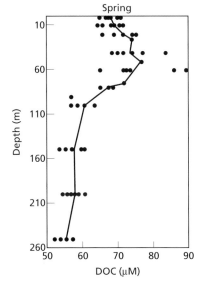
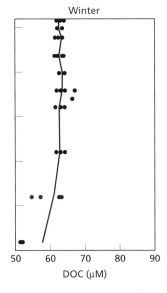

**Fig. 9.8** Seasonal variation in DOC in the oceanic water column (after Carlson *et al.*, 1994). In the oligotrophic Sargasso Sea DOC depth profiles in the upper 250 m are strongly influenced by factors such as convective mixing and vernal stratification. In the spring this gives rise to an accumulation of DOC from an increase in primary production. This accumulated DOC is partly consumed in the summer and autumn, and the remainder is exported in winter, resulting in a homogeneous distribution of DOC throughout the upper water column during this season. The winter export of DOC is thought to be equal to, or greater than, the particle flux in the area.

with a turnover rate of days, could account for a large fraction (~20–40%) of the bulk DOC when production *outstripped* consumption in the spring phytoplankton bloom and the concentrations of DOC were relatively high.

## 9.3 The marine organic carbon cycle

The concentrations of both POC and DOC are higher, and more variable, in surface than in deep waters. Deep and surface waters therefore may be considered to be two distinct organic carbon reservoirs, and the main features in the marine organic carbon cycle can be interpreted in terms of a **surface water → deep water → sediment** global ocean journey, involving fluvial, atmospheric and primary production carbon sources, transport down the water column and incorporation into the sedimentary sink; changes that occur in the sediment sink itself are considered in Sections 14.1–14.4. The data for organic carbon in the oceans are continually being refined, and two data sets relating to the organic carbon journey are given in Table 9.3. The movements of organic carbon between, and within, the reservoirs are discussed below.

It can be seen from the data in Table 9.3 that the photosynthetic fixation of $CO_2$ is by far the most important input of organic carbon to the oceans. At first sight, this may seem surprising because the living biomass itself makes up only a small portion of the total organic carbon in the oceans. However, most of the photosynthetic organic carbon is recycled in the upper water column (the surface reservoir) by grazing within the food web, and for many years it was thought that the organic carbon cycle in the sea followed reasonably simple pathways, which could be summarized as follows.

1 Carbon dioxide is fixed by phytoplankton during primary production.

2 Secondary production, involving the consumption of phytoplankton by zooplankton, etc., and tertiary production move the fixed $CO_2$ along the food web.

3 The oxidative destruction and remineralization of dead organisms at depth restores $CO_2$ and nutrients back into the water column, thus closing the cycle.

In fact, the cycle is not fully closed, and some of the organic carbon escapes to be added to both the DOC and POC oceanic pools. This occurs in two principal ways.

1 Remineralization results in the recycling of much of the phytosynthetically fixed carbon in the upper ocean, and only ~0.2% of primary production ultimately is trapped in marine sediments. Some of the photosynthetic fixed carbon, however, escapes as POC (e.g. faecal material and 'marine snow') during the tissue destruction and remineralization stage and is transported to deep water; it is this fraction (export production), together with escaping DOC, which needs to be replaced by new nutrients (new production—see Section 9.2.2).

2 In addition to the POC that directly escapes the 'nutrient loop' and is transported to depth in the water column, organic material can escape recycling in other ways. For example, there is an additional source to the DOC pool from processes such as exudation by phytoplankton, excretion by zooplankton and post-death decay processes. This source has been estimated to yield ~10% of the total primary production; i.e. ~$3.6 \times 10^{15}$–$5.0 \times 10^{15}\,g\,C\,yr^{-1}$ (see Table 9.3). Further, the remineralization stage in the 'nutrient loop' leaves behind a refractory residual organic material, which has been estimated by Williams (1975) to be ~$1.8 \times 10^{15}\,g\,C\,yr^{-1}$.

Around 95% of the down-column flux of sinking POC is associated with large faecal-type aggregates, although they account for only ~5% of the total POC present (see Section 10.4). It is probable, therefore, that there are at least two fractions of POC in ocean water.

1 A **large-sized fraction**, consisting of faecal material and 'marine snow', some of which can be utilized in the food web, which is responsible for a continuous POC flux from the surface waters to bottom sediments. This down-column POC flux varies both geographically and seasonally, and can be related to annual cycles of primary production in the surface waters—see Section 12.1.1.

2 A **small-sized fraction**, consisting mainly of residual refractory organic matter produced in surface waters, which has a homogeneous distribution in the water column below the euphotic zone.

The composition of the sinking POC has been studied by a number of workers, and it has become apparent that the material that finally reaches the sediment surface bears little relationship to the compounds originally produced via primary production in surface waters (Lee & Wakeham, 1992). For example, it has been shown that <1% of the total

**Table 9.4** Estimates of the annual export POC flux (new production) normalized in the base of the euphotic zone for a series of oceanographic provinces; units, mg C m$^{-2}$ day$^{-1}$.

| Oceanic province | Location* | Water depth (m) | POC export flux (mg C m$^{-2}$ day$^{-1}$) | Data source |
|---|---|---|---|---|
| Upwelling regions | Off California | 100 | 230 | Martin et al. (1987) |
| | Off Namibia | 100 | 63–173 | Wefer & Fisher (1993) |
| | Off West Africa | 100 | 90–96 | Wefer & Fisher (1993) |
| Open-ocean | Equatorial Atlantic | 100 | 16–49 | Wefer & Fisher (1993) |
| | North Atlantic, spring bloom | 150 | 118 | Lochte et al. (1993) |
| | North Atlantic, Sargasso Sea | 150 | 26 | Lohrenz et al. (1992) |
| | Central Pacific, subtropical gyre | 100 | 40 | Martin et al. (1987) |
| | North Pacific, subtropical gyre | 150 | 29 | Karl et al. (1996) |
| | North Pacific, open-ocean composite | 100 | 50 | Martin et al. (1987) |
| | Eastern equatorial Pacific | 80–100 | 59–143 | Murray et al. (1989) |
| | Equatorial Pacific† | 100 | 24–60 | Buesseler et al. (1995) |
| | Equatorial Pacific†: El Niño | 120 | 23 | Bacon et al. (1996) |
| |            Non-El Niño | 120 | 29 | Bacon et al. (1996) |
| | Equatorial Pacific†: El Niño | 120 | 20–76 | Murray et al. (1996) |
| |            Non-El Niño | 120 | 43–234 | Murray et al. (1996) |
| | Equatorial Pacific†: El Niño | 120 | 7.2–16 | Luo et al. (1995) |
| |            Non-El Niño | 100 | 18–60 | Luo et al. (1995) |

* These are general oceanic locations; for details of how POC export fluxes vary in a Pacific region (the equatorial Pacific) see Fig. 9.9 and text.
† Indicates that POC export fluxes were estimated using Th isotopes.

sterols and fatty acids (Gagosian *et al.*, 1982) and <10% of the amino acids (Lee & Cronin, 1982) produced in the euphotic zone reach the sea bed.

The POC (export production flux) escaping the surface waters undergoes considerable degradation as it continues to sink, and as a result estimates of POC down-column flux rates are strongly dependent on water depth. To overcome this, a number of workers have used flux–depth formulations of the type devised by Martin *et al.* (1987) to normalize their POC flux data to specific depths in the water column; e.g. ~100 m for export from the euphotic zone ('new' production), and ~4000 m for export to the deep ocean interior. A summary of various data for the export flux of POC from the euphotic zone is given in Table 9.4, from which it can be seen that the highest POC fluxes are found in areas of upwelling and the lowest in the central oceanic gyres. The relationship between the POC export flux and new production, however, is not always clear cut. Although the sinking of POC from the surface layer has been thought to be equivalent to the new production arising from nutrients that make up the loss which

escapes the 'biological loop', in some instances there can be a decoupling between primary production and export production, which affects the efficiency at which the 'biological pump' operates (see e.g. Karl *et al.*, 1996). For example, according to Luo *et al.* (1995) export production at the base of the euphotic zone in the equatorial Pacific represents only a small fraction of the new production, indicating that much of the organic matter must be removed from the euphotic zone in the form of DOC (see also, Bacon *et al.*, 1996). Murray *et al.* (1996) also considered the relationship between new and export production in the central equatorial Pacific and found it to be somewhat complicated. The authors compared data for (i) new production and (ii) the POC export flux at the base of the euphotic layer (120 m), at a series of stations on a 140°W transect. Data for two surveys (Survey I, El Niño and Survey II, non-El Niño) are illustrated in Fig. 9.9, which also indicates how the POC export flux varies with location in the equatorial Pacific. Both new and export production were higher during Survey II (non-El Niño) than Survey I (El Niño), favouring the hypothesis that the fluxes would

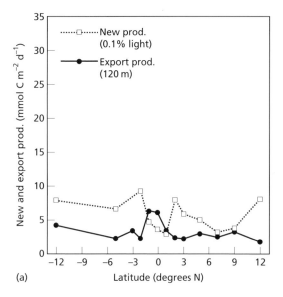

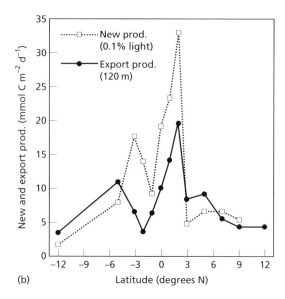

**Fig. 9.9** Comparison between new production and POC export in the equatorial Pacific (from Murray *et al.*, 1996). (a) Survey I, El Niño; (b) Survey II, non-El Niño. New production values are for the 0.1% light-level depth and are expressed in carbon units calculated assuming a carbon : nitrogen fixation ratio of 6.6. The uncertainty for the new production is assumed to be 30%. POC fluxes are for a depth of 120 m.

be lower during warmer El Niño periods because temperature, nutrient and mixing conditions force a shift in the biological community towards smaller phytoplankton. During Survey I, new production was higher than export production and the difference was 'highly significant'. During Survey II, however, new production was higher than export production on a 'highly significant' basis only over the region from 2°N to 3°S. Overall, however, in both surveys the POC export flux was lower than the new production, and the authors suggested that the new production not exported in a particulate (POC) form is potentially exported in a dissolved form (DOC), with the difference between new production and POC export flux being the 'potential DOC export flux'. The highest potential DOC fluxes for Survey I were in off-equatorial zones, with a significant potential DOC production in the equatorial zone from 2°N to 3°S in Survey II, the latter being consistent with the hypothesis that there is DOC export from the equatorial zone of the central Pacific. Despite this, annual average rates of new production (0.47 Gt C yr$^{-1}$) and POC export production (0.42 Gt C yr$^{-1}$)

were in good agreement on a regional equatorial Pacific scale (10°N–10°S: 90°W–180°E) scale. This suggests that on an annual basis, significant export of DOC need not be invoked to balance new and export production in the equatorial Pacific.

Using data obtained mainly from sediment traps it is generally accepted now that the flux of POC sinking through the upper ~100 m of the water column is in the order of ~$6 \times 10^{15}$ g C yr$^{-1}$, which accounts for ~12% of the primary production in the surface layers (see e.g. Martin *et al.*, 1987). This flux decreases as the POC continues to sink, and at a depth of ~4000 m it falls to ~1% of the surface primary production. For example, Martin *et al.* (1987) carried out a study as part of the VERTEX programme in the northeast Pacific. The authors estimated that around 50% of the carbon leaving the surface waters is rapidly regenerated in the upper 300 m of the water column, ~75% is regenerated by 500 m and ~90% by 1500 m. On this basis, therefore, only a few per cent of the POC escaping the euphotic zone reaches the sediment surface.

The down-column transport of the export POC that escapes the 'nutrient loop' is an extremely important oceanic process and drives the '**global carbon flux**', which is the principal mechanism for the transport of material from the surface layer to deep water and the sediment sink; this process is considered in detail in Section 10.4. With respect to the oceanic organic carbon budget, however, it should be stressed that only a very small fraction of the organic matter actually reaching the sea bed is ultimately stored in the sediments. For example, Santos *et al.* (1994) gave data on the organic carbon budget in sediments from the Porcupine Abyssal plain in the northeast Atlantic Ocean. Combining their data with those from other workers, they determined the following organic carbon fluxes in the region:

**1** the deep-water (3500 m) annual carbon flux was ~1.0 g C m$^{-2}$ yr$^{-1}$, which corresponded to a surface productivity of ~70 g C m$^{-2}$ yr$^{-1}$;

**2** the burial flux of organic carbon in the sediments was ~0.12 g C m$^{-2}$ yr$^{-1}$. Thus the burial flux of organic carbon into the sediments was ~0.2% of the surface production, with ~10% of the down-column flux actually reaching the sediment.

Santos *et al.* (1994) reported that initially the down-column flux material arrives fairly evenly distributed over the sea bed, but then within a matter of hours it can become highly 'patchy' as it is redistributed by near-bottom currents to become concentrated in depressions and grooves in the sediment surface. Smith *et al.* (1996) gave data on the deposition of fresh phytoplankton detritus (**phytodetritus**) on the sea floor in the central equatorial Pacific following seasonal phytoplankton blooms. Greenish flocculent material ('**fluff**'—see Section 12.2) formed a layer, up to ~5 mm thick in places, covering ~90% of the sea floor between ~5°S and 5°N of the Equator, which was grazed by bottom-feeding organisms, such as holothurians. The 'fluff' contained 1–12.5% organic carbon, which was between 5 and 39 times richer than the bottom sediments, and had microbial activity rates that were five times higher than those in the sediments. In addition, $^{234}$Th activities indicated that the green 'fluff' had been transported down the water column within the previous 100 days. The authors estimated that between ~5°S and 5°N of the Equator the standing stock of the phytodetritus made up ~3% of the annual flux of organic carbon to the sea bed.

The magnitude of the POC flux to the benthic layer varies considerably in both time and space, and it had been assumed that the coupling of the deep-sea floor environment to surface waters is the result of the rapid transport of POC through the water column. However, a number of difficulties have become apparent when attempts are made to reconcile the POC flux with its mineralization to $CO_2$ in the bottom sediments, as measured by sediment commu-

nity oxygen consumption (SCOC) rates. According to Smith *et al.* (1992) POC fluxes in the North Atlantic are sufficient to meet the SCOC. In contrast, in the North Pacific it was thought initially that the supply of POC into the benthic layer fell short of meeting the SCOC by as much as 97%. Smith *et al.* (1992), however, reconciled the POC flux and the SCOC at a station in the North Pacific by including previously undetected *episodic* inputs of POC to the benthic layer. The overall question of the manner in which POC fluxes to the benthic layer and sediment oxygen consumption are related still remains unresolved. It was thought originally that variations in POC fluxes to the sea floor are accompanied by temporal variability in the sediment oxygen demand, but this was questioned by Sayles *et al.* (1994), who reported that in the oligotrophic Atlantic sediment oxygen consumption does not vary significantly despite large seasonal variations in POC fluxes to the benthic layer. According to these authors, this implies that large areas of the sea floor may be characterized by seasonally *invariant* sediment oxygen demand.

It may be concluded, therefore, that only a small fraction of the carbon phytosynthetically fixed in the surface layers escapes from the euphotic zone. However, it can be seen from the data in Table 9.3 that there are several problems associated with the **surface water → deep water → sediment** organic carbon journey, and therefore with any attempts to close the oceanic organic carbon budget. A number of these problems were highlighted by Hedges (1992).

**1** The carbon burial rate in marine sediments (~0.1 × 10$^{15}$ g C yr$^{-1}$) is very low compared with the magnitude of the oceanic input fluxes. Most of the oceanic input arises from internal primary production (~50 × 10$^{15}$ g C yr$^{-1}$), and this input exceeds carbon burial in sediments by a factor of about 500. This is not unexpected because the plankton are composed mainly of labile components, which are recycled, and only the material that escapes the 'nutrient loop' is transported out of the upper waters. Even if the input from primary production is ignored, however, the total external inputs (fluvial and atmospheric) of DOC (~0.5 × 10$^{15}$ g C yr$^{-1}$) still exceed the sediment burial rate by a factor of five. Overall, therefore, even if all the DOC pool was to be remineralized, less than ~0.2% of primary production and only ~20% of the terrestrial inputs would be preserved in sediments.

**2** The flux of DOC out of the oceans (the turnover rate of $\sim0.1 \times 10^{15}\,g\,C\,yr^{-1}$) alone is similar to the sediment burial rate of $(\sim0.1 \times 10^{15}\,g\,C\,yr^{-1})$. However, fluvial DOC is imported to the oceans at a rate $(\sim0.2 \times 10^{15}\,g\,C\,yr^{-1})$ that is around two times higher than the DOC turnover rate. Further, there is little evidence that most of the organic matter in sediments is derived from the DOC pool.

**3** It would appear, therefore, that DOC is destroyed (consumed) somewhere in the oceans. In this context, Smith & Mackenzie (1987) proposed that on a global-scale the oceans are heterotrophic; i.e. they consume more organic carbon than they produce. The processes that consume the DOC are not clear, but Hedges (1992) has suggested that photolysis may play an important role in removing DOM from sea water; e.g. by the photoproduction of LMW compounds from DOC, which can be utilized rapidly by bacteria.

*To summarize.* The principal feature in the oceanic organic carbon cycle is the phytosynthetic production of carbon in the surface waters and its recycling at relatively shallow depths. This biologically mediated loop in the recycling of organic carbon in the upper water column is related intimately to the fate of nutrient elements and involves the internal oceanic formation, and subsequent destruction, of POC. However, the biological loop is not 100% efficient and a small fraction, perhaps $\sim5\%$, of the POC formed in surface layers escapes to deep water as large-sized faecal material and 'marine snow'. The large-sized organic aggregates are the principal driving force behind the **global carbon flux**. They also act as carrier phases for trace elements that have been removed from solution, and their down-column transport provides a mechanism for the delivery of particulate-associated constituents to the surface of the sediment reservoir. This mechanism operates on an ocean-wide scale, and it is now recognized that the global carbon flux is one of the major processes controlling both the throughput of material in the ocean system and the trace-element composition of sea water.

### 9.4 Organic matter in the oceans: summary

**1** Organic carbon in the oceans consists of a particulate phase (POC), which makes up $\sim5\%$ of the total organic carbon (TOC), and a dissolved phase (DOC), which constitutes $\sim95\%$ of the TOC.

**2** The POC and DOC pools receive contributions from both internal (autochthonous) and external (allochthonous) sources. However, there is considerable dispute over the origin of the DOC in the oceanic pool.

**3** The principal feature in the oceanic distributions of POC and DOC is that the concentrations of both components are higher, and more variable, in surface than in deep waters.

**4** There is an organic carbon cycle in the oceans. Essentially, this involves the recycling of phytoplankton-fixed organic carbon within the biological system, mainly in the euphotic layer, from which some POC and DOC escapes by a variety of biological and non-biological mechanisms. The POC that is exported from the euphotic layer makes up $\sim5\%$ of the photosynthetically fixed carbon, and is transported down the water column as the global carbon flux.

**5** As the POC of the global carbon flux sinks down the water column it undergoes further degradation, and $\sim90\%$ of the material leaving the surface waters is regenerated by the time it reaches the sediment surface.

**6** Degradation of the POC continues in the sedimentary environment, and less than $\sim10\%$ of the POC flux reaching the sea bed is ultimately buried in the sediments.

Particulate matter plays a critical role in marine geochemistry, in removing and transporting dissolved elements, especially trace elements, via the global carbon flux. The total suspended matter (TSM) in sea water does not consist only of organic material, however, and before reviewing the factors that control the distribution of trace elements in sea water, the overall nature of this TSM will be described in the following chapter.

### References

Amon, R.M.W. & Benner, R. (1994) Rapid cycling of high-molecular-weight dissolved organic matter in the oceans. *Nature*, **369**, 549–52.

Armstrong, F.A.J. (1965) Silicon. In *Chemical Oceanography*, 1st edn, J.P. & G. Skirrow (eds), Vol. 1, 409–32. London: Academic Press.

Bacon, M.P., Cochran, J.K., Hirschberg, D., Hammar, T.R. & Fleer, A.P. (1996) Export flux of carbon at the equator during the EqPac time-series cruises estimated from [234]Th measurements. *Deep Sea Res.*, 43, 1133–53.

Barber, R.T. & Chavez, F.P. (1991) Regulation of primary productivity rate in the equatorial Pacific. In *What Controls Phytoplankton Production in Nutrient-rich Areas of the Open Sea?* S.W. Chisholm & F.M.M. Morel (eds). *Limnol. Oceanogr.*, 36, 1803–15.

Blumer, M. (1970) Dissolved organic compounds in sea water: saturated and olefinic hydrocarbons and singly branched fatty acids. In *Organic Matter in Natural Waters*, D.W. Hood (ed.), 153–67. Institute of Marine Science, Alaska, Occasional Publication No. 1.

Broecker, W.S. (1974) *Chemical Oceanography*. New York: Harcourt Brace Jovanovich.

Broecker, W.S. & Peng, T.H. (1982) *Tracers in the Sea*. Palisades, NY: Lamont-Doherty Geological Observatory.

Bruland, K.W. (1980) Oceanographic distributions of cadmium, zinc, nickel and copper in the North Pacific. *Earth Planet. Sci. Lett.*, 47, 176–98.

Buesseler, K.O., Andrews, J.A., Kartman, M.C., Belastock, R. & Chai, F. (1995) Regional estimates of the export flux of particulate organic carbon from thorium-234 during the JGOFS EqPac Program. *Deep Sea Res.*, 42, 777–804.

Bushaw, K.L., Zepp, R.G., Tarr, M.A., *et al.* (1996) Photochemical release of biologically available nitrogen from aquatic dissolved organic matter. *Nature*, 381, 404–7.

Campbell, J.W. & Aarup, T. (1992) New production in the North Atlantic derived from seasonal patterns of surface chlorophyll. *Deep Sea Res.*, 39, 1669–94.

Carlson, D.J. & Ducklow, H.W. (1995) Dissolved organic carbon in the upper ocean of the central equatorial Pacific Ocean, 1992: daily and finescale vertical variations. *Deep Sea Res.*, 42, 639–50.

Carlson, D.J., Brann, M.L., Mague, T.H. & Mayer, M.L. (1985) Molecular weight distribution of dissolved organic materials in seawater determined by ultrafiltration: a re-examination. *Mar. Chem.*, 16, 155–71.

Carlson, D.J., Ducklow, H.W. & Michaels, A.F. (1994) Annual flux of dissolved organic carbon from the euphotic zone in the northwestern Sargasso Sea. *Nature*, 371, 405–408.

Cawet, G. (1981) Non-living particulate matter. In *Marine Organic Chemistry*, E.K. Duursma & R. Dawson (eds), 71–89. Amsterdam: Elsevier.

Chavez, F.P. & Barber, R.T. (1987) An estimate of new production in the equatorial Pacific. *Deep Sea Res.*, 34, 1229–43.

Chavez, F.P., Buck, K.R. & Coale, K.H. (1991) In *What Controls Phytoplankton Production in Nutrient-rich Areas of the Open Sea?* S.W. Chisholm & F.M.M. Morel (eds). *Limnol. Oceanogr.*, 36, 1816–33.

Chavez, F.P., Buck, K.R., Service, S.K., Newton, J. & Barber, R.T. (1996) Phytoplankton variability in the central and eastern tropical Pacific. *Deep Sea Res.*, 43, 835–70.

Chisholm, S.W. & Morel, F.M.M. (eds) (1991a) *What Controls Phytoplankton Production in Nutrient-rich Areas of the Open Sea? Limnol. Oceanogr.*, 36.

Chisholm, S.W. & Morel, F.M.M. (1991b) Preface. In *What Controls Phytoplankton Production in Nutrient-rich Areas of the Open Sea?* S.W. Chisholm & F.M.M. Morel (eds). *Limnol. Oceanogr.*, 36.

Chrost, R.H. & Faust, M.A. (1983) Organic carbon release by phytoplankton: its composition and utilisation by bacterioplankton. *J. Plankton Res.*, 5, 477–93.

Coale, K.H., Fitzwater, S.E., Gordon, R.M., Johnson, K.S. & Barber, R.T. (1996a) Control of community growth and export production by upwelled iron in the equatorial Pacific Ocean. *Nature*, 379, 621–4.

Coale, K.H., Johnson, K.S., Fitzwater, S.E., *et al.* (1996b) A massive phytoplankton bloom induced by an ecosystem-scale iron fertilisation experiment in the equatorial Pacific Ocean. *Nature*, 383, 495–501.

Cooper, D.J., Watson, A.J. & Nightingale, P.D. (1996) Large decrease in ocean-surface $CO_2$ fugacity in response to *in situ* iron fertilisation. *Nature*, 383, 511–13.

Cullen, J.J. (1991) Hypothesis to explain high-nutrient conditions in the open sea. In *What Controls Phytoplankton Production in Nutrient-rich Areas of the Open Sea?* S.W. Chisholm & F.M.M. Morel (eds). *Limnol. Oceanogr.*, 36, 1578–99.

Cullen, J.J., Lewis, M.R., Davies, C.O. & Barber, R.T. (1992) Photosynthetic characteristics and estimated growth rates indicate grazing is the proximate control of primary production in the equatorial Pacific. *J. Geophys. Res.*, 971, 639–54.

De Baar, H.J.W., de Jong, J.T.M., Bakker, D.C.E., *et al.* (1995) Importance of iron for plankton blooms and carbon dioxide drawdown in the Southern Ocean. *Nature*, 373, 412–15.

Druffel, E.R.M., Williams, P.M. & Suzuki, Y. (1989) Concentrations and radiocarbon signatures of dissolved organic matter in the Pacific Ocean. *Geophys. Res. Lett.*, 16, 991–4.

Druffel, E.R.M., Williams, P.M., Bauer, J.E. & Ertel, J.R. (1992) Cycling of dissolved and particulate organic matter in the open ocean. *J. Geophys. Res.*, 97, 15639–59.

Duce, R.A. (1986) The impact of atmospheric nitrogen, phosphorous, and iron species on marine biological productivity. In *The Role of Air–Sea Exchange in Geochemical Cycling*, P. Buat-Menard (ed.), 479–529. Dordrecht: Reidel.

Duce, R.A. & Tindale, N.W. (1991) Chemistry and biology of iron and other trace metals. *Limnol. Oceanogr.*, 36, 1715–26.

Dugdale, R.C. & Goering, J.J. (1967) Uptake of new and regenerated forms of nitrogen in primary productivity. *Limnol. Oceanogr.*, 12, 196–206.

Dugdale, R.C. & Wilkerson, F.P. (1992) Nutrient limitation of new production. In *Primary Productivity and Biogeochemical Cycles in the Sea*, P.G. Falkowski & A.D. Woodhead (eds), 107–22. New York: Plenum Press.

Dugdale, R.C., Wilkerson, F.P. & Minas, H.J. (1995) The role of a silicate pump in driving new production. *Deep Sea Res.*, **42**, 697–19.

Eadie, B.J., Jeffrey, L.M. & Sackett, W.M. (1978) Some observations on the stable carbon isotope composition of dissolved and particulate organic carbon in the marine environment. *Geochim. Cosmochim. Acta*, **42**, 1265–9.

Eglinton, G. & Hamilton, R.J. (1963) The distribution of alkanes. In *Chemical Plant Taxonomy*, T. Swain (ed.), 187–218. New York: Academic Press.

Eppley, R.W. & Peterson, B.J. (1979) Particulate organic matter flux and planktonic new production in the deep ocean. *Nature*, **282**, 677–80.

Ehrardt, M., Osterroht, C. & Petrick, G. (1980) Fatty-acid methyl esters dissolved in seawater and associated with suspended particulate material. *Mar. Chem.*, **10**, 67–76.

Fellows, D.A., Karl, D.M. & Knauer, G.A. (1981) Large particle fluxes and the vertical transport of living carbon in the upper 1500 m of the northeast Pacific Ocean. *Deep Sea Res.*, **28**, 921–36.

Fenical, W. (1981) Natural halogenated organics. In *Marine Organic Chemistry*, E.K. Duursma & R. Dawson (eds), 375–93. Amsterdam: Elsevier.

Finden, D.A.S., Tipping, E., Jaworski, G.H.M. & Renolds, C.S. (1984) Light-induced reduction of natural iron (III) oxide and its relevance to phytoplankton. *Nature*, **309**, 783–4.

Frost, B.W. (1991) The role of grazing in nutrient-rich areas of the open sea. In *What Controls Phytoplankton Production in Nutrient-rich Areas of the Open Sea?* S.W. Chisholm & F.M.M. Morel (eds). *Limnol. Oceanogr.*, **36**, 1616–30.

Fuhrman, J.A. & Capone, D.G. (1991) Possible biological consequences of ocean fertilisation. In *What Controls Phytoplankton Production in Nutrient-rich Areas of the Open Sea?* S.W. Chisholm & F.M.M. Morel (eds). *Limnol. Oceanogr.*, **36**, 1951–9.

Gagosian, R.B., Smith, S.O. & Nigrelli, G.E. (1982) Vertical transport of steroid alcohols and ketones measured in a sediment trap experiment in the equatorial Atlantic Ocean. *Geochim. Cosmochim. Acta*, **46**, 1163–72.

Hedges, J.I. (1978) The formation and clay mineral reactions of melanoidins. *Geochim. Cosmochim. Acta*, **42**, 69–76.

Hedges, J.I. (1992) Global biogeochemical cycles: progress and problems. *Mar. Chem.*, **39**, 67–93.

Hedges, J.I. & Keil, R.G. (1995) Sedimentary organic matter preservation: an assessment and speculative synthesis. *Mar. Chem.*, **49**, 81–115.

Hudson, R.J.M. & Morel, F.M.M. (1990) Iron transport in marine phytoplankton: kinetics of cellular and medium coordination reactions. *Limnol. Oceanogr.*, **35**, 1002–20.

Jackson, G.A. (1988) Implications of high dissolved organic matter concentrations for oceanic properties and processes. *Oceanography*, November, 621–3.

Jackson, G.A. & Williams, P.M. (1985) Importance of dissolved organic nitrogen and phosphorus to biological nutrient cycling. *Deep Sea Res.*, **32**, 223–35.

Jenkins, W.J. & Goldman, J.C. (1985) Seasonal oxygen cycling and primary production in the Sargasso Sea. *J. Mar. Res.*, **43**, 465–91.

Jickells, T.D., Blackburn, T.H., Blanton, J.O., *et al.* (1991) What determines the fate of materials within ocean margins? In *Ocean Margin Processes in Global Change*, R.F.C. Mantoura, J.-M. Martin & R. Wollost (eds). Chichester: J. Wiley & Sons.

Johnson, K.S., Coale, K.H., Elrod, V.A. & Tindale, N.W. (1994) Iron photochemistry in seawater from the equatorial Pacific. *Mar. Chem.*, **46**, 319–34.

Kamykowski, D. & Zentara, S.J. (1986) Predicting plant nutrient concentrations from temperature and sigma-*t* in the upper kilometer of the world ocean. *Deep Sea Res.*, **33**, 89–105.

Karl, D.M., Christian, J.R., Dore, J.E., *et al.* (1996) Seasonal and interannual variability in primary production and particle flux at station ALOHA. *Deep Sea Res.*, **43**, 539–68.

Kattner, G.G., Gercken, G. & Hammer, K.D. (1983) Development of lipids during a spring plankton bloom in the northern North Sea. *Mar. Chem.*, **14**, 163–73.

Kawase, M. & Sarmiento, J.L. (1985) Nutrients in the Atlantic thermocline. *J. Geophys. Res.*, **90**, 8961–79.

Kennicutt, M.C. & Jeffrey, L.M. (1981) Chemical and GC-MS characterisation of marine dissolved lipids. *Mar. Chem.*, **19**, 367–87.

Kieber, D.J., McDaniel, J. & Mopper, K. (1989) Photochemical source of biological substances in seawater: implications for carbon cycling. *Nature*, **341**, 637–9.

Kirchman, D.L., Suzuki, Y., Garside, C. & Ducklow, H.W. (1991) High turnover rates of dissolved organic carbon during a spring phytoplankton bloom. *Nature*, **352**, 612–14.

Koblentz-Mishke, O.J., Volkovinsky, V.V. & Kabanova, Y.G. (1970) Plankton primary production in the world ocean. In *Scientific Exploration of the South Pacific*, W.S. Wooster (ed.), 183–93. Washington, DC: National Academy of Science.

Krom, M.D., Brenner, S., Kress, N. & Gordon, L.I. (1991) Phosphorus limitation of primary productivity in the Eastern Mediterranean Sea. *Limnol. Oceanogr.*, **36**, 424–32.

Kumar, N., Anderson, R.F., Mortlock, R.A., *et al.* (1995) Increased biological productivity and export production in the glacial Southern Ocean. *Nature*, **378**, 675–80.

Lalli, C. & Parsons, T.R. (1993) *Biological Oceanography*. Oxford: Pergamon Press.

Lee, C. & Bada, J.L. (1977) Dissolved amino acids in the equatorial Pacific, Sargasso Sea and Biscayne Bay. *Limnol. Oceanogr.*, **22**, 502–10.

Lee, C. & Cronin, C. (1982) The vertical flux of particulate organic nitrogen in the sea: decomposition of amino acids in the Peru upwelling area and equatorial Atlantic. *J. Mar. Res.*, **40**, 227–51.

Lee, C. & Henrichs, S.M. (1993) How the nature of dissolved organic matter might affect the analysis of dissolved organic carbon. *Mar. Chem.*, **41**, 105–120.

Lee, C. & Wakeham, S.G. (1992) Organic matter in the water column: future research challenges. *Mar. Chem.*, **39**, 95–118.

Liebeziet, G., Bolter, M., Brown, I.F. & Dawson, R. (1980) Dissolved free amino acids and carbohydrates at pycnocline boundaries in the Sargasso Sea and related microbial activity. *Oceanol. Acta*, **3**, 357–62.

Lochte, K., Ducklow, H.W., Fasham, M.J.R. & Stienen, C. (1993) Plankton succession and carbon cycling at 47°N 20°W during the JGOFS North Atlantic Bloom Experiment. *Deep Sea Res.*, **40**, 91–114.

Lohrenz, S.E., Knauer, G.A., Asper, V.L., Tuel, M., Michaels, A.F. & Knapp, A.H. (1992) Seasonal variability in primary production and particle flux in the northwestern Sargasso Sea: U.S. JGOFS Bermuda Atlantic Time-series Study. *Deep Sea Res.*, **39**, 1373–91.

Luo, S., Ku, T.-L., Kusakabe, M., James, K., Bishop, B. & Yang, J.-L. (1995) Tracing particle cycling in the upper ocean with $^{230}$Th: an investigation in the equatorial Pacific along 140°W. *Deep Sea Res.*, **42**, 805–29.

Mantoura, R.F.C. & Woodward, E.M.S. (1983) Conservative behaviour of riverine dissolved organic carbon in the Severn Estuary: chemical and geochemical implications. *Geochim. Cosmochim. Acta*, **47**, 1293–309.

Martin, J.H. (1990) Glacial–interglacial $CO_2$ change: the iron hypothesis. *Paleoceanography*, **5**, 1–13.

Martin, J.H. (1992) Iron as a limiting factor in oceanic productivity. In *Primary Productivity and Biogeochemical Cycles in the Sea*, P.G. Falkowski & A.D. Woodhead (eds), 123–37. New York: Plenum Press.

Martin, J.H. & Gordon, R.M. (1988) Northeast Pacific iron distributions in relation to phytoplankton productivity. *Deep Sea Res.*, **35**, 177–96.

Martin, J.H., Knauer, G.A., Karl, D.M. & Broenkow, W.W. (1987) VERTEX: carbon cycling in the northeast Pacific. *Deep Sea Res.*, **34**, 267–85.

Martin, J.H., Coale, K.H., Johnson, K.S., *et al.* (1994) Testing the iron hypothesis in ecosystems of the equatorial Pacific Ocean. *Nature*, **371**, 123–9.

Meyers-Schulte, K.C. & Hedges, J.I. (1986) Molecular evidence for a terrestrial component of organic matter dissolved in ocean water. *Nature*, **321**, 61–3.

Miller, C.B., Frost, B.W., Wheller, P.A., Landry, M.R., Welschmeyer, N. & Powell, T.M. (1991) Ecological dynamics in the subarctic Pacific, a possible iron-limited ecosystem. In *What Controls Phytoplankton Production in Nutrient-rich Areas of the Open Sea?* S.W. Chisholm & F.M.M. Morel (eds). *Limnol. Oceanogr.*, **36**, 1600–15.

Minas, J.M. & Minas, M. (1992) Net community production in 'high nutrient–low chlorophyll' waters of the tropical and Antarctic Oceans: grazing versus iron hypothesis. *Oceanol. Acta*, **15**, 145–62.

Mitchell, G.B., Brody, E.A., Holm-Hanson, O., McClain, C. & Bishop, J. (1991) Light limitation of phytoplankton biomass and macronutrient utilisation in the Southern Ocean. In *What Controls Phytoplankton Production in Nutrient-rich Areas of the Open Sea?* S.W. Chisholm & F.M.M. Morel (eds). *Limnol. Oceanogr.*, **36**, 1662–77.

Moore, R.M., Milley, J.E. & Chatt, A. (1984) The potential for biological mobilisation of trace elements from aeolian dust in the oceans and its importance in the case of iron. *Oceanol. Acta*, **7**, 221–8.

Mopper, K., Zhou, X., Kieber, R.J., Kieber, D.J., Sikorski, R.J. & Jones, R.D. (1991) Photochemical degradation of dissolved organic carbon and its impact on the oceanic carbon cycle. *Nature*, **353**, 60–2.

Murray, J.W., Downs, J.N., Strom, S., Wei, C.-L. & Jannasch, H.W. (1989) Nutrient assimilation, export production and $^{234}$Th scavenging in the eastern equatorial Pacific. *Deep Sea Res.*, **36**, 1471–89.

Murray, J.W., Young, J., Newton, J., *et al.* (1996) Export flux of particulate organic carbon from the central equatorial Pacific determined using a combined drifting trap $^{234}$Th approach. *Deep Sea Res.*, **43**, 1095–1132.

Neinhuis, P.H. (1981) Distribution of organic matter in living marine organisms. In *Marine Organic Chemistry*, E.K. Duursma & R. Dawson (eds), 31–69. Amsterdam: Elsevier.

Olson, J.S., Garrels, R.M., Berner, R.A., Armentano, T.V., Dyer, M.I. & Taalon, D.H. (1985) The natural carbon cycle. In *Atmospheric Carbon Dioxide and the Global Carbon Cycle*, J.R. Trabalka (ed.), 175–213. Washington, DC: U.S. Department of Energy.

Owens, N.J.P., Rees, A.P., Woodward, E.M.S. & Mantoura, R.F.C. (1989) Size-fractionated primary production and nitrogen assimilation in the northwest Mediterranean Sea during January 1989. *Water Pollut. Rep.*, **13**, 126–35.

Parrish, C.C. & Wangersky, P.J. (1988) Iatroscan-measured profiles of dissolved and particulate marine lipids classes over the Scotian Slope in the Bedford Basin. *Mar. Chem.*, **23**, 1–15.

Parsons, T.R. (1975) Particulate organic carbon in the sea. In *Chemical Oceanography*, J.P. Riley & G. Skirrow (eds), Vol. 2, 365–83. London: Academic Press.

Peltzer, E.T. & Gagosian, R.B. (1989) Oceanic geochemistry of aerosols over the Pacific Ocean. In *Chemical Oceanography*, J.P. Riley & R. Chester (eds), Vol. 10, 282–338. London, Academic Press.

Peltzer, E.T. & Hayward, N.A. (1996) Spatial and temporal variability of total organic carbon along 140°W in the equatorial Pacific in 1992. *Deep Sea Res.*, **43**, 1155–80.

Platt, T., Sathyendranth, S., Ulloa, O., Harrison, W.G., Hoepffner, N. & Goes, J. (1992) Nutrient control of phytoplankton photosynthesis in the Western North Atlantic. *Nature*, **356**, 229–31.

Redfield, A.C. (1934) On the proportion of organic derivatives in sea water and their relation to the composition of plankton. In *James Johnstone Memorial Volume*, 177–92. Liverpool: Liverpool University Press.

Redfield, A.C. (1958) The biological control of chemical factors in the environment. *Am. J. Sci.*, **46**, 205–21.

Reid, T.R., Live, D.H., Faulkner, D.J. & Butler, A. (1993) A siderophore from a marine bacterium with an exceptional ferric ion affinity constant. *Nature*, **366**, 455–8.

Saliot, A. (1981) Natural hydrocarbons in sea water. In *Marine Organic Chemistry*, E.K. Duursma & R. Dawson (eds), 327–74. Amsterdam: Elsevier.

Sambrotto, R.N., Savidge, G., Robinson, C., *et al.* (1993) Elevated consumption of carbon relative to nitrogen in the surface ocean. *Nature*, **363**, 248–50.

Santos, V., Billett, D.S.M., Rice, A.L. & Wolff, G.A. (1994) Organic matter in deep-sea sediments from the Porcupine Abyssal Plain in the north-east Atlantic Ocean. I—Lipids. *Deep Sea Res.*, **41**, 787–819.

Santschi, P.H., Guo, L., Baskaran, M., *et al.* (1995) Isotopic evidence for the contemporary origin of high-molecular weight organic matter in oceanic environments. *Geochim. Cosmochim. Acta*, **59**, 625–31.

Sathyendranath, S., Longhurst, A., Caverhill, C.M. & Platt, T. (1995) Regionally and seasonally differentiated primary production in the North Atlantic. *Deep Sea Res.*, **42**, 1773–1802.

Sayles, F.L., Martin, W.R. & Deuser, W.G. (1994) Response of benthic oxygen demand to particulate organic carbon supply in the deep sea near Bermuda. *Nature*, **371**, 686–9.

Sharp, J.H. (1983) The distribution of inorganic nitrogen and dissolved and particulate organic nitrogen in the sea. In *Nitrogen in the Marine Environment*, E.J. Carpenter & D.G. Capone (eds), 1–35. New York: Academic Press.

Sharp, J.H., Benner, R., Bennett, L., *et al.* (1995) Analyses of dissolved organic carbon in seawater: the JGOFS EqPac methods comparison. *Mar. Chem.*, **48**, 91–108.

Smith, C.R., Hoover, D.J., Doan, S.E., *et al.* (1996) *Deep Sea Res.*, **43**, 1309–38.

Smith, K.L., Baldwin, R.J. & Williams, P.M. (1992) Reconciling particulate organic carbon flux and sediment community oxygen consumption in the deep North Pacific. *Nature*, **359**, 313–16.

Smith, S.V. & Mackenzie, F.T. (1987) The ocean as a net heterotrophic system: implications from the carbon biogeochemical cycle. *Global Biogeochem. Cycles*, **1**, 187–98.

Strickland, J.D.H. (1965) Production of organic matter in the primary stage of the marine food chain. In *Chemical Oceanography*, 1st edn, J.P. Riley & G. Skirrow (eds), Vol. 1, 477–610. London: Academic Press.

Sugimura, Y. & Suzuki, Y. (1988) A high temperature catalytic oxidation method for the determination of non-volatile dissolved organic carbon in seawater by direct injection of a liquid sample. *Mar. Chem.*, **24**, 105–31.

Sugugawa, H., Handa, N. & Ohta, K. (1985) Isolation and characterisation of low molecular weight carbohydrates in seawater. *Mar. Chem.*, **17**, 341–62.

Summerhayes, C.P., Prell, W.L. & Emeis, K.C. (eds) (1992) *Upwelling Systems: Evolution since the Early Miocene.* Bath: Geological Society of London, Special Publication No. 64.

Sunda, W.G., Swift, D.G. & Huntsman, S.A. (1991) Low iron requirement in oceanic phytoplankton. *Nature*, **351**, 55–7.

Sverdrup, H.U., Johnson, M.W. & Fleming, R.H. (1942) *The Oceans.* New York: Prentice Hall.

Suzuki, Y. (1993) On the measurement of DOC and DON in seawater. *Mar. Chem.*, **41**, 287–8.

Suzuki, Y., Sugimura, Y. & Itoh, T. (1985) A catalytic oxidation method for the determination of total nitrogen dissolved in seawater. *Mar. Chem.*, **16**, 83–97.

Takahasi, T., Broecker, W.S. & Langer, S. (1985) Redfield ratio based on chemical data from isopycnal surfaces. *J. Geophys. Res.*, **90**, 6907–24.

Townsend, D.W., Keller, M.D., Sieracki, M.E. & Ackieson, S.G. (1992) Spring phytoplankton blooms in the absence of vertical water column stratification. *Nature*, **360**, 59–62.

Traganza, E.D., Silva, V.M., Austin, D.M., Hanson, W.L. & Bronsink, S.H. (1983) Nutrient mapping and recurrence of coastal upwelling by satellite remote sensing: its implication to primary production and the sediment record. In *Coastal Upwelling*, E. Suess & J. Thiede (eds), 61–83. New York: Plenum Press.

Tsuchiya, M. (1985) The subthermocline phosphate distribution and circulation in the far eastern equatorial Pacific Ocean. *Deep Sea Res.*, **32**, 299–313.

Turner, S.M., Nightingale, P.D., Spokes, L.J., Liddicoat, M.I. & Liss, P.S. (1996) Increased dimethyl sulphide concentrations in sea water from *in situ* iron enrichment. *Nature*, **383**, 513–16.

Walsh, J.J. (1976) Herbivory as a factor in patterns of nutrient utilisation in the sea. *Limnol. Oceanogr.*, **21**, 1–13.

Watson, A.J. & Whitfield, M. (1985) Composition of particles in the global ocean. *Deep Sea Res.*, **32**, 1023–39.

Wefer, G. & Fisher, G. (1993) Seasonal patterns of vertical particle flux in equatorial and coastal upwelling areas of the eastern Atlantic. *Deep Sea Res.*, **40**, 1613–45.

Williams, P.J. (1975) Biological and chemical aspects of dissolved organic material in sea water. In *Chemical Oceanography*, J.P. Riley & G. Skirrow (eds), Vol. 2, 301–63. London: Academic Press.

Williams, P.M. (1971) The distribution and cycling of organic matter in the ocean. In *Organic Compounds in Aquatic Environments*, S.D. Faust & J.V. Hunter (eds), 145–63. New York: Marcel Dekker.

Williams, P.M. & Druffel, E.R.M. (1987) Radiocarbon in dissolved organic matter in the central North Pacific Ocean. *Nature*, **330**, 246–8.

Williams, P.M. & Gordon, L.I. (1970) Carbon-13:carbon-12 ratios in dissolved and particulate organic matter in the sea. *Deep Sea Res.*, **17**, 19–27.

Williams, P.M., Oeschger, H. & Kinney, P. (1969) Natural radiocarbon activity of dissolved organic carbon in the North-East Pacific Ocean. *Nature*, **224**, 256–9.

Van den Berg, C.M.G. (1995) Evidence for organic complexation of iron in sea water. *Mar. Chem.*, **50**, 139–57.

Villareal, T.A., Altabet, M.A. & Culver-Rymsza, K. (1993) Nitrogen transport by vertically migrating diatom mats in the North Pacific Ocean. *Nature*, **363**, 709–12.

Zafiriou, O.C. (1983) Natural water photochemistry. In *Chemical Oceanography*, J.P. Riley & R. Chester (eds), Vol. 8, 339–70. London: Academic Press.

Zafiriou, O.C. (1986) Atmospheric, oceanic and interfacial photochemistry as factors influencing air–sea exchange fluxes and processes. In *The Role of Air–Sea Exchange in Geochemical Cycling*, P. Buat-Menard (ed.), 185–207. Dordrecht: Reidel.

Zhuang, R.G. & Duce, R.A. (1993) The adsorption of dissolved iron on marine aerosol particles in surface waters of the open ocean. *Deep Sea Res.*, **40**, 1413–29.

Zhuang, R.G., Duce, R.A. & Brown, P.R. (1992) Link between iron and sulphur cycles suggested by detection of Fe(II) in marine aerosols. *Nature*, **355**, 537–9.

# 10      Particulate material in the oceans

Lal (1977) estimated that the total mass of suspended material in the oceans is $\sim 10^{16}$ g, which is equivalent to an average seawater concentration of only $\sim 10$–$20$ ng l$^{-1}$. This suspended material moves through the ocean system, but the journey it undertakes is a dynamic one and its concentration and composition are subject to continuous change as a result of processes such as aggregation, disaggregation, zooplankton scavenging, decomposition and dissolution (Gardner *et al.*, 1985). As it undertakes this journey, the particle microcosm plays a vital role in regulating the chemical composition of sea water via the removal of dissolved constituents (e.g. trace elements and nutrients) from solution and their down-column and lateral transport to the bottom-sediment sink. Indeed, such is the extent to which the behaviour of dissolved trace metals is dominated by suspended solids that Turekian (1977) referred to the phenomenon as the *great particle conspiracy*. Here, then, perhaps lies the key to Forchhammer's (1865) 'facility with which the elements in sea water are made insoluble'. In the present chapter attention will be paid to the sources, distribution and composition of the *total suspended material* (TSM) in the sea, and in following this route an attempt will be made to decipher the role that the TSM plays in the major oceanic biogeochemical cycles.

## 10.1 The measurement and collection of oceanic total suspended matter

A number of techniques have been used to measure the concentrations of TSM in sea water. Some of these are indirect, i.e. the concentrations of TSM are measured *in situ* in the water column, but no actual samples are collected; these techniques include those based on optical phenomena, such as light absorption (transmissometry) and light scattering (nephelometry). Other techniques involve the

direct collection, and subsequent analysis, of samples of TSM, e.g. by filtration or centrifugation. In addition, one of the most important recent advances in the direct collection of TSM has come through the introduction of the sediment trap, a device that collects material as it sinks down the water column. An important advantage of this type of device is that it can retain the large-sized, relatively rare, particles that dominate the vertical TSM flux (see Section 12.1).

## 10.2 The distribution of total suspended matter in the oceans

Jerlov (1953) used light-scattering measurements to make one of the first major surveys of oceanic TSM. From the data obtained he was able to identify a number of overall trends in the distribution of the TSM, including the recognition of what were subsequently termed **nepheloid layers**, i.e. layers of relatively turbid water that can extend hundreds of metres above the sea bed. This work was followed by a series of investigations made by Russian scientists in the 1950s, and the results of these have been summarized by Lisitzin (1972). The next principal development came in the 1960s from work carried out by American groups, mainly at the Lamont-Doherty Geological Observatory, who used optical methods to study the oceanic distribution of TSM. One of the major findings to emerge from this work was the confirmation of the presence of the nepheloid layers in many regions of the World Ocean (see e.g. Ewing & Thorndike, 1965; Connary & Ewing, 1972; Eittreim *et al.*, 1976). More recent studies on the global distribution of oceanic TSM have been carried out as part of the GEOSECS and HEBBLE programmes, and down-column TSM fluxes, which have been measured by the deployment of sediment traps in many locations, have been assessed on an ocean-

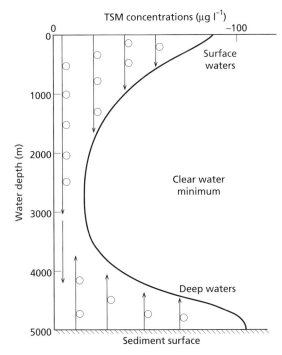

**Fig. 10.1** Typical oceanic TSM profile for a water column with a well-developed nepheloid layer (from Chester, 1982; after Biscaye & Eittreim, 1977). The arrows indicate that vertical particle settling can be affected by horizontal advective processes.

wide scale in the GOFS and VERTEX programmes (see Chapter 12).

Much of the indirect data on the distribution of oceanic TSM were obtained in the form of optical parameters. Subsequently, however, Biscaye & Eittreim (1977) converted nephelometer data from the Atlantic Ocean water into units of absolute TSM concentrations. In this way, they were able to identify a number of distinctive features in the vertical and horizontal distributions of TSM in the water column, and these are illustrated in Fig. 10.1 in the form of a generalized vertical profile. From this figure it can be seen that the distribution of oceanic TSM can be described in terms of a **three-layer model** in which the main features are: (i) a surface-water layer; (ii) a clear-water minimum layer; and (iii) a deep-water layer. This three-layer model offers a convenient framework within which to describe the distribution of TSM in the World Ocean.

### 10.2.1 The surface-water layer

In surface waters, TSM concentrations are higher, and more variable, in coastal and estuarine regions than they are in the open ocean. This results largely from the combined effects of (1) an input of externally produced particulates via river run-off and atmospheric transport, and (2) the internal generation of particulates from primary production, both of which have their strongest signals in coastal waters.

1 Although on a global scale ~90% of river suspended material (RSM) is retained in estuaries (see Section 3.2.7.1), the material that does escape has a higher concentration in coastal waters. Further, some estuaries can transport **plumes** of suspended material for considerable distances out to sea in the surface-water layer. For example, Gibbs (1974) reported that in periods of high discharge, suspended material transported by the Amazon formed a plume, in which concentrations of TSM could exceed $5000 \, \mu g \, l^{-1}$, extending seawards as far as ~100 km.

2 Particulate organic material, including both living and dead fractions, makes up a major proportion of TSM in many surface waters of the oceans, and most of it results from the photosynthetic fixation of $CO_2$ during primary production. The principal features in the global distribution of this production are the 'ring' of high-fertility zones around the edges of the ocean basins, and the generally low fertility in the central gyres, which leads to enhanced concentrations in coastal surface waters (see Section 9.1.3.3).

In these regions, however, the actual concentrations of TSM exhibit large temporal and spatial variations. For example, Chester & Stoner (1972) reported a range of $<100->3000 \, \mu g \, l^{-1}$ of TSM in a variety of coastal waters from the World Ocean. Away from the coastal regions, the lowest TSM concentrations are found in open-ocean areas of low productivity (oligotrophic zones), especially at the centres of the central gyre systems, where they can fall to values as low as $<10 \, \mu g \, l^{-1}$. There are also inter-oceanic variations in the concentrations of TSM in the mixed layer. For example, Chester & Stoner (1972) reported an average TSM concentration of ~75 $\mu g \, l^{-1}$ for open-ocean surface waters of the Atlantic and Indian Oceans, and according to Lal (1977) concentrations in the Pacific are about two or three times lower than this.

## 10.2.2  The clear-water minimum

This was the term used by Biscaye & Eittreim (1977) to describe the subsurface region in which there is a decrease in the concentrations of TSM. The decrease is caused by the destruction of particulate material as it sinks from surface waters, or is moved out by advective processes, and the depth at which the minimum is reached varies from one area of the ocean to another. The decrease in particulate concentrations with depth is largely the result of processes that affect the organic matter (oxidative destruction) and shell fractions (dissolution) of the TSM. The overall result of these processes is a decrease in the concentrations of TSM with depth below the surface. Apart from the fact that the concentrations are lower, however, the distribution of TSM around the clear-water minimum parallels that in surface waters. Thus, TSM concentrations at the clear-water minimum represent the particle contribution resulting from the downward transport of material from surface waters and therefore reflect a balance between surface primary production, variable rates of particle setting and down-column particle destruction–dissolution processes. The distribution of TSM at the clear-water minimum in the Atlantic Ocean is illustrated in Fig. 10.2(a).

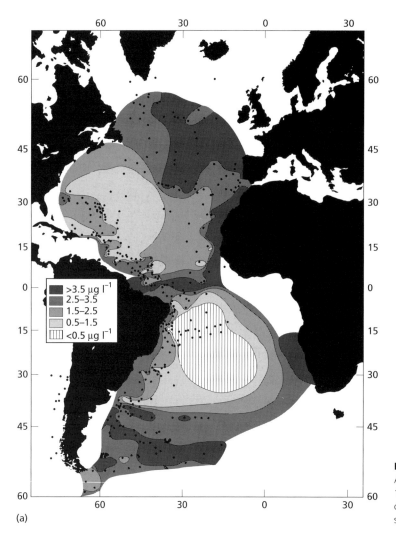

(a)

**Fig. 10.2** Distribution of TSM in the Atlantic Ocean (from Biscaye & Eittreim, 1977). (a) Concentration of TSM at the clear-water minimum. (b) Net particulate standing crop in the deep water.

### 10.2.3 The deep-water layer

If the sinking of particulate matter from the surface reservoir were the only source of TSM to deep waters, then the concentrations of TSM would be expected to continue to fall all the way down the water column as the destructive processes continued, and the clear-water minimum would not be developed. In fact, it is now known that in many oceanic areas there is an increase in TSM concentrations below the minimum, and this therefore requires an additional (i.e. non-surface water) source of particulate matter. Much of this additional source comes from the resuspension of bottom sediments into turbid nepheloid layers. In these layers the concentration of particles decreases upwards away from the sediment source, to fall away completely at the clear-water minimum. However, even in deep waters there is a contribution to the TSM from particles sinking through the minimum. In order, therefore, to assess the spatial distribution of material in the nepheloid layers, Biscaye & Eittreim (1977) identified two particle populations below the clear-water minimum.

*The gross particulate standing crop.* This was identified as the total TSM below the clear-water minimum. There are two main features in the distribution of the gross particulate standing crop.

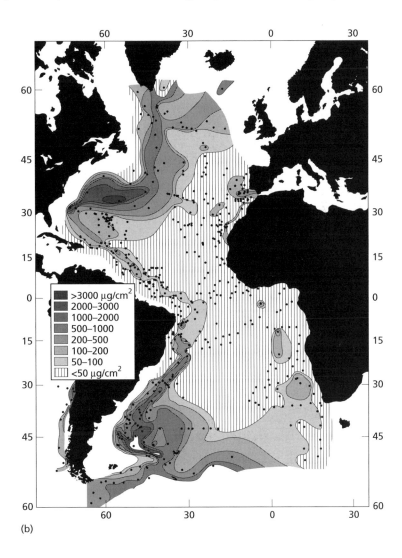

**Fig. 10.2** *Continued*

(b)

1 Relatively low concentrations are found in the central portions of both the North and South Atlantic, reflecting the low surface-water TSM concentrations in the main gyres.

2 The concentrations in the western Atlantic basins are higher than those in the eastern basins, sometimes by as much as an order of magnitude. In the western basins, the maximum TSM concentrations are coincident with the axes of the deep-water boundary currents (see Section 7.3.3).

The distribution of the gross particulate standing crop therefore reflects the effects of two different kinds of processes: those that control the surface-water distribution of TSM and those that regulate the additional supply of particulates necessary to sustain the three-layer water-column model.

*The net particulate standing crop.* This was defined as the amount of TSM below the clear-water minimum that is in excess of the clear-water concentration itself. Thus, the authors were able to obtain a much clearer assessment of the deep-water, or abyssal, particulate signal by excluding 'noise' from the surface-derived source. The distribution of the net particulate standing crop in Atlantic deep-waters is illustrated in Fig. 10.2(b). On the basis of these data Biscaye & Eittreim (1977) concluded that the abyssal signal results from processes that raise particles into near-bottom waters and maintain them in suspension. These processes include supply from the continental shelves by turbidity currents and advective transport, and the direct resuspension of sediment by bottom currents. It was shown in Section 7.3.3 that a characteristic feature of deep-water circulation is the existence of strong boundary currents on the western sides of the oceans, and it is these erosive boundary currents that are the most effective agents in sediment resuspension, and thus in the formation of nepheloid layers; these layers have particle concentrations that are usually in the range $\sim 200$–$500 \mu g\, l^{-1}$. One problem associated with the development of nepheloid layers, however, has been that currents with enough energy to erode the bottom, and to maintain the particles in suspension, have not been found concurrently with high concentrations of particles in the layers themselves. It has been known for some time that bottom-current velocities can vary, and can include local episodes of high-speed currents. Further, data obtained from the High Energy Benthic Bound-ary Layer Experiment (HEBBLE) have shown that there is a high temporal and spatial variability in nepheloid-layer particle concentrations, leading to the inference that strong episodic sediment-modifying events, or 'storms', have occurred on the sea bed. Such deep-sea sediment transport storms have, in fact, been identified by Gross *et al.* (1988) in the Nova Scotia Rise area of the North Atlantic. Four storms of high kinetic energy and near-bed flow were observed over a 1-yr period. These storms were associated with particle concentrations $>2000 \mu g\, l^{-1}$, and the authors suggested that the occurrence of a few of these large episodic events per year could account for most of the suspended load in the deep-sea nepheloid layer.

It may be concluded, therefore, that the spatial and temporal distributions of TSM in the oceans are controlled by the classic oceanographic parameters such as the transport of material from the continents, primary production and the major water circulation patterns. The vertical distribution of TSM in the water column can be described in terms of a three-layer model in which the main particulate sources are the surface ocean, from which particles sink downwards, and the bottom sediments, from which particles are lifted by erosive currents and moved upwards. An intermediate, or clear-water minimum, layer is developed at that part of the water column where these two principal supply mechanisms have their smallest effects, the surface source trailing out downwards and the bottom source falling off upwards. In addition to these mainly vertical supply processes, the lateral advection of particles at mid-depths, e.g. from shelf sediments, can modify the simple three-layer down-column TSM profile (see Section 10.2).

## 10.3 The composition of oceanic total suspended matter

Oceanic TSM consists of a mixture of components, some of which have external sources and some of which are produced internally. In the present section attention is confined largely to the surface-water TSM, and the composition of the down-column flux is considered in Chapter 12.

### 10.3.1 Externally produced components of TSM

These are formed on the continents, in estuaries, or in the air, and are brought to the oceans mainly via river run-off, atmospheric deposition and locally via ice transport.

Crustal weathering products are a ubiquitous, although often minor, component of surface-water TSM, the principal crust-derived minerals being aluminosilicates (e.g. clay minerals and feldspars) and quartz. Chester *et al.* (1976) gave data on the distributions of aluminosilicates in surface waters from a number of contrasting oceanic regions. These data showed that in the China Sea, which receives a relatively large fluvial input, aluminosilicate material made up ~60% of the TSM. Off the coast of West Africa, where there is a relatively large aeolian input arising from dust pulses originating in the Sahara desert (see Section 4.1.4.1), aluminosilicates accounted for ~10% of the total suspended solids. Krishnaswami & Sarin (1976) reported relatively high concentrations of aluminosilicates in high-latitude North and South Atlantic waters, where they probably resulted from glacial weathering. In open-ocean waters, however, aluminosilicates generally make up less than ~5% of the TSM. The overall pattern to emerge from studies such as these is that the highest concentrations of aluminosilicates are found in regions where the terrestrial source signals are strongest, and the lowest concentrations are found in the open-ocean central gyres remote from the land masses.

In addition to crustal weathering products externally produced material in oceanic TSM includes organic suspensions, flocculated metal-organic colloids and precipitated iron and manganese oxides formed in fluvial and estuarine environments.

### 10.3.2 Internally produced oceanic TSM components

These include biological material (tissues and shells) and inorganic precipitates, together with various resuspended sediment components.

Oceanic TSM components that are produced in the biosphere include both organic matter and shell material. The factors controlling the distribution of organic matter in the oceans have been described in Section 9.2, and the highest concentrations are usually associated with regions of high primary production found in a ring around the major ocean basins. Particulate organic matter (POM), which is composed of living organisms (phytoplankton, zoo-

plankton, etc.) and detritus (dead organisms, faecal debris, etc.) makes up a large fraction of the TSM found in many surface waters of the World Ocean, usually greater than ~50% of the total solids (see e.g. Chester & Stonier, 1972; Krishnaswami & Sarin, 1976; Masuzawa *et al.*, 1989). The shell material secreted by marine organisms consists largely of either calcium carbonate or opal (see Section 15.2.1). Together, these shell components make up a major fraction of surface-water TSM, but there are geographical trends in their distributions; e.g. opaline shells can dominate in some high-latitude regions, whereas carbonate shells can dominate, particularly in the central gyres, at lower latitudes.

Some non-biogenic components of TSM are also produced within the ocean itself from components dissolved in sea water; these include material such as non-biogenic barite, a number of carbonates, and iron and manganese oxyhydroxides. In the lower and mid-water column, hydrothermal precipitates consisting of minerals such as chalcopyrite, sphalerite and hydrous iron and manganese oxides are also added to the TSM population (see Section 15.3.6).

In addition to the components described above, various anthropogenic materials can be found in oceanic TSM. These include sewage products, nuclear components, petroleum hydrocarbons, pesticides, PCBs and other synthetic organics, and tars. The distributions and fates of these anthropogenic components in oceanic TSM have been reviewed by Preston (1989).

Data for the elemental compositions of a number of the principal components that contribute to the oceanic particulate population are given in Table 10.1(a). These are:

1 marine organisms, which are representative of the internally produced TSM biomass population;

2 faecal pellets, which offer an estimate of the composition of large-sized aggregates that leave surface waters (see Section 10.4);

3 river and atmospheric particulate material, which represent the composition of continentally derived, i.e. external, mainly natural, material transported to the ocean system.

It must be stressed, however, that these are average values, and that wide variations can be found in the elemental compositions of the individual components themselves. Various examples of the elemental composition of bulk TSM are listed in Table 10.1(b).

**Table 10.1** The elemental composition of oceanic total suspended material (TSM); units, $\mu g\,g^{-1}$.

(a) Some principal components of oceanic TSM.

| Element | Microplankton: North Pacific* | Phytoplankton: Monteray Bay, California* | Zooplankton: Monteray Bay, California* | Zooplankton: North Pacific* | Zooplankton: North Atlantic† | Bulk plankton, average‡ | Bulk plankton: marine organisms, average§ | Faecal pellets¶ | Faecal pellets‖ | River particulate matter** | Soil-sized aerosols†† |
|---|---|---|---|---|---|---|---|---|---|---|---|
| Al | 72–108 | 7–2850 | <8–313 | 9–31 | — | 202 | 159 | 20800 | — | 94000 | — |
| Fe | 1030–4000 | 49–3120 | 54–1070 | 90–1720 | 567–1467 | 306 | 862 | 21600 | 24000 | 48000 | 52000 |
| Mn | 3.4–32 | 2.1–30 | 2.2–12 | 2.9–7.1 | 0–23 | 7.7 | 9.3 | 2110 | 243 | 1050 | 1312 |
| Cu | 40–104 | 1.3–45 | 4.4–23 | 6.2–58 | 10–90 | 14 | 27 | 650 | 226 | 100 | 157 |
| Ni | 11–12 | <0.5–13 | <0.5–13 | 5–13 | 15–77 | 12 | 17 | — | 20 | 90 | 91 |
| Co | <1 | <1 | <1 | <1 | 8–20 | — | <1 | 15 | 3.5 | 20 | 9 |
| Cr | <4 | <21 | <1 | <1 | — | — | <1 | — | 38 | 100 | 85 |
| V | <3 | <3 | <3 | <3 | — | — | <3 | 76 | — | 170 | 145 |
| Ba | 51–70 | 5–500 | 4–257 | 51–70 | — | 55 | 60 | 192 | — | 600 | 487 |
| Sr | 6800–9650 | 53–3934 | 83–810 | 380–3000 | 57–520 | — | 862 | 1430 | 78 | 150 | 101 |
| Pb | 17–39 | <1–47 | <1–12 | 22–14 | 0–123 | — | 20 | — | 34 | 100 | 465 |
| Zn | 285–4190 | 3–703 | 53–279 | 60–750 | 120–400 | 131 | 257 | <20 | 950 | 250 | 683 |
| Cd | 1.0–2.2 | 0.4–6 | 0.8–10 | 1.9–3.5 | 2–9 | 22 | 4.6 | — | 19.6 | 1 | — |
| Hg | 0.11–0.53 | 0.10–0.59 | 0.07–0.16 | 0.04–0.45 | — | — | 0.16 | — | — | — | — |

* Martin & Knauer (1973). † Martin (1970). ‡ Collier & Edmond (1983). § Chester & Aston (1976). ¶ Spencer *et al.* (1978). ‖ Fowler (1977). ** Martin & Whitfield (1983). †† Chester & Stoner (1974).

(b) Oceanic TSM.

| Element | Tropical North Atlantic TSM: surface water* | Tropical North Atlantic TSM: deep water* | North Atlantic TSM: surface water† | South Atlantic TSM: surface water† | Indian Ocean TSM: surface water† | Sargasso Sea TSM: surface water (10 m)‡ | Sargasso Sea TSM: deep water (2000 m)‡ | Northeast Pacific TSM: surface water (30 m)‡ | Northeast Pacific TSM: deep water (2872 m)‡ | East Pacific TSM: upwelling zone, surface water (0 m)§ | East Pacific TSM: oligotrophic zone, surface water (0 m)§ |
|---|---|---|---|---|---|---|---|---|---|---|---|
| Al | 3000 | 7500 | — | — | — | — | — | — | — | — | — |
| Fe | 8800 | 15000 | — | — | — | — | — | — | — | — | — |
| Mn | 140 | 320 | 145 | 85 | 385 | 126 | 1581 | 27 | 7855 | 10.5 | 18 |
| Cu | 145 | 200 | 74 | 52 | 202 | 17 | 245 | 8.9 | 202 | 83 | 3.7 |
| Ni | 70 | 130 | — | — | — | 11 | 57 | 8.8 | 148 | 5.6 | 4.4 |
| Co | 5 | 16 | 11 | 16 | 14 | — | — | — | — | — | — |
| Cr | 125 | 170 | — | — | — | — | — | — | — | — | — |
| V | 22 | — | 38 | 69 | 60 | — | — | — | — | — | — |
| Pb | 180 | 570 | 52 | 72 | 44 | — | — | — | — | — | — |
| Zn | 640 | 1000 | 159 | 260 | 231 | 11 | 98 | 43 | 145 | 89 | 42 |
| Cd | — | — | — | — | — | 0.81 | 0.70 | 15 | 0.91 | — | — |
| Hg | 16 | 36 | — | — | — | — | — | — | — | — | — |
| Sc | 0.5 | 1.3 | — | — | — | — | — | — | — | — | — |
| Se | 8 | 17 | — | — | — | — | — | — | — | — | — |
| Ag | 4 | 9.3 | — | — | — | — | — | — | — | — | — |
| Sb | 7 | 10 | — | — | — | — | — | — | — | — | — |
| Au | 0.4 | — | — | — | — | — | — | — | — | — | — |

* Buat-Menard & Chesselet (1979). † Chester & Stoner (1975). ‡ Sherrell (1989); quoted in Saarger (1994). § Jannish (1990); quoted in Saarger (1994).

**Table 10.2** The concentration of some particulate elements in surface seawaters from various regions; units, ng l$^{-1}$.

| Element | China Sea: strong river input* | Northeast Atlantic, off coast of West Africa: upwelling zone and strong aeolian input* | Gulf of Mexico: coastal† | Tropical North Atlantic‡ | Atlantic Ocean: GEOSECS‡ | Sargasso Sea§ | Northwest Atlantic¶ | North + South Atlantic‖ | South Atlantic: open-ocean** | Indian Ocean: open-ocean** | Weddell Sea†† |
|---|---|---|---|---|---|---|---|---|---|---|---|
| Al | 45 559 | 3755 | — | 78 | 118 | 32 | 160 | 164 | 88 | 107 | — |
| Fe | — | — | — | 222 | 341 | 18 | 121 | 223 | — | — | — |
| Mn | 778 | 54 | 1110 | 3.5 | 7.6 | 1.9 | 11 | 7.0 | 5.6 | 2.1 | — |
| Ni | — | 43 | — | 1.8 | — | 0.34 | 1.1 | 5.2 | — | 3.0 | 3.9 |
| Cr | — | — | — | 3.1 | 9.6 | — | 3.1 | — | — | — | — |
| V | 41 | — | — | — | — | — | — | — | 9.0 | 4.0 | — |
| Cu | 74 | 14 | 790 | 3.6 | 25 | 0.40 | 3.4 | 9.1 | 6.6 | 3.5 | 5.4 |
| Pb | 47 | 19 | — | 4.5 | — | 0.23 | 2.8 | — | 9.0 | 3.0 | — |
| Zn | 149 | 85 | 1200 | 16 | 23 | 0.34 | 11 | — | 18 | 9.3 | 142 |
| Cd | — | — | — | — | — | 0.03 | — | 0.04 | 1.1 | 0.17 | 1.8 |

* Chester (1982). † Slowey & Hood (1971). ‡ Buat-Menard & Chesselet (1979). § Sherrell & Boyle (1992). ¶ Wallace *et al.* (1977). ‖ Krishnaswami & Sarin (1976). ** Stoner (1974) & Griffiths (1978). †† Westerlund & Ohman (1991).

A number of authors have provided data on the concentrations of particulate elements in surface sea water, and a compilation of some of these is given in Table 10.2. The overall chemical composition of bulk oceanic TSM is controlled by the proportions in which the various components described above are present, and in a very general sense this is apparent from the particulate element concentration data given in Table 10.2. Thus, the concentrations are highest in surface waters from the China Sea (large river input), intermediate in the eastern margins of the North Atlantic (large aeolian input, intense primary production), and lowest in the open-ocean regions of the Atlantic and Indian Oceans (low external inputs, low primary production).

## 10.4 Total-suspended-matter fluxes in the oceans

In terms of the three-layer distribution model described in Section 10.2, it is apparent that TSM sinks through the water column from the surface layer to provide a *downward* particulate signal. This signal decreases in strength with depth, but below the clear-water minimum it encounters the outriders of an *upward* particulate signal from the

sea bed. In addition to these vertical signals, the water-column TSM profiles can be modified by *laterally* advected particulate signals. These various signals interact, both to govern the throughput of particulate material in the ocean system and to control the net output of the material to the sediment sink. This **particulate throughput** is the key mechanism controlling the rates at which many dissolved trace metals are removed from the ocean reservoir, but before attempting to understand the way in which this 'great particle conspiracy' operates it is necessary to understand the nature of the particulate fluxes involved.

McCave (1984) made a detailed study of the size spectra of suspended particles in the oceans, which can be related to the three-layer model described in Section 10.2. Various authors (see e.g. Brun-Cottan, 1986) have described the particle-size distributions of the total particle population of oceanic TSM using the log-normal law. McCave (1984), however, showed that in terms of particle-size spectra in ocean waters, the particle *number* data can nearly always be fitted by a power-law distribution over a large part of the measured range. Expressed as a cumulative number $N$ as a function of particle diameter $d$, the size distribution may be described by

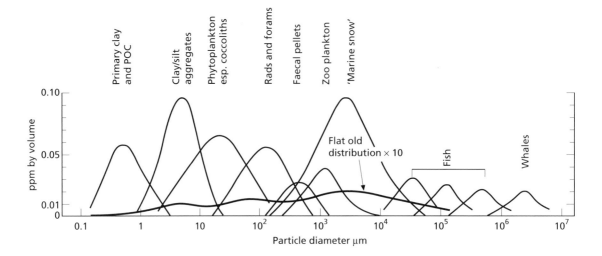

**Fig. 10.3** Hypothetical wide-spectrum oceanic TSM particle-size distribution by volume, with suggested distributions of component particle types (from McCave, 1984).

$$N = kd^{-\beta} \tag{10.1}$$

where $N$ is the number of particles with diameters $>d$. A value of $\beta = 3$ indicates equal particle *volumes* in logarithmically increasing size grades. This situation is found in mid-depth regions (i.e. the clear-water minimum) where $\beta = 3$, with $d$ in the range 1–100 μm; here, therefore, approximately equal volumes of material are found in logarithmically increasing size grades and the particle-size distributions by volume are flat. Close to the sources of particles, however, i.e. the surface ocean (surface layer) and the sea bed (deep-water layer), the size distributions in particle volume space are not flat, but instead show a number of peaks, some of which are characteristic of the source (e.g. clay and foraminifera) and some of which are related to particle aggregation–see Fig. 10.3.

Most particle production, and modification of the particle size, takes place in the **upper ocean**. The overall distribution of small particles (as sensed by nephelometers) and large particles (as sensed by sediment traps) show a maximum in surface and bottom regions, with a broad minimum in between (cf. the three-layer model). The observed loss of peaks from the suspensions to yield flat distributions requires aggregation of material as the fine particles settle slowly down the water column; this aggregation takes place by processes such as the shear-controlled coagulation between particles of similar size, and the capture of small particles by larger biological aggregates, such as 'marine snow', which escape degradation in the upper layers of the ocean. Thus, the breakdown of larger particles as they fall from the surface waters, leading to the production of smaller particles, is matched by the formation of the larger units. The overall result of this is to produce a particle-size spectrum in the oceans that is flattened in mid-water column locations. Thus, the ocean particle size-distribution by volume tends to be flat at mid-depths (equivalent to a cumulative particle number distribution of $\beta = 3$) but is peaked in the nepheloid and surface layers.

The large biologically produced particles that fall from the surface zone undergo destruction in deeper waters, with the release of fine, more refractory, particles, which is counterbalanced by their aggregation into larger units. This continuous cycle of aggregation and disaggregation modifies initial size distributions and produces a particle-size spectrum over much of the oceans that is flattened, with no peaks that can be identified with specific particle populations.

According to McCave (1984) the size distribution of suspended particulates is a function of a number of variables, which include the source and nature of the particles, the physical and biological processes that cause their aggregation, and the age of the suspension. In practice, most of the particles suspended in sea water have diameters $<2$ μm (McCave, 1975), and those with diameters >20 μm are rare (Honjo, 1980). Particles are transported out of the mixed or surface layer down the water column by gravitational and Stokes settling. However, there is a relatively rapid transport of particles from the surface layers which is in excess of that predicted from Stokes' law for particles having diameters <2 μm. Evidence for the accelerated sinking of particles has come from a number of sources. These include the measured, or inferred, down-column transit times of minerals, biogenic components, stable trace metals and radionuclides, all of which indicate that relatively large-diameter particles are required to sediment the phases out of sea water. As most TSM is composed of particles with diameters <2 μm, it must be assumed that this removal is preceded by the aggregation of the particles into larger units. This requirement for a 'fast' particle settling mechanism has been considered by various authors. In a benchmark paper, McCave (1975) made a theoretical study of the transport of TSM in the oceans and concluded that the principal feature in the vertical down-column flux of particles is that relatively rare, rapidly sinking, large particles contribute most of the **flux**, whereas more common smaller particles contribute most of the **concentration**. Brun-Cottan (1976) also concluded that small particles (<5 μm diameter) might behave conservatively in the water column and undergo lateral movement in a given water mass, but that it is mainly the larger (>50 μm diameter) particles, or aggregates, formed in surface waters that fall directly to the sea bed. The large particles are formed by aggregation in the surface layers, and it has become apparent that the generation of faecal material by organisms is an important route by which smaller particles can undergo such aggregation into larger units, and faecal material does in fact make up a high proportion of the large-particle population. It must be stressed, however, that there is a continual break-up of the larger particles, and Cho & Azam (1988) have suggested that it is the action of free-living bacteria that gives rise to the large-scale production of the small particles at the expense of the large aggregates.

It was pointed out above that the continual cycle of aggregation and disaggregation leads to the production of an oceanic particle population that has no specific size-class populations within it. Nonetheless, in the context of both biogeochemical reactivity (e.g. the removal of dissolved elements by active biological uptake and passive scavenging) and particle flux transport (which is dominated by large-sized material), it is convenient to envisage the microcosm of particles as being divided into two general populations.

1 A common, small-sized population (probably <5 μm diameter), termed **fine particulate matter** (FPM) by Lal (1977), which undergoes large-scale horizontal transport; i.e. a suspended population (concentration ~20–50 μg l⁻¹, sinking rate ≪1 m day⁻¹), which dominates the standing stock of TSM.

2 A rare, large-sized population, referred to as **coarse particulate matter** (CPM) by Lal (1977), which consists mainly of particle aggregates >50 μm in diameter that undergo vertical settling; i.e. a sinking population, which is the prime carrier in down-column transport from the surface water to the ocean depths.

It is now widely recognized that *faecal material* (pellets and debris) makes up a significant fraction of the rain of particles from the surface layer to deep water. Faecal material, however, is not the only type of biogenic particulate material to fall from the surface. For example, **marine snow** is a term that has been applied to relatively large amorphous aggregates of biological origin. To some extent this may be analogous to faecal material, because faecal pellets are known to be an important component of this marine snow (see e.g. Shanks & Trent, 1980). However, marine snow also includes other components, and Honjo *et al.* (1984) prefer the term **large amorphous aggregates** (LAA) to describe the large-sized biogenic particles. In this terminology, therefore, LAA consist of the debris of phytoplankton and zooplankton, including faecal material, which is hydrated into a matrix of aggregates to which are attached other components, such as micro-organisms and clay particles. Shanks & Trent (1980) concluded that, like faecal pellets, marine snow can accelerate the transport of material to the deep ocean and may in fact be the main package in which the vertical flux takes place. It is apparent, therefore, that the main flux of particulate material from the surface ocean to the sediment surface is driven by large organic aggregates, which form the main carriers of the global carbon flux; i.e. the organic matter that escapes recycling and is replaced in 'new' production (see Section

9.1.1). As they descend through the water column, these large aggregates drag down small-sized inorganic particles. Some of these inorganic particles are lithogenous in origin, and although they make up only a small fraction (usually less than ~5%) of the total TSM in surface water they form the main contributors of material to lithogenous pelagic clays. McCave (1984) therefore suggested the intriguing idea that 'the processes of pelagic sedimentation of lithogenic matter may be viewed as a side effect of excretion and disposal of other waste products of the ocean's biological system'.

The concept of a two-particle population in sea water has considerably advanced our understanding of how TSM is transported through the ocean reservoir. However, it has become apparent that the picture is more complicated than this simple twofold particle classification would suggest. For example, Lal (1980) demonstrated that, although the measured sinking rate of particles based on the removal of Pu, Pb, Th and Fe from sea water is at least an order of magnitude greater than that based on the sinking of the FPM, it is also two or three orders of magnitude smaller than that derived from the sinking of the CPM. To explain this, Lal (1980) suggested that a piggy-back particle mechanism operates in the water column. In this mechanism, small particles (<1 µm diameter), which are very active in trace metal and radionuclide scavenging, undergo impaction with the larger particles and adhere to them, probably via organic matter coatings. The large particles then carry the smaller ones down the water column by an on-and-off piggy-back mechanism. McCave (1984) also concluded that there is a scavenging of small particles by larger units in the surface and nepheloid regions of the water column and that this is an important particle transport process.

The down-column flux is driven by CPM, which is composed mainly of biogenic aggregates. As a result, primary production in the euphotic layer exerts a fundamental control on the initiation of this down-column flux, via export production. An important thrust in marine biogeochemistry over the past few years has been to model the relationship between primary production at the surface and the flux of material through the interior of the ocean, and this is considered in Chapter 12. At present, it is sufficient to point out that it does appear as if the primary down-column flux is closely related to surface productivity,

and that the flux often varies in magnitude in relation to seasonal changes in photosynthetic activity in the euphotic zone. Further, it is generally recognized that around 95% of the organic matter produced in the surface ocean is recycled in the upper few hundred metres of the water column. However, the 5% or so that does escape from surface waters, i.e. the export production that is replaced by new production (see Section 9.1), is composed largely of aggregates and it is these that dominate the down-column material flux. For example, Bishop *et al.* (1977) demonstrated that ~95% of the down-column flux of particulate material at a site in the equatorial Atlantic is associated with large faecal aggregates, although these only account for <5% of the total POC. These authors estimated that the transit times for the faecal material through a 4-km water column would be ~10–15 days. During a transit of this relatively short duration, lateral displacement resulting from deep-ocean advection in the region would be only ~40 km. As a result, material deposited in the underlying sediments would reflect the oceanic variability in the surface water TSM, thus offering an explanation for the source-related clay mineral distribution patterns found in deep-sea sediments (see Section 15.1.2). It has generally been considered that the *primary* flux of particulate material through the water column is in a downward direction; however, Smith *et al.* (1989) showed that upward fluxes of particulate organic matter composed of positively buoyant particles can also be important.

## 10.5 Down-column changes in the composition of oceanic TSM and the three-layer distribution model

The vertical transport of TSM in the water column is driven mainly by the CPM flux, and as the particulates settle they undergo considerable modification via processes such as decomposition, aggregation–disaggregation, zooplankton grazing and dissolution.

The major components of the sinking particulate mass flux are biogenic (organic matter, calcium carbonate and opal) and lithogenic (see Section 12.1.1.2). The proportions of these components change as the particles descend the water column. The most notable change is that affecting organic matter, which undergoes carbon recycling in subsurface waters. For ex-

**Table 10.3** Changes in the composition of the particle mass flux composition with depth (data from Masuzawa et al., 1989).

| Trap depth (m) | Mass flux (mg m$^{-2}$ day$^{-1}$) | Flux composition (%) | | | |
|---|---|---|---|---|---|
| | | CaCO$_3$ | Opal | Organic matter | Lithogenic matter |
| 890 | 139 | 8.7 | 41.0 | 34.3 | 16.0 |
| 1100 | 116 | 8.2 | 41.0 | 33.2 | 17.6 |
| 1870 | 50.4 | 9.6 | 36.8 | 23.8 | 29.8 |
| 2720 | 49.4 | 8.3 | 37.4 | 9.4 | 44.8 |
| 3420 | 60.0 | 9.8 | 27.2 | 6.0 | 57.0 |

ample, Martin et al. (1987) concluded that 50% of the organic carbon removed from surface waters is regenerated at depths <300 m, 75% is regenerated by 500 m, and 90% by 1500 m (see Section 12.1). The biogenic shell components change less dramatically with depth, but the overall effect is to increase the proportions of lithogenic material in the sinking mass flux. For example, Masuzawa et al. (1989) presented data on the mass flux composition of material collected by five sediment traps deployed between 890 m and 3420 m in the Japan Sea. A summary of the data from the traps is given in Table 10.3, from which it is apparent that the contribution of organic matter to the total flux decreases from ~35% at 890 m to ~5% in the deepest trap. In contrast, the contribution made by lithogenic material to the total flux increases from ~15% at 890 m to ~57% at 3420 m.

In general, therefore, it may be concluded that internally produced biogenic components decrease, and land-derived (lithogenic) components increase, in importance as the CPM flux carries TSM down the water column. There are a number of reasons for this that can be related to the three-layer TSM water-column distribution model, which combines downward and upward particulate signals, but which also can be affected by lateral signals.

*The downward signal.* Part of the decrease in the proportion of the biogenic components results from their loss as a result of the oxidative decomposition of POM and the dissolution of shell material, processes that occur both in the euphotic zone and at depth during the descent of the TSM through the water column. It is these processes that lead to the setting up of the clear-water minimum zone, and they have two effects on the bulk TSM.

**1** Because biogenic components make up a large percentage of the surface water TSM, their destruction leads to a decrease in the absolute concentration of particles towards the clear-water minimum zone.

**2** As these components are removed, there is an increase in the relative proportions of the non-biogenic material that survives to reach mid-water depths.

*The upward signal.* The organic carbon content of most deep-sea sediments is generally quite low (<5%; see Section 14.2), with the result that the resuspended particulate population is relatively rich in aluminosilicates and sometimes also in shell material. The *in situ*, i.e. non-basin-boundary, resuspended flux of bottom sediments, which results in the production of nepheloid layers, therefore will increase the proportions of non-biogenic material in bottom water TSM.

*The lateral signal.* The mixing of bottom sediment into the water column in the erosive boundary currents on the western edges of the ocean basins, and its advective transport, is now known to be a major pathway for the introduction of small-sized refractory particles into the basin interiors (see e.g. Brewer et al., 1980; Honjo et al., 1982; Spencer, 1984). The increase in the proportions of non-biogenic components at mid-water depths below the clear-water minimum can therefore also result from a direct addition of fine aluminosilicate material that is transported laterally from the continental margins and is then transferred into the vertical CPM flux.

The general relationships in the processes that drive the down-column TSM flux are illustrated schematically in Fig. 10.4.

Up to this point we have concentrated on TSM, and the distinction between TSM and dissolved components is usually operationally defined by a ~0.45 μm cut-off, using a membrane filter. There is,

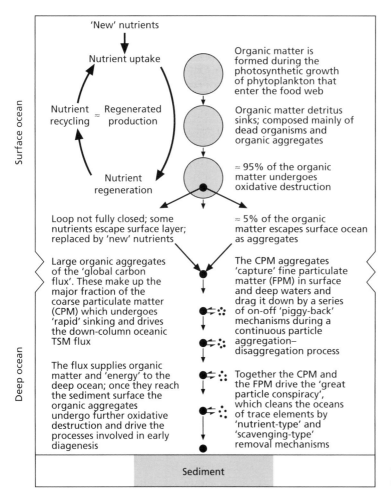

**Fig. 10.4** A schematic representation of the processes that drive the down-column flux of oceanic TSM.

however, a population of material that bridges the 'dissolved–particulate' classification, but which is usually described as 'dissolved' by conventional filtration techniques. This population consists of **colloids**, i.e. submicron particles between ~1 nm and ~1 μm. Isao *et al.* (1990) identified colloids in the size range 0.38–1 μm in the North Pacific. They reported that 95% of the colloidal particles were **non-living** and occurred in the upper 50 m of the water column at concentrations in the order of $5 \times 10^7$–$8 \times 10^7$ particles ml$^{-1}$. The concentrations decreased to $2 \times 10^6$ ml$^{-1}$ at 200 m. In the upper 200 m, ~95% of the colloidal material measured was <0.6 μm. Wells & Goldberg (1991) presented data on the concentration and vertical distribution of colloids (<120 nm) in waters ~15 km off-shore in the Santa Monica Basin.

They found that the colloids were highly stratified, i.e. (i) concentrations were below detection (<$10^4$ particles ml$^{-1}$) at the surface; (ii) they increased sharply near the lower thermocline (~40–100 m) to concentrations of >$10^9$ particles ml$^{-1}$; and (iii) decreased again to <$10^4$ particles ml$^{-1}$ below the thermocline. This distribution is different from that of the larger colloids (0.38–1 μm) described by Isao *et al.* (1990).

Organic matter appears to be the most important constituent of the colloids reported in both studies, although Wells & Goldberg (1992) also identified inorganic material such as iron colloids and clay minerals, together with trace metals, in oceanic colloids. The mainly organic nature of the colloids suggests that they are derived primarily from biological processes, and Isao *et al.* (1990) concluded that a significant proportion (perhaps even as much as *c.* 10%) of the material defined by conventional techniques as dissolved organic matter (DOM) in the upper ocean may be in the form of these colloids. According to Wells & Goldberg (1991) the vertical stratification of the colloids they described indicates that they are reactive and have short residence times in sea water. In this context, Moran & Buesseler (1992) estimated that the residence time of colloidal $^{234}$Th ($<0.2\,\mu m$) with respect to aggregation into small particles ($>0.2–53\,\mu m$) in the upper ocean was *c.* 10 days. The $^{234}$Th activity of the colloidal particles was the same, on average, as that of small particles ($>0.2–53\,\mu m$), implying that the colloidal matter may be as important as traditionally defined TSM in the cycling of Th, and possibly other reactive elements, between dissolved and particulate forms in sea water (see Section 11.6.3.1).

It may be concluded, therefore, that macromolecular colloidal matter in the upper ocean has a short residence time and a rapid turnover. Colloids may act as a reactive intermediary in the marine geochemistry of trace metals (see Chapter 12), and a biologically labile pool of colloidal matter would also affect carbon cycling in the oceans.

## 10.6 Particulate material in the oceans: summary

1 The vertical distribution of TSM in the water column can be described in terms of a three-layer model in which a surface layer, a clear-water minimum layer and a deep-water layer are distinguished. The principal mechanisms that contribute to the setting up of this model are a primary downward signal from the ocean surface, transporting mainly biogenic aggregates, and a secondary (or resuspended) upward signal from the bottom sediment, carrying mainly aluminosilicates. The two end-members in the system are therefore the **surface ocean** and the **surface sediment**. The surface ocean is the zone in which biological cycles are especially active, and it is here that biogenic solids interact with the dissolved and particulate elements transported by river run-off, atmospheric deposition and vertical mixing, and so enable them to enter the oceanic biogeochemical cycles.

2 Oceanic TSM consists of a variety of components from both external sources (e.g. aluminosilicates and quartz) and internal sources (e.g. biological material, such as tissues and shells).

3 There is a continual cycle of aggregation and dis-aggregation of oceanic TSM, which leads to the production of a particle population that has no specific size classes within it. From the point of view of biogeochemical reactivity, however, it is convenient to distinguish between fine particulate matter (FPM, diameters $<5–10\,\mu m$) and coarse particulate matter (CPM, diameters $>50\,\mu m$).

4 Much of the particulate flux to deep waters and sediments is driven by the CPM, which consists mainly of large organic aggregates. The CPM can also carry FPM to deep waters by 'on–off' piggy-back-type mechanisms.

In the present chapter we have described the distribution, composition and sinking characteristics of the oceanic TSM population. It is this microcosm of particles that plays a vital role in regulating the chemical composition of sea water via **particulate–dissolved** equilibria. In the next chapter the distributions of trace elements in the oceans will be discussed, and in doing this the TSM story will be used as a background to assess the factors that control the down-column transport of these elements. It is this vertical transport to the benthic boundary layer that eventually results in the ultimate removal of the elements into the sediment reservoir, thus initiating the final, but still complex, act in the 'great particle conspiracy', which was used as the starting point in the present chapter.

## References

Biscaye, P.E. & Eittreim, S.T. (1977) Suspended particulate loads and transports in the nepheloid layer of the abyssal Atlantic Ocean. *Mar. Geol.*, **23**, 155–72.

Bishop, J.K.B., Edmond, J.M., Kellen, D.R., Bacon, M.P. & Silker, W.B. (1977) The chemistry, biology, and vertical flux of particulate matter from the upper 400 m of the equatorial Atlantic Ocean. *Deep Sea Res.*, **24**, 511–48.

Brewer, P.G., Nozaki, Y., Spencer, D.W. & Fleer, A.P. (1980) Sediment trap experiments in the deep North Atlantic: isotopic and elemental fluxes. *J. Mar. Res.*, **38**, 703–28.

Brun-Cottan, J.C. (1976) Stokes settling and dissolution rate model for marine particles as a function of size distribution. *J. Geophys. Res.*, **81**, 1601–5.

Brun-Cottan, J.C. (1986) Vertical transport of particles within the ocean. In *The Role of Air–Sea Exchange in Geochemical Cycling*, P. Buat-Menard (ed.), 83–111. Dordrecht: Reidel.

Buat-Menard, P. & Chesselet, R. (1979) Variable influence of the atmospheric flux on the trace metal chemistry of oceanic suspended matter. *Earth Planet. Sci. Lett.*, **42**, 399–411.

Chester, R. (1982) The concentration, mineralogy, and chemistry of total suspended matter in sea water. In *Pollutant Transfer and Transport in the Sea*, G. Kullenberg (ed.), 67–99. Boca Raton: CRC Press.

Chester, R. & Aston, S.R. (1976) The geochemistry of deep-sea sediments. In *Chemical Oceanography*, J.P. Riley & R. Chester (eds), Vol. 6, 281–390. London: Academic Press.

Chester, R. & Stoner, J.H. (1972) Concentration of suspended particulate matter in surface seawater. *Nature*, **240**, 552–3.

Chester, R. & Stoner, J.H. (1974) The distribution of Mn, Fe, Cu, Ni, Co, Ga, Cr, V, Ba, Sr, Sn, Zn and Pb in some soil-sized particulates from the lower troposphere over the World Ocean. *Mar. Chem.*, **2**, 157–88.

Chester, R. & Stoner, J.H. (1975) Trace elements in total particulate material from suface sea water. *Nature*, **255**, 50–51.

Chester, R., Cross, D., Griffiths, A.G. & Stoner J.H. (1976) The concentrations of 'aluminosilicates' in particulates from some surface waters of the World Ocean. *Mar. Geol.*, **22**, M59–67.

Cho, B.C. & Azam, F. (1988) Major role of bacteria in biogeochemical fluxes in the ocean's interior. *Nature*, **322**, 441–3.

Collier, R.W. & Edmond, J.M. (1983) Plankton compositions and trace element fluxes from the surface ocean. In *Trace Metals in Sea Water*, C.S. Wong, E. Boyle, K.W. Bruland, J.D. Burton & E.D. Goldberg (eds), 789–809. New York: Plenum.

Connary, S.C. & Ewing, M. (1972) The nepheloid layer and bottom circulation in the Guinea and Angola Basins. In *Studies in Physical Oceanography*, A.L. Gordon (ed.), 169–84. London: Gordon & Breach.

Eittreim, S., Thorndike, E.M. & Sullivan, L. (1976) Turbidity distribution in the Atlantic Ocean. *Deep Sea Res.*, **23**, 1115–27.

Ewing, M. & Thorndike, E. (1965) Suspended matter in deep ocean water. *Science*, **147**, 1291–4.

Forchhammer, G. (1865) On the composition of sea water in the different parts of the ocean. *Philos. Trans. R. Soc. London*, **155**, 203–62.

Fowler, S.W. (1977) Trace elements in zooplankton particulate products. *Nature*, **269**, 51–3.

Gardner, W.D., Southard, J.B. & Hollister, C.D. (1985) Sedimentation, resuspension and chemistry of particles in the northwest Atlantic. *Mar. Geol.*, **65**, 199–242.

Gibbs, R.J. (1974) The suspended material of the Amazon shelf and tropical Atlantic Ocean. In *Suspended Solids in Water*, R.J. Gibbs (ed.), 203–10. New York: Plenum.

Griffiths, A.H. (1978) *Elemental chemistry of surface sea water particulates and eolian dusts*. PhD Thesis, University of Liverpool.

Gross, T.F., Williams, A.J. & Nowell, A.R.M. (1988) A deep-sea sediment transport storm. *Nature*, **331**, 518–21.

Honjo, S. (1980) Material fluxes and modes of sedimentation in the mesopelagic and bathypelagic zones. *J. Mar. Res.* **38**, 53–97.

Honjo, S., Spencer, D.W. & Farrington, J.W. (1982) Deep advective transport of lithogenic particles in Panama Basin. *Science*, **216**, 516–8.

Honjo, S., Doherty, K.W., Agrawal, Y.C. & Asper, V.L. (1984) Direct optical assessment of large amorphous aggregates (marine snow) in the deep ocean. *Deep Sea Res.*, **31**, 67–76.

Isao, K., Hara, S., Terauchi, K. & Kogure, K. (1990) Role of sub-micrometre particles in the ocean. *Nature*, **345**, 242–44.

Jerlov, N.G. (1953) Particle distribution in the ocean. *Rep. Swed. Deep Sea Exped.*, **3**, 71–97.

Krishnaswami, S. & Sarin, M.M. (1976) Atlantic surface particulate: composition, settling rates and dissolution in the deep-sea. *Earth Planet. Sci. Lett.*, **32**, 430–40.

Lal, D. (1977) The oceanic microcosm of particles. *Science*, **198**, 997–1009.

Lal, D. (1980) Comments on some aspects of particulate transport in the oceans. *Earth Planet. Sci. Lett.*, **49**, 520–7.

Lisitzin, A.P. (1972) *Sedimentation in the World Ocean*. Tulsa, OK: Society of Economic Paleontologists and Mineralogists, Special Publication 17.

McCave, I.N. (1975) Vertical flux of particles in the ocean. *Deep Sea Res.*, **22**, 491–502.

McCave, I.N. (1984) Size spectra and aggregation of suspended particles in the deep ocean. *Deep Sea Res.*, **31**, 329–52.

Martin, J.H. (1970) The possible transport of trace metals via moulted copepod exoskeletons. *Limnol. Oceanogr.*, **15**, 756–61.

Martin, J.H. & Knauer, G.A. (1973) The elemental composition of plankton. *Geochim. Cosmochim. Acta*, **37**, 1639–53.

Martin, J.M. & Whitfield, M. (1983) The significance of the river input of chemical elements to the ocean. In *Trace Elements in Sea Water*, C.S. Wong, E. Boyle, K.W. Bruland, J.D. Burton & E.D. Goldberg (eds), 265–96. New York: Plenum.

Martin, J.H., Knauer, G.A., Karl, D.M. & Broenkow, W.W. (1987) VERTEX: carbon cycling in the northeast Pacific. *Deep Sea Res.*, **34**, 267–86.

Masuzawa, T., Noriki, S., Kurosaki, T. & Tsunogai, S. (1989) Compositional changes of settling particles with water depth in the Japan Sea. *Mar. Chem.*, **27**, 61–78.

Moran, S.B. & Buesseler, K.O. (1992) Short residence time of colloids in the upper ocean estimated from [238]U–[234]Th disequilibria. *Nature*, **359**, 221–3.

Preston, M.R. (1989) Marine pollution. In *Chemical Oceanography*, J.P. Riley (ed.), Vol. 9, 53–196. London: Academic Press.

Saarger, M.S. (1994) *On the relationships between dissolved trace metals and nutrients in seawater.* Doctoral Thesis, Vrije Universiteit Amsterdam.

Shanks, A.L. & Trent, J.D. (1980) Marine snow: sinking rates and potential role in vertical flux. *Deep Sea Res.*, **27**, 137–43.

Sherrell, R.M. & Boyle, E.A. (1992) The trace metal composition of suspended particles in the oceanic water column near Bermuda. *Earth Planet. Sci. Lett.*, **111**, 155–74.

Slowey, J.F. & Hood, D.W. (1971) Copper, manganese and zinc concentrations in Gulf of Mexico waters. *Geochim. Cosmochim. Acta*, **35**, 121–38.

Smith, K.L., Williams, P.M. & Druffel, F.R.M. (1989) Upward fluxes of particulate organic matter in the deep North Pacific. *Nature*, **337**, 724–6.

Spencer, D.W. (1984) Aluminium concentrations and fluxes in the ocean. In *Global Ocean Flux Study*, 206–20. Washington, DC: National Academy Press.

Spencer, D.W., Brewer, P.G., Fleer, A., Honjo, S., Krishnaswami, S. & Nozaki, Y. (1978) Chemical fluxes from a sediment trap experiment in the deep Sargasso Sea. *J. Mar. Res.*, **36**, 493–523.

Stoner, J.H. (1974) *Trace element geochemistry of particulates and waters from the marine environment.* PhD Thesis, University of Liverpool.

Turekian, K.K. (1977) The fate of metals in the oceans. *Geochim. Cosmochim. Acta*, **41**, 1139–44.

Wallace, G.T., Hoffman, G.L. & Duce, R.A. (1997) The influence of organic matter and atmospheric decomposition on the particulate trace metal concentration in northwest Atlantic surface seawater. *Mar. Chem.*, **5**, 143–70.

Wells, M.L. & Goldberg, E.D. (1991) Occurrence of small colloids in sea water. *Nature*, **353**, 342–44.

Wells, M.L. & Goldberg, E.D. (1992) Marine submicron particles. *Mar. Chem.*, **40**, 5–18.

Westerlund, S. & Ohman, P. (1991) Cadmium, copper, cobalt, nickel, lead and zinc in the water column of the Weddell Sea, Antarctica. *Geochim. Cosmochim. Acta*, **55**, 2127–46.

# 11 Trace elements in the oceans

Trace elements* are present in sea water at concentrations that range down to picomoles per litre ($pmol\,l^{-1}$) and even lower. Such small concentrations pose extreme analytical problems, and it is only relatively recently that these have been fully overcome. It is now known, for example, that contamination and the lack of sufficiently precise analytical techniques have led to reported concentration data that for some trace elements were too high by factors as much as $10^3$. Because of this, real trends in the trace-element data were sometimes totally masked by noise in the system, creating what Chester (1985) described as a 'frustration barrier', which prevented marine chemists from being able to relate trace-metal distribution patterns to a consistent oceanographic framework. In a keynote review, however, Bruland (1983) pointed out that the mid-1970s had seen a quantum leap in our knowledge of the oceanic distributions of trace elements. This leap had become possible as a result of major improvements in both analytical and collection techniques, especially with regard to the elimination of sample contamination, which allowed the noise to be filtered out of the data. A selection of the new trace-element concentration data in sea water is given in Table 11.1. At the same time that trace-element *concentration* data were being refined, there were also advances in our understanding of the *speciation* of the elements in sea water.

## 11.1 Introduction

One of these key steps in the story of the new (post-

* With a few exceptions, e.g. carbon and the nutrients N, P and Si, when the term 'trace element' is used throughout this volume it refers to elements which are, to differing degrees, metallic. In terms of accepted convention, therefore, for most purposes the term **trace metal** may be considered to be interchangeable with that of **trace element**, and unless otherwise stated it is employed in this way in the text.

1975) dissolved trace-element data was the setting up of GEOSECS (Geochemical Ocean Sections Study), which was designed to provide a framework of hydrographic and geochemical measurements that could be used in the study of oceanic circulation and mixing processes. A number of oceanic water-column sections were carefully selected and vertical concentration profiles of parameters such as salinity, temperature, TSM, dissolved gases and a range of radioactive and stable trace elements were measured. As the new, low-concentration, dissolved trace-element data began to emerge, however, they presented marine chemists with a serious dilemma when attempts were made to interpret them. The reason for this, as Boyle *et al.* (1977) pointed out, was quite simply that neither the fact that extreme precautions had been taken during the collection and analysis of the samples, nor the finding that the results were lower than previous ones, in themselves gave validity to the new data. Boyle *et al.* (1977) concluded, therefore, that, rather than accepting low trace-element concentrations at face value, the validation of the data must rest on three primary criteria.

1 The new trace element concentrations must be confirmed by interlaboratory agreement; this was an integral part of GEOSECS.

2 The vertical distribution profiles obtained from the new data must show smooth variations that can be related to hydrographic and chemical features displayed by conventionally measured properties, which themselves have well established distributions.

3 The regional variations derived on the basis of the new data should be compatible with the large-scale physical and chemical circulation patterns known to operate in the ocean system. In other words, the new trace element data must be *oceanographically consistent*.

The problem therefore revolves around the overall approach that should be adopted when attempts are

**Table 11.1** A selection of 'new' data on the speciations, concentrations and types of vertical distributions of trace elements in sea water (from Bruland, 1983).

| Element | Probable main species in oxygenated sea water | Range and average concentration at 35‰ salinity* | Type of distribution |
|---|---|---|---|
| Li | $Li^+$ | $25\,\mu mol\,kg^{-1}$ | Conservative |
| Be | $BeOH^+$, $Be(OH)_2^0$ | $4–30\,pmol\,kg^{-1}$; $20\,pmol\,kg^{-1}$ | Nutrient-type and scavenging |
| B | $H_3BO_3$ | $0.416\,mmol\,kg^{-1}$ | Conservative |
| C | $HCO_3^-$, $CO_3^{2-}$ | $2.0–2.5\,mmol\,kg^{-1}$; $2.3\,mmol\,kg^{-1}$ | Nutrient-type |
| N | $NO_3^-$ (also as $N_2$) | $<0.1–45\,\mu mol\,kg^{-1}$; $30\,\mu mol\,kg^{-1}$ | Nutrient-type |
| O | $O_2$ (also as $H_2O$) | $0–300\,\mu mol\,kg^{-1}$ | Mirror image of nutrient-type |
| F | $F^-$, $MgF^+$ | $68\,\mu mol\,kg^{-1}$ | Conservative |
| Na | $Na^+$ | $0.468\,mol\,kg^{-1}$ | Conservative |
| Mg | $Mg^{2+}$ | $53.2\,mmol\,kg^{-1}$ | Conservative |
| Al | $Al(OH)_4^-$, $Al(OH)_3^0$ | $(5–40\,nmol\,kg^{-1}$; $20\,nmol\,kg^{-1})$ | Mid-depth minima |
| Si | $H_4SiO_4$ | $<1–180\,\mu mol\,kg^{-1}$; $100\,\mu mol\,kg^{-1}$ | Nutrient-type |
| P | $HPO_4^{2-}$, $NaHPO_4^-$, $MgHPO_4^0$ | $<1–3.5\,\mu mol\,kg^{-1}$; $2.3\,\mu mol\,kg^{-1}$ | Nutrient-type |
| S | $SO_4^{2-}$, $NaSO_4^-$, $MgSO_4^0$ | $28.2\,mmol\,kg^{-1}$ | Conservative |
| Cl | $Cl^-$ | $0.546\,mol\,kg^{-1}$ | Conservative |
| K | $K^+$ | $10.2\,mmol\,kg^{-1}$ | Conservative |
| Ca | $Ca^{2+}$ | $10.3\,mmol\,kg^{-1}$ | Slight surface depletion |
| Sc | $Sc(OH)_3^0$ | $8–20\,pmol\,kg^{-1}$; $15\,pmol\,kg^{-1}$ | Surface depletion |
| Ti | $Ti(OH)_4^0$ | $(<20\,nmol\,kg^{-1})$ | ? |
| V | $HVO_4^{2-}$, $H_2VO_4^-$, $NaHVO_4^-$ | $20–35\,nmol\,kg^{-1}$; $30\,nmol\,kg^{-1}$ | Slight surface depletion |
| Cr | $CrO_4^{2-}$, $NaCrO_4^-$ | $2–5\,nmol\,kg^{-1}$; $4\,nmol\,kg^{-1}$ | Nutrient-type |
| Mn | $Mn^{2+}$, $MnCl^+$ | $0.2–3\,nmol\,kg^{-1}$; $0.5\,nmol\,kg^{-1}$ | Depletion at depth |
| Fe | $Fe(OH)_3^0$ | $0.1–2.5\,nmol\,kg^{-1}$; $1\,nmol\,kg^{-1}$ | Surface depletion, depletion at depth |
| Co | $Co^{2+}$, $CoCO_3^0$, $CoCl^+$ | $(0.01–0.1\,nmol\,kg^{-1}$; $0.02\,nmol\,kg^{-1})$ | Surface depletion, depletion at depth |
| Ni | $Ni^{2+}$, $NiCO_3^0$, $NiCl^+$ | $2–12\,nmol\,kg^{-1}$; $8\,nmol\,kg^{-1}$ | Nutrient-type |
| Cu | $CuCO_3^0$, $CuOH^+$, $Cu^{2+}$ | $0.5–6\,nmol\,kg^{-1}$; $4\,nmol\,kg^{-1}$ | Nutrient-type and scavenging |
| Zn | $Zn^{2+}$, $ZnOH^+$, $ZnCO_3^0$, $ZnCl^+$ | $0.05–9\,nmol\,kg^{-1}$; $6\,nmol\,kg^{-1}$ | Nutrient-type |
| Ga | $Ga(OH)_4^-$ | $(0.3\,nmol\,kg^{-1})$ | ? |
| Ge | $H_4GeO_4$, $H_3GeO_4^-$ | $\leqslant7–115\,pmol\,kg^{-1}$; $70\,pmol\,kg^{-1}$ | Nutrient-type |
| As | $HAsO_4^{2-}$ | $15–25\,nmol\,kg^{-1}$; $23\,nmol\,kg^{-1}$ | Nutrient-type |
| Se | $SeO_4^{2-}$, $SeO_3^{2-}$, $HSeO_3^-$ | $0.5–2.3\,nmol\,kg^{-1}$; $1.7\,nmol\,kg^{-1}$ | Nutrient-type |
| Br | $Br^-$ | $0.84\,mmol\,kg^{-1}$ | Conservative |
| Rb | $Rb^+$ | $1.4\,\mu mol\,kg^{-1}$ | Conservative |
| Sr | $Sr^{2+}$ | $90\,\mu mol\,kg^{-1}$ | Slight surface depletion |
| Y | $YCO_3^+$, $YOH^{2+}$, $Y^{3+}$ | $(0.15\,nmol\,kg^{-1})$ | ? |
| Zr | $Zr(OH)_4^0$, $Zr(OH)_5^-$ | $(0.3\,nmol\,kg^{-1})$ | ? |
| Nb | $Nb(OH)_6^-$, $Nb(OH)_5^0$ | $(\leqslant50\,pmol\,kg^{-1})$ | ? |
| Mo | $MoO_4^{2-}$ | $0.11\,\mu mol\,kg^{-1}$ | Conservative |
| (Tc) | $TcO_4^-$ | No stable isotope | — |
| Ru | ? | ? | ? |
| Rh | ? | ? | ? |
| Pd | ? | ? | ? |
| Ag | $AgCl_2^-$ | $(0.5–35\,pmol\,kg^{-1}$; $25\,pmol\,kg^{-1})$ | Nutrient-type |
| Cd | $CdCl_2^0$ | $0.001–1.1\,nmol\,kg^{-1}$; $0.7\,nmol\,kg^{-1}$ | Nutrient-type |
| In | $In(OH)_3^0$ | $(1\,pmol\,kg^{-1})$ | ? |
| Sn | $SnO(OH)_3^-$ | $(1–12, \sim4\,pmol\,kg^{-1})$ | High in surface waters |
| Sb | $Sb(OH)_6^-$ | $(1.2\,nmol\,kg^{-1})$ | ? |
| Te | $TeO_3^{2-}$, $HTeO_3^-$ | ? | ? |
| I | $IO_3^-$ | $0.2–0.5\,\mu mol\,kg^{-1}$; $0.4\,\mu mol\,kg^{-1}$ | Nutrient-type |
| Cs | $Cs^+$ | $2.2\,nmol\,kg^{-1}$ | Conservative |
| Ba | $Ba^{2+}$ | $32–150\,nmol\,kg^{-1}$; $100\,nmol\,kg^{-1}$ | Nutrient-type |

*Continued on p. 260*

**Table 11.1** *Continued*

| Element | Probable main species in oxygenated sea-water | Range and average concentration at 35‰ salinity* | Type of distribution |
|---|---|---|---|
| La | $La^{3+}$, $LaCO_3^+$, $LaCl^{2+}$ | $13-37\,pmol\,kg^{-1}$; $30\,pmol\,kg^{-1}$ | Surface depletion |
| Ce | $CeCO_3^+$, $Ce^{3+}$, $CeCl^{2+}$ | $16-26\,pmol\,kg^{-1}$; $20\,pmol\,kg^{-1}$ | Surface depletion |
| Pr | $PrCO_3^+$, $Pr^{3+}$, $PrSO_4^+$ | ($4\,pmol\,kg^{-1}$) | Surface depletion |
| Nd | $NdCO_3^+$, $Nd^{3+}$, $NdSO_4^+$ | $12-25\,pmol\,kg^{-1}$; $20\,pmol\,kg^{-1}$ | Surface depletion |
| Sm | $SmCO_3^+$, $Sm^{3+}$, $SmSO_4^+$ | $2.7-4.8\,pmol\,kg^{-1}$; $4\,pmol\,kg^{-1}$ | Surface depletion |
| Eu | $EuCO_3^+$, $Eu^{3+}$, $EuOH^{2+}$ | $0.6-1.0\,pmol\,kg^{-1}$; $0.9\,pmol\,kg^{-1}$ | Surface depletion |
| Gd | $GdCO_3^+$, $Gd^{3+}$ | $3.4-7.2\,pmol\,kg^{-1}$; $6\,pmol\,kg^{-1}$ | Surface depletion |
| Tb | $TbCO_3^+$, $Tb^{3+}$, $TbOH^{2+}$ | ($0.9\,pmol\,kg^{-1}$) | Surface depletion |
| Dy | $DyCO_3^+$, $Dy^{3+}$, $DyOH^{2+}$ | ($4.8-6.1\,pmol\,kg^{-1}$; $6\,pmol\,kg^{-1}$) | Surface depletion |
| Ho | $HoCO_3^+$, $Ho^{3+}$, $HoOH^{2+}$ | ($1.9\,pmol\,kg^{-1}$) | Surface depletion |
| Er | $ErCO_3^+$, $ErOH^{2+}$, $Er^{3+}$ | $4.1-5.8\,pmol\,kg^{-1}$; $5\,pmol\,kg^{-1}$ | Surface depletion |
| Tm | $TmCO_3^+$, $TmOH^{2+}$, $Tm^{3+}$ | ($0.8\,pmol\,kg^{-1}$) | Surface depletion |
| Yb | $YbCO_3^+$, $YbOH^{2+}$ | $3.5-5.4\,pmol\,kg^{-1}$; $5\,pmol\,kg^{-1}$ | Surface depletion |
| Lu | $LuCO_3^+$, $LuOH^{2+}$ | ($0.9\,pmol\,kg^{-1}$) | Surface depletion |
| Hf | $Hf(OH)_4^0$, $Hf(OH)_5^-$ | ($<40\,pmol\,kg^{-1}$) | ? |
| Ta | $Ta(OH)_5^0$ | ($<14\,pmol\,kg^{-1}$) | ? |
| W | $WO_4^{2-}$ | $0.5\,nmol\,kg^{-1}$ | ? |
| Re | $ReO_4^-$ | ($14-30\,pmol\,kg^{-1}$; $20\,pmol\,kg^{-1}$) | ? |
| Os | ? | ? | ? |
| Ir | ? | ? | ? |
| Pt | ? | ? | ? |
| Au | $AuCl_2^-$ | ($25\,pmol\,kg^{-1}$) | ? |
| Hg | $HgCl_4^{2-}$ | ($2-10\,pmol\,kg^{-1}$; $5\,pmol\,kg^{-1}$) | ? |
| Tl | $Tl^+$, $TlCl^0$; or $Tl(OH)_3^0$ | $60\,pmol\,kg^{-1}$ | Conservative |
| Pb | $PbCO_3^0$, $Pb(CO_3)_2^{2-}$, $PbCl^+$ | $5-175\,pmol\,kg^{-1}$; $10\,pmol\,kg^{-1}$ | High in surface waters, depleted at depth |
| Bi | $BiO^+$, $Bi(OH)_2^+$ | $\leqslant 0.015-0.24\,pmol\,kg^{-1}$ | Depletion at depth |

* Parentheses indicate uncertainty about the accuracy or range of concentration given.

made to interpret the new trace-metal data. According to Edmond *et al.* (1979), the local water-column distributions of dissolved constituents in open-ocean waters reflect oceanic circulation patterns, with a large proportion of the chemical signal being determined by long-distance advection and the mixing of water masses of different end-member compositions. It is necessary, therefore, to extract chemical information from this advective background. This was pointed out initially by Chan *et al.* (1977). These authors presented data on the concentration of Ba in Atlantic Ocean profiles sampled during GEOSECS, and showed that the physical circulation is the dominant factor affecting the distribution of the element in the Atlantic basins. In order to extract chemical information from this background, the authors adopted the approach of relating the distribution of Ba to that of other species with distributions in the water column that are well understood. In this way, they were able to demonstrate that Ba was involved in a deep-water regeneration cycle similar to those of the refractory nutrients, silicate (opal) and calcite—see Worksheet 7.3. This kind of comparison is developed more fully in Section 11.5. For the moment, however, it is important to understand that the North Atlantic is a conservative ocean with respect to *unreactive* elements that spend a relatively long time in sea water. For example, Measures *et al.* (1984) showed that this was the case for Be. However, these authors also pointed out that for *reactive* species, the North Atlantic will not be a conservative ocean. It is apparent, therefore, that the time a dissolved trace element spends in the oceanic water column will exert a basic control on its large-scale distribution patterns. Before attempting to understand the factors that control the distributions of trace elements in sea water, it is therefore necessary to introduce the concept of oceanic residence times.

## 11.2 Oceanic residence times

The residence time of an element in the oceans is the average time it spends in the sea before being removed into the sediment sink. In a *steady-state* system it is assumed that the input of an element (mainly via river run-off, atmospheric deposition and hydrothermal exhalations) per unit time is balanced by its output (mainly via sedimentation). There are a number of ways in which equations can be written to describe this relationship. For example, the residence times ($\tau$) of an element is often calculated from the equation

$$\tau = A/(dA/dt) \tag{11.1}$$

where $A$ is the total amount of the element in suspension or solution in sea water, and $dA/dt$ is the amount introduced or removed per unit time. In making this type of calculation it is assumed that the element is completely mixed in the system in a time that is short compared with its residence time, and that neither $A$ nor $dA/dt$ change appreciably in three to four times this period.

Attempts have been made to estimate the residence times of elements in the oceans using the approach outlined above on the basis of *input* (e.g. Barth, 1952) and *output* (e.g. Goldberg & Arrhenius, 1958) data. Residence-time estimates based on the two techniques are given in Table 11.2. Goldberg (1965) concluded that in spite of the drastic oversimplifications involved there is a remarkable degree of agreement between the residence-time estimates derived from the two techniques, and he was able to draw a number of general conclusions regarding the residence times of elements in the oceans.

1 The values of $\tau$ span a range of six orders of magnitude, e.g. from Na ($2.6 \times 10^8$ yr) to Al (100 yr), which reflect the variations in the reactivities of the elements in sea water.

2 The longest residence times are found for the lower atomic number alkali metals and alkaline earths (excluding Be), which are characterized by a general lack of reactivity of their aqueous ions (mainly simple hydrated cations) in sea water.

3 Intermediate residence times ($\sim 10^3$–$10^4$ yr) are found for trace metals such as Zn, Mn, Co and Cu.

4 The shortest residence times ($\sim 10$–$10^3$ yr) have been calculated for elements such as Al, Ti, Cr and Fe. For these elements the residence times are less than the mixing times for oceanic water masses.

It is of course obvious that any calculation of a residence time carried out using eqn (11.1) is extremely sensitive to the values used for the input and output mechanisms. This especially is true for data on the input of trace metals from their primary sources. Most calculations used only river inputs in the estimation of oceanic residence times, and even now our knowledge of these inputs for trace metals is sparse (see Section 6.1). Further, many residence-time calculations assume that the river inputs are delivered to the open ocean, i.e. they take no account of estuarine processes. Bewers & Yeats (1977) attempted to overcome this problem by using new trace metal data for *net* fluvial fluxes (see Section 6.1.4.2), which they combined with theories on the removal of trace metals in the coastal zone, to update the oceanic residence times of a series of trace metals. A summary of the data provided by these authors is included in Table 11.2; in general, these residence-time estimates tend to be smaller than previous ones, especially for Fe and Mn.

Whitfield (1979) introduced the concept of a **mean oceanic residence time** (MORT), which is defined as the total quantity of an element present in the oceans divided by its input rate (from rivers) or its output rate (to the sediment). The MORT values, which assume the whole ocean to be a well-stirred system, are only approximate quantities because they ignore a number of important input (e.g. atmospheric deposition, hydrothermal venting) and output (e.g. atmospheric exchange, the rock sink) terms. However, they do offer an overview of the reactivities, i.e. the intensity of the particle–water interactions, of elements in the ocean system. Values for MORT are listed with other residence time data in Table 11.2.

The wider problems involved in the overall concept of oceanic residence times have been discussed by a number of authors.

Bruland (1980) attempted to re-evaluate the residence time concept. He pointed out that for elements that are conservative in sea water the residence time can be defined as outlined above, i.e. $\tau = A/(dA/dt)$. These conservative elements usually have residence times that are $\gtrsim 10^6$ yr. The concept of oceanic residence times, however, becomes more complex

**Table 11.2** Residence times of elements in sea water; units, yr.

| Element | Goldberg (1965) River input | Goldberg (1965) Sedimentation | Brewer (1975) | Others | MORT values§ |
|---|---|---|---|---|---|
| Li | $1.2 \times 10^7$ | $1.9 \times 10^7$ | $2.3 \times 10^6$ | — | $5.5 \times 10^5$ |
| B | — | — | $1.3 \times 10^7$ | — | $0.9 \times 10^7$ |
| F | — | — | $5.2 \times 10^5$ | — | $4.8 \times 10^5$ |
| Na | $2.1 \times 10^8$ | $2.6 \times 10^8$ | $6.8 \times 10^7$ | — | $7.4 \times 10^7$ |
| Mg | $2.2 \times 10^7$ | $4.5 \times 10^7$ | $1.2 \times 10^7$ | — | $1.5 \times 10^7$ |
| Al | $3.1 \times 10^3$ | $1.0 \times 10^2$ | $1.0 \times 10^2$ | — | $3.7 \times 10^2$ |
| Si | $3.5 \times 10^4$ | $1.0 \times 10^4$ | $1.8 \times 10^4$ | — | $1.4 \times 10^4$ |
| P | — | — | $1.8 \times 10^5$ | — | $1.9 \times 10^4$ |
| Cl | — | — | $1 \times 10^8$ | — | $1.1 \times 10^8$ |
| K | $1 \times 10^7$ | $1.1 \times 10^7$ | $7 \times 10^6$ | — | $9.2 \times 10^6$ |
| Ca | $1 \times 10^6$ | $8 \times 10^6$ | $1 \times 10^6$ | — | $1.1 \times 10^6$ |
| Sc | — | — | $4 \times 10^4$ | — | $5.5 \times 10^3$ |
| Ti | — | — | $1.3 \times 10^4$ | — | $3.7 \times 10^3$ |
| V | — | — | $8 \times 10^4$ | — | $9.2 \times 10^4$ |
| Cr | — | — | $6 \times 10^3$ | — | $1.1 \times 10^4$ |
| Mn | — | — | $1 \times 10^4$ | 39–53* | $8.9 \times 10^2$ |
| Fe | — | — | $2 \times 10^2$ | 27–30* | $1.8 \times 10^3$ |
| Co | — | — | $3 \times 10^4$ | — | $9.2 \times 10^3$ |
| Ni | $1.5 \times 10^4$ | $1.8 \times 10^4$ | $9 \times 10^4$ | $1.8–3.6 \times 10^3$*, $6 \times 10^3$† | $1.4 \times 10^4$ |
| Cu | $4.3 \times 10^4$ | $5 \times 10^4$ | $2 \times 10^4$ | $4.1–6.4 \times 10^3$*, $5 \times 10^3$‡ | $2.4 \times 10^3$ |
| Zn | — | — | $2 \times 10^4$ | $0.78–1.8 \times 10^4$* | $1.2 \times 10^2$ |

*Continued*

for the non-conservative elements. Dissolved elements that undergo passive particle scavenging tend to have relatively short residence times with respect to the mixing time of the ocean, and their distributions are controlled largely by their *external* inputs. For example, recent data indicate that the residence time of dissolved Al with respect to its atmospheric input, which is the major open-ocean source of the element, is ~100–200 yr. However, the nutrient-type elements (see Section 11.5.2), which are involved in an active biological removal mechanism, undergo a surface depletion and a subsurface regeneration, and this regeneration at depth implies an oceanic residence time that is long with respect to the oceanic mixing cycle (~1600 yr); thus, the residence times of phosphate and silicate have been estimated to be $180 \times 10^3$ and $18 \times 10^3$ yr, respectively. These elements therefore undergo numerous internal cycles within sea water prior to their final removal in the sediments. The nutrient-type trace metals involved in these cycles take part in upwelling, down-column transport via organic carriers and regeneration at depth. As a result, their distributions in the oceans are virtually independent of their external points of entry, and their internal exit points are controlled mainly by *internal* oceanic processes. Bruland (1980) therefore concluded that the nutrient-type elements must have an oceanic residence time that is long with respect to the time-scale of oceanic mixing, and set a lower limit of $\tau$ ~$5 \times 10^3$ yr for such trace metals, e.g. Cd, Zn and Ni. Bruland (1980) then compared this $\tau_{min}$ with some of the more recent residence-time estimates given in the literature. To this end, he cited the following $\tau$ values: Cd, $50 \times 10^3$ yr (input data, Boyle *et al.*, 1976); Ni, $40 \times 10^3$ yr (input data, Sclater *et al.*, 1976); and Ni, $6 \times 10^3$ yr (output data, Sclater *et al.*, 1976). The $\tau$ for Ni based on output data is therefore close to the minimum permitted by the theory, but those estimated from input data are considerably in excess of it. Compared with the nutrient-type elements, the scavenging-type elements will have shorter residence times in sea water, and this nutrient–scavenging difference has important implications for the transport of trace

**Table 11.2** *Continued*

| Element | Goldberg (1965) | | Brewer (1975) | Others | MORT values§ |
| --- | --- | --- | --- | --- | --- |
| | River input | Sedimentation | | | |
| Ga | — | — | $1 \times 10^4$ | — | $1.2 \times 10^4$ |
| As | — | — | $5 \times 10^4$ | — | $3.2 \times 10^4$ |
| Se | — | — | $2 \times 10^4$ | — | $3.7 \times 10^4$ |
| Br | — | — | $1 \times 10^8$ | — | $1.2 \times 10^8$ |
| Rb | $6.1 \times 10^6$ | $2.7 \times 10^5$ | $4 \times 10^6$ | — | $2.9 \times 10^6$ |
| Sr | $1.0 \times 10^4$ | $1.9 \times 10^7$ | $4 \times 10^6$ | — | $4.9 \times 10^6$ |
| Mo | $2.1 \times 10^6$ | $5 \times 10^5$ | $2 \times 10^5$ | — | $7.3 \times 10^5$ |
| Ag | $2.5 \times 10^5$ | $2.1 \times 10^6$ | $4 \times 10^4$ | — | $4.9 \times 10^3$ |
| Cd | — | — | — | $7.7–9.2 \times 10^3$* | $1.8 \times 10^4$ |
| Sb | — | — | $7 \times 10^3$ | — | $8.8 \times 10^3$ |
| I | — | — | $4 \times 10^5$ | — | $3.1 \times 10^5$ |
| Cs | — | — | $6 \times 10^5$ | — | $4.2 \times 10^5$ |
| Ba | $5 \times 10^4$ | $8.4 \times 10^4$ | $4 \times 10^4$ | — | $1.2 \times 10^4$ |
| La | — | — | $6 \times 10^{12}$ | — | $2.2 \times 10^3$ |
| W | — | — | $1.2 \times 10^5$ | — | $1.2 \times 10^5$ |
| Au | — | — | $2 \times 10^5$ | — | $7.3 \times 10^4$ |
| Hg | — | — | $8 \times 10^4$ | $3.5 \times 10^2$‡ | $1.8 \times 10^4$ |
| Pb | $5.6 \times 10^2$ | $2 \times 10^3$ | $4 \times 10^2$ | — | $1.1 \times 10^3$ |
| Th | — | — | $2 \times 10^2$ | — | $3.7 \times 10^3$ |
| U | — | — | $3 \times 10^6$ | — | $4.9 \times 10^5$ |

* Bewers & Yeats, 1977; † Martin & Whitfield, 1983; ‡ Gill & Fitzgerald, 1988; § Martin & Whitfield, 1983.

elements in the ocean system. For example, the nutrient-type elements, which undergo recycling and have a longer residence time, can take part in inter-ocean transport and will tend to be enriched in the older intermediate waters of the North Pacific at the end of the oceanic 'grand tour'; thus, there is a fivefold inter-oceanic enrichment of dissolved Zn in the North Pacific compared with the North Atlantic. In contrast, scavenging-type, or particle-reactive, elements tend to be depleted in the North Pacific relative to the North Atlantic, for which the external inputs are greater; dissolved Al is an example of this, showing a 40-fold depletion in the North Pacific (Orians & Bruland, 1985). This kind of inter-ocean fractionation is important in the authigenic flux of trace elements to sediments, and this is discussed in Section 16.5.

It may be concluded, therefore, that in view of the very considerable difficulties involved, the overall *trends* in the residence time characteristics of the various elements given in Table 11.2 hold in a general sense, but the actual residence time values themselves should still be regarded as being no more than speculative.

## 11.3 An oceanic trace-metal framework

When a box-model approach is adopted it is convenient to divide the ocean into a surface and a deep reservoir (see Section 7.5), and if a system of oceanographic consistency is the principal aim of marine chemistry it must apply to trace-element distributions in both reservoirs.

*The surface-water reservoir.* The most important sources for the input of trace elements to the surface ocean reservoir are: (i) river run-off at the ocean margins, (ii) atmospheric deposition, which operates to varying degrees over the entire surface ocean, (iii) diffusion from shelf sediments, and (iv) upwelling, which has its greatest effects in certain well-defined areas of high primary production. The distributions of trace elements in the surface ocean therefore will reflect the magnitudes of the external, i.e. river and atmospheric, source-term signals and the effects of surface circulation patterns, upon which are superimposed the results of internal (i.e. oceanic) processes involved in the major biogeochemical cycles.

*The deep-water reservoir.* Trace elements are supplied to this reservoir from: (i) the surface ocean, by processes such as downwelling and the settling of TSM; and (ii) the deep ocean itself, from mid-depth hydrothermal sources and diffusion from bottom sediments. Deep-water trace-element distributions therefore will be governed by the strengths of the source-term signals and the effects of deep-water circulation patterns, upon which are superimposed the involvement of the elements in the internal biogeochemical processes.

In general, therefore, there are three parameters that must be considered when any attempt is made to describe the distributions of trace elements (or other components) within the oceans. These are: **source terms**, which supply the elements; large-scale **circulation processes**, which govern the transport of both dissolved and particulate elements within the oceans; and internal **biogeochemical processes**, which involve particulate–dissolved interactions and lead, sometimes via recycling stages, to the eventual removal of the trace elements from the system. In any oceanographically consistent framework, the distributions of the trace elements must be related to all three of these parameters. In order to set up such a framework within which to evaluate the new trace-element data, it is convenient to consider the *geographical* (or lateral) and the *vertical* distributions of the element separately.

In recent years new data have been reported on the concentrations of a wide range of trace elements in waters from all the major oceans, and the list of elements studied is growing all the time. A compilation of the ranges of concentrations and their averages is shown in Table 11.1. However, to attempt a 'periodic table' element-by-element description of all the trace constituents in sea water is clearly beyond the scope of this volume. Rather, the treatment adopted will concentrate on a limited number of *process-illustrating* elements (e.g. Cd, Zn, Ni, Pb, Cu, Al and Mn), which will serve as examples to illustrate the recently developed theories on the factors that control the distributions of trace elements in sea water. Perhaps the two most important features to

emerge from the 'new' trace-element data are: (i) the concentrations are much lower than was previously thought, and (ii) there are dramatic **horizontal** and **vertical** concentration gradients in the oceans.

In some trace-element studies the seawater samples are filtered prior to analysis, whereas in others bulk water samples are analysed; however, as the dissolved fraction is dominant for many trace elements, the data given in the present chapter will generally refer to the dissolved species unless stated otherwise. It must be stressed, however, that evidence is beginning to emerge on the importance of colloids in the particulate–dissolved speciation of trace elements in sea water, especially in estuarine and coastal waters. For example, Moran *et al.* (1996) showed that a significant fraction of the 'dissolved' ($<0.2\,\mu m$) Pb, Fe and Hg in productive shelf waters is associated with colloids. Benoit *et al.* (1994) reported that colloidal Fe, Al, Pb and Zn (but not Cu) in six Texas estuaries accounted for most of the concentrations of these trace elements passing through a $0.4\,\mu m$ filter (see also, Muller, 1996). Martin *et al.* (1995) found that in the Venice Lagoon (Italy) 18% of the Ni, 34% of the Cd, 46% of the Cu, 54% of the Mn, 58% of the Pb, and 87% of the Fe passing through a $0.4\,\mu m$ filter, i.e. which would previously have been classified as 'dissolved', were in fact associated with colloidal material. The study of the role played by colloids in oceanic chemistry is still in its infancy but probably will be one of the major research 'growth areas' in the coming years, especially with respect to all forms of trace-element speciation.

## 11.4 Geographical variations in the distributions of trace elements in surface ocean waters: coastal–open-ocean horizontal gradients

In Section 11.3 it was pointed out that the distributions of trace elements in the surface ocean reservoir are controlled by the magnitude of the external source signals and the effects of surface circulation, upon which are superimposed the results of internal (i.e. oceanic) processes involved in the major biogeochemical cycles. These parameters combine together to maintain surface-water trace-element concentration patterns in both coastal and open-ocean environments.

Coastal seas are regions of importance for both

**Table 11.3** Trace-metal fluxes to the North Sea; units, t yr$^{-1}$ (after Chester *et al.*, 1994).

| Trace metal | Atmospheric fluxes | Fluvial fluxes | Discharges and dumping fluxes |
|---|---|---|---|
| Ni | 869 | 240–270 | 903 |
| Cr | 277 | 590–630 | 3382 |
| Cu | 544–1067 | 1290–1330 | 1580 |
| Zn | 3891–4145 | 7360–7370 | 9850 |
| Pb | 1058–2042 | 920–980 | 2470 |
| Cd | 49 | 46–52 | 43 |

their economic (e.g. mineral and oil exploration, fishing, waste disposal) and social (e.g. leisure activities, wildlife preservation) utilization. The external input signals of the principal transport mechanisms (fluvial run-off and atmospheric deposition) that supply trace metals to oceanic surface waters from continental sources are stronger to coastal than open-ocean waters. Further, there are a number of additional pollutant inputs to the coastal zone. These include: (i) **disposal dumping**—e.g. sewage sludge, radioactive waste, dredge spoils, military hardware, off-shore structures; (ii) **deliberate discharges**—e.g. from sewage works, power stations, industrial plants, oil refineries, radioactive reprocessing plants; and (iii) **accidental discharges**—e.g. from oil tanker wrecks. An indication of the importance of these additional pollutant sources to a coastal zone is given by the data in Table 11.3, from which it can be seen that for the North Sea, inputs from discharges and dumping exceed those from fluvial and atmospheric sources for some trace metals. It may be concluded that shallow coastal seas are a major receiving zone for both natural and pollutant inputs to the ocean, and the health of coastal waters has given rise to much concern over recent years.

One effect of the enhanced inputs to the coastal zone is that the surface-water concentrations of some trace elements are higher in coastal and shelf waters than they are in open-ocean waters. Examples of this have been provided for a number of individual trace metals.

**1** *Iron.* Martin & Gordon (1988) reported a 300-fold decrease in surface-water total Fe concentrations on a transect from the North Pacific central gyre (~0.3 nmol l$^{-1}$) to the California coast (~100 nmol l$^{-1}$), and Symes & Kester (1985)

found a 100-fold decrease in total Fe from offshore (~3 nmol l$^{-1}$) to inshore waters (~300 nmol l$^{-1}$) on a transect across the USA Atlantic continental shelf.

**2** *Manganese.* Landing & Bruland (1987) measured a 10-fold decrease in dissolved Mn in surface waters on a California Current (~10 nmol l$^{-1}$) to North Pacific central gyre (~1 nmol l$^{-1}$) transect.

**3** *Zinc.* Bruland & Franks (1983) reported a 40-fold decrease in Zn concentrations in Atlantic surface water from the USA east coast (~2.6 nmol l$^{-1}$) to the Sargasso Sea (~0.06 nmol l$^{-1}$).

In the coastal zone trace elements are delivered to surface or, for some pollutant inputs, subsurface waters and the coastal → open-ocean trace-metal concentration decreases reported above largely reflect differences in external source inputs between the two environments. However, the coastal zone is not simply a body of water that acts as the receiving reservoir for the **external** inputs of trace elements. Within it, the trace elements are distributed between the water column, biota and sediment reservoirs, and the zone provides an environment of intense **internal** reactivity involving a complex interplay between physical, chemical and biological processes (see Fig. 11.1). Together, these processes constrain the fates of trace elements, and may result in their permanent, or temporary, retention within the coastal zone, or their export out of it.

The export of trace elements out of the coastal zone involves the physical transfer of either dissolved or particulate trace elements by advective water transport. The retention of trace elements within the zone occurs via a number of often interlinked physical, biological and chemical processes.

### 11.4.1 Permanent retention

**1** Physical processes. Kremling (1983, 1985) provided data on the dissolved concentrations of Cd, Cu, Ni and Mn on an 'open-ocean North Atlantic to northern Scottish coast to European North Sea coast' transect. The most striking feature in the surface-water distributions of the metals was a sharp decrease in concentration as the shelf–open-ocean Atlantic boundary was crossed—see Fig. 11.2. The shelf edge is characterized by strong horizontal salinity gradients at **fronts**, which mark the boundary between Scottish coastal and open-ocean water regimes. The data thus offer an example of how trace metals can be

(a)

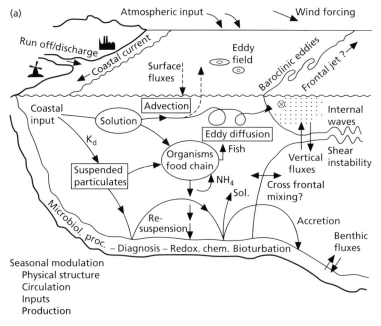

Seasonal modulation
  Physical structure
  Circulation
  Inputs
  Production

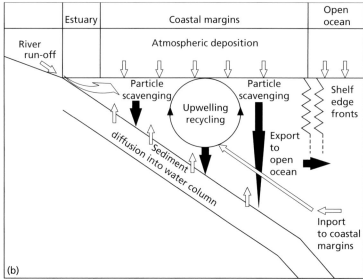

(b)

**Fig. 11.1** Trace metals in the coastal zone. (a) General processes influencing the fates of trace metals and pollutants in coastal margins (from Simpson, 1994). (b) Schematic representation of trace-metal pathways in the coastal margins.

retained in the coastal zone by the physical circulation regime.

**2** Chemical processes. The permanent retention of dissolved trace elements in the coastal zone via chemical processes requires their scavenging transfer to the particulate state and the incorporation of the particulates into bottom sediments. However, incorporation into sediments need not be a 'one-step' trap-

ping process. Coastal and shelf sediments have a relatively high concentration of organic matter (see Section 14.2), and are thus susceptible to suboxic as well as oxic diagenesis (see Section 14.1). As a result, elements that have been incorporated into reducing coastal sediments can be released by diagenetically mediated processes (see Section 14.3.2), and some fraction of these can escape back into the overlying

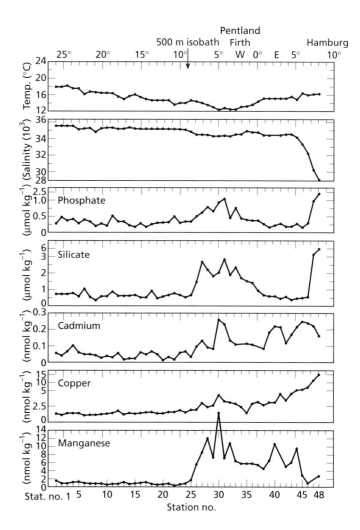

**Fig. 11.2** Surface-water distributions of Cd, Cu, Ni, Mn and Al on an open-ocean Atlantic–European coast transect (from Kremling, 1983): the shelf edge is indicated by the 500 m isobath.

waters; e.g. from direct diffusion across the sediment–water interface, or by the sweeping out of interstitial waters during sediment resuspension.

### 11.4.2 Temporary retention

This can occur, for example, during upwelling associated with coastal primary production. During primary production there is an uptake of 'nutrient-type' trace metals by biota. Following the death of the biota and the sinking down the water column of their remains, nutrients and nutrient-type trace metals are released from degraded particulate matter into solution in subsurface waters and transferred back to the surface as part of the 'nutrient loop' (see Section

9.1.2). This recycling works to retain the trace metals in the water column of the coastal zone for longer than they would otherwise have remained there, but it is not a permanent retention; for example, water brought to surface by coastal upwelling can be advected into the central gyres along isopycnals (see e.g. Collier & Edmond, 1984).

It was shown above that the strength of the source signals to coastal waters can result in them having higher trace element concentrations than open-ocean waters. This can impose 'fingerprints' on the concentration patterns, which sometimes can be traced out into the open-ocean. However, because of the complex interrelated biogeochemical processes occurring in the coastal zone, the coastal →

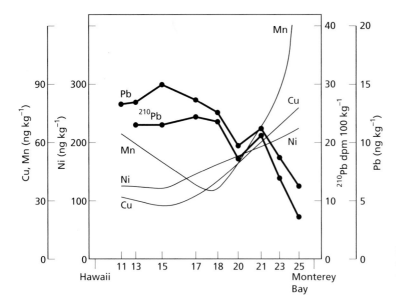

**Fig. 11.3** Trace-element surface-water distribution patterns on a North Pacific transect (from Schaule & Patterson, 1981).

open-ocean decreases in concentration reported above for Fe, Mn and Zn are not necessarily found for all trace elements. For example, if the trapping of an element can involve its permanent removal from solution in the coastal zone, and if it has a relatively strong atmospheric source direct to open-ocean surface waters, its dissolved concentration may increase away from the coastal zones despite the fact that it has a stronger source signal to nearshore waters.

### 11.4.3 Coastal–open-ocean horizontal gradients

Surface-water distribution patterns have been used to unscramble coastal and open-ocean trace-metal input signals. This type of approach was adopted by Schaule & Patterson (1981) who reported data for the distributions of a series of trace metals in surface ocean waters of the North Pacific on a transect that extended from off the coast of California out to the central gyre. The transect passed through a variety of oceanic environments, which included: (i) the nutrient-rich waters of the biologically productive outer shelf region; (ii) an intermediate open-ocean region; and (iii) the biologically non-productive centre of the North Pacific gyre. Generalized surface-water distribution profiles of the trace metals determined in the investigation are illustrated in Fig. 11.3.

On the basis of their distributions along this transect the trace metals can be divided into four broad groups.

### 1 *Copper and nickel*

The distributions of Cu and Ni exhibit a negative, i.e. decreasing, horizontal surface-concentration gradient away from the coastal upwelling area out towards the open-ocean, and reach their lowest values around the central gyre. This type of negative coastal → open-ocean surface-water concentration profile is to be expected:

1 if trace metals are supplied to the coastal receiving zones by processes such as river run-off, atmospheric deposition, diffusion from sediments and shallow-depth coastal upwelling;
2 if only a fraction of the elements from these inputs subsequently escapes the coastal zone and is transported horizontally to the open ocean by advection–diffusion processes.

### 2 *Lead*

The concentrations of Pb in the Pacific transect exhibit a gradient that is the opposite to that shown by both Cu and Ni, i.e. the lowest Pb concentrations are found in the biologically active waters of the outer shelf region and there is a general increase out towards the central gyre, in which the surface waters have about three times as much dissolved Pb as those of the shelf region. The concentrations of dissolved Pb at depth in the water column are lower than those in surface waters (see Section 11.5.3), which precludes the possibility that vertical redistribution by upwelling can be the dominant supply mechanism for Pb in the mixed layer. Schaule & Patterson (1981) concluded, therefore, that the surface-water distribution of dissolved Pb had resulted from strong *external* inputs to the surface waters. However, the surface-water Pb distribution pattern, with lower values in the coastal areas and higher values in the central gyre, has a concentration gradient which means that the external input into coastal waters cannot all subsequently undergo off-shelf transport by advection–diffusion to the open-ocean. The distribution of common Pb along the Pacific transect is very similar to that of $^{210}$Pb, which has a predominantly aeolian input to the ocean. Both forms of Pb are subject to removal processes from surface waters, and Schaule & Patterson (1981) proposed that the distribution of Pb in the surface waters of the Pacific transect was maintained by the element having a strong atmospheric input to both coastal and open-ocean areas, coupled with a higher rate of removal from surface waters in the coastal areas, where the rates of biological activity, and so the concentrations of TSM, are higher than they are in the central gyres. The predominantly atmospheric input of Pb to the oceans is entirely consistent with the elemental source strengths discussed in Section 6.4.

## 3 Manganese

The surface-water distribution of Mn along the Pacific transect exhibits a pattern that is intermediate between those of Cu and Ni on the one hand, and Pb on the other. That is, the highest concentrations of Mn are found in the coastal waters and there is a general decrease out into the open ocean; thus, Mn exhibits a negative concentration gradient, which is similar to those of Cu and Ni—see Fig. 11.3. This type of profile suggests that there must be a significant input of Mn to the coastal receiving zone (e.g. by river run-off or diffusion from bottom sediments), where much of it is removed from the surface waters, so leading to a diminished lateral transport of the

element to the open ocean. In the coastal zone, therefore, it would appear that the dominant input to Mn is by fluvial transport and/or sediment diffusion, whereas that for Pb is largely atmospheric. In the open-ocean waters, however, the surface distribution of Mn differs from those of Cu and Ni in that its concentrations increase to reach higher values in the central gyre. As with Pb, this type of distribution may indicate an atmospheric input to the open-ocean North Pacific, with the lowest rate of removal from surface waters occurring in the biologically inactive central gyre. Unlike Pb, however, the input of Mn from river run-off and/or sediment diffusion in the coastal regions overshadows that from the atmosphere (see also Landing & Bruland, 1980; Bruland, 1983). There is no doubt, however, that in some open-ocean regions the surface waters can retain 'fingerprints' of an atmospheric input of Mn.

## 4 Other trace metals

Surface-water profiles are also available for elements that were not studied by Schaule & Patterson (1981). From the point of view of understanding the involvement of trace metals in the biogeochemical oceanic cycles, Cd is one of the most important of these elements. Boyle *et al.* (1981) and Boyle & Huested (1983) gave data on the distribution of this element in surface waters of the North Atlantic and North Pacific oceans, and reported that the highest Cd concentrations were found in areas of equatorial upwelling (80 pmol l$^{-1}$) and the lowest in the non-upwelling open-ocean (<10 pmol l$^{-1}$). Kremling (1985) also found that Cd was enriched in high-latitude, nutrient-rich, waters of the Atlantic relative to the oligotrophic subtropical waters. Another element that is important from the process-orientated viewpoint is Al. Orians & Bruland (1986) presented results which showed that in the North Pacific the concentrations of dissolved Al are lowest in the eutrophic California Current (0.3–1 nmol l$^{-1}$) and increase westwards into the subtropical North Pacific Gyre (~5 nmol l$^{-1}$), with values in the South Pacific Gyre being fivefold lower than those in the North Pacific. The authors suggested that much of the fluvial input of dissolved Al is taken out of solution by intense scavenging in the particle-rich estuarine and highly productive coastal zones, and concluded that the increasing coastal → open-ocean horizontal concentration trends indicate that the primary source of dissolved Al to the surface ocean is via solubilization from aeolian dust following atmospheric deposition. Surface values for dissolved Al therefore will reflect the atmospheric source strength; thus, concentrations decrease in the order North Atlantic > North Pacific > South Pacific (see also Orians & Bruland, 1985).

### 11.4.4 Geographical variations in trace elements: summary

The general patterns in the surface-water distributions of a number of process-orientated trace elements in the oceans have been described above, and these can be summarized as follows.

The concentrations of many trace elements are usually higher, sometimes by orders of magnitude, in the surface waters of the coastal receiving zones than they are in open-ocean waters. This reflects the inputs from two major *primary* external sources, i.e. fluvial run-off and atmospheric deposition, both of which have stronger trace-metal signals in the coastal environment. The coastal receiving zone is defined by a series of intensively active boundaries, which include the river–ocean, the air–sea, the sediment–seawater and the shelf–slope–open-ocean interfaces. Boundary processes occurring across these interfaces make the zone a region of high trace-element reactivity. Once they have reached the coastal zone, trace elements can be affected by a number of processes, the most important of which are summarized below.

1 They can be removed into bottom sediments in association with inorganic and/or organic (biological) particulate matter.

2 They can undergo recycling. This can take place in the water column, e.g. within the biological removal–upwelling cycle, or across the sediment–water or air–sea interfaces.

3 They can be exported directly out of the coastal zone to the open ocean.

Trace elements are transported to open-ocean surface waters from the coastal receiving zones by advection and diffusion. As admixture between shelf, slope and open-ocean waters takes place there is a negative trace-element concentration gradient away from the boundaries of the continents, leading to progressively smaller surface-water concentrations in the more remote oceanic areas. However, this overall lateral, physically transported, negative coastal → open-ocean concentration gradient can be modified in two principal ways.

1 The gradient will be sharpened if the trace element has a predominantly nearshore supply and undergoes very strong removal in the coastal zone by processes such as enhanced scavenging, nutrient trapping and retention at fronts.

2 The direct open-ocean supply of trace elements can significantly modify the coastal → open-ocean negative concentration gradients, sometimes to the extent of totally reversing the trend. For example, an atmospheric supply coupled with a strong scavenging removal in coastal waters results in the surface-water concentrations of Pb in the North Pacific being highest in the remote unproductive regions of the central gyre.

These elemental distribution patterns raise the critical question of the status of the central oligotrophic oceanic gyres in the marine cycles of the trace metals, and open up the possibility of using interoceanic differences in trace-element concentrations as source indicators. According to Bruland & Franks (1983) it is to be expected that trace element concentrations in the subtropical gyres of the North Atlantic and the North Pacific, i.e. oceanographically equivalent areas, should reflect any differences in the magnitudes of the source terms in the two regions. The North Atlantic has stronger river and atmospheric input signals that does the North Pacific (see Table 6.21), with the result that the two principal external sources of trace elements to surface waters deliver greater quantities of the metals to the North Atlantic. The average concentrations of a series of trace metals in the surface waters of the subtropical gyres in the North Atlantic and North Pacific are listed in Table 11.4. From this table it can be seen that the concentrations of Mn, Cu and Pb in the surface waters of the North Atlantic Gyre are about twice as high as those in the North Pacific Gyre, thus reflecting the magnitudes of their external inputs to the two regions. The concentrations of Cd, Zn and Ni are essentially the same in surface waters from both gyre systems, however, suggesting that a source, or sources, in addition to these arising from river and atmospheric transport must be operating to maintain this surface-water similarity. It was shown above that some elements (e.g. Cd, Ni) appear to have higher surface-water concentrations in regions of upwelling. Thus, the transport of water from depth in the oceanic column would

**Table 11.4** Concentrations of some trace metals in surface and deep waters of the subtropical central oceanic gyres; units, nmol l$^{-1}$.

| Element | North Atlantic Gyre | | North Pacific Gyre | |
|---|---|---|---|---|
| | Surface water* | Deep water† | Surface water* | Deep water† |
| Cd | 0.0020 | 0.29 | 0.0015 | 0.87 |
| Cu | 1–1.5 | 2.0 | 0.5 | 4.0 |
| Ni | 2.3 | 6.0 | 2.1 | 10 |
| Zn | 0.06 | 1.7 | 0.07 | 8.5 |
| Pb | 0.17 | 0.025 | 0.075 | 0.005 |
| Mn | 2.4 | 0.6 | 1.0 | 0.2 |

\* Data from Bruland & Franks (1983).
† Data from Bruland & Franks (1983), Schaule & Patterson (1981) and Bruland (1983). Some data have been obtained from distribution profiles and are only approximate.

seem to be the most probable source for these elements to the mixed layer, even in the central gyres where the rate of upwelling is relatively slow.

It may be concluded, therefore, that trace-element distributions in the surface ocean reflect:

1 their source strengths, i.e. the magnitudes of their input mechanisms from both external and internal sources;

2 their removal strengths, i.e. the magnitude of their output mechanisms, which must take account of the upwelling recycling of the nutrient-type trace elements; and

3 surface-water circulation patterns, the effects of which are superimposed on those of both the input and output mechanisms.

## 11.5 The vertical distribution of trace elements in the water column

Surface-water distributions can be extremely useful in identifying the effects that source strengths have on 'fingerprinting' trace-element distributions in the mixed layer. Both the surface and the deep ocean reservoirs are zones of trace-metal reactivity, however, and in order to assess the effects that internal oceanic processes have on trace-metal distributions it is necessary to obtain data on their vertical as well as their lateral profiles. The most common approach that has been adopted to aid the interpretation of vertical profiles is to relate the distributions of individual elements to those of other species with distributions that are relatively well understood. These species include the following:

1 the *nutrients*, which can be used to assess the involvement of trace metals in both the labile and refractory stages of the oceanic biogeochemical cycles;

2 *dissolved oxygen*, to which redox-mediated reactions can be related; and

3 *particulate matter*, to which scavenging reactions and sediment resuspension can be related.

In his review paper, Bruland (1983) utilized the new data to distinguish between a number of types of vertical trace-element distribution profiles in the oceans. These distributions can be summarized as follows.

### 11.5.1 Conservative-type vertical trace-metal profiles

Elements with conservative profiles have a constant concentration relative to salinity, which results from their generally low reactivity in sea water. Trace elements with this type of profile include the hydrated cations of $Rb^+$ and $Cs^+$, and the molybdate oxyanion $MoO_4^{2-}$.

### 11.5.2 Nutrient-type (recycled), surface-depletion–depth-enrichment vertical trace-metal profiles

The characteristic features of the vertical concentration profiles of trace metals with nutrient-type profiles are (i) a depletion in surface waters, and (ii) an enrichment at some depth within the water column. These features arise from the involvement of the elements in the oceanic biogeochemical cycles (for a description of the nutrient cycles see Section 9.1). Phytoplankton utilize nutrients (phosphate, nitrate, silicate) in the euphotic zone and as they grow they extract trace elements from the water, thus leading to their depletion in these surface waters. As the organisms die and sink down the water column they undergo oxidative decay, during which there is a regeneration of the nutrients, and the associated trace elements, back into solution. There also is a net flux of organic debris out of the euphotic zone (the CPM—see Section 10.4), which carries nutrients and trace metals to deeper waters where decomposition results in a further release of material back into solution as the organic carbon is oxidized. Bruland (1983) identified three types of nutrient-related vertical trace-metal profiles.

*Labile nutrient-type vertical trace-metal profiles.* The labile nutrients (phosphate, nitrate and organic carbon) are associated with the soft tissue phases of organisms, and undergo rapid regeneration in the *upper* water column. Trace metals having labile nutrient-type profiles thus have vertical distributions similar to those of phosphate and nitrate, i.e. they exhibit a surface depletion followed by a shallow-water regeneration, which leads to a mid-depth concentration maximum. This type of profile is shown by dissolved Cd—see Fig. 11.4(a).

The oceanic distribution of dissolved Cd has been studied by a number of workers in the post-1975 period, and concentrations are reported to be in the range $\sim$1–2 pmol l$^{-1}$ in surface waters, rising to $\sim$1 nmol l$^{-1}$ at depth (see e.g. Boyle *et al.*, 1976; Martin *et al.*, 1976; Bruland, 1980; Bruland & Franks, 1983; Danielsson & Westerlund, 1983; Moore, 1983; Boyle & Huested, 1983; Burton *et al.*, 1983; Boulegue, 1983; Spivack *et al.*, 1983; Kremling, 1983, 1985). From these various studies a reasonably detailed picture of the oceanic distribution of Cd has emerged. The predominant feature in the vertical profile of dissolved Cd is a shallow-water regeneration cycle similar to those of phosphate and nitrate. In the North Pacific, for example, dissolved Cd and phosphate are linearly correlated, at phosphate concentrations in excess of 0.2 $\mu$mol l$^{-1}$, by the regression:

$$[Cd] = (0.347 \pm 0.007)[P] - (0.068 \pm 0.017)$$
$$(\text{mean} \pm SD; \ r = 0.992)$$

where [Cd] is in units of nmol l$^{-1}$ and [P] is in units of $\mu$mol l$^{-1}$. However, there are distinct regional changes in Cd–nutrient correlations (see e.g. Boyle & Huested, 1983).

The similarity between the dissolved Cd and phosphate profiles (see Fig. 11.4a) suggests that there is an organic tissue Cd carrier phase produced by phytoplankton in the euphotic layer, which sinks and is decomposed mainly in the upper waters, thus releasing the Cd back into solution; the carrier-phase association is considered in more detail in Section 11.6.3.2. This type of vertical Cd distribution profile appears to be characteristic of the upper water column of all the major oceans. Although the vertical profiles of dissolved Cd are generally similar, however, the deep-water concentrations of the element vary considerably between the major oceans, decreasing in the order North Pacific ($\sim$0.8 nmol l$^{-1}$) > Indian Ocean ($\sim$0.5 nmol l$^{-1}$) > North Atlantic ($\sim$0.3 nmol l$^{-1}$). According to Bruland & Franks (1983), this is a consequence of the deep-water circulation pattern in which the North Pacific lies at the end of the 'global grand tour' Atlantic $\rightarrow$ Indian $\rightarrow$ Pacific deep-water circulation path, and so is older than the deep-water in the other oceans.

*Refractory nutrient-type vertical trace-metal profiles.* The refractory nutrients are associated with the hard skeletal parts of organisms and have a deep-water regeneration. Trace metals having refractory nutrient-type profiles have vertical distributions similar to those of silicate, i.e. a surface depletion followed by a deep-water regeneration, which leads to a deep-water concentration maximum. Zinc is an example of a trace metal having a refractory nutrient-type profile.

Data on the vertical distribution of Zn in the oceanic water column have been provided by various workers, and concentrations seem to range between $\sim$0.05 nmol l$^{-1}$ in surface waters and $\sim$9 nmol l$^{-1}$ at depth (see e.g. Bruland *et al.*, 1978; Bruland, 1980; Danielsson, 1980; Bruland & Franks, 1983; Danielsson & Westerlund, 1983, Magnusson & Westerlund, 1983). In general, the vertical distribution of dissolved Zn is highly correlated with that of silicate, thus showing a strong surface depletion and a deep-water enrichment; a typical dissolved Zn profile, together with that for silicate, is illustrated in Fig. 11.4(b). The correlation between Zn and silicate in North Pacific waters was assessed by Bruland (1980), who reported the following relationship:

$$[Zn] = (0.0535 \pm 0.0008)[Si] - (0.02 \pm 0.09)$$
$$(\text{mean} \pm SD; \ r = 0.996)$$

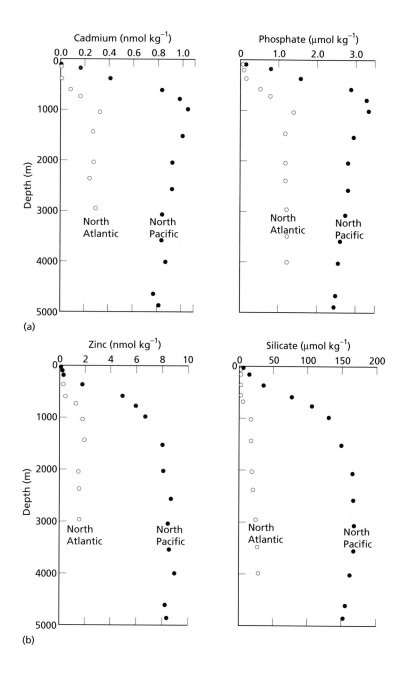

**Fig. 11.4** (a–e) Vertical profiles of dissolved trace elements in the oceans (from Bruland, 1983; which lists the original data sources).

where [Zn] is in units of nmol l⁻¹ and [Si] is in units of μmol l⁻¹. From this type of vertical profile, in which the distribution of Zn mirrors that of silicate, it would appear that the Zn is involved in a deep regeneration cycle of the kind that affects opal (silicate) and calcium carbonate. Like those of Cd, the deep-water concentrations of Zn are controlled by oceanic circulation patterns, with the highest values being found in the North Pacific deep-waters at the end of the global circulation path.

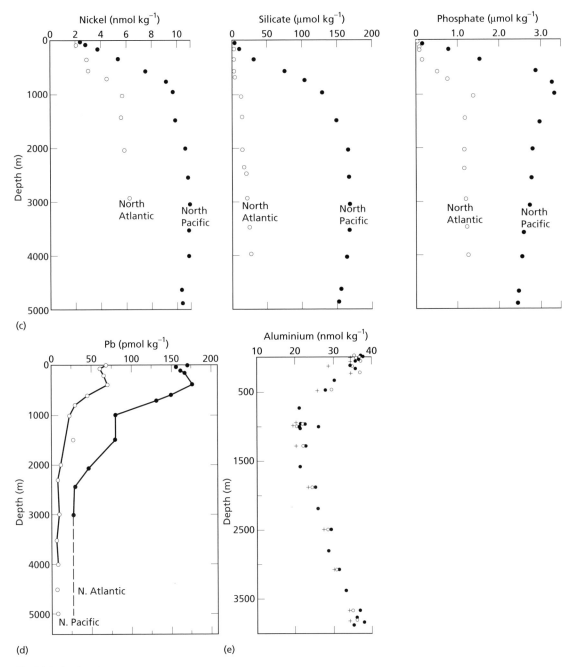

(c)

(d)

(e)

**Fig. 11.4** *Continued*

*Combined labile–refractory nutrient-type vertical trace-metal profiles.* Trace metals with this type of vertical profile have both a shallow-water and a deep-water regeneration cycle. Nickel is an example of such an element.

The vertical distribution of Ni in oceanic waters has been widely reported and concentrations appear to range from ~2–3 nmol l$^{-1}$ in surface waters to ~12 nmol l$^{-1}$ in the deep-waters of the North Pacific (see e.g. Sclater *et al.*, 1976; Bruland, 1980; Danielsson, 1980; Boyle *et al.*, 1981; Bruland & Franks, 1983; Boyle & Huested, 1983; Danielsson & Westerlund, 1983; Magnusson & Westerlund, 1983; Spivack *et al.*, 1983). Although the vertical distribution of Ni in the water column displays a nutrient-type profile, the relationship between the element and the nutrients is not as clear-cut as those for either Cd (labile-type) or Zn (refractory-type). This was demonstrated by Sclater *et al.* (1976), who showed that vertical profiles of dissolved Ni in both the Atlantic and Pacific Oceans could be correlated with either phosphate (labile-type nutrient) or silicate (refractory-type nutrient)—see Fig. 11.4(c). The authors concluded, therefore, that the vertical distribution of Ni in the water column could be interpreted in terms of the incorporation of the element into both the soft tissue parts (phosphate analogue) and the hard skeletal parts (silicate analogue) of organisms, which would lead to both a shallow- and a deep-water regeneration stage, with each having a dissolved Ni concentration maximum. Like all nutrient-type elements, Ni has higher concentrations in the deep-waters of the North Pacific than in those of the North Atlantic.

The surface depletion of the nutrient-type trace metals, together with that of the nutrients themselves, indicates an association with plankton. Some metals such as Cu, Ni and Zn are essential for the growth of phytoplankton and are thought to be involved in metabolic processes. In contrast, the physiological function of Cd, and therefore its phytoplankton requirement, is not apparent, although it has been suggested that it might fulfil the biochemical role of Zn in Zn-depleted waters. Biota are selective with respect to the form of a trace metal taken up from sea water, and this is considered in relation to metal speciation in Section 11.6.2.3.

There is no doubt that the involvement of trace metals such as Cd, Ni, Zn and Cu in the oceanic bio-geochemical cycles leads to them having nutrient-type, or for Cu (see below) mixed nutrient-type–scavenging-type, distributions in the water column. However, the relationships between the trace metals and the nutrients are not always simple. For example, although the correlations between some trace metals and their nutrient analogues can be described by linear equations, there is evidence that the coefficients of these equations are not globally unique and can in fact vary regionally (see e.g. Boyle *et al.*, 1981; Spivack *et al.*, 1983).

### 11.5.3 Scavenging-type, surface-enrichment–depth-depletion vertical trace-metal profiles

Two prime requirements are necessary in order for a dissolved trace element to maintain this type of vertical oceanic profile: (i) it must be introduced mainly into surface waters; and (ii) it must be removed rapidly from solution before it can be transported down the water column, i.e. its residence time must be short relative to the mixing time of the ocean. Trace elements can have a direct input to surface waters either via atmospheric transport (e.g. Pb) or by horizontal mixing into surface waters following delivery from rivers or release from shelf sediments (e.g. Mn). In addition, Bruland (1983) identified a further surface water input in which selected oxidation states, or specific chemical forms, of an element can undergo *in situ* production in surface waters followed by changes in deeper waters. For example, the production of arsenite from arsenate in the euphotic zone is a result of biologically mediated redox reactions, and this is followed by oxidation of the arsenite to arsenate at depth.

Lead is perhaps the most widely studied of the trace metals that exhibit a surface-enrichment–depth-depletion vertical oceanic profile. Much of our knowledge of the distribution of Pb in the marine environment has come from the pathfinding work carried out by Clair Patterson and his group at the California Institute of Technology, Pasadena. Thanks to this work, the principal features in the oceanic distribution of Pb are now known in some considerable detail, and they may be summarized as follows.

1 The concentrations of Pb in surface waters are higher in the North Atlantic than in the North Pacific, and in both localities they increase from the ocean

margins out towards the central gyres (see Section 11.4.3).

**2** Surface-water concentrations of Pb are higher than those in deep-waters. This effect is most pronounced in the North Atlantic—see Fig. 11.4(d).

**3** Deep-water concentrations of Pb in the North Atlantic, i.e. at the start of the global deep-water 'grand tour', are higher than those in the North Pacific, i.e. at the end of the deep-water circulation path.

This is in contrast to the distributions of the nutrient-type trace metals, which have their highest deep-water concentrations in the North Pacific (see above). The factors that produce these features in the oceanic distribution of Pb have been interpreted within a general framework in which the principal input of Pb to the oceans is via the atmosphere into surface waters. The Pb is then removed relatively rapidly throughout the whole water column by scavenging reactions involving small-sized TSM. However, the atmospheric input is of a sufficient magnitude to maintain the enhanced surface-water concentrations. In deep-waters the residence time of dissolved Pb is short compared with the deep-water circulation cycle, which, together with a stronger Atlantic than Pacific surface input signal, results in deep-water Pb concentrations in the North Atlantic being higher than those in the North Pacific.

Recent evidence suggests that, although it has a more complex oceanic chemistry, Hg shares some of the distribution features common to Pb. For example, in the Atlantic Ocean surface-water ($\sim$4 pmol l$^{-1}$) and deep-water ($\sim$4 pmol l$^{-1}$) concentrations are higher than those in the Pacific Ocean: $\sim$1.7 pmol l$^{-1}$ and $\sim$1 pmol l$^{-1}$, respectively (see e.g. Gill & Fitzgerald, 1988; Fitzgerald, 1989). The major input of Hg to the oceans is atmospheric deposition, and this inter-oceanic fractionation is consistent with a stronger atmospheric input of the element to the Atlantic than to the Pacific. In the northwest Atlantic (Sargasso Sea) the vertical profile of Hg in the water column exhibits a maximum within the main thermocline region, which may be an advective feature. Below this maximum the concentrations fall off, indicating that rapid scavenging occurs and maintains the low concentrations found in deep waters; in the northeast Pacific, however, there is apparently a bottom source of Hg, which enhances concentrations in deep waters below $\sim$3000 m. As a result of its

oceanic reactivity, Hg has a relatively short residence time in sea water, which Gill & Fitzgerald (1988) estimated to be $\sim$350 yr; this is less than oceanic mixing times, and indicates that Hg has a high biogeochemical reactivity as well as a rapid removal from the water column. The distribution of Hg is of particular interest because it can be used to illustrate the importance that advection can have on the transport of elements in the oceans. According to Gill & Fitzgerald (1988) Hg is advected to the open-ocean following its diagenetic release from coastal sediments. The authors also suggested, however, that the element can be redistributed within the open-ocean regions themselves through advective water transport via mixing along isopycnals, from areas where there is an enhanced atmospheric input to areas where the input is less strong. Thus, the authors suggested that the Hg maximum in the main thermocline in the Sargasso Sea could have originated from the isopycnal transport of Hg-enriched waters from surface outcrops at higher latitudes, thus demonstrating that horizontal transport can affect the vertical distribution of Hg in the water column. The sources, speciation and fluxes of Hg, and the atmospheric and oceanic cycling of the element, have been reviewed by Fitzgerald (1989).

Some trace metals have vertical distribution profiles in the oceanic water column that combine features of both the nutrient-type (surface-depletion–depth-enrichment) and the scavenging-type (surface-enrichment–depth-depletion) profiles. For example, Cu has a 'mixed nutrient-type–scavenging-type' vertical distribution profile in open-ocean waters, in which the metal is depleted in surface waters but increases in a nearly linear manner throughout the water column. This linear increase with depth shows a departure from the nutrient-type profile and results from the *in situ* scavenging of Cu in deep waters superimposed on the nutrient-type surface assimilation, depth regeneration profile. The conditions necessary to maintain the open-ocean Cu profile therefore are: (i) a primary input to surface waters; (ii) removal in the upper water layer; (iii) intermediate- and deep-water scavenging; and (iv) an increase in concentration towards the sea bed as a result of a bottom source following the release of the metal from the sediment–interstitial water complex.

Iron is another trace element that is now known to have a vertical oceanic profile influenced by both nutrient-type and scavenging-type processes; this is discussed in Section 11.6.2.3.

### 11.5.4 Mid-depth-minimum-type vertical trace-metal profiles

Two examples of these profiles can be identified: those associated with a surface- and a deep-water source, and those associated with suboxic waters.

#### 11.5.4.1 Profiles associated with a surface- and a deep-water source

These are associated with a surface-water source, scavenging throughout the water column and a bottom-water source. The surface-enrichment–depth-depletion type of profile (see above) will be modified significantly if there is also a source signal from sediments at the base of the water column; the three-layer model for the vertical distribution of TSM in the oceanic waters is an analogue for this—see Section 10.2.

Aluminium has been cited as an example of an element that can have a mid-depth-minimum-type profile, which can be maintained by both surface-water and deep-water inputs. Al(III) has a strong tendency to hydrolyse in sea water to form particle-reactive dissolved species, and its concentration apparently is controlled by scavenging processes. Hydes (1979) derived a depth–concentration profile for dissolved Al in the North Atlantic and reported a gradual decrease in concentration from the surface waters ($\sim$38 nmol l$^{-1}$) down to $\sim$1000 m ($\sim$22 nmol l$^{-1}$), followed by a steady increase to $\sim$4000 m ($\sim$38 nmol l$^{-1}$)—see Fig. 11.4(e). To account for this type of vertical distribution, Hydes (1979) outlined the following sequence of events.

1 Al is solubilized from the atmospheric material deposited to the North Atlantic surface-water layers (see Section 6.2).

2 Once the Al is brought into solution, it is scavenged by siliceous shells, leading to a decrease in dissolved Al concentrations below the surface layer. Orians & Bruland (1985) suggested that in addition to passive adsorption on to particles, e.g. in the open-ocean, there is evidence that dissolved Al can be removed from solution via an active biological uptake mechanism, e.g. in confined basins such as the Mediterranean Sea.

3 Unlike the situation for Pb (see above), the decrease in dissolved Al is, in this type of profile, balanced by an input from deep-water sources, e.g. by dissolution from suspended sediment particles.

The vertical water-column distribution of Al, however, can be modified by advective as well as by local source-scavenging relationships. This has been demonstrated for the Atlantic by Measures et al. (1986). One of the interesting features in the vertical distribution of Al in this ocean is that whereas in the northwest Atlantic the profiles exhibit a surface maximum (from aeolian deposition), a mid-depth minimum and an increase in deep water, those in the northeast Atlantic do not show the deep-water increase. Measures et al. (1986) related the different deep-water Al signatures in these and other areas to the distribution of water masses in the vertical oceanic sections. Thus, they suggested that European shelf waters, which are enriched in Al having a fluvial origin, are advected into the southern Greenland Sea where they participate in the formation of Antarctic Intermediate Water, which is the major contributor to the Greenland–Scotland overflows. Thus, North Atlantic Deep Water at high latitudes in both the eastern and western basins is enriched in Al. Although this enrichment extends south to at least 30°N in the western basin, it is not found in the eastern basin, where the circulation is more sluggish at this latitude; this therefore offers a possible explanation for the difference in the deep-water Al profiles in the northwest and northeast Atlantic. According to this concept, therefore, the advection of Al-rich coastal waters to regions of convective water-mass formation can result in the injection of Al-rich waters into the deep interior of the ocean, where they modify the vertical Al profile. The surface enrichment type of dissolved Al water-column profile is evidently typical of areas in which there is a significant atmospheric mineral aerosol signal to the surface waters. However, the concentrations of dissolved Al differ from ocean to ocean, depending on local input source strength signals. For example, Orians & Bruland (1985, 1986) presented data which showed that although the vertical distribution features for dissolved Al are similar in the North Atlantic and the North Pacific, the concentrations are 8–40 times lower in the central North Pacific than in the central North Atlantic. This inter-oceanic fractionation of dissolved Al can be explained by geographical variations in the atmospheric Al sources, which are much stronger in the central North Atlantic. In some regions, however, variations in atmospheric source strength can actually modify the vertical dissolved Al profile. For example, at high latitudes the atmospheric input may not be sufficient to maintain the surface maximum, and here the maximum may in fact be entirely absent, although there is still an input of dissolved Al from the bottom sediments (see e.g. Olafsson, 1983). In general, however, the vertical distribution of Al in sea water is controlled by a surface atmospheric input, a deep-water source (e.g. from the advection of Al-rich waters or via sediment resuspension) and intense scavenging throughout the water column.

### 11.5.4.2. Profiles associated with suboxic waters

According to Bruland (1983) this type of mid-depth minimum, which results from solubilization–precipitation reactions associated with redox changes, can be associated with the presence of subsurface, oxygen-depleted (or suboxic) water layers, which are found in some oceanic areas, such as the eastern tropical Pacific and the northern Indian Ocean (see Section 8.3). Under these conditions, mid-depth concentration minima in elemental profiles can be established when the reduced form of an element is relatively insoluble, or when it is removed from solution in association with particulate phases. Bruland (1983) used Cr(III) as an example of the reduced form of a dissolved element that is stable under suboxic conditions but is rapidly scavenged from the water column by particulate matter.

### 11.5.5 Mid-depth-maxima-type vertical trace-metal profiles

Two profiles of this type were described by Bruland (1983): those associated with mid-depth sources, and those associated with suboxic waters.

#### 11.5.5.1 Profiles associated with mid-depth trace-metal sources

It was shown in Section 6.3 that hydrothermal activity at the spreading ridge crests can introduce major quantities of some elements directly into the water column at mid-depths, thus providing concen-

tration maxima in vertical trace-metal profiles. For example, hydrothermal inputs can cause a dramatic increase in the concentration of Mn in intermediate and deep waters; a profile of this type, taken in the vicinity of the Mid-Atlantic Ridge, is illustrated in Fig. 11.5(e).

#### 11.5.5.2 Profiles associated with suboxic waters

Mid-depth trace-metal concentration maxima can result from in situ redox processes, and occur when the reduced form of an element is relatively soluble in comparison with its oxidized form. Reduced species behaving in this manner include Mn(II) and Fe(II). For example, suboxic Mn concentration maxima associated with an oxygen minimum zone have been found in the Pacific Ocean (see e.g. Klinkhammer & Bender, 1980; Murray et al., 1983; Landing & Bruland, 1987).

### 11.5.6 Anoxic-water-type vertical trace-metal profiles

Anoxic waters can be formed in a variety of marine environments, including:
1 areas where the water circulation is restricted, e.g. in coastal inlets having a fjord type of circulation and in deep-sea trenches;
2 at the surface exit of marine hydrothermal systems;
3 within sediment interstitial waters (see e.g. Emerson et al., 1983).
In regions of restricted water circulation, reducing conditions can be set up as a result of the redox couple $SO_4^{2-}$–$H_2S$. Then, as in the suboxic waters described above, trace-metal maxima can occur when the reduced form of the element is more soluble than the oxidized form (e.g. the Mn(II) and Fe(II) forms of manganese and iron, respectively) and minima result when the reduced form is less soluble or when it is heavily scavenged (e.g. Cr(III)). In some regions where the water column is stratified, with anoxic conditions being developed in subsurface waters, there is a recycling of components between oxic and anoxic layers; e.g. in the Black Sea, the world's largest stable anoxic marine basin. As there are dramatic changes in the solubilities of many trace metals at the oxic–anoxic boundary layer (or **redox front**) at the $O_2$–$H_2S$ interface, this can lead to distinctive vertical trace-metal water-column profiling associated with both redox

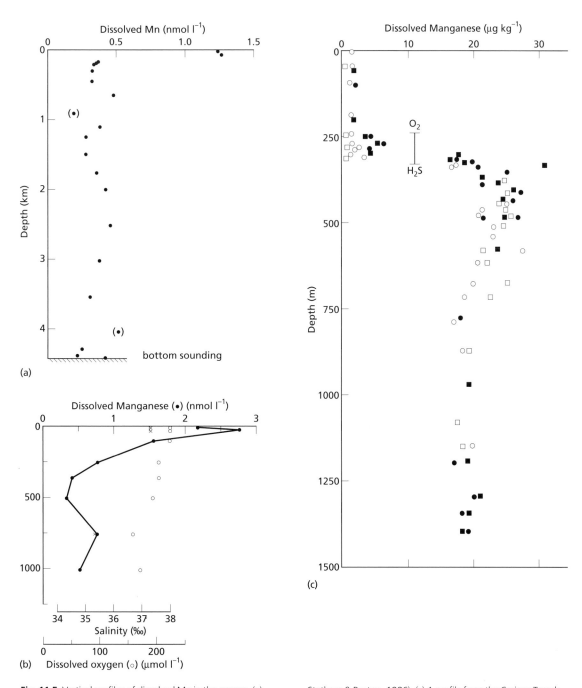

(a)

(b)

(c)

**Fig. 11.5** Vertical profiles of dissolved Mn in the oceans. (a) A profile from the eastern North Atlantic showing a typical distribution of dissolved Mn in the absence of intermediate- and deep-water sources; the principal feature is the high concentration of dissolved Mn in the mixed layer as the result of a surface input (from Statham & Burton, 1986). (b) A profile from the eastern North Atlantic showing a dissolved Mn maximum in the mixed-layer high-concentration region (from Statham & Burton, 1986). (c) A profile from the Cariaco Trench showing an increase in dissolved Mn across the $O_2$–$H_2S$ redox boundary (from Bacon et al., 1980). (d) A profile from the Pacific Ocean showing an increase in dissolved Mn in the region of the oxygen minimum (from Klinkhammer & Bender, 1980). (e) A profile from the Mid-Atlantic Ridge (26°N) showing the effect of hydrothermal inputs (from Klinkhammer et al., 1986).

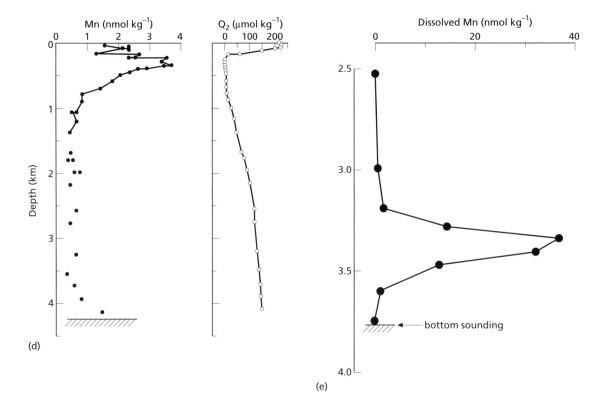

(d)

(e)

**Fig. 11.5** *Continued*

cycling across the interface, e.g. for Mn, and sulphide precipitation in the underlying sulphidic water, e.g. for Cu, Zn and Cd, for which anoxic basins may act as traps (see e.g. Spencer & Brewer, 1971; Emerson *et al.*, 1983; Kremling, 1983; Jacobs *et al.*, 1985; Lewis & Landing, 1991). A number of aspects of the chemistry of the Black Sea have been reviewed in the volume edited by Izdar & Murray (1991), and in *Deep-Sea Research*, Vol. 38 (Supplement 2A).

Various types of trace-element profiles in the oceanic water column have now been identified, and the distributions of a series of process-orientated elements have been described in terms of analogue constituents that have well-established distributions in sea water. In this manner it has been shown that vertical profiles can provide information on the factors that control the seawater chemistries of individual trace elements. For this purpose the characteristic profile shapes of a wide range of elements are identified in Table 11.1. Vertical profiles are therefore

useful as process indicators. However, some elements can exhibit 'multishape' profiles because more than one process has affected their seawater chemistries. This can be illustrated with respect to the vertical distribution of dissolved Mn in the oceanic water column.

Dissolved Mn is present in sea water in the +2 oxidation state, mainly as free hydrated $Mn^{2+}$, and in oxygenated sea water the Mn(II) is thermodynamically unstable with respect to oxidation to insoluble manganese oxides (Bruland, 1983). According to Burton & Statham (1988) the dominant factor that affects the aquatic geochemistry of Mn is the change of oxidation state between reducing and oxidizing environments, coupled with a marked difference in the reactivity of the metal between the two oxidation states; the Mn(II) form, which is the stable state under reducing conditions, is more soluble and geochemically mobile than the Mn(IV) state.

Dissolved Mn(II) ($Mn_d$) is supplied to the *surface* ocean mainly via river run-off and diffusion from shelf sediments at the ocean basin margins, and by dissolution from atmospheric particulates over all regions. The $Mn_d$ is extremely particle reactive, and originally it was thought that the uptake on to suspended particles occurred via microbial oxidation (see e.g. Sunda & Huntsman, 1988). According to Moffett (1997), however, although microbial oxidation is found in the Sargasso Sea, a non-oxidative, biologically mediative uptake occurs in the equatorial Pacific; i.e. the geochemical cycling of Mn is different in the two environments. Relatively high concentrations of $Mn_d$ are maintained in surface waters by atmospheric deposition and *in situ* photochemical reduction and photoinhibition of microbial oxidation, processes that do not occur in the lower euphotic layer, where an increase in particulate Mn sometimes can be found. Dissolved Mn is scavenged throughout the water column on a time-scale of $\sim 10$–$100\,yr$, and once scavenged, this form of Mn is regenerated only under reducing conditions. As a result of these input and removal processes, vertical profiles of $Mn_d$ in a 'typical' oceanic water column show a generally consistent overall pattern in which the concentrations are highest in the surface waters and decrease rapidly to low values, which usually are maintained to the bottom. An example of such a 'surface-input–water-column-scavenging' vertical $Mn_d$ profile is illustrated in Fig. 11.5(a). Subsurface $Mn_d$ maxima are sometimes found within the overall high mixed-layer concentration pattern. An example of such a maximum is shown in Fig. 11.5(b) and, according to Statham & Burton (1986), features of this kind arise as a net result of input, removal, physical mixing, cycling processes (such as photoreduction) and possible involvement of the metal in biological systems. The 'typical' $Mn_d$ vertical profile, however, can be severely perturbed by subsurface sources, and a number of examples of this are described below.

1 Mn redox recycling across the $O_2$–$H_2S$ boundary, leading to an increase in $Mn_d$, can occur in nearshore waters with restricted circulation (e.g. the Black Sea) and in deep-sea trenches (e.g. the Cariaco Trench). An example of a vertical $Mn_d$ profile of this type is given in Fig. 11.5(c).

2 Intermediate-depth $Mn_d$ maxima can be found in some areas at depths close to the oxygen minima (see Section 8.3), especially in the Pacific Ocean. Two principal processes may contribute to the setting up of this type of $Mn_d$ maxima. One of these is the redox-mediated release of pre-scavenged $Mn_d$, which is solubilized from particles sinking from the surface as they reach the oxygen minima, and/or release via the decomposition of organic matter. Various authors (see e.g. Martin & Knauer, 1984; Landing & Bruland, 1987), however, have shown that the release from particles was insufficient to account for the observed increase in $Mn_d$ in the maxima, and have invoked the lateral off-shelf advection of $Mn_d$, which may have originated via redox-mediated diffusion from shelf sediments. An example of a vertical water-column $Mn_d$ profile showing an intermediate-depth concentration maximum is illustrated in Fig. 11.5(d).

3 Major perturbations to the typical down-column $Mn_d$ profiles can be found in regions of hydrothermal activity. Here, the hydrothermal inputs can cause a dramatic increase in the concentrations of $Mn_d$ in intermediate-depth and deep waters; a profile of this type is illustrated in Fig. 11.5(e). Hydrothermal inputs of $Mn_d$ have been found in both the Atlantic and the Pacific Oceans. However, Burton & Statham (1988) have pointed out that the dominant contrast between the two oceans is that in the Atlantic the ridge topography restricts the dispersal of the hydrothermal Mn plume to the rift valley. In contrast, in the Pacific hydrothermal Mn can be transported for $\sim 2000\,km$ from the ridge-crest venting sources (see Section 6.3).

By using Mn as an example, it is therefore apparent that although the various types of vertical water-column distribution profiles are useful for identifying individual processes, the profiles of some dissolved elements can be influenced by more than one biogeochemical process. In addition, it must be stressed that although biogeochemical processes exert a strong influence on the vertical profiles of trace elements in the water column, physical advective transport also can play an important role in maintaining the shapes of the profiles.

## 11.6 Processes controlling the removal of trace elements from sea water

### 11.6.1 Introduction

If it is assumed that an element has been continually added to the oceans over geological time, then either its concentration will build up in sea water or it will be removed from the aqueous phase. The World

Ocean is now thought to be in a steady-state, and although their residence times in sea water vary considerably, the trace elements are not building up to reach concentrations equal to their long-time inputs. In fact, some of the trace elements are being removed relatively fast in times of less than a hundred to a few thousand years, i.e. Forchhammer's 'facility with which the elements in sea water are made insoluble' has been operative. The question that must now be addressed is 'What are the processes that drive the facility which makes the elements insoluble'?

One of the first attempts to evaluate the processes involved in the removal of trace elements from sea water was made by Krauskopf (1956), who assessed the effects of: (i) saturation precipitation; (ii) adsorption on to particulate matter; and (iii) interactions with the biosphere on the removal of a series of elements from the ocean. His main conclusions may be summarized as follows.

1 Saturation precipitation could only act as a potential control on the removal of Sr and Ba.

2 Adsorption on to particulate material such as ferric hydroxide, manganese dioxide, the clay mineral montmorillonite, dried plankton and peat moss was a possible control on the removal of Zn, Cd, Pb, Bi, Cu, Hg, Ag and Mo.

3 Interactions with the biosphere may regulate the concentrations of V, Ni and possibly also Co, W and Mo in sea water.

Krauskopf's study therefore clearly indicated the potential importance of inorganic particle adsorption and interaction with the biosphere in the removal of trace elements from sea water. Since then, many advances have been made in our understanding of both the concentrations of trace elements in sea water and the factors that control their distributions. The picture is still by no means totally clear, but the new data offer a better position from which to review the mechanisms that remove trace elements from the oceans, especially with respect to the evaluation of the relative importance of inorganic scavenging and biological uptake in the removal processes.

In the two preceding sections we have reviewed the lateral and vertical distributions of trace elements in the ocean system. On the basis of their lateral distributions it was possible to make a distinction between two general types of trace metals.

1 **Type I** was exemplified by Pb. The main trend in the surface-water distribution of Pb is an increase in dissolved concentrations away from the coastal regions towards the central gyres, where it reached its highest concentration. Outside the coastal waters the concentrations of Mn also increased in the central gyres, but the overall distribution of this element was complicated by a strong nearshore source.

2 **Type II** was exemplified by Cd and Ni, which had their highest surface-water concentrations in areas of upwelling and their lowest concentrations in the oligotrophic central gyres.

The reasons for the differences between the type I and type II elements became apparent when their vertical water-column profiles were examined. These showed that the concentrations of Pb and Mn decreased rapidly with depth, whereas those of Cd and Ni increased with depth. Thus, if specialized conditions, such as those found in suboxic and anoxic waters, are ignored, then the vertical profiles of type I and type II trace metals may be considered to be essentially maintained by two basically different biogeochemical processes, upon which the results of advective transport can be superimposed. The type I distribution is controlled by a surface input and a rapid scavenging by particulate phases throughout the water column, on to which may be superimposed the effects of other processes, such as those associated with mid-depth and bottom trace-metal sources; i.e. the type I element distributions are controlled largely by **external cycling processes**. In contrast, the type II distribution is essentially the result of the involvement of the trace metals in the major biological (nutrient) cycles, which results in a surface depletion and a subsurface enrichment in dissolved concentrations; i.e. the type II element distributions are controlled mainly by **internal cycling processes**. This classification is by no means all-embracing and some trace elements have intermediate-type distributions (e.g. Cu). In a very general sense, however, the vertical distributions of many trace metals may be considered to be controlled by either a **scavenging-type** or a **nutrient-type** mechanism, or by a combination of both, involving the removal of dissolved species by some form of particulate matter.

The distinction between the scavenging-type and the nutrient-type reactions not only draws attention to the central role played by particulate matter in the removal of trace elements from sea water, but also highlights the importance of the histories of the individual particle types; thus, the internally produced biotic particles (nutrient-type) are involved in different trace-metal cycles than are the externally produced non-biotic particles (scavenging-type). However, it will be shown later in the text that in reality the situation is not as clear-cut as this; for example, the scavenging component of marine particulate matter has an organic nature (see Section 11.6.3.1). Further, it now appears that biota-generated particles may act as passive substrates for the uptake of surface-active trace elements (see Section 11.6.3.1). Nonetheless, for the purpose of a working hypothesis, it is still useful at this stage to categorize trace element removal mechanisms into the '**scavenging**' and '**nutrient-type**' reactions.

The importance of particulate material in the removal of trace metals from sea water has been widely acknowledged for many years, to the extent that Turekian (1977) referred to the overall process as the 'great particle conspiracy' (see Chapter 10). However, marine geochemists are still only at a relatively early stage in their understanding of how this conspiracy works.

Particulate matter can act as either a source or a sink for trace elements dissolved in sea water. The *in situ* removal of dissolved elements on to particulate matter (i.e. their consumption) has been termed the **J-efflux**, and their *in situ* release from particulate matter (i.e. their production) has been termed the **J-flux**. It was shown in Section 10.4 that oceanic TSM can be divided into two general populations:

1 a small-sized ($\leqslant$5–10 µm diameter) fraction, termed the FPM;

2 a large-sized ($\geqslant$50 µm diameter) fraction, referred to as the CPM.

This size classification is critical in understanding how the 'great particle conspiracy' operates. For example, the scavenging-type and nutrient-type mechanisms can be referred to the FPM and CPM particulate matter populations, respectively. Thus, involvement in the biological cycles introduces the nutrient-type trace elements into the large aggregated CPM population, whereas the adsorptive or scavenging removal of trace metals from sea water is dependent upon the surface area and concentration of particles, and so is dominated by the small-sized FPM population. However, this FPM sinks at very slow rates and it is the large-sized, fast-sinking, particles that deliver most of the vertical material flux in the oceans. Therefore, in order to satisfy the residence-time requirements of some trace metals in the water column it is necessary to achieve a balance between the slow- and fast-sinking particles. The direct coupling between the two populations is the transfer of small particles to the large particle flux by mechanisms such as 'piggy-backing' or packaging into faecal-type material by organisms at surface and mid-depths (see Section 10.4), but these are not one-step processes and are best described in terms of an aggregation–disaggregation particle continuum.

It may be concluded that to a first approximation trace metals can be removed from sea water by scavenging-type and nutrient-type mechanisms, and in the following sections the reactions controlling these two mechanisms will be considered in more detail. Before attempting to do this, however, it is necessary to understand something of the speciation of the trace metals in sea water, because this plays a vital role in controlling the pathway that a specific trace metal will follow in its removal from the water column.

### 11.6.2 Trace-element speciation

The speciation of components in multi-electrolyte solutions such as sea water is notoriously difficult to assess, and a full treatment of the subject will not be attempted here. Instead, attention will be focused on two key questions.

1 What is speciation?

2 Why is speciation important to our understanding of the chemical dynamics of trace elements in sea water?

For a general discussion of speciation in the environment, the reader is referred to the volume edited by Ure & Davidson (1995).

### 11.6.2.1 *What is speciation?*

**Speciation** can be defined as the individual physico-chemical forms of an element that together make up its total concentration. The full water-column speciation of an element therefore involves its distribution between free ions, ion pairs, complexes (both inorganic and organic), colloids and particles. According to Andreae (1986) the major processes that affect chemical speciation between these components in sea water can be related to a number of exchange reactions. These are: *solid–aqueous phase exchange, elec-*

*tron exchange* (redox chemistry), *proton exchange* (acid–base chemistry) and *ligand exchange* (complex chemistry). This author also points out that the processes that determine the species distribution of an element can be described within the context of fundamental interactions in the bonding environment around the atom, i.e. the formation of covalent bonds, electron exchange in redox reactions, and various types of ligand exchange (complex formation, acid–base chemistry and surface interactions). A number of generalities can be identified with respect to these bonding environments.

1 For a few elements, mainly the non-metallic and metalloid elements, the formation of covalent bonds is important in controlling species distribution in sea water.

2 Coordination bonding chemistry (acid–base reactions, precipitation–dissolution, complex formation) dominates the species distribution of the metallic elements.

3 Redox reactions lead to changes in electron configuration that are reflected in both covalent and coordination bonding characteristics.

Because both proton exchange (acid–base reactions) and electron exchange (redox) processes are extremely important in controlling speciation distribution, pH and redox potential (Pe) are uniquely important parameters in aqueous chemistry, and are often termed the 'master variables' of the system.

The basic concepts underlying trace-element speciation in sea water have been described by Stumm & Brauner (1975). Atoms, molecules and ions will tend to increase the stability of their outer-shell electron configurations by undergoing changes in their co-ordinative relationships, e.g. by acid–base, precipitation and complex formation reactions. Any combination of cations (the **central atom**) with molecules or anions (the **ligand**) containing unshared electron pairs (**bases**) is termed coordination (or complex formation); this can be either electrostatic, covalent, or a mixture of both. In an aqueous solution, cations are coordinated with water molecules, and Stumm & Brauner (1975) distinguish between two types of complex species in sea water.

1 In an **ion pair** (or outer-sphere species), the metal ion, the ligand or both retain the coordinated water when the complex is formed, so that the metal ion and the ligand are separated by water molecules. In an ion pair, the association between the cation and anion is largely the result of long-range electrostatic attraction.

2 In a **complex** (or inner-sphere species), the interacting ligand is immediately next to the metal cation, and a dehydration step must precede the association reaction. In this type of species, short-range or covalent forces contribute towards the bonding. The number of linkages attaching ligands to a central group is known as the coordination number. When a base contains more than one ligand, and can thus occupy more than one coordination position in a complex, it is termed a multidentate (as opposed to a unidentate) ligand; complex formation with these ligands is termed chelation, and the complexes are referred to as **chelates**. Chelates are usually much more stable relative to the corresponding complexes with unidentate ligands. The order of stability of the transition metal chelates formed with ligands (especially organic ligands) follows the so-called Irving–Williams order $Mn^{2+} < Fe^{2+} < Co^{2+} < Ni^{2+} < Cu^{2+} < Zn^{2+}$.

### 11.6.2.2 *Important trace-element species in sea water*

*Ligand exchange; complex chemistry.* Considerable advances in trace-metal speciation chemistry have been made using electrochemical techniques, and it is convenient to use electrochemical concepts to describe ligand complex formation in sea water. Following Bailey *et al.* (1978) the formation of a complex in solution can be represented by a number of steps, each of which can be described by an **equilibrium constant** (formation constant). The complexes usually undergo continuous breaking and remaking of the metal–ligand bonds, and this introduces a fundamental concept in speciation chemistry. If the bond-breaking step is rapid, the ligand exchange reactions are also rapid, and the complex is classified as being **labile**. In contrast, if the reactions are slow, the complex is classified as **non-labile**, or inert.

According to Nurnberg & Valenta (1983) the inorganic ligands normally present in sea water usually form labile mononuclear complexes with trace metals. Because of factors such as their high rate constants of formation and disassociation, these complexes are very mobile and will undergo reversible electrode processes in electrochemical determinations. This also is the situation for certain weak complexes formed between trace metals and some ligands of the dissolved organic matter (DOM) in sea water. In contrast, other DOM ligands form more stable non-labile complexes with trace metals. The reduction of these species requires a considerable overvoltage and the electrode process therefore becomes irreversible.

In terms of their association–disassociation rate constants it is therefore convenient from a biogeochemical viewpoint to consider some aspects of both the inorganic and organic trace-metal complex formation in sea water within the framework of a classification that distinguishes between two general species types.

**1** Those that form kinetically very **labile** complexes. These species can be considered to be in thermodynamic equilibrium, and species distributions can be predicted from mathematical models.

**2** Those that are involved in the formation of very **non-labile** (or inert) complexes. For these, species distribution can be identified using experimental techniques, which can isolate the inert complexes.

A schematic representation of this labile–non-labile concept is illustrated in Fig. 11.6 in terms of the role

played in the oceans in the biogeochemical cycles of trace metals.

**1** *Speciation involving labile complexes.* This form of trace-metal speciation can occur with inorganic, and also some organic, ligands. The determination of the labile complexes uses theoretical equilibrium models based on thermodynamic compositions. There are a number of problems involved in both the setting up of these models and in the manner in which the data they generate are interpreted; e.g. factors such as the use of different suites of stability constants often yield different speciation pictures. In sea water, the abundant inorganic ligands that are significant for trace-metal speciation are $Cl^-$, $OH^-$, $CO_3^{2-}$ and $SO_4^{2-}$. The processes involved in this speciation can be influenced by the formation of ion pairs (electrostatic attraction only) between cations and anions, and by mixed ligand complexes. Other effects that must be considered include side reactions and mixed ligand complexes. A number of workers have attempted to calculate the equilibrium speciation of dissolved components with the inorganic ligands present in sea water (see e.g. Sillen, 1961; Garrels & Thompson, 1962; Zarino & Yamamoto, 1972; Dyrssen &

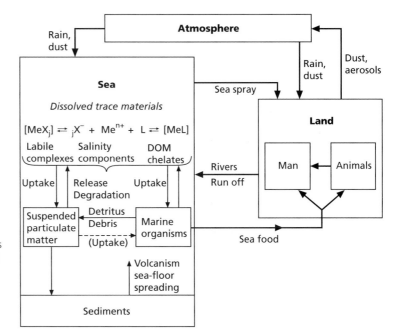

**Fig. 11.6** Schematic representation of the biogeochemical cycle of heavy metals (from Nurnberg & Valenta, 1983). [*MeX$_j$*] represents labile metal complexes in which *Me* is the metal and *X* is the inorganic ligand, and [*MeL*] represents non-labile metal chelates in which *Me* is the metal and *L* the organic ligand.

Wedborg, 1974; Stumm & Brauner, 1975; Florence & Batley, 1976; Turner *et al.*, 1981), and the speciation data obtained are usually expressed as percentages of the total metal concentration.

A list of predicted trace-metal species in oxygenated sea water is given in Table 11.1; the species, which are derived mainly from data provided by Stumm & Morgan (1981) and Turner *et al.* (1981), are taken from Bruland (1983). It must be stressed, however, that the speciation signatures given in the literature are often at variance with each other, e.g. as a result of different stability constant data used in the models. Despite this, two main features can be identified from the data in Table 11.1.

(a) The alkali and alkaline earth metals do not form strong complexes, and even their tendency to form ion pairs is limited, with the result that these elements exist in sea water largely as *simple cations*.

(b) For the trace elements, the species are distributed between the *free ion* and *complexes* with the various ligands.

It must be pointed out, however, that the data in the table disregard the formation of complexes with organic ligands.

*2 Speciation involving non-labile complexes with organic matter.* Some organic ligands can form stable complexes, or even chelates, with trace metals in sea water, even though there is side-reaction competition from salinity components such as Ca and Mg (see e.g. Nurnberg & Valenta, 1983).

A number of electrochemical techniques have been used to investigate trace-metal–organic complexation in sea water. Three of the most common of these are: (i) differential pulse anodic stripping voltammetry (DPASV); (ii) differential pulse cathodic stripping voltammetry (DPCSV); and (iii) complexation capacity titration; i.e. the ability of the water sample to remove added trace metal from the free-ion pool. For a detailed description of electrochemical techniques, the reader is referred to Whitfield & Jagner (1981) and van den Berg (1989).

*Electron exchange; redox chemistry.* Redox speciation affects some elements (e.g. Mn, Fe, I, Cr, As and Se) that can exist in variable oxidation states in sea water. For example, the thermodynamically unstable Mn(II) is soluble in sea water but oxidizes to form insoluble Mn(II) oxyhydroxides. As a result, Mn is

solubilized by reduction and precipitated by oxidation, and cycling between the two species of the metal can occur throughout the water column and in bottom sediments. In surface waters redox changes affecting dissolved Mn can occur on a diel cycle, involving night-time removal of dissolved Mn(II) by manganese-oxidizing bacteria and daytime regeneration resulting from photoreduction and photoinhibition of the Mn-oxidizing bacteria. In subsurface waters Mn oxidation changes can occur at the $O_2$–$H_2S$ boundary and in the dissolved oxygen minimum zone (see Section 11.5), e.g. in the eastern tropical Pacific and northern Indian Ocean. In the sediment–interstitial-water complex, redox reactions driven by the utilization of manganese oxides as secondary oxidants can set up a dissolved-phase–solid-phase Mn cycle (see Section 14.6.2). Chromium is another trace metal that has a highly soluble redox state (Cr(VI)) and a very insoluble state (Cr(III)) in sea water; in contrast to Mn, however, Cr is solubilized by oxidation and precipitated by reduction. As a result, a maximum in dissolved Mn (Mn(II)) and a minimum in dissolved Cr (Cr(VI)) can occur in the dissolved-oxygen minimum zone. For example, Landing & Bruland (1987) reported the presence of a dissolved-Mn maximum coincident with the oxygen minimum zone in the eastern tropical North Pacific water column. Redox changes therefore can have a strong influence on the dissolved water-column, and sometimes also on the sediment, distributions of the redox-affected trace metals such as Mn.

### 11.6.2.3 Why is speciation important to our understanding of the chemical dynamics of trace elements in sea water?

In the preceding discussion it was shown that trace elements dissolved in sea water can exist in a variety of different forms, which include **free hydrated ions, inorganic complexes**, and **organic complexes**, and the answer to the question is that the different forms of the elements undergo different geochemical and biological interactions. This is especially important with respect to two fundamental processes that underpin trace-metal cycles in sea water; dissolved $\leftrightarrow$ particulate reactivity and bioactivity.

**1 Dissolved $\leftrightarrow$ particulate reactivity**, i.e. surface interactions. Speciation involves an interactive competition between various components in natural waters to 'capture' dissolved trace metals. One of these components is particulate matter, and inorganic complex formation strongly influences adsorption equilibria; hydroxy, sulphato, carbonato and uncharged inorganic complexes tend to be sorbed much more strongly at an interface than free metal ions (Stumm & Brauner, 1975). Perhaps more important from a biogeochemical viewpoint, the complexing of dissolved trace metals by organic ligands stabilizes the metals in solution and so can reduce their particle reactivity (scavenging).

2 **Bioactivity.** Speciation is extremely important in determining the behaviour of a trace metal in the biosphere with respect to its uptake by phytoplankton, i.e. at the entry point into marine biological cycles. During the uptake of trace metals from sea water it is generally assumed that phytoplankton discriminate against strongly complexed species, preferring to take up the fraction that is present in a free ionic form. Further, reducing the free-ion concentration by competitive organic complexing can reduce the toxicity of trace metals. The **bioavailability** of trace metals therefore is strongly influenced by their speciation in sea water. For a detailed discussion of the concepts associated with the bioavailability of elements in sea water, see Sunda *et al.* (1981), Sunda & Huntsman (1983), Turner (1987), Morel (1986), Hudson & Morel (1990, 1993), Morel *et al.* (1991) and the volume edited by Tessier & Turner (1995).

The way in which different forms of an element take part in different biogeochemical interactions is illustrated below with respect to organically complexed trace metals, the hydride elements and methylation.

*Organic complexation; trace-metal–biota relationships.* Although our understanding of the organic complexing of trace metals in sea water is still in its relative infancy, advances have been made in recent years. These advances are summarized below for a number of individual metals and are related, where appropriate, to particle scavenging and bioactivity.

1 **Copper:** a mixed nutrient–scavenging type trace metal. The importance of organically complexed Cu in sea water has been demonstrated by several authors (see e.g. Huizenga & Kester, 1983; Wood *et al.*, 1983; Mills & Quinn, 1984; van den Berg, 1984; Buckley & van den Berg, 1986; Kramer, 1986;

Moffett & Zika, 1987; Coale & Bruland, 1988, 1990; Hanson *et al.*, 1988).

The complexation of Cu by natural ligands is most pronounced in surface and euphotic-zone waters where primary production occurs. For example, van den Berg (1984) reported that between 94% and 98% of the total dissolved Cu in Irish Sea surface waters was present as organic complexes. Coale & Bruland (1988) showed that > 99.7% of the total dissolved Cu(II) in the surface waters (upper 100 m) of the central northeast Pacific was bound in strong organic complexes.

There is evidence that speciation can play an important role in the particle scavenging of Cu. In this context, Mills & Quinn (1984) reported that organically complexed Cu species in surface waters from Narragansett Bay ranged from 14% to 70% of the total Cu. The authors concluded that the organic Cu mixes conservatively, so that organic complexing decreases the extent to which the total dissolved Cu is available for particle scavenging: the process that induces non-conservative behaviour. Organic complexing therefore may be an important factor in controlling the transport of dissolved Cu out of the estuarine environment. This control also can extend beyond estuaries. For example, Huizenga & Kester (1983) analysed total and labile (non-organically complexed) Cu in the northwestern Atlantic and reported that on a shelf transect there was a linear relationship (i.e. conservative behaviour) between total Cu and salinity, with < 5% of the total Cu being in a labile form. The authors suggested that Cu may have behaved conservatively on the shelf transect because the non-labile (mainly complexed) Cu, which made up >95% of the total Cu, is less reactive to particle scavenging than free Cu. It is apparent, therefore, that the **inorganic/organic speciation** of Cu (and other trace metals) will play an important role in controlling the amounts of the metals that are sedimented from the estuarine, coastal and open-ocean water columns by association with particulate matter (see also Wangersky, 1986).

Copper speciation is also important from the point of view of bioactivity. For example, the data reported by Coale & Bruland (1988) showed that >99.7% of the total dissolved Cu(II) in the surface waters (upper 100 m) of the central northeast Pacific is bound in strong organic complexes mainly by the stronger $L_1$ (concentration ~1.8 nM in the upper 100 m;

$1 \, nM = 1 \, \mu mol \, l^{-1}$) of two Cu-complexing ligands (or classes of ligands), which probably have a biological source; e.g. from prokaryotic cyanobacteria (picoplankton). The concentration of $L_1$ decreased with depth, but its excess concentration to that of total dissolved Cu ($\sim 0.5 \, nM$ at the surface, increasing to $\sim 1.5 \, nM$ at 500 m) caused the high degree of organic complexation found in the upper water column. This organic complexation reduced the fraction of inorganic Cu species to <0.3% of the total dissolved Cu, and free hydrated Cu amounted to only $\sim 4\%$ of this inorganic fraction. The possibility therefore arises that via the production of complexing organic ligands some cyanobacteria may detoxify Cu by lowering its free-ion activity to levels that would not inhibit their growth.

**2 Zinc:** a nutrient-type trace metal. Several workers have reported studies on the speciation of dissolved Zn in sea water (see e.g. van den Berg & Dharmvanij, 1984; van den Berg, 1985; Bruland, 1989; Donat & Bruland, 1990).

The data provided by van den Berg (1985) suggested that zinc–organic complexes dominate Zn speciation in Irish Sea waters. The author identified two classes of ligands, which had concentrations of 26 and 62 nM. Van den Berg & Dharmvanij (1984) concluded that 24% of the total dissolved Zn in a surface seawater sample from the South Atlantic was bound in complexes by organic ligands, which were present at a concentration of 30 nM. Bruland (1989) found that >98% of the total dissolved Zn in the upper 200 m of the central North Pacific Gyre was bound in organic complexes, which were stronger than those reported by van den Berg (1985) and van den Berg & Dharmvanij (1984), by Zn-specific organic ligands, which were present in low concentrations (1.2 nM). The concentration of the ligand exceeded that of dissolved Zn from surface waters to a depth of $\sim 350 \, m$, over which it had a conservative vertical distribution. The complexation of Zn by natural ligands, like that for Cu, is most pronounced in surface and euphotic-zone waters, where primary production occurs.

Evidence in the literature suggests that the reproductive rates of neritic (inshore shelf water, <200 m depth) phytoplankton species are limited by free $[Zn^{2+}]$ concentrations $<10^{-11.5} \, M$, whereas those of oceanic (offshore water, >200 m depth) species were either not limited, or only slightly limited, at concentrations of $\sim 10^{-12.5} \, M$. This lead Bruland et al. (1991) to suggest the possibility that the speciation of Zn may affect phytoplankton communities in a 'habitat-related' (neritic versus oceanic) pattern. The Zn-limiting data also infer that the oceanic surface-water free $[Zn^{2+}]$ concentration of $10^{-11.8} \, M$, such as reported by Bruland (1989) for the North Pacific central gyre, are neither high enough to be biotoxic nor low enough to be biolimiting with respect to the growth rates of the oceanic phytoplankton species. It is not therefore readily apparent that the high degree of organic complexation of the essential, and potentially biolimiting, Zn in open-ocean surface waters is biologically advantageous, although it may help to keep the metal in solution by preventing it being scavenged by particles. For example, Bruland (1989) showed that in the upper 600 m of the water column in the North Pacific, zinc-complexing ligands strongly buffered the free Zn ion activity with respect to perturbations in total dissolved Zn, and suggested that the complexation of the Zn by organic ligands may decrease the scavenging of the metal by particulate material.

**3 Cadmium:** a 'nutrient-type' trace metal. Bruland (1992) studied the complexation of Cd by organic ligands in the upper 600 m of the central North Pacific. He reported that $\sim 70\%$ of the total dissolved Cd in the surface waters was bound in strong complexes by relatively Cd-specific organic ligands present at low concentrations ($\sim 0.1 \, nM$). The ligand class was found only in the upper 175 m of the water column, with maximum concentrations between 40 and 100 m, and in intermediate and deep waters, complexation with inorganic chloride ligands dominated the Cd speciation. Generally similar findings were reported by Sakamoto-Arnold et al. (1987) for the speciation of Cd in a Gulf Stream warm-core ring in the western North Atlantic. Clearly, therefore the organic complexation of Cd is important in oceanic surface waters.

**4 Lead:** a 'scavenging-type' trace metal. Lead is highly toxic to biota, but relatively little is known of its organic complexation in sea water. Capodaglio et al. (1990) determined Pb complexation in surface waters from 11 locations in the eastern North Pacific. The total dissolved Pb concentrations varied between 17 and 49 nM and exhibited little spatial structure. The authors found that the organically complexed Pb fraction contributed a significant proportion

(~50%) of the total Pb. The data were consistent with one class of organic ligands, present at low concentrations (0.2–0.5 nM). The presence of this ligand, together with various inorganic ligands in sea water, gave rise to a concentration of free ionic Pb of ~0.4 pM. This study highlighted the importance of complexation with organic ligands for trace metals such as Pb, which are not involved in naturally occurring biological processes.

5 **Cobalt** (surface depletion, depletion at depth, potentially biolimiting) and **nickel** (nutrient-type). Relatively few studies have been carried out on the complexation of these two metals with organic ligands in open-ocean sea waters. Zhang *et al.* (1990) found that a variable fraction (45–100%; mean, 73%) of dissolved Co was strongly complexed in the Irish Sea and the Sheltd River estuary. Donat & Bruland (1988) reported that coastal and open-ocean seawater samples had ~50% of their total dissolved Co in association with strong organic complexes. Van den Berg & Nimmo (1987) reported that ~30% to ~50% of the dissolved Ni in UK coastal waters was present as very strong organic complexes.

6 **Iron**: a mixed nutrient–scavenging type trace metal. Iron is probably the most important bioac-

tive trace metal and has been invoked as a limiting micronutrient to primary production in HNLP regions (see Section 9.1.3.2). Both the general oceanic chemistry of Fe, and its speciation, are not well understood, but a recent review by Johnson *et al.* (1997) provides the best available data. These authors reviewed the concentrations of dissolved iron in the World Ocean and identified a number of important features in the distribution of the metal.

(a)  The dissolved Fe concentrations had a uniform shape, displaying a 'nutrient-type' profile (see e.g. Fig. 11.7b) at all the stations studied. It should be noted, however, that in some regions where there is a strong external source of the metal, dissolved Fe can exhibit a surface-enhanced scavenged-type profile (see Fig. 11.7a). The surface concentrations of dissolved Fe reported by Johnson *et al.* (1997) were <0.2 nmol l$^{-1}$ (average, 0.07 nmol l$^{-1}$), with an average of 0.76 nmol l$^{-1}$ below 500 m. The concentration of dissolved Fe was relatively constant below 1000 m.

(b)  The residence time for dissolved Fe was estimated to be 100–200 yr, indicating a rapid removal from the water column. Other trace elements with short residence times are characterized by vertical

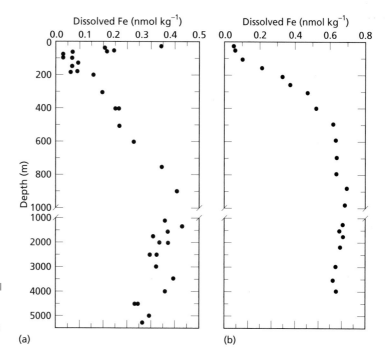

**Fig. 11.7** Vertical water-column profiles of dissolved iron. (a) A profile from the North Pacific central gyre, typical of a scavenged-type trace metal having a significant external source (after Bruland *et al.*, 1991). (b) A profile from a remote subarctic Pacific HNLP region with the lack of a strong external iron source (after Martin *et al.*, 1989).

profiles that show: (i) a surface-enhancement and depth-depletion; (ii) a decrease with depth; and (iii) deep-water concentrations that decrease with age as the water passes from the Atlantic to the Pacific. For Fe, however, there is no decrease below 1000 m and no inter-oceanic fractionation.

(c) The major source of Fe to the oceans is aeolian transport, but overall the surface-water integrated concentrations of Fe (0–500 m) were not well correlated with the aeolian fluxes.

Johnson *et al.* (1997) concluded that the nutrient-type Fe profiles are maintained by a mechanism that reduces the scavenging rate of dissolved Fe at concentrations <0.6 nmol l$^{-1}$; i.e. in the deep water. This mechanism may be either an equilibrium between dissolved and particulate iron, or binding by organic ligands. It was not possible to identify which mechanism was operative, but the authors suggested that the complexation of Fe by strong organic ligands, which have been found in the Pacific and Atlantic at concentrations of ~0.6 nmol l$^{-1}$ (see e.g. Rue & Bruland, 1995; Wu & Luther, 1995), could explain the water-column profiles of iron. Thus, by keeping the Fe soluble, this mechanism would allow a nutrient-type profile to develop before scavenging begins to remove the metal, and would therefore diminish inter-ocean fractionation. According to Johnson *et al.* (1997), therefore, Fe would be taken up by phytoplankton in surface waters and reminer-alized at depth, like the nutrient elements, and then held in solution in subsurface waters by the organic ligand solubilizing mechanism. As the Fe concentrations increase and exceed those of the ligands the metal would then be free to undergo scavenging. Without the ligands, therefore, the oceans probably would be far more depleted in Fe, and as the ligands are derived from biota this is a large-scale feedback mechanism. Several aspects of the oceanic Fe chemistry were questioned by Boyle (1997), Sunda (1997) and Luther & Wu (1997).

A number of general conclusions can be drawn from the trace-metal speciation data presented above.

1 Large fractions of the total dissolved concentrations of Cu (~99%) and Zn (~95%), and smaller but still significant fractions of Cd (~70%) and Pb (~50%), in open-ocean surface waters are bound in organic complexes by ligands, which are often metal-specific.

2 The degree of complexation of Cu, Zn, Cd and Pb is highest in the euphotic zone and decreases sharply in subsurface waters. There is evidence to suggest that the ligands are produced by phytoplankton, although bacteria and fungi may also produce complexing ligands.

3 The metal-specific ligands are present at different concentrations in open-ocean North Pacific surface waters, the Cu-complexing ligands being present at ~2 nM, the Zn-complexing ligands at ~1.2 nM, the Pb-complexing ligands at ~0.37 nM, and the Cd-complexing ligands at ~0.1 nM.

4 The organic complexation of the metals reduces their free ion concentrations to low values. For example, in the open-ocean North Pacific the free-ion concentrations have been estimated to be as follows: $Cu^{2+}$ ~$10^{-13.2}$ M, $Zn^{2+}$ ~$10^{-12}$ M, $Cd^{2+}$ ~$10^{-13.6}$ M and $Pb^{2+}$ ~$10^{-12.3}$ M.

The relationship between trace metals and biota in sea water involves a complex web of interrelated factors, and because the organic ligands that complex trace metals originate from biota it is now clear that speciation changes are an intimate part of this web. As a consequence, as Bruland (1992) pointed out, it is apparent that oceanic **biology** is directly, or indirectly, influencing oceanic **chemistry**: in particular, the chemical speciation and free-ion concentrations of trace metals such as Cu, Zn, Cd and Pb. Because of this 'chemical–biological' interplay there is a need to review trace-metal–biota interactions in sea water in the light of recent findings.

Although the **external inputs** of many trace metals and nutrients to the oceans are dominated by physical and geochemical processes, the **internal cycles** of these metals are driven largely by biochemical processes, e.g. assimilation during photosynthesis. The availability of trace metals to biota is dependent not simply on their total concentrations in sea water but also on their speciation chemistries, both of which are important constraints on trace-metal–plankton interactions. The availability of trace metals to plankton is related to their free-ion concentrations and is controlled by the concentrations required to facilitate growth, i.e. those between **biolimiting** and **biotoxic** concentrations. It is, therefore, important to remember that the free-metal-ion concentrations must remain between these limiting and toxic levels.

Biological processes can strongly influence the oceanic chemistries of trace metals and, in turn, the trace metals themselves can influence primary productivity and plankton community structure. Bruland *et al.* (1991) termed these '**interactive influences**', and pointed out that they are particularly important for Mn, Fe, Co, Ni, Cu and Zn. These metals are required by phytoplankton for a variety of metabolic functions; i.e. they are 'bioactive' and can, under certain circumstances, have the potential to be either biolimiting or biotoxic.

The bioactive trace metals can have nutrient-type (e.g. Zn), scavenging-type (e.g. Mn) or mixed-type (e.g. Cu) vertical water-column profiles. Trace metals are actively assimilated or passively adsorbed (scavenged) on to reactive surface sites of plankton, or other biological particles, and transferred vertically out of the surface layers with the sinking of these particles (see Section 10.4). As a result, whatever the type of vertical metal-distribution profile, that proportion of the metal which is associated with the biota will be released into the water column at subsurface depths, together with the macronutrients, following degradative destruction of the organic matter. This subsurface water can be brought to the surface during upwelling. During this process the nutrient-type and the scavenging-type trace metals can behave differently, leading to inter-metal fractionations. Thus, the metals that have relatively short deep-water scavenging times (see Table 11.5), e.g. Fe (40–77 yr), Mn (51–65 yr) and Co (40 yr), tend to be stripped from subsurface waters by scavenging following their release from biota. This stripping process is less important for Cu (385–650 yr), and even less important for Zn,

Ni (15 850 yr) and Cd (177 800 yr), all of which have residence times governed by biogeochemical recycling.

Recently, a new picture has begun to emerge of the role played by the bioactive trace metals, in which it is suggested that it is not the free-ion concentrations of a single metal, but rather the **free-ion ratios** between metals that constrain their biological effects. For example, phytoplankton limitation by multiple metals may be more severe than by any one metal alone. Competitive interactions can occur between essential nutrient metals and competing inhibitory metals at cell-surface uptake sites, or at intracellular metabolic sites, and only when too much of the 'wrong' metal is bound to the sites do toxic effects become apparent. According to Bruland *et al.* (1991) these competitive interactions between metals are termed **metal antagonisms**, and have been reported for Mn and Cu, Zn and Cu, Fe and Cu, Cd and Fe, Mn and Fe and Mn and Zn.

Trace-metal antagonisms between Cu and the potentially biolimiting metals Fe, Mn and Zn may prove especially important in the oceans. For example, it has been shown that elevated free $[Mn^{2+}]$ concentrations can alleviate the biologically toxic effects of free $[Cu^{2+}]$; i.e. phytoplankton growth inhibition is a function of free-ion ratios of the two metals, and only when the free $[Cu^{2+}]$ is increased relative to free $[Mn^{2+}]$ do the inhibiting effects appear. As a result, it seems that when free $[Cu^{2+}]$ exceeds a critical value, the values of the $[Mn^{2+}]$:$[Cu^{2+}]$ ratio may be more important than the absolute concentrations of either metal. The $[Mn^{2+}]$:$[Cu^{2+}]$ ratio can vary under different oceanic conditions, and Bruland *et al.* (1991) cited the waters of the North Pacific central gyre as an example of this. In areas of maximum growth rates in the surface waters the $[Mn^{2+}]$:$[Cu^{2+}]$ ratio is ~6300. In contrast, in water upwelled from 200 to 400 m, assumed to be the source of freshly upwelled water, the ratio could be as low as 2 and complete inhibition of growth could be expected. For upwelled waters to sustain phytoplankton growth, the organisms therefore must maintain suitable free $[Mn^{2+}]$:$[Cu^{2+}]$ ratios, either by reducing free $[Cu^{2+}]$ or increasing free $[Mn^{2+}]$. One way of doing this is for the plankton to reduce free $[Cu^{2+}]$ by producing strong Cu-complexing organic ligands, which could reduce the $[Mn^{2+}]$:$[Cu^{2+}]$ ratios to levels favourable for growth. Another consequence

**Table 11.5** The deep-water scavenging residence times of some trace elements in the oceans (data from Balistrieri *et al.*, 1981; Orians & Bruland, 1986; Whitfield & Turner, 1987).

| Element | Scavenging residence time (years) | Element | Scavenging residence time (years) |
|---|---|---|---|
| Sn | 10 | Mn | 51–65 |
| Th | 22–33 | Al | 50–150 |
| Fe | 40–77 | Sc | 230 |
| Co | 40 | Cu | 385–650 |
| Po | 27–40 | Be | 3700 |
| Ce | 50 | Ni | 15850 |
| Pa | 31–67 | Cd | 177800 |
| Pb | 47–54 | Particles | 0.365 |

of trace-metal antagonisms may be that the 'iron limitation' concept, which has been invoked to explain low productivity in high-nutrient, low-productivity regions (see Section 9.1.3.2), may not in fact result from Fe alone, but from the antagonistic effects of Fe and Cu. A note of caution should perhaps be added here, because this view is not universally held. For example, Scharek *et al.* (1997) carried out experiments to simulate the response of Southern Ocean phytoplankton to the addition of trace metals. They reported that responses of various Antarctic oceanic communities to the addition of trace metals differed, but that Fe was the most important bioactive trace metal, with Mn, Co and Zn being only of minor importance. Further, they found that the accumulation of cellular Fe was neither enhanced, nor suppressed, by Mn, Co and Zn: i.e. co-limitation by these metals did *not* occur.

Trace metals exhibit spatial, as well as vertical, variations in concentrations; e.g. decreases from coastal to open-ocean waters (see Section 11.4). Free-ion requirements of phytoplankton for the bioactive metals may be related to these spatial variations with, in general, oceanic phytoplankton species that live in environments having lower metal concentrations being able to tolerate lower free-ion concentrations before showing reduced reproductive rates. Two speculative conclusions may be drawn from an overview of recent studies.

**1** It is probable that a combination of macronutrients and bioactive trace metals may be simultaneously controlling biological production in the oceans. We are only now beginning to understand the role played by the trace metals, but it is apparent that some may be toxic and some essential, and that they may act individually or by antagonistic interactions between multiple metals.

**2** Cycling of the trace metals may be largely controlled by biological processes.

The feedback system identified above between **biological** and **chemical** systems in the oceans may be of the utmost importance in controlling and maintaining productivity in the HNLP areas of the open-ocean remote from terrestrial inputs. In these areas biology strongly influences the distributions and chemical speciation of the bioactive trace metals, and the speciation of the metals themselves influences primary production, species composition and trophic structure.

*The hydride elements and methylation.* Another area in which trace-element speciation is important has been highlighted by the work on the 'hydride' elements in sea water. Andreae (1983) has pointed out that a number of metals and metalloids in the fourth, fifth and sixth main groups of the periodic table form volatile hydrides on reduction with sodium borohydride. These elements include Ge, Sn, Pb, As, Sb and Se, and an interesting feature of their marine chemistry is that some of them can be present in sea water in organometallic forms. For example, Froelich *et al.* (1983, 1985) have identified at least three forms of Ge in marine and estuarine waters, inorganic germanic acid ($Ge_i$), monomethylgermanium (MMGe) and dimethylgermanium (DMGe), and the authors showed that a large fraction of the oceanic Ge apparently exists in the methyl forms. The geochemical importance of these findings is that the different forms of Ge follow different geochemical pathways in the estuarine–marine environment. In general, $Ge_i$ behaves like silica; it has a high concentration in river water, where its source is the weathering of crustal rocks, and low concentrations in sea water. In estuaries, the behaviour of $Ge_i$ follows that of silica, and it can exhibit non-conservative behaviour following uptake by diatoms. In the oceans, the horizontal and vertical distributions of $Ge_i$ also mimic those of silica and reflect uptake on to, and dissolution from, siliceous organisms; i.e. it is a nutrient-type element displaying a non-conservative behaviour pattern. In contrast, both MMGe and DMGe behave in a conservative manner in estuaries and in the open ocean; both species are barely detectable in river water, and in estuaries they probably have an oceanic source.

Selenium is another element that has organic and inorganic species that exhibit different biogeochemical behaviour patterns in the oceans. Cutter & Bruland (1984) reviewed the oceanic chemistry of the element, and demonstrated the existence of three dissolved Se species: selenite (Se(IV)), selenate (Se(VI)) and an operationally defined organic selenide. The water-column distribution of these species is illustrated in Fig. 11.8(a). In surface waters of the North and South Pacific, organic selenide comprised ~80% of the total dissolved Se, with the maximum being associated with primary production. In deep-waters, however, organic selenide was undetectable. In contrast to the organic Se, the selenite and selenate

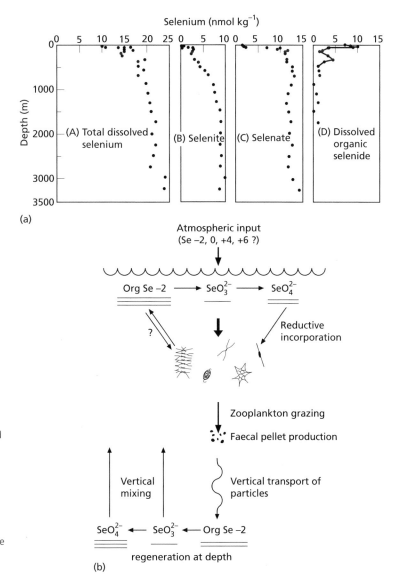

**Fig. 11.8** Selenium in sea water. (a) Vertical water-column profiles of total dissolved selenium, selenite, selenate and organic selenide in the Pacific Ocean (from Cutter & Bruland, 1984). (b) Schematic representation of the marine biogeochemical cycle of selenium (from Cutter & Bruland, 1984). The underlining indicates the relative concentrations of selenium species in surface and deep waters. The preferential uptake of selenite in surface waters is shown by the largest of the dissolved-to-particulate arrows.

species exhibited nutrient-type vertical distributions and were enriched in deep-waters. On the basis of their new data, the authors re-evaluated the marine biogeochemical cycle of Se. This cycle, which is illustrated in Fig. 11.8(b), involves the selective uptake of the element, its reductive incorporation into biogenic material, its delivery to the deep sea as particulate organic selenide via sinking detritus, and a regeneration back into the dissolved state.

Methylation is important in the marine chemistries of As, Sb, Sn and Hg. Arsenic is present in sea water as As(V), as As(III) and methylated As species. As(V) (arsenate) is the predominant dissolved As species, especially in deep-water. In the surface layer, however, arsenate is taken up by phytoplankton, together with phosphate, and is excreted as As(III) (arsenite) and as the methylated species methylarsenate and dimethylarsenate, the methylated species making up ~10% of the total As in the euphotic zone in some oceanic regions (Andreae, 1986). Thus, plankton (and bacteria) can influence the speciation of As in sea water, and the speciation transformations in the euphotic zone lead to As(V) having a nutrient-type distribution in the water column. In some respects, the distribution of Sb in the water column resembles that of As, with Sb(V) predominating over Sb(III), and methylated Sb making up ~10% of the total antimony in surface water. Methylated species of Sn have been identified in polluted estuarine and coastal waters; however, they are at very low concentrations in the open ocean. Toxic butyltin, like methyltin, is produced in industrial processes and it is also used as an antifouling agent, and these can be sources of the compound to coastal waters. According to Andreae (1986), Hg has very high stability constants with organic ligands, with the result that, in addition to forming a variety of inorganic complexes, it can form true organometallic compounds, which have been identified in sea water. Methylmercury has been identified in some polluted coastal waters, but there is uncertainty over its concentration in the open ocean.

*Summary.* It may be concluded that speciation is important because different species of the same element can behave in different ways with respect to their entry into marine biogeochemical cycles. In particular, the speciation of a trace element has a strong influence on its **bioavailability**, i.e. uptake into biota, and its **particle reactivity**, i.e. uptake on to inorganic particles. Some speciation forms, especially those involving organic complexation, can stabilize metals in solution. However, it is the transition from the dissolved to the particulate state that ultimately controls the residence time of an element in sea water and mediates its delivery to the sediment sink; i.e. its journey out of the seawater reservoir.

The speciation of a trace element exerts an important influence on the processes involved in dissolved → particulate transitions, and these processes are discussed in the following section in the wider context of the incorporation of trace elements into the oceanic biogeochemical cycles that drive them on their journey through the water column.

### 11.6.3 The principal routes for the removal of trace elements from sea water

It was suggested in Section 11.6.1 that the mechanisms by which trace metals are removed from sea water can be classified in terms of either scavenging-type or nutrient-type processes, and each of these is considered individually below.

#### 11.6.3.1 Scavenging-type mechanisms

The adsorptive removal of trace metals occurs on to particles that sink down the water column, and Goldberg (1954) gave this removal process the general name of **scavenging**. The dissolved down-column concentration curves for scavenged elements are concave in deep water when plotted against a conservative tracer, indicating loss from the water column. The extent of this scavenging removal can be expressed in terms of a deep-water scavenging residence time; a selection of these residence times is given in Table 11.5. Much of our knowledge of the scavenging of dissolved elements from sea water has been derived from the unique chemistry of the radionuclide elements, and their distributions in the oceans have allowed scavenging models to be constructed. In this respect, members of three natural radioactive decay series have proved to be useful tools for the interpretation of scavenging reactions; these are the $^{238}$U series, the $^{232}$Th series and the $^{235}$U series. The fundamental concept behind the use of these radionuclides is the parent–daughter relationship, and the underlying rationale is that the main input of the daughter to the sea is via the *in situ* decay of its immediate radioactive parent. Thus, as Broecker & Peng (1982) have pointed out, by measuring the concentration of the parents of adsorption-prone daughters, it is possible to calculate what the production rate of the daughter should be, and by determining the actual *in situ* concentration of the daughter it is possible to establish if it is being taken out of solution by particulate matter and also, if this is the case, at what rate the process occurs.

Following Bacon (1984), the rate of removal ($R_d$) of the daughter from a parcel of water is proportional to the difference in activity between the parent ($A_p$) and the daughter ($A_d$):

$$R_d = \lambda_d (A_p - A_d) \tag{11.2}$$

where $\lambda_d$ is the radioactive decay constant of the daughter.

A considerable amount of work has been carried out on Th isotopes in sea water. Three isotopes of Th have been used as marine tracers: $^{230}$Th (half-life 7.52

$\times 10^4$ yr), and the two shorter lived species $^{228}$Th (half-life 1.91 yr) and $^{234}$Th (half-life 24.1 days). Bacon & Anderson (1982) measured the three Th isotopes in sea water and found that for all of them most activity was in the dissolved form. For example, only ~17% of the $^{230}$Th was associated with particulate matter; however, both dissolved and particulate $^{230}$Th *increased* with depth in the water column.

In recent years many workers have used radioactive disequilibrium data to interpret scavenging processes in the oceans, and a variety of scavenging models have appeared in the literature. Bacon & Anderson (1982) divided these models into three general types and then used them to identify the processes that control the scavenging of $^{230}$Th in the water column; these processes have to take account of the increase in both dissolved and particulate $^{230}$Th with depth in the water column. The approach adopted by Bacon & Anderson (1982) provides a useful framework within which to describe the theoretical concepts underlying trace-metal scavenging in the oceans. In all three models, it is implicit that Th isotopes are supplied only by their radioactive parents dissolved in sea water, and that they are transferred to particles by adsorption, a process that is assumed to be first-order with respect to the dissolved Th concentration. In the present context, the term 'particulate' Th refers to particles ≲30 μm in diameter, which are assumed to sink at velocities of a few hundred metres per year or less.

*Scavenging model I: irreversible uptake.* In this type of model it is assumed that there is an irreversible binding of the daughter radionuclide to particle surfaces, which is followed by a slow sinking of the mass of particles to the sea bed. This type of irreversible uptake also assumes that the reaction site is so far out of equilibrium that the reverse, i.e. desorption, reactions are insignificant. If the suspended particulate matter has a long enough residence time in sea water, then for any isotope that is removed by scavenging a steady state is reached in which loss by decay is balanced by gain from uptake. A number of authors have applied this type of model, in association with advection–diffusion models, to both radioactive and stable trace-metal distributions in sea water (see e.g. Craig, 1974; Brewer, 1975; Boyle *et al.*, 1977).

Following Simpson (1982) the one-dimensional vertical advection–diffusion models that assume irreversible uptake for radionuclides can be described by the following equations.

1 Dissolved species:

$$\frac{\delta C_d}{\delta t} = K_z \frac{\delta^2 C_d}{\delta z^2} - w \frac{\delta C_d}{\delta z} + P_d - (\lambda + k_1)C_d \qquad (11.3)$$

2 Particulate species:

$$\frac{\delta C_p}{\delta t} = K_z \frac{\delta^2 C_p}{\delta z^2} - (U_s + w)\frac{\delta C_p}{\delta z} + k_1 C_d - \lambda C_p \qquad (11.4)$$

where $C_d$ is the dissolved concentration of the nuclide, $C_p$ is the particulate concentration of the nuclide, $K_z$ is the vertical eddy diffusion coefficient, $w$ is the vertical advective velocity, $U_s$ is the particle settling velocity, $\lambda$ is the radioactive decay constant, $P_d$ is the rate of production of the daughter from the dissolved parent, $z$ is depth, $t$ is time and $k_1$ is the removal rate constant from solution to the particulate phase (adsorption rate coefficient).

In practice, however, such models proved to be inadequate when applied to the distribution of thorium isotopes in the water column. In terms of these Th isotopes, irreversible uptake can be described by the relationship

$$\text{Th(dissolved)} \xrightarrow{k_1} \text{Th(particulate)} \qquad (11.5)$$
$$\downarrow$$
$$S$$

where $k_1$ is the **scavenging rate constant** and S mediates slow particle sinking.

Bacon & Anderson (1982) used a mathematical treatment to apply their Th isotope data to this model and rejected it as a basis for describing the Th scavenging on the evidence of the thorium particulate/thorium dissolved ($C_p/C_d$) distribution. In their treatment they allowed $k_1$ to be fixed by the observed deep-water $^{234}$Th distribution. Under this constraint, it is required that the $C_p/C_d$ value for $^{230}$Th increase very sharply with depth to the extent that, for reasonable sinking velocities, well over half the $^{230}$Th would be found in the particulate form. As this was clearly not the case (see above), the authors concluded that the simple irreversible uptake model cannot simultaneously satisfy the observed distributions of both the $^{230}$Th and the $^{234}$Th isotopes; i.e. because the actual particulate Th is less than the predicted value, an additional mechanism is required to account for the loss of particulate Th from the water column. Bacon &

Anderson (1982) therefore looked at other models in which the distribution of $C_p/C_d$ is governed not only by $k_1$ but also by an additional transfer coefficient, which represents a loss of the particulate Th.

*Scavenging model II: irreversible uptake with fast particle removal.* In this model, the additional transfer coefficient for the loss of particulate Th is represented by the incorporation of rapidly sinking particles ($k_2$) into the down-column flux. Here, therefore, it is the small-sized particulate matter itself that is lost from the water column, e.g. by 'piggy-backing'. Thus

$$\text{Th(dissolved)} \xrightarrow{k_1} \text{Th(particulate)} \xrightarrow{k_2}$$
$$\downarrow S$$
$$\xrightarrow{k_2} \text{Th(fast particle)} \qquad (11.6)$$

This is similar to the 'piggy-backing' model suggested by Lal (1980) (see Section 10.4). However, this type of small-particle–large-particle 'piggy-backing' does not result in an *in situ* addition of trace metals to deep waters and so cannot account for the loss of particulate Th at depth in the water column. This was confirmed by Bacon & Anderson (1982), who showed that although model II predicted an increase in particulate Th with depth, it did not at the same time predict an increase in dissolved $^{230}$Th with depth.

*Scavenging model III: reversible exchange.* In this model, the additional transfer coefficient for the loss of particulate Th is by desorption from particles. Thus

$$\text{Th(dissolved)} \underset{k_{-1}}{\overset{k_1}{\rightleftarrows}} \text{Th(particulate)} \qquad (11.7)$$
$$\downarrow S$$

Bacon & Anderson (1982) concluded that this model is the only one that successfully predicts that there will be an increase in both dissolved and particulate $^{230}$Th with depth in the water column.

From the mathematical treatment of their data Bacon & Anderson (1982) therefore concluded that although both models I and III predicted linear increases in particulate $^{230}$Th with depth, only model III was able to predict that the dissolved $^{230}$Th also increases in this manner.

It was pointed out in Section 10.4 that the TSM in the oceans consisted of a small-sized (non-sinking)

and a large-sized (fast-sinking) population. It is assumed that it is the fine suspended particulate population that controls the adsorption of radionuclides, and stable trace metals, from sea water. To bring about downward transport, particle aggregation into the large, fast-sinking, particle population is necessary. However, in order to satisfy residence time requirements, Bacon *et al.* (1985) have pointed out that it is also necessary for disaggregation to occur. The authors therefore envisaged a continuous exchange of material between the small- and large-sized particle populations, so that before reaching the sea floor a particle may be exchanged several times between the non-sinking small population and the fast-sinking aggregated population, with the overall result that fine particles work their way down the water column at an average speed of ~300–1000 m yr$^{-1}$. The aggregation may result from small particles 'piggy-backing' on to the larger ones (see Section 10.4), but the mechanism of disaggregation in the deep sea was not known at the time. In this context, Cho & Azam (1988) have suggested that non-sinking particles may be produced at the expense of the larger aggregates via decomposition by free-swimming bacteria at depth in the water column, thus providing a biological explanation for the down-column radionuclide transport models. The assumption that the scavenging of trace metals from sea water is essentially confined to the fine, largely inorganic, suspended particulate population, however, has been questioned following the discovery that a correlation can be found between dissolved Th removal and the *biological* production of particles. This is discussed in Section 11.6.3.3, but at this stage it should be pointed out that Clegg & Whitfield (1991) drew attention to the fact that any model which assumes reversible Th exchange (sorption) only on to small particles is a simplification of the real situation and that uptake on to material such as faecal pellets and phytoplankton must be considered. The authors outlined a generalized model for the scavenging of thorium from sea water, and concluded that: (i) the return of sorbed Th to the dissolved state is controlled by remineralization in the upper 100–200 m of the water column; and (ii) that the scavenged Th flux is correlated with new production.

Perhaps the most important conclusion that can be drawn from the study carried out by Bacon & Anderson (1982) is that, at least to a first approximation, the suspended particulate matter in deep water exists in a state of equilibrium with respect to the exchange of Th by adsorption–desorption reactions. This can be described in terms of an equilibrium model. In the framework of models of this type, the residence time of an element with respect to its removal from the oceans by scavenging is controlled by its equilibrium partitioning between dissolved and adsorbed forms, and by the residence time of the particles with which it reacts. Thus, a knowledge of the equilibrium distribution of the element, combined with an independently determined value for the residence time of the particulate material, will yield an estimate of the scavenging rate of the element itself. This relationship can be expressed in the following manner:

$$\frac{1}{\tau_{Me}} = \frac{1}{\tau_P} \frac{[Me]_P}{[Me]_t} \qquad (11.8)$$

where $\tau_{Me}$ is the residence time of the element (metal), $\tau_P$ is the residence time of the particulate matter, and $[Me]_P$ and $[Me]_t$ are the particulate (adsorbed) and total concentrations of the element, respectively.

The quantity $[Me]_P/[Me]_t$ can be calculated from direct field measurements, thus allowing the scavenging residence time of an element to be determined. This approach was adopted by Bacon & Anderson (1982), who concluded that on the basis of their equilibrium model the removal of Mn, Cu, Pb, Th and Pa from sea water appears to be controlled by a single population of particles with a residence time of *c.* 5–10 yr in the water column. It was shown in Section 10.4 that most of the vertical flux of particulate matter from the surface to deep water is controlled by large aggregates (CPM; diameter $\geqslant 50\,\mu m$), which have much faster sinking times than 5–10 yr. However, scavenging is dominated by the small-sized population (FPM; diameter $\ll 10\,\mu m$), which has a long residence time but can undergo accelerated settling by 'piggy-backing' on to and off the large aggregates. This, combined with reversible scavenging, yields scavenging residence times for trace metals and isotopes that lie between the residence times of the individual small- and large-particle populations. The deep-water scavenging residence times of a number of metals in sea water are listed in Table 11.5.

The concept that dissolved–particulate reactions in the water column may be controlled by an eq-

uilibrium mechanism, and so lead to the possibility that marine chemists may be able to describe the reactions involved in terms of classic surface chemistry theory, was not a new one. Equilibrium models, i.e. those involving reversible uptake, had in fact been proposed by other workers to explain trace-metal scavenging within the water column. For example, Balistrieri *et al.* (1981) had defined an equilibrium scavenging model, based on the principles of surface chemistry, that could be combined with field data to determine the scavenging fate of trace metals in the oceans. To do this, the quantity $[Me]_P/[Me]_t$ was calculated on the basis of equilibrium constants. Such an approach is intellectually appealing because it allows the oceanic residence times of elements to be predicted without a prior knowledge of their seawater concentrations. The results obtained by Balistrieri *et al.* (1981) strongly indicated that the scavenging component of marine particulate matter has an *organic* nature, i.e. the trace-metal scavenging is controlled by organic coatings. The approach adopted by Balistrieri *et al.* (1981) is open to criticism because it may have oversimplified particle–metal interactions in the ocean. Nonetheless, it represented an important step towards the stage when trace-metal scavenging in sea water could be modelled adequately.

It is now apparent that for many of the scavenging-type elements the reactions in vertical sections of the water column can be described in terms of some kind of reversible exchange mechanism. However, it has also become evident that the vertical seawater profiles of some trace metals can be perturbed by the effects of processes that occur at the sediment–water interface in both coastal and open-ocean areas. For example, Bacon & Anderson (1982) showed that there was a sharp decrease in the concentration of dissolved [230]Th towards the sea bed, which they interpreted as resulting from an accelerated uptake of the isotope from solution on to resuspended sediment particles, which act as a sink for dissolved elements. Similar perturbations to the dissolved down-column profiles also have been reported for [210]Pb (see e.g. Bacon *et al.*, 1976). In addition to this accelerated uptake in the benthic boundary layer (see Section 12.2), the dissolved forms of some trace metals are known to be removed preferentially at the *ocean boundaries* as a result of intensified or accelerated scavenging. Most of the non-biogenic (mainly clay) material in deep-sea sediments is small-sized, with ~60–70% being ≤2 μm in diameter (see Section 13.1). The resuspension of this material at the sediment–water interface is a common feature in the distribution of oceanic TSM (see the three-layer TSM model described in Section 10.2), especially along the erosive western boundary currents where nepheloid layers are strongly developed. The particles resuspended here, which are small-sized, can then be transported laterally, thus providing a horizontal supply of TSM to supplement the vertical flux.

The generation of resuspended particles at the ocean boundaries also can have an important effect on the distributions of dissolved elements in sea water. This arises because the higher concentrations of fine particles at the boundaries lead to zones of **enhanced scavenging** (cf. turbidity maxima in estuaries—see Section 3.2.4). Thus, processes at the ocean boundaries may act as a *source* for fine particles and as a *sink* for some dissolved elements via enhanced scavenging. Although the vertical transport of elements down the water column by incorporation into the sinking particle flux, and reference to scavenging-type and nutrient-type processes, offers a useful insight into how trace elements are removed from sea water it is evident that lateral transport into and out of the boundary regions (e.g. along isopycnals) must also be considered in assessing the processes that control the marine cycles of the trace elements.

To summarize, the scavenging of trace elements from sea water is a complex process. The scavenging removal itself is dominated by the fine-particle population, but is then complicated by an association of the fine particles with the large-particle flux by reversible physical mechanisms (e.g. 'piggy-backing') and by reversible chemical adsorption that may tend towards an equilbrium state.

### 11.6.3.2 Nutrient-type mechanisms

Nutrient-type distributions, i.e. surface-depletion–subsurface-enrichment, are displayed by those trace elements which are involved in the major oceanic biological cycles. This involvement results in the trace elements being removed from solution in the surface waters and then being transported to depth by biogenic carriers. A large fraction of the carrier material undergoes oxidative destruction below the surface layer, thus resulting in the regeneration and recycling of the dissolved species. Elements can become associated with the biogenic carriers either because they are specific to biological requirements or because they are taken up by analogy with essential elements; this can result from a lack of discrimination in the biological mechanisms or from 'mistaken identity' (see e.g. Whitfield & Turner, 1987). The biogenic carriers themselves are of two types: (i) those composed of soft tissue material (**labile carriers**); and (ii) those composed of hard skeletal parts (**refractory carriers**). The relative importance of these two types of biogenic trace-element carriers has been assessed by Collier & Edmond (1983, 1984). These authors collected plankton samples from a variety of marine environments and subjected them to a series of leaching–decomposition experiments in order to identify the major and trace element compositions of the principal carrier phases. The elements investigated included C, N, P, Ca, Si, Fe, Mn, Ni, Cu, Cd, Al, Ba and Zn. Although the study may be criticized because the plankton collection employed nets that discriminated against small particles, the results have led to considerable advances in our understanding of the processes involved in the removal of the nutrient-type elements from sea water.

A number of carrier phases, and types of association, are possible between trace elements and marine particulates. According to Collier & Edmond (1983) these trace-element associations include: (i) those with terrigenous material scavenged by biogenic particles; (ii) those involving specific biochemical processes related to metabolism; (iii) those with structural skeletal materials such as calcite, opal and celestite; and (iv) those with hydrous metal oxide precipitates, or organics, via active surface scavenging. The authors assessed the relative importance of these various associations, and some of the more important results of the study are summarized below.

**1** Calcium carbonate and opal were not found to be significant carriers for any of the elements studied.

**2** A phase containing Al and Fe in terrigenous ratios was present in all the plankton samples. However, this non-biogenic carrier phase made an insignificant contribution to the concentrations of the trace elements studied.

**3** The majority of the trace elements were associated directly with the non-skeletal organic phases of the plankton in three types of host associations. In order of their general ease of metal release to sea water via remineralization, these were a **very labile** association (which included the nutrient P and major fractions of the total amounts of Cd, Mn, Ni and Cu), a **moderately refractory** association (which included significant fractions of the total amounts of Cu, Ni, Cd, Ba, Mn and Zn), and a **strongly refractory** association (which included large fractions of the total amounts of Si, Al, Fe and Zn).

4 The concentrations of trace elements in surface ocean biogenic particulate matter are not fixed in simple proportions to their surface water concentration ratios to the nutrients. To demonstrate this the authors showed that, although there are considerable variations in the surface-water dissolved-element : nutrient ratios (e.g. between nutrient-rich upwelling and oligotrophic waters), the particulate trace-metal : carrier-phase ratios are relatively constant. Thus, the trace-element composition of the particulate material appears to be fixed, and limited, mainly by the properties of the organic materials involved and the metabolism of the plankton, rather than by the dissolved concentrations of the elements in the surface waters. In this context, Bruland *et al.* (1991) suggested that the ratios of a series of 'bioactive' trace metals to P and C in plankton samples yield an approximate 'Redfield-type' elemental composition of plankton organic tissue of: $C:N:P:Fe:Zn:Cu$, Mn, Ni, $Cd = 106:16:1:0.005:0.002:0.0004$.

One of the principal aims of the investigation carried out by Collier & Edmond (1983, 1984) was to quantify the biogenic fluxes of trace elements out of the surface layer. The particulate trace-element : carrier-phase ratio constancy allowed the biogenic flux of each element to be predicted in terms of the major oceanic biological cycles, and Collier & Edmond (1983) used a **carrier model** to make their flux estimates. These were then compared with flux values derived from a **two-box vertical model**, which describes the dissolved distributions of the trace elements. Both of these illustrate how models can be used to elucidate the factors that control the distributions of trace elements in the ocean system, and for this reason they will be described in detail.

*The carrier model.* In this model, independently derived estimates of the fluxes of major biologically cycled elements, such as C, N, P, Ca (carbonate) and Si (opal), were used, and the trace elements were coupled to these fluxes through their ratios to the

major elements, which represent the biogenic carrier phases. As extensive data were available for P, this element was used in the model to represent the **primary flux** of organic matter. Collier & Edmond (1984) then used the following relationship to compare elemental ratios in the plankton with those in the surrounding sea water:

$$\alpha = \frac{(\text{metal}/P)_{\text{plankton}}}{(\text{metal}/P)_{\text{water}}} \tag{11.9}$$

Differences in the regeneration of the nutrients and trace elements in the surface waters will result in different elemental ratios in the sinking particulates, which are given by the relationship:

$$\beta = \frac{(\text{metal}/P)_{\text{sinking}}}{(\text{metal}/P)_{\text{plankton}}} \tag{11.10}$$

In these equations $\alpha$ represents particulate formation by primary producers, and $\beta$ represents the various processes that modify the ratio before the particulate matter sinks out of the surface layer. Thus, for regeneration that takes place within the surface layer, the product $\alpha\beta$ offers a comparison between the relative enrichment, or depletion, of the particulate trace-element flux with respect to the dissolved ratios. The authors used their leaching experiment data to obtain a first-order estimate of the magnitude of $\beta$, which was calculated as the percentage of the total element *not* released by leaching to sea water, divided by the same percentage for P. Thus, $\beta$ is a relative enrichment factor for the residual trace-element/P ratio in the particles, and the product of the plankton sample composition and $\beta$ gives an estimate of the composition of the material settling out of the surface layer. In the carrier model, the net, or 'new production', P flux out of the surface layer is coupled with the residual trace-element/P ratios to calculate the down-column biogenic carrier flux. The carrier model is illustrated in Fig. 11.9(a).

*The two-box model.* The two reservoirs in this model are the **surface ocean** and the **deep ocean** (see Section 7.5.2). The parameters used to define the model are the dissolved distributions of the elements, the mixing rates of water between the reservoirs, and the primary input and output fluxes of the elements. The particulate flux out of the surface layer

($P$) is then calculated to balance the sum of the other input–output fluxes; thus

$$P = V_M(C_D - C_S) + V_R C_R + A \qquad (11.11)$$

where $V_M$ (3.5 m yr$^{-1}$) is the exchange rate between the surface and deep boxes of water with concentra-tions $C_S$ and $C_D$, respectively; $V_R C_R$ represents the river input; and $A$ represents the atmospheric input. The two-box model is shown diagrammatically in Fig. 11.9(b). Further details on both models are given in Collier & Edmond (1984).

The fluxes obtained from the two-box model may be regarded as estimates of those required to maintain

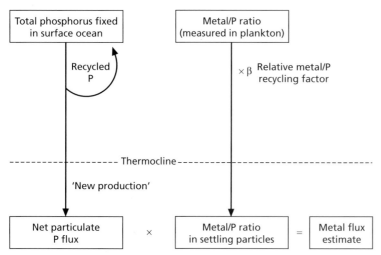

(a)

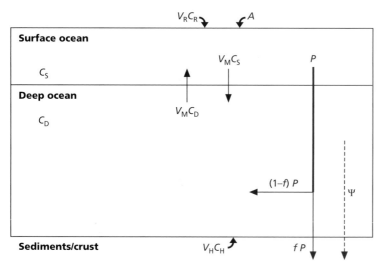

(b) $\qquad P = V_M(C_D - C_S) + V_R C_R + A$

**Table 11.6** Down-column flux estimates of trace metals using an organic carrier phase model and a two-box model; units, ng cm$^{-2}$ yr$^{-1}$ (after Collier & Edmond, 1984).

| Element | Organic-carrier estimate | Box-model estimate |
|---------|--------------------------|---------------------|
| Cd | 56–112 | 38 |
| Ni | 70–135 | 152 |
| Cu | 95–114 | 76–95 |
| Mn | 33–44 | 110 |
| Ba | 961–1374 | 5494 |
| Zn | 654–1635 | 163–196 |
| Fe | 1396–4468 | 1117–3909 |

the vertical gradients in the dissolved distributions of the elements. Thus, by comparing the *carrier*-phase fluxes with these *total* fluxes it was possible to make first-order assessments of the importance of biological particulates in down-column trace-element fluxes. The flux estimates obtained using the two models are given in Table 11.6, and the data can be summarized as follows.

1 The fluxes derived from the two models agree, within a factor of 2, for Cd, Ni and Cu, indicating that the data used in the carrier model are representative of the major particulate flux that maintains the vertical flux of these elements.

2 Mn is incorporated in plankton and released during regeneration processes. However, comparison of the magnitudes of the fluxes predicted from the models indicates that the biogenic carrier phases transport considerably less Mn out of the surface

layer than do the non-biogenic carriers. Thus, the association of Mn with plankton should not have a significant effect on the dissolved distribution of the metal.

3 The biogenic carrier flux for Ba is only sufficient to account for ~20–30% of the box model flux. This biogenic carrier flux for Ba includes associations with organic matter, calcium carbonate, opal and celestite, and Collier & Edmond (1983, 1984) suggested that a secondary process resulting in the formation of Ba-rich particles (such as barite) probably occurs below the surface zone of nutrient depletion.

4 The flux predicted by the carrier model for Zn is an order of magnitude higher than that calculated by applying the dissolved distributions of the element to the two-box model. This may be the result of sample contamination in any of the values used in either model, and as a result it is not yet possible to identify with any certainty the Zn carrier phase in the oceans.

### 11.6.3.3 Scavenging-type versus nutrient removal mechanisms

The division of trace-element uptake mechanisms into scavenging-type (passive surface adsorption) and nutrient-type (active biological uptake) classes is a useful way of describing the vertical distributions and down-column transport of some elements. Honeyman *et al.* (1988), however, took a wider view of the processes that remove trace elements from sea water, and attempted to bring together recent advances in our knowledge of particle scavenging

**Fig. 11.9** Models for the down-column transport of trace elements in the oceans (from Collier & Edmond, 1984). (a) The carrier model. This makes use of independent predictions of the fluxes of major organically cycled elements. Trace elements are coupled to these biogenic fluxes through their ratios to the major elements that represent the biogenic carrier phase. In the model illustrated, *P* is used to represent the primary flux of organic matter out of the surface layer. Independent estimates of the *P* flux are derived from total surface ocean carbon fixation rates. The fraction of the 'new' carbon production, i.e. that which is not recycled in the surface layer, but which actually constitutes a net flux to the deep ocean (see Section 9.1.1), is estimated from nitrogen flux data. The *P* flux associated with this net out-of-surface carbon flux is estimated using a Redfield C : P ratio of 106 : 1 (see Section 9.2.3.1). The resulting *P* flux (~1 μmol P cm$^{-2}$ yr$^{-1}$) is coupled to out-of-surface fluxes of the minor element concentrations in sinking particles using β (see text) as

an estimate of the residual (i.e. non-rapidly recycled) metal : P ratios, and the product of β and the plankton sample composition is used to give an estimate of the composition of the material settling out of the surface layer. This is then applied to the 'new' production estimate of the net *P* flux to calculate the biogenic carrier flux out of the surface ocean for each trace element. (b) The two-box model: *P*, particulate matter flux out of the surface ocean; $V_R$, $C_R$, volume and concentration of an element in river water; *A*, atmospheric input; $C_S$, $C_D$, dissolved concentrations of an element in the surface and deep reservoirs; *f*, fraction of particulate matter preserved in sediments; ψ, additional particulate matter flux resulting from scavenging within the deep ocean; $V_H$, $C_H$, volume and concentration of an element in hydrothermal solutions. The particulate matter flux (*P*) is calculated by the mass balance of all other inputs and outputs to the surface ocean reservoir.

mechanisms in terms of the removal of dissolved Th from sea water.

The solids in natural aqueous systems contain surface functional groups, which can form surface complexes with solutes by coordination reactions that follow the principles of coordination chemistry. Various authors have used **surface complexation models** to describe the uptake of trace elements by solid surfaces. For example, a surface complexation model is described below for the ultimate removal of elements from the oceans. For a detailed review of the application of coordination chemistry to the adsorption and transformation of trace-element species at particle–water interfaces, the reader is referred to Leckie (1986). Honeyman *et al.* (1988) pointed out that two general hypotheses have been produced to account for oceanic scavenging, both of which involve marine particles. Using Th as an example, these scavenging reactions were characterized as (i) control by biological removal, and (ii) control by equilibrium exchange reactions. Thermodynamic scavenging models assume complete equilibrium in the system between the dissolved and particulate phases, and the equilibrium values are assumed to be constant for similar conditions regardless of reaction time relative to particle residence time (Jannasch *et al.*, 1988). If the particle residence time is short compared with the adsorption time, however, an equilibrium condition may not be attained, and the system may depend on kinetic factors (see e.g. Nyffeler *et al.*, 1984; Jannasch *et al.*, 1988).

It is now known that the rate of transfer of dissolved Th to the particulate state is a function of particle concentration or particle residence time (see e.g. Nyffeler *et al.*, 1984), i.e. the rate constants of adsorption vary with the particle concentration. Further, one of the most significant findings in recent years was that there is a correlation between dissolved Th removal rate constants and the production of particles via primary productivity (see e.g. Coale & Bruland 1985, 1987). The key to understanding the removal processes, therefore, lies in the fact that the adsorption of Th varies with the concentration of particles in the system. Honeyman *et al.* (1988) pointed out that this has important implications on Th fractionation because in near-shore surface ocean waters the particle residence time is equal to, or less than, the residence time of the

dissolved Th. As particles are leaving the system the fraction of Th in the particulate phase will be less than it would be if the particles had an infinite residence time. Under these conditions, therefore, the system should not be viewed as an equilibrium system.

Honeyman *et al.* (1988) derived a series of equations of describe Th uptake and then plotted the log value for the forward Th sorption rate constant, i.e. uptake, against the log value of the particle concentration, using data for a wide variety of oceanic environments. These included the highly productive California coastal zone, with medium particle concentrations; the low productive China shelf zone, with high particle concentrations; and the deep ocean, which is low in both biological productivity and particle concentrations. The plot is reproduced in Fig. 11.10, and shows that there is a strong linear relationship between the log value of the forward sorption rate constant and the log value of the particle concentration. This particle dependence of the scavenging rate holds over seven orders of magnitude variation in particle concentration. Linear regression analysis of the data (excluding that for the Amazon) gave a slope of 0.58, and the authors described the partitioning of Th between dissolved and particulate phases by the equation:

$$\frac{A_{Th, part}}{A_{Th, diss}} = \frac{R_f (C_p)^{0.58}}{R_r + \lambda_{Th} + \lambda_{part}} \quad (11.12)$$

where $A_{Th, part}$ and $A_{Th, diss}$ are the activities of particulate and dissolved Th, respectively; $R_f$ ($= 0.079 \, l^{-0.58} \, mg^{-0.58} \, day^{-1}$) is the rate constant for the transfer of Th from the dissolved to the particulate phase; $R_r$ ($= 0.007 \, day^{-1}$) is the rate constant for the reverse reaction; $C_p$ is the mass concentration of particles (expressed in units of mg $l^{-1}$); and $\lambda_{Th}$ and $\lambda_{part}$ are the decay constant for the thorium isotope (day$^{-1}$) and the rate constant for particle removal (day$^{-1}$), respectively. Thus, the partitioning of Th depends on an interplay between the reaction rates, the particle concentration, the decay characteristics of the particular Th isotope, and the residence time of the particles in the system, which depends on the water column height, the particle concentration and the particle flux.

Trace-metal removal times based on adsorption reactions are thought to be rapid, i.e. of the order of

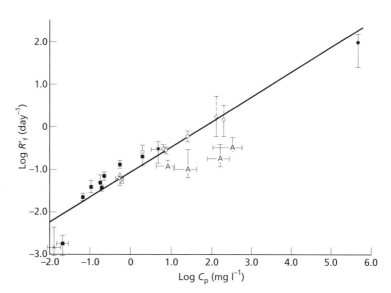

**Fig. 11.10** A plot of the forward sorption rate constant ($\log R_f$) for Th versus particle concentration using field data from various oceanic regions (from Honeyman *et al.*, 1988). The oceanic regions cover a wide variety of particle concentrations, particle types and primary productivity intensities: ($\square$ and $\times$) deep sea; ($\blacksquare$) surface water, California Current; ($\bigcirc$) coastal waters, Yangtze River; (A) coastal waters, Amazon River; ($\triangle$) surface waters, Funka Bay; ($\blacktriangledown$) surface waters, Narragansett Bay; ($\bullet$) sediment porewaters, Buzzards Bay. The full line represents the linear regression excluding the Amazon data.

seconds, but in natural systems they appear to be of the same order as physical mixing processes, i.e. days to weeks. In order, therefore, to combine all the various features of the Th removal processes, Honeyman *et al.* (1988) suggested that the aggregation or coagulation of colloids, which also is a function of particle concentration, may play a part in controlling these slow rates. They proposed, therefore, that the constancy of the 'Th removal rate constant–particle concentration' relationship can be attributed to a Th removal mechanism that involves a combination of surface coordination reactions (based on surface coordination chemistry involving surface complex formation—see above) and colloidal particle-to-particle aggregation. This surface co-ordination–colloidal aggregation model implies that although particles in a highly productive system may be of biological origin and may be supplied at a rate that is controlled by primary production, they simply serve as **new adsorptive passive substrates** for the surface-active Th. It must be stressed, however, that Honeyman *et al.* (1988) based their model on Th, and other metals may have different controlling properties. Thus, the extent to which Th can act as an analogue for other surface-active metals requires further research. Nonetheless, the model proposed by Honeyman *et al.* (1988) links microlevel adsorption reactions to macrolevel oceanic scavenging processes, and offers an explanation for the correlations found

between dissolved Th removal rate constants and biological productivity.

Despite the fact that biologically produced particles may act as passive substrates for the uptake of surface-active trace metals, it is still useful to view down-column trace-metal transport in terms of scavenging-type uptake, which involves inorganic (perhaps organic-coated) FPM, and nutrient-type uptake, which involves organic CPM aggregates. It is the CPM organic aggregates that dominate the down-column transport of material to the sediment sink, pulling down the FPM with them. Once they reach the sea bed, however, a large fraction of the organic carrier material is destroyed in open-ocean regions during oxic diagenesis, releasing the associated trace elements back into solution, and the fraction that is buried in the sediment will continue to be degraded during suboxic diagenesis (see Section 14.3). As a result of processes such as this, Li (1981) has suggested that the most important mechanism for the *ultimate* removal of elements from the deep sea is adsorption on to the surfaces of fine particles. These include aluminosilicates, managanese oxides and iron oxides, and from data based on factor analyses Li (1981) showed that the following relationships exist in both ferromanganese modules and pelagic clays.

1 Elements such as K, Rb, Cs, Be, Sc and Ga are associated mainly with the aluminosilicate detritus.

2 The elements associated with iron oxide phases form anions or oxyanions (e.g. P, S, Se, Te, As, B, Sn, I, Br, F, U, Pb and Hg) and hydroxide complexes of tri- and tetravalent cations (e.g. Ti, Ge, Zr, Hf, Th, Y, In, Pd and Cr).

3 The elements associated with manganese oxides are mono- and divalent cations (e.g. Mg, Ca, Ba, Tl, Co, Ni, Cu, Zn, Bi, Ag and Cd) and oxyanions (e.g. Mo, W and Sn).

Li (1981) attempted to rationalize these relations in terms of theoretical adsorption models.

The uptake of elements on to the surfaces of particulate matter involves a distribution ratio ($K_d$), which is the ratio of the concentration of the element in the solid phase to that in the aqueous phase. A variety of theoretical models have been proposed to describe the inorganic adsorption of trace elements on to particulate surfaces. However, the surface complexation model (see e.g. Schindler, 1975) has been most commonly used in marine chemistry, and this will serve to illustrate the general mechanisms that are thought to be involved in the ultimate removal of elements reaching the sediment surface. In this model, it is assumed that metal ions are removed by adsorption on to inorganic particulate matter at oxide surfaces covered with OH groups, which act as ligands. The hydrolysis of the oxide surfaces produces hydrous oxide surface groups such as $\equiv Si-OH$, $\equiv Mn-OH$ and $=Fe-OH$. Protonation or deprotonation results in a surface charge on the oxide surfaces, which Li (1981) represents as:

$$\equiv MeOH_2^+ \rightleftharpoons \equiv MeOH + H^+ \qquad (11.13)$$

$$K_{a1}^s = \frac{\{MeOH\}[H^+]}{\{MeOH_2^+\}} \qquad (11.14)$$

$$\equiv MeOH \rightleftharpoons MeO^- + H^+ \qquad (11.15)$$

$$K_{a2}^s = \frac{\{MeO^-\}[H^+]}{\{MeOH\}} \qquad (11.16)$$

where Me is the metal (e.g. Mn, Fe, Si, Al) of the solid oxides, [ ] is the concentration of species in the aqueous phase, { } is the concentration of surface species on the solid oxides, and $K_{a1}^s$ and $K_{a2}^s$ are acidity constants.

Li (1981) applied the surface complexation model to the observed partitioning of a series of elements between sea water and pelagic clays, and between river water and river suspended particulates. To do this, the adsorption of a cation ($M^{z+}$) on to a surface hydroxyl group was represented as:

$$M^{z+} + \equiv Me-OH \rightleftharpoons \equiv Me-O-M^{(z-1)+} + H^+ \qquad (11.17)$$

$$*K_{1(app)}^s = \frac{[H^+]\{Me-O-M^{(z-1)+}\}}{[M^{z+}]\{MeOH\}} \qquad (11.18)$$

where the units are as for eqns (11.13)–(11.16), and $*K_{1(app)}^s$ is the apparent equilibrium constant. The bond strength of a metal–oxygen bond (O–M) for various cations on silica gel has been shown to be linearly correlated with their $\log *K_1$, where $*K_1$ is the first hydrolysis constant for the reaction:

$$M^{z+} + H_2O \rightleftharpoons H-O-M^{(z-1)+} + H^+ \qquad (11.19)$$

$$*K_1 = \frac{[H^+][HOM^{(z-1)+}]}{[M^{z+}][H_2O]} \qquad (11.20)$$

Schindler (1975) demonstrated that $\log *K_{1(app)}$ of silica gel is also linearly correlated with $\log *K_1$. Li (1981) suggested therefore that the partitioning of various *cations* between solid and liquid phases in rivers and the oceans should be correlated positively with the relative bond strength between the cations and the oxygen of the hydrous oxide surface, or with $\log *K_1$. This relationship is illustrated in Fig. 11.11(a, c), in which the partitioning of the cations between liquid and solid phases in the ocean is plotted against $\log *K_1$ values.

From this figure it can be seen that there is a general positive correlation between the two parameters for most mono- and divalent cations, i.e. the higher the $\log *K_1$ values, the more the cations are partitioned towards the solid phase (surface adsorption). The tri- and tetravalent cations, which usually form hydroxyl complexes in natural water, plot as a broad maximum. These usually are associated with iron oxide phases in pelagic deposits and the fact that their partitioning coefficient ratios do not increase with $*K_1$ may be a function of the adsorption dynamics on to the iron oxide surfaces.

For undissociated acids, monovalent anions and divalent anions there is an inverse relationship between $\log K_1$ (first dissociation constant) or $\log K_2$ (second dissociation constant) and the partitioning coefficients for river water and sea water. This is shown in Fig. 11.11(b, d), and indicates that the higher the $\log K_1$ or $\log K_2$ values, the more the species

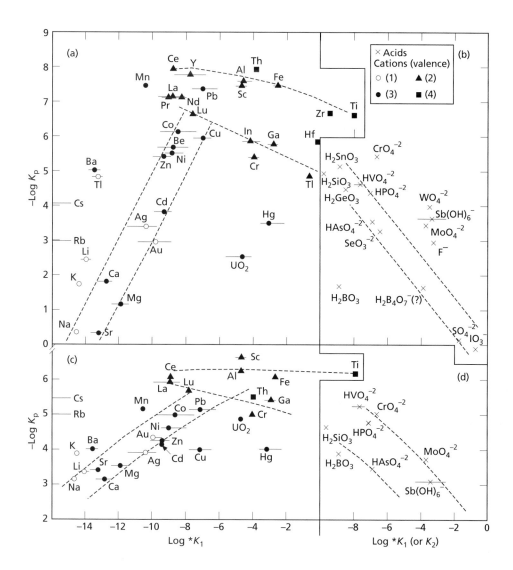

**Fig. 11.11** Plots of the relationship between (a, b) $\log K_p = \log C_{op}/C_{sw}$ (the concentration ratio of an element in oceanic pelagic clay and sea water) and (c, d) $\log K_p = \log C_p/C_r$ (the concentration ratio of an element in river-suspended particulates and river water) versus $\log *K_1$ (the first hydrolysis constant) or $\log K_2$ (the second dissociation constant of acids) (after Li, 1981).

is partitioned to the liquid phase (less adsorption). A number of species fall outside the main correlation. These include (i) $MoO_4^{2-}$, $WO_4^{2-}$ and $SbO_3^-$, which may be taken up in authigenic manganese phases, and (ii) $F^-$, which may enter apatite.

Overall, therefore, Li (1981) concluded that with the exception of those which form authigenic phases,

the observed partitioning of most elements between the solid and liquid phases of river water and sea water can be explained adequately in terms of the surface complexation model, and that adsorption on to the surfaces of iron oxides, manganese oxides and clay minerals was the most important mechanism for the ultimate removal of elements from the oceans. Whitfield & Turner (1982), however, criticized the surface complexation model on a number of grounds, and proposed instead that the partitioning of elements between the solid and liquid phases in river and sea water could be better rationalized using a more simple electrostatic model (see Section 17.3).

## 11.7 Trace elements in sea water: summary

**1** Trace-element distributions in sea water are controlled by a complex interaction between their source strengths, their removal strengths and water circulation patterns.

**2** The vertical distributions of trace elements in the water column can be related to a number of analogues. On this basis, the profiles can be divided into the following classes: (i) conservative type (unreactive); (ii) nutrient surface-depletion–depth-enrichment type; (iii) scavenging surface-enrichment–depth-depletion type; (iv) mid-depth-minimum type; (v) mid-depth-maximum type; and (vi) anoxic-water type.

**3** Dissolved trace elements are removed from sea water by the oceanic microcosm of particles, via scavenging-type and nutrient-type mechanisms, and the subsequent settling of these particles down the water column, either directly from the surface water or indirectly following subsurface lateral advection.

**4** Relatively large fractions (>50%) of the dissolved concentrations of some trace metals (e.g. Cu, Zn, Cd, Pb) in open-ocean surface waters are bound in organic complexes by ligands, which are often metal-specific. This organic speciation plays an important role in the biogeochemical cycles of the metals.

The processes involved in the down-column transport of the trace elements are part of the 'great particle conspiracy', and in the next chapter an attempt will be made to assess the rate at which the conspiracy operates.

## References

Andreae, M.O. (1983) The determination of the chemical species of some of the 'hydride elements' (arsenic, antimony, tin and germanium) in seawater: methodology and results. In *Trace Metals in Sea Water*, C.S. Wong, E. Boyle, K.W. Bruland, J.D. Burton & E.D. Goldberg (eds), 1–19. New York: Plenum.

Andreae, M.O. (1986) Chemical species in seawater and marine particulates. In *The Importance of Chemical 'Speciation' in Environmental Processes*, M. Bernhard, F.E. Brinckman & P.J. Sadler (eds), 301–35. Berlin: Springer-Verlag.

Bacon, M.P. (1984) Radionuclide fluxes in the ocean interior. In *Global Ocean Flux Study*, 180–205. Washington, DC: National Academy Press.

Bacon, M.P. & Anderson, R.F. (1982) Distribution of thorium isotopes between dissolved and particulate forms in the deep sea. *J. Geophys. Res.*, 87, 2045–56.

Bacon, M.P., Spencer, D.W. & Brewer, P.G. (1976) $^{210}Pb/^{226}Ra$ and $^{210}Po/^{210}Pb$ disequilibria in seawater and suspended particulate matter. *Earth Planet. Sci. Lett.*, 32, 277–96.

Bacon, M.P., Brewer, P.G., Spencer, D.W., Murray, J.W. & Goddard, J. (1980) Lead-210, polonium-210, manganese and iron from the Cariaco Trench. *Deep Sea Res.*, 27, 119–35.

Bacon, M.P., Huh, C.-A., Fleer, A.P. & Deuser, W.G. (1985) Seasonality in the flux of natural radionuclides and plutonium in the deep Sargasso Sea. *Deep Sea Res.*, 32, 273–86.

Bailey, R.A., Clarke, H.M., Ferris, J.P., Krause, S. & Strong, R.L. (1978) *Chemistry of the Environment*. New York: Academic Press.

Balistrieri, L., Brewer, P.G. & Murray, J.W. (1981) Scavenging residence time of trace metals and surface chemistry of sinking particles in the deep ocean. *Deep Sea Res.*, 28, 101–21.

Barth, T.W. (1952) *Theoretical Petrology*. New York: Wiley.

Benoit, G., Oktay-Marshall, S.D., Cantu, A., *et al.* (1994) Partitioning of Cu, Pb, Ag, Zn, Fe, Al and Mn between filter-retained particles, colloids, and solution of six Texas estuaries. *Mar. Chem.*, 45, 307–36.

Bewers, J.M. & Yeats, P.A. (1977) Oceanic residence times of trace metals. *Nature*, 268, 595–8.

Boulegue, J. (1983) Trace metals (Fe, Zn, Cd) in anoxic environments. In *Trace Metals in Sea Water*, C.S. Wong, E. Boyle, K.W. Bruland, J.D. Burton & E.D. Goldberg (eds), 563–77. New York: Plenum.

Boyle, E. (1997) What controls dissolved iron concentrations in the world ocean?—a comment. *Mar. Chem.*, 57, 163–67.

Boyle, E.A. & Huested, S. (1983) Aspects of the surface distributions of copper, nickel, cadmium and lead in the North Atlantic and North Pacific. In *Trace Metals in Sea Water*, C.S. Wong, E. Boyle, K.W. Bruland, J.D. Burton & E.D. Goldberg (eds), 379–94. New York: Plenum.

Boyle, E.A., Sclater, F. & Edmond, J.M. (1976) On the marine geochemistry of cadmium. *Nature*, 263, 42–4.

Boyle, E.A., Sclater, F. & Edmond, J.M. (1977) The distribution of dissolved copper in the Pacific. *Earth Planet. Sci. Lett.*, 37, 38–54.

Boyle, E.A., Huested, S.S. & Jones, S.P. (1981) On the distribution of copper, nickel and cadmium in the surface waters of the North Atlantic and North Pacific Ocean. *J. Geophys. Res.*, 86, 8048–66.

Brewer, P.G. (1974) Minor elements in sea water. In *Chemical Oceanography*, J.P. Riley & G. Skirrow (eds), Vol. 1, 415–96. London: Academic Press.

Broecker, W.S. & Peng, T.-H. (1982) *Tracers in the Sea*. Palisades, NY: Lamont-Doherty Geological Observatory.

Bruland, K.W. (1980) Oceanographic distributions of cadmium, zinc, nickel and copper in the North Pacific. *Earth Planet. Sci. Lett.*, 47, 176–98.

Bruland, K.W. (1983) Trace elements in sea water. In *Chem-*

*ical Oceanography*, J.P. Riley & R. Chester (eds), Vol. 8, 157–220. London: Academic Press.

Bruland, K.W. (1989) Complexation of zinc by natural organic ligands in the central North Pacific. *Limnol. Oceanogr.*, **34**, 269–85.

Bruland, K.W. (1992) Complexation of cadmium by natural organic ligands in the central North Pacific. *Limnol. Oceanogr.*, **37**, 1008–17.

Bruland, K.W. & Franks, R.P. (1983) Mn, Ni, Zn and Cd in the western North Atlantic. In *Trace Metals in Sea Water*, C.S. Wong, E. Boyle, K.W. Bruland, J.D. Burton & E.D. Goldberg (eds), 395–414. New York: Plenum.

Bruland, K.W., Knauer, G.A. & Martin, J.H. (1978) Cadmium in northeast Pacific waters. *Limnol. Oceanogr.*, **23**, 618–25.

Bruland, K.W., Donat, J.R. & Hutchins, D.A. (1991) Interactive influences of bioactive trace metals on biological production in oceanic waters. *Limnol. Oceanogr.*, **36**, 1555–77.

Buckley, P.J.M. & van den Berg, C.M.G. (1986) Copper complexation profiles in the Atlantic Ocean. A comparative study using electrochemical and ion exchange techniques. *Mar. Chem.*, **19**, 281–96.

Burton, J.D. & Statham, P.J. (1988) Trace metals as tracers in the ocean. *Philos. Trans. R. Soc. London*, **325**, 127–45.

Burton, J.D., Maher, W.A. & Statham, P.J. (1983) Some recent measurements of trace metals in Atlantic Ocean waters. In *Trace Metals in Sea Water*, C.S. Wong, E. Boyle, K.W. Bruland, J.D. Burton & E.D. Goldberg (eds), 415–26. New York: Plenum.

Capodaglio, G., Coale, K.H. & Bruland, K.W. (1990) Lead speciation in surface waters of the eastern North Pacific. *Mar. Chem.*, **29**, 221–33.

Chan, L.H., Drummond, D., Edmond, J.M. & Grant, B. (1977) On the barium data from the Atlantic GEOSECS Expedition. *Deep Sea Res.*, **24**, 613–49.

Chester, R. (1985) Book reviews. *Chem. Geol.*, **51**, 150–1.

Chester, R., Bradshaw, G.F., Ottley, C.J., *et al.* (1994) The atmospheric distributions of trace metals, trace organics and nitrogen species over the North Sea. In *Understanding the North Sea System*, H. Charnock, K.R. Dyer, J.M. Huthnance, P.S. Liss, J.H. Simpson & P.B. Tett (eds). *Philos. Trans. R. Soc. London, Ser. A*, **343**, 545–56.

Cho, B.C. & Azam, F. (1988) Major role of bacteria in biogeochemical fluxes in the ocean's interior. *Nature*, **332**, 441–3.

Clegg, S.L. & Whitfield, M. (1991) A generalised model for the scavenging of trace metals in the open ocean—II. Thorium scavenging. *Deep Sea Res.*, **38**, 91–120.

Coale, K.H. & Bruland, K.W. (1985) [234]Th : [238]U disequilibria within the California Current. *Limnol. Oceanogr.*, **30**, 22–33.

Coale, K.H. & Bruland, K.W. (1987) Oceanic stratified euphotic zone as elucidated by [234]Th : [238]U disequilibria. *Limnol. Oceanogr.*, **32**, 189–200.

Coale, K.H. & Bruland, K.W. (1988) Copper complexation in the northeast Pacific. *Limnol. Oceanogr.*, **33**, 1084–1101.

Coale, K.H. & Bruland, K.W. (1990) Spatial and temporal variability in copper complexation in the North Pacific. *Deep Sea Res.*, **47**, 317–36.

Collier, R.W. & Edmond, J.M. (1983) Plankton compositions and trace element fluxes from the surface ocean. In *Trace Metals in Sea Water*, C.S. Wong, E. Boyle, K.W. Bruland, J.D. Burton & E.D. Goldberg (eds), 789–809. New York: Plenum.

Collier, R.W. & Edmond, J.M. (1984) The trace element geochemistry of marine biogenic particulate matter. *Progr. Oceanogr.*, **13**, 113–99.

Craig, H. (1974) A scavenging model for trace elements in the deep sea. *Earth Planet. Sci. Lett.*, **23**, 393–402.

Cutter, G.A. & Bruland, K.W. (1984) The marine biogeochemistry of selenium: a re-evaluation. *Limnol. Oceanogr.*, **29**, 1179–92.

Danielsson, L.G. (1980) Cadmium, cobalt, iron, lead, nickel and zinc in Indian Ocean water. *Mar. Chem.*, **8**, 199–215.

Danielsson, L.G. & Westerlund, S. (1983) Trace metals in the Arctic Ocean. In *Trace Metals in Sea Water*, C.S. Wong, E. Boyle, K.W. Bruland, J.D. Burton & E.D. Goldberg (eds), 85–95. New York: Plenum.

Donat, J.R. & Bruland, K.W. (1988) Simultaneous determination of dissolved cobalt and nickel in seawater by cathodic stripping voltammetry preceded by collection of cyclohexane 1,2-dione dioxime complexes. *Anal. Chem.*, **60**, 240–44.

Donat, J.R. & Bruland, K.W. (1990) A comparison of two voltammetric techniques for determining zinc speciation in northeast Pacific Ocean waters. *Mar. Chem.*, **28**, 301–23.

Dyrssen, D. & Wedborg, M. (1974) Equilibrium calculations of the speciation of the elements in sea water. In *The Sea*, E.D. Goldberg (ed.), Vol. 5, 181–95. New York: Interscience.

Edmond, J.M., Jacobs, S.S., Gordon, A.L., Mantyla, A.W. & Weiss, R.F. (1979) Water column anomalies in dissolved silica over opaline pelagic sediments and the origin of the deep silica maximum. *J. Geophys. Res.*, **84**, 7809–26.

Emerson, S., Jacobs, L. & Tebo, B. (1983) The behaviour of trace metals in marine anoxic waters: solubilities at the oxygen–hydrogen sulphide interface. In *Trace Metals in Sea Water*, C.S. Wong, E. Boyle, K.W. Bruland, J.D. Burton & E.D. Goldberg (eds), 579–608. New York: Plenum.

Fitzgerald, W.F. (1989) Atmospheric and oceanic cycling of mercury. In *Chemical Oceanography*, J.P. Riley, R. Chester & R.A. Duce (eds), Vol. 10, 151–186. London: Academic Press.

Florence, T.M. & Batley, G.E. (1976) Trace metal species in seawater. *Talanta*, **23**, 179–86.

Froelich, P.N., Hambrick, G.A. & Andreae, M.O. (1983)

Geochemistry of inorganic and methyl germanium species in three estuaries. *Eos*, **45**, 715.

Froelich, P.N., Hambrick, G.A., Kaul, L.W., Byrd, J.T. & Lecointe, O. (1985) Geochemical behaviour of inorganic germanium in an unperturbed estuary. *Geochim. Cosmochim. Acta*, **49**, 519–24.

Garrels, R.M. & Thompson, M. (1962) A chemical model for sea water at 25°C and one atmosphere total pressure. *Am. J. Sci.*, **260**, 57–60.

Gill, A.G. & Fitzgerald, W.F. (1988) Vertical mercury distribution in the oceans. *Geochim. Cosmochim. Acta*, **52**, 1719–28.

Goldberg, E.D. (1954) Marine geochemistry. I. Chemical scavengers of the sea. *J. Geol.*, **62**, 249–65.

Goldberg, E.D. (1965) Minor elements in sea water. In *Chemical Oceanography*, 1st edn, J.P. Riley & G. Skirrow (eds), Vol. 1, 181–95. London: Academic Press.

Goldberg, E.D. & Arrhenius, G.O.S. (1958) Chemistry of Pacific pelagic sediments. *Geochim. Cosmochim. Acta*, **13**, 153–212.

Hanson, A.K., Sakamoto-Arnold, C.M., Huizenga, D.L. & Kester, D.R. (1988) Copper complexation in Sargasso Sea and Gulf Stream warm-core ring waters. *Mar. Chem.*, **23**, 181–203.

Honeyman, B.D., Balistrieri, L.S. & Murray, J.W. (1988) Oceanic trace metal scavenging: the importance of particle concentration. *Deep Sea Res.*, **35**, 227–46.

Hudson, R.J.M. & Morel, F.M.M. (1990) Iron transport in marine phytoplankton: kinetics of cellular and medium coordination reactions. *Limnol. Oceanogr.*, **35**, 1002–20.

Hudson, R.J.M. & Morel, F.M.M. (1993) Trace metal transport by marine microorganisms: implications of metal coordination kinetics. *Deep Sea Res.*, **40**, 129–150.

Huizenga, D.I. & Kester, D.R. (1983) The distribution of total and electrochemically available copper in the northwestern Atlantic Ocean. *Mar. Chem.*, **13**, 281–91.

Hydes, D.J. (1979) Aluminium in sea water: control by inorganic processes. *Science*, **205**, 1260–2.

Izdar, E. & Murray, J.W. (eds) (1991) *Black Sea Oceanography*. Dordrecht: Kluwer.

Jacobs, L., Emerson, S. & Skei, J. (1985) Partitioning and transport of metals across the $O_2/H_2S$ interface in a permanently anoxic basin: Framvaren Fjord, Norway. *Geochim. Cosmochim. Acta*, **49**, 1433–44.

Jannasch, H., Honeyman, B.D., Balistrieri, L.S. & Murray, J.W. (1988) Kinetics of trace element uptake by marine particles. *Geochim. Cosmochim. Acta*, **52**, 567–77.

Johnson, K.S., Gordon, R.M. & Coale, K.H. (1997) What controls dissolved iron concentrations in the world ocean?—a comment. *Mar. Chem.*, **57**, 137–61.

Klinkhammer, G.P. & Bender, M.L. (1980) The distribution of manganese in the Pacific Ocean. *Geochim. Cosmochim. Acta*, **46**, 361–84.

Klinkhammer, G., Elderfield, H., Greaves, M., Rona, P. &

Nelson, T. (1986) Manganese geochemistry near high temperature vents in the Mid-Atlantic Rift valley. *Earth Planet. Sci. Lett.*, **80**, 230–40.

Kramer, C.J.M. (1986) Apparent copper complexation capacity and conditional stability constants in North Atlantic waters. *Mar. Chem.*, **18**, 335–49.

Krauskopf, K.B. (1956) Factors controlling the concentrations of thirteen rare metals in sea-water. *Geochim. Cosmochim. Acta*, **12**, 61–84.

Kremling, K. (1983) Trace metal fronts in European shelf waters. *Nature*, **303**, 225–7.

Kremling, K. (1985) The distribution of cadmium, copper nickel, manganese and aluminium in surface waters of the open Atlantic and European shelf area. *Deep Sea Res.*, **32**, 531–55.

Lal, D. (1980) Comments on some aspects of particulate transport in the oceans. *Earth Planet. Sci. Lett.*, **49**, 520–7.

Landing, W.M. & Bruland, K.W. (1980) Manganese in the North Atlantic. *Earth Planet. Sci. Lett.*, **49**, 45–56.

Landing, W.M. & Bruland, K.W. (1987) The contrasting biogeochemistry of iron and manganese in the Pacific Ocean. *Geochim. Cosmochim. Acta*, **51**, 29–43.

Leckie, J.O. (1986) Adsorption and transformation of trace element species at sediment/water interface. In *The Importance of Chemical 'Speciation' in Environmental Processes*, M. Bernhard, F.E. Brinckman & P.J. Sadler (eds), 237–54. Berlin: Springer-Verlag.

Lewis, B.L. & Landing, W.M. (1991) The biogeochemistry of manganese and iron in the Black Sea. *Deep Sea Res.*, **38**, S773–S803.

Li, Y.-H. (1981) Ultimate removal mechanisms of elements from the ocean. *Geochim. Cosmochim. Acta*, **45**, 1659–64.

Luther, G.W. & Wu, J. (1997) What controls dissolved iron concentrations in the world ocean?—a comment. *Mar. Chem.*, **57**, 173–79.

Magnusson, B. & Westerlund, S. (1983) Trace metals in the Skagerrak and Kategat. In *Trace Metals in Sea Water*, C.S. Wong, E. Boyle, K.W. Bruland, J.D. Burton & E.D. Goldberg (eds), 467–73. New York: Plenum.

Martin, J.H. & Gordon, R.M. (1988) Northeast Pacific iron distributions in relation to phytoplankton productivity. *Deep Sea Res.*, **35**, 177–96.

Martin, J.H. & Knauer, G.A. (1984) VERTEX: manganese transport through the oxygen minima. *Earth Planet. Sci. Lett.*, **67**, 35–47.

Martin, J.H., Bruland, K.W. & Broenkow, W.W. (1976) Cadmium transport in the California Current. In *Marine Pollutant Transfer*, H.L. Windom & R.A. Duce (eds), 159–84. Lexington, MA: Lexington Books.

Martin, J.H., Gordon, R.M. & Broenkow, W.W. (1989) VERTEX: phytoplankton/iron studies in the Gulf of Alaska. *Deep Sea Res.*, **37**, 1639–53.

Martin, J.-M. & Whitfield, M. (1983) The significance of the river input of chemical elements to the ocean. In *Trace*

*Metals in Sea Water*, C.S. Wong, E. Boyle, K.W. Bruland, J.D. Burton & E.D. Goldberg (eds), 265–96. New York: Plenum.

Martin, J.-M., Dai, M.-H. & Cauwet, G. (1995) Significance of colloids in the biogeochemical cycling of organic carbon and trace metals in the Venice Lagoon (Italy). *Limnol. Oceangr.*, **40**, 119–31.

Measures, C.I., Grant, B., Khadem, M., Lee, D.S. & Edmond, J.M. (1984) Distribution of Be, Al, Se and Bi in the surface waters of the western North Atlantic and Caribbean. *Earth Planet. Sci. Lett.*, **71**, 1–12.

Measures, C.I., Edmond, J.M & Jickells, T.D. (1986) Aluminium in the northwest Atlantic. *Geochim. Cosmochim. Acta*, **50**, 1423–9.

Mills, G.L. & Quinn, J.G. (1984) Dissolved copper and copper–organic complexes in the Narragansett Bay Estuary. *Mar. Chem.*, **15**, 151–72.

Moffett, J.W. (1997) The importance of microbial Mn oxidation in the upper ocean: a comparison of the Sargasso Sea and equatorial Pacific. *Deep Sea Res.*, **44**, 1277–91.

Moffett, J.W. & Zika, R.G. (1987) Solvent extraction of copper acetylacetonate in studies of Cu(II) speciation in seawater. *Mar. Chem.*, **21**, 301–13.

Moore, R.M. (1983) The relationship between distributions of dissolved cadmium, iron and aluminium and hydrography in the central Arctic Ocean. In *Trace Metals in Sea Water*, C.S. Wong, E. Boyle, K.W. Bruland, J.D. Burton & E.D. Goldberg (eds), 131–42. New York: Plenum.

Moran, S.B., Yeats, P.A. & Balls, P.W. (1996) On the role of colloids in trace metal solid–solution partitioning in continental shelf waters: a comparison of model results and field data. *Continental Shelf Res.*, **16**, 397–408.

Morel, F.M.M. (1986) Trace metals–phytoplankton interactions: an overview. In *Biogeochemical Processes at the Land–Sea Boundary*, P. Lasserre & J.-M. Martin (eds), 177–89. Amsterdam: Elsevier.

Morel, F.M.M., Hudson, R.J.M. & Price, N.M. (1991) Limitation of productivity by trace metals in the sea. *Limnol. Oceanogr.*, **36**, 1742–55.

Muller, F.L.L. (1996) Interactions of copper, lead and cadmium with the dissolved, colloidal and particulate components of estuarine and coastal waters. *Mar. Chem.*, **52**, 245–68.

Murray, J.W., Spell, B. & Paul, B. (1983) The contrasting geochemistry of manganese and chromium in the eastern tropical Pacific Ocean. In *Trace Metals in Sea Water*, C.S. Wong, E. Boyle, K.W. Bruland, J.D. Burton & E.D. Goldberg (eds), 643–69. New York: Plenum.

Nurnberg, H.W. & Valenta, P. (1983) Potentialities and applications of voltammetry in chemical speciation of trace metals in the sea. In *Trace Metals in Sea Water*, C.S. Wong, E. Boyle, K.W. Bruland, J.D. Burton & E.D. Goldberg (eds), 671–97. New York: Plenum.

Nyffeler, U.P., Li, Y.-H. & Santschi, P.S. (1984) A kinetic approach to describe trace element distribution between particles and solution in natural aquatic systems. *Geochim. Cosmochim. Acta*, **48**, 1513–22.

Olafsson, J. (1983) Mercury concentrations in the North Atlantic in relation to cadmium, aluminium and oceanographic parameters. In *Trace Metals in Sea Water*, C.S. Wong, E. Boyle, K.W. Bruland, J.D. Burton & E.D. Goldberg (eds), 475–85. New York: Plenum.

Orians, K.J. & Bruland, K.W. (1985) Dissolved aluminium in the central North Pacific. *Nature*, **316**, 427–9.

Orians, K.J. & Bruland, K.W. (1986) The biogeochemistry of aluminium in the Pacific Ocean. *Earth Planet. Sci. Lett.*, **78**, 397–410.

Preston, M.R. & Chester, R. (1996) Chemistry and pollution of the marine environment. In *Pollution: Causes, Effects and Control*, R.M. Harrison (ed.), 26–51. Cambridge: Royal Society of Chemistry.

Rue, E.L. & Bruland, K.W. (1995) Complexation of iron(III) by natural organic ligands in the Central North Pacific as determined by a new competitive ligand equilibration/adsorptive cathodic stripping voltammetric method. *Mar. Chem.*, **50**, 117–38.

Sakamoto-Arnold, C.M., Hanson, A.K., Huizenga, D.L. & Kester, D.R. (1987) Spatial and temporal variability of cadmium in Gulf Stream warm-core rings and associated waters. *J. Mar. Res.*, **45**, 201–30.

Scharek, R., Van Leeuwe, M.A. & De Baar, H.J.W. (1997) Responses of Southern Ocean phytoplankton to the addition of trace metals. *Deep Sea Res.*, **44**, 209–27.

Schaule, B.K. & Patterson, C.C. (1981) Lead concentrations in the Northeast Pacific: evidence for global anthropogenic perturbations. *Earth Planet. Sci. Lett.*, **54**, 97–116.

Schindler, P.W. (1975) Removal of trace metals from the oceans: a zero order model. *Thalassia Jugoslav.*, **11**, 101–11.

Sclater, F.R., Boyle, E. & Edmond, J.M. (1976) On the marine geochemistry of nickel. *Earth Planet. Sci. Lett.*, **31**, 119–28.

Sillen, L.G. (1961) The physical chemistry of sea water. In *Oceanography*, M. Sears (ed.), 549–81. Washington, DC: American Association for the Advancement of Science.

Simpson, J.H. (1994) Introduction to the North Sea Project. In *Understanding the North Sea System*, H. Charnock, K.R. Dyer, J.M. Huthnance, P.S. Liss, J.H. Simpson & P.B. Tett (eds). *Philos. Trans. R. Soc. London, Ser. A*, **343**, 1–4.

Simpson, W.R. (1982) Particulate matter in the oceans— sampling methods, concentration and particle dynamics. *Oceanogr. Mar. Biol. Annu. Rev.*, **20**, 119–72.

Spencer, D.W. & Brewer, P.B. (1971) Vertical advection diffusion and redox potentials as controls on the distribution of manganese and other trace metals dissolved in waters of the Black Sea. *J. Geophys. Res.*, **76**, 5877–92.

Spivack, J., Huested, S.S. & Boyle, E.A. (1983) Copper, nickel and cadmium in the surface waters of the Mediterranean. In *Trace Metals in Sea Water*, C.S. Wong,

E. Boyle, K.W. Bruland, J.D. Burton & E.D. Goldberg (eds), 505–12. New York: Plenum.

Statham, P.J. & Burton, J.D. (1986) Dissolved manganese in the North Atlantic Ocean, 0–35°N. *Earth Planet. Sci. Lett.*, **79**, 56–65.

Stumm, W. & Brauner, P.A. (1975) Chemical speciation. In *Chemical Oceanography*, J.P. Riley & G. Skirrow (eds), Vol. 1, 173–239. London: Academic Press.

Stumm, W. & Morgan, J.J. (1981) *Aquatic Chemistry*. New York: Wiley.

Sunda, W.G. (1997) What controls dissolved iron concentrations in the world ocean: a comment. *Mar. Chem.*, **57**, 169–72.

Sunda, W.G. & Huntsman, S.A. (1983) Effect of competitive interactions between manganese and copper on cellular manganese and growth in estuarine and oceanic species of the diatom. *Thalassiosira. Limnol. Oceanogr.*, **28**, 924–34.

Sunda, W.G. & Huntsman, S.A. (1988) Effect of sunlight on redox cycles of manganese in the southwestern Sargasso Sea. *Deep Sea Res.*, **35**, 1297–1317.

Sunda, W.G., Barber, R.T. & Huntsman, S.A. (1981) Phytoplankton growth in nutrient rich seawater: importance of copper–manganese cellular interactions. *J. Mar. Res.*, **39**, 567–86.

Symes, J.L. & Kester, D.R. (1985) The distribution of iron in the northwest Atlantic. *Mar. Chem.*, **17**, 57–74.

Tessier, A. & Turner, D.R. (eds) (1995) *Metal Speciation and Bioavailability in Aquatic Systems*. Chichester: John Wiley.

Turekian, K.K. (1977) The fate of metals in the ocean. *Geochim. Cosmochim. Acta*, **41**, 1139–44.

Turner, D.R. (1987) Speciation and cycling of arsenic, cadmium, lead and mercury in natural waters. In *Lead, Mercury, Cadmium and Arsenic in the Environment*, SCOPE, T.C. Hutchinson & K.M. Meema (eds), 175–86. New York: Wiley.

Turner, D.R., Whitfield, M. & Dickson, A.G. (1981) The equilibrium speciation of dissolved components in freshwater and seawater at 25°C and 1 atm pressure. *Geochim. Cosmochim. Acta*, **45**, 855–81.

Ure, A.M. & Davidson, C.M. (eds) (1995) *Chemical Speciation in the Environment*. London: Chapman & Hall.

Van den Berg, C.M.G. (1984) Organic and inorganic speciation of copper in the Irish Sea. *Mar. Chem.*, **14**, 201–12.

Van den Berg, C.M.G. (1985) Determination of the zinc complexing capacity in seawater by cathodic stripping voltammetry of zinc–APDC complex ions. *Mar. Chem.*, **16**, 121–30.

Van den Berg, C.M.G. (1989) Electrochemical chemistry of sea-water. In *Chemical Oceanography*, J.P. Riley (ed.), Vol. 9, 197–245. London: Academic Press.

Van den Berg, C.M.G. & Dharmvanij, S. (1984) Organic complexation of zinc in estuarine interstitial and surface water samples. *Limnol. Oceanogr.*, **29**, 1025–36.

Van den Berg, C.M.G. & Nimmo, M. (1987) Determinations of interactions of nickel with dissolved organic material in seawater using cathodic stripping voltammetry. *Sci. Total Environ.*, **60**, 185–95.

Wangersky, P.J. (1986) Biological control of trace metal residence time and speciation: a review and synthesis. *Mar. Chem.*, **18**, 269–97.

Whitfield, M. (1979) The mean oceanic residence time (MORT) concept, a rationalization. *Mar. Chem.*, **8**, 101–23.

Whitfield, M. & Jagner, D. (eds) (1981) *Marine Electrochemistry*. New York: Wiley.

Whitfield, M. & Turner, D.R. (1982) Ultimate removal mechanisms of elements from the ocean—a comment. *Geochim. Cosmochim. Acta*, **46**, 1989–92.

Whitfield, M. & Turner, D.R. (1987) The role of particles in regulating the composition of sea water. In *Aquatic Surface Chemistry Chemical Processes at the Particle–Water Interface*, W. Stumm (ed.), 457–93. New York: Wiley.

Wood, A.M., Evans, D.W. & Alberts, J.J. (1983) Use of an ion exchange technique to measure copper complexing capacity on the continental shelf of the southeastern United States and in the Sargasso Sea. *Mar. Chem.*, **13**, 305–26.

Wu, J. & Luther, G.W. (1995) Complexation of Fe(III) by natural organic ligands in the Northwest Atlantic Ocean by a competitive ligand equilibration method and kinetic approach. *Mar. Chem.*, **50**, 159–77.

Zarino, A. & Yamamoto, T. (1972) A pH dependent model for the chemical speciation of copper, zinc and lead in seawater. *Limnol. Oceanogr.*, **17**, 661–71.

Zhang, H., van den Berg, C.M.G. & Wollast, R. (1990) The determination of interactions of cobalt(II) with organic compounds in seawater using cathodic stripping voltammetry. *Mar. Chem.*, **28**, 285–300.

# 12 Down-column fluxes and the benthic boundary layer

In the last two chapters we have described the distributions of particulate matter and dissolved trace elements in the oceans, and have discussed the manner in which they interact in the water column to exert a control on the elemental composition of sea water. The principal mechanism underlying this control is the removal of dissolved trace elements by the oceanic microcosm of particles via scavenging-type and nutrient-type carrier-phase associations, and the subsequent settling of these particles down the water column. These transport processes are part of the *great particle conspiracy*, and in the present chapter an attempt will be made to estimate the rate at which the conspiracy operates by assessing the magnitude of the fluxes of elements to deep waters via the particulate carriers. Following this, the journey undertaken by the particulate matter will be taken a stage further by tracking the sedimenting solids to the sediment–water interface.

## 12.1 Down-column fluxes

It was shown in Section 10.4 that the total suspended material in the oceans can be divided conveniently into two general populations: (i) a small-sized population, termed fine particulate matter (**FPM**), which sinks at a slow rate; and (ii) a large-sized population (**CPM**), which consists mainly of biogenic particle aggregates that undergo relatively fast vertical settling. The FPM sink mainly by association with the CPM, the coupling between the two particulate populations occurring by 'piggy-backing'. The total suspended matter (TSM) flux sinking vertically down the water column from the surface therefore can be defined as:

$$\text{total flux} = \text{FPM}_f + \text{CPM}_f \tag{12.1}$$

where $\text{FPM}_f$ is the flux associated with the fine material and $\text{CPM}_f$ is that associated with the coarse material.

The treatment given below will focus on the total, or particle mass, flux down the water column. The FPM is especially important for trace metals, however, and an attempt will be made to decouple it from the CPM when the down-column metal fluxes are considered. Most of the mass flux of material is carried down the water column in association with large-sized aggregates, such as faecal pellets and 'marine snow'. Thus, the vertical transport is dominated by the **global carbon flux**, i.e. the export flux that escapes the 'nutrient loop' and is replaced by new production. In general, the strength of the down-column flux varies with the degree of primary production in surface waters, although there are exceptions to this (see below). The magnitude of the flux has been estimated by both sediment-trap and radioisotope techniques. The general use of radioisotopes in dissolved–particulate equilibrium studies has been reviewed in Section 11.6.3.1, and attention at this stage will be confined largely to sediment-trap techniques.

Much of our knowledge of the strengths of down-column, or more strictly down-column plus laterally advected, fluxes has come from the use of sediment traps, which according to Brewer *et al.* (1986) provide direct evidence of 'sediments in the making'. The down-column fluxes that carry the material forming the sediments are driven by particulate matter and in many of the field experiments the sediment traps have been deployed in relation to the three-layer model for the distribution of TSM in the water column (see Section 10.2); i.e. samples have been collected in the surface layer, the clear-water minimum and the deep-water layer. This relationship between particle collection and the three-layer model is very important because it identifies a fundamental constraint that must be placed on the interpretation

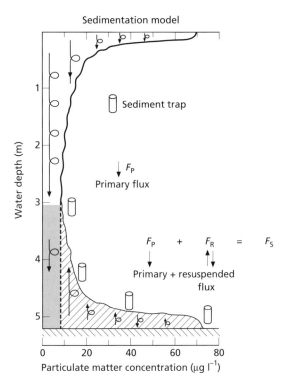

**Fig. 12.1** The relationship between the down-column location of sediment traps and the three-layer oceanic TSM flux model (from Gardner *et al.*, 1985).

of sediment-trap data. That is, when attempts are made to estimate the strengths of down-column particulate fluxes it is necessary to make a distinction between the two sources of TSM to the water column, i.e. a primary *downward* flux from the surface layer and a resuspended *upward* flux from the sea bed. The overall relationship between the three-layer TSM model and the various particle fluxes is illustrated in Fig. 12.1.

There are a number of difficulties associated with the design, calibration and water-column location of sediment traps; e.g. on the basis of $^{234}$Th data Buesseler (1991) concluded that traps located in the upper ocean may not provide an accurate measure of particle fluxes. A detailed treatment of the hydrodynamics of sediment traps is outside the scope of this volume, and to gain an understanding of the topic the reader is referred to Gust *et al.* (1994). The main advantage of the traps is that they are capable of collecting both small and large-sized sinking material,

and despite the inherent problems they have played a vital role in mass-flux studies. Over the past decade or so, sediment-trap experiments have provided a considerable body of data on the magnitude and composition of the down-column particle fluxes to the interior of the ocean.

### 12.1.1 The magnitude and major component composition of the particle mass flux to the interior of the ocean

The experiments have been carried out over a range of oceanic environments and a summary of the data from a series of mostly long-term (at least 1 yr) trap deployments is given in Table 12.1. Although the sediment traps have been deployed at a variety of depths, a number of features of the particle mass flux to the interior of the ocean can be identified when the data in Table 12.1 are combined with other findings reported in the literature. These features are summarized below.

#### 12.1.1.1 The magnitude of the particle mass flux

Mass fluxes to the interior of the ocean appear to lie in the range $\sim$20–250 mg m$^{-2}$ day$^{-1}$. The overall geometric average for the sample set in Table 12.1 is 82 mg m$^{-2}$ day$^{-1}$; however, within this there are considerable variations between the oceanic regions. The highest values, greater than $\sim$100 mg m$^{-2}$ day$^{-1}$, are found in areas of upwelling; e.g. off the coasts of West Africa in the North Atlantic, and Namibia in the South Atlantic, mass fluxes lie in the range 102–245 mg m$^{-2}$ day$^{-1}$. In the Arabian Sea a mass flux of 153 mg m$^{-2}$ day$^{-1}$ was recorded under the influence of monsoonal upwelling at the 16°N, 60°E site. Relatively high values are also found in continental shelf and slope marginal regions; e.g. mass fluxes on the shelf off North America are as high as $\sim$350 mg m$^{-2}$ day$^{-1}$. The mass flux data for the Atlantic shelf–slope areas quoted in Table 12.1 are for the clear-water minimum region (see Fig. 12.1), and as such represent a measure of the primary down-column flux. The down-column particle mass flux profiles in many shelf–slope locations, however, are extremely complex as a result of laterally transported and resuspended inputs in these dynamic environments. A number of examples of the particle regimes in such environments have been described in the literature.

**Table 12.1** Magnitude and composition of the mass particle flux to the interior of the ocean.

(a) Atlantic Ocean.

| Region | Location Latitude | Longitude | Trap depth (m) | Mass flux (mg m$^{-2}$ day$^{-1}$) | Flux composition (%) CaCO$_3$ | Opal | Organic matter | Lithogenic matter | Data source* |
|---|---|---|---|---|---|---|---|---|---|
| Open-ocean | 75°N | 11°E | 1650 | 77.5 | 23.5 | 7 | 18.5 | 51 | 1 |
| | 48°N | 21°W | 3750 | 72 | 59 | 21.5 | 7.5 | 12 | 2 |
| | 48°N | 19°W | 3100 | 61 | 52 | 18.5 | 17 | 12.5 | 3 |
| | 34°N | 21°W | 4500 | 58 | 60 | 9.5 | 8 | 22.5 | 2 |
| | 32°N | 64°W | 3200 | ≈40 | — | — | — | — | 4 |
| | 32°N | 71°W | 5400 | 54 | 55 | 9 | 7.5 | 28.5 | 5 |
| | 31°N | 55°W | 5206 | 13 | — | — | — | — | 6 |
| | 31°N | 25°W | 4440 | 33.5 | 72 | — | 8.5 | <19.5 | 7 |
| | 28°N | 22°W | 3600 | 27.5 | 61 | 5 | 10 | 24 | 3 |
| | 24°N | 23°W | 3870 | 41.5 | 49 | 7 | 9.5 | 34.5 | 3 |
| | 23°N | 64°W | 5847 | 32 | 40.5 | 9 | 8 | 42.5 | 5 |
| | 22°N | 25°W | 4120 | 51 | — | — | — | 36 | 8 |
| | 13°N | 57°W | 4250 | 80 | — | — | — | — | 10 |
| | 13°N | 54°W | 5068 | 47 | — | — | — | 31 | 6 |
| | 01°N | 11°W | 3921 | 102 | 54 | 12 | 12 | 22 | 9 |
| | 02°S | 10°W | 696 | 26 | 56 | 7.5 | 23.5 | 13 | 9 |
| Off-coast upwelling | 21°N | 20°W | 2195 | 183 | 43 | 8 | 8 | 41 | 9 |
| | 21°N | 22°W | 3502 | 152 | 52 | 4 | 6 | 38 | 9 |
| | 19°N | 20°W | 2190 | 245 | 45 | 7 | 17.5 | 30.5 | 3 |
| | 20°S | 09°E | 1640 | 162 | 56 | 14 | 20 | 10 | 9 |
| | 20°S | 09°E | 599 | 134 | 63 | 8.5 | 21 | 7.5 | 9 |
| | 20°S | 09°E | 1648 | 102 | 65 | 8 | 20 | 7 | 9 |
| Shelf–slope | 27°41'N | 78°54'W | 675 | 220 | 70 | — | 13 | — | 14 |
| | 33°30'N | 76°15'W | 1345 | 280 | 49 | — | 21 | — | 14 |
| | 38°23'N | 65°45'W | 3515 | 205 | 16 | — | 15 | — | 14 |
| | 38°50'N | 72°31'W | 2162 | 342 | 51 | — | 10 | — | 13 |
| | 38°28'N | 72°02'W | 2316 | 115 | 33.5 | — | 11 | — | 13 |
| | 38°19'N | 69°37'W | 3059 | 242 | 35 | — | 9.5 | — | 13 |

(b) Indian Ocean.

| Region | Location Latitude | Longitude | Trap depth (m) | Mass flux (mg m$^{-2}$ day$^{-1}$) | Flux composition (%) CaCO$_3$ | Opal | Organic matter | Lithogenic matter | Data source* |
|---|---|---|---|---|---|---|---|---|---|
| Arabian Sea | 16°N | 60°E | 3024 | 153 | 53 | 24 | 12 | 11 | 11 |
| | 14°N | 65°E | 2919 | 86 | 63 | 11 | 12 | 14 | 11 |
| | 15°N | 69°E | 2787 | 91.5 | 50 | 16 | 12 | 22 | 11 |
| Outer shelf, western margin of India | ~15°N | ~73°E | 30 | 400 | 66 | — | 13 | — | 15 |
| | ~15°N | ~73°E | 70 | 178 | 44.5 | — | 22 | — | 15 |
| | ~15°N | ~73°E | 103 | 70 | 59 | — | 15.5 | — | 15 |

*Continued on p. 314*

**Table 12.1** *Continued*

(c) Pacific Ocean.

| Region | Location Latitude | Location Longitude | Trap depth (m) | Mass flux (mg m⁻² day⁻¹) | Flux composition (%) CaCO₃ | Opal | Organic matter | Lithogenic matter | Data source* |
|---|---|---|---|---|---|---|---|---|---|
| | 58°N | 179°E | 3137 | 144 | 11 | 58 | 7 | 24 | 12 |
| | 53°N | 149°E | 1061 | 129 | 9 | 53 | 7 | 31 | 12 |
| | 50°N | 145°W | 3800 | 153$^{(82/83)}$ | <1 | 53 | 8 | 38 | 12 |
| | 50°N | 145°W | 3800 | 47$^{(83/84)}$ | 52 | 38 | 10 | <1 | 12 |
| | 50°N | 145°W | 3800 | 61$^{(84/85)}$ | 48 | 44 | 6 | <2 | 12 |
| | 48°N | 138°W | 3500 | 70.5 | 48 | 42 | 9 | <1 | 12 |
| | 48°N | 128°W | 2200 | 67.5 | 47 | 29 | 9 | 15 | 5 |
| | 42°N | 125°W | 2829 | 513 | 8 | 25 | 5 | 62 | 5 |
| | 42°N | 127°W | 2830 | 163 | 15 | 27.5 | 7 | 50.5 | 5 |
| | 42°N | 132°W | 3664 | 28 | 41 | 32 | 17.5 | 9.5 | 5 |
| | 40°N | 128°W | 4230 | 107 | 14 | 21 | 8 | 57 | 5 |
| | 15°N | 151°W | 5582 | 11 | — | — | — | 15 | 6 |
| | 11°N | 140°W | 3400 | 38 | 54 | 30 | 8.5 | 7.5 | 12 |
| | 11°N | 140°W | 4628 | 28 | 53 | 33.5 | 6 | 7.5 | 5 |
| | 9°N | 140°W | 2250 | 22 | 60 | 25 | 12 | 3 | 12 |
| | 5°N | 81°W | 3791 | 180 | — | — | — | 41 | 6 |
| | 5°N | 140°W | 2100 | 75 | 66 | 23 | 11 | <1 | 12 |
| | 2°N | 140°W | 2200 | 73.5 | 67 | 23 | 10 | <1 | 12 |
| | 1°N | 140°W | 3495 | 61 | 56 | 31 | 11 | 2 | 12 |
| | 1°N | 140°W | 3495 | 110 | 54.5 | 37 | 9.5 | <1 | 12 |
| | 1°N | 139°W | 4445 | 100 | 56.5 | 29 | 7 | 7.5 | 5 |
| | Equator | 140°W | 3618 | 95 | 66 | 21 | 7 | 6 | 12 |
| | 2°S | 140°W | 3593 | 85.5 | 57 | 22 | 9 | 12 | 12 |
| | 5°S | 140°W | 2209 | 61 | 66 | 21 | 9 | 4 | 12 |
| | 12°S | 140°W | 3594 | 21 | 79.5 | 10 | 8 | 2.5 | 12 |

(d) Japan Sea.

| Region | Location Latitude | Location Longitude | Trap depth (m) | Mass flux (mg m⁻² day⁻¹) | Flux composition (%) CaCO₃ | Opal | Organic matter | Lithogenic matter | Data source* |
|---|---|---|---|---|---|---|---|---|---|
| | 41°N | 139°E | 2720 | 49.4 | 9 | 37 | 9 | 45 | 16 |

* 1, Honjo *et al.* (1988); 2, Honjo & Manganani (1993); 3, Jickells *et al.* (1966); 4, Deuser (1987); 5, Dymond & Lyle (1994); 6, Honjo *et al.* (1982); 7, Lampitt *et al.* (1992); 8, Kremling & Streu (1993); 9, Wefer & Fischer (1993); 10, Jickells *et al.* (1990); 11, Haake *et al.* (1993); 12, Honjo *et al.* (1995); 13, Gardner *et al.* (1985); 14, Hinga *et al.* (1979); 15, Ramaswamy (1987); 16, Masuzawa *et al.* (1989).

These include those for: (i) the Southern Middle Atlantic Bight, studied during the SEEP II Program (see e.g. Biscaye & Anderson, 1994); (ii) the North Sea (see e.g. Kempe & Jennerjahn, 1988); (iii) the Gulf of Lions in the Western Mediterranean Sea, investigated as part of the ECOMARGE Programme (see e.g. Monaco *et al.*, 1990); and (iv) The Black Sea (see e.g. Izdar *et al.*, 1987). In contrast to upwelling and coastal regions, the lowest mass-flux values are found in the central ocean gyres, where mass particle fluxes appear to lie in the range ~25–65 mg m⁻² day⁻¹. In general, therefore, the mass

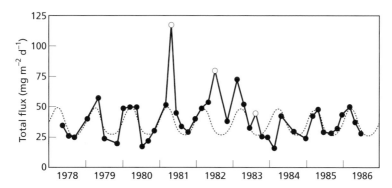

**Fig. 12.2** Temporal variations in particle fluxes in the deep ocean (from Deuser, 1987). An eight-year record of variations in total particle flux to a sediment trap located at 3200 m in the Sargasso Sea is shown. Points represent average fluxes over two-month trap deployment periods, plotted at the period mid-points. The dotted curve represents the average timing of the winter–spring flux maximum: the three measurements indcated by open circles were not included in the calculation of the average curve.

particle fluxes to the ocean interior mirror TSM concentrations in surface waters (see Section 10.2), and reflect the magnitude of primary production in the surface waters.

### 12.1.1.2 *The composition of the particle mass flux*

The average composition of TSM transported to the interior of the ocean as derived from the data in Table 12.1 is: calcium carbonate 50.5%, opal 19.5%, organic matter 11% and lithogenic* material 19%. Within these average values there are inter-oceanic differences in the composition of the particle mass flux. Overall, the mass flux for the Atlantic Ocean has the following average composition: calcium carbonate 52.5%, opal 9.5%, organic matter 13% and lithogenous material 25%. In contrast, in the Pacific Ocean the average flux composition is: calcium carbonate 45%, opal 31.5%, organic matter 9% and lithogenous material <14.5%. In addition, there are latitudinal differences in the flux composition in the individual oceans; for example, in the Pacific Ocean opal dominates the total flux at the high-latitude northern regions, whereas calcium carbonate dominates at lower latitudes. The average values therefore will be biased by the number of samples from individual oceanic environments included in the database,

but it may be concluded that over most oceanic regions the mass particle flux to the interior of the ocean is dominated by calcium carbonate.

### 12.1.1.3 *The mass particle flux and primary production in surface waters*

*Seasonality in the mass particle flux.* Various authors have identified seasonal signals in mass flux data obtained from sediment traps. Deuser (1987) reported data on particle flux measurements taken over an 8-yr period from a sediment trap deployed at 3200 m in the Sargasso Sea. The data are illustrated in Fig. 12.2 and show evidence of both seasonal and annual variability in the magnitude of the particle fluxes, the seasonal variations exhibiting a winter–spring maximum and a summer–autumn minimum. The seasonal correlation is consistent with the concept that in this region it is the mixed-layer depth, or conversely the degree of stratification, that controls primary production by regulating the flow of nutrients into the euphotic zone; i.e. the magnitude of the particle flux, the new production, is related to, but may be off-set from, primary production in the overlying waters. Honjo & Manganani (1993) showed that the spring bloom in the North Atlantic (the NABE—see Section 8.4.6.1) strongly influenced mass fluxes to the deep ocean at two sites located at 34°N, 21°W and 48°N, 21°W. At the 34°N site the bloom persisted for ~140 days (38% of the year) but pro-

---

* The term lithogenic refers mainly to land-derived minerals and has the same meaning as lithogenous (see Section 15.1).

vided over 60% of the annual mass flux. At the 48°N station the bloom persisted for 102 days (28% of the year) but contributed ~50% of the annual biogenic particle flux. Wefer & Fischer (1993) found a strong seasonality at a sediment trap site off Namibia, where a distinct bimodal flux pattern occurred with peaks in the austral autumn (May–June), which coincided with a non-upwelling off-shore bloom, and the austral spring (October–November), which was associated with upwelling events. Haake *et al.* (1993) reported that 58% of the annual biogenic particle flux in a deep sediment trap in the western Arabian Sea was recorded during a 4-month southwest monsoonal upwelling-associated period.

*Decoupling of surface production from the export of particulate material out of the euphotic zone.* It is generally accepted that the POC, which dominates the flux out of the euphotic layer, is equivalent to the new production which replaces the nutrients that escape the 'nutrient loop'. However, the 'surface-water primary-production–particle-mass-flux' relationship is not always a simple one and in some oceanic regions the particle flux may be decoupled from surface-water primary production. A number of explanations have been offered to account for this decoupling.

**1** Grazing of phytoplankton. For example, Honjo *et al.* (1988) found a number of seasonal phases in the particle mass flux to deep water in the Norwegian Sea. Within this, the smallest particle flux was associated with the spring bloom, and it was proposed that during the bloom extensive grazing had left behind a relatively small particle mass for transportation to deep water.

**2** Lateral particle transport. For example, Honjo *et al.* (1995) reported a decoupling of the particle mass flux and surface-water primary production in the equatorial Pacific, and suggested that one explanation for this may be that there was a poleward transport of particulate matter from these locations before it could reach the interior traps.

**3** Export as DOC. Some of the organic carbon that escapes the 'nutrient loop' may do so in the form of DOC (see Section 9.3), and it has been proposed that this occurs over parts of the equatorial Pacific. For example, Bacon *et al.* (1996) calculated that the flux of POC to the base of the euphotic zone at the Equator in the Pacific amounted to only ~2% of the primary production and was insufficient to balance the new production, which was estimated to be ~17% of the primary production. The authors concluded that this was consistent with the hypothesis that a major fraction of the new production was exported in the form of DOC (see also Buesseler *et al.*, 1995).

**4** More efficient phytoplankton recycling of nutrients in the euphotic zone in permanently stratified water columns than in systems that are subject to deep winter mixing, leading to a more efficient 'biological pump' under the latter conditions (see e.g. Jickells *et al.*, 1996).

### 12.1.1.4 Particle sinking rates

Particle sinking rates depend on a number of factors, which include the origin, composition, size, morphology and density of the particulate matter. As a result, different components of the flux sink at different speeds. Attention here, however, is focused on data derived from sediment traps that integrate the sinking rates of all captured material. Deuser (1986) estimated that the bulk particulate material reached a trap located at 3200 m in the Sargasso Sea in around 1 month, which yields an average sinking time of ~100 m day$^{-1}$. Pfannkuche (1993) reported an average sinking speed of 75–110 m day$^{-1}$ to the interior of the ocean at 47°N, 20°W in the North Atlantic. Honjo & Manganini (1993) showed that the surface bloom in the North Atlantic (34°N, 21°W) penetrated into the ocean interior with a lag-time of a few weeks, at a sinking speed of ~46 m day$^{-1}$ to ~1800 m and an accelerated rate of ~178 m day$^{-1}$ at deeper layers. Honjo *et al.* (1995) estimated a minimum settling velocity of 180 m day$^{-1}$ for material collected at a series of traps in the equatorial Pacific.

The association of biogenic with lithogenic components (see below), the so-called '*mineral ballast*', can increase the sinking speeds of the oceanic mass particle flux. Despite such a coupling, however, Honjo *et al.* (1995) pointed out that the sinking rate of the particle mass flux in the equatorial Pacific, where there was little 'mineral ballast', was faster (180 m day$^{-1}$) than that in the North Atlantic at a station (34°N, 21°W) which received an input of Saharan aeolian dust (46 m day$^{-1}$). The authors suggested that the higher rates of particle sinking in the equatorial Pacific may have arisen because large planktonic

Foraminifera play a greater role there than in the North Atlantic.

It was pointed out in Section 9.3 that both the magnitude and composition of the particle mass flux can change significantly with depth in the water column as a result of organic matter degradation and, to a lesser extent, mineral dissolution. For this reason, problems can arise when direct comparisons are made between mass fluxes derived from sediment traps located at different depths in the water column. To overcome this, various authors (see e.g. Martin *et al.*, 1987; Bishop, 1989) have attempted to quantitatively model the decrease in mass fluxes with depth. In this way sediment-trap data can be normalized to specific depths, thus allowing direct intertrap comparisons to be made. Using this approach, Jickells *et al.* (1996) combined data from a series of investigations in order to assess the trends in particle fluxes to the interior of the ocean on a regional scale. Information from 10 sediment trap moorings on a 20°W transect extending between 19° and 48°N in the northeast Atlantic was assessed. Particle flux data for the various sites, normalized to 4000 m using the formulation of Martin *et al.* (1987), are listed in Table 12.2. The region of the northeast Atlantic covered by the sediment-trap transect can be divided into a number of biogeochemical provinces, or zones, on the basis of a variety of classification schemes. Essentially, when taken together, the various schemes yield an overall picture in which the interaction of phytoplankton communities with the surface ocean dynamics results in a north to south gradient from temperate (North Atlantic Drift Region) to subtropical (Subtropical Gyre East), involving (i) a province with a deep winter mixed layer, (ii) a province with permanent stratification, and (iii) an upwelling province. The results of the study are summarized below.

*Particle mass flux magnitude.* A clear trend was apparent in the mass flux data at the various sites (see Table 12.2a), with the average fluxes decreasing southwards from 48°N, 20°W to 28°N, 22°W, then increasing again under the influence of the North African upwelling regime, particularly at the 21°N, 21°W, 21°N, 20°W and the 19°N, 20°W sites. These particle fluxes to the interior of the ocean are entirely consistent with the large-scale oceanographic parameters. Because productivity is closely coupled with

**Table 12.2** Particle mass fluxes to the interior of the northeast Atlantic derived from sediment-trap deployments: data normalized to 4000 m for the loss of POC (data after Jickells *et al.*, 1996).

(a) Particle mass flux and total flux composition; $mg\,m^{-2}\,day^{-1}$.

| Site | Mass flux | $CaCO_3$ | Opal | Organic matter | Lithogenic matter |
|---|---|---|---|---|---|
| 48°N 21°W | 71.5 | 42.4 | 15.4 | 5.1 | 8.6 |
| 48°N 20°W | 59.0 | 31.7 | 11.3 | 8.4 | 7.6 |
| 34°N 21°W | 58.7 | 34.9 | 5.5 | 5.3 | 13.0 |
| 31°N 24°W | 33.6 | 24.0 | — | 3.1 | <6.5 |
| 28°N 22°W | 27.7 | 16.7 | 1.4 | 3.0 | 6.6 |
| 24°N 23°W | 41.1 | 20.3 | 2.9 | 3.9 | 14.3 |
| 21°N 20°W | 177.0 | 78.7 | 14.8 | 9.0 | 74.5 |
| 21°N 21°W | 150.9 | 79.0 | 6.5 | 8.1 | 57.3 |
| 19°N 20°W | 227.7 | 110.2 | 16.9 | 25.4 | 75.2 |
| 22°N 25°W | 51 | — | — | — | 18.2 |

(b) Percentage total flux composition.

| Site | $CaCO_3$ | Opal | Organic matter | Lithogenic matter |
|---|---|---|---|---|
| 48°N 21°W | 59 | 21.5 | 7 | 12 |
| 48°N 20°W | 54 | 19 | 14 | 13 |
| 34°N 21°W | 59 | 9 | 9 | 22 |
| 31°N 24°W | 71 | — | 9 | <19 |
| 28°N 22°W | 60 | 5 | 11 | 24 |
| 24°N 23°W | 49 | 7 | 9 | 34 |
| 21°N 20°W | 45 | 8 | 5 | 42 |
| 21°N 21°W | 52 | 4 | 5 | 38 |
| 19°N 20°W | 48 | 7 | 11 | 33 |
| 22°N 25°W | — | — | — | 36 |

(c) Percentage biogenic flux composition.

| Site | $CaCO_3$ | Opal | Organic matter | Biogenic flux ($mg\,m^{-2}\,day^{-1}$) |
|---|---|---|---|---|
| 48°N 21°W | 67 | 24 | 8 | 62.9 |
| 48°N 20°W | 62 | 22 | 16 | 51.4 |
| 34°N 21°W | 76 | 12 | 12 | 45.7 |
| 31°N 24°W | 71–89 | 0–9.3 | 9–11 | 27.1–33.6 |
| 28°N 22°W | 79 | 7 | 14 | 21.1 |
| 24°N 23°W | 75 | 11 | 14 | 27.1 |
| 21°N 20°W | 77 | 14 | 9 | 102.5 |
| 21°N 21°W | 84 | 7 | 9 | 93.6 |
| 19°N 20°W | 72 | 11 | 17 | 152.5 |

surface ocean dynamics, e.g. stratification of the water column, the north to south progression has a number of consequences. For example, there is a southward reduction in winter deep mixing rates, and a shallowing of the winter mixed layer, until permanent stratification develops and nutrient availability decreases. In regions of decreased nutrient availability there is a more efficient cycling of carbon within the phytoplankton web, and consequently less 'new' production. As a result, particle mass fluxes, which are related to 'new' production, decrease towards the south and then increase again towards the North African upwelling region. There are also variations in the seasonality of the mass flux, which is greater at the more northern sites. At high latitudes, the temporal distribution of primary production is sinusoidal, whereas at lower latitudes it is more constant throughout the year. Further, seasonal spring blooms occur earlier in the year to the south of 48°N, and are less pronounced, as a result of the earlier stabilization of the water column. These trends are reflected in the mass flux data, which exhibit a seasonality at the 48°N site that was almost five times as great as that at the 28°N, 22°W site and 10 times that at the 19°N, 20°W site.

*Mass flux composition.* Data on the composition of the mass flux at the various sites is given in Table 12.2(b). The **lithogenic** flux increases from <6.5–14 mg m$^{-2}$ day$^{-1}$ for the sites lying between 48° and 24°N to 57–75 mg m$^{-2}$ day$^{-1}$ for the three southernmost traps. In addition, the percentage contribution of the lithogenic material to the total mass flux also increases southwards, from 12% to 34% of the total flux at the northernmost sites to 33% to 42% at the three southernmost sites. The authors attributed this increase in the proportion of lithogenic material to the increasing influence of the Saharan dust plume at the more southern sites and possibly to a more efficient packaging of the dusts under the higher biogenic flux conditions associated with upwelling. To assess changes in the biogenic flux, Jickells *et al.* (1996) recalculated their data on a lithogenic-free basis, and the results are listed in Table 12.2(c). At all sites the mass particle flux is dominated by coccoliths, and the **carbonate** flux makes up between 62% and 89% of the biogenic flux. It is apparent from the data in Table 12.2(c) that the biogenic flux at the northern sites 48°N, 21°W and 48°N, 20°W has much higher percentages of **opal** (average of ~23% on a

lithogenic-free basis) than do the fluxes of the other sites (average of less than ~10% on a lithogenic-free basis). This decrease follows primary productivity characteristics in the surface waters. The northern sites have a strong spring bloom dominated by diatoms, and there is a decrease in siliceous phytoplankton species southwards, some of which are absent from the southern sites. However, there was no evidence that the percentage opal increased in the upwelling region off West Africa. This may be the result of a low supply of silicate during the upwelling; however, the opal mass flux increased at the 19°N, 20°W site (see Table 12.2b). The biogenic flux to the sediment traps decreases from north to south by a factor of about two or three, being ~52 mg m$^{-2}$ day$^{-1}$ at 48°N, 20°W and ~27 mg m$^{-2}$ day$^{-1}$ at 24°N, 23°W. The ratio of the POC flux at 48°N, 20°W to that at 28°N, 22°W is ~2.8. Primary production maps indicate that gradients over this region are substantially less than a factor of two. This implies that the transport of biogenic material derived from primary production is more efficient at the higher latitudes, supporting the suggestion that phytoplankton ecosystems in permanently stratified water columns, such as at 24°N, 23°W and 28°N, 22°W, evolve to recycle nutrients within the euphotic zone more efficiently than high-latitude systems that are subject to deep winter mixing.

The most important result of the study carried out by Jickells *et al.* (1996) was that the authors were able to demonstrate that mass particle fluxes to the deep ocean reflect different regional-scale surface water biogeochemical provinces; i.e. they were able to inject 'oceanographic consistency' into the sediment-trap data.

### 12.1.2 The elemental composition of the particle mass flux to the interior of the ocean

Data for the elemental composition of the major components that contribute to oceanic TSM, and for some examples of TSM itself, have been given in Table 10.1, and data for the concentrations of particulate trace metals in surface waters have been given in Table 10.2. In surface waters biological productivity is the most important factor controlling the distribution and composition of particulate matter, and this results in regional-scale variability in particulate trace metal and nutrient concentrations (Saager, 1994). Further, surface-water TSM is not representative of that in the entire water column because particulate matter undergoes considerable modification as it sinks, which affects its trace-metal composition as well as that of the major components.

It was pointed out above (eqn 12.1) that the total down-column TSM flux can be expressed as:

total flux = $FPM_f + CPM_f$

where $FPM_f$ is the flux associated with the fine material and $CPM_f$ is that associated with the coarse material. With respect to elemental composition, the FPM and total TSM will be described separately, and an attempt will be made to decouple the signals from the FPM and CPM particulate populations in deep waters.

### 12.1.2.1 The trace-metal composition of the FPM down-column flux component

Few data are available on the trace-metal composition of the FPM component of the total down-column flux. However, the studies reported by Westerlund & Ohman (1991) and Sherrell & Boyle (1992) provide useful information.

Westerlund & Ohman (1991) used bottle casts deployed at a series of depths to collect water in the Weddell Sea, a trace metal and nutrient-rich polar environment. Both dissolved and particulate matter (extracted using 0.4 μm Nucleopore filters) were determined and the authors reported the following trends in their data.

1 In this region the dissolved concentrations of Cd, Cu and Zn were not linked to the bioproduction cycle, as they are in many other oceanic waters.
2 Particulate Cd had higher concentrations in the surface waters, with an invariant very low concentration profile below ~500 m. In the surface waters the high particulate Cd was probably generated by biota.
3 Particulate Zn, Cu and Ni generally showed the same concentration ranges, with only minor variations, throughout the water column.

Sherrell & Boyle (1992) determined the trace-metal composition of suspended particulates filtered from water collected using an *in situ* pump, with a two-stage filter holder fitted with a 53 μm pre-filter and a 1 μm Nucleopore primary filter, deployed at various depths in the Sargasso Sea: an open-ocean oligotrophic site. The major trends in their data can be summarized as follows.
1 The particulate lithogenic trace metals Al and

Fe exhibited parallel variations through the water column, displaying low concentrations in near-surface waters, an increase with depth in the upper thermocline, followed by a uniform distribution below ~1000 m before increasing again in the bottom nepheloid layer as the result of sediment resuspension; the down-column profile of particulate Al is illustrated in Fig. 12.3(a).
2 Particulate Mn, Co, Pb, Zn, Cu and Ni exhibited relatively low concentrations near the surface, which increased to maxima at ~500 m, then decreased before increasing again as a result of sediment resuspension, although Pb showed no bottom increase; the down-column profiles of particulate Mn and Zn are shown in Fig. 12.3((b) and (c) respectively.
3 Particulate Cu increased from the surface waters to a maximum in the thermocline, but was then relatively constant at all depths below ~500 m, probably as the result of deep-water scavenging, which is not found for the other trace metals; the down-column profile of particulate Cu is illustrated in Fig. 12.3(d).
4 The authors suggested that the overall similarity in the vertical profiles of Mn, Co, Pb, Zn, Cu and Ni may be the result of their uptake from the dissolved pool by Mn oxides, although other host phases such as organic matter also may be important in this respect.
5 Particulate Cd was the only trace metal to exhibit a maximum in the near-surface waters—see Fig. 12.3(e).

Saager (1994) reviewed the data on the vertical distributions of a series of particulate trace metals in sea water and drew the following conclusions.
1 Profiles of particulate Cd exhibit a surface-water maximum and a sharp decrease with depth. The maxima is most pronounced in productive waters, supporting the hypothesis that the distributions of both particulate and dissolved Cd are controlled by involvement in the biological cycle.
2 The vertical profiles of particulate Ni, Cu and Zn do not follow their dissolved profiles, which exhibit nutrient, or 'mixed'-type, water-column distributions (see Section 11.5). Faecal pellets and marine snow can effectively scavenge trace metals from sea water and Fisher *et al.* (1991) showed that Zn and Cd are remineralized from this type of material most rapidly in surface waters. However, according to Saager (1994), it is apparent from the particulate profiles of Ni, Cu and Zn that, with the exception of the situation in

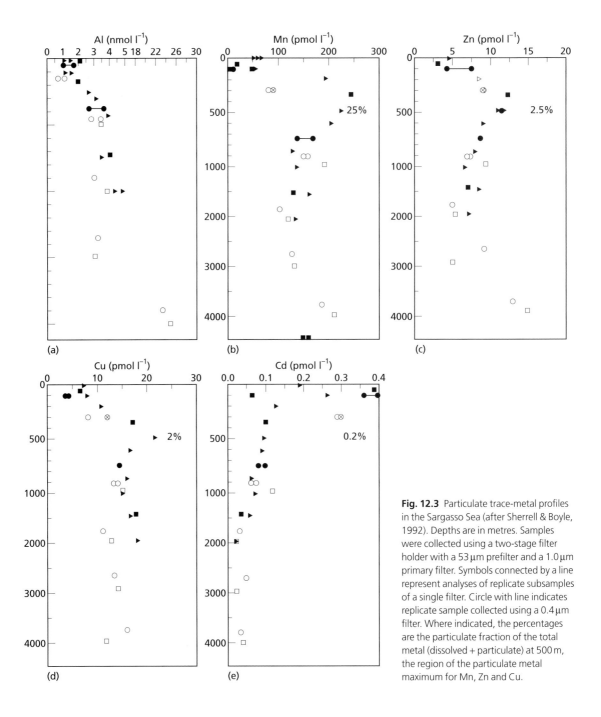

**Fig. 12.3** Particulate trace-metal profiles in the Sargasso Sea (after Sherrell & Boyle, 1992). Depths are in metres. Samples were collected using a two-stage filter holder with a 53 μm prefilter and a 1.0 μm primary filter. Symbols connected by a line represent analyses of replicate subsamples of a single filter. Circle with line indicates replicate sample collected using a 0.4 μm filter. Where indicated, the percentages are the particulate fraction of the total metal (dissolved + particulate) at 500 m, the region of the particulate metal maximum for Mn, Zn and Cu.

very productive waters, there is relatively little *in situ* metal regeneration in deep waters. This poses a problem because if the dissolved–nutrient relationships in deep waters are not maintained by *in situ* remineralization then some other explanation for them must be found. Saager (1994) proposed such an alternative explanation and suggested that dissolved deep-water distributions of Ni, Cu and Zn may be related to diagenetic processes, which affect the small amount of organic matter that reaches the sediment surface. Much of this organic matter undergoes degradation at, or near, the sediment surface (see Section 14.1), and can lead to out-of-sediment pore-water fluxes of trace metals and nutrients to the water column (see Section 14.6). Thus, the remineralization stage may be deferred to the sediment reservoir, rather than actually taking place in the deep-water column itself. Saager (1994) concluded that although the diagenetic effect is difficult to quantify, abyssal pore waters that have received dissolved trace metals and nutrients from degraded organic matter may contribute to the relationship between the nutrients and Ni, Cu and Zn in deep waters by decoupling the relationship between the metals and nutrients in the settling particulates. In addition, advective transport from upwelling regions, where a shallow regeneration of some trace metals can take place, may have a strong effect on the regeneration trends of both dissolved trace metals and nutrients in deep-waters.

### 12.1.2.2 *The FPM—CPM coupling in down-column trace-metal transport*

We have considered the trace-metal composition of the FPM. We now move forward to consider how the FPM is coupled to the CPM to provide the total (i.e. FPM + CPM) trace-metal flux to the interior of the ocean, and the model proposed by Sherrell & Boyle (1992) provides an excellent description of how this coupling operates. In setting up the model it is assumed that the CPM 'sinking' population is formed in surface waters by biological activity and that the FPM 'suspended' population becomes associated with it by 'piggy-backing'. In this context, the scavenging of trace metals can be regarded as the sum of two processes: (i) incorporation into the surface-water particulates which form the CPM that sinks rapidly to the deep ocean; and (ii) association with deep suspended particles of the FPM.

The 'Sherrell–Boyle' model uses a one-dimensional two-box flux approach, and the particulates are divided into CPM and FPM populations. The CPM is free-sinking but the suspended FPM, which makes up the bulk of the TSM, sinks only by association with the CPM, and is maintained by disaggregation of the larger particles (CPM). Thus, the large particulate

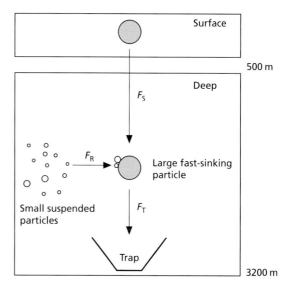

**Fig. 12.4** The 'Sherrell–Boyle' model to describe the coupling between suspended (FPM) and sinking (CPM) particulates in the down-column transport of trace metals in the Sargasso Sea (from Sherrell & Boyle, 1992). $F_s$ is the sinking flux formed in surface waters by the association of FPM with CPM. $F_R$ is the 'repackaging' flux associated with the removal of suspended (FPM) particulate metals. $F_T$ is the total flux, i.e. $F_T = F_s + F_R$. The flux arrows indicate only the processes of interest in the model and are not intended to demonstrate all fluxes in the system. For a full description of the 'Sherrell–Boyle' model, see text.

flux regulates the removal of the suspended particulates, but it is only the suspended FPM which, because of its greater concentration, larger surface areas and longer residence times, is assumed to exchange metals with the dissolved pool.

The 'Sherrell–Boyle' model is illustrated in Fig. 12.4. The total metal flux leaving the deep box is equal to the sum of (i) the surface-derived flux ($Fs$) and (ii) the flux associated with the removal of suspended particulate material, i.e. the 'repackaging flux' ($F_R$); thus

$$F_T = F_S + F_R \qquad (12.2)$$

where $F_T$ was taken as the mean flux determined from sediment-trap data, and the 'repackaging' flux was calculated from the suspended metal data, where

$$F_R = (Me_p D)/\tau_p \qquad (12.3)$$

where $Me_p$ is the mean deep-ocean suspended metal concentration per volume of sea water, $D$ is the depth

**Table 12.3** The 'Sherrell–Boyle' down-column particulate flux model applied to the Sargasso Sea water column; units, ng cm$^{-2}$ yr$^{-1}$ (data from Sherrell & Boyle, 1992). For explanation of the model—see text.

| Trace metal | 'Repackaging' flux ($F_R$) | Total flux ($F_T$) | $\dfrac{F_R}{F_T}$ |
|---|---|---|---|
| Al | 4045 | 17 530 | 0.23 |
| Fe | 2792 | 9494 | 0.30 |
| Mn | 335 | 1044 | 0.32 |
| Ni | 14.7 | 45 | 0.32 |
| Cu | 39.5 | 95 | 0.41 |
| Zn | 20 | 189.5 | 0.11 |
| Cd | 0.225 | 0.337 | 0.67 |
| Pb | 20 | 66 | 0.30 |

of the deep box and $\tau_p$ is the estimated suspended particle residence time. The value of $\tau_p$ was estimated by Sherrell & Boyle (1992) using measurements of particulate $^{230}$Th in the region. The model requires a knowledge of the relative contributions made to the total flux by surface scavenging and deep-water removal at a single location, and Sherrell & Boyle (1992) used data for both contributions at the Sargasso Sea site, utilizing sediment-trap data and their own suspended-particulate data.

Using the model, Sherrell & Boyle (1992) demonstrated that the repacking flux ($F_R$) provided less than about half (29–54%) of the total flux ($F_T$) for Al, Fe, Mn, Ni, Cu, Zn and Cd in the Sargasso Sea region. Further, the fraction yielded by the repackaged flux was remarkably constant—see Table 12.3, column 4. Thus, the repackaging and vertical removal of deep suspended particles plays a relatively minor role in governing the total flux of trace metals in the deep ocean, and processes taking place in the surface waters (<500 m) contribute about two-thirds of the deep flux.

The overall scenario to emerge from applying the 'Sherrell–Boyle' model to the Sargasso Sea depicts large particles (CPM) that, having interacted with small particles (FPM) in the upper water column, sink through the deep ocean exchanging, on average, only a minor portion of their mass and trace-metal content with the suspended particulates (FPM).

If these Sargasso Sea results can be applied to other central ocean gyres, it becomes apparent that the removal of suspended particles in the deep ocean interior is not the dominant sink for trace metals from the ocean. Rather, as suggested by a growing body of evi-

dence, the control is exerted at the **ocean boundaries**: i.e. (i) the surface ocean, and (ii) the continental margins. Thus, the whole oceanic residence times of trace metals may be strongly influenced by processes occurring at the ocean boundaries rather than in dissolved–particulate fractionation in deep waters.

### 12.1.2.3 The total particulate down-column trace-metal flux

It was pointed out above that there is a coupling between the packaging of biogenic and lithogenic components in the mass particle flux. This has important implications for the trace-metal composition of the material transported into the ocean interior and subsequently to the sediment surface. Deuser et al. (1983) showed that there was a strong correlation between the organic carbon (biogenic) and aluminium (lithogenic) fluxes into a sediment trap located at a depth of 3200 m in the Sargasso Sea. The mass particle flux appeared to follow an annual cycle similar to primary production in the surface waters, and although particulate Al varied randomly with respect to this cycle, the removal of the element to deep water was intimately linked to the rapid down-column transport of the organic matter. This biogenic–lithogenic association can be extended to include trace metals initially in solution in the surface waters. In this context, Coale & Bruland (1985) demonstrated surface-layer couplings (i) between the rates of removal of dissolved elements by particulate adsorption and the rate of primary production, and (ii) between the residence times of the particulates and the formation of faecal pellets. The biogenic–lithogenic coupling, however, is not a universal feature of the down-column transport of trace metals at all locations and three studies carried out at sites in different regions of the North Atlantic, which receive different trace-metal inputs, will serve to illustrate this. The three sites are identified below.

1 The western tropical North Atlantic (WTNA), a site relatively close to the South American mainland (Jickells et al., 1990). The sediment trap was located at 4250 m, with a non-continuous deployment time of 25 months. The total mass particle flux of material <37 μm had an average of 53.5 mg m$^{-2}$ day$^{-1}$, with no evidence of a strong seasonality.

2 The Sargasso Sea (SS), an open-ocean site (Jickells et al., 1984). The sediment trap was located at

3200 m, with a deployment time of 24 months. The total mass particle flux of material <37 μm had an average of 36.5 mg m$^{-2}$ day$^{-1}$, and showed strong seasonality.

**3** The eastern tropical North Atlantic (ETNA), a site off the coast of West Africa (Kremling & Streu, 1993). Two sediment traps were deployed and data here are given for the one located at 4120 m, with a deployment time of 5 months. The total mass particle flux of material <200 μm had an average of 49 mg m$^{-2}$ day$^{-1}$.

Data for the average trace-metal concentrations, $EF_{crust}$ values (see Section 4.2.1.1), flux rates and correlations with biogenic (organic carbon) or lithogenic (Al) material for TSM from the three sites are given in Table 12.4. These data are discussed below in terms of the sources, deep-ocean transport mechanisms and fluxes of the trace metals.

*Trace-metal sources to the surface waters.* The trace-metal composition of the deep ocean particulates at the three sites differ (see Table 12.4a), but a direct comparison between them is complicated because temporal variations can bias the average values. The principal sources of trace metals to surface ocean waters are via fluvial and atmospheric transport, with atmospheric sources dominating in many open-ocean regions. To evaluate the composition of the particulate populations at the three sites in relation to their sources their $EF_{crust}$ values are given in Table 12.4(b), from which it can be seen that those for Fe, Mn, Ni, Co and V are <10 for samples from all the populations. These metals are usually non-enriched elements (NEEs) in the marine aerosol, and their sources are crust-controlled (see Section 4.2.1.1). In contrast, the $EF_{crust}$ values of Cu, Zn, Pb and Cd, which are often anomalously enriched elements (AEEs) in the marine aerosol, vary between the three sites, being highest at the SS site, intermediate at the WTNA site and lowest at the ETNA site. At the ETNA site the $EF_{crust}$ values of all the metals, including Cu, Pb, Zn and Cd, are ≪10 and are similar to those in Saharan dusts, indicating that aeolian material deposited from the northeast trades is the major contributor of both the NEEs and AEEs to the sediment trap (Kremling & Streu, 1993). Jickells *et al.* (1984) concluded that aeolian sources also dominate the supply of trace metals to the SS site, and that the relatively high $EF_{crust}$ values for Cu, Zn

and Pb may reflect an anthropogenic perturbation by atmospheric material originating on the North American continent. At the WTNA site, however, the situation is more complex and the sediment-trap material is derived from a number of sources, which include aeolian deposition and an input from South American rivers to the surface waters, and lateral ocean transport to subsurface waters (Jickells *et al.*, 1990).

*Trace-metal transport to the deep ocean.* At both the SS and the WTNA sites the deep-ocean fluxes of Al, representative of lithogenic material, and other trace metals all correlate significantly with the flux of organic matter. This suggests that the metals are transported to depth in association with sinking organic debris (see above). Further, the Al : organic-carbon ratio at the WTNA site is twice as high as at the SS site, implying that during the 'packaging' process the biota in the surface waters can cope with the higher concentrations of lithogenic material found at the WTNA site. In contrast, at the ETNA site the fluxes of most trace metals through the water column were closely associated with Al (lithogenic material), but none showed any correlation with P, which was used as an indicator of organic matter. Kremling & Streu (1993) concluded that the decoupling of the lithogenic and biogenic components arose because the relatively large aeolian input delivered to this site from the northeast trades dilutes the biogenic components. The authors suggested that this 'dilution mechanism' is to be expected in oceanic regions that have relatively low primary production, but which receive large injections of crustal dust: i.e. the 'biological control' on the removal of lithogenic particulates may be overridden when sufficient concentrations of lithogenic material are delivered to the sea surface. However, this is difficult to reconcile with the fact that although the flux of lithogenic material at the WTNA site is higher than that at the SS site, organic matter regulates the down-column fluxes at both locations. An alternative explanation may be that the input of crustal components to the ETNA site is largely in the form of non-continuous dust 'pulses' (see Section 4.1.4.1), which deliver relatively large quantities of material to the sea surface over relatively short periods of time and so can intermittently perturb the steady-state conditions in the surface waters.

**Table 12.4** Trace-metal concentrations and fluxes derived from sediment-trap deployments at three sites in the North Atlantic.

(a) Average trace-metal concentrations; units, $\mu g\,g^{-1}$.

| Trace metal | Western tropical North Atlantic*: trap depth, 4250 m, <37 μm fraction | Sargasso Sea†: trap depth, 3200 m, <37 μm fraction | Eastern tropical North Atlantic‡: trap depth, 4120 m, <200 μm fraction |
|---|---|---|---|
| Al | 47768 | 18500 | 37143 |
| Fe | 23417 | 9200 | 19836 |
| Mn | 512 | 1104 | 408 |
| Ni | 59 | 46 | 27 |
| Co | 11 | — | 8.8 |
| V | 72 | 40 | 52 |
| Cu | 53 | 100 | 32 |
| Pb | 98 | 68 | 18 |
| Zn | 201 | 187 | 72 |
| Cd | — | 0.91 | 0.46 |
| P | 689 | 682 | 4129 |
| POC (%) | 5.0 | 4.19 | — |

(b) Average trace-metal $EF_{crust}$ values.

| Trace metal | Western tropical North Atlantic*: trap depth, 4250 m, <37 μm fraction | Sargasso Sea†: trap depth, 3200 m, <37 μm fraction | Eastern tropical North Atlantic‡: trap depth, 4120 m, <200 μm fraction |
|---|---|---|---|
| Al | 1.0 | 1.0 | 1.0 |
| Fe | 0.71 | 0.72 | 0.78 |
| Mn | 0.92 | 5.1 | 0.94 |
| Ni | 1.4 | 2.8 | 0.81 |
| Co | 0.76 | — | 0.79 |
| V | 0.94 | 1.4 | 0.87 |
| Cu | 1.6 | 7.7 | 1.2 |
| Pb | 14 | 24.5 | 3.2 |
| Zn | 4.9 | 12 | 2.3 |
| Cd | — | 20.5 | 5.2 |

(c) Average trace metal fluxes and correlations; units, $\mu g\,m^{-2}\,day^{-1}$.

| Trace metal | Western tropical North Atlantic: trap depth, 4250 m, <37 μm fraction | | Sargasso Sea: trap depth, 3200 m, <37 μm fraction | | Eastern tropical North Atlantic: trap depth, 4120 m, <200 μm fraction | |
|---|---|---|---|---|---|---|
| | Flux | Correlation with POC | Flux | Correlation with POC | Flux | Correlation with Al |
| Al | 2400 | 0.94 | — | — | 1455 | — |
| Fe | 1300 | 0.89 | 327 | 0.98 | 767 | 0.99 |
| Mn | 26 | 0.73 | 37 | 0.96 | 16 | 0.79 |
| Ni | 3.4 | 0.95 | 1.6 | 0.95 | 1.1 | 0.49 |
| Co | 0.56 | 0.84 | — | — | 0.32 | 0.85 |
| V | 3.8 | 0.92 | 1.3 | 0.95 | 2.0 | 0.99 |
| Cu | 3.1 | 0.93 | 3.5 | 0.93 | 1.2 | 0.92 |
| Pb | 5.1 | 0.90 | 2.4 | 0.94 | 0.68 | 0.96 |
| Zn | 7.8 | 0.93 | 6.8 | 0.91 | 2.7 | 0.65 |
| Cd | — | — | <0.03 | — | 0.039 | — |

* Data after Jickells *et al.* (1990). † Data after Jickells *et al.* (1984). ‡ Data after Kremling & Streu (1993).

*Trace-metal fluxes to the deep ocean* (Table 12.4c). Using Fe as an indicator, the terrigenous flux at the three sites decreases in the order WTNA ($1300\,\mu g\,m^{-2}\,day^{-1}$) > ETNA ($767\,\mu g\,m^{-2}\,day^{-1}$) > SS ($327\,\mu g\,m^{-2}\,day^{-1}$). Using P as an indicator, the biogenic flux at the ETNA ($19\,\mu g\,m^{-2}\,day^{-1}$) and SS ($24\,\mu g\,m^{-2}\,day^{-1}$) sites is similar, but is higher at the WTNA site ($35\,\mu g\,m^{-2}\,day^{-1}$). Thus, both the lithogenic and biogenous fluxes are highest at the WTNA site, possibly as a result of patches becoming separated from the Orinoco and Amazon plumes invading the region and adding to the aeolian input. The average trace-metal fluxes at the three sites vary but, with the exception of Pb (see below), they all agree to within factors which range between ~2 and ~4. On an individual site basis, the fluxes of Cu, Zn and Pb are two to three times higher at the SS that at the ETNA site, despite the higher terrigenous flux at the eastern location, probably resulting from the elevated concentrations of these metals in aerosols carried from North America relative to those transported in the Atlantic northeast trades. The importance of the aeolian inputs at the two sites is confirmed by the fact that atmospheric fluxes to the Sargasso Sea and the northeastern Atlantic are similar to those found in the sediment traps. The major difference between the WTNA and the other two sites is the elevated Pb flux at the WTNA. At the WTNA site deposition from the atmosphere is an order of magnitude less than that to the sediment trap, and Jickells *et al.* (1990) suggested that the elevated Pb flux may result from an enhanced scavenging of the metal owing to a relatively high concentration of total particulate matter at this location. The WTNA site is the closest of the three sites to the ocean margins, where enhanced Pb removal by scavenging can occur. It is not known if these scavenging processes also affect the fluxes of Cu and Zn at the WTNA site.

Material collected in the sediment traps at the three locations provide a number of insights into the processes which constrain trace-metal fluxes to the interior of the ocean. In addition, variations in the concentrations of elements with depth in the water column can provide evidence on the processes which control the chemical composition of the particle mass flux to the interior of the ocean. For example, Brewer *et al.* (1980) provided data on the fluxes of a range of elements to the deep ocean from sediment traps located at two North Atlantic sites, one in the

Sargasso Sea and one off Barbados. Aluminium increased in both concentration and flux down the water column at both sites, and the authors used this element to normalize the concentration of trace metals in the sediment-trap material. In this manner they were able to identify the factors that controlled the chemical composition of the flux, and they divided a series of elements into three groups.

1 A **terrigenous group**, in which element : Al ratios were constant with depth and were close to those in crustal material. The elements in this group were K, Ti, La, Co and $^{232}$Th.

2 A **biogenous group**, in which element : Al ratios decreased with depth. The elements in this group, i.e. Ca, Sr, Mg, Si, Ba, U, I and $^{226}$Ra, are all known to be incorporated, at least to some extent, into biogenic particulates such as organic matter, calcium carbonate and opal.

3 A **scavenged group**, in which the elements were characterized by increasing element : Al ratios with depth. Manganese, Cu, Fe, Sc and $^{230}$Th were placed in this group. These metals are all known to be highly surface-active in sea water, as indicated by their relatively short oceanic residence times.

Masuzawa *et al.* (1989) collected material using a vertical series of sediment traps at a single site in the Sea of Japan and gave data on the compositional changes in the particulates with depth in the water column. On the basis of both elemental concentrations and element : Al ratios the authors identified four groups of elements.

1 A **terrigenous group**, in which the elemental concentrations increased with depth and the element : Al ratios remained constant. This group contained Sc, La, Th, Hf, V, Ta, K, Rb and Cs.

2 A **biogenous group**, in which both the elemental concentrations and the element : Al ratios decreased with depth. The elements in this group were Ba, Ca and Sr.

3 A **scavenged group**, in which both the elemental concentrations and the element : Al ratios increased with depth. This group included Mn, Fe and Co.

These three groupings were identical to those proposed by Brewer *et al.* (1980). However, by including variations in both elemental concentrations and element : Al ratios, Masuzawa *et al.* (1989) were able to identify an additional group.

4 A **biogenic-scavenged group**, in which the elemental concentrations remained almost constant, or

increased slightly, with depth, but in which the element/Al ratios decreased with depth. Arsenic, Zn, Se and Ag were found in this group. The decrease in element/Al ratios indicates that these elements are released from settling particles through the decomposition of biogenic components, but the relative constancy of the elemental concentrations with depth suggests that they are also accumulated by particulate material as it settles down the water column.

With the exception of those for the WTNA site, the various trace-metal fluxes discussed above are derived mainly from sediment traps located in open-ocean locations. An example of how the fluxes operate in environments close to the continents has been provided by Gardner *et al.* (1985). In this study, sediment traps were used to measure both the primary and resuspended fluxes of trace metals at two sites; one on the continental slope and one on the continental rise in the northwest Atlantic. The data derived from this work showed that the primary

trace-metal fluxes in the slope–rise areas are considerably higher than those found for the open-ocean. The results for these sites are included in Table 12.5, which summarizes data on the down-column fluxes of elements to a number of oceanic locations.

### 12.1.3 Organic components of the particle mass flux to the interior of the ocean

Organic carbon decreases in concentration as the particle mass flux sinks from surface waters, and several authors have provided data on the fates of individual organic compounds, such as lipids (including hydrocarbons), amino acids and carbohydrates, during this process.

Matsuda & Handa (1986) gave data on the hydrocarbons collected in sediment traps at three sites in the eastern North Atlantic. The hydrocarbons included *n*-alkanes in the range $n$-$C_{15}$ to $n$-$C_{32}$, with the major constituent being $n$-$C_{17}$ which was

**Table 12.5** Down-column fluxes of some elements to the interior of the ocean; all data except Buat-Menard & Chesselet (1979) from sediment-trap deployments; units, $\mu g\, m^{-2}\, day^{-1}$.

| Location | Depth above bottom (m) | Ca | Mg | Si | Al | Fe |
|---|---|---|---|---|---|---|
| Northwest Atlantic: continental slope–rise*: | | | | | | |
|   Site KN | 500 | 14658 | 1315 | — | 3178 | — |
| | 13 | 32329 | 4192 | — | 10575 | — |
|   Site DOS | 518 | 25918 | 3781 | — | 10164 | — |
| | 118 | 38685 | 6877 | — | 20658 | — |
| Sargasso Sea†: | 5367 | | | | | |
|   primary flux | | 1529 | 52 | — | 137 | 142 |
|   resuspended flux | | 2389 | 879 | — | 2140 | 1556 |
| Sargasso Sea‡: | | | | | | |
|   total flux | 976 | 1882 | 38 | 419 | 60 | 38 |
|   total flux | 3964 | 5370 | 126 | 1479 | 444 | 153 |
| Off Barbados‡ | 389 | 11233 | 315 | 4466 | 392 | 176 |
| | 988 | 9041 | 251 | 4849 | 578 | 35.5 |
| | 3755 | 11205 | 414 | 7288 | 1548 | 847 |
| | 5086 | 9068 | 416 | 4370 | 1802 | 945 |
| Sargasso Sea§ | 3200 | — | — | — | — | 327 |
| Western tropical North Atlantic¶ | 4250 | 5800 | 760 | 51 | 2400 | 1300 |
| Eastern tropical North Atlantic‖ | 4120 | — | — | — | 1455 | 767 |
| Tropical North Atlantic** calculated primary down-column flux | | — | — | — | 142 | 392 |

*Continued on p. 327*

thought to originate from phytoplankton. The vertical flux of the hydrocarbons tended to decrease exponentially with depth at all three sites. In addition, there were significant changes in the composition of the hydrocarbons at two of the stations, the most striking feature being a rapid decrease in the relative abundance of $n$-$C_{17}$ as a result of biological degradation.

Carbohydrates are a major component of cycling organic matter in the oceans, making up ~13% of sinking particulate organic carbon (see e.g. Hernes *et al.*, 1996). The down-column fluxes of carbohydrates appear to decrease with depth (see e.g. Wefer *et al.*, 1982; Ittekkot *et al.*, 1984; Liebezeit, 1987). For example, Liebezeit (1987) showed that in the Drake Passage the total carbohydrate fluxes decreased from $2.6\,\mathrm{mg\,m^{-2}\,day^{-1}}$ at the surface to $1.2\,\mathrm{mg\,m^{-2}\,day^{-1}}$ at 2540 m. There were also a number of down-column trends in the relative proportions of the individual carbohydrates. Both fucose and glucose decreased in

proportion with depth, indicating that they were present in relatively easily degradable forms. In contrast, galactose, xylose, arabinose and rhamnose increased in proportion with depth, probably as a result of their incorporation into less degradable structural compounds. Further, despite the overall decrease in the total carbohydrate fluxes with depth, the relative contribution of the carbohydrates to the total organic flux actually increased down the water column, indicating that other compounds in the POC pool are degraded at faster rates than the carbohydrates.

Ittekkot *et al.* (1984) gave data on the fluxes of amino acids (dominated by aspartic acid and glycine) and amino sugars (dominated by glucose, galactose, fucose and rhamnose) to the deep ocean in the Panama Basin, and showed that there was an overall decrease in the acids and sugars with depth. The fluxes varied seasonally, however, with peaks that were associated with primary production in the

**Table 12.5** *Continued*

| Ti | Mn | Sc | Ba | Sr | Ni | Co | V | Cu | Pb | Zn |
|---|---|---|---|---|---|---|---|---|---|---|
| — | 88 | — | 77 | 134 | — | — | 6.8 | 77 | — | — |
| — | 41 | — | 140 | 274 | — | — | 21 | 93 | — | — |
| — | 162 | — | 88 | 217 | — | — | 19 | 148 | — | — |
| — | 384 | — | 162 | 274 | — | — | 38 | 192 | — | — |
| 8.2 | 14 | 0.26 | 1.3 | 9.4 | — | 0.03 | 0.49 | 4.4 | — | <0.14 |
| 126 | 30 | 0.46 | 14 | 12 | — | 0.52 | 4.1 | 2.2 | — | <20 |
| 4.3 | 0.36 | 0.008 | 5.3 | 13 | — | 0.06 | 0.13 | 1.5 | — | 1.7 |
| 18.5 | 16 | 0.04 | 13 | 36.5 | — | 0.27 | 0.68 | 5.7 | — | 7.0 |
| 28 | 2.9 | 0.05 | — | 70 | — | — | 0.58 | — | — | 31 |
| 41 | 3.5 | 0.11 | 29.5 | 67 | — | 0.11 | 1.6 | 1.3 | — | 19 |
| 86.5 | 17.5 | 0.26 | 35 | 85 | — | 0.41 | 2.8 | 4.3 | — | 18 |
| 108 | 23 | 0.33 | 253 | 62 | — | 0.55 | 3.3 | 6.4 | — | 23 |
| — | 37 | — | — | — | 1.6 | — | 1.3 | 3.5 | 2.4 | 6.8 |
| 130 | 26 | — | 36 | 51 | 3.4 | 0.56 | 3.8 | 3.1 | 5.1 | 7.8 |
| — | 16 | — | — | — | 1.1 | 0.32 | 2.0 | 1.2 | 0.68 | 2.7 |
| — | 6.6 | — | — | — | 3.2 | 0.25 | 1.0 | 6.4 | 9.0 | 28.5 |

* Gardner *et al.* (1985). † Spencer *et al.* (1978). ‡ Brewer *et al.* (1980). § Jickells *et al.* (1984). ¶ Jickells *et al.* (1990). ‖ Kremling & Streu (1993). ** Buat-Menard & Chesselet (1979).

surface waters. The authors also found that the distributions of sugars and amino acids in the deep-water zone (>3000 m) was different from that observed in the bottom sediments, leading to the conclusion that significant biochemical activity takes place in the benthic transition layer between the sediment surface and the deepest traps.

Hernes *et al.* (1996) carried out a detailed study of the neutral carbohydrate geochemistry in the waters of the central equatorial Pacific, using data acquired from plankton tows, floating sediment traps (105 m depth), moored sediment traps (~1000 m and ~4000 m depth) and bottom sediments. The principal neutral carbohydrates in the tows and sediment traps decreased in the order: glucose (~20–50 wt% total neutral carbohydrates expressed as total aldoses) > mannose (~11–25 wt%) > galactose (~8–25 wt%) > fucose (~2–13 wt%) > rhamnose (~2–12 wt%). Total neutral carbohydrate fluxes showed a large decrease with depth in the central equatorial Pacific that followed the organic carbon fluxes, and could be explained by the remineralization of settling POC. The accumulation of total neutral carbohydrates in the sediments averaged <1% of the total flux found in the floating sediment traps and <0.05% of the total neutral carbohydrate primary production. The carbohydrate fluxes in the equatorial Pacific were similar to those in other open-ocean regions; i.e. generally in the range <0.1 to <2 mg m$^{-2}$ day$^{-1}$. However, higher fluxes can be found in coastal regions—up to ~50 mg$^{-2}$ day$^{-1}$. Overall, the data showed that selective degradation occurs in the equatorial Pacific and the following trends were identified. There was: (i) a preferential loss of glucose and ribose in the upper water column; (ii) a loss of glucose throughout the water column and into the sediments; (iii) a relative increase of mannose, xylose, rhamnose and fucose in the sediments; and (iv) mid-water maximum in galactose, which appears to resist degradation in the water column but is lost in the bottom sediments.

Hernes *et al.* (1996) subdivided the carbohydrates into **non-structural**, or storage (e.g. energy source) types, and **structural** (e.g. cell wall) types, and concluded that the carbohydrates in the central equatorial Pacific consist of three components:

1 a relatively labile, preferentially lost, non-structural component, rich in glucose and ribose;

2 an intermediate labile structural component, containing galactose and perhaps arabinose;

3 a more refractory, preferentially preserved, structural component enriched in rhamnose, fucose, xylose and mannose.

Thus, the neutral carbohydrate compositions appear to be driven by the preferential loss of storage glucose and ribose, and the preferential preservation of structural components. An important implication of this is that the compositional signatures of the neutral carbohydrates in sediments may be influenced more by their planktonic source than by changes that affect organic matter during its passage through the water column, i.e. the diagenetic pathway. The carbohydrates can also indicate the speed with which the mass organic carbon flux descends the water column. It was shown above that ribose can be lost preferentially from the carbohydrates, but Ittekkot *et al.* (1984) showed that ribose also can be preserved at all water depths and so by escaping degradation can provide evidence of a rapid down-column transport of organic material from surface to deep water.

## 12.2 The benthic boundary layer: the sediment–water interface

The data given in Table 12.5 offer some of the best estimates presently available of the down-column fluxes of elements in a variety of oceanic regions. The point therefore has been reached at which particulate material in the oceans leaves sea water and enters the sediment reservoir, a step that involves crossing the sediment–water interface. This interface is the site of a number of biogeochemical reactions, which can mediate the dissolved–particulate recycling of some of the elements in the 'young' material that has been transported down the water column as part of the TSM flux. However, instead of dealing here with a single interface, i.e. the sediment–water interface, it is more useful to relate the changes that take place in bottom waters to the wider concept of the **benthic boundary layer** (BBL), because by adopting this unified approach the bottom reaction zone is extended above the top of the sediment layer and taken back into the water column itself. The BBL is an extremely dynamic ocean environment in which the 'great particle conspiracy' continues to operate. Before following the (sea water → sediment–seawater interface → sediment) TSM transport sequence to the sediment stage itself, we therefore will consider the processes that occur in the benthic boundary layer.

The benthic boundary layer is a zone in which frictional forces result from the interaction of the deep oceanic circulation with the ocean bottom, evidenced for example in the formation of nepheloid layers (see Section 10.2.3). In addition to these physical forces the BBL, which includes the sediment–water interface, is the site of relatively large gradients in chemical and biological properties, and the processes associated with these gradients are involved in the active cycling of elements between the water and sediment reservoirs. Overall, therefore, the BBL is one of the most dynamic environments in the entire ocean system.

The down-column transport of TSM transfers nutrients and organic carbon from the surface ocean to deep waters, and the remineralization of the carbon provides energy to the bottom benthos. Burial in the sediments removes bioactive material from the system and therefore regulates oceanic fertility (Brewer *et al.*, 1986). In addition, the composition of the TSM and the magnitude of its flux play an important role in controlling the chemical environment in the bottom sediments (Dymond, 1984). The reason for this is that organic carbon is the determining factor controlling the type of redox conditions that are set up in the sediments (see Section 14.4). Because these redox conditions largely control the diagenetic reactions at both the sediment–seawater and the sediment–interstitial-water interfaces, the POC flux to the sediments may be regarded as the driving force behind most of the diagenetic reactions occurring in marine sediments (Emmerson & Dymond, 1984).

In Section 10.5 it was pointed out that in order to assess the net downward flux to the BBL it is necessary to distinguish between the primary and the resuspended fluxes that operate in this deep-ocean reservoir. According to Dymond & Lyle (1994) it also is necessary to distinguish between **distal** and **local** resuspension fluxes. Distal resuspension, e.g. by nepheloid transport or turbidity currents, can be part of the primary, or net, flux of material to a sediment site. However, local resuspension represents an artificial flux as it merely recycles material already deposited. Trace-metal data for both the primary and resuspended fluxes are given in Table 12.5, and in order to move one step further along the TSM transport sequence it is necessary to establish how much of

the net down-column flux is actually buried in the bottom sediment.

The mass particle flux that descends to the sea bed supplies material to marine sediments. This flux may be termed the **rain rate,** and the extent to which the flux material is incorporated into bottom sediments may be referred to as the **burial rate.** The difference between the rain rate (material supply) and the burial rate (material preservation) is then a measure of the recycling of the flux material that reaches the sea floor: i.e. the **benthic regeneration flux.** This flux indicates the extent of the exchange of material between the sediment and the overlying water column and is a key component in the process by which the particulate rain is transformed into the sedimentary record. In its simplest form Dymond & Lyle (1994) expressed this 'benthic regeneration equation' as:

the rain rate of any particulate component =
 the burial rate of that component +
 benthic regeneration flux $\qquad$ (12.4)

Dymond (1984) proposed an elegant conceptual model to assess how much of the net mass particle flux is buried in the underlying sediments. The model is illustrated in Fig. 12.5 and offers an excellent framework within which to address the problem of the fate of down-column transported elements once they reach the BBL. In designing the model, Dymond (1984) made a distinction between (i) **refractory** elements (e.g. Al, Fe, Ti), and (ii) **labile** elements (e.g. organic C, P, N). The primary particulate flux of the refractory elements can increase with depth as a result of either particle scavenging of dissolved elements or additions from horizontal advection (e.g. from the continental margins and distal resuspension). Close to the sea bed the refractory elements can also exhibit a sharp increase in flux from the resuspended flux, because the sediments are enriched in refractory material relative to the sinking flux. In contrast, the particulate flux of the labile elements will decrease with depth, owing to processes such as particle decomposition, dissolution and disaggregation.

Local resuspension of sedimentary material does not appear to be significant at elevations a few hundred metres above the sea bed (the resuspension zone), and in the Dymond (1984) model net, or primary, particulate flux measurements are extrapolated downwards through the resuspension zone

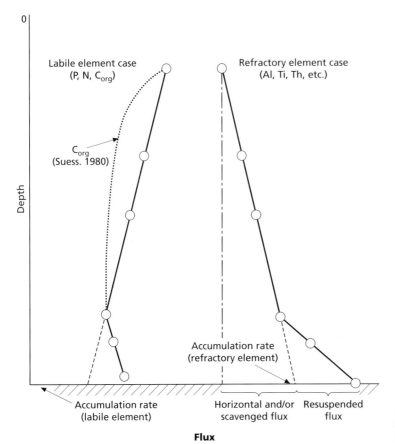

**Fig. 12.5** Conceptual model of particle-associated fluxes to the benthic boundary layer (from Dymond, 1984).

from measurements taken above the zone itself—see Fig. 12.5. Dymond & Lyle (1994) used this model to compare the primary particle **rain rate**, determined from sediment-trap data, to the sediment **burial rate** at 11 sites in the Atlantic and Pacific oceans that covered a number of different biogeochemical oceanic environments. These were:

1 Four sites (H, M, S and C) in the equatorial Pacific.
2 Five sites off the west coast of the USA (MFZ, NS, MW, G and JDF). Site MFZ is in the California Current. Sites NS, MW and G form a California Current transect from very productive waters influenced by upwelling (NS) to the relatively unproductive central gyre (G). Site JDF is located above the Juan de Fuca Ridge; this is a site of intense present-day hydrothermal activity, but trap material was collected above the hydrothermal effluent plume.
3 The Hatteras Abyssal Plain (HAP), a site ~300 km off the east coast of the USA.

4 The Nares Abyssal Plain (NAP), a deep-water site in the western Atlantic.

Chemical data for the various sites are given in Table 12.6, and are discussed below in terms of the refractory and labile elements. It must be stressed, however, that the sedimentation at the HAP site may have been influenced by bottom-transported turbidite inputs (i.e. distal resuspension), which would not have been picked up in the sediment traps. It is included here in order to illustrate this effect.

*Refractory elements; terrigenous rain.* For all sites, with the exception of HAP, the burial rates of Al, Fe and Ti are within a factor of two of their rain rates, and the differences are probably the result of the comparison of a single year's sediment trap rain-rate data to the $10^3$–$10^5$ yr time-scale that is integrated into the sediment accumulation rate measurements. At the HAP site the burial rates are considerably in excess of the rain rates, probably as a result of bottom-transported inputs, which are not sensed by the sediment trap data (see above). Excluding the HAP site, the rain rate of particulate Al varies by a factor of 500 between the various sites, generally being higher in the nearshore locations. The NS–MW–G transect provides an example of the decrease in the refractory terrigenous flux with increasing distance from the continental margin, the Al rain rates varying by a factor of about 50 across the transect in the order: NS (940 µg cm² yr⁻¹) > MW (200 µg cm² yr⁻¹) > G (20 µg cm² yr⁻¹) in an off-margin sequence. The generally good overall agreement between the rain rate and burial rate of the refractory elements indicates that sediment-trap studies provide a valid approach for studying the recycling of biogenic elements at the sea floor.

*Labile elements; biogenic rain.* At most sites biogenic debris, consisting mainly of organic carbon, calcium carbonate and opal, dominates the particle mass flux. Organic carbon preservation at the pelagic sites (>500 km from shore; NAP, G, H, M, S and C) is relatively low, and less than ~5% of the down-column rain is buried in the sediments. In contrast, at the sites closer to the continental margins there is a higher preservation of organic carbon, and the burial rate ranges between ~10% and ~30% of the rain rate. This is consistent with the overall preservation of organic matter in marine sediments and indicates the importance of the continental margins as major sites for the removal of organic carbon from the ocean; this topic is discussed in Sections 14.2 and 14.4. Calcium carbonate preservation in marine sediments is a function of water depth (see Section 15.2). However, the data in Table 12.6 indicate that in most of the sediments less than ~30%, and in many <5%, of the carbonate rain is preserved. Overall, the percentage of opal preserved in the sediments ranges from ~1 to ~20% of the rain rate. The preservation of opal in marine sediments is more complex than that of calcium carbonate (see Section 15.2), but the rain rate of opal provides a constraint on its dissolution. The data provided by Dymond & Lyle (1994) indicate that there is a relationship between the magnitude of the opal rain rate and its sediment burial rate, with, in general, the highest rain rate resulting in a higher burial rate in the sediments.

Dymond (1984) also provided data on the particle-flux–sedimentation-rate relationships for a wide range of trace metals at the H and M sites. The author concluded that a number of particle reactive elements are accumulated in the sediments at about the same rates at which they are delivered down the water column—see Fig. 12.6. Lead was an exception to this, with only less than ~20% of its flux to the BBL being preserved in the depositing sediment; however, this may have been because much of the Pb in the water column is of a recent anthropogenic origin, and so has not yet been integrated into the sediments, which accumulate only at slow rates.

The reactivity of a particulate element in aqueous environments is determined largely by the manner in which it is partitioned between the components of the particulate carriers (see Section 3.1.4), and Dymond (1984) gave data on the partitioning of a series of elements among four operationally defined, sequentially leached, particulate fractions. In this way the elements were classified as: (i) carbonate and exchangeable cations; (ii) organically bound cations; (iii) amorphous hydroxide-associated cations; and (iv) refractory-associated cations. The partitioning data were illustrated with respect to the down-column particulate fluxes of Mn and Cu. The data for Mn showed that in the water column above the resuspension zone the element is present almost exclusively in **refractory** and **carbonate–exchangeable** forms. The proportions of the two associations remained essentially constant, indicating that there was a continuous scavenging of Mn down the water column and a continuous horizontal addition of refractory Mn to the site. In the resuspension zone, however, ~35% of the total Mn was associated with **amorphous hydroxides**; this is the dominant association for Mn in deep-sea deposits (see Section 16.5) and results here from sediment resuspension. For Cu the situation was different. This element had a major fraction of its flux associated with the organic host phase and the proportion of total Cu in this association increased with depth, thus highlighting the scavenging of Cu by an organic carrier phase throughout the water column. According to Dymond (1984) it is presumably this form of Cu that is regenerated at the sea floor. Data to support this were provided by Chester *et al.* (1988), who investigated the solid-state speciation of Cu in surface water particulates and sediments from the Atlantic Ocean. These authors showed that, whereas on average ~50% of the total Cu in the surface-water

**Table 12.6** Rain rates (primary down-column fluxes obtained from sediment-trap deployments) and sediment burial rates in a number of oceanic locations; units, µg cm² yr⁻¹ (data from Dymond & Lyle, 1994).

| Site | Latitude N | Longitude W | Trap depth (m) | Al | | Fe | |
|---|---|---|---|---|---|---|---|
| | | | | Rain rate | Burial rate | Rain rate | Burial rate |
| Atlantic: | | | | | | | |
| HAP | 32.73 | 70.82 | 5400 | 47 | 312 | 25 | 684 |
| NAP | 23.20 | 63.98 | 5847 | 52.4 | 59.8 | 29.5 | 33.1 |
| Pacific: | | | | | | | |
| MFZ | 39.49 | 127.69 | 4230 | 220 | 254 | 135 | 162 |
| JDF | 47.97 | 128.10 | 2200 | 30.6 | 29.1 | 19.3 | 44.7 |
| NS | 42.09 | 125.77 | 2829 | 940 | 746 | — | — |
| MW | 42.19 | 127.58 | 2830 | 200 | 308 | — | — |
| G | 41.55 | 132.00 | 3664 | 20 | 34 | — | — |
| H | 6.57 | 92.77 | 3565 | 7.0 | 6.4 | 3.75 | 5.4 |
| M | 8.83 | 103.98 | 3150 | 15.1 | 14.2 | 15.7 | 20.3 |
| S | 11.06 | 140.14 | 4620 | 1.8 | 2.1 | 1.01 | 1.3 |
| C | 1.04 | 138.94 | 4445 | 5.41 | 4.8 | 2.61 | 3.2 |

*Continued*

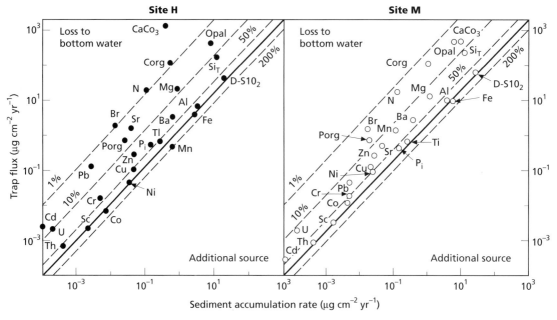

**Fig. 12.6** Comparison between sediment accumulation rates and measured particle fluxes at two sites in the eastern equatorial Pacific (from Dymond, 1984). The heavy line indicates equal particulate and burial fluxes; the broken lines indicate the percentage of the particle flux preserved in the sediment. $P_i$ is inorganic P, Si is total Si, D-SiO$_2$ is detrital silica.

**Table 12.6** *Continued*

| Ti | | Organic carbon | | CaCO$_3$ | | Opal | |
|---|---|---|---|---|---|---|---|
| Rain rate | Burial rate | Rain rate | Burial rate | Rain rate | Burial rate | Rain rate | Burial rate |
| 2.28 | 18.51 | 76.4 | 18.7 | 1105 | 948 | 180 | 135 |
| 3.01 | 3.43 | 48.3 | 1.93 | 482 | 36.6 | 106 | 1 |
| 12 | 15 | 127 | 25 | 550 | 57 | 850 | 102 |
| — | — | 109 | 12.8 | 1170 | 185 | 728 | 74 |
| 46 | 52 | 490 | 130 | 1490 | 80 | 4772 | 850 |
| 11 | 30 | 220 | 50 | 910 | 100 | 1663 | 200 |
| — | — | 90 | 3.5 | 420 | 20 | 330 | 22 |
| 0.74 | 0.40 | 89 | 0.83 | 659 | 1 | 385 | 16 |
| 0.85 | 0.97 | 138 | 3.8 | 2093 | 42.2 | 711 | 30 |
| — | — | 30 | 0.15 | 542 | 0.3 | 345 | 6 |
| — | — | 130 | 2.59 | 2093 | 64.2 | 1069 | 113 |

particulates was present in an organically associated form, this fell to less than ~10% in the surface sediments (see Fig. 12.7); i.e. the organic form of Cu had undergone regeneration on deposition.

The resuspension of the bottom sediment has an important effect on the down-column trace-metal flux profiles. This resuspended trace-metal flux can be generated in a number of ways. For example, it can result from the addition of metal-rich sediment particles directly to the water column. However, the resuspended particles can themselves scavenge additional dissolved metals from the bottom-water column, thus leading to a region of 'enhanced scavenging' in the resuspension zone. This could arise, for example, as metals released by the regeneration of organic particles are made available for scavenging by the resuspended material. An alternative mechanism to explain the increases in deep-water particulate fluxes was postulated by Dymond (1984), who introduced the concept of a **rebound flux**. This rebound flux was thought to result from the resuspension not of well-deposited sediment components, but rather of recently laid down primary flux material (see e.g. Billet *et al.*, 1983). Dymond (1984) suggested that the rebound material may be similar to the organic-rich 'fluff' found over deep-sea sediments. The concept was later modified, however, by Walsh *et al.* (1988) to exclude 'fluff' and in order to distinguish it from the resuspension of surface sediment, the rebound flux was defined as particles that have reached the sediment surface by settling out of the water column but which have not yet become incorporated into the sediment itself. Using the modified terminology, therefore, the material in the BBL can be divided into: (i) a local primary down-column flux; (ii) a local rebound flux; and (iii) a local, or distal, resuspended sediment flux.

## 12.3 Down-column fluxes and the benthic boundary layer: summary

1 There is a 'rain' of particulate material to the sea floor over the World Ocean, with sinking speeds in the range ~50–200 m day$^{-1}$. The magnitude of the particle mass flux varies in both space and time but it appears to be in the range ~20–250 mg m$^{-2}$ day$^{-1}$, with the highest values being found in areas of coastal upwelling and the smallest in the central ocean gyres.

2 The particle mass flux to the interior of the ocean is composed primarily of organic carbon, calcium carbonate, opal and lithogenic material, with calcium carbonate dominating in most open-ocean regions. As it sinks down the water column the relative proportions of the major components change, and the percentage of organic matter decreases from as much as greater than ~90% in surface waters to usually less than ~10% at depth.

3 The down-column fluxes, which are driven by

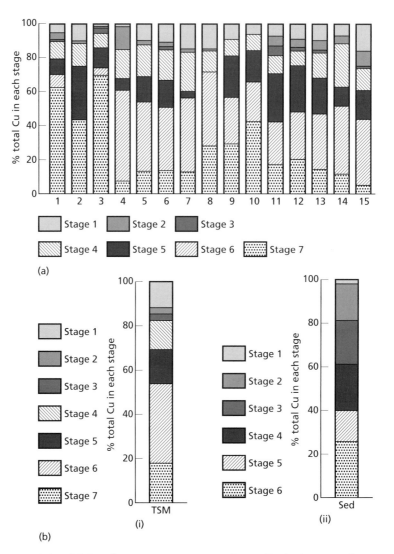

**Fig. 12.7** The solid-state speciation of Cu in surface water particulates (TSM) and deep-sea sediments (from Chester *et al.*, 1988). (a) The average partitioning signatures for ΣCu in Atlantic surface seawater particulates. Samples were collected along a north–south Atlantic transect. The Cu host associations for the various stages are as follows: stage 1, loosely held or exchangeable associations; stage 2, carbonate and surface oxide associations; stage 3, easily reducible associations, mainly with 'new' oxides and oxyhydroxides of manganese and amorphous iron oxides; stage 4, humic organic associations; stage 5, moderately reducible associations, mainly with 'aged' manganese oxides and crystalline iron oxides; stage 6, refractory organic associations; stage 7, detrital or residual associations. (b) The average partitioning signatures for ΣCu in Atlantic deep-

sea sediments. For the deep-sea sediment samples the humic and refractory organic associations were combined into one organic-associated stage. The Cu host associations are as follows: stage 1, loosely held or exchangeable associations; stage 2, carbonate and surface oxide associations; stage 3, easily reducible associations, mainly with 'new' oxides and oxyhydroxides of manganese and amorphous iron oxides; stage 4, moderately reducible associations, mainly with 'aged' manganese oxides and crystalline iron oxides; stage 5, organic associations; stage 6, detrital or residual associations. Note the differences in the solid-state speciation of Cu between the open-ocean surface water TSM (samples 4–15), in which ~50% of the ΣCu is held in some form of organic association, and the deep-sea sediments, in which ≤10% of the ΣCu is associated with organic hosts.

export production from the euphotic zone, are related to primary production in the surface waters and often show seasonal variations. Thus, the down-column mass particle flux can be related to major oceanographic parameters in surface waters and 'oceanographic consistency' can be injected into the system.

4 'Slippages' in the system can occur, however, when mass particle fluxes are decoupled from the surface production. For example, in some HNLP regions the nutrients are not exhausted during primary production and only a relatively small fraction of the new production is exported in the form of POC, the export being largely via DOC. As a result, the 'biological pump' that draws particulate carbon down the water column does not operate in some regions as efficiently as it would otherwise.

5 The down-column fluxes of trace metals generally vary in relation to the total particle mass fluxes, being highest in coastal and marginal areas, and lowest in the open-ocean.

6 The total down-column flux of trace metals involves both the CPM (the sinking population) and the FPM (the suspended population), but the FPM can sink only in association with the CPM. The total particulate down-column transport in most regions involves a coupling between biogenic and lithogenic components, with the result that most lithogenic elements are transported to depth in association with organic components. The lithogenic material, however, can be decoupled from the biogenic material, and sink down the water column independently, when there is a relatively large input of terrigenous components: e.g. during the sporadic delivery of aeolian material in the form of dust 'pulses', which perturb the steady-state conditions in surface waters.

7 As it approaches the sea bed the sinking particle mass flux enters the benthic boundary layer (BBL). This layer is an oceanic region of considerable reactivity through which down-column and laterally transported TSM must pass, and perhaps be recycled, before entering the sediment reservoir.

8 The difference between the down-column rain rate and the sediment burial rate is a measure of the extent to which elements undergo benthic recycling. In open-ocean regions in which the bottom deposits are not affected by distal transport (e.g. turbidity currents) the rain rates and burial rates of refractory elements, such as Al, Fe, Ti, are generally in good agreement. In contrast, for labile elements, such as C and P, rain rates exceed burial rates, indicating their recycling in the BBL. Some trace metals can also be recycled across the BBL.

9 It has become clear, therefore, that although particulate matter is the driving force behind the reactions that control the elemental composition of sea water, the movement of both particulate and dissolved elements through the ocean system is not a simple one-stage journey, but rather involves a series of complex interactive recycling stages.

This complicated journey has now been followed down to the sediment–seawater interface. It must be stressed, however, that even when the material has crossed this interface and has been incorporated into the bottom deposits it has not simply entered a static reservoir for the permanent storage of components extracted from the water column. On the contrary, diagenetic reactions occur within the sediment–interstitial-water complex, and these can severely modify the mineral and elemental compositions of the sediments themselves. Before looking at the reactions that occur in the sediment reservoir, however, it is necessary to know something of the fundamental characteristics of the oceanic sediments themselves and these are described in the next chapter.

## References

Bacon, M.P., Cochran, J.K., Hirschberg, D., Hammar, T.R. & Fleer, A.P. (1996) Export flux of carbon at the equator during the EqPac time-series cruises estimated from [234]Th measurements. *Deep Sea Res.*, **43**, 1133–53.

Billet, D.S.M., Lampitt, R.S., Rice, A.L. & Mantoura, R.F.C. (1983) Seasonal sedimentation of phytoplankton to the deep-sea benthos. *Nature*, **302**, 520–2.

Bishop, J.K.B. (1989) Regional extremes in particulate matter composition and flux: effects on the chemistry of the ocean interior. In *Productivity of the Ocean: Present and Past*, W.H. Berger, V.S. Smetack & G. Wefer (eds), 117–37. Chichester: J. Wiley & Sons.

Biscaye, P.E. & Anderson, R.F. (1994) Fluxes of particulate matter on the slope of the southern Middle Atlantic Bight: SEEP II. *Deep Sea Res.*, **41**, 459–509.

Brewer, P.G., Nozaki, Y., Spencer, D.W. & Fleer, A.P. (1980) Sediment trap experiments in the deep North Atlantic: isotopic and chemical fluxes. *J. Mar. Res.*, **38**, 703–28.

Brewer, P.G., Bruland, K.W., Eppley, R.W. & McCarthy, J.J. (1986) The Global Ocean Flux Study (GOFS): status of the U.S. GOFS Program. *Eos*, 827–32.

Buat-Menard, P. & Chesselet, R. (1979) Variable influences of the atmospheric flux on the trace metal chemistry of oceanic suspended matter. *Earth Planet. Sci. Lett.*, **42**, 399–411.

Buesseler, K.O. (1991) Do upper-ocean sediment traps provide an accurate record of the particle flux. *Nature*, **353**, 420–23.

Buesseler, K.O., Andrews, J.A., Hartman, M.C., Belastock, R. & Chai, F. (1995) Regional estimates of the export flux of particulate organic carbon derived from thorium-234 during the JGOFS EqPac Program 1995. *Deep Sea Res.*, 777–804.

Chester, R., Thomas, A., Lin, F.J., Basaham, A.S. & Jacinto, G. (1988) The solid state speciation of copper in surface water particulates and oceanic sediments. *Mar. Chem.*, **24**, 261–92.

Coale, K.H. & Bruland, K.W. (1985) $^{234}$Th:$^{238}$U disequilibria within the California Current. *Limnol. Oceanogr.*, **30**, 22–33.

Deuser, W.G. (1986) Seasonal and interannual variations in deep-water particulate fluxes in the Sargasso Sea and their relation to surface hydrography. *Deep Sea Res.*, **33**, 225–91.

Deuser, W.G. (1987) Variability of hydrography and particle flux: transient and long-term relationships. In *Particle Flux in the Ocean*, E. Izdar & S. Honjo (eds), 179–93. Hamburg: Mitt. Geol.-Palont. Inst. Univ. Hamburg, SCOPE/UNEP, Sonderband 62.

Deuser, W.G., Brewer, P.G., Jickells, T.D. & Commeau, R.F. (1983) Biological control of the removal of biogenic particles from the surface ocean. *Science*, **219**, 388–91.

Dymond, J. (1984) Sediment traps, particle fluxes, and benthic boundary layer processes. In *Global Ocean Flux Study*, 261–84. Washington, DC: National Academy Press.

Dymond, J. & Lyle, M. (1994) Particle fluxes in the ocean and implications for sources and preservation of ocean sediments. In *Material Fluxes on the Surface of the Earth*, 125–42. Washington, DC: National Academy Press.

Emmerson, S. & Dymond, J. (1984) Benthic organic carbon cycles: toward a balance of fluxes from particle settling and pore water gradients. In *Global Ocean Flux Study*, 205–384. Washington, DC: National Academy Press.

Fisher, N.S., Nolan, C.V. & Fowler, S.W. (1991) Scavenging and retention of metals by zooplankton faecal pellets and marine snow. *Deep Sea Res.*, **38**, 1261–75.

Gardner, W.D., Souchard, J.B. & Hollister, C.D. (1985) Sedimentation, resuspension and chemistry of particles in the northwest Atlantic. *Mar. Geol.*, **65**, 199–242.

Gust, G., Michaels, A.F., Johnson, R., Dueser, W.G. & Bowles, W. (1994) Mooring line motions and sediment trap hydrodynamic: *in situ* intercomparison of three common deployment designs. *Deep Sea Res.*, **41**, 831–57.

Haake, B., Ittekkot, V., Rixen, T., Ramaswamy, V., Nair, R.R. & Curry, W.B. (1993) Seasonal and interannual variability of particle fluxes to the deep Arabian sea. *Deep Sea Res.*, **40**, 1323–44.

Hernes, P.J., Hedges, J.I., Peterson, M.I., Wakeham, S.G. & Lee, C. (1996) Neutral carbohydrate geochemistry of particulate material in the central equatorial Pacific. *Deep Sea Res.*, **43**, 1181–204.

Hinga, K.R., Sieburth, J.M. & Ross Heath, G. (1979) The supply and use of organic material at the deep-sea floor. *J. Mar. Res.*, **37**, 557–79.

Honjo, S. & Manganini, S.J. (1993) Annual biogenic particle fluxes to the interior of the North Atlantic Ocean: studies at 34°N 21°W and 48°N 21°W. *Deep Sea Res.*, **35**, 1223–34.

Honjo, S., Manganini, S.J. & Coale, J.J. (1982) Sedimentation of biogenic matter in the deep-ocean. *Deep Sea Res.*, **29**, 609–25.

Honjo, S., Manganini, S.J. & Wefer, G. (1988) Annual particle flux and a winter outburst of sedimentation in the northern Norwegian Sea. *Deep Sea Res.*, **35**, 1223–34.

Honjo, S., Dymond, J., Collier, R. & Manganini, S.J. (1995) Export production of particles to the interior of the equatorial Pacific Ocean during the 1992 EqPac experiment. *Deep Sea Res.*, **42**, 831–70.

Ittekkot, V., Degens, E.T. & Honjo, S. (1984) Seasonality in the fluxes of sugars, amino acids and amino sugars to the deep ocean: Panama Basin. *Deep Sea Res.*, **31**, 1071–83.

Izdar, E., Konuk, T., Ittekkot, V., Kempe, S. & Degens, E.T. (1987) Particle flux in the Black Sea: nature of the organic matter. In *Particle Flux in the Ocean*, E. Izdar & S. Honjo (eds), 179–93. Hamburg: Mitt. Geol.-Palont. Inst. Univ. Hamburg, SCOPE/UNEP, Sonderband 62.

Jickells, T.D., Deuser, W.G. & Knap, A.H. (1984) The sedimentation rates of trace elements in the Sargasso Sea measured by sediment trap. *Deep Sea Res.*, **31**, 1169–78.

Jickells, T.D., Deuser, W.D., Fleer, A. & Hembleben, C. (1990) Variability of some elemental fluxes in the western tropical Atlantic Ocean. *Oceanologica Acta*, **13**, 291–98.

Jickells, T.D., Newton, P.P., King, P., Lampitt, R.S. & Boutle, C. (1996) A comparison of sediment trap records of particle fluxes from 19 to 48°N in the northeast Atlantic and their relation to surface water productivity. *Deep Sea Res.*, **43**, 971–86.

Kempe, S. & Jennerjahn, T. (1988) The vertical particle flux in the northern North Sea, its seasonality and composition. In *Biogeochemistry and Distribution of Suspended Matter in the North Sea and Implications to Fisheries Biology*, S. Kempe, V. Dethlefsen, G. Liebezeit & U. Harms (eds), 229–68. Hamburg: Mitt. Geol.-Palont. Inst. Univ. Hamburg, SCOPE/UNEP, Sonderband 65.

Kremling, K. & Streu, P. (1993) Saharan dust influenced trace element fluxes in deep North Atlantic subtropical waters. *Deep Sea Res.*, **40**, 1155–68.

Lampitt, R.S. (1992) The contribution of deep-sea macroplankton to organic remineralisation: results from sediment trap and zooplankton studies over the Madeira Abyssal Plain. *Deep Sea Res.*, **39**, 221–33.

Liebezeit, G. (1987) Particulate carbohydrate fluxes in the Bransfield Strait and Drake passage. *Mar. Chem.*, **20**, 255–64.

Martin, J.H., Knauer, G.A., Karl, D.M. & Broenkow, W.W. (1987) VERTEX: carbon cycling in the northeast Pacific. *Deep Sea Res.*, **34**, 267–86.

Masuzawa, T., Noriki, S., Kurosaki, T. & Tsunogai, S. (1989) Compositional changes of settling particles with water depths in the Japan Sea. *Mar. Chem.*, **27**, 61–78.

Matsuda, H. & Handa, N. (1986) Vertical flux of hydrocarbons as measured in sediment traps in the eastern North Pacific Ocean. *Mar. Chem.*, **20**, 179–95.

Monaco, T., Courp, T., Heussner, S., Carbonne, J., Fowler, S.W. & Denjaux, B. (1990) Seasonality and composition of particle fluxes during ECOMARGE-I, western Gulf of Lions. *Continental Shelf Res.*, **10**, 959–87.

Pfannkuche, O. (1993) Benthic response to the sedimentation of particulate organic matter at the BIOTRANS station, 4°N, 20°W. *Deep Sea Res.*, **40**, 135–49.

Ramaswamy, V. (1987) Particle flux during the southwest monsoon on the western margin of India. In *Particle Flux in the Ocean*, E. Izdar & S. Honjo (eds), 233–42. Hamburg: Mitt. Geol.-Palont. Inst. Univ. Hamburg, SCOPE/UPEND, Sonderband 62.

Saager, M.S. (1994) *On the relationships between dissolved trace metals and nutrients in seawater*. Doctoral Thesis, Vrije Universiteit Amsterdam.

Sherrell, R.M. & Boyle, E.A. (1992) The trace metal composition of suspended particles in the oceanic water column near Burmuda. *Earth Planet. Sci. Lett.*, **111**, 155–74.

Spenser, D.W., Brewer, P.G., Fleer, A., Honjo, S., Krishnaswami, S. & Nozaki, Y. (1978) Chemical fluxes from a sediment trap experiment in the deep Sargasso Sea. *J. Mar. Res.*, **36**, 493–523.

Walsh, I., Fischer, K., Murray, D. & Dymond, J. (1988) Evidence for resuspension of rebound particles from near-bottom sediment traps. *Deep Sea Res.*, **35**, 59–70.

Wefer, G. & Fischer, G. (1993) Seasonal patterns of vertical particle flux in equatorial and coastal upwelling areas of the eastern Atlantic. *Deep Sea Res.*, **40**, 1613–45.

Wefer, G., Suess, E., Balzer, W., *et al.* (1982) Fluxes of biogenic compounds from sediment trap deployments in circumpolar waters of the Drake Passage. *Nature*, **299**, 145–7.

Westerlund, S. & Ohman, P. (1991) Cadmium, copper, cobalt, nickel, lead and zinc in the water column of the Weddell Sea, Antarctica. *Geochim. Cosmochim. Acta*, **55**, 2127–46.

# Part III
# The Global Journey: Material Sinks

# 13 Marine sediments

Marine sediments represent the major sink for material that leaves the seawater reservoir. Before attempting to understand how this sink operates, however, it is necessary briefly to describe the sediments within a global ocean context.

## 13.1 Introduction

The sea floor can be divided into three major topological regions: the continental margins, the ocean-basin floor and the mid-ocean ridge system.

*The continental margins.* These include the continental shelf, the continental slope and the continental rise. The continental shelf is the seaward extension of the land masses, and its outer limit is defined by the **shelf edge**, or break, beyond which there is usually a sharp change in gradient as the continental slope is encountered. The continental rise lies at the base of this slope. In many parts of the world both the continental slope and rise are cut by **submarine canyons**. These are steep-sided, V-shaped valleys, which are extremely important features from the point of view of the transport of material from the continents to the oceans because they act as conduits for the passage of terrigenous sediment from the shelves to deep-sea regions by processes such as turbidity flows. **Trenches** are found at the edges of all the major oceans, but are concentrated mainly in the Pacific, where they form an interrupted ring around the edges of some of the ocean basins. The trenches are long (up to ~4500 km in length), narrow (usually <100 km wide) features that form the deepest parts of the oceans and are often associated with island arcs; both features are related to the tectonic generation of the oceans. The trenches are important in the oceanic sedimentary regime because they can act as traps for material carried down the continental shelf. In the absence of trenches, however, much of the bottom-transported material is carried away from the continental rise into the deep sea.

*The ocean-basin floor.* The ocean-basin floor lies beyond the continental margins. In the Atlantic, Indian and northeast Pacific Oceans **abyssal plains** cover a major part of the deep-sea floor. These plains are generally flat, almost featureless expanses of sea bottom composed of thick (>1000 m) layers of sediment, and have been formed by the transport of material from the continental margins by turbidity currents, which spread their loads out on the deep-sea floor to form thick turbidite sequences. Thus, large amounts of sediment transported from the continental margins are laid down in these abyssal plains, which fringe the ocean margins in the so-called 'hemipelagic' deep-sea areas. The plains are found in all the major oceans, but because the Pacific is partially ringed by a trench belt that acts as a sediment trap, they are more common in the Atlantic and Indian Oceans. Because the Pacific has fewer abyssal plains than the other major oceans, **abyssal hills** are more common here, covering up to ~80% of the deep-sea floor in some areas. **Seamounts** are volcanic hills rising above the sea floor, which may be present either as individual features or in chains. Seamounts are especially abundant in the Pacific Ocean.

*The mid-ocean ridge system.* This ridge system is one of the major topographic features on the surface of the planet. It is an essentially continuous feature, which extends through the Atlantic, Antarctic, Indian and Pacific Oceans for more than ~60 000 km, and the 'mountains' forming it rise to over 3000 m above the sea bed in the crestal areas. The topography of the ridge system is complicated by a series of large semi-parallel fracture zones, which cut across it in many areas. It is usual to divide the ridge system into crestal and flank regions. The flanks lead away from ocean basins, with a general increase in height as the crestal areas are approached. In the Mid-Atlantic Ridge the crestal regions have an extremely rugged topography with a central rift valley (~1–2 km deep) that is surrounded by rift mountains.

The distribution of these various topographic features is illustrated for the North Atlantic in Fig. 13.1 and for the World Ocean in Fig. 13.2(a). The way in which these sea-bed features were formed can be related to the tectonic history of the oceans in terms of the theory of **sea-floor spreading**. In essence, this theory can be summarized as follows. The mid-ocean

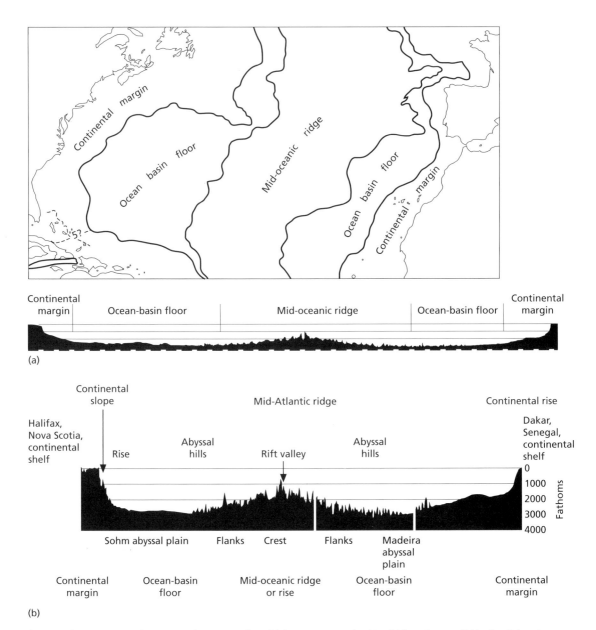

**Fig. 13.1** The topography of the North Atlantic ocean floor ((a) from Heezen *et al.*, 1959; (b) from Kennet, 1982; after Holcombe, 1977).

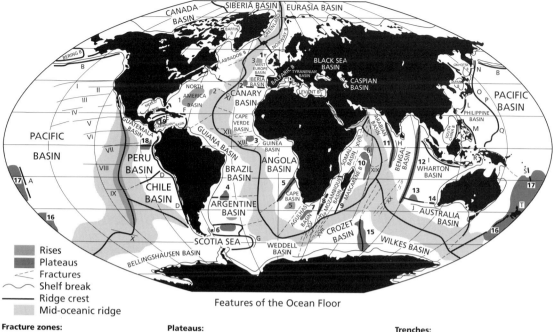

Features of the Ocean Floor

**Fracture zones:**

| | | | |
|---|---|---|---|
| I. | Mendocino | XI. | Atlantis |
| II. | Pioneer | XII. | Vema |
| III. | Murray | XIII. | Romanche |
| IV. | Molokai | XIV. | Chain |
| V. | Clarion | XV. | Mozambique |
| VI. | Clipperton | XVI. | Prince Edward |
| VII. | Galapagos | XVII. | Malagasy |
| VIII. | Marquesas | XVIII. | Owen |
| IX. | Easter | XIX. | Rodriguez |
| X. | Eltanin | XX. | Amsterdam |

**Plateaus:**

| | |
|---|---|
| **1.** Rockall Plateau | **11.** Chagos-Laccadive Plateau |
| **2.** Azores Plateau | **12.** Ninetyeast Ridge |
| **3.** Sierra Leone Plateau | **13.** Broken Ridge |
| **4.** Rio Grande Plateau | **14.** Naturaliste Plateau |
| **5.** Walfisch Ridge | **15.** Kerguelen Plateau |
| **6.** Falkland Ridge | **16.** New Zealand Plateau |
| **7.** Agulhas Plateau | **17.** Melanesia Plateau |
| **8.** Mozambique Ridge | **18.** Galapagos Plateau |
| **9.** Madagascar Ridge | **19.** Jamaica Plateau |
| **10.** Mascarene Plateau | |

**Trenches:**

| | | | |
|---|---|---|---|
| A. | Kermedec-Tonga | K. | Manila |
| B. | Aleutian | L. | Nansei-Shoto |
| C. | Middle America | M. | Philippine |
| D. | Peru-Chile | N. | Kurile-Kamchatka |
| E. | Cayman | O. | Japan |
| F. | Puerto Rico | P. | Bonin |
| G. | South Sandwich | Q. | Mariana |
| H. | Chagos | R. | Banda |
| I. | Java | S. | New Hebrides |
| J. | Diamantina | T. | Hikurangi |

Legend:
- Rises
- Plateaus
- Fractures
- Shelf break
- Ridge crest
- Mid-oceanic ridge

**Rises**

1. Bermuda Rise
2. Corner Rise
3. Rockall Rise
4. Argentine Rise
5. Schmidt-Ott Rise
6. Madingley Rise

(a)

**Fig. 13.2** Topographic features in the oceans. (a) The distribution of the major oceans and their principal topographic features (from Heezen & Hollister, 1971). (b) The distribution of the principal lithospheric plates (from Parsons & Richter, 1981). The principal plates are identified; the abbreviations CO = Cocos, CAR = Caribbean, AR = Arabian, PH = Philippine. Absolute plate velocities (cm yr$^{-1}$) are indicated at selected points. Convergent boundaries are indicated by arrow-heads that point from the subducting plate towards the non-subducting plate.

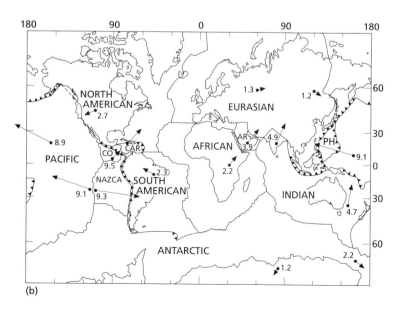

(b)

ridges are associated with the rising limbs of convection cells in the mantle and the sea floor 'cracks apart' at the crest regions. New crust is formed here, and as it is generated the previous crust is moved aside on either side of the crests and is lost at the edges of the oceans under zones of trenches, island arcs and young mountains associated with the descending limbs of convection cells where the ocean floor is carried back down into the mantle. Thus, the ocean floor is continually being created and destroyed in response to convection in the mantle. According to the evidence available it would appear that the spreading takes place at rates of a few centimetres per year. However, some ridges spread faster than others, allowing fast and slow spreading centres to be identified, although the spreading rates do not appear to have been constant with time. The slowest rates are found along parts of the Mid-Atlantic and Mid-Indian Ocean Ridges ($\sim$1–2.5 cm yr$^{-1}$), and the highest around the East Pacific Rise ($\sim$5 cm yr$^{-1}$).

One of the most important findings to emerge from these tectonic studies is that the ocean basins are relatively young on the geological time-scale, and magnetic evidence, combined with sediment age data from the Deep Sea Drilling Project (DSDP), indicates that the oldest oceanic crust is around Middle Jurassic in age, i.e. it was formed $\sim$170 Ma ago. The concept of a dynamic ocean floor that is being generated at the ridge crests and resorbed into the mantle at the ocean edges is intimately related to the overall pattern of global tectonics. The unifying theory which brings together the various aspects of modern thinking on global structure is the **plate tectonic** concept. Put simply, it is now thought that the surface of the Earth consists of a series of thin ($\sim$100–150 km) plates that are continuously in motion. Six (or sometimes seven) major plates were originally identified, but additional plates have since been included (see Fig. 13.2b). The plates have boundaries between them, and Jones (1978) has classified these into three principal kinds.

1 **Conservative boundaries**, at which the plates slide past each other without any creation or destruction of oceanic crust.

2 **Constructive boundaries**, which are marked by the crests of the active spreading oceanic ridges where new crust is formed as the plates diverge. Thus, constructive boundaries are crust *sources*.

3 **Destructive boundaries**, at which the plates converge and one plate moves beneath another one. These are marked by the presence of trenches, island arcs or young mountain belts. Thus, destructive margins are crust *sinks*, and the crust is lost by subduction under the trenches, where it eventually becomes sorbed back into the mantle. This is an extremely important process, because when the crust is sorbed back into the mantle it takes with it the sediment that has accumulated on the ocean floor. This has important implications for marine geochemistry because it means that the sediments deposited at the bottom of the sea, which are a principal sink for much of the material poured into sea water, are eventually returned back into the global geochemical cycle.

A large part of the floor of the World Ocean is covered by sediments. The thickness of this sediment blanket varies from place to place; for example, in the Atlantic it averages >1 km, in the Pacific it is <1 km, and over the whole ocean it averages $\sim$500 m. The sediments forming this blanket are the ultimate marine sink for all the particulate components that survive destruction in the ocean reservoir. The sediments are also the major sink for dissolved elements, although for some of these elements basement rocks can also act as a sink (see Chapter 5). Marine sediments therefore form one of the principal oceanic reservoirs.

In terms of a very simplistic model, the initial sediment in a newly formed ocean basin will be deposited on to the basalt basement as fresh crust is generated at the ridge spreading centres. This initial sediment is a hydrothermal deposit composed of metal-rich precipitates (see Sections 15.3.6 & 16.6.2), and as the ridge lies above the carbonate compensation depth (CCD; see Section 15.2.4.1) carbonate sediments begin to accumulate. As the basin continues to open up, the sediment surface away from the ridge falls below the CCD and clays are deposited (see Fig. 16.7). As sedimentation continues, the composition of the deposits is controlled by a variety of interacting chemical, physical and biological factors. Because of changes in the relative importance of these various factors as the plate on which the sediment is deposited moves through different oceanic environments (e.g. high- and low-productivity zones), the nature of the deposits forming the marine 'sediment blanket' has varied through geological times as the ocean basins evolved to their present state in response to sea-floor spreading. As a result of this **sea-floor spreading**, or continental drift, changes have occurred in the sizes, shapes, latitudinal distributions, water circulation patterns and depths of the ocean basins, all of which affected the patterns of sedimentation. Much of our current knowledge of the deeper parts of the oceanic sediment column has come from material collected during the DSDP. Detailed examination of the DSDP cores has shown, for example, that there have been gross changes in both the compositions and rates of accumulation of marine sediments over the past 120 million yr (Davies & Gorsline, 1976). During this time the depositional environments have passed through a variety of conditions leading to the preferential preservation of carbonate-rich, silica-rich, or terrigenous-rich formations, and have sometimes given rise to periods of hiatuses in which removal exceeds deposition and sediment is absent. Such hiatuses, which are found in all the oceans, probably reflect changes in current strengths and circulation patterns, and are most pronounced on the western edges of the oceans where the boundary currents are strongest (see Section 7.3.3). In addition to those associated with alternations in the shapes of the oceanic basins following the tectonic evolution of the oceans, changes in sediment distribution patterns are also brought about by global climatic variations, which can affect both the strengths of the external input mechanisms, which deliver material to the seawater reservoir, and the internal oceanic conditions (e.g. sea level, water temperature, current movements, primary productivity). For example, large-scale climatic changes attendant on glacial–interglacial transitions can modify sedimentation patterns considerably. The Holocene–Pleistocene transition, which marks the last transition from glacial to interglacial conditions, is often marked in oceanic sediments by a decrease in carbonate concentrations, as a result of the rising of the CCD (see Section 15.2.4.1) during interglacials and its lowering during glacials, and changes in the assemblages of planktonic microfossils, which occurred mainly in response to variations in temperatures in the euphotic productive water layers.

The importance of changes in the type of deposit with depth in the oceanic sediment column will be identified where necessary in the text. In the present volume, however, we are concerned chiefly with the role played by the sediments as marine sinks for components that have flowed through the seawater reservoir. Diagenetic changes, which have their most immediate effect on the composition of sea water, take place in the upper few metres of the sediment column, i.e. during early diagenesis (see Section 14.1), and for this reason attention will be focused mainly on the present-day sediment distribution patterns, which reflect contemporary trends in climatic conditions, oceanic current patterns and sea-floor topography.

Marine sediments are deposited under a wide variety of depositional environments. However, it is useful at this stage to make a fundamental distinction between nearshore and deep-sea deposits, a distinction that recognizes the importance of the shelf break in dividing two very different oceanic depositional regimes.

**Nearshore sediments** are deposited mainly on the shelf regions under a wide variety of regimes that are strongly influenced by the adjacent land masses. As a result, physical, chemical and biological conditions in nearshore areas are much more variable than in deep-sea regions. Nearshore depositional environments include estuaries, fjords, bays, lagoons, deltas, tidal flats, the continental terrace and marginal basins.

**Deep-sea sediments** are usually deposited in depths of water >500 m, and factors such as remoteness from the land-mass sources, reactivity between particulate and dissolved components within the oceanic water column, and the presence of a distinctive biomass lead to the setting up of a deep-sea environment that is unique on the planet. Because of this, deep-sea sediments, which cover more than 50% of the surface of the Earth, have very different characteristics from those found in continental or nearshore environments. Two of the most distinctive characteristics of these deep-sea sediments are (i) the particle size, and (ii) the rate of accumulation, of their land-derived non-biogenic components.

1 The land-derived fractions of deep-sea sediments are dominated by clay-sized, i.e. <2 μm diameter, components, which usually account for ~60–70% of the non-biogenic material in them. In contrast, material having a wide variety of particle sizes is found in nearshore sediments, and in general clay-sized components constitute a much smaller fraction of the land-derived solids—see Table 13.1.

2 Various techniques are available for the measurement of the accumulation rates of marine sediments. These include dating by magnetic reversals, fossil

**Table 13.1** The average content of the <2 μm fraction in marine sediments and suspended river particulates (data from Griffin *et al.*, 1968).

| Sediment type | Location | Wt% <2 μm fraction |
|---|---|---|
| Pelagic sediment | Atlantic Ocean | 58 |
| | Pacific Ocean | 61 |
| | Indian Ocean | 64 |
| Shelf sediment | USA Atlantic coast | 2 |
| | Gulf of Mexico | 27 |
| | Gulf of California | 19 |
| | Sahul Shelf, northwest Australia | 72 |
| Suspended river particulates | 33 USA rivers | 37 |

**Table 13.2** Accumulation rate of land-derived fractions in deep-sea sediments; units, mm $10^3\,yr^{-1}$.

| Oceanic region | Accumulation rate (mm $10^3\,yr^{-1}$) | |
|---|---|---|
| | A* | B† |
| South Pacific | 0.45 | 1.0 |
| North Pacific | 1.5 | 5.8 |
| South Atlantic | 1.9 | 6.0 |
| North Atlantic | 1.8 | 5.7 |
| Indian Ocean | — | 4.4 |

* Data from Goldberg & Koide (1962), Goldberg *et al.* (1963) and Goldberg & Griffin (1964).
† Data from Ku *et al.* (1968).

assemblages and the decay of radionuclides. The radionuclide decay methods are the most commonly used, and normally involve either $^{14}C$ or members of the uranium, thorium and actino-uranium decay series. There are differences between the accumulation rates derived by the various techniques, and sometimes even between those obtained using the same technique. Because of this two data sets are given in Table 13.2 for the accumulation rates of the land-derived material in deep-sea sediments. Although there are differences between the two data sets, however, it is apparent that the land-derived fractions of deep-sea sediments are accumulating at a rate of the order of a few millimetres (usually <10 mm

in pelagic sediments) per 1000 yr. There are, however, variations in the accumulation rates of this land-derived material within the oceans. An example of how the accumulation rates of land-derived material in deep-sea sediments vary with the environment of deposition was provided by Griffin *et al.* (1968). These authors showed that the highest accumulation rates for this material were found adjacent to the continents (**hemi-pelagic areas**), where they could reach values >10 mm $10^3\,yr^{-1}$. Away from the influence of the continents, the rates decreased to reach minimum values (~0.5 mm $10^3\,yr^{-1}$) in remote open-ocean (**pelagic**) areas, e.g. around the Mid-Atlantic Ridge. However, Ku *et al.* (1968) reported somewhat higher values (~1 mm $10^3\,yr^{-1}$) for these remote regions, and estimated the average accumulation rate for the non-carbonate fractions of deep-sea sediments to be ~2 mm $10^3\,yr^{-1}$. The trends in the accumulation rate data derived by Griffin *et al.* (1968) are illustrated in Fig. 13.3(a). Bostrom *et al.* (1973) made a compilation of the accumulation rates of the non-carbonate fractions of Indo-Pacific deep-sea sediments, and also demonstrated that there is an overall decrease in the rates away from the land margins towards the open-ocean regions; these trends are illustrated in Fig. 13.3(b). Oceanic carbonate oozes usually accumulate at rates in the range 1–3 cm $10^3\,yr^{-1}$.

It is apparent, therefore, that under present-day conditions, land-derived material is accumulating in deep-sea sediments at a rate of the order of a few millimetres per 1000 yr. In contrast to deep-sea deposits, those deposited in nearshore areas have land-derived fractions that can accumulate at rates much greater than a few millimetres per year.

## 13.2 The formation of deep-sea sediments

The processes that are involved in the formation of deep-sea sediments can be linked to the manner in which material is transported to, and distributed within, the World Ocean, and by considerably simplifying the situation it is possible to distinguish between five general types of sediment transport mechanisms that operate in the deep sea.

*Gravity currents.* Gravity currents transport material to the deep sea from the shelf regions by slides, slumps and gravity flows. Of the various types of gravity flows, it is the **turbidity currents** that have the

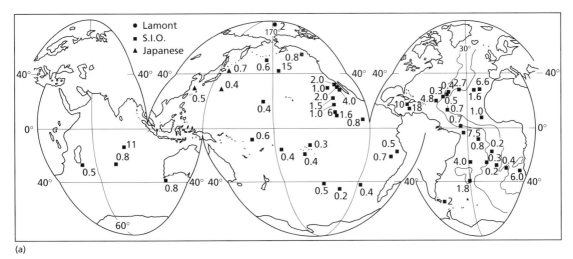

(a)

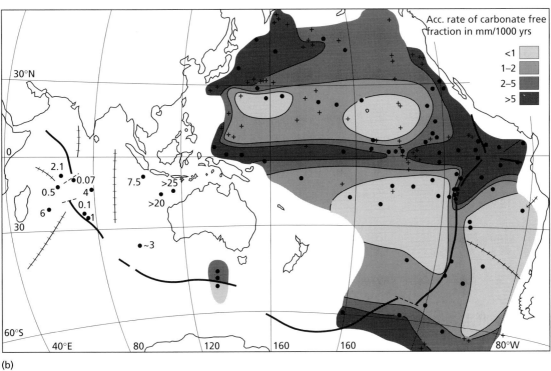

(b)

**Fig. 13.3** Accumulation rates of non-carbonate material in deep-sea sediments. (a) Accumulation rate trends, showing low values around the Mid-Atlantic Ridge (from Griffin *et al.*, 1968). (b) Accumulation rates in the Indo-Pacific Ocean (from Bostrom *et al.*, 1973).

greatest influence on the movement of material from the shelf regions to the open ocean. These turbidity currents are short-lived, high-velocity, density currents that can carry vast quantities of suspended sediment off the shelves, often via submarine canyon conduits. The deep-sea deposits generated by these currents are termed **turbidites**, and usually they are made up of sand layers interbedded with pelagic deposits of smaller grain size. *Proximal* turbidites have been deposited relatively close to the source of the transported sediment, whereas *distal* types have been carried for much greater distances. Turbidite deposition is an extremely important marine sedimentary process and is thought to be responsible for the formation of features such as submarine fans on the continental rise, and the abyssal plains on the deep-sea floor.

*Geostrophic deep-ocean or bottom currents.* Bottom currents have a significant influence on the distribution of sediment on the deep-sea floor. These currents transport material that has a finer grain size than that carried by turbidity currents, and result in the formation of features such as sediment piles and ridges. Bottom currents are most strongly developed along the western boundaries of the oceans (see Section 7.3.3), and it is here that nepheloid layers (layers of suspended sediment) are best developed.

*Mid-depth currents.* Various kinds of TSM can be transported by advection via mid-depth oceanic circulation patterns. These include material released directly from nearshore sediments, material resuspended from deep-sea sediments at basin edges, and hydrothermal components. As well as transporting material away from the basin margins, mid-depth currents can also carry material laterally from the centres to the edges of the oceanic basins.

*Surface and near-surface currents.* Surface current movements, which are dominated by the gyre circulation patterns (see Section 7.3.2), transport the oceanic biomass, together with fine-grained land-derived material introduced to the surface ocean by river run-off, atmospheric deposition and glacial transport.

*Vertical or down-column transport.* This is the great ocean-wide carbon-driven transport process (or global ocean flux), which carries material from the surface ocean to the sea bed. During the process, the material becomes incorporated into the major oceanic biogeochemical cycles, which are involved in the down-column flux of particulate material, and play such a vital role in controlling the chemistry of the oceans (see Sections 11.6.3.1 & 11.6.3.2). The

down-column flux can also incorporate the advective mid-depth flux (see Section 10.5).

On the basis of these different mechanisms, important distinctions therefore can be made between the transport vectors involved in: (i) *lateral* off-shelf movements; (ii) *lateral* sea-bed movements; (iii) *lateral* mid-depth movements; (iv) *lateral* sea surface movements; and (v) *vertical* down-column movements.

Each of these transport vectors has a material flux associated with it that contributes material to deep-sea sediments, and a study reported by Grousset & Chesselet (1986) can be used both to illustrate how these transport vectors operate on a quasi-global scale and to assess the extent of any coupling between them. These authors carried out an investigation into the Holocene (10 000 yr BP) mid-ocean ridge sedimentary regime that prevailed in the North Atlantic between ~45°N and ~65°N. The major sources of sediment to the region in the Holocene were the North American mainland and Iceland, with local mid-ocean ridge sources being generally of only minor importance. By using a number of tracers to identify the sediment source materials and to elucidate the mechanisms by which they were transported, the authors were able to construct a first-order model for the Holocene sedimentary regime in the region. This model is illustrated in Fig. 13.4, and involves the following source–flux relationships.

**1 Source 1** was the North American mainland from which material was supplied by surface currents ($\Phi_s$), turbidity currents ($\Phi_t$), and aeolian transport ($\Phi_e$). Material from this source decreased in the underlying sediments in a west-to-east direction, and it was suggested that a coupling of the aeolian transport flux into the down-column flux ($\Phi_v$) was the principal mechanism responsible for driving this gradient.

**2 Source 2** was Iceland, from which material was supplied via the turbidity current flux ($\Phi_t$) and the geostrophic current flux ($\Phi_a$). Material from this source decreased in concentration in a north-to-south direction in the underlying sediments, and it was proposed that this gradient arose from a coupling of the two fluxes that transported material from Iceland.

**3 Source 3** was the Mid-Atlantic Ridge, but this supplied only a minor amount of sedimentary material to the region.

The two thick arrows in the flux model (Fig. 13.4),

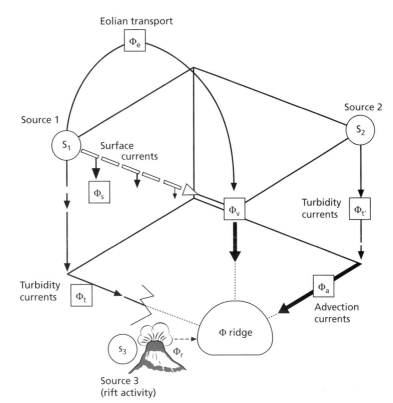

**Fig. 13.4** A first-order model for the mid-ocean ridge Holocene sedimentary regime in the North Atlantic (from Grousset & Chesselet, 1986). The three major sediment sources are $S_1$, $S_2$ and $S_3$, which supply the following specific flux materials: $\Phi_t$ and $\Phi_{t'}$, which are fluxes associated with downslope processes; $\Phi_s$, which is the surface current flux; $\Phi_a$ which is the advective bottom flux; $\Phi_e$ which is the aeolian flux; $\Phi_r$, which is the rift flux. These are the main components of $\Phi_v$ (the vertical flux) and $\Phi_a$ (the advective flux)—see text.

representing the down-column ($\Phi_v$) and the geostrophic ($\Phi_a$) fluxes, are the main pathways by which continental material was transported from the North America and Icelandic sources, respectively, and these combine to form the ridge flux ($\Phi_{ridge}$), the hydrothermal flux being negligible.

The study reported by Grousset & Chesselet (1986) clearly demonstrates how the various mechanisms that transport land-derived material to the ocean reservoir interact on a quasi-global oceanic scale. This kind of interaction is extremely important in controlling the chemical composition of the bottom sediments. There are a number of reasons for this, and three of the more important of these are identified below.

1 Solids derived from different continental sources may have different chemical compositions, and the extent to which they are mixed will affect the overall composition of a sediment to which they contribute.

2 Material can be transported at varying rates by the different transport mechanisms; for example, turbidity currents deposit material at a much faster rate than down-column settling. The accumulation rate of a sediment determines the length of time its surface is in contact with the overlying water column, and therefore influences both the degree to which the sediment components react with material dissolved in sea water and the extent to which the diagenetic sequence proceeds within the sediment column.

3 The relative magnitudes of the **sea bottom** and the **vertical down-column particle fluxes** are extremely important in constraining the chemical composition of non-biogenous deep-sea sediments. There are two main reasons for this. First, the material that is initially dumped on the shelf regions has a coarser particle size, and is poorer in trace metals, than the material that escapes the coastal zone via the surface ocean; this concept was developed in Section 3.1.4. Secondly, the material that undergoes down-column transport is involved in the dissolved–particulate reactions that remove trace elements from sea water and so becomes a sink for additional trace elements, which are not picked up by bottom-transported material. This bottom-transport–down-column

vertical transport trace-metal fractionation is considered in more detail in Section 16.5.

## 13.3  A general scheme for the classification of marine sediments

A number of the parameters discussed above can be combined together to outline a general scheme for the classification of marine sediments. It must be stressed, however, that this is by no means a rigorous sediment classification, and several attempts have been made to produce much more detailed and lithologically consistent classifications of deep-sea sediments. None the less, the simplified classification will serve its purpose, which is merely to act as a framework within which to describe the geochemistry of marine sediments. This general scheme is outlined below.

### 13.3.1  Nearshore sediments

Nearshore or coastal sediments are deposited on the margins of the continents under a wide variety of conditions in chemical environments that range from oxic to fully anoxic in character. The sediments include gravels, sands, silts and muds and they are composed of mixtures of terrigenous, authigenic and biogenous components; the latter is mainly shell material, but relatively high concentrations of organic carbon are found in sediments deposited under reducing conditions. Nearshore sediments contain material having a wide variety of grain sizes, fine-grained material being found in low-energy environments and sands in high-energy environments. Over many shelves terrigenous sediments (e.g. terrigenous muds, which accumulate at relatively high rates, greater than several millimetres per year) are the prevalent type of deposit, but on some shelf regions carbonate deposition can predominate, e.g. on broad shallow shelves where the supply of terrigenous debris is small.

### 13.3.2  Deep-sea sediments

On the basis of the major transport mechanisms that supply the sediment-forming material, deep-sea deposits can be subdivided into hemi-pelagic and pelagic types.

#### 13.3.2.1  Hemi-pelagic deep-sea sediments

These sediments are deposited in areas that fringe the continents, e.g. on the abyssal plains. The land-derived material in hemi-pelagic sediments has been transported mainly by bottom processes, i.e. by the turbidity current and geostrophic bottom-current fluxes described above, and much of it originated on the shelf regions. The inorganic hemi-pelagic sediments include **lithogenous clays** (or muds), **glacial marine sediments**, **turbidites** and **mineral sands**. All of these sediments can contain varying proportions of biogenous shell material. The rates of deposition of the land-derived material in hemi-pelagic deep-sea sediments can be $\geq 10\,mm\ 10^3\,yr^{-1}$, and they often contain as much as $\sim 1{-}5\%$ organic carbon. The preservation of this organic carbon is a function of the extent to which the diagenetic sequence has proceeded (see Sections 14.3 & 14.4), and hemi-pelagic clays are often grey-green in colour, indicating reducing conditions, below a thin oxidized red layer.

#### 13.3.2.2  Pelagic deep-sea sediments

These sediments accumulate in open-ocean areas. They are generally deposited in the absence of effective bottom currents, and Davies & Laughton (1972) have defined pelagic sediments as those 'laid down in deep-water under quiet current conditions'. The bulk of the material forming these deposits has settled down the water column to blanket the bottom topography with a sediment cover. Thus, pelagic sediments are formed via the **vertical down-column flux** identified above, and so can be differentiated from hemi-pelagic deposits, which have been formed largely from **bottom transport processes**. It is common practice to subdivide pelagic (and hemi-pelagic) deep-sea sediments into inorganic and biogenous categories.

*Inorganic pelagic deep-sea sediments.* In the past these have been defined as containing <30% biogenous skeletal remains, and as having a large fraction ($\geq 60\%$) of their non-biogenous material in the <2 μm (clay) size class. Traditionally, therefore, it is these sediments that have been known as the pelagic clays, or simply the **deep-sea clays**. The land-derived material in all types of pelagic sediments has been in suspension for relatively long periods in the water column and is deposited at slow rates, i.e. usually a few millimetres per 1000 yr. As a result, much of the organic carbon reaching the sediment surface is destroyed in the early stages of the diagenetic sequence (see Section 14.1), and the sediments usually contain only ~0.1–0.2% organic carbon. Pelagic clays are therefore oxidizing to a considerable depth, and often have a red colour owing to the presence of ferric iron. Because of this, they have often been termed **red clays**. Some authors have further subdivided pelagic clays on the basis of the origin of their principal components, e.g. into **lithogenous** and **hydrogenous** types; for a description of these categories, see Chapter 15.

*Biogenous pelagic deep-sea sediments.* These sediments have been defined traditionally as containing >30% biogenous shell remains. They are usually referred to by the term **oozes**, and are subdivided into calcareous and siliceous types.

The **calcareous oozes** contain >30% skeletal carbonates, and are classified on the basis of the predominant organisms present into **foraminiferal ooze** (sometimes termed *Globigerina* ooze after the most common of the forams), **nanofossil ooze** (or coccolith ooze) and **pteropod ooze**.

The **siliceous oozes** contain >30% opaline silica skeletal remains, and are subdivided into **diatom oozes** and **radiolarian oozes**, depending on the principal silica-secreting organism present.

## 13.4 The distribution of marine sediments

The distribution of sediments in the World Ocean is illustrated in Fig. 13.5. In Fig. 13.5(a), the sediments are classified on a general basis in which individual types of deep-sea clay are not specified. In contrast, Fig. 13.5(b) presents an example of a classification in which the deep-sea clays are subdivided into lithogenous and hydrogenous types. A number of the principal features in the distributions of deep-sea sediments can be identified from these diagrams.
1 Deep-sea clays and calcareous oozes are the predominant type of deep-sea deposit.
2 The calcareous oozes cover large tracts of the open-ocean floor at water depths <3–4 km (see Sections 13.7 and 15.2.4.1).

3 The siliceous oozes form a ring around the high-latitude ocean margins in the Antarctic and North Pacific (both diatom oozes), and are also found in a band in the equatorial Pacific (radiolarian oozes).
4 Extensive deposits of glacial marine sediments are confined to a band fringing Antarctica, and to the high-latitude North Atlantic.
5 Lithogenous clays cover a large area of the North Pacific, whereas the clays in the South Pacific are mainly hydrogenous in character.

## 13.5 The chemical composition of marine sediments

The material that is finally deposited in the marine sediment reservoir has undergone a complex journey before reaching its sea-bed sink. The sediments forming this reservoir respresent, if not the ultimate end-point, at least a major geological time-scale halt in the global mobilization–transportation cycle. A knowledge of the chemical composition of this sediment reservoir is therefore important for an understanding of the global cycles of many elements. However, before attempting to synthesize the various factors that act together to control the chemistry of these sediments it is necessary to establish a compositional database from which to work. For this purpose, three data compilations have been prepared.
1 The first compilation is given in Table 13.3, and lists the **overall elemental compositions** of the principal types of marine sediments, i.e. nearshore muds, deep-sea clays and deep-sea carbonates. The table also includes chemical data for the continental crust, soils, suspended river particulates and crustal aerosols, which are given so that the compositions of marine sediments can be evaluated within the global context of their principal terrestrial feeder materials.
2 The second compilation is given in Table 13.4, and lists the **major element compositions** of the main types of deep-sea sediments.
3 The third compilation is given in Table 13.5, and presents data showing a number of **overall trends** in the elemental compositions of a wide range of deep-sea deposits. The aim of this table is to permit compositional differences in the deposits to be evaluated within an ocean-wide framework.

From these combined databases a number of gross features in the geochemistry of marine sediments can

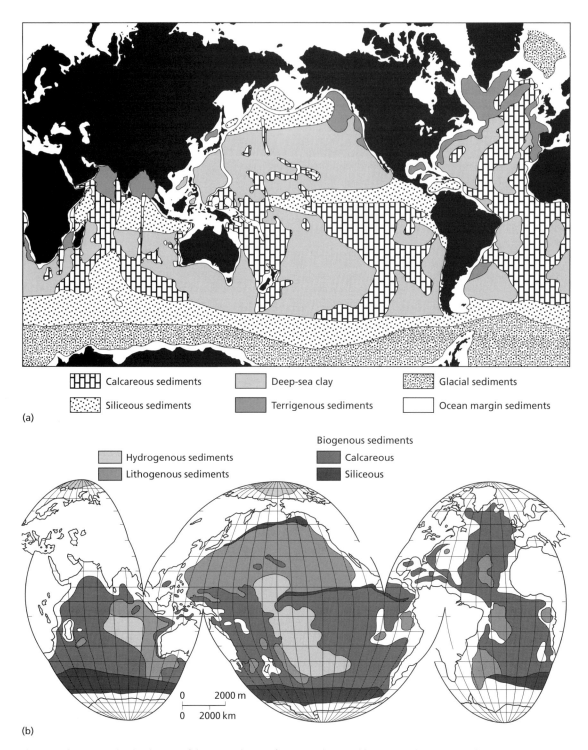

(a)

(b)

**Fig. 13.5** The present-day distributions of the principal types of marine sediments. (a) The general distribution of marine sediments (from Davies & Gorsline, 1976). (b) The distribution of deep-sea sediments classified on the basis of their components (from Riley & Chester, 1971).

**Table 13.3** Elemental composition of marine sediments and some continental material (units, $\mu g\,g^{-1}$). Estimated values are given in brackets.

| Element | Continental crust* | Continental soil* | River particulate material* | Crustal dust† | Near-shore mud‡ | Deep-sea clay* | Deep-sea carbonate§ |
|---|---|---|---|---|---|---|---|
| Ag | 0.07 | 0.05 | 0.07 | — | — | 0.1 | 0.0X |
| Al | 69 300 | 71 000 | 94 000 | 82 000 | 84 000 | 95 000 | 20 000 |
| As | 7.9 | 6 | 5 | — | 5 | 13 | 1.0 |
| Au | 0.01 | 0.001 | 0.05 | — | — | 0.003 | 0.00X |
| B | 65 | 10 | 70 | — | — | 220 | 55 |
| Ba | 445 | 500 | 600 | — | — | 1500 | 190 |
| Br | 4 | 10 | 5 | — | — | 100 | 70 |
| Ca | 45 000 | 15 000 | 21 500 | — | 29 000 | 10 000 | 312 400 |
| Cd | 0.2 | 0.35 | (1) | — | — | 0.23 | 0.23 |
| Ce | 86 | 50 | 95 | — | — | 100 | 35 |
| Co | 13 | 8 | 20 | 23 | 13 | 55 | 7 |
| Cr | 71 | 70 | 100 | 79 | 60 | 100 | 11 |
| Cs | 3.6 | 4 | 6 | — | — | 5 | 0.4 |
| Cu | 32 | 30 | 100 | 47 | 56 | 200 | 50 |
| Er | 3.7 | 2 | (3) | — | — | 2.7 | 1.5 |
| Eu | 1.2 | 1 | 1.5 | — | — | 1.5 | 0.6 |
| Fe | 35 000 | 40 000 | 48 000 | 48 000 | 65 000 | 60 000 | 9000 |
| Ga | 16 | 20 | 25 | — | — | 20 | 13 |
| Ge | 1.5 | X | — | — | — | 1.2 | 0.1 |
| Gd | 6.5 | 4 | (5) | — | — | 7.8 | 3.8 |
| Hf | 5 | — | 6 | — | — | 4.5 | 0.41 |
| Ho | 1.6 | 0.6 | (1) | — | — | 1.0 | 0.8 |
| In | 0.1 | — | — | — | — | 0.08 | 0.02 |
| K | 24 000 | 14 000 | 20 000 | 20 000 | 25 000 | 28 000 | 2900 |
| La | 41 | 40 | 45 | — | — | 45 | 10 |
| Li | 42 | 25 | 25 | — | 79 | 45 | 5 |
| Lu | 0.45 | 0.40 | 0.50 | — | — | 0.50 | 0.50 |
| Mg | 16 400 | 5000 | 11 800 | — | 21 000 | 18 000 | — |
| Mn | 720 | 1000 | 1050 | 865 | 850 | 6000 | 1000 |
| Mo | 1.7 | 1.2 | 3 | 1.8 | 1 | 8 | 3 |
| Na | 14 200 | 5000 | 5300 | 5100 | 40 000 | 40 000 | 20 000 |
| Nd | 37 | 35 | 35 | — | — | 40 | 14 |
| Ni | 49 | 50 | 90 | 73 | 35 | 200 | 35 |
| P | 610 | 800 | 1150 | — | 550 | 1400 | 350 |
| Pb | 16 | 35 | 100 | 52 | 22 | 200 | 9 |
| Pr | 9.6 | — | (8) | — | — | 9 | 3.3 |
| Rb | 112 | 150 | 100 | — | — | 110 | 10 |
| Se | 0.05 | 0.01 | — | — | — | 4.5 | 0.5 |
| Sb | 0.9 | 1 | 2.5 | — | — | 0.8 | 0.15 |
| Sc | 10 | 7 | 18 | — | 12 | 20 | 2 |
| Si | 275 000 | 330 000 | 285 000 | — | 250 000 | 283 000 | 32 000 |
| Sm | 7.1 | 4.5 | 7.0 | — | — | 7.0 | 3.8 |
| Sn | 2 | (0.1) | — | — | 2 | 1.5 | 0.X |
| Sr | 278 | 250 | 150 | — | 160 | 250 | 2000 |
| Ta | 0.8 | 2 | 1.2 | — | — | 1 | 0.0X |
| Tb | 1.05 | 0.7 | 1 | — | — | 1 | 0.6 |
| Th | 9.3 | 9 | 14 | — | — | 10 | — |
| Ti | 3800 | 5000 | 5600 | 5700 | 5000 | 5700 | 770 |
| Tm | 0.5 | 0.6 | (0.4) | — | — | 0.4 | 0.1 |
| U | 3 | 2 | 3 | — | — | 2 | — |
| V | 97 | 90 | 170 | 120 | 145 | 150 | 20 |
| Y | 33 | 40 | 30 | — | — | 32 | 42 |
| Yb | 3.5 | — | 3.5 | — | — | 3 | 1.5 |
| Zn | 127 | 90 | 250 | 75 | 92 | 120 | 35 |
| Zr | 165 | 300 | — | — | 240 | 150 | 20 |

X, order of magnitude.

* Data for Ge, In, Se, Sn and Zr from a variety of sources; data for all other elements from Martin & Whitfield (1983).

† A Saharan dust population from the Atlantic Northeast Trades; data from Murphy (1985).

‡ Data from Wedepohl (1960).

§ Turekian & Wedepohl (1961).

be identified, and for this purpose it is convenient to treat the major and trace elements separately.

*Major elements.* The major element composition of marine sediments is largely controlled by the relative proportions of the sediment-forming minerals. The principal minerals are the clays, and biogenous carbonates and opal. On the basis of the mutual proportions of these minerals the sediments can be divided into three broad types, i.e. clays, carbonate oozes and siliceous oozes. Average major element analyses for the three sediment types are given in Table 13.4, from which it can be seen that aluminium is concentrated in the clays, calcium in the carbonates and silicon in the siliceous oozes. Manganese and Fe are included in Table 13.4, and although they are usually present as major elements they will also be included with the trace elements because both Fe and Mn phases can act as scavenging agents for dissolved metals.

*Trace elements.* From the data given in Table 13.5, it is possible to establish a number of trends in the distributions of some trace elements in deep-sea sediments (see e.g. Chester & Aston, 1976). These trends can be summarized as follows.

**1** In general, deep-sea carbonates are impoverished in most trace elements relative to deep-sea clays, Sr being an exception to this.

**2** Certain trace elements, e.g. Cr, V and Ga, have similar concentrations in both nearshore muds and deep-sea clays (DSC).

**3** In contrast, other trace elements, e.g. Mn, Cu, Ni, Co and Pb, are enhanced in the DSC relative to nearshore muds. Thus, a fundamental oceanic fractionation between nearshore muds and deep-sea clays can be introduced for these **enriched** or **excess elements.**

**4** The excess trace elements are enhanced to a

**Table 13.4** The major element composition of the principal types of deep-sea sediments* (units, wt% oxides).

| Major element | Calcareous | Lithogenous clay | Siliceous | Oceanic† average |
|---|---|---|---|---|
| $SiO_2$ | 26.96 | 55.34 | 63.91 | 42.72 |
| $TiO_2$ | 0.38 | 0.84 | 0.65 | 0.59 |
| $Al_2O_3$ | 7.97 | 17.84 | 13.30 | 12.29 |
| $Fe_2O_3$ | 3.00 | 7.04 | 5.66 | 4.89 |
| FeO | 0.87 | 1.13 | 0.67 | 0.94 |
| MnO | 0.33 | 0.48 | 0.50 | 0.41 |
| CaO | 0.30 | 0.93 | 0.75 | 0.60 |
| MgO | 1.29 | 3.42 | 1.95 | 2.18 |
| $Na_2O$ | 0.80 | 1.53 | 0.94 | 1.10 |
| $K_2O$ | 1.48 | 3.26 | 1.90 | 2.10 |
| $P_2O_5$ | 0.15 | 0.14 | 0.27 | 0.16 |
| $H_2O$ | 3.91 | 6.54 | 7.13 | 5.35 |
| $CaCO_3$ | 50.09 | 0.79 | 1.09 | 24.87 |
| $MgCO_3$ | 2.16 | 0.83 | 1.04 | 1.51 |
| Organic C | 0.31 | 0.24 | 0.22 | 0.27 |
| Organic N | — | 0.016 | 0.016 | 0.015 |
| Total | 100.0 | 100.0 | 100.0 | 100.0 |
| Total $Fe_2O_3$ | 3.89 | 8.23 | 6.42 | — |

* Data from El Wakeel & Riley (1961).
† Weighted mean calculated on the basis of the areal coverage of the sea floor by each type of sediment: calcareous, 48.7%; lithogenous, 37.8%; siliceous, 13.5%.

**Table 13.5** The concentrations of some trace elements in deep-sea deposits* (units, $\mu g\,g^{-1}$).

| Trace element | Nearshore muds | Deep-sea carbonate | Atlantic deep-sea clay | Pacific deep-sea clay | Active ridge sediment | Ferromanganese nodules |
|---|---|---|---|---|---|---|
| Cr | 100 | 11 | 86 | 77 | 55 | 10 |
| V | 130 | 20 | 140 | 130 | 450 | 590 |
| Ga | 19 | 13 | 21 | 19 | — | 17 |
| Cu | 48 | 30 | 130 | 570 | 730 | 3300 |
| Ni | 55 | 30 | 79 | 293 | 430 | 5700 |
| Co | 13 | 7 | 38 | 116 | 105 | 3400 |
| Pb | 20 | 9 | 45 | 162 | — | 1500 |
| Zn | 95 | 35 | 130 | — | 380 | 3500 |
| Mn | 850 | 1000 | 4000 | 12500 | 60000 | 220000 |
| Fe | 69900 | 9000 | 82000 | 65000 | 180000 | 140580 |

* From Chester & Aston (1976).

greater extent in Pacific than in Atlantic deep-sea clays.

5 Ferromanganese nodules have particularly high concentrations of the elements that are enhanced in deep-sea clays ((3) above), but only small concentrations of the elements that are not enriched in DSC ((2) above).

6 Metalliferous, active ridge, sediments are enhanced in elements such as Fe, Mn, Cu, Zn, Ni and Co, relative to normal DSC.

7 Although it is not apparent from the data given in Table 13.5, some of the early surveys carried out on the distributions of trace elements in deep-sea sediments revealed that the excess trace elements reached their highest values in deposits that had accumulated at very slow rates in areas remote from the land masses.

These various trends offer a skeleton around which to build a discussion of the factors that combine together to control the elemental composition of marine sediments. Even at this stage a number of elemental fractionations can be identified in the marine sediment complex. For example, there is a major fractionation of some trace metals between nearshore muds and deep-sea clays. In addition, further fractionation stages occur within the various kinds of deep-sea deposits themselves, e.g. between the ridge-crest metalliferous deposits and the deep-sea clays. Thus, it begins to appear as if some kind of **sequential enhancement** is occurring for certain elements within marine sediments. The following chapters will be devoted to an attempt to understand how these elemental fractionations, and sequential enrichments, may have arisen. To provide a framework for this, further use will be made of the concept of chemical signals.

## 13.6 Chemical signals to marine sediments

Marine sediments may be thought of as having received a variety of chemical signals, or fluxes, which, in a number of combinations, have resulted in them acquiring their present composition. Two principal questions therefore must be asked in order to understand how the sediments attained this composition.

1 What is the chemical composition of the dissolved and particulate material carried by the signals themselves?

2 How can the effects of the individual signals be unscrambled in order to provide a reasonably coher-

ent explanation for the geochemical characteristics of individual sediments, or sediment suites?

One way of addressing these questions is to identify the individual components that combine together to form marine sediments, and then to establish whether or not the *processes* by which they are formed can be related to specific **individual chemical signals**, i.e. to adopt a process-orientated approach to the problem of describing the chemical compositions of the sediments. It is possible to classify the components of marine sediments into a series of genetically different types, and a number of schemes have been proposed for this purpose. The scheme adopted in the present volume is a modification of that outlined by Goldberg (1954), which classifies the components in terms of their geospheres of origin. In the original scheme the sediment components were subdivided into a single aqueous phase, i.e. **interstitial waters**, and four solid phases, which were classified according to the origin of their component elements as **lithogenous, biogenous, hydrogenous** and **cosmogenous**. In the modified scheme used here, however, the hydrogenous material will be subdivided into a number of different types (see Section 15.3.1).

## 13.7 Marine sediments: summary

1 A large part of the floor of the World Ocean is covered by a blanket of sediment, which has an average thickness of ~500 m.

2 Nearshore sediments are deposited on the shelf region, under a wide variety of depositional environments. Deep-sea sediments are deposited seaward of the shelf, under conditions of slow accumulation, and cover more than 50% of the surface of the Earth. The deep-sea sediments are subdivided into **hemi-pelagic** types, which are deposited in areas fringing the continents mainly via bottom transport mechanisms, and **pelagic** types, which are deposited in open-ocean areas mainly via down-column transport mechanisms.

3 Calcareous oozes cover large tracts of the open-ocean floor at water depths ≪3–4 km. Siliceous oozes from a ring around the high-latitude ocean margins.

4 The components of marine sediments can be subdivided into an aqueous phase (interstitial water) and four solid phases, which, on the basis of their geospheres of origin, are classed as lithogenous, hydrogenous, biogenous and cosmogenous.

5 Relative to nearshore muds, deep-sea clays contain enhanced concentrations of some trace elements, e.g. Mn, Cu, Ni, Co and Pb; these are often referred to as **excess** elements.

The solid sediment-forming components that make up marine sediments can be thought of as building blocks, which are stacked together in various proportions to form an individual sediment, or a suite of sediment types. However, at this stage an extremely important concept must be introduced; this is that sediments are *not* an inert reservoir, and as a result the building blocks are not simply stacked together in a way that retains their original compositions. Rather, they can be subjected to a series of *diagenetic* reactions following their deposition. Further, it is important to understand that these diagenetic reactions, which take place mainly via the medium of the interstitial waters, not only modify the compositions of pre-existing building blocks but also can supply elements that result in the formation of new blocks. In order, therefore, to be able to evaluate fully the processes involved in the formation of the sediment components, diagenesis will be described before the components themselves are considered. To do this, diagenetic reactions will be discussed in terms of the aqueous, i.e. the interstitial water, sediment phase. This will be followed by a description of the individual sediment-forming components themselves, and finally an attempt will be made to identify, and unscramble, the chemical signals that are transmitted to marine sediments.

## References

Bostrom, K., Kraemer, T. & Gartner, S. (1973) Provenance and accumulation rates of opaline silica, Al, Ti, Fe, Mn, Cu, Ni and Co in Pacific pelagic sediments. *Chem. Geol.*, **11**, 123–48.

Chester, R. & Aston, S.R. (1976) The geochemistry of deep-sea sediments. In *Chemical Oceanography*, J.P. Riley & R. Chester (eds), Vol. 6, 281–390. London: Academic Press.

Davies, T.A. & Gorsline, D.S. (1976) Oceanic sediments and sedimentary processes. In *Chemical Oceanography*, J.P. Riley & R. Chester (eds), Vol. 5, 1–80. London: Academic Press.

Davies, T.A. & Laughton, A.S. (1972) Sedimentary processes in the North Atlantic. In *Initial Reports of the Deep Sea Drilling Project*, Vol. 12, 905–34. Washington, DC: US Government Printing Office.

El Wakeel, S.K. & Riley, J.P. (1961) Chemical and mineralogical studies of deep-sea sediments. *Geochim. Cosmochim. Acta*, **25**, 110–46.

Goldberg, E.D. (1954) Marine geochemistry. Chemical scavengers of the sea. *J. Geol.*, **62**, 249–55.

Goldberg, E.D. & Griffin, J.J. (1964) Sedimentation rates and mineralogy in the South Atlantic. *J. Geophys. Res.*, **69**, 4293–309.

Goldberg, E.D. & Koide, M. (1962) Geochronological studies of deep-sea sediments by the ionium/thorium method. *Geochim. Cosmochim. Acta*, **26**, 417–50.

Goldberg, E.D., Koide, M., Griffin, J.J. & Peterson, M.N.A. (1963) A geochronological and sedimentary profile across the North Atlantic Ocean. In *Isotope and Cosmic Chemistry*, H. Craig, S.L. MIller & G.J. Wasserburg (eds), 211–32. Amsterdam: North-Holland.

Griffin, J.J., Windom, H. & Goldberg, E.D. (1968) The distribution of clay minerals in the World Ocean. *Deep Sea Res.*, **15**, 433–59.

Grousset, F.E. & Chesselet, R. (1986) The Holocene sedimentary regime in the northern Mid-Atlantic Ridge region. *Earth Planet. Sci. Lett.*, **78**, 271–87.

Heezen, B.C. & Hollister, C.D. (1971) *The Face of the Deep*. New York: Oxford University Press.

Heezen, B.C., Tharp, M. & Ewing, M. (1959) The floors of the oceans. *Geol. Soc. Am. Spec. Pap.*, **65**, 1–122.

Holcombe, T.L. (1977) Ocean bottom features—terminology and nomenclature. *Geojournal*, **6**, 25–48.

Jones, E.J.W. (1978) Sea-floor spreading and the evolution of the ocean basins. In *Chemical Oceanography*, J.P. Riley & R. Chester (eds), Vol. 7, 1–74. London: Academic Press.

Kennet, J.P. (1982) *Marine Geology*. Englewood Cliffs, NJ: Prentice Hall.

Ku, T.L., Broecker, W.S. & Opdyke, N. (1968) Comparison of sedimentation rates measured by paleomagnetic and ionium methods of age determination. *Earth Planet. Sci. Lett.*, **4**, 1–16.

Martin, J.-M. & Whitfield, M. (1983) The significance of the river input of chemical elements to the ocean. In *Trace Metals in Sea Water*, C.S. Wong, E.A. Boyle, K.W. Bruland, J.D. Burton & E.D. Goldberg (eds), 265–96. New York: Plenum.

Murphy, K.J.T. (1985) *The trace metal chemistry of the Atlantic aerosol*. PhD thesis, University of Liverpool.

Parsons, B. & Richter, F.M. (1981) Mantle convection and the oceanic lithosphere. In *The Sea*, Vol. 7, C. Emiliani (ed.), 73–117. New York: Interscience.

Riley, J.P. & Chester, R. (1971) *Introduction to Marine Chemistry*. London: Academic Press.

Turekian, K.K. & Wedepohl, K.H. (1961) Distribution of the elements in some major units of the Earth's crust. *Bull. Geol. Soc. Amer.*, **72**, 175–92.

Wedepohl, K.H. (1960) Spurenanalytische Untersuchungen an Tiefseetonen aus dem Atlantik. *Geochim. Cosmochim. Acta*, **18**, 200–31.

# 14 Sediment interstitial waters and diagenesis

Interstitial waters are aqueous solutions that occupy the pore spaces between particles in rocks and sediments. In some nearshore deposits groundwater seepages can occur, and around the ridge-crest areas circulating hydrothermal solutions (i.e. modified sea water) can enter the sediment column. For most marine sediments, however, the interstitial fluids originated as sea water trapped from the overlying water column. The interstitial-water–sediment complex is a site of intense chemical, physical and biological reactions, which can lead both to the formation of new and altered mineral phases and to changes in the composition of waters themselves. These changes may be grouped together under the term *diagenesis*, which has been defined by Berner (1980) as 'the sum total of processes that bring about changes in a sediment or sedimentary rock subsequent to its deposition in water'. Many of the important diagenetic changes that affect marine sediments take place during *early diagenesis*, which occurs during the burial of the deposits to a depth of a few hundred metres.

## 14.1 Early diagenesis: the diagenetic sequence and redox environments

### 14.1.1 Introduction

Many of the chemical changes that take place during early diagenesis are redox-mediated, i.e. they depend on the redox environment in the sediment–interstitial-water–seawater system. In turn, this redox environment is largely controlled by the degree to which organic carbon is preserved, or undergoes decomposition, in the sediment complex. Most marine environments are oxic, and as a result $\geq 90\%$ of the organic carbon that reaches the deep-sea floor via the vertical particle flux is oxidized close to the sediment–water interface. This process occurs largely via catabolic microbial reactions that are involved in

the breakdown of organic molecules to simple molecules or inorganic species. Berner (1980) has derived equations to describe diagenesis and these are given in Worksheet 14.4 in the context of the diagenesis of Mn in marine sediments.

### 14.1.2 The diagenetic sequence

Diagenetic processes in sediments are driven by redox reactions that are mediated by the decomposition of organic carbon, and some of the basic concepts involved in sedimentary redox processes are described in Worksheet 14.1.

It is now generally recognized that there is a diagenetic sequence of catabolic processes in sediments, the nature of which depends on the particular oxidizing agent that 'burns' the organic matter. As sedimentary organic matter is metabolized it donates electrons to several oxidized components in the interstitial-water–sediment complex, and when oxygen is present it is the preferred electron acceptor. During the diagenetic sequence, however, the terminal electron-accepting species alter as the oxidants are consumed in order of decreasing energy production per mole of organic carbon oxidized. Thus, as oxygen is exhausted, microbial organisms switch to a succession of alternative terminal electron acceptors in order of decreasing thermodynamic advantage (see e.g. Froelich *et al.*, 1979; Galoway & Bender, 1982; Wilson *et al.*, 1985). Using the schemes outlined by, among others, Froelich *et al.* (1979) and Berner (1980), the general **diagenetic sequence** in marine sediments can be outlined in the following general way.

*Aerobic metabolism.* Aerobic organisms can use **dissolved oxygen** from the overlying or interstitial waters to 'burn' organic matter. The organic matter that undergoes early diagenesis can be considered

**Worksheet 14.1: Redox reactions in sediments**

Aqueous solutions do not contain free protons and free electrons. However, according to Stumm & Morgan (1981) it is possible to define the relative proton and electron activities in these solutions. Acid–base processes involve the transfer of protons, and pH, which can be written

$$pH = -\log_{10} a_{H+} \qquad (1)$$

measures the relative tendency of a solution to accept or transfer protons. The activity of a hypothetical hydrogen ion is high at low pH and low at high pH, and pH is a master variable in acid–base equilibria.

In a similar manner, it is also possible to define a convenient parameter to describe redox intensity. Redox reactions involve the transfer of electrons, and $p\varepsilon$, which can be written

$$p\varepsilon = -\log_{10} a_{e-} \qquad (2)$$

measures the relative tendency of a solution to accept or transfer electrons. A high $p\varepsilon$ indicates a relatively high tendency for oxidation, and pe is a master variable in redox equilibria.

An oxidation–reduction reaction is termed a *redox* reaction, and can be written as two half-reactions in which a reduction is accompanied by an oxidation in terms of a redox couple. To illustrate this, Drever (1982) used the reduction of $Fe^{3+}$ by organic matter, represented by (C). Thus:

$$4Fe^{3+} + (C) + 2H_2O \rightarrow 4Fe^{2+} + CO_2 + 4H^+ \qquad (3)$$

In this equation neither molecular oxygen nor electrons are shown explicitly. The equation can be broken down into two half-reactions, one involving only Fe and the other only C. Thus:

$$4Fe^{3+} + 4e^- \rightarrow 4Fe^{2+} \qquad (4)$$

in which $Fe^{3+}$ undergoes reduction to $Fe^{2+}$, and

$$(C) + 2H_2O \rightarrow CO_2 + 4H^+ + 4e^- \qquad (5)$$

in which the organic matter undergoes oxidative destruction to yield $CO_2$. It must be remembered, however, that these do not represent complete chemical reactions because aqueous solutions do not contain free electrons.

The half-reaction concept can be related to measurements in electrochemical half-cells, and allows another parameter to be introduced into redox chemistry. This parameter is $E_h$, in which the electron activity is expressed in volts, the h subscript indicating that the $E_h$ value is expressed relative to the standard hydrogen electrode, which is used as a zero reference. The relative activity of electrons in a solution can therefore be expressed in units of electron activity ($p\varepsilon$), which is a

*continued*

dimensionless quantity, or in volts ($E_h$), and the relation between pe and $E_h$ is given by:

$$p\varepsilon = \frac{F}{2.3RT} E_h \tag{6}$$

where $F$ is Faraday's constant, $R$ is the gas constant and $T$ is the absolute temperature; at 25°C $E_h = 0.059p\varepsilon$. Electrode-measured $E_h$ values in oxidizing natural waters are difficult to relate to a specific redox pair, and both Stumm & Morgan (1981) and Drever (1982) have pointed out that it is important to distinguish between electrode-measured $E_h$ and $E_h$ calculated from the activities of a redox pair.

In the present text the general concept of *redox conditions* will be used, in which positive $E_h$ (redox potential) values indicate oxidizing conditions and negative values indicate reducing conditions; i.e. half-reactions of high $E_h$ are oxidizing, and those of low $E_h$ are reducing. Thus, a half-reaction with a lower $E_h$ will undergo oxidation when combined with a half-reaction of higher $E_h$. This reaction combination can be used to describe redox-mediated diagenetic reactions in sediments.

$E_h$ conditions in sediments are controlled mainly by the decomposition of photosynthetically produced organic matter by non-photosynthetic bacteria, and are constrained by the rate of supply of the organic matter (primary production) and the rate at which it accumulates (sedimentation rate). This bacterial decomposition of organic matter is driven by a sequence of reactions that switch to a successive series of oxidants, or electron acceptors, which represent lower $p\varepsilon$ levels. During the reaction sequence, in which the organic matter is decomposed by micro-organisms, the organisms acquire energy for their metabolic requirements.

Only a relatively few elements (C, N, O, S, Fe and Mn) are predominant participants in aquatic redox processes. The overall relationships that involve these elements in the microbially mediated redox sequence have been summarized diagrammatically by Stumm & Morgan (1981); their scheme is reproduced in Fig. (i) in which the energy yields associated with the various processes in the diagenetic sequence are given in the form of reaction combinations that are initiated at various $E_h$ and $p\varepsilon$ values. For example, the first stage in the sequence involves the oxidation of organic matter by dissolved oxygen (A + L), with successive reactions following the decreased $p\varepsilon$ and $E_h$ levels. The full 'diagenetic sequence', and the sedimentary environments associated with the various stages in the sequence, are discussed in detail in the text. Examples of the diagenetic succession are given in the box, in Fig. (i), from which it can be seen, for example, that there is a tendency for the more energy-yielding reactions to take precedence over those that are less energy-yielding. Thus, the sequence begins with aerobic respiration (A + L), followed by denitrification (B + L), etc.

*continued on p. 360*

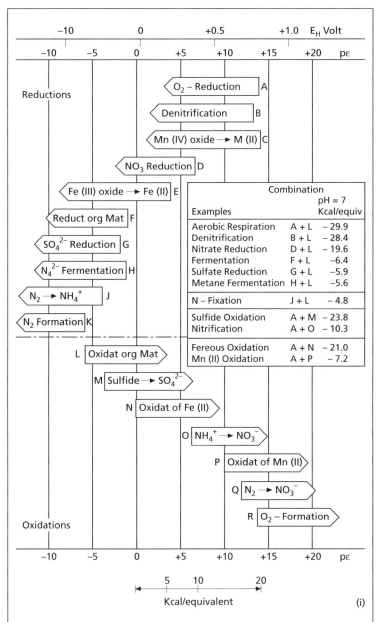

**Fig. (i)** The microbially mediated diagenetic sequence in sediments (from Stumm & Morgan, 1981).

to have the Redfield composition of $(CH_2O)_{106}(NH_3)_{16}(H_3PO_4)$ (see Section 9.2.3.1). The oxidation of organic matter by aerobic organisms therefore can be represented by a general equation such as that proposed by Galoway & Bender (1982):

$$5(CH_2O)_{106}(HN_3)_{16}(H_3PO_4) + 690O_2 \rightarrow$$
$$530CO_2 + 80HNO_3 + 5H_3PO_4 + 610H_2O \quad (14.1)$$

The $CO_2$ released during this reaction can lead to carbonate dissolution, and the ammonia can be oxidized to nitrate, a process termed **nitrification**. Under oxic conditions most of the remains of dead animals and plankton are apparently destroyed at this stage in the diagenetic sequence. For example, according to Bender & Heggie (1984), >90% of the organic carbon that reaches the deep-sea floor is oxidized by $O_2$. Oxygen therefore may be regarded as the *primary* oxidant involved in the destruction of organic matter, and in a closed system reaction (14.1) (**oxic diagenesis**) will continue until sufficient oxygen has been consumed to drive the redox potential low enough to favour the next most efficient oxidant. Thus, as dissolved oxygen becomes depleted, organic matter decomposition can continue using oxygen from *secondary* oxidant sources (**suboxic diagenesis**).

*Anaerobic metabolism.* Anaerobic metabolism takes over when the content of dissolved oxygen falls to very low levels, or becomes entirely exhausted, and a series of secondary oxidants are utilized. These secondary oxidants include nitrate, $MnO_2$, $Fe_2O_3$ and sulphate.

**1 Nitrate.** According to Berner (1980), when the dissolved oxygen levels fall to ~5% of their concentration in aerated waters the decomposition of organic matter can occur using oxygen from nitrate, a reaction that can be represented as follows:

$$5(CH_2O)_{106}(NH_3)_{16}(H_3PO_4) + 472HNO_3 \rightarrow$$
$$276N_2 + 520CO_2 + 5H_3PO_4 + 886H_2O \qquad (14.2a)$$

This process is termed **dentrification**. In the reaction given above it is assumed that all organic nitrogen released is in the form of ammonia, which is then oxidized to molecular nitrogen by the reaction:

$$5NH_3 + 3HNO_3 \rightarrow 4N_2 + 9H_2O \qquad (14.2b)$$

However, this is not the only possible pathway, and Froelich *et al.* (1979) have pointed out that the fate of the nitrogen has important consequences in diagenesis with respect to the sequence in which the secondary oxidants are used. These authors suggested that if all the nitrogen goes to $N_2$ then the use of nitrate as a secondary oxidant overlaps with that of $MnO_2$, but that if the nitrogen is released as ammonia and is not oxidized to $N_2$ then $MnO_2$ is apparently reduced before nitrate.

The use of the other secondary oxidants can be illustrated by reactions of the type described by Froelich *et al.* (1979).

**2 Manganese oxides**

$$(CH_2O)_{106}(NH_3)_{16}(H_3PO_4) + 236MnO_2 + 472H^+$$
$$\rightarrow 236Mn^{2+} + 106CO_2 + 8N_2 + H_3PO_4 + 336H_2O$$
$$(14.3)$$

**3 Iron oxides**

$$(CH_2O)_{106}(NH_3)_{16}(H_3PO_4) + 212Fe_2O_3 + 848H^+$$
$$\rightarrow 424Fe^{2+} + 106CO_2 + 16NH_3 + H_3PO_4 + 530H_2$$
$$(14.4)$$

**4 Sulphate**

$$(CH_2O)_{106}(NH_3)_{16}(H_3PO_4) + 55SO_4 \rightarrow$$
$$106CO_2 + 16NH_3 + 55S^{2-} + H_3PO_4 + 106H_2O$$
$$(14.5)$$

**5** Following sulphate reduction, biogenic methane can be formed by two possible reaction pathways, which may be generalized as:

$$CH_3COOH \rightarrow CH_4 + CO_2 \quad \text{acetic acid formation}$$
$$(14.6a)$$

or

$$CO_2 + 8H_2 \rightarrow CH_4 + 2H_2O \quad CO_2 \text{ reduction} \quad (14.6b)$$

Thus, diagenesis proceeds in a general sequence in which the oxidants are utilized in the order: oxygen > nitrate ≥ manganese oxides > iron oxides > sulphate. In this diagenetic sequence it is assumed that in marine sediments $O_2$, $NO_3^-$, $MnO_2$, $Fe_2O_3$ (or FeOOH) and $SO_4^{2-}$ are the only electron acceptors, and that organic matter (represented by the Redfield composition) is the only electron donor. Furthermore, it is assumed that the oxidants are limiting, i.e. each reaction proceeds to completion before the next one starts. However, the diagenetic processes are not always sequential; for example, although it is usually thought that sulphate reduction precedes methane formation, Oremland & Taylor (1978) have suggested that the two processes can occur simultaneously, i.e. they are not mutually exclusive.

### 14.1.3 Diagenetic and redox environments

As diagenesis proceeds a number of end-member sedimentary environments are set up, which Berner (1981) was able to relate to a **diagenetic zone** sequence.

**1 Oxic** environments are those in which the interstitial waters of the sediments contain measurable dissolved oxygen, and diagenesis occurs via aerobic metabolism. Under these conditions little organic matter is preserved in the sediments, and in terms of the reactions given above the diagenetic sequence has proceeded only to reaction (14.1) in oxic environments.

**2 Anoxic** environments are those in which the sediment interstitial waters contain no measurable dissolved oxygen, i.e. diagenesis here has to proceed via the secondary oxidants through anaerobic metabolism. The anoxic environments were subdivided into a number of types.

**3 Non-sulphidic post-oxic** environments. These environments, which contain no measurable dissolved sulphides, are common in many deep-sea sediments, and are perhaps more often referred to in the literature as **suboxic** environments. The condition necessary to set up this type of sedimentary environment is a supply of organic carbon sufficient that diagenesis can proceed beyond the oxic stage. Under these conditions, nitrate, manganese oxides and iron oxides are used as secondary oxidants, but the sequence does not reach the stage at which sulphate is utilized for this purpose. In suboxic sediments, therefore, there is a relatively large, but still limited, supply of metabolizable organic matter, and the diagenetic sequence has proceeded to reactions (14.2)–(14.4) given above.

**4 Sulphidic** environments. These result when the diagenetic sequence has reached the stage at which the bacterial reduction of dissolved sulphate takes place with the production of $H_2S$ and $HS^-$. If a sufficient supply of metabolizable organic matter is available, sulphate reduction can be a common feature in marine sediments as a result of the relatively high concentration of sulphate in both sea water and marine sediment interstitial waters. In practice, however, constraints on the supply and preservation of organic matter mean that sulphate reduction is largely restricted to nearshore sediments. In sulphate

environments the diagenetic sequence has now proceeded to reaction (14.5).

**5 Non-sulphidic methanic** environments. In some sediments that contain a relatively large amount of metabolizable organic matter the diagenetic reactions can pass through the stage at which oxygen, nitrate, manganese oxides, iron oxides and sulphate are sequentially utilized. Continued decomposition of organic matter results in the formation of dissolved methane, e.g. by reactions (14.6a) and (14.6b)*.

The general diagenetic sequence outlined earlier, in which the various oxidants are consumed in the order oxygen > nitrate ≥ manganese oxides > iron oxides > sulphate leads to the setting up of a series of diagenetic zones in sediments. These zones give rise to an environmental succession in the processes of organic matter decomposition, which involves oxygen consumption (respiration), nitrate (and/or manganese and iron) reduction, sulphate reduction and methane formation, and depending on the amount of available organic matter, any sediment can pass through each of these zones during deposition and burial. Thus a **vertical** diagenetic zone sequence can be set-up in a sediment.

Truly anoxic waters, where sediments are *initially* deposited under anoxic conditions, prevail over only a small area of the oceans (see Section 8.3). The vast majority of environments at the sea floor are therefore oxidizing, and there is usually a layer of oxic material at the sediment surface. As a result of the consumption of dissolved oxygen in the interstitial waters, however, the sediments can become reducing, and ultimately anoxic, at depth as the diagenetic sequence proceeds. The depth at which the oxic–anoxic change occurs depends largely on a combination of the magnitude of the down-column carbon flux, by which the carbon is *supplied*, and the sediment accumulation rate, by which it is *buried*. For example, according to Muller & Mangini (1980) a bulk sedimentation rate of ≤1–4 cm $10^3$ yr$^{-1}$ is necessary for the deposition of an oxygenated sedimentary column. Thus, the thickness of the surface sediment oxic layer will tend to increase from nearshore to pelagic regions as the accumulation rate decreases. An example of how the thickness of the oxic surface layer in deep-sea sediments varies has been provided by Lyle (1983) for a series of hemi-pelagic deposits from the eastern Pacific. The **redox boundary**, which is indicative of the change between oxidizing (oxic–positive redox potential) and reducing

---

* Methane bubbles, generated by biogeochemical processes, are common in organic-rich, muddy near-shore sediments and can escape into the overlying waters from the sea bed. For a description of these 'gassy' sediments, the reader is referred to 'Modelling Gassy Sediment Structure and Behaviour' (*Cont. Shelf Res.*, **18**, 1998; guest eds M.D. Richardson & A.M. Davies).

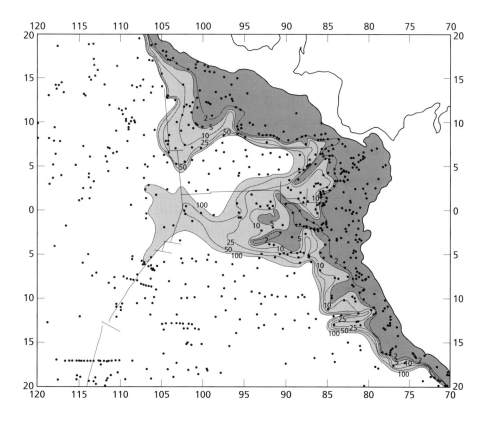

**Fig. 14.1** Variations in the thickness (in cm) of the surface oxic layer in sediments from the eastern Pacific (from Lyle, 1983).

(anoxic–negative redox potential) conditions in sediments is often accompanied by a colour change, which generally is from red-brown (oxidized) to grey-green (reduced). The depth of the colour transition in the sediments from the eastern tropical Pacific is illustrated in Fig. 14.1, and Lyle (1983) was able to identify a number of general features in the distribution of the brown oxic surface layer that can be related qualitatively to primary productivity in the overlying euphotic layer. For example, a very thin brown layer (<2 cm) is found: (i) on the continental margin, where upwelling results in a high surface productivity; and (ii) in a westward-extending tongue around the Equator, which lies below the zone of equatorial upwelling. In contrast, the tongue of sediment having a brown layer >1 m in thickness at ~5°N results from the transport of low productive surface water into the area by the North Equatorial Countercurrent.

It is apparent, therefore, that there are a range of redox environments in marine sediments, which can

be expressed, on the basis of the increasing thickness of the surface oxic layer, in the following sequence.

**1 Anoxic sediments.** These are usually found in coastal areas, or in isolated basins and deep-sea trenches. They have organic carbon contents in the range ~5–≥10%, and are reducing throughout the sediment column if the redox boundary is found in the overlying waters.

**2 Nearshore sediments.** These sediments, which usually have organic carbon contents of ≤5%, accumulate at a relatively fast rate and become anoxic at shallow depths so that the brown oxic layer is usually no more than a few centimetres in thickness.

**3 Hemi-pelagic sediments.** These have intermediate sedimentation rates, and organic carbon contents that are typically around 2%. The thickness of the oxic layer in these deposits ranges from a few centimetres up to around a metre.

**4 Pelagic sediments.** These are deposited at very slow rates and have organic carbon contents that usually are only ~0.1–0.2%. In these sediments the oxic layer extends to depths well below 1 m, and often to several tens of metres.

It was pointed out above that early diagenesis in marine sediments follows a **vertical** zone sequence. It is also apparent that there is a **lateral**, i.e. nearshore → hemi-pelagic → pelagic, diagenetic zone sequence in the oceanic environment.

The concentration of organic matter in a sediment is a critical parameter in determining how far the diagenetic sequence progresses, and the factors controlling the distribution, protection and preservation of organic matter in marine sediments are discussed in the following sections.

## 14.2 Organic matter in marine sediments

### 14.2.1 Introduction

The organic matter in marine sediments is important, not only because the sediments provide a significant reservoir in the global carbon cycle, but also because organic matter drives early diagenesis and thus plays a major role in the chemistry of the oceans.

### 14.2.2 The sources and distribution of organic matter in marine sediments

The organic matter in marine sediments is derived from terrestrial, marine and anthropogenic sources. Hedges & Keil (1995) re-evaluated the evidence relating to the distribution of organic carbon in marine sediments, and their revised budget is given in Table 14.1. The total burial rate of organic carbon is $\sim 0.16 \times 10^{15}\,g\,C\,yr^{-1}$, of which $\sim 90\%$ is deposited in deltas and the continental shelves and upper slopes. It has been estimated that $\sim 1\%$ of terrestrial productivity ($\sim 60 \times 10^{15}\,g\,C\,yr^{-1}$) is transported to the oceans by rivers. This yields a total fluvial flux of $\sim 0.4 \times 10^{15}\,g$ C yr$^{-1}$, which is made up of about equal proportions of DOM and POM (see Table 9.3). Because the aeolian contribution is relatively small (less than $\sim 0.1 \times 10^{15}\,g\,C\,yr^{-1}$), rivers are the main **external**, allochthonous, source for the transport of organic carbon to the oceans, and much of the organic matter that accumulates in estuarine and coastal sediments is land-derived (e.g. from higher plants). However, atmospheric transport is also thought to provide an important mechanism for the transport of terrestrial organic matter to deep-sea regions. For example, from data obtained in the SEAREX Programme, Gagosian (1986) was able to show that at remote North and South Pacific sites the hydrocarbons in aerosols had a clear terrestrial (plant wax) signature, indicating long-range transport from the land masses. On the basis of flux data, Gagosian *et al.* (1987) concluded that atmospheric transport could have a major impact on the terrestrially derived lipid material found in deep-sea sediments.

Primary production, largely by phytoplankton in the open-sea, is the principal **internal**, autochthonous, source of organic carbon in the oceans, and has been estimated to generate $\sim 50 \times 10^{15}\,g\,C\,yr^{-1}$. Estimates of the amount of this primary production that reaches the ocean floor vary (see Section 9.3) but if $\sim 1\%$ is assumed to reach the sea bed at depths of water in excess of 4000 m, this would give a flux of $\sim 0.5 \times 10^{15}\,g\,C\,yr^{-1}$, which is comparable with the fluvial flux of $\sim 0.4 \times 10^{15}\,g\,C\,yr^{-1}$; i.e. the flux of plankton remains to the deep ocean is comparable to the export rate of organic carbon from land to sea.

Various **source markers** have been used to distinguish marine from terrestrial inputs; these include *n*-alkanes, polycyclic hydrocarbons, *n*-alcohols, sterols, diterpenoids, lignins and bulk $^{13}C/^{12}C$ ratios.

| Sediment type | Deltaic | Shelf | Slope | Pelagic | Total |
|---|---|---|---|---|---|
| Deltaic sediments | 70(44) | 0 | 0 | 0 | 70 |
| Shelves and upper slope | 0 | 68(42) | 0 | 0 | 68 |
| Biogenous sediments; high productivity zones | 0 | 0 | 7(4) | 3(2) | 10 |
| Shallow water carbonates | 0 | 6(4) | 0 | 0 | 6 |
| Pelagic sediments; low productivity zones | 0 | 0 | 0 | 5(3) | 5 |
| Anoxic basins | 0 | 1(0.5) | | 0 | 1 |
| Total organic carbon burial | | | | | 160 |

**Table 14.1** Organic carbon burial rates, together with the percentages buried, in marine sediments deposited under different environmental regimes; units, $10^{12}\,g\,C\,yr^{-1}$; units in parentheses, per cent of total burial (data from Hedges & Keil, 1995; recalculated from Berner, 1989).

In addition to the sources identified above, both petroleum and natural gas can reach the marine environment by seepages.

The extent to which organic matter is preserved in a sediment is critical in determining how far the diagenetic sequence progresses. There are considerable variations in the organic matter content of marine sediments. On the basis of the information summarized above, however, it is apparent that two major marine sedimentary organic matter reservoirs can be identified, and it is important to distinguish between them because they have very different preservation characteristics.

*Nearshore sediments.* Fluvial inputs are delivered initially to nearshore regions, which are also the sites of much of the oceanic primary productivity. Nearshore deposits, which accumulate at relatively fast rates, usually contain ~1–5% organic carbon, but the concentrations can be considerably higher in sediments deposited in some anoxic basins and under areas of high primary production; for example, Calvert & Price (1970) reported that organic-rich diatomaceous muds on the Namibian shelf contained up to ~25% organic carbon.

*Deep-sea sediments.* Heath *et al.* (1977) carried out a survey of the distribution of organic carbon in deep-sea sediments, and showed that its concentrations vary from ~5% in reduced hemi-pelagic sediments deposited close to the continental margins to ≪1% in most oxic pelagic clays. The authors reported that there is a general correlation between the organic carbon content and the accumulation rate of a sediment, and a regression line obtained from a log–log plot of organic carbon accumulation rate versus sedimentation rate yielded the relationship (organic carbon accumulation rate) = 0.01 (sedimentation rate).

Sedimentation rate is an environmental parameter often thought to have an effect on the reactivity and preservation of organic matter because bacterial activities are most active close to the sediment surface and fast deposition removes the organic matter down through this diagenetically active zone. Such 'fast' burial also helps to 'cap' the accumulating sediments from the input of dissolved oxidizing agents, such as oxygen, nitrate and sulphate. Together, these various factors lead to the concept of 'greater organic matter

preservation with faster burial'. However, variations of percentage organic matter with sediment accumulation are subject to 'dilution effects' by coarse minerals, and to overcome this it has been the practice to use organic matter **burial efficiency** as a preservation indicator. The burial efficiency is defined as the accumulation of organic matter below the diagenetically active surface sediment divided by the organic flux to the sea floor. Burial efficiency rates vary from <1% in deep-sea sediments to ~80% for some shallow-water deposits and correlate directly with sedimentation rate. None the less, burial rates are only mechanistically meaningful if the preserved organic matter is deposited fresh for the first time, i.e. from primary production in the water column. The **exclusive** deposition of fresh plankton remains to nearshore sediments, where most organic matter preservation occurs, is unlikely because these sediments will contain an appreciable fraction of at least partially degraded fluvial organic matter. Further, reworked sediments are more common in the coastal zone and should contain a large component of old reworked organic matter, including planktonic remains, that may have suffered numerous 'deposition–resuspension' cycles.

### 14.2.3 The classification of organic matter in marine sediments

The organic matter in marine sediments is composed of a wide range of more than 1000 different organic molecules (Degens & Mopper, 1976). According to Simoniet (1978) the classes of compounds identified include: hydrocarbons, e.g. alkanes, cycloalkanes and iso-prenoids; fatty acids; fatty alcohols, ketones and wax esters; steroids; triterpenoids; tetraterpenoids; pigments; amino acids and peptides; purines and pyrimidines; carbohydrates (sugars); aromatic hydrocarbons, e.g. polynuclear aromatic hydrocarbons (PAH); natural polymers, e.g. chitin, cellulose, cutin and lignin; branched and cyclic hydrocarbons of the 'hump'; and kerogen and humates.

The organic matter exists in a number of forms, which according to Ertel & Hedges (1985) include: (i) discrete organic particles, e.g. vascular plant debris and planktonic tissues; (ii) surface films on inorganic substrates, e.g. humic-clay complexes; and (iii) integral components of inorganic matrices, e.g. kerogen. However, ~90% of the organic matter cannot be

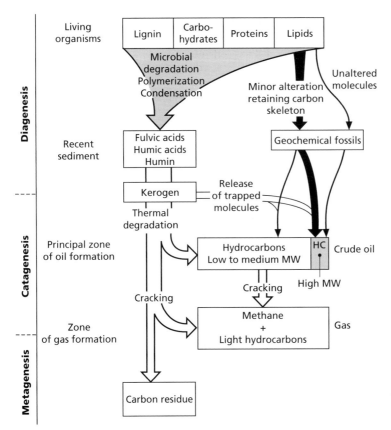

**Fig. 14.2** Stages in the diagenesis, catagenesis and metagenesis of organic matter in sediments (from Tissot & Welte, 1984).

physically separated from its mineral matrix (see Section 14.3.5). A number of organic compounds are also found in the interstitial waters of marine sediments. Thus, a variety of organic matter products originating in the biosphere cross the sediment–seawater interface to enter the geosphere, and once there they undergo a complex series of reactions in a sequence that follows the depth of burial of the sediment (see Fig. 14.2). The full sequence is in the order: diagenesis (down to ~1000 m) → catagenesis (several kilometres depth, with an increase in temperature and pressure) → metagenesis or metamorphism (Tissot & Welte, 1984).

### 14.2.4 The overall behaviour of organic matter during diagenesis

The organic matter in the upper portions of marine sediments can be divided into two general categories:
**1 hydrolysable** components (in this context,

'hydrolysable' material is that which can be converted by hydrolysis into water-soluble substances);
**2 non-hydrolysable** material, which makes up the bulk of the organic matter and is composed of 'humic-like' components.

The formation of these two categories of organic material can be described in terms of a series of transformation reactions, which act upon the organic components. The organic material reaching the surface of marine sediments is composed largely of biogenic macromolecules, such as proteins, carbohydrates, lipids (including hydrocarbons) and lignins, together with various uncharacterized substances (see Sections 9.2.3 & 12.1.3). According to Tissot & Welte (1984), the most striking difference between the compositions of the organic matter in living organisms and that in surface, and near-surface, sediments is that a large proportion of the biogenic macromolecules have been lost from the sedimentary deposits. This is largely the result of bacterial activity, during which the biogenic polymers (**biopolymers**), such as proteins and carbohydrates, undergo destruction. During this process water-soluble complexes containing amino acids and sugars are formed, which, together with hydrocarbons and fatty acids, are found in the upper portions of marine sediments. It is these components that comprise the free or hydrolysable content of the sediments. Most of these hydrolysable compounds are destroyed or modified at a shallow depth, especially in oxic environments, and the non-degradable residues become part of polycondensed structures (**geopolymers**), such as humic and fulvic acids, which subsequently undergo insolubilization to form 'humin'. A significant fraction of this humin is hydrolysable in young sediments, but with burial this decreases and eventually kerogen is formed. The term **kerogen** refers to the condensed macromolecular fraction of the organic matter in sediments that is insoluble in organic solvents. Kerogen can originate from either marine (plankton) or terrestrial (higher plant waxes) sources, or from a combination of both, and three different types of kerogen have been identified.

The general reaction pathway for the modification of organic matter during the complete diagenetic stage in marine sediments therefore is to convert *biopolymers* into *geopolymers*. In this context, therefore, the overall sequence for the full diagenesis of organic matter in sediments can be written: degraded cellular material → water-soluble complexes containing amino acids, lipids and carbohydrates → fulvic acids → humic acids → kerogen (see e.g. Nissenbaum & Kaplan, 1972). Under some extremely reducing conditions, however, the sequence can bypass the fulvic and humic acid stages. Some of the humic material in marine sediments can have a terrestrial source, although this type appears to be restricted largely to estuarine and coastal sea deposits.

The manner in which the sequence operates up to the stage at which fulvic and humic acids are formed can be illustrated with respect to the diagenesis of amino acids in the upper sections of marine sediments. Various lines of evidence suggest that the concentrations of amino acids in marine sediments decrease with depth. For example, Burdige & Martens (1988) gave data on the concentrations of total hydrolysable amino acids (THAA) in anoxic rapidly accumulating sediments in an organic-rich coastal marine basin (Cape Lookout Bight, USA), and

showed that they exhibited an exponential decrease with depth similar to, but at a faster rate than, that for total organic carbon. The most abundant amino acids in the sediments were aspartic acid, glutamic acid, glycine and alanine, which is a similar abundance to that in the two major sources of organic matter to the region, i.e. vascular saltmarsh plants (terrestrial) and plankton (marine). Kinetic modelling of the Cape Lookout Bight data indicated that *c*. 45% of the input of amino acids to the surface sediment is remineralized in the upper *c*. 40 cm, which amounted to ~27% of the regeneration of total organic carbon.

On the basis of their data, Burdige & Martens (1988) proposed a general model to describe the major processes involved in the early diagenesis of amino acids in *anoxic* sediments. The principal features in this model, which is illustrated in Fig. 14.3, can be summarized as follows.

1 Amino acids initially deposited in marine sediments usually can be divided into two fractions, a labile (or metabolizable) fraction, contained in proteinaceous material, and a refractory fraction, which is degraded at rates significantly slower than those at which the labile amino acids are remineralized.

2 The decrease in the concentrations of THAA with depth is assumed to be related to the microbial utilization of the labile amino acid fraction, to produce pore-water dissolved free and dissolved combined amino acids.

3 Individual dissolved free amino acids are metabolized to yield ammonium, methane and/or $\Sigma CO_2$ by sulphate reducers, methanogens and/or fermentative bacteria.

4 Although the pore-water amino acids may act as intermediates in the remineralization of hydrolysable amino acids in anoxic sediments, they also can be involved in non-biological reactions (e.g. adsorption), which may result in the reincorporation of the amino acids back into the sediments as humic and/or fulvic acids, i.e. the process of geopolymerization (or humification) identified above.

Conditions can exist, however, under which the hydrolysable organic fraction does not decrease significantly with depth in the sediments. For example, Steinberg *et al.* (1987) determined amino acids and carbohydrates in recent sediments in a 'mud patch' region on the continental margin off the coast of southern New England (USA) as part of the Shelf Edge Exchange Program (SEEP)—for a full descrip-

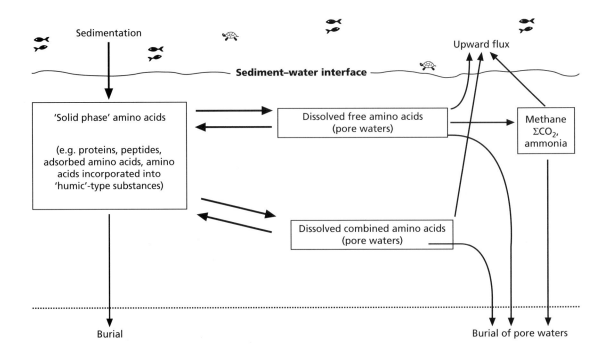

**Fig. 14.3** A model for the diagenesis of amino acids in anoxic marine sediments (from Burdige & Martens, 1988).

tion of the various aspects of this programme see *Continental Shelf Research*, Vol. 8, 1988. An example of the amino acids found in the upper portion of one of the SEEP sediments is illustrated in Fig. 14.4, showing the dominance of glycine in the sample. The total amino acid composition of the five 'mud patch' samples was similar, and there was very little decrease in their concentrations with depth in the cores. The authors suggested that this lack of a depth gradient could arise because only refractory material has reached the sediment and/or because bioturbation and resuspension have homogenized the upper sections of the cores.

The overall effect of diagenesis in many marine sediments is to transform protein, carbohydrate and lignin biopolymers into geopolymers, the end-product being the formation of kerogen. However, in addition to kerogen, sediments contain an organic matter fraction that has suffered only minor degradation. This fraction originates from high-molecular-weight lipids, including hydrocarbons, and because

these molecules retain their carbon skeleton they are often termed **geochemical fossils** or **biological markers**. The *n*-alkanes are among the least reactive of the organic compounds, and will serve as an example of a geochemical fossil.

The *n*-alkanes make up the major fraction of the lipid hydrocarbons in surficial sediments, and they can have a number of origins. These include (i) direct inputs from terrestrial, anthropogenic or marine sources, (ii) diagenetic formation (e.g. from the post-depositional reduction of fatty acids), and (iii) migration from deep sources. There are a number of ways in which the sources of the *n*-alkanes can be evaluated. The *n*-alkane distribution pattern in immature marine sediments is commonly characterized by the predominance of odd-carbon-number, high-molecular-weight homologues ($C_{25}$–$C_{33}$), which are typical of higher plant waxes, over light alkanes in the range $C_{15}$–$C_{23}$, which are typical of marine autochthonous production, thus indicating a significant terrestrial contribution. In addition, Grimalt & Albaiges (1987) have shown that the occurrence of $C_{12}$–$C_{22}$ homologues, with a strong even-carbon-number preference, is widespread in freshwater and marine sediments formed under both oxic and anoxic conditions.

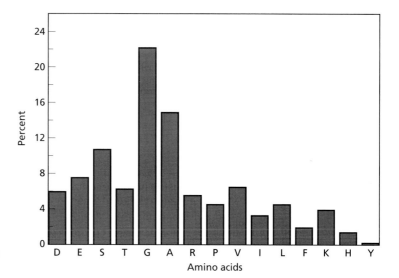

**Fig. 14.4** The amino acid composition of a surface sediment from the continental margin off southern New England, USA (from Steinberg *et al.*, 1987). D, aspartic acid; E, glutamic acid; S, serine; T, threonine; G, glycine; A, alanine; R, arginine; P, proline; V, valine; L, leucine; I, isoleucine; F, phenylalanine; K, lysine; H, histidine; Y, tyrosine. Note the dominance of glycine in the sample.

The authors concluded that these *n*-alkanes are most likely to have originated from a variety of autochthonous biological sources (e.g. micro-organisms), rather than from diagenetic processes.

According to Simoniet (1978), marine sediments found close to the continents contain *n*-alkanes from both terrestrial and marine sources (with terrestrial components being dominant where the continental input is strong, and marine components dominating under areas of intense primary production), whereas sediments from deep-sea regions contain *n*-alkanes derived predominantly from marine sources.

It may be concluded, therefore, that the diagenetic '*preservation–destruction*' of organic matter in marine sediments proceeds along two general pathways, which lead to the formation of two organic fractions:

1 protein, carbohydrate and lignin biopolymers are transformed into geopolymers, the end-product being the formation of the kerogen fraction;
2 a fraction that originates from high-molecular-weight lipids, including hydrocarbons, which are retained in the sediment with only minor modification.

However, the distinction between two such fractions in sediments is far from clear, and it is apparent that, for example, lipids are incorporated into kerogen at an early stage of diagenesis (G. Wolff, personal communication).

## 14.3 Early diagenesis in marine sediments

### 14.3.1 Introduction

It was pointed out in Section 14.1.3 that on the basis of the premise that the total metabolic activity within the upper layers of marine sediments depends on the supply of organic carbon and the rate at which it is preserved, there is a *lateral*, i.e. a nearshore → hemipelagic → pelagic, diagenetic zone sequence. Diagenesis in the various sedimentary environments found in this lateral sequence is discussed below.

### 14.3.2 Diagenesis in nearshore sediments

The vast majority of the organic matter in oceanic sediments is located in (i) deltaic, and (ii) shelf and upper slope sediments, which contain 44% and 42%, respectively, of the total organic matter in marine deposits (see Table 14.1). In terms of the lateral sequence, the extreme nearshore diagenetic environment is represented by the anoxic sediment end-member. Although anoxic waters are rare, they are a very special diagenetic environment because the sediments associated with them are actually deposited under anoxic conditions, rather than becoming anoxic at shallow depths in the sediment column itself, and the oxic–anoxic boundary can occur in the water column above the sediment surface (e.g. in parts of the Black Sea), which sets up a specialized water chemistry. Anoxic conditions can also develop in deep-sea trenches and fjords where the water circulation is restricted. In the sediments deposited in these anoxic environments diagenesis can reach the methanic stage. For example, methane, which is a terminal product of the anaerobic degradation of organic matter, can be generated in anoxic sediments found in both deep-sea trenches (e.g. the Cariaco Trench) and in nearshore regions (e.g. in the Sannich Inlet, British Columbia).

In other types of nearshore sediments that are not deposited under an anoxic, or an intermittently anoxic, water column, the extent to which the reactions involved in early diagenesis proceed is dependent on local environmental conditions. For example, diagenesis can be affected by factors such as bioturbation and bottom stirring, which tend to be unimportant in anoxic regions, where there is usually an absence of both benthic fauna and erosive currents.

A characteristic feature of many nearshore muds is that the surface oxic layer is very thin, and in the anoxic sediments below diagenesis often progresses to the sulphate reduction stage. However, there is something of a paradox here. The nearshore environment is the receiving region for fluvial inputs of continental organic matter, and is also the region where most primary production occurs; hence the nearshore environment has a large supply of organic matter. Nearshore sediments are deposited at relatively fast rates, which results in the rapid burial of organic matter below the thin surface oxic layer. Within nearshore sediments diagenesis can proceed to the sulphate reduction stage, and yet nearshore deposits store relatively large quantities of organic matter. It is almost as if some protective mechanism operates in the shelf and upper slope sediments to preserve organic matter, and we shall return to this paradox later.

Both anoxic and normal nearshore sediments contain a wide variety of authigenic minerals. These include: iron sulphides, such as the polysulphide pyrite ($FeS_2$) and its diamorph marcasite ($FeS_2$); the metastable iron sulphides mackenawite and greigite; glauconite; chamosite; various carbonates (including those of iron and manganese); and iron and manganese oxides. Manganese minerals, e.g. mixed manganese carbonates, manganese sulphide (MnS) and manganese phosphate ($Mn_3(PO_4)_2$), are sometimes prominent among the authigenic phases. For a detailed description of these minerals the reader is referred to Calvert (1976).

### 14.3.3 Diagenesis in hemi-pelagic sediments

As the thickness of the surface oxic layer increases out into open-ocean areas the extent to which the diagenetic sequence progresses decreases, and this can be traced through hemi-pelagic (oxic–suboxic) to truly pelagic (oxic) environments. Froelich *et al.* (1979) made a study of early diagenesis in suboxic hemi-pelagic sediments from the eastern equatorial Atlantic. The cores investigated typically had a light tan-coloured surface layer $\sim 35\,cm$ in thickness, with a low organic carbon content (0.2–0.5%), underlain by a dark olive green terrigenous sediment, which had a higher organic carbon content ($\sim 0.5$ to >1%). The authors gave data on a number of constituents in the interstitial waters of the sediments, and their findings can be summarized as follows.

1 Dissolved nitrate concentrations increased from those of the ambient bottom water to a maximum, then decreased linearly to approach zero at approximately the depth of the tan–olive green lithological transition.

2 Dissolved $Mn^{2+}$ concentrations were very low at the surface but began to increase at a depth lying between the nitrate maximum and the nitrate zero.

3 Dissolved $Fe^{2+}$ concentrations were under the detection limit to a depth below the nitrate zero, then began to increase.

4 Sulphate concentrations never differed detectably from those of the ambient bottom water, i.e. there was no indication that the sediments had entered the sulphate reduction zone.

Froelich *et al.* (1979) interpreted their data in terms of the general vertical-depth-zone diagenetic model, which is related to the sequential use of oxidants for the destruction of organic carbon, and they were able to identify a number of distinct zones in the sediments. The sequence is illustrated in Fig. 14.5, and the individual zones are described below.

*Zone 1.* This is the interval over which oxygen is being consumed during the destruction of organic matter (diagenetic reaction 14.1). During this process the ammonia is oxidized to nitrate.

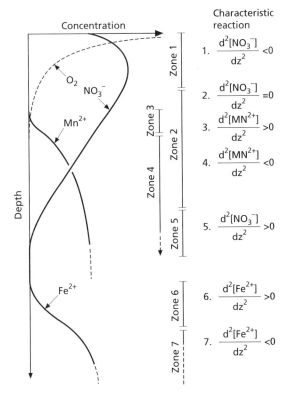

Concentration

Characteristic
reaction

1. $\dfrac{d^2[NO_3^-]}{dz^2} < 0$

2. $\dfrac{d^2[NO_3^-]}{dz^2} = 0$

3. $\dfrac{d^2[MN^{2+}]}{dz^2} > 0$

4. $\dfrac{d^2[MN^{2+}]}{dz^2} < 0$

5. $\dfrac{d^2[NO_3^-]}{dz^2} > 0$

6. $\dfrac{d^2[Fe^{2+}]}{dz^2} > 0$

7. $\dfrac{d^2[Fe^{2+}]}{dz^2} < 0$

**Fig. 14.5** Schematic representation of diagenetic zones and trends in interstitial-water profiles during the suboxic diagenesis of marine sediments (from Froelich *et al.*, 1979).

*Zones 2 and 5.* Below the nitrate maximum, nitrate diffuses downwards, the linearity of the gradient suggesting that nitrate is neither consumed nor produced in this zone. The downward diffusing nitrate is reduced by denitrification (diagenetic reaction 14.2) at the depth of the nitrate zero in zone 5.

*Zones 3 and 4.* These overlap with zones 2 and 5. Zone 4 is the interval over which organic carbon is oxidized by manganese oxides (diagenetic reaction 14.3) to release dissolved $Mn^{2+}$ into the interstitial waters. This then diffuses upwards to be oxidatively converted to solid $MnO_2$ at the top of the diffusion gradient in zone 3. This reduction–oxidation cycling results in the setting up of a 'sedimentary manganese trap' (see Section 14.6.3 and Worksheet 14.4).

*Zones 6 and 7.* Zone 7 is the region over which organic carbon oxidation takes place via oxygen

formed during the reduction of ferric oxides (diagenetic reaction 14.4). $Fe^{2+}$ is released into solution, and diffuses upwards to be consumed near the top of zone 7 and in zone 6.

This keynote investigation therefore provided clear evidence of how the diagenetic sequence operates in hemi-pelagic sediments, and demonstrates that the oxidants are, in fact, consumed in the predicted sequence, i.e. oxygen > nitrate ≅ manganese oxides > iron oxides > sulphate (although the sulphate reduction stage was not reached in these sediments).

### 14.3.4 Diagenesis in pelagic sediments

As the thickness of the surface oxidizing layer continues to increase, the diagenetic sequence can be traced as far as the least extreme oceanic end-member. This is represented by pelagic sediments, in which the surface oxic layer can extend downwards for several tens of metres. Wilson *et al.* (1985) reported data on the interstitial water constituents of a **pelagic** deep-sea sediment (station 10552) from the northeastern Atlantic, which had an accumulation rate of ~0.42 cm $10^3$ yr$^{-1}$. The interstitial water profiles for the sediment showed that: (i) dissolved oxygen was present throughout the interval sampled (0–2 m); (ii) nitrate concentrations increased downwards, i.e. there was no evidence of the consumption of nitrate as a secondary oxidant; (iii) no detectable dissolved $Mn^{2+}$ was present; and (iv) there was no evidence of sulphate reduction. The diagenetic sequence in this sediment had therefore not progressed beyond the oxic stage at which oxygen is consumed for the degradation of organic matter (diagenetic reaction 14.1), and there was little mobilization of the redox-sensitive elements.

Wilson *et al.* (1985) also provided information on the interstitial water chemistry of a mixed-layer **pelagic–turbidite** sediment. This sediment (station 10554) consisted of a thin (≤10 cm) upper light brownish grey layer of pelagic material, underlain by a turbidite sequence made up of an upper light brownish grey unit and a lower green unit. The sediment had an overall accumulation rate of ~10 cm $10^3$ yr$^{-1}$. The organic carbon content was ~0.6% in the surface layer and rose to ~1.6% around the top of the green layer. The interstitial water chemistry of the sediment at this site was considerably more complex than that of the exclusively pelagic deposit at Site 10552, the major trends being as follows.

1 Dissolved oxygen fell off rapidly with depth in a linear manner, to approach zero close to the surface of the green layer.

2 Nitrate concentrations also decreased in a continuous linear fashion, from a near-surface maximum to approach zero at a depth slightly greater than that for oxygen, although traces of both oxygen and nitrate were found in the green layer.

3 Dissolved $Mn^{2+}$ was found in solution below the top of the green layer.

4 There was no evidence of sulphate reduction.

It was apparent, therefore, that in the turbidite layer of the mixed sediment the diagenetic sequence had progressed considerably further than in the pure pelagic sediment, and had in fact reached the nitrate $\cong$ manganese oxide reduction stage. The main point of interest, however, was that there were two distinct colour units *within* the turbidite layer itself, the upper brownish grey layer, which appears to be oxic now, and the lower green layer, which still remained anoxic. On the basis of the linearity of the dissolved interstitial water profiles of both oxygen and nitrate, and the fact that they approach zero around the top of the green layer (which is coincident with a sharp rise in the concentration of organic carbon in the sediment), Wilson *et al.* (1985) concluded that the colour change represents a **downward-moving oxidation front**, the rate of movement of which is controlled by the rates of diffusion of oxygen and nitrate from the bottom water to the front itself. In terms of this concept, therefore, the upper brown turbidite layer is a zone in which the metabolizable organic carbon has already been oxidized as the front moved downwards. Under normal steady-state diagenesis, the oxic–anoxic colour change (or redox) boundary migrates *upwards* at the same rate as the sediment accumulates, so as to remain at a *constant* depth below the sediment–water interface. However, the progressive *downward* migration of the turbidite oxidation front is not a steady-state process; rather it implies that there is a continuing readjustment of the redox profile *after* the deposition of an organic-rich unit, which initially was at the original pelagic sediment surface. Almost all the diagenetic changes in the upper part of the mixed pelagic–turbidite sediment occur in two zones, one close to the sediment–water interface and one close to the top of the green layer (the migrating oxidation front). The concept of a secondary subsurface maximum in metabolic activity has important consequences, implying that even a relatively infrequent deposition of organic-rich units can have a major effect on the redox state of the host sediment column. Thus, the differences in the redox profiles of the pelagic and mixed pelagic–turbidite deep-sea sediments are not primarily the result of differences in total metabolic activity, but rather occur in response to differences in how the activity is distributed within the sediment column.

Thin trace-metal-rich layers are found in a number of deep-sea sediments at, or just below, the last glacial–Holocene transition, where there was a change from relatively rapid glacial sediment accumulation with reducing conditions at shallow depths to a more slowly accumulating Holocene depositional regime. The elements Mn, Ni, Co, Fe, P, V, Cu, Zn and U can be enriched in the vicinity of the layers. Wilson *et al.* (1986a,b) and Wallace *et al.* (1988) have proposed that the locus of formation of the layers is coincident with the present-day oxic–suboxic redox boundary, and can be explained in terms of the concept of a progressive oxidation front moving into the sediments following the change from glacial to interglacial conditions, with the depth of the oxidant penetration from sea water hovering around a near-fixed depth and being balanced by the reductant counterflux of $Mn^{2+}$ and $Fe^{2+}$ from the reducing sediments below.

Although attention in this volume is concentrated mainly on the role played by early diagenesis in present-day sediments the identification of downward-migrating non-steady-state diagenetic fronts associated with the deposition of turbidite sequences also has wider implications for seawater–sediment chemistry, especially from the point of view of the response of the oceans to climatic changes. For example, turbidite deposition on the Madeira Abyssal Plain may reflect global climatic changes associated with glaciations and sea-level fluctuations (see e.g. Weaver & Kuijpers, 1983), and thus their modification by downward-migrating diagenetic fronts provides an example of how the present oceans respond to the past climatic changes. In this respect, the Atlantic Ocean may still be recovering from late Quaternary climatic changes (Wallace *et al.*, 1988; T.R.S. Wilson, personal communication).

## 14.4 New concepts in organic matter preservation in marine sediments

It was pointed out in Section 14.3.2 that large quantities of organic matter are preserved in shelf and upper slope sediments, and that it appears as if some protective mechanism operates in these nearshore environments to preserve organic matter, a mechanism that is apparently 'switched off' in the deep-sea sedimentary environment. Hedges & Keil (1995) addressed this paradox, and proposed a theory which challenged many of the traditional concepts that have surrounded the preservation of organic matter in the marine environment.

Hedges & Keil (1995) pointed out that because fluvial organic matter has been subjected to severe microbial attack in soils and aquifers it might be expected to be more resistant than recently biosynthesized organic matter derived from primary production. In fact, the evidence available indicates that most organic matter in marine sediments is autochthonous, and the total burial of organic matter in marine sediments ($\sim 0.16 \times 10^{15}\,g\,C\,yr^{-1}$) is in any case only around one-third of the fluvial discharge ($\sim 0.4 \times 10^{15}\,g\,C\,yr^{-1}$). Thus, the indications are that fluvial organic matter is rapidly and extensively mineralized within either sea water or surficial marine sediments. This poses another paradox in the global carbon cycle; i.e. whereas the extremely recalcitrant organic substances such as kerogen, soil humus and fluvial organic matter are extensively oxidized back to $CO_2$ in relatively short time periods in the oceans, seemingly reactive remains of recently living marine organisms (autochthonous material) are preserved, especially in some nearshore sediments.

Organic matter degrades across a range of time-scales that vary over ten orders of magnitude from minutes for the breakdown of biochemicals in animal guts to $10^6\,yr$ for organic matter mineralization in deep-sea sediments. Further, the low concentrations of organic matter in deep-sea sediments underlying all but the most productive waters clearly demonstrate the potential for essentially complete mineralization. However, organic matter can be found in much higher concentrations in shelf sediments, leading to a wide-scale preservation of organic matter which is unique to these environments. Two conclusions can be drawn from this:

1 the organic matter that ultimately survives degradation to be preserved mainly in deposits on the shelves and upper slopes must be highly unusual in either structure or circumstances of burial;
2 the processes that preserve it must occur along all the continental margins and switch off in the deep-sea.

In addressing this, Hedges & Keil (1995) stressed that degradation and preservation are **opposite** processes, and concluded that although (i) productivity in the surface waters, (ii) sediment accumulation rate, (iii) bottom-water oxicity, and (iv) organic-matter source can be accepted as key variables, the mechanisms governing the preservation of organic matter in sediments are complex and unclear. In this context, at least two essential features of the organic matter regime in marine sediments require explanation:

1 conditions of essentially zero organic matter preservation prevail throughout the open-ocean;
2 landward of this, some protective organic matter process must widely occur.

To explain features such as these, Hedges & Keil (1995) proposed their theory, building on the pioneering work of Mayer (see e.g. Mayer, 1994), to explain the preservation of organic matter in marine sediments. The theory utilizes the widely accepted observation that fine-grained sediments are invariably associated with higher concentrations of organic matter than coarse-grained sediments, and attempts to unravel this **textural association** by suggesting that although the factors that determine organic matter preservation vary with the depositional regime, they have in common a critical interaction between **organic** and **inorganic** materials which occurs over locally variable time-scales.

More than $\sim 90\%$ of the organic matter from a wide variety of sedimentary environments cannot be physically separated from its mineral matrix. This organic matter varies directly in concentration with sediment surface area, and so appears to be sorbed on to mineral grains. Further, it is concentrated in fine-grained deposits along continental margins, in which the minerals have relatively large surface areas. According to Hedges & Keil (1995), if deltaic deposits are excluded, sediments accumulating along continental shelves and upper slopes characteristically exhibit mineral 'surface-area–organic-matter loadings' which are equivalent to a single monolayer coating: i.e. 'monolayer equivalent' coatings. It is significant that these monolayer coatings include a fraction of organic molecules that are labile, but which resist mineralization as they pass rapidly through oxygenated surface sediments and are **preserved** in the underlying anoxic deposits. The implication therefore is that the monolayers bind organic matter in a **protective coating**; however, these initially preserved labile organics can be degraded via oxic–suboxic diagenesis when shelf sediments are transported to the deep-sea as distal turbidites—see below, and Section 14.3.4.

Not all sediments in the marine environment bind organic matter by 'monolayer' equivalents. For example, deltaic sediments, which account for another ~45% of oceanic organic carbon, often show less than monolayer coatings: i.e. '**submonolayer**' coatings. Such 'submonolayer' coatings are also found in continental rise and open-ocean abyssal plain (hemipelagic) and pelagic clays, where slower accumulation rates and deeper oxygen penetration result in increased oxygen exposure times and little organic matter preservation: concentrations of less than ~1% organic carbon, which amount to only ~5% of the total global ocean organic matter (see Table 14.1). In contrast, in organic-rich sediments underlying highly productive, low-oxygen, waters, e.g. those off Peru, there is a direct relationship between organic matter content and mineral surface area, but at organic loadings two to five times a 'monolayer equivalent', i.e. '**multilayer equivalent**' coatings. The burial of organic matter, and its relationship to organic-matter–mineral-surface coatings, is illustrated in Fig. 14.6.

There is evidence for the widespread sorptive

**Fig. 14.6** Idealized diagram depicting estimates of organic matter burial (expressed as a percentage of the total sediment burial), and organic-matter–mineral-surface layer loadings, in a number of marine sediment types (after Hedges & Keil, 1995).

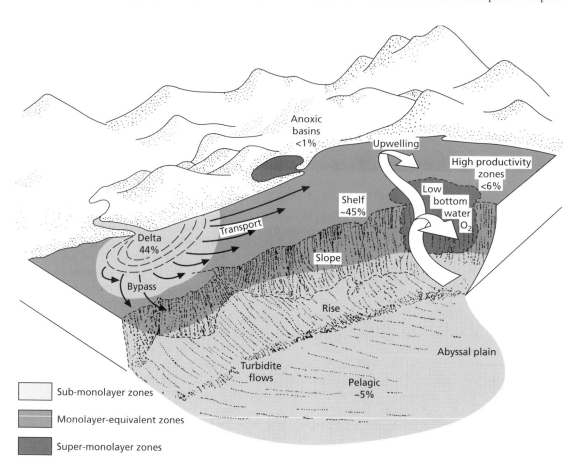

preservation of sedimentary organic matter in mono-layers along continental shelves and upper slopes, and a slow oxic–suboxic degradation in seaward sediments. This appears to produce a 'redoxcline' which manifests itself as a 'monolayer–submonolayer organic-matter coating transition' on the lower continental slope, below which little organic matter is preserved to be buried in the sediments. A transition zone between **sorptive organic matter protection** along coastal margins and **oxic–suboxic degradation** within deep-sea sediments therefore should be expected to lie somewhere on the lower continental margins if the inferred competition between these opposing forces of 'sorptive protection' and 'oxic–suboxic degradation' actually exists in nature. Hedges & Keil (1995) maintain that evidence for such a transition has been found in the fact that there is a strong correlation between organic matter content and sediment grain size in some upper shelf, but not lower shelf and continental rise, sediments.

The 'transition' may result, in part, from the nature of the organic matter itself. The margin sediments contain monolayers which include recycled fluvial organic matter that has already been extensively degraded on land and this recycled material should, on average, be more refractory than recently biosynthesized organic matter raining directly out of the water column from primary production. Whatever the cause, the 'monolayer–submonolayer' organic coating transition, or boundary, can be thought of as representing a '**preservation threshold**', with the monolayer organic matter being most susceptible to preservation. As a result, the overall preservation of organic matter in marine sediments therefore may be controlled largely by competition between **sorption** at different protective thresholds and **oxic–suboxic degradation**.

The inference that almost all sedimentary organic matter is sorbed on to mineral surfaces, often as monolayers, has a number of implications. One of the most important of these is that the organic molecules involved were once in a dissolved state, or otherwise they would not have spread so uniformly, or thinly, over the surfaces of sedimentary mineral surfaces. Because most fresh organic matter is particulate, and less than c. 10% of the organic matter in marine sediments is present as discrete particles, then appreciable degradation must have been necessary to dissolve the organic matter prior to sorption. Further, most

minerals in soils and river suspensions carry mono-layer-equivalent organic coatings. However, the extensive presence of submonolayer coatings in deltaic sediments, and decreases in the fraction of terrestrial organic matter within the monolayer coatings of shelf sediments, indicate that an exchange of organic molecules both on to and off the mineral surfaces has occurred. It therefore is implicit in the Hedges & Keil (1995) theory that the monolayer-equivalent coatings include a fraction of reversibly bound organic molecules, and experimental data indicated that intrinsically labile organic matter is preserved by **easily reversible** sorption on to mineral matter. This reversible partitioning of organic molecules between particle surfaces and ambient waters is a potentially important process because it links sedimentary POM, DOM and minerals into an intricate network of interactions.

In the Hedges & Keil (1995) theory organic-matter coatings on mineral surfaces are therefore: (i) partially reversible; and (ii) strongly protective of potentially labile organic substances that are susceptible to appreciable degradation only under oxic–suboxic conditions in the open-ocean environment. These organic-matter coatings vary in thickness from monolayers to multilayers, with the monolayers being protected to a greater extent than the multilayers. The theory proposes that the delivery of **mineral surface area** is therefore the primary control on organic matter preservation in non-deltaic coastal margin regions. However, a direct cause for the stability of the 'monolayer equivalents' of organic molecules in these coatings has not been established.

Deep-sea deposits demonstrate that some degradative mechanism must consistently overpower 'surface-area–sorption' phenomena to produce extremely organic-matter-poor sediments covering open-ocean regions. This raises the question of the importance of '**protective sorption versus oxic–suboxic degradation**' as determinants in the preservation of organic matter in marine sediments. In some deep-sea sediments the organic-matter-poor conditions occur across the complete particle-size spectrum, despite the fact that the inputs of organic matter exceed those necessary to produce a 'monolayer equivalent' by factors of 10 to 100 times. Hedges & Keil (1995) suggest that the factors controlling the destruction of organic matter in deep-sea deposits are the opposite to those leading to its preservation in continental margin sediments; i.e. oxic–suboxic degradation overpowers sorptive preservation.

The most direct explanation of the organic-matter-poor conditions in deep-sea sediments comes from the long-term exposure of the deposits to oxygen and other electron acceptors, such as manganese and iron oxides (oxic–suboxic diagenesis—see Section 14.1.2), which may be referred to collectively as the 'oxygen effect'. In assessing the effects of oxic–suboxic degradation three types of organic matter were recognized by Hedges & Keil (1995): (i) totally refractive organic matter; (ii) oxygen-sensitive organic matter (OSOM), which degrades slowly in the presence of oxygen and/or metal oxides under oxic and suboxic, but not anoxic, conditions; and (iii) hydrolysable organic matter that is completely mineralized regardless of conditions (see also Section 14.2.4). In slowly deposited pelagic open-ocean sediments, in which oxygen has penetrated extensively, hydrolysable and OSOM organic matter will have had time to degrade completely, leaving only refractory components behind. As sedimentation rates increase, e.g. in the pelagic → hemi-pelagic → nearshore sediment sequence (see Section 14.3), oxygen penetration depths become shallower giving rise to briefer oxygen exposure times, and OSOM degradation will be effectively 'switched off' at progressively shallower depths and the 'protective sorption' effect will take over.

One of the most dramatic examples of the organic matter 'nearshore preservation' versus 'open-ocean degradation' syndrome is provided by the fate of organic matter in shelf sediments that subsequently have been deposited as distal turbidites in abyssal plain regions. When marginal fine-grained, organic-matter-rich sediments slump into the deep-sea, the organic matter at the top of the resulting turbidite is exposed to high concentrations of bottom-water oxygen and low sedimentation rates. As a result, molecular oxygen diffuses into the top of the turbidite where it reacts with the organic matter at the 'redox fronts' (see Section 14.3.4). In one example quoted by Hedges & Keil (1995), ~80% of the organic matter that had been stable for ~140 000 yr in the presence of sulphate below the redox front in an emplaced turbidite was degraded completely within ~10 000 yr following deposition as a result of prolonged exposure to molecular oxygen and secondary oxidants.

That is, exposure to the 'oxidation effect' alone was sufficient to decrease the organic matter concentration from values typical of nearshore sediments (~1%) to values characteristic of deep-sea deposits (~0.15–0.20%). These turbidites therefore provide clear evidence that the factors which preserve organic matter below the oxic layer in nearshore sediments can be overridden by long-term reaction with oxygen.

The Hedges & Keil (1995) theory brings a number of exciting insights into the factors controlling the fate of organic matter in marine sediments. In particular, it offers an explanation for the large-scale preservation of organic matter in shelf and upper slope sediments, and its almost total degradation in many open-ocean deposits, by linking into a 'preservation versus degradation' framework. In terms of the theory, much of the organic matter in the shelf and upper slope sediments is protected by forming a monolayer coating on mineral surfaces (**organic matter preservation**) and continues to be protected in the absence of deep oxygen penetration in the fast accumulating deposits. This preservation mechanism is switched off in the deep-sea, where there is deeper oxygen penetration in slowly accumulating sediments, and oxic–suboxic diagenesis (**organic matter degradation**) becomes predominant; i.e. the organic-matter coatings are destroyed. There is, therefore, competition between 'sorptive preservation' (shelf and upper slope sediments) and 'diagenetic degradation' (deep-sea sediments), with the threshold boundary between the two processes lying somewhere at the base of the continental margins.

A number of criticisms have been levelled at the Hedges & Keil (1995) 'mineral–organic-matter sorption model'.

Pedersen (1995) suggested that the problem of what controls the accumulation and preservation of organic matter in marine sediments is, in fact, a multivariate problem that defies a 'unifying theory'. For example, he points out that the local settling flux of organic matter correlates well with primary productivity in the overlying waters and may offer a 'better fit' to organic matter burial in sediments than the 'mineral–organic-matter sorption' model. However, the 'primary productivity' model also has its problems because, for example, some productive continental shelves host little organic matter in the underlying sediments, apparently because of the removal of fine-grained particles by hydrodynamic processes—e.g. on the Oman Margin (see e.g. Pedersen *et al.*, 1992). Pedersen (1995) concluded that although the theory advanced by Hedges & Keil (1995) can explain many fundamental aspects of the 'textural relationship', two aspects give rise to concern:

1 The assumption that mono- or multilayer coatings account for all organic matter in sediments from a wide range of environments requires the sorbed organic matter to pass through a 'solubilization step', as indeed Hedges & Keil (1995) point out themselves. If this is the case, Pedersen (1995) raises the question 'how then do the various individual organic matter compounds, e.g. fatty acids and pigments, which are present in both the precursor undecomposed organic matter **and** the sediments, retain their integrity through the degradation step?'

2 The concept that particle surface area plus long exposure to the non-sulphate oxidant suite (NSOS) provide the ultimate control on the organic matter content of deep-ocean sediments does not fit the data on a broad scale, and the Pacific Ocean was cited as an example to illustrate this. Eastern Pacific equatorial sediments have a relatively low surface area (mainly sand-sized forams and calcite fragments) but a high organic matter content. In contrast, sediments to the west have similar grain-sized characteristics, but much lower contents of organic matter. Further, pelagic clays from the Pacific central gyres have a high surface area, but low organic-matter contents. According to Pedersen (1995), the reason for this is that the magnitude of primary production provides the grand overall, first-order, control on the organic matter contents of Pacific sediments.

Berner (1995) drew attention to one of the principal advances of the Hedges & Keil (1995) theory, i.e. the proposal that the sorption of organic matter has a **stabilizing** effect on its microbial decomposition, and concluded that this may be an important process for the preservation of organic matter in sediments. The author also pointed out, however, that the theory neglects the ingestion of sediment by the micro-benthos. Essentially all the fine-grained sediment deposited under oxygenated bottom-waters passes through the gut of worms and other bottom-dwelling organisms. This will result in the 'coating' of the sediment particles by organic matter and bacteria, and it may be that the organic-matter layers arise within animal guts and not through sorption from solution.

Henrichs (1995) offered a serious major criticism of the Hedges & Keil (1995) theory by proposing that **easily reversible** adsorption on to mineral surfaces cannot preserve labile organic compounds, and that only **strong adsorption** can prevent the diffusive loss of organic molecules from sediments, regardless of whether or not they can be decomposed by bacteria. The model Henrichs (1995) put forward predicts that only very strongly adsorbed molecules will be preserved, and that these are produced in sediments during the decay of the more labile organic matter. The intensity of the adsorption required for preservation is inversely proportional to the rate constant for the decomposition of organic matter, as a result of which the sorptive preservation of very labile compounds is much less likely than the preservation of compounds that are inherently refractory. The lack of preservation of organic matter in oxic deep-sea sediments can be explained if **very strong** adsorption does not occur in this environment. The essence of the Henrichs' theory therefore is that sorbed organic matter must either be intrinsically refractory and/or strongly bound to a protective surface to have a high probability of preservation.

The overall conclusion that may be drawn from these various studies is that although a major proportion of the organic matter in marine sediments may be either strongly or weakly sorbed on to mineral surfaces, the balance between 'sorption at different protective thresholds' and 'oxic–suboxic degradation' has not yet been established. For example, it is possible that the organic-matter 'preservation versus degradation' conundrum could be explained without the need to invoke any kind of protective coating in terms of:

1 **protection** in fast accumulating nearshore sediments with a relatively high input of organic matter and little oxygen penetration, in which the organic matter is buried quickly;

2 oxic–suboxic **degradation** in slowly accumulating deep-sea sediments with a relatively low input of organic matter and extensive oxygen penetration.

In this case, the diagenesis of organic matter in turbidites derived from nearshore regions and deposited in the deep-sea would simply be an effect of transporting them to a different environment in which more organic matter diagenesis is possible. Thus, the preservation of organic matter in marine sediments would be a function of the rate of supply of the organic matter, the rate of burial of the organic matter, and the availability of oxygen; i.e. a protective organic-matter coating would not be a necessary requirement of the system. Despite this, there is considerable evidence that organic matter protective coatings are found in marine sediments, and the conundrum is yet to be solved.

## 14.5 Diagenesis: summary

1 Early diagenesis in marine sediments follows a general pattern in which a series of oxidants are utilized for the destruction of organic carbon in the following general sequence: oxygen > nitrate $\cong$ manganese oxides > iron oxides > sulphate.

2 The diagenetic sequence passes through each of the oxidant utilization stages successively in a down-column direction, thus setting up a *vertical* gradient.

3 The degree to which organic matter suffers degradation in marine sediments may be regarded as the result of a competition between 'sorptive protection' and 'oxic–suboxic diagenetic degradation'. The extent to which the diagenetic sequence itself proceeds depends largely on the rate of supply of organic matter to the sediment surface and the rate at which it is buried. Both of these parameters are linked into the 'sorptive protection' versus 'oxic–suboxic diagenetic degradation' competition, and both decrease away from the continental margins towards the open-ocean, thus setting up a *lateral* gradient in the diagenetic sequence. As a consequence of this, the diagenetic stage reached drops progressively from (i) nearshore anoxic sediments (methanic stage), to (ii) nearshore sediments having a thin oxic layer of a few millimetres (sulphate reduction stage), to (iii) hemi-pelagic deep-sea sediments having an oxic layer of a few centimetres (nitrate $\cong$ manganese oxide, iron oxide reduction stages), to (iv) truly pelagic deep-sea sediments, in which diagenesis does not progress beyond the oxic stage at which oxygen consumes virtually all the organic carbon brought to the sediment surface. In shelf and slope sediments organic matter can be protected by a monolayer coating and can survive down to the sulphate reduction stage. In the open-ocean, in contrast, long-term exposure to oxygen overcomes sorptive protection and little organic matter survives in the sediments.

During diagenesis elements are mobilized into solution and so can migrate through the interstitial waters. Some of these elements (together with those already present in the interstitial waters) are incorporated into newly formed or altered minerals. However, a fraction of the elements can escape capture in this way, and so be released into the overlying sea water. In the next section, therefore, the wider aspects of diagenetic mobilization will be considered in relation both to the depletion-enrichment of elements in interstitial waters and to the potential fluxes of the elements to the oceanic water column.

## 14.6 Interstitial water inputs to the oceans

### 14.6.1 Introduction

Early diagenesis involving the oxidative destruction of organic matter can play an important role in the interstitial water chemistries and oceanic cycling of a number of components. These include the following.

1 The *bioactive*, or *labile*, elements such as C, N, P, together with Ca (calcium carbonate) and Si (opal). A characteristic of these elements is that only a relatively small fraction of their down-column rain rates are preserved in marine sediments, with most of them being recycled; for example, Martin & Sayles (1994) estimated that around 90% of the organic carbon, between 30 and 80% of the calcium carbonate and between ~50 and ~90% of the opaline silica reaching the sea floor is recycled (see also Section 12.2).

2 The *oxidants* used to destroy organic matter: oxygen, nitrate, Mn and Fe oxides, and sulphate (see Section 14.1.2). For example, the net rate of loss of sulphate from sea water as a consequence of sulphate reduction during organic-matter diagenesis is $\sim 0.5 \times 10^{12}\,mol\,yr^{-1}$, which is smaller than the input rate of sulphate from rivers ($3.7 \times 10^{12}$) but is similar to removal by hydrothermal activity ($0.84 \times 10^{12}$) (see e.g. Martin & Sayles, 1994).

3 *Trace metals.* Diagenetic processes involved in the oxidative destruction of organic matter are intimately related to the interstitial water chemistries of many trace metals, including those transported down the water column by organic carriers and those associated with the secondary oxidants (e.g. Mn oxides). For metals in associations such as these, the diagenetic destruction of organic matter acts as a **recycling** term.

4 *Refractory elements.* In addition, changes occur during early diagenesis that are not related to the

oxidative destruction of organic matter. These changes affect the *non-bioactive*, or refractory, elements (e.g. Na, K, Kg). For these elements the down-column rain rates and sediment burial rates are generally in good agreement (see Section 12.2), and in the interstitial-water–sediment complex their reactivity is associated mainly with aluminosilicate mineral alteration rather than with the destruction of organic matter. This mineral alteration can act as either a seawater **source** or **sink** term.

The transport of dissolved material in interstitial waters takes place by convection and diffusion. When the sediment thickness is less than ~150 m, convective processes can move the water through the deposits, but as the sediment thickness increases diffusive transport becomes dominant. The external boundary conditions acting on the interstitial-water–sediment 'sandwich' are therefore set by: (i) the basalt basement below; and (ii) the water column above, which supplies the fluids that are trapped in the depositing sediments. The signals transmitted through interstitial waters therefore reflect the compositions of the overlying sea water and the underlying basalt, and are modified by *in situ* diagenetic processes within the sediment sandwich itself. The elemental composition of interstitial waters therefore is controlled by a number of interrelated factors, which include:

1 the nature of the original trapped fluid, usually sea water;

2 the nature of the transport processes, i.e. convection or diffusion;

3 reactions in the underlying basement, including both high- and low-temperature basalt–seawater interactions;

4 reactions in the sediment column;

5 reactions across the sediment–seawater interface.

Together, the reactions taking place in the '**seawater–sediment-sandwich–basalt**' complex can involve either the release or the uptake of dissolved components. As a result, changes can be produced in the composition of the interstitial waters relative to the parent sea water, and **diffusion gradients** can be set up under which the components will migrate from high- to low-concentration regions. In contrast, under some conditions the compositions of the interstitial water and sea water will not differ significantly and concentration gradients will be absent.

## 14.6.2 Major elements

Diagenesis associated with the destruction of organic matter, involving C, N and P, has been described in Sections 14.1 to 14.5, and with respect to the **major** interstitial water constituents attention in the present section will be confined largely to the non-biogenic elements.

Analyses of interstitial waters were carried out as long ago as the last century (see e.g. Murray & Irvine, 1895). Until recently, however, data for the chemistry of interstitial water have suffered from a number of major uncertainties. According to Sayles (1979) these arose from:

1 sampling procedures, such as temperature-induced artefacts inherent in the water extraction techniques;

2 imprecise analytical techniques (especially for trace elements); and

3 a lack of detail close to the sediment–seawater interface, a region where a number of important reactions take place.

Because of factors such as these, much of the early interstitial-water data must be regarded as being unreliable. In order to rectify some of these uncertainties, Sayles (1979) carried out a study of the composition of interstitial waters collected using *in situ* techniques from the upper 1–2 m of a series of sediments from the North and South Atlantic on a marginal–central ocean transect. A number of trends could be identified from the data obtained on this transect.

1 The interstitial waters were almost always *enriched* in $Na^+$, $Ca^{2+}$ and $HCO_3^-$, and *depleted* in $K^+$ and $Mg^{2+}$, relative to sea water. In addition, $SO_4^{2-}$ was slightly enriched at most stations.

2 The extent of these depletions or enrichments varied from element to element. For example, the enrichments in $Na^+$ were relatively small, and although the concentration increased with depth the gradient was only gradual. In contrast, the other major cations had interstitial-water distribution profiles that were characterized by sharp gradients in the upper 15–30 cm of the sediments with only a limited change at greater depths.

3 The concentrations of $Mg^{2+}$, $K^+$, $Ca^{2+}$ and $HCO_3^-$ in the interstitial waters all exhibited a pronounced geographical variability, with the highest concentrations being found in waters from the marginal

sediments and the lowest in those from the central ocean areas.

It may be concluded therefore that, relative to sea water, the interstitial fluids of the upper 1–2 m of oceanic sediments are generally enriched in calcium, sodium and bicarbonate and are depleted in potas-

sium and magnesium. Some of the processes causing these interstitial-water enrichments and depletions are discussed below, and a number of the basic concepts relating to the behaviour of chemical species in interstitial water are discussed in Worksheet 14.2.

The formation of interstitial-water gradients can be

---

**Worksheet 14.2: Some basic concepts relating to the behaviour of chemical species in the sediment–interstitial-water complex**

Interstitial waters are the medium through which elements migrate during diagenetic reactions. In sediments the interstitial-water properties change very much more rapidly in the vertical than in the horizontal direction, with the result that the changes can often be described by one-dimensional models. The transport of solutes through interstitial waters takes place by convection, advection and diffusion. In the context used here *advection* refers to transport by the physical movement of the water phase, and *diffusion* refers to migration of a chemical species throught the water as a result of a gradient in its concentration (or chemical potential). Diffusion in an aqueous solution can be described mathematically by Fick's laws, which, for one dimension, may be written as follows (see e.g. Berner, 1980):

**1 First law**

$$J_i = -D_i \frac{\partial c_i}{\partial x} \tag{1}$$

**2 Second law**

$$\frac{\partial c_i}{\partial t} = D_i \frac{\partial^2 c_i}{\partial x^2} \tag{2}$$

Here $J_i$ is the diffusion flux of component $i$ in mass per unit area per unit time, $c_i$ is the concentration of component $i$ in mass per unit volume, $D_i$ is the diffusion coefficient of $i$ in area per unit time, and $x$ is the direction of maximum concentration gradient; the minus sign in the first law indicates that the flux is in the opposite direction to the concentration gradient. Fick's first law is applied to calculations that involve steady-state systems, and the second law is applied to non-steady-state systems. Before Fick's laws can be applied directly to sediments, however, it is necessary to take account of the nature of the sediment–interstitial-water complex. Interstitial waters are dispersed throughout a sediment and the rate of diffusion of a solute through them is less than that in water alone, i.e. as predicted by Fick's laws, because of the solids present (the porosity effect) and because the diffusion path has to move around the grains; the term *tortuosity* ($\theta$) is used

*continued*

to describe the ratio of the length of the sinuous diffusion path to its straight-line distance (Berner, 1980). Tortuosity is usually determined indirectly from measurements of electrical resistivity of sediments and of the interstitial waters separated from them, using the relationship

$$\theta^2 = \phi F \tag{3}$$

where $\phi$ is the porosity and $F$ is the *formation factor* ($F = R/R_0$, where $R$ is the electrical resistivity of the sediment and $R_0$ is the resistivity of the interstitial water alone). Formation factors in marine sediments usually appear to lie in the range *c.* 1 to *c.* 10, so that the *effective diffusion coefficient* ($D'$) in a sediment will be less than in the solution alone by a factor of up to *c.* 10. The effective diffusion coefficient can be calculated from the relationship

$$D' = \frac{D_\phi}{\theta^2} \tag{4}$$

where $D$ is the diffusion coefficient in solution, $\phi$ is the porosity and $\theta$ is the tortuosity.

A detailed treatment of how to apply Fick's laws directly to sediments is given in Berner (1980), and for a comprehensive mathematical treatment of migrational processes and chemical reactions in interstitial waters the reader is referred to the 'benchmark' publication by Lerman (1977).

*Examples in the text*
Gieskes (1983) has pointed out that if it is assumed that only vertical transport through interstitial waters is important, then the flux of a chemical constituent can be described by the equation

$$J_b = -p D_b \frac{\partial c}{\partial z} + puc \tag{5}$$

where $J_b$ is the mass flux, $p$ is the porosity, $z$ is the depth coordinate in centimetres (positive downwards), $u$ is the interstitial-water velocity relative to the sediment–water interface in $cm_b s^{-1}$ (i.e. the advection rate), $c$ is the mass concentration in $mol\,cm_p^{-3}$ and $D_b$ is the diffusion coefficient in the bulk sediment (the subscript b indicates that concentrations and distances are measured over the bulk sediment (i.e. solids and interstitial waters) and the subscript p indicates the interstitial water phase only).

The mass balance of the solute is given by

$$\frac{\partial pc}{\partial t} = \frac{\partial}{\partial z}(J_b) + R \tag{6}$$

where $R$ is a chemical source–sink term, i.e. the reaction rate ($mol\,cm_b^{-3}\,s^{-1}$).

*continued on p. 382*

If the interstitial-water density and the solid density do not change in a given depth horizon, then

$$\frac{\partial p}{\partial t} = \frac{\partial pu}{\partial z} \tag{7}$$

and eqn (6) becomes

$$p\frac{\partial c}{\partial t} = \frac{\partial}{\partial z}\left(pD_b \frac{\partial c}{\partial z}\right) - pu\frac{\partial c}{\partial z} + R \tag{8}$$

and when steady state exists, this becomes

$$0 = \frac{\partial}{\partial z}\left(pD_b \frac{\partial c}{\partial z}\right) - pu\frac{\partial c}{\partial z} + R \tag{9}$$

According to Gieskes (1983) if conditions (e.g. sedimentation rates, temperature gradients) have been stable during relatively recent times (the last 10–12 Ma), then the steady-state assumption is valid for pelagic sediments, which have accumulation rates of $\sim$20 m Ma$^{-1}$. The author then considered how a concentration–depth gradient, such as that illustrated in Fig. (i), could be explained.

Gieskes (1983) considered the factors that might control the concentration–depth relationship in the dissolved Ca profile illustrated in Fig. (i) and related them to changes in three variables. These vari-

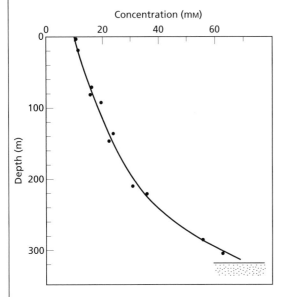

**Fig. (i)** Concentration–depth profile of dissolved Ca in interstitial waters of a DSDP core (from Gieskes, 1983).

*continued*

ables were diffusion ($D_b$), reaction rate ($R$) and advection ($u$); i.e. the profiles were interpreted within a *diffusion–advection–reaction* framework. To illustrate this approach, three cases were considered.

*Case 1*, in which the rate of diffusion varies; i.e. $R = 0$, $D_b = f(z)$ and $u = 0$. Under these conditions, there is no reaction and no significant contribution from advection. Thus, only a gradual decrease in $D_b$ with depth could then explain the increased curvature with depth in the otherwise conservative profile.

*Case 2*, in which the reaction rate varies; i.e. $R \neq 0$, $D_b$ is constant and $u = 0$. Thus, the profile implies a removal of calcium from solution, notwithstanding the significant source term for dissolved Ca at the lower boundary.

*Case 3*, in which the advection rate is not zero; i.e. $R = 0$, $D_b$ is constant and $u \neq 0$. Thus, under these conditions of no reaction and constant diffusion coefficient, the curvature in the profile would be caused by the relatively large advective term.

Gieskes (1983) then considered two types of Ca–Mg interstitial-water concentration–depth profiles. In the first type, there are linear correlations between $\Delta$Ca and $\Delta$Mg, i.e. $R = 0$. Using data that included information on porosity, and diffusion coefficients (evaluated from a knowledge of formation factors; see above), a solution of eqn (9) assuming $R = 0$ indicated that the depth profiles of Ca and Mg could be explained in terms of conservative behaviour (i.e. transport through the interstitial water column alone and no reaction), with the boundary conditions being fixed by concentrations in the underlying basalts and the overlying sea water. In the second type, there are non-liner correlations between $\Delta$Ca and $\Delta$Mg, which implies reaction in the sediment column; i.e. $R \neq 0$. Under these conditions, derivatives of eqn (9) must be evaluated geometrically from the concentration–depth profiles in order to model the data. These two types of Ca–Mg profiles are described in the text.

illustrated with respect to calcium and magnesium. Concentration gradients, showing increases in calcium and decreases in magnesium with depth, have been reported in the interstitial waters of many deep-sea sediments. The theories advanced to explain the existence of these gradients include:

1 the formation of dolomite or high-magnesium calcite during the dissolution and recrystallization of shell carbonates, which would account for the interstitial-water gains in calcium and losses in magnesium;

2 the adsorption of magnesium on to opal phases, which would result in a decrease in magnesium in the interstitial waters of siliceous sediments.

However, in addition to changes in magnesium and calcium there is often a decrease in $\delta^{18}O$ values in the interstitial waters, and according to Lawrence *et al.* (1975) this cannot be explained by reactions involving either carbonate or opaline diagenesis. Instead, these authors proposed that the changes in $\delta^{18}O$ values result from reactions taking place during the alteration either of the basalts of the basement or of volcanic material dispersed throughout the sediment column. The problem was addressed by Gieskes

(1983), who identified two types of calcium–magnesium profiles in the interstitial waters from 'long-core' oceanic sediments and linked them to reactions involving interstitial waters and volcanic rocks, either in the basalt basement or in the sediment column itself, or in both.

*Reactions involving basement rocks.* In the interstitial waters of some sediments there is a linear $\Delta$Ca versus $\Delta$Mg correlation down to the base of the sediment column, i.e. gains in Ca are matched by losses in Mg. Gieskes (1983) classified the behaviour of the two elements under these conditions as *conservative*, and suggested that their distributions in the interstitial waters are controlled largely by their transport through the sediment column following reactions in the basement basalts, which act as a sink for magnesium and a source for calcium; i.e. the sediment sandwich itself is mainly unreactive, at least with respect to calcium and magnesium. An example of a DSDP core in which calcium and magnesium behave conservatively in the interstitial waters is given in Fig. 14.7(a). It can be seen from this figure that although calcium and magnesium have not been affected by reactions in the sediment layer, such diagenetic reactions have affected the interstitial-water profiles of other constituents in a manner that reflects the types of processes involved. For example:

**1** the maximum in the strontium concentrations probably results from the recrystallization of carbonates in the nanofossil ooze, during which new mineral phases are formed that have lower strontium concentrations than the parent material;

**2** potassium concentrations decrease with depth, probably as a result of the uptake of the element during the formation of potassium feldspar;

**3** the sulphate profile shows a decrease with depth, indicating that there has been sulphate reduction within the sediment column.

*Reactions involving both basement rocks and dispersed volcanic material.* In the interstitial waters of some deep-sea sediments there is a non-linear correlation between $\Delta$Ca and $\Delta$Mg, and the behaviour of the elements is classified as *non-conservative*. Under these conditions, Gieskes (1983) assumed that reactions controlling the interstitial-water distributions of the two elements had occurred both in the underlying basalt basement and in the sediment column. The

sediment reactions were thought to have taken place mainly between the interstitial waters and dispersed volcanic material, and to have led to the addition of calcium and the removal of magnesium from the fluids. An example of the non-conservative behaviour of calcium and magnesium in the interstitial waters of sediments from a DSDP site is illustrated in Fig. 14.7(b), from which it can be seen that other constituents have also suffered diagenetic modifications. For example, variations in the $\delta^{18}$O (see e.g. Lawrence *et al.*, 1975) and the $^{87}$Sr:$^{86}$Sr ratios (see e.g. Hawkesworth & Elderfield, 1978) in the interstitial waters indicate that exchange has taken place between volcanic material and the fluids. This DSDP site is located on the Bellinghausen Abyssal Plain, and is of particular interest because it contains a **silicification front** in the sediment column. At this front the conversion of biogenic opal-A to opal-CT (porcellanite) occurs, together with other diagenetic changes in the sediment–interstitial-water complex that can affect mineral phases such as carbonates and volcanic material. In addition to being zones of diagenetic reaction, silicification fronts, at which the silica minerals are recrystallized, cause a decrease in the porosity of the sediment, thus acting as what Gieskes (1983) termed 'diffusion barriers'.

Other types of specialized reactions that bring about changes in the composition of interstitial waters, such as those associated with evaporite deposits and high-temperature hydrothermal activity, have been reviewed by Manheim (1976) and Gieskes (1983).

Data provided by Sayles (1979) showed that the reactions that control the fluxes of $Mg^{2+}$, $Ca^{2+}$, $K^+$ and $HCO_3^-$ across the sediment–sea-water interface occur mainly in the top $\sim$30 cm of the sediment column, accounting for between 70 and 90% of the total fluxes. Only sodium had a deep sediment source. It may be concluded, therefore, that fluxes across the sediment–water interface that are based on data from the uppermost portions of the sediment column should give the most realistic assessment of those actually occurring in nature. Sayles (1979) calculated total diffusive fluxes of a number of elements across the interface and found that they varied geographically, being higher in the marginal than in the central areas of the Atlantic. To derive global fluxes the results were therefore weighted for high- and low-flux areas. The data were subsequently refined by

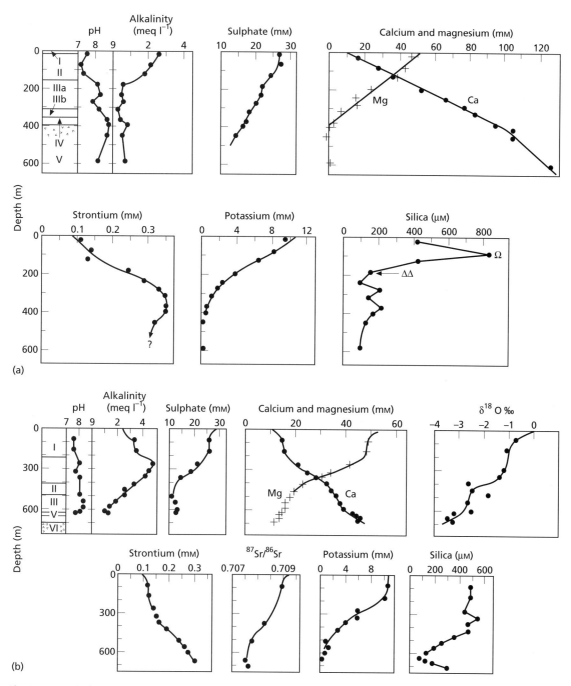

**Fig. 14.7** Interstitial water profiles at DSDP sites. (a) Concentration–depth profiles at DSDP site 446 (24°42′N, 132°47′E). Lithology: I, brown terrigenous mud and clay; II, pelagic clay and ash, siliceous; IIIa, mudstone, clay stone, siltstone, sandstone; IIIb, calcareous clay and mudstones,
turbidites; IV, calcareous claystones, glauconite, mudstones; V, basalt sills and intrusions. (b) Concentration–depth profiles at DSDP site 323 (63°41′S, 98°00′W). Lithology: I, clay, silt, diatom ooze; II, chert and claystone; III, claystone; IV, nanno chalk; V, zeolitic clay; VI, basalt. (From Gieskes 1983.)

**Table 14.2** Diffusive fluxes of a number of components across the sediment–water interface; units, $10^{18}\,\mu eq\,yr^{-1}$ (after Martin & Sayles, 1994).

| Component | Interstitial water flux* | | | |
| | Ocean margin | Central ocean | Total* | River flux |
|---|---|---|---|---|
| Na | +3.6 | +4.3 | +7.9 | 8.3 |
| Mg | −5.4 | −5.2 | −10.6 | 11.7 |
| Ca | +7.5 | +6.9 | +14.4 | 27.2 |
| K | −5.8 | −6.6 | −1.1 | 1.3 |
| $HCO_3^-$ | +8.1 | +10.4 | +18.5 | 33.3 |

\* + = release to sea water; interstitial water source. − = loss from sea water; interstitial water sink.

Martin & Sayles (1994) and are listed in Table 14.2. To put the magnitude of these interstitial water fluxes in an oceanographic context fluvial fluxes are also included in Table 14.2, and a number of conclusions can be drawn from the data.

**1** For magnesium and potassium the interstitial waters act as a *sink*, with ~90% of the magnesium, and ~85% of the potassium river inputs being balanced by diagenetic uptake in the sediments. On this basis, therefore, early diagenetic reactions remove magnesium and potassium from sea water at rates that are nearly equal to their river inputs.

**2** For calcium, sodium and bicarbonate, the interstitial waters act as a *source* to sea water and augment the river supply by ~53% for calcium, ~95% for sodium and ~55% for bicarbonate. On this basis, therefore, sodium is added to sea water via early diagenesis at a rate comparable to its river input.

There is, however, considerable disagreement in the literature on the magnitude of interstitial water fluxes to the oceans. For example, Gieskes (1983) summarized a series of DSDP data and concluded that on the basis of concentration–depth profiles, calculated over greater depths than those used by Sayles (1979), the calcium flux out of interstitial waters is <1% of the river flux, that for magnesium is ~2% and that for potassium is ~10%. Gieskes (1983) concluded therefore that the diffusion of calcium, magnesium and potassium, and also sodium, from sediments does not play a significant role in their overall geochemical mass balances: a conclusion which is very different from that reached by Martin & Sayles (1994). According to Sayles (1979) the uptake of magnesium

and potassium calculated from the fluxes across the sediment–seawater interface from his data would result in their concentrations being considerably in excess of those actually found in the sediments. He suggested, therefore, that the interstitial water fluxes for magnesium and potassium may have been overestimated by as much as 50%. However, this would still leave their removal from sea water, relative to fluvial inputs, considerably greater than that predicted by Gieskes (1983). Further, it has been proposed that hydrothermal activity at the ridge centres acts as a sink for ~45% of fluvial magnesium (see e.g. Thompson, 1983), which if correct would not require the very large interstitial water sink proposed by Martin & Sayles (1994).

### 14.6.3 Trace elements

'New', i.e. post-1975, data can be used to assess the fates of trace elements in the interstitial-water–sediment system. The fate of a component following deposition in a sediment is constrained by its post-depositional mobility, since in order to be added to the interstitial water from solid sediment phases it must first be solubilized to the dissolved state. For many trace metals this solubilization is intimately related to the oxidative destruction of organic matter during early diagenesis. The trace metals that are released in this process to become concentrated in interstitial waters relative to overlying sea water can follow one of two general pathways:

**1** they can be released (i.e. recycled) back into sea water via upward migration across the sediment–water interface;

**2** they can be reincorporated into sediment components following upward or downward migration through the interstitial waters.

Thus, it is necessary to introduce the concept of a *net* interstitial water flux, which in the present context is the flux that escapes reaccumulation and is added directly to sea water.

Several studies can throw light on the extent to which these net, or out-of-sediment, fluxes operate. In one of these, Hartmann & Muller (1982) gave data on the mean concentrations of Mn, Cu, Zn and Ni in the interstitial waters of a series of oxic pelagic sediments (siliceous oozes, clays and calcareous oozes) from the central Pacific—a summary of their data is given in Table 14.3. All the trace metals were enriched in the interstitial waters relative to ambient sea bottom-water and, with the exception of Zn, they had their highest concentrations near to the sediment surface (0–2 cm depth). As a result of the interstitial-water > seawater concentration differences, diffusive fluxes will be set up to transport the dissolved metals across the sediment–water interface. Hartmann & Muller (1982) made lower limit estimates of the magnitudes of these fluxes from the upper 2 cm of the sediments and these are listed in Table 14.3, together with estimates of the rates at which the elements accumulate in the deposits.

A similar type of study was carried out by Callender & Bowser (1980) on the interstitial water chemistry of Mn and Cu in nodule-rich pelagic sediments from the northeastern equatorial Pacific. The distributions of Mn and Cu in the interstitial waters varied with depth, that of Cu exhibiting a regular pattern, with the highest concentrations being in the waters from the upper 10 cm. The authors used their data to calculate diffusive interstitial fluxes and the findings are given in Table 14.4. Callender & Bowser (1980) concluded that Cu is transported to the sediments mainly in association with a biogenic carrier phase (see also Sawlan & Murray (1983) and Section 11.6.3.2), and that the interstitial-water profiles are

**Table 14.3** Average concentrations of Mn, Cu, Zn and Ni in interstitial waters from central Pacific clays, siliceous oozes and carbonate oozes, together with enrichments relative to ambient sea water and their fluxes into the sediments (data from Hartmann & Muller, 1982).

| Element | Depth within sediment column (cm) | Mean concentration in interstitial waters ($\mu$g l$^{-1}$) | Mean enrichment factor relative to sea water | Diffusive flux from sediments ($\mu$g cm$^{-2}$ yr$^{-1}$) | Rates of accumulation to sediments ($\mu$g cm$^{-2}$ yr$^{-1}$) |
|---|---|---|---|---|---|
| Mn | 0–2 | 3.7 ± 2.2 | 18.6 | 0.18 | 0.3–0.7 |
| | 2–38 | 3.6 ± 3.2 | 15.2 | | |
| | 38–973 | 2.8 ± 2.4 | 12.2 | | |
| Cu | 0–2 | 4.8 ± 2.4 | 28.4 | 0.23 | 0.01–0.06 |
| | 2–38 | 2.1 ± 1.5 | 12.4 | | |
| | 38–973 | 0.9 ± 0.6 | 5.4 | | |
| Zn | 0–2 | 17.3 ± 20.7 | 3.2 | 0.58 | 0.01–0.02 |
| | 2–38 | 19.9 ± 20.3 | 3.7 | | |
| | 38–973 | 19.6 ± 17.2 | 3.6 | | |
| Ni | 0–2 | 2.2 ± 2.2 | 4.6 | 0.08 | 0.01–0.02 |
| | 2–38 | 1.4 ± 1.1 | 3.0 | | |
| | 38–973 | 1.1 ± 0.9 | 2.3 | | |

**Table 14.4** Interstitial water concentration, diffusive fluxes and accumulation rates of Mn and Cu in northeast equatorial Pacific Ocean sediments (data from Callender & Bowser, 1980).

| Sediment | Manganese | | | Copper | | |
|---|---|---|---|---|---|---|
| | Concentration 0–2 cm ($\mu$g l$^{-1}$) | Mean diffusion flux from sediments ($\mu$g cm$^{-2}$ yr$^{-1}$) | Mean rate of accumulation in sediments ($\mu$g cm$^{-2}$ yr$^{-1}$) | Concentration 0–2 cm ($\mu$g l$^{-1}$) | Mean diffusion flux from sediments ($\mu$g cm$^{-2}$ yr$^{-1}$) | Mean rate of accumulation in sediments ($\mu$g cm$^{-2}$ yr$^{-1}$) |
| Terrigenous mud | 1.5 | 0.28 | 2.55 | 11.5 | 0.72 | 0.038 |
| Pelagic clay | 1.2–11 | 0.015 | 1.06 | 5–9 | 0.31 | 0.06 |
| Siliceous ooze | 0.07–<0.7 | 0.025 | 0.18 | 1–3 | 0.17 | 0.015 |
| Calcareous ooze | 3.0 | 0.12 | 0.20 | 14 | 0.78 | 0.012 |

maintained by its rapid release in surficial sediments and its uptake on to solid phases at depth. The rapid release from surficial sediments results in a diagenetic flux of Cu across the sediment–water interface. The authors used the calculated diffusive flux rates and net sediment accumulation rates to estimate that the regeneration of Cu from these deposits is >90%; i.e. most of the sedimentary flux of Cu to the sea floor is returned as dissolved Cu via a diagenetic flux across the sediment–water interface. This was in contrast to the behaviour of Mn. Manganese is transported down the water column following particle scavenging reactions (see Section 11.6.3.1), and although it undergoes diagenetic remobilization much of the resultant diffusive flux is taken up by sediment components and so becomes trapped before it can cross the sediment–seawater interface in a dissolved form. For example, Callender & Bowser (1980) estimated that <10% of the Mn in the pelagic clays and siliceous oozes which they studied undergoes diagenetic transport into the overlying sea water. Unlike Cu, therefore, most of the remobilized Mn is trapped with the sediments.

The remobilization and recycling of trace elements in the sediment–interstitial-water–seawater complex is intimately related to the diagenetic environment under which the sediment is deposited, and this can be summarized in terms of the findings of a number of investigations.

*Pelagic sediments.* These are oxic to considerable depths (see Section 14.3.4), and surface enrichments do not appear to be a characteristic feature of the distributions of Mn, Fe, Ni and Cd in their interstitial waters. Dissolved Cu, however, does show a concentration maximum at or near the sediment–seawater interface. This can be illustrated by data provided by Klinkhammer *et al.* (1982) for sediments collected at MANOP site S in the central equatorial Pacific. These sediments were oxic down to 30 cm, and their interstitial-water profiles are shown in Fig. 14.8(a). The authors derived a model in which it was assumed that in oxic sediments the destruction of organic matter is accomplished by dissolved oxygen, i.e. early diagenesis is analogous to oxidation in the overlying water column and interstitial water near the sediment–seawater interface is a continuum of ambient bottom waters. Nickel and Cd are nutrient-type elements that have reasonably well-defined

metal–nutrient relationships in sea water (see Section 11.5.2), and their interstitial-water concentrations in the oxic sediments at site S could be predicted adequately by the model from the interstitial-water nutrient concentrations. The most striking feature of the dissolved Cu profile reported by Klinkhammer *et al.* (1982), and confirmed in other pelagic cores by Sawlan & Murray (1983), is the presence of a concentration maximum near to the sediment–seawater interface (see Fig. 14.8a). The diagenetic model could not, however, be applied to the concentrations of Cu in the interstitial waters because the amount of this element released at the interface was considerably in excess of that predicted by the decomposition of organic debris on the basis of the metal–nutrient data; a second model was therefore constructed to describe the regeneration of Cu in the sediments. The models used by Klinkhammer *et al.* (1982) are described in Worksheet 14.3.

*Hemi-pelagic sediments.* These are in a reducing condition below a relatively thin oxic surface layer (see Section 14.3.3), and their interstitial waters tend to show large concentration gradients in the distributions of Mn, Ni and Cu. This can be illustrated by the data provided by Klinkhammer *et al.* (1982) for sediments collected at MANOP site C in the central equatorial Pacific (see Fig. 14.8b); see also the variety of profiles given by Sawlan & Murray (1983). At site C there is a sharp increase in the dissolved concentrations of Mn in the interstitial waters below 9 cm. This is characteristic of *suboxic* interstitial waters and results from the utilization and solubilization of Mn oxides in the diagenetic sequence. Above 9 cm, however, the dissolved Mn concentrations are lower and relatively constant, i.e. the concentration maximum does not extend through the oxic layer to the sediment–water interface. These upper oxidized portions of deep-sea sediments often exhibit an increase in solid-phase manganese, and it is generally accepted that this arises from the reprecipitation trapping of the solubilized Mn under the oxic conditions. The distribution of Ni in the interstitial waters generally follows that of Mn, i.e. there is no surface maximum, and the subsurface maximum occurs at a similar depth to that of Mn, probably as a result of its release from the Mn oxides. There is, however, a surface maximum in the interstitial-water Cu profiles, with the concentrations at the surface usually being

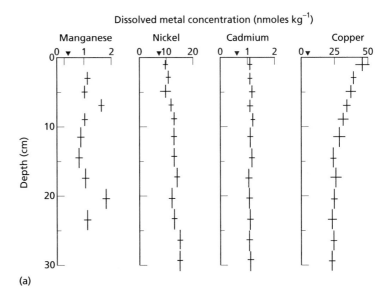

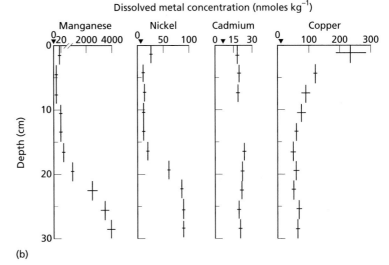

**Fig. 14.8** Interstitial-water profiles of dissolved trace metals (from Klinkhammer *et al.*, 1982). (a) Profiles from a sediment that was in an oxic condition down to 30 cm: MANOP site S, central equatorial Pacific. (b) Profiles from a sediment that was in a suboxic condition below an oxic surface layer: MANOP site C, central equatorial Pacific. Arrowheads indicate bottom-water concentrations.

considerably higher than those in pelagic sediment interstitial waters. This surface maximum leads to a diffusive flux out of the sediment—see below.

*Highly reducing shelf sediments.* In these sediments the upper oxic zone is either confined to the top few centimetres or is absent altogether. Sawlan & Murray (1983) gave data on the distributions of a number of elements in the interstitial waters from shelf sediments of this kind. They reported that in these

suboxic environments elevated dissolved Mn concentrations can be found at or near the sediment–water interface. These sediments, with their high core-top dissolved Mn concentrations, represent a window to the bottom water through which Mn can diffuse, thus reinforcing the status of shelf sediments for the supply of dissolved Mn to sea water (see Section 11.4). In contrast, Cu did not show the surface enrichments reported for the pelagic and hemi-pelagic sediment interstitial waters, and the concentrations were

**Worksheet 14.3: Models for the regeneration of trace metals in oxic sediments**

Klinkhammer *et al.* (1982) reported data on the concentrations of Ni, Cd and Cu in bottom waters and oxic sediment interstitial waters at MANOP site S in the central equatorial Pacific, and used these to set up diagenetic models. A characteristic feature of the metal interstitial-water profiles at the oxic site (S) is that the steepest concentration gradient is found across the sediment–seawater interface, which implies that most metal regeneration in these oxic sediments takes place across this interface. The concentration–depth profiles are maintained by a combination of the regeneration and the interaction between the dissolved metals and the sediment below the interface. Nickel and Cd exhibit little tendency to react with sediment components under these conditions, with the result that their interstitial water profiles are monotonic at site S and are generated by the burial of surficial pore water. In contrast, Cu is readily taken up by sediment components, which leads to an exponential decrease in interstitial-water dissolved Cu concentrations with depth — see Fig. (i).

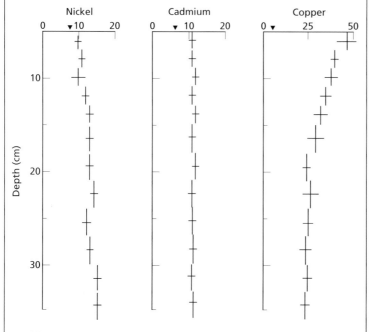

**Fig. (i)** Dissolved metal concentrations in the interstitial waters of oxic sediments at MANOP site S (nmol l⁻¹). Arrow heads indicate bottom-water concentrations.

*continued*

*Ni and Cd regeneration*

Under conditions of oxic diagenesis the degradation of organic matter in surface sediments utilizes dissolved oxygen, and Klinkhammer *et al.* (1982) assumed that in the simplest case early oxic diagenesis on the sea floor is analogous to oxidation in the overlying water column, i.e. a 'continuum model'. Nickel and Cd are nutrient-type elements in sea water so that under oxic conditions it should be possible to predict the Ni and Cd concentrations in the surficial interstitial waters from the interstitial nutrient concentrations. To set up a model for this, Klinkhammer *et al.* (1982) assumed that both the metals and the nutrients are lost from the sediment–seawater interface by diffusion only, so that the flux of metal $M$ across the interface is related to the corresponding flux of a nutrient $N$ by a proportionality constant $M/N$. Thus

$$D_M \left( \frac{dM}{dz} \right)_{z=0} = \frac{M}{N} D_N \left( \frac{dN}{dz} \right)_{z=0} \tag{1}$$

By assuming a linear concentration gradient across the interface, eqn (1) reduces to the relationship

$$\frac{M}{N} = \frac{(M_0 - M_{BW})D_M}{(N_0 - N_{BW})D_N} \tag{2}$$

In the water column Ni mimics silica and Cd is related to nitrate, and using data from the literature the appropriate diffusion coefficient ratios ($D$) were calculated to be $D_{Cd}/D_{NO_3} = 0.36$ and $D_{Ni}/D_{Si} = 0.68$. Variables $M_0$ and $M_{BW}$ are metal concentrations in the top interstitial-water interval and bottom sea water, and $N_0$ and $N_{BW}$ are the corresponding nutrient concentrations. A comparison of the results obtained from eqn (2) with the ratios found in sea water is a test of continuity. Klinkhammer *et al.* (1982) made such a comparison and the data from site S are reproduced in Table (i), and strongly support the 'continuum model'.

*Cu regeneration*

Copper is readily taken up by sediment components, which leads to an exponential decrease in interstitial-water dissolved Cu concentrations

**Table (i)** Metal : nutrient ratios (Ni : Si and Cd : NO₃) at site S calculated from eqn (2) compared with those observed in general sea water and bottom water at the site (from Klinkhammer *et al.*, 1982).

| Element | $(M:N)_{\text{site S}} \times 10^5$ | $(M:N)_{\text{SW}} \times 10^5$ | $(M:N)_{\text{BW}} \times 10^5$ |
|---------|--------------------------------------|----------------------------------|----------------------------------|
| Ni | 3.2 | 3.3 | 5.3 |
| Cd | 3.3 | 2.3 | 1.8 |

*continued on p. 392*

with depth. In addition, dissolved Cu is released into sea water at the interface. The amount of Cu released in this way, however, is considerably greater than that which would be predicted from the decomposition of organic material consisting of plankton. Thus, the model developed for Ni and Cd is inappropriate for Cu. Klinkhammer *et al.* (1982) therefore assumed that the interstitial-water dissolved Cu profiles are sustained by scavenging from the water column combined with vigorous recycling at the interface and uptake into the sediment. The authors then attempted to model the interstitial-water dissolved Cu profile in the following manner.

The shape of the interstitial-water dissolved Cu profile at site S (see Fig. (i)) shows a concentration gradient that indicates diffusion upwards (into sea water) and downwards (into the sediment) from the interface, and the negative curve suggests an uptake by the sediment at depth. The authors concluded that the simplest model consistent with this type of profile was a steady-sate approach in which the Cu is controlled by diffusion and first-order removal, which can be represented by the equation:

$$D\frac{\partial^2 C}{\partial z^2} - kC = 0. \tag{3}$$

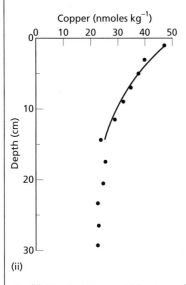

(ii)

**Fig. (ii)** Dissolved Cu interstitial-water profiles in the oxic sediments at site S. Filled circles are the measured concentrations in the interstitial waters. The full curve indicates the curve derived from eqn (4) using the following parameters: $C_1 = 47\,\text{nmol}\,\text{kg}^{-1}$, $C_2 = 26\,\text{nmol}\,\text{kg}^{-1}$, $h = 13\,\text{cm}$ and $k = 0.17\,\text{yr}^{-1}$.

*continued*

The solution of eqn (3) for a one-layer sediment is given by

$$C = \frac{C_1 \sinh[R(h-z)] + C_2 \sinh(Rz)}{\sinh(Rh)} \quad (4)$$

where $R = (k/D)^{1/2}$, $C$ is the concentration predicted at depth $z$, $C_1$ and $C_2$ are the concentrations at the upper ($z = 0$) and lower ($z = h$) boundaries, $D$ is the diffusion coefficient corrected for porosity, tortuosity and temperature ($D = 1.4 \times 10^{-6}\,cm\,s^{-1}$), and $k$ is the reaction constant in $s^{-1}$. Values for $k$ were calculated using a best-fit approach. The authors modelled their data to a depth of $c$. 15 cm in the sediment, below which there is little variation. The results for site S are illustrated in Fig. (ii). It is apparent from Fig. (ii) that there is very good agreement between the predicted and observed dissolved Cu interstitial-water profiles, i.e. the profiles are adequately explained by the remobilization of Cu at the interface and its removal into the sediment below.

**Table 14.5** Diffusive fluxes of some elements from deep-sea sediments.

| Sediment type | Diffusive flux ($\mu g\,cm^{-2}\,yr^{-1}$) | | | | Data source |
| | Cu | Mn | Ni | Cd | |
|---|---|---|---|---|---|
| Oxic pelagic | 0.12 | 0.0008 | −0.002 | −0.002 | Klinkhammer *et al.*, 1982 |
| Pelagic red clay | 0.18 | — | — | — | Sawlan & Murray, 1983 |
| Hemi-pelagic | 0.38 | — | — | — | Sawlan & Murray, 1983 |

generally similar to those in ambient sea water, i.e. there was no concentration gradient to drive an out-of-sediment flux, such as was found for the pelagic and hemi-pelagic sediments.

A summary of some of the flux data provided by both Klinkhammer *et al.* (1982) and Sawlan & Murray (1983) is given in Table 14.5.

It may be concluded that the fluxes of some trace metals, such as Mn, Ni and Cu, through the interstitial waters are driven by reactions involving the destruction of organic carbon, and so are linked to the diagenetic sequence. As a result, some trace metals are released close to the sediment surface during *oxic diagenesis* when organic carbon is destroyed by dissolved oxygen, whereas others are released at depth during *suboxic* diagenesis, when secondary oxidants, such as Mn oxides, are utilized for the destruction of the organic matter. The elements released into the interstitial waters can then migrate in either upward or downward directions,

and the shapes of the concentration profiles are often diagnostic of the flux direction. This can be illustrated with respect to the interstitial-water distributions of Cu and Mn.

*Copper.* In the interstitial waters of both pelagic and hemi-pelagic sediments dissolved Cu can have its maximum concentrations at or near the sediment surface, and the shapes of the dissolved Cu profiles reported by Klinkhammer *et al.* (1982) (see Fig. 14.8) indicate both upward and downward diffusion from the sediment–seawater interface.

**1 Downward diffusion.** The decrease in dissolved Cu below the interface indicates downward diffusion and the removal of the element into the solid sediment phases, i.e. the downward-transported Cu is trapped in solid sediment material either at or near the sediment surface or at depth within the sediment column. According to various estimates, from at least half (Sawlan & Murray, 1983) to most

(Klinkhammer *et al.*, 1982) of the Cu in hemi-pelagic and pelagic deep-sea sediments has an oxic diagenetic origin and is supplied by downward diffusion.

**2 Upward diffusion.** Because of the position of the dissolved Cu maximum at the sediment–seawater interface, the dissolved Cu transported upwards will drive a diffusive flux into the overlying sea water. Klinkhammer *et al.* (1982) estimated this flux to be $\sim 1800\,nmol\,cm^{-2}\,10^3\,yr^{-1}$ at MANOP site S and $\sim 6600\,nmol\,cm^{-2}\,10^3\,yr^{-1}$ at MANOP site C. The authors concluded that the Cu must be released from a very thin layer at the top of the sediment, which they estimated should have a Cu content of $\sim 2500–5000\,\mu g\,g^{-1}$. As concentrations of this kind are not picked up in sediment analysis, the thin layer is probably present as 'fluff' (see Section 12.3), from which it was suggested that the Cu is recycled to sea water around nine times before it is eventually buried in the sediment. The out-of-sediment diffusive fluxes derived by Klinkhammer *et al.* (1982) are given in Table 14.4. Sawlan & Murray (1983) modelled their interstitial-water data and estimated that the diffusive flux of dissolved Cu into sea water represented around 75% of the total Cu delivered to the sediments. They also calculated that at least 50% of the total Cu in hemi-pelagic sediments had a diagenetic origin from the into-sediment flux. It may be concluded, therefore, that a large fraction of the Cu remobilized in sediments can, under some conditions, diffuse into the overlying sea water. This out-of-sediment diffusive Cu flux can impose a 'fingerprint' on the water-column profile of dissolved Cu. For example, Boyle *et al.* (1977) calculated that on the basis of water-column profiles in the Pacific there must be a strong general flux of dissolved Cu from the sediment surface into the water column; this flux will, of course, include Cu taken into solution in the benthic boundary layer, as well as that which is remobilized in the sediment and then escapes across the interface. Boyle *et al.* (1977) applied a diffusion–advection model to estimate that the out-of-sediment dissolved Cu flux should have a magnitude of *c.* 0.16 $\mu g\,cm^{-2}\,yr^{-1}$, which is equivalent to an ocean-wide flux of $\sim 0.51 \times 10^{12}\,g\,yr^{-1}$. Using an approach based on the down-column transport of Cu in the North Atlantic, and the rate at which it accumulates in the underlying sediments, Chester (1981) showed that there was a regeneration of Cu at the depositional surface which would amount to a flux of $\sim 0.17\,\mu g$ $cm^{-2}\,yr^{-1}$, or to an ocean-wide flux of $\sim 0.55 \times 10^{12}\,g$ $yr^{-1}$ (see Table 14.6). Both of the latter flux values are close to those derived from interstitial-water concentration gradients (see Table 14.6), and so offer independent verification of the order-of-magnitude release of dissolved Cu from pelagic sediments.

*Manganese.* In strongly reducing shelf sediments, with the suboxic zone lying near the surface, the dissolved Mn concentration maxima are found close to the sediment–seawater interface, and the Mn can escape into sea water through these windows. In hemi-pelagic sediments, however, dissolved interstitial-water Mn usually has its highest concentrations at depth in the sediments as a result of the utilization of Mn oxides for the destruction of organic carbon in the suboxic zone; other elements, such as Ni, Co, Cu and Zn, which are associated with the oxides, also can be released in this zone. The shape of the dissolved Mn profiles in the interstitial waters (see e.g. Fig. 14.8b) characteristically exhibits a decrease in concentration in the oxic zone, indicating a removal of the element into the sediment phases. This is confirmed by the distribution of solid-phase Mn in the sediments, and it has been known for many years that Mn profiles in deep-sea sediments often display higher concentrations in the oxic layers close to the surface than at depth.

The reactions involved in the diagenesis of Mn can be related to interstitial-water-phase–solid-phase changes, and can be expressed in a simple form as:

$$Mn^{2+} \text{(soluble)} \underset{\text{reduction}}{\overset{\text{oxidation}}{\rightleftharpoons}} MnO_x \text{(solid hydrous oxide)}$$

$$(14.7)$$

where *x* generally is less than 2. This reaction governs the diagenetic mobility of Mn in sediments, and the general conditions that control both the solid-phase and dissolved Mn in sediments have been described by a number of models.

**1** The *Lynn–Bonatti model.* One of the initial models was that proposed by Lynn and Bonatti (1965), and in essence it can be summarized as follows. Manganese oxides (**first generation**) are deposited at the sediment surface and are subsequently buried below the redox boundary where they are reduced. This results in the production of dissolved $Mn^{2+}$, which then diffuses upwards, along a concentration gradient in the interstitial water, and is oxidized and precipitated in the upper sediment layer as hydrous Mn oxides (**second generation**). As sedimentation proceeds, the second-generation oxides are again carried down into the reducing zone and the cycle starts again. The overall result of the Mn recycling is the trapping of solid-phase Mn in a narrow zone at the redox boundary.

2 The *Froelich model*. The general pattern of Mn behaviour was refined by Froelich *et al.* (1979) in their classic paper on the diagenetic sequence in deep-sea sediments. In this model it is proposed that when the sequence reaches the stage at which manganese oxides are utilized to provide dissolved oxygen for the oxidation of organic matter, the reduction, mobilization and upward diffusion of $Mn^{2+}$ are initiated. Thus, the reduction–mobilization provides a mechanism for the stripping of Mn from the sediments as they accumulate, and the upward diffusion provides a mechanism for transporting it back into oxic layers, where it is subsequently redeposited following oxidation to $MnO_2$. With further sediment accumulation the new $MnO_2$ passes into the zone where it is used as an oxidant and is again remobilized. This leads to the setting up of the sedimentary Mn trap, which can give rise to Mn-rich bands, or **spikes**, in some types of sediment (see Fig. 14.9). Froelich *et al.* (1979) proposed that the depth of the Mn spike is controlled by the balance of oxygen diffusing downwards and $Mn^{2+}$ diffusing upwards. In a steady-state system the concentration of Mn in the spike will increase until the sedimentary input of reactive Mn is balanced by the efficiency of reduction and remobilization. Thus, such a steady-state system would display the highly concentrated Mn spike near the top of the dissolved $Mn^{2+}$ gradient. The Froelich model is illustrated diagrammatically in Fig. 14.9(a).

3 The *Burdige-Gieskes model*. The general type of steady-state Mn diagenesis has also been described by Burdige & Gieskes (1983). In their model the sedimentary column is divided into four distinct zones.

(a) **Oxidized zone.** This is the zone in which first-generation Mn oxides accumulate, and the concentration of dissolved $Mn^{2+}$ is essentially zero.

(b) **Manganese oxidation zone.** In this zone the dissolved $Mn^{2+}$ profile increases with depth and is concave upwards as a result of the diffusion of $Mn^{2+}$ from below across the redox boundary and its consumption by oxidation in the zone. The oxidation product is a solid hydrous Mn oxide, so that

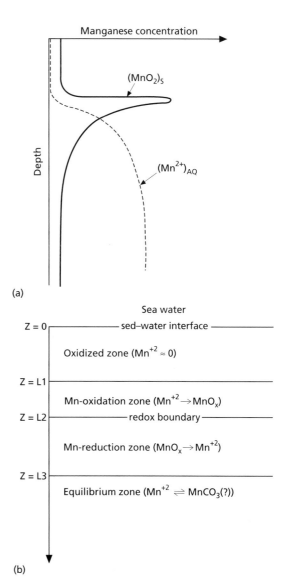

**Fig. 14.9** Manganese diagenesis in sediments under steady-state conditions. (a) Schematic representation of dissolved (AQ) and solid-phase (S) Mn profiles in a hypothetical steady-state sediment system (from Froelich *et al.*, 1979). (b) Schematic representation of the zonation of marine sediments with respect to Mn diagenesis (from Burdige & Gieskes, 1983).

there is an increase in this component with depth. The **redox boundary** separates the Mn oxidation and Mn reduction zones. This boundary is the depth in the sediment below which $Mn^{2+}$ is favoured over solid Mn oxide phases.

(c) **Manganese reduction zone.** As solid Mn oxides are buried below the redox boundary they undergo reduction, e.g. as the diagenetic sequence switches to the secondary oxidants. $Mn^{2+}$ is released into the interstitial waters and the $Mn^{2+}$ profiles are concave downwards. Thus, dissolved $Mn^{2+}$ increases with depth, whereas solid Mn decreases.

(d) **Equilibrium zone.** Here dissolved Mn reaches a maximum and may in fact decrease with depth perhaps owing to the formation of some kind of Mn carbonate.

The main features in the Burdige–Gieskes model are illustrated diagrammatically in Fig. 14.9(b) and a mathematical treatment of the model is given in Worksheet 14.4.

---

**Worksheet 14.4: The Burdige-Gieskes model for the steady-state diagenesis of Mn in marine sediments**

Burdige & Gieskes (1983) outlined a steady-state pore-water–solid-phase diagenetic model for Mn in marine sediments. The model has been described qualitatively in the text, and also can be used to illustrate how *diagenetic equations* are used. The Burdige–Gieskes steady-state diagenetic model, which involves a series of Mn reaction zones, is illustrated in Fig. (i).

The derivation of the diagenetic equations used to translate this into a quantitative mathematical model is summarized below.

Burdige & Gieskes (1983) presented a mathematical treatment of their Mn 'zonation model' using the steady-state diagenetic equations given by Berner (1980):

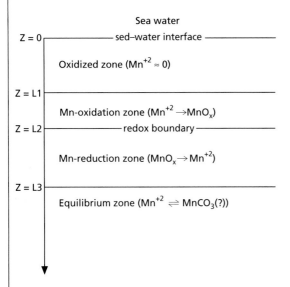

**Fig. (i)** Manganese steady-state diagenetic zonation model (from Burdige & Gieskes, 1983).

*continued*

$$D_b \frac{\partial^2 C_p}{\partial z^2} - w \frac{\partial C_p}{\partial z} + R(z) = 0 \tag{1}$$

$$-w \frac{\partial C_s}{\partial z} - \frac{\phi}{1 - \phi} R(z) = 0 \tag{2}$$

where $D_b$ is the bulk sediment diffusion coefficient (identical with Berner's $D_s$ term), $C_p$ is the concentration of Mn in the pore waters, $C_s$ is the concentration of Mn in the solid phase, $w$ is the sedimentation rate, $\phi$ is the porosity and $R(z)$ is a rate expression for either oxidation or reduction. The term $\phi/(1 - \phi)$, which has units $cm^3_{pore\ water}/cm^3_{dry\ sed}$, is used to convert Mn concentrations between the solid and liquid phases.

With respect to the Mn model the following assumptions are implicit in the diagenetic equations.

1 Steady-state diagenesis operates.

2 Vertical gradients are much more important than horizontal gradients, to the extent that the latter can be ignored.

3 Diffusion in pore waters occurs via molecular processes, i.e. the diffusion follows Fick's laws.

4 Porosity and diffusion coefficients are constant with depth.

5 Advection is constant and equal to the sedimentation rate.

6 Solid-phase diffusion can be neglected.

7 The supply of solid Mn to the sediment surface is constant with time.

8 Adsorption of $Mn^{2+}$ can be neglected.

9 Bioturbation can be neglected because in most marine sediments it will be above the oxidation (or reducing) zone.

The authors also drew attention to two other factors, which are detailed below.

Variable $D_b$ differs from the free-ion diffusion in sea water $(D)$ because of tortuosity effects resulting from the presence of sediment particles, and $D_b$ can be related to $D$ by the following equation:

$$D_b = D/\phi F \tag{3}$$

where $F$ is a 'formation factor', which is measured as the ratio of the bulk sediment resistivity to the pore-water resistivity (see Worksheet 14.3).

In eqns (2) and (3), $R(x)$ is the rate expression for either oxidation or reduction. In the presence of an abundant surface area (such as that of a sediment), Mn precipitation–oxidation is a pseudo-first-order process, so that

$$R_{ox}(z) = k_{ox} C_p \tag{4}$$

and eqn (5) therefore should be an appropriate rate expression for oxidation under these conditions, in which the reaction product is

*continued on p. 398*

assumed to be a hydrous oxide ($MnO_x$—see text). The authors assumed that whatever the mechanism involved, the rate of Mn reduction will be proportional to the amount of solid Mn available (which is presumed to be all hydrous oxide). Thus, $R_{red}(z)$ can be expressed as

$$R_{red}(z) = k_{red}C_s. \tag{5}$$

Combining the rate expressions with eqns (1) and (2), the following set of equations was obtained:

**1** For the oxidizing zone ($L_1 \leq z \leq L_2$)

$$D_b \frac{\partial^2 C_p^{ox}}{\partial z^2} - w \frac{\partial C_p^{ox}}{\partial z} - k_{ox}C_p^{ox} = 0 \tag{6}$$

$$-w \frac{\partial C_s^{ox}}{\partial z} + \frac{\phi}{1-\phi} k_{ox}C_p^{ox} = 0 \tag{7}$$

**2** For the reducing zone ($L_2 \leq z \leq L_3$)

$$D_b \frac{\partial^2 C_p^{red}}{\partial z^2} - w \frac{\partial C_p^{red}}{\partial z} + \frac{1-\phi}{\phi} k_{red}C_s^{red} = 0 \tag{8}$$

$$-w \frac{\partial C_s^{red}}{\partial z} - k_{red}C_s^{red} = 0 \tag{9}$$

By defining a series of boundary conditions the solutions to eqns (6)–(9) were given as

$$C_p^{ox} = A\sinh[\alpha(z - L_1)] \tag{10}$$

$$C_p^{red} = G - \frac{1-\phi}{\phi} \frac{Ew^2}{k_{red}D_b} \exp[-\beta(z - L_2)] \tag{11}$$

$$C_s^{ox} = C_s^o \cosh[\alpha(z - L_1)] \tag{12}$$

$$C_s^{red} = E\exp[-\beta(z - L_2)] \tag{13}$$

where

$$\alpha = (k_{ox}/D_b)^{1/2} \tag{14}$$

(assuming $4D_b k_{ox} \gg w^2$, i.e. advection in pore waters is negligible compared with diffusion) and

$$\beta = k_{red}/w \tag{15}$$

$$L_{ox} = L_2 - L_1 \tag{16}$$

$$A = \frac{C_s^o(1-\phi)w}{\phi\alpha D_b} \tag{17}$$

$$E = C_s^o \cosh(\alpha L_{ox}) \tag{18}$$

*continued*

$$G = A[\sinh(\alpha L_{\mathrm{ox}}) + \alpha/\beta \cosh(\alpha L_{\mathrm{ox}})] \qquad (19)$$

There are two unknowns ($k_{\mathrm{ox}}$ and $k_{\mathrm{red}}$) in the equation, which can be used for a least-squares fit of the equations to actual data.

By inserting typical parameters for marine sediments into the equations the authors derived model profiles for the diagenesis of dissolved and solid-phase Mn. These are illustrated in Fig. (iia), and show close agreement with the diagrammatic representation suggested by Froelich *et al.* (1979)—see Fig. (iib).

The Mn post-depositional migration associated with this model may be summarized as follows (see text for detailed explanation). As the sediments are buried below the redox boundary, solid Mn oxides are reduced to $Mn^{2+}$, which diffuses upwards along the concentration gra-

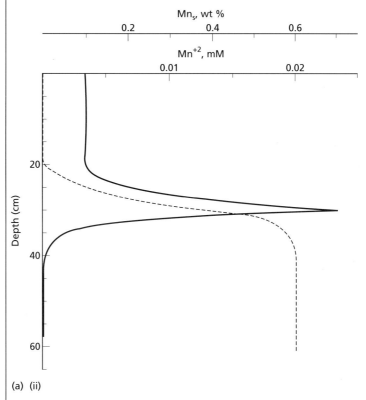

**(a) (ii)**

**Fig. (ii)** Theoretical models for Mn diagenesis in marine sediments. (a) Model for porewater (broken curve) and solid-phase (full curve) Mn profiles (from Burdige & Gieskes, 1983). Profiles predicted from eqns (10) to (19) using the following parameters: $D_b = 71.6\,\mathrm{cm^2\,yr^{-1}}$, $w = 3\,\mathrm{cm\,10^3\,yr^{-1}}$, $\phi = 0.8$, $\rho = 2.6\,\mathrm{cm^{-3}}$ of sediment, $L_1 = 20\,\mathrm{cm}$, $L_2 = 30\,\mathrm{cm}$, $C_s^o = 0.1\,\mathrm{wt\%}$, $k_{\mathrm{ox}} = 5\,\mathrm{yr^{-1}}$, and $k_{\mathrm{red}} = 1.50 \times 10^{-3}\,\mathrm{yr^{-1}}$. (b) Schematic representation of model predicted by Froelich *et al.* (1979).

*continued on p. 400*

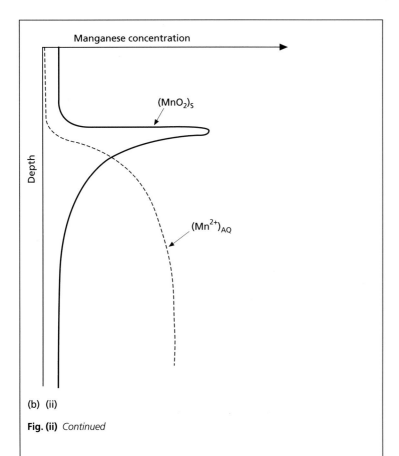

Manganese concentration

Depth

$(MnO_2)_S$

$(Mn^{2+})_{AQ}$

(b) (ii)

**Fig. (ii)** *Continued*

dient to be oxidized and reprecipitated. With further sedimentation, these reprecipitated oxides are again brought to the reducing zone to be redissolved, with the net result being the Mn trap at the narrow zone at the redox boundary.

Burdige & Gieskes (1983) also applied their equations to solid and porewater data from two cores in the eastern Equatorial Atlantic and obtained very good fits to the data.

4 The *Pedersen model*. Both Froelich *et al.* (1979) and Burdige & Gieskes (1983) sucessfully applied their steady-state Mn diagenesis models to field situations and were able to identify the presence of Mn spikes in deep-sea sediments displaying an upper oxic zone separated from a lower reducing zone by a redox boundary. However, Pedersen *et al.* (1986) showed that the diagenesis of Mn in such sediments need not always take place under steady-state conditions. These authors gave data on the distribu-

tion of solid-phase and interstitial-water Mn in a hemi-pelagic core from the East Pacific Rise. The interstitial-water $Mn^{2+}$ profile showed a negative concave-upward form and there was an enrichment of solid-phase Mn in the near-surface layer, two features characteristic of hemi-pelagic sediments. However, the solid-phase and water-phase Mn were in disequilibrium, indicated by the fact that there was no concentration gradient for dissolved Mn in the upper 5 cm of the core in the region over which the solid-phase Mn increased in concentration. The authors concluded that the increase in solid-phase Mn in the top 5 cm therefore could not be supplied by an upward diffusion of dissolved $Mn^{2+}$ and its precipitation close to the sediment surface. Instead, in the Pedersen non-steady-state model it was proposed that precipitation of dissolved Mn takes place below the enriched horizon. This was thought to be due to non-steady-state diagenesis brought about by a decrease in primary productivity in the overlying water, which lessened the amount of organic carbon reaching the sediment and so lowered the oxygen demand at or near the interface and set up a downward migrating oxidation front (see Section 14.4.2) which eroded the $Mn^{2+}$ interstitial water profile.

From these various examples of Mn diagenesis it is apparent that much of the Mn remobilized in deep-sea sediments is trapped in the upper oxic layers and does not cross the sediment–seawater interface to yield a strong out-of-sediment flux, and any Mn that does get this far (e.g. in the absence of an oxic layer) is often incorporated into ferromanganese nodules (see Section 15.3.4.6). In general, therefore, Mn redox recycling in deep-sea sediments that have an upper oxic layer leads to a concentration of the element in the layer. Sawlan & Murray (1983) therefore identified two types of *upward* metal fluxes in oceanic sediments:

1 those to the base of the Mn oxidation zone, where they can become trapped, which are termed the **oxidation zone fluxes**;

2 those to the sediment–seawater interface, from which they can diffuse into the overlying sea water, which are referred to as the **benthic fluxes**.

In addition, **downward** metal fluxes can transport metals (e.g. Cu) to depth in the sediment.

A summary of the more recent data on the shallow depth diffusive fluxes of Mn, Cu and Zn is given in Table 14.6. For Cu and Mn there is a reasonable amount of data available, and for these two elements there is a remarkable degree of agreement on the magnitude of the fluxes that have been obtained from interstitial-water concentration gradients in *pelagic* sediments, those for Mn ranging from $\sim0.14 \times 10^{12}$ to $\sim0.56 \times 10^{12}\,g\,yr^{-1}$ and those for Cu being in the range $\sim0.31 \times 10^{12}$–$1.1 \times 10^{12}\,g\,yr^{-1}$.

**Table 14.6** Data on global shallow (0–50 cm) depth diffusive fluxes of Mn, Cu and Zn from marine sediments; units, $10^{12}\,g\,yr^{-1}$.

| Element | Sediment type | Diffusive flux* | Original data source |
|---|---|---|---|
| Mn | Estuarine | 189 | Elderfield & Hepworth, 1975 |
| | Hemi-pelagic mud | 0.88 | Callender & Bowser, 1980 |
| | Pelagic sediments | 0.14 | Callender & Bowser, 1980 |
| | | 0.56 | Hartmann & Muller, 1982 |
| | | 0.22 | Elderfield, 1976 |
| | | 0.14 | Manheim, 1976 |
| | Pelagic sediment range | 0.1–0.6 | |
| Cu | Estuarine | 11 | Elderfield & Hepworth, 1975 |
| | Hemi-pelagic mud | 2.3 | Callender & Bowser, 1980 |
| | Pelagic sediments | 1.1 | Callender & Bowser, 1980 |
| | | 0.73 | Hartmann & Muller, 1982 |
| | | 0.51 | Boyle *et al.*, 1977 |
| | | 0.55 | Chester, 1981 |
| | | 0.31 | Dymond (in Callender & Bowser, 1980) |
| | Pelagic sediment range | 0.3–1 | |
| Zn | Estuarine | 3.4 | Elderfield & Hepworth, 1975 |
| | Pelagic sediments | 1.8 | Hartmann & Muller, 1982 |
| | | 3.0 | |
| | Pelagic sediment range | 2–3 | |

* All values are approximate.

## 14.7 Interstitial water inputs to the oceans: summary

1 The interstitial waters of most oceanic sediments originate as trapped sea water.

2 The external boundary conditions for the interstitial-water–sediment complex are set by the basalt basement below and the water column above.

3 Reactions take place in the sediment–interstitial-water sandwich, during which changes are produced in the composition of the interstitial waters relative to the parent sea water, and diffusion gradients are set up via which elements migrate from high- to low-concentration regions.

4 The most intense reactions usually take place in the upper ~30–50 cm of the sediment column. These reactions take place in response to the diagenetic environment, the intensity of the reactions generally decreasing in the order highly reducing shelf sediments > hemi-pelagic sediments > pelagic sediments.

5 Data are now available on the diffusive interstitial water fluxes of a number of major and trace components. For some of these components interstitial waters act as sinks for their removal from sea water; for others they act as sources for their addition to sea water.

6 When considering interstitial water sources it is extremely important to make a distinction between primary and secondary fluxes. Interactions that take place between the basalt basement and interstitial waters, and between some types of dispersed volcanic material and interstitial waters, can lead to dissolved components being introduced into sea water for the first time. These are the **primary fluxes** in the same way as hydrothermal fluxes are primary in nature. In contrast, a large fraction of the trace metals that are mobilized into interstitial waters from particulate material via the diagenetic sequence has already been taken out of solution in sea water in the biogeochemical cycles that operate in the water column, and so is simply being returned to the dissolved state. These are the recycled or **secondary fluxes**. The dissolved trace metals associated with these fluxes can sometimes impose 'fingerprints' on the water column if they escape recapture at the sediment surface. However, the important point is that they have been recycled *within* the ocean system, and as such do not represent an additional primary external source. Despite this, the diagenetic remobilization and release of elements that are initially removed from solution by incorporation into estuarine and coastal sediments can sometimes reverse the processes that occur in estuaries, and Bruland (1983) has suggested that for these elements such a diagenetic release can be regarded as an 'additional dissolved river input' to the oceans.

The diagenetic reactions that take place both within the sediment–interstitial-water complex and at the seawater–sediment interface can strongly modify the composition of the material that reaches the sediment reservoir. The nature of these reactions has now been described, and in the following chapter we will consider the composition of the components that form the oceanic sediments in terms of their genetic histories, which, for some material, includes diagenetic modification.

## References

Bender, M.L. & Heggie, D.T. (1984) Fate of organic carbon reaching the deep sea flow: a status report. *Geochim. Cosmochim. Acta*, **48**, 977–86.

Berner, R.A. (1980) *Early Diagenesis: a Theoretical Approach*. Princeton, NJ: Princeton University Press.

Berner, R.A. (1981) A new geochemical classification of sedimentary environments. *J. Sed. Petrol.*, **51**, 359–65.

Berner, R.A. (1989) Biogeochemical cycles of carbon and sulfur in the modern ocean and their effect on atmospheric oxygen over Phanerozoic time. *Palaeogeogr. Palaeoclimatol. Palaeoecol.*, **73**, 97–122.

Berner, R.A. (1995) Sedimentary organic matter preservation: an assessment and speculative synthesis. *Mar. Chem.*, **49**, 121–2.

Boyle, E.A., Sclater, F.R. & Edmond, J.M. (1977) The distribution of copper in the Pacific. *Earth Planet. Sci. Lett.*, **37**, 38–54.

Bruland, K.W. (1983) Trace elements in sea water. In *Chemical Oceanography*, J.P. Riley & R. Chester (eds), Vol. 8, 156–220. London: Academic Press.

Burdige, D.J. & Gieskes, J.M. (1983) A pore water/solid phase diagenetic model for manganese in marine sediments. *Am. J. Sci.*, **283**, 29–47.

Burdige, D.J. & Martens, C.S. (1988) Biogeochemical cycling in an organic-rich coastal marine basin: 10. The role of amino acids in sedimentary carbon and nitrogen cycling. *Geochim. Cosmochim. Acta*, **52**, 1571–84.

Callender, E. & Bowser, C.J. (1980) Manganese and copper geochemistry of interstitial fluids from manganese nodule-rich pelagic sediments of the northeastern equatorial Pacific Ocean. *Am. J. Sci.*, **280**, 1063–96.

Calvert, S.E. (1976) The mineralogy and geochemistry of near-shore sediments. In *Chemical Oceanography*, J.P.

Riley & R. Chester (eds), Vol. 6, 187–280. London: Academic Press.

Calvert, S.E. & Price, N.B. (1970) Minor metal contents of Recent organic-rich sediments off South West Africa. *Nature*, **227**, 593–5.

Chester, R. (1981) Regional trends in the distribution and sources of aluminosilicates and trace metals in recent North Atlantic deep-sea sediments. *Bull. Inst. Geol. Bassin Aquitaine*, **31**, 325–35.

Degens, E.T. & Mopper, K. (1976) Factors controlling the distribution and early diagenesis of organic materials in marine sediments. In *Chemical Oceanography*, J.P. Riley & R. Chester (eds), Vol. 6, 59–113. London: Academic Press.

Drever, J.I. (1982) *The Geochemistry of Natural Waters*. Englewood Cliffs, NJ: Prentice-Hall.

Elderfield, H. (1976) Manganese fluxes to the oceans. *Mar. Chem.*, **4**, 103–32.

Elderfield, H. & Hepworth, A. (1975) Diagenesis, metals and pollution in estuaries. *Mar. Pollut. Bull.*, **6**, 85–7.

Ertel, J.R. & Hedges, J.I. (1985) Sources of sedimentary humic substances: vascular plant debris. *Geochim. Cosmochim. Acta*, **49**, 2097–107.

Froelich, P.N., Klinkhammer, G.P., Bender, M.L., *et al.* (1979) Early oxidation of organic matter in pelagic sediments of the eastern equatorial Atlantic: suboxic diagenesis. *Geochim. Cosmochim. Acta*, **43**, 1075–90.

Gagosian, R.B. (1986) The air–sea exchange of particulate organic matter: the sources and long-range transport of lipids in aerosols. In *The Role of Air–Sea Exchange in Geochemical Cycling*, P. Buat-Menard (ed.), 409–42. Dordrecht: Reidel.

Gagosian, R.B., Peltzer, E.T. & Merrill, J.T. (1987) Long-range transport of terrestrially derived lipids in aerosols from the South Pacific. *Nature*, **325**, 800–3.

Galoway, F. & Bender, M. (1982) Diagenetic models of interstitial nitrate profiles in deep-sea suboxic sediments. *Limnol. Oceanogr.*, **27**, 624–38.

Gieskes, J.M. (1983) The chemistry of interstitial waters of deep sea sediments: interpretations of deep-sea drilling data. In *Chemical Oceanography*, J.P. Riley & R. Chester (eds), Vol. 8, 221–69. London: Academic Press.

Grimalt, J. & Albaiges, J. (1987) Sources and occurrence of $C_{12}$–$C_{22}$ *n*-alkanes distributions with even carbon-nuclear preference in sedimentary environments. *Geochim. Cosmochim. Acta*, **51**, 1379–84.

Hartmann, M. & Muller, P.J. (1982) Trace metals in interstitial waters from central Pacific Ocean sediments. In *The Dynamic Environment of the Ocean Floor*, K.A. Fanning & F. Manheim (eds). Lexington, MA: Lexington Books.

Hawkesworth, C.J. & Elderfield, H. (1978) The strontium isotopic composition of interstitial waters from Sites 245 and 336 to the Deep Sea Drilling Project. *Earth Planet. Sci. Lett.*, **40**, 423–32.

Heath, G.R., Moore, T.C. & Dauphin, J.P. (1977) Organic carbon in deep-sea sediments. In *The Fate of Fossil Fuel CO$_2$ in the Oceans*, N.R. Anderson & A. Malahoff (eds), 605–25. New York: Plenum.

Hedges, J.I. & Keil, R.G. (1995) Sedimentary organic matter preservation: an assessment and speculative synthesis. *Mar. Chem.*, **49**, 81–115.

Henrichs, S. (1995) Sedimentary organic matter preservation: an assessment and speculative synthesis—a comment. *Mar. Chem.*, **49**, 127–36.

Klinkhammer, G., Heggie, D.T. & Graham, D.W. (1982) Metal diagenesis in oxic marine sediments. *Earth Planet. Sci. Lett.*, **61**, 211–19.

Lawrence, J.R., Gieskes, J.M. & Broecker, W.S. (1975) Oxygen isotope and carbon composition of DSDP pore waters and the alteration of Layer II basalts. *Earth Planet. Sci. Lett.*, **27**, 1–10.

Lerman, A. (1977) Migrational processes and chemical reactions in interstitial waters. In *The Sea*, E.D. Goldberg, I.N. McCave, J.J. O'Brien & J.H. Steele (eds), Vol. 6, 695–738. New York: Interscience.

Lyle, M. (1983) The brown-green color transition in marine sediments: a marker of the Fe(II)–Fe(III) redox boundary. *Limnol. Oceanogr.*, **28**, 1026–33.

Lynn, D.C. & Bonatti, E. (1965) Mobility of manganese in diagenesis of deep-sea sediments. *Mar. Geol.*, **3**, 457–74.

Manheim, F.T. (1976) Interstitial waters of marine sediments. In *Chemical Oceanography*, J.P. Riley & R. Chester (eds), Vol. 6, 115–86. London: Academic Press.

Martin, W.R. & Sayles, F.L. (1994) Seafloor diagenetic fluxes. In *Material Fluxes on the Surface of the Earth*, 143–63. Washington, DC: National Academy Press.

Mayer, L.M. (1994) Surface area control of organic carbon accumulation in continental shelf sediments. *Geochim. Cosmochim. Acta*, **58**, 1271–84.

Muller, P.J. & Mangini, A. (1980) Organic carbon decomposition ratio in sediments of the Pacific manganese nodule belt dated by $^{230}$Th and $^{231}$Pa. *Earth Planet. Sci. Lett.*, **51**, 96–114.

Murray, J. & Irvine, R. (1895) On the chemical changes which take place in the composition of the seawater associated with blue muds on the floor of the ocean. *Trans. R. Soc. Edinburgh*, **37**, 481–507.

Nissenbaum, A. & Kaplan, I.R. (1972) Chemical and isotopic evidence for the *in situ* origin of marine substances. *Limnol. Oceanogr.*, **17**, 570–82.

Oremland, R.S. & Taylor, B.F. (1978) Sulfate reduction and methanogenesis in marine sediments. *Geochim. Cosmochim. Acta*, **42**, 209–14.

Pedersen, T.F. (1995) Sedimentary organic matter preservation: an assessment and speculative synthesis—a comment. *Mar. Chem.*, **49**, 117–19.

Pedersen, T.F., Vogel, J.S. & Southon, J.R. (1986) Copper and manganese in hemipelagic sediments at 21°N, East Pacific Rise: diagenetic contrasts. *Geochim. Cosmochim. Acta*, **50**, 2019–31.

Pedersen, T.F., Shimmield, G.B. & Price, N.B. (1992) Lack of enhanced preservation of organic matter in sediments under the oxygen minimum in the Oman Margin. *Geochim. Cosmochim Acta*, **56**, 545–51.

Sawlan, J.J. & Murray, J.W. (1983) Trace metal remobilization in the interstitial waters of red clay and hemipelagic marine sediments. *Earth Planet. Sci. Lett.*, **64**, 213–30.

Sayles, F.L. (1979) The composition and diagenesis of interstitial solutions—I. Fluxes across the seawater–sediment interface in the Atlantic Ocean. *Geochim. Cosmochim. Acta*, **43**, 527–45.

Simoniet, B.R.T. (1978) The organic chemistry of marine sediments. In *Chemical Oceanography*, J.P. Riley & R. Chester (eds), Vol. 7, 233–311. London: Academic Press.

Steinberg, S.M., Venkatesan, M.I. & Kaplan, I.R. (1987) Organic geochemistry of sediments from the continental margin off southern New England, U.S.A.—Part I. Amino acids, carbohydrates and lignin. *Mar. Chem.*, **21**, 249–65.

Stumm, W. & Morgan, J.J. (1981) *Aquatic Chemistry*. New York: Wiley.

Thompson, G. (1983) Hydrothermal fluxes in the oceans. In *Chemical Oceanography*, J.R. Riley & R. Chester (eds), Vol. 8, 270–337. London: Academic Press.

Tissot, B.P. & Welte, D.H. (1984) *Petroleum Formation and Occurrence*. Berlin: Springer-Verlag.

Wallace, H.E., Thomson, J., Wilson, T.R.S., Weaver, P.P.E., Higgs, N.C. & Hydes, J.D. (1988) Active diagenetic formation of metal-rich layers in N.E. Atlantic sediments. *Geochim. Cosmochim. Acta*, **52**, 1557–69.

Weaver, P.P.E. & Kuijpers, A. (1983) Climatic control of turbidite deposition on the Madeira Abyssal Plain. *Nature*, **306**, 360–3.

Wilson, T.R.S., Thompson, J., Colley, S., Hydes, D.J. & Higgs, N. (1985) Early organic diagenesis: the significance of progressive subsurface oxidation fronts in pelagic sediments. *Geochim. Cosmochim. Acta*, **49**, 811–22.

Wilson, T.R.S., Thomson, J., Hydes, J.D., Colley, S., Culkin, F. & Sørensen, J. (1986a) Oxidation fronts in pelagic sediments: diagenetic formation of metal-rich layers. *Science*, **232**, 927–75.

Wilson, T.R.S., Thomson, J., Hydes, D.J., Colley, S., Culkin, F. & Sørensen, J. (1986b) Metal-rich layers in pelagic sediments: reply. *Science*, **234**, 1129.

# 15    The components of marine sediments

A number of the components that form the sediment building blocks will be described in the present chapter. However, no attempt will be made to adopt an all-embracing catalogue approach to the subject; rather, a limited number of individual components have been selected on the basis that they can provide information on the *processes* that control the mineralogical and chemical compositions of oceanic sediments. These sediment building-block components are described below in terms of a modification of the 'geosphere of origin' classification proposed by Goldberg (1954).

## 15.1 Lithogenous components

### 15.1.1 Definition

Following Goldberg (1954), lithogenous components are defined as those which arise from land erosion, from submarine volcanoes or from underwater weathering where the solid phase undergoes no major change during its residence in sea water.

A wide variety of solid products are mobilized on the continents and transported to the oceans via river run-off, atmospheric deposition and glacial transport. These solid products have a wide range of composition and there is no reason why any type of continental mineral should not be found, albeit it at small concentrations, in oceanic sediments. Quantitatively, the most important land-derived minerals in the sediments are the clay minerals and quartz, together with smaller amounts of other minerals such as the feldspars. Some of these lithogenous minerals are deposited in nearshore sediments, but the finest fractions can reach open-ocean areas. The distributions of the fine-fraction lithogenous minerals in deep-sea sediments can therefore offer an insight into how material derived from the continents is dispersed throughout the marine environment. This dispersion process is illustrated below with respect to the distribution of the clay minerals.

### 15.1.2 The distributions of the principal clay minerals in deep-sea sediments

The structure of the clay minerals consists of a combination of tetrahedral layers, which are made up of $SiO_4$ tetrahedra linked to form a sheet, and octahedral layers, which are made up of two layers of oxygen atoms or hydroxyl ions with cations of Al, Fe or Mg between them. Riley & Chester (1971) divided the principal clay minerals into two general types. In **type I** clays, the basic structure consists of a two-layer sheet of one tetrahedral and one octahedral layer (1:1 clays); in **type II** clays, the basic unit is a sandwich of two tetrahedral layers (in which there is usually some substitution of $Si^{4+}$ by $Al^{3+}$ in the $SiO_4$ tetrahedra) with one octahedral layer between them (2:1 clays). The clay minerals can be formed in a number of ways; for example some are stable weathering residues, some are metamorphic, some are hydrothermal and some are reconstituted during reverse weathering processes. A range of these clay minerals have been found in marine sediments, but the most common varieties belong to the **kaolinite, chlorite, illite** and **montmorillonite** groups. These clay minerals make up a large proportion of the <2 μm land-derived (carbonate-free) fraction of deep-sea sediments, and as the members of the four major groups are sometimes formed under different geological conditions they are especially attractive for the study of the sources and dispersal patterns of solids in the oceans.

The studies reported by Biscaye (1965) and Griffin *et al.* (1968), on the <2 μm carbonate-free sediment fractions, provided a database that allowed major trends in the oceanic distributions of the clays to be established. The trends identified in these two studies are outlined below in terms of the principal clay mineral groups, and the distributions of the individual clays are illustrated in Fig. 15.1. It is well documented that most clay minerals can be formed under a variety of geological conditions. However, in looking for major trends in the distribution patterns of the clays only their most important general sources will be identified, and although this will of necessity produce an oversimplified picture it is still a useful approach to adopt on an ocean-wide scale.

*Kaolinite.* The formation of kaolinite, which is a type I clay, is characteristic of intense tropical and desert weathering, and it is clearly apparent from Fig. 15.1(a) that the highest concentrations of the mineral

**Fig. 15.1** (a–d) The concentration of clay minerals in the <2 μm size fractions of deep-sea sediments (from Griffin *et al.*, 1968); values expressed as percentages of a 100% clay sample.

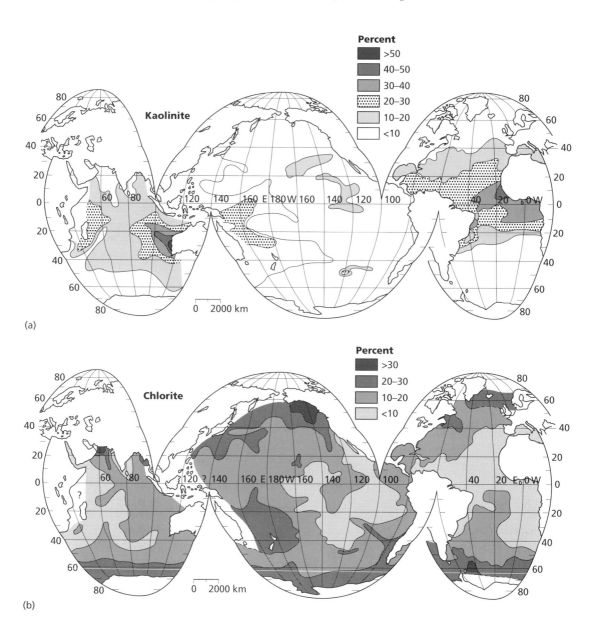

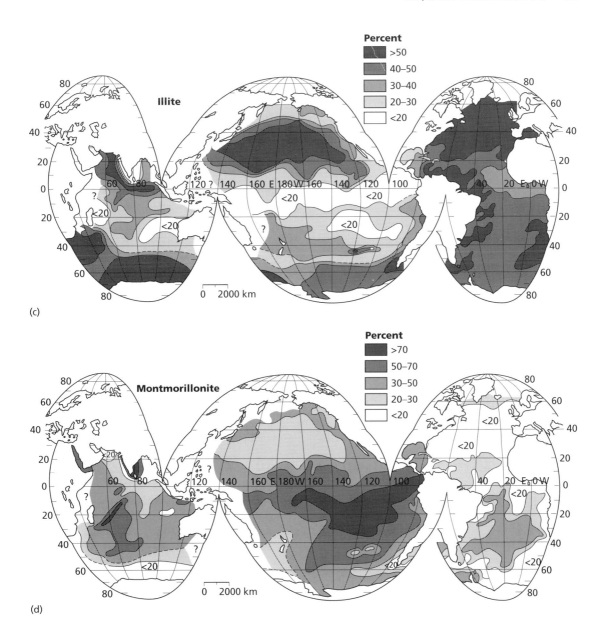

**Fig. 15.1** *Continued*

are found in deep-sea sediments from equatorial regions. The mineral is transported to these regions from the surrounding arid and semi-arid land masses via fluvial and atmospheric pathways, and because of its distribution, Griffin *et al.* (1968) referred to kaolinite as the *low-latitude* clay mineral.

*Chlorite.* Chlorite is a type II clay mineral and has a brucite layer between the basic three-layer sandwich. It can be seen from Fig. 15.1(b) that chlorite has its highest concentrations in sediments from the polar seas. The mineral is found in metamorphic and sedimentary rocks of the Arctic and Antarctic regions from which, in the general absence of chemical weathering, it is released by mechanical processes and is subsequently dispersed by ice rafting; the latter is confirmed by the sharp break-off in concentrations around 50°N and 50°S, which corresponds roughly to the iceberg transport limit. Thus, this chlorite is a primary mineral and not a weathering residue. Chlorite is also formed under a variety of other geological–climatic conditions, but because of its high concentrations in sediments from polar regions, Griffin *et al.* (1968) termed it the *high-latitude* clay mineral. In general, the overall distribution of chlorite in deep-sea sediments is therefore the inverse of that of the low-latitude clay mineral kaolinite.

*Illite.* Members of the illite group are type II clays, which have $K^+$ ions lying between the three-layer basic sandwich. Illites are the most common of the clay minerals in deep-sea sediments. They are formed under a wide variety of geological conditions and, unlike kaolinites and chlorites, they are not confined to particular latitudinal bands. Most of the illites in deep-sea sediments are land-derived and their distributions are controlled by (i) the amount of land surrounding an oceanic area, and (ii) the extent to which they are diluted by clays that have specific source regions (e.g. kaolinite and chlorite). As a result of a combination of these two factors, the highest concentrations of illite are found in deep-sea sediments from *mid-latitudes* (where there is less dilution from chlorite and kaolinite) in the Northern Hemisphere (which has a higher proportion of surrounding land areas than the Southern Hemisphere). This is particularly apparent in the North Pacific where the sediments have an average of 40% illite, compared with only 26% in the South Pacific—see Table 15.1. The distribution of illite in deep-sea sediments is illustrated in Fig. 15.1(c).

*Montmorillonite.* This is sometimes termed smectite or 'expanding-lattice' clay. Montmorillonites are type II clays, in which water molecules are found lying between the basic three-layer sandwich units. According to Riley & Chester (1971) montmorillonites in deep-sea clays can be formed in at least three different ways:

1 as primary detrital weathering residues formed on the continents;

2 as secondary detrital degraded lattices of the illite or chlorite type which have had their intersheet $K^+$ ions replaced by water molecules;

3 as *in situ* products of the submarine weathering of volcanic material.

It is the latter form that imposes specific patterns on the distribution of montmorillonite in deep-sea sediments. As a result, the highest concentrations of the clay are found in areas that have a plentiful supply of parent volcanic debris, and in which the sedimentation rates are low enough to allow the transformation reactions to occur before the volcanic debris is buried. This is reflected in the sequence of increasing montmorillonite concentrations in deep-sea sediments, which are in the order: North Atlantic (16%) < South Atlantic (26%) < North Pacific (35%) < Indian Ocean (41%) < South Pacific (53%)—see Table 15.1. Thus, the deep-sea sediments from the South Pacific are richer in montmorillonite than those from the other oceanic areas. According to Griffin *et al.* (1968), the highest concentrations of montmorillonite in the South Pacific itself are found in sediments from mid-ocean areas, where the mineral is found in association with volcanic material, with concentration gradients decreasing towards the continents. This led the authors to propose that much of the montmorillonite in the mid-ocean sediments has an *in*

| Oceanic area | Average % clay minerals* | | | |
| --- | --- | --- | --- | --- |
| | Chlorite | Montmorillonite | Kaolinite | Illite |
| North Atlantic | 10 | 16 | 20 | 55 |
| South Atlantic | 11 | 26 | 17 | 47 |
| North Pacific | 18 | 35 | 8 | 40 |
| South Pacific | 13 | 53 | 8 | 26 |
| Indian | 12 | 41 | 17 | 33 |

**Table 15.1** Average concentrations of the principal clay minerals in the <2 μm carbonate-free fractions of sediments from the major oceans (data from Griffin *et al.*, 1968).

* Individual clay mineral percentages are expressed in terms of a 100% clay sample.

*situ* origin from the alteration of volcanic material, and on this basis they characterized montmorillonite as being indicative of a *volcanic regime*. The distribution of montmorillonite in deep-sea sediments is illustrated in Fig. 15.1(d).

The North Pacific has more rivers draining into it, and a greater area of surrounding land, than the South Pacific, and this is reflected in the distributions of illite and montmorillonite in the deep-sea sediments from these two regions. That is, in the North Pacific, which receives a relatively large contribution of land-derived material, the sediments have some of the highest concentrations of the ubiquitous detrital clay mineral illite, whereas those in the South Pacific have the highest concentrations of the authigenic clay mineral montmorillonite. Griffin & Goldberg (1963) suggested therefore that the North Pacific is an area of mainly detrital deposition, but that in the South Pacific authigenic deposition is more important.

### 15.1.3 The chemical significance of the clay minerals in marine processes

Once they are brought into contact with saline waters the clay minerals can take part in ion-exchange reactions of the type

$$\text{cation A (clay)} + \text{cation B}^x \text{ (solution)} =$$
$$\text{cation B (clay)} + \text{cation A}^x \text{ (solution)} \quad (15.1)$$

where $x$ is the charge on the cation. In addition to this normal ion exchange, the clay minerals can remove major cations from sea water during the reconstitution of degraded clay minerals, i.e. those that have had cations stripped from intersheet positions.

A number of workers have attempted to model the composition of sea water on the basis of thermodynamic equilibria between components dissolved in the water and the mineral phases of marine sediments. This type of modelling was given a great impetus in the 1960s by Sillen. In his models, Sillen proposed that the composition of sea water was controlled by equilibria between individual minerals, which would fix the ratios of the cations in solution. The removal of major ions supplied to the oceans from river run-off was then assumed to take place by the *transformation* of one clay mineral phase into another, and the equilibria were considered to operate between phases such as seawater–quartz–kaolinite–illite–chlorite–montmorillonite–calcite–

zeolites as part of the overall atmosphere–seawater–sediment system (see e.g. Sillen, 1961, 1963, 1965, 1967). In terms of these models, the coexistence of the various phases would fix the mutual ratios of all the cations, so that only the sum of their concentrations was variable, and this was assumed to be fixed by the chloride concentration. The Sillen models offered a potentially attractive explanation for the composition of sea water. However, subsequent data (see above) showed that many of the clay minerals in deep-sea sediments are in fact land-derived in origin and are generally unreactive, at least to the extent that their source-controlled distribution patterns indicate that they do not appear to undergo significant intermineral transformations. In an attempt to overcome this particular objection to the equilibria model theory, Mackenzie & Garrels (1966) proposed that, rather than transformations between existing clay minerals, it was the transformation of either amorphous aluminosilicate material (formed during weathering) or degraded clays into new crystalline clay phases that controlled the reactions required by the equilibrium models. Thus

$$\text{X-ray amorphous Al silicate} + SiO_2 + HCO_3^- +$$
$$\text{cations} = \text{cation Al silicate} + CO_2 + H_2O \quad (15.2)$$

The normal weathering of aluminosilicates involves the uptake of $CO_2$ from the atmosphere and the release of bicarbonate. Thus

$$\text{silicate} + CO_2 + H_2O = \text{clay mineral} + HCO_3^- +$$
$$H_4SiO_4 + \text{cations} \quad (15.3)$$

In effect, therefore, the processes occurring in the sea (i.e. in which cations, dissolved silica and bicarbonate are removed from solution) are the reverse of the weathering process on the continents; hence, they are often referred to as **reverse weathering**. There is no doubt that reverse weathering does take place within marine sediments: for example, in the formation of mixed-layer clays, zeolites, etc. However, according to Drever (1982) uptake during the formation of authigenic minerals of this type can account for only a small amount of the major ions supplied to the oceans via river input. To be effective, therefore, the clays must undergo alteration before burial in the sediments: e.g. on the initial mixing of river and sea water, as suggested by Mackenzie & Garrels (1966).

The status of equilibria reactions between the clay minerals (and other sediment phases) in fixing the ionic composition of sea water remains a grey area. There is a tendency towards equilibria between the sediment phases and sea water, and although it is now accepted that the clay mineral transformations suggested by the Sillen equilibrium models do not take place, efforts have been made to replace them with reverse-weathering-type reactions. Other workers have suggested that the thermodynamic equilibria in the oceans cannot be approached fast enough to negate kinetic effects; for example, Broecker (1971) outlined a kinetic model to explain the composition of sea water. There is also another criticism that can be directed against the equilibrium model approach. The models involved propose that the cationic ratios in sea water are controlled by equilibria among the major solid phases present in marine sediments. It is now known, however, that basalt–seawater reactions can act as a sink for species such as Mg, and can in fact account for ~50% of the river flux of this element (see Section 17.2). The importance of these basalt–seawater reactions in controlling the chemistry of the oceans has been recognized by Holland (1978), who revised an earlier scheme presented by Mackenzie & Garrels (1966) for the mass-balance removal of river-derived constituents from sea water to take account of these reactions. The volume written by Holland (1978) is strongly recommended as a keynote work for students who wish to gain a deeper insight into the factors controlling the chemical compositions of the oceans and the atmosphere.

### 15.1.4 Clay minerals in deep-sea sediments: summary

The average concentrations of the four principal clay minerals in deep-sea sediments from the major oceans are listed in Table 15.1, and their origin–distribution trends may be summarized as follows.

1 Illite, chlorite and kaolinite are largely land-derived, i.e. lithogenous, clays.

2 Chlorite and kaolinite tend to be concentrated in surficial rocks and soils from particular latitudinal belts, and this is reflected in their distributions in oceanic sediments, chlorite being the *low-latitude* clay and kaolinite the *high-latitude* clay.

3 Illite is also mainly land-derived, but because it is formed under a wide variety of weathering conditions it is not diagnostic of any particular supply region.

4 The main features in the distribution of montmorillonite in deep-sea sediments arise from its transformation from volcanic debris. Thus, montmorillonite is characteristic of volcanic regimes.

## 15.2 Biogenous components

### 15.2.1 Definition

In the classification suggested by Goldberg (1954) biogenous components are defined as those produced in the biosphere, and as such include both organic matter and inorganic shell material. Organic matter in marine sediments has been discussed in Sections 14.1–14.4, and will not be considered here. Other biogenous components include phosphates (e.g. skeletal apatite) and sulphates (e.g. barite). However, these usually make up only a few per cent of marine sediments, and in the present section attention will be focused on the major sediment-forming biogenic components, i.e. carbonate and opaline silica shell material.

### 15.2.2 Carbonate and opaline skeletal material in marine sediments

The most important carbonate-secreting organisms in the oceans are foraminifera, coccolithophorids and pteropods (see Section 9.1.3.1). **Forams** are grazers. The planktonic forams have chambered shells (or tests) composed of calcite that range in size from ~30 μm to ~1 mm. These planktonic forams, which are classified into a superfamily (the Globigerinacea), constitute the most common biogenous components in deep-sea sediments, evidenced by the fact that *Globigerina* oozes are the dominant type of pelagic sediment. **Coccolithophorids** are nanoplankton (plants) secreting calcite shells (generally *c.* 10 μm in size), which after the death of the organism tend to break up into individual plates termed coccoliths. These coccoliths are a major component of many calcareous deep-sea sediments. **Pteropods** are pelagic molluscs that secrete large (millimetre-sized) shells of aragonite. These shells undergo dissolution at shallower depths than those composed of calcite (see below), and pteropod oozes have only a limited occurrence on the ocean floor.

The principal opal-secreting organisms in the marine environment are diatoms and radiolaria, with minor amounts of silica being produced by the silicoflagellates and the sponges. **Diatoms** are plants with frustules, which range in size from a few micrometres to around 2 mm. **Radiolarians** are grazers, with tests ranging in size from a few tens to a few hundred micrometres.

Both carbonate- and opal-secreting organisms contribute biogenous components to deep-sea sediments, calcareous oozes (which contain >30% skeletal carbonate) covering ~50% of the deep-sea floor and siliceous oozes (which contain >30% skeletal opal) covering ~15%. However, the oceanic distributions of the calcareous and siliceous oozes are very different. This is illustrated in Fig. 15.2, from which it can be seen that, whereas the calcareous oozes are found mainly on topographic highs in mid-ocean areas, the siliceous oozes tend to be restricted to regions underlying coastal and equatorial upwelling, i.e. they are correlated with high primary production in the surface waters. There are a number of reasons for this overall difference in the distributions of calcareous and siliceous oozes, and these generally can be related to the factors that control the output (production) of the organisms and the dissolution of their skeletal remains.

### 15.2.3 Shell production

The production of shells depends on the fertility of the ocean, which is measured by primary productivity. The two most striking features in the global distribution of primary production are: (i) a ring of high-fertility zones in coastal areas around the edges of the ocean basins; and (ii) a generally low fertility in the central gyres (see Section 9.1.3.3). Berger (1976) has identified a fundamental **biogeographical dichotomy** in the distribution of planktonic populations in the oceans between these two extreme fertility end-members. This dichotomy is represented by a coastal high-fertility regime, within a few hundred miles of the shore, which is characterized mainly by *diatoms*, and an oceanic low-fertility regime associated with a deep permanent thermocline (e.g. in the central gyres), which is characterized by *coccoliths*. This coastal–oceanic fertility dichotomy produces a geographical separation between opal-secreting and calcite-secreting organisms, and so imparts a major fingerprint on their distributions in deep-sea sediments.

### 15.2.4 Shell dissolution–preservation processes

After sinking from the surface layer of the ocean both calcareous and opaline skeletal remains undergo dissolution processes in response to physicochemical parameters. Sea water varies in the extent to which it is undersaturated with respect to carbonate, but it is always undersaturated with respect to opaline silica. As a result, dissolution affects carbonate and siliceous organisms to varying extents, and tends to strengthen the difference produced by the fertility dichotomy. For this reason it is convenient to treat the dissolution of calcareous and siliceous skeletal remains separately.

#### 15.2.4.1 Carbonate shells

From the time scientists first became interested in the oceans it was realized that the carbonate content of deep-sea sediments varies considerably from one location to another. Further, it soon became apparent that this was related to the depth of water under which the deposits had accumulated, with the higher carbonate contents being found in sediments located on topographic highs. There is, therefore, a first-order **depth control** on the occurrence of carbonate sediments in the World Ocean. In addition, there is a general depth that forms a boundary between the deposition of carbonate and non-carbonate sediments (or at least those containing only a few per cent of carbonate). This is termed the **calcium carbonate compensation depth**. Above this depth, carbonate, shells will accumulate in the sediments; at the depth itself, the rate of supply equals the rate of dissolution; at greater depths, dissolution exceeds supply and there is no *net* accumulation of significant amounts of carbonate material. The compensation depth differs for calcite and aragonite. For example, the calcite compensation depth (CCD) appears to be at ~4.5–5 km over much of the Atlantic, at ~5 km in the tropical Indian Ocean, and at ~3–5 km in the Pacific—see Fig. 15.3. In contrast, the aragonite compensation depth (ACD) is much shallower, ranging from ~3 km in the western tropical North Atlantic, to between ~1 and ~2 km in the western tropical Pacific, and as low as a few hundred metres in the tropical North Pacific (Berger, 1976). It is also important to recognize that in addition to the carbonate/non-carbonate deposition represented by the CCD boundary, there is

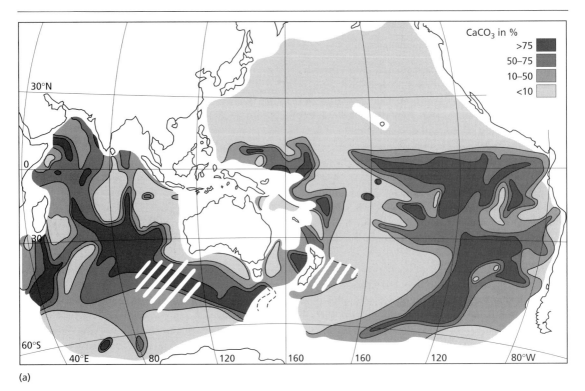

(a)

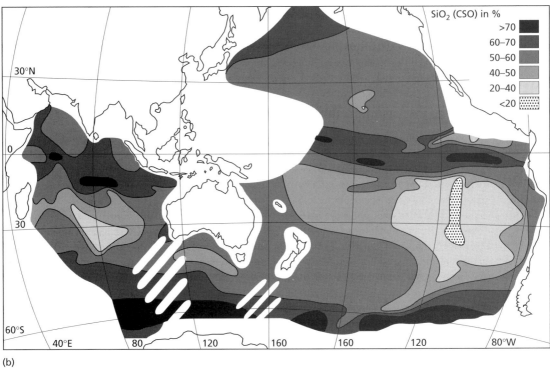

(b)

**Fig. 15.2** The distributions of calcium carbonate and opaline silica in deep-sea sediments. (a) The distribution of calcium carbonate in Indo-Pacific sediments (from Bostrom *et al.*, 1973). (b) The distribution of opaline silica in Indo-Pacific sediments (from Bostrom *et al.*, 1973); data on a carbonate-, salt- and organic-matter-free basis.

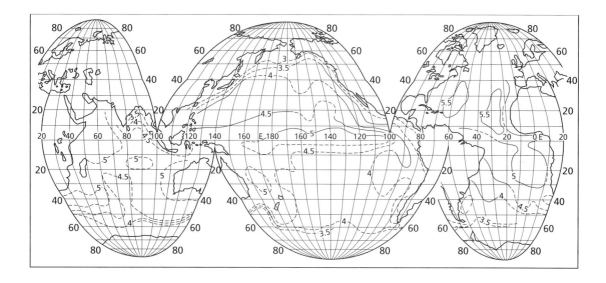

**Fig. 15.3** The distribution of the calcium carbonate compensation depth in the World Ocean (from Berger & Winterer, 1974). This is the depth (km) of the level at which the carbonate content of deep-sea sediments decreases to a few per cent.

dissolution of shell material above this depth in the sediments. This leads to variations in the relative degrees to which carbonate shells have undergone dissolution, and there is a horizon in the sediment that separates well-preserved from poorly preserved shell assemblages. Berger (1968) referred to this dissolution facies boundary, above which species are preserved more or less intact, as the sedimentary **lysocline**. In areas of low productivity the lysocline will be close to the level of the CCD, but there can be considerable separation of the two boundaries below fertile regions, where there is an enhanced supply of carbonate. Berger (1976) pointed out that the lysocline surfaces are usually found at shallower depths in the Pacific than in the Atlantic, and suggested that this is the fundamental reason for the difference in the extent to which carbonates cover the sea floor in the two oceans, i.e. ~67.5% in the Atlantic as opposed to only ~36% in the Pacific.

The main feature in the distribution of calcium carbonate in deep-sea sediments therefore is its common occurrence on topographic highs. One of the keys necessary to understand how this depth control operates lies in the fact that the degree to which sea

water is undersaturated with respect to calcium carbonate varies down the water column. The upper waters are always supersaturated with respect to calcium carbonate; however, the degree of saturation decreases by a factor of up to 10 with depth, leading to undersaturation at intermediate depths. It is this saturation–undersaturation depth pattern that can be regarded as being the driving force behind carbonate dissolution. Both calcite and aragonite are significantly soluble in sea water. The solubility increases with increasing pressure (and so depth) and decreasing temperature, with the result that carbonate shells are susceptible to dissolution both during their transit down the water column and when they reach the bottom sediment.

Much of the controversy over the nature of the factors that control the CCD has revolved around the central issue of whether they are constrained by thermodynamic or kinetic functions (see e.g. Drever, 1982).

The degree to which the water column is undersaturated with respect to calcium carbonate should exert a fundamental control on carbonate shell preservation. Because this undersaturation increases with depth, it has been suggested that **thermodynamic equilibrium** controls apply to the system. Originally it was thought that sea water was supersaturated with respect to the calcite at depth above the CCD, although some doubt had been cast on this.

However, Li *et al.* (1969) used measured values for the partial pressure of CO₂ and the total content of dissolved inorganic CO₂ to calculate the degree of saturation of calcium carbonate in the water column. Their results indicated that Atlantic waters become undersaturated with respect to calcite at depths of ~4–5 km and with respect to aragonite at depths of ~1–2.5 km, whereas those of the Pacific become undersaturated with respect to calcite at ~1.5–3.0 km and with respect to aragonite at ~0.3 km. The authors concluded that the CCD in both Atlantic sediments (~4.5 km) and Pacific sediments (~3.5 km) therefore does reflect a transition from saturation to undersaturation in the overlying waters, i.e. a saturation horizon, and that the differences in depth of the CCD in the two oceans is a function of the differences in the dissolved CO₂ in the two water masses.

Broecker & Peng (1982) also considered the controls on the CCD and related them to reactions occurring in the seawater carbonate system (see Section 8.4.2), with special reference to a **saturation horizon**. When CaCO₃ undergoes dissolution, calcium ions and carbonate ions are formed, and as the calcium content of sea water is approximately constant Broecker & Peng (1982) simplified the degree of calcium carbonate saturation by expressing it in terms of carbonate ions only. Thus

$$D = \frac{[CO_3^{2-}] \text{ sea water}}{[CO_3^{2-}] \text{ saturated sea water}} \qquad (15.4)$$

where $D$ is the ratio of the carbonate ion content of a seawater sample to that of one saturated in the laboratory under the same temperature and pressure conditions. Thus, $D$ is a measure of the degree to which the seawater sample is saturated with respect to calcite or aragonite, and so indicates the strength of the driving force for carbonate dissolution. The authors gave data on the distribution of carbonate ion in the oceans, and compared the values to those of the saturation carbonate ion content for calcite and aragonite. They then used the relationship $[CO_3^{2-}]_{in\ situ}-[CO_3^{2-}]_{sat}$ as a measure of the tendency for carbonate to dissolve, positive values indicating supersaturation and negative values undersaturation. In this terminology, the zero value is termed the *saturation horizon*. Broecker & Peng (1982) found that for calcite the saturation horizon was deepest in the western Atlantic Ocean (~4.5 km), intermediate in the western Indian Ocean (~3.5 km) and least deep in

the North Pacific (<3 km). The saturation horizons for aragonite were much shallower than those for calcite, being ~3.5 km in the Atlantic and actually falling within the main thermocline in the Indian and Pacific Oceans. For calcite, the depth of the saturation horizon was about the same as that of the sedimentary lysocline, and it is here that the driving force for calcite dissolution is zero.

Morse & Berner (1972) proposed a *kinetic* model in which it was suggested that the destruction rate of carbonate sediments is controlled by chemical dissolution kinetics associated with an abrupt change in dissolution rate, which is related to a critical pH; i.e. there is a *critical kinetic horizon* rather than a saturation horizon.

Edmond (1974) critically examined the contrasting thermodynamic and kinetic hypotheses. The author demonstrated that neither the **saturation horizon** in the thermodynamic model, nor the **critical kinetic horizon** in the kinetic model, was coincident with the CCD or lysocline, and that both were in fact found at depths which were shallower than the CCD over parts of the ocean. Edmond (1974) looked closely at the analogy between carbonate and silicate sedimentation in the deep ocean. Although sea water is always undersaturated to some degree with respect to opaline silica, siliceous sediments are preserved at all water depths, their major concentration at the ocean margins being dictated by the patterns of productivity. Thus, as Edmond (1974) points out, a steady-state situation operates for opaline shells, with dissolution being sufficiently slow to allow a significant fraction of the tests to survive long enough to be protected within the sediment. Edmond (1974) also demonstrated that there is a similarity in the distributions of alkalinity and silica in the oceans, and thus in the sites of dissolution of both calcite and opal. However, the solution chemistries of the two phases are very different, which led the author to suggest that the dissolution of calcite and opal is not particularly sensitive to equilibrium effects. He concluded, therefore, that the down-column transit of both opaline and calcareous tests through the water column is fast enough to make the contrasting influences of the two chemical regimes unimportant, and that the calcite compensation depth is best described by a steady-state *physical* model, in which calcite dissolution is accelerated by bottom water motions, rather than by a thermodynamic or a kinetic chemical model. At present, the controversy over the nature of the factors controlling the CCD remains unresolved. Despite this, the CCD can play an important role in oceanic chemical dynamics and this can be illustrated by the study reported by Broecker (1971).

Broecker (1971) postulated a kinetic model for the composition of sea water in which it was suggested that the amount of carbon leaving the ocean is fixed by the phosphorus cycle, but that the presence of large amounts of calcium carbonate in deep-sea sediments proves that this organic removal is not adequate to cope with all the incoming carbon. The excess carbon is therefore precipitated by organisms as $CaCO_3$, but the rate at which this happens exceeds the rate at which the carbonate is removed into sediments, witnessed by the fact that shell dissolution, rather than preservation, occurs in deep waters. Broecker (1971) suggests, therefore, that organisms are generating more $CaCO_3$ than is necessary for the removal of carbon. This has caused the carbonate ion content of the sea to drop to a level low enough so that a large area of the sea floor is covered by waters that are corrosive to $CaCO_3$; this area has a size sufficient to ensure that the overproduction of $CaCO_3$ is just balanced by deep-water solution. The position of the CCD (or the lysocline, or the saturation horizon) is critical to this balance. For example, if it were too deep, more carbonate would accumulate and the removal of carbon would exceed its supply. As a result, the carbonate ion content of sea water would decrease, forcing the CCD upwards until the balance was again achieved. Broecker (1971) concludes that the carbonate ion content of sea water is therefore fixed by the requirement that carbon flows smoothly through the ocean system. The relationship between the rate of production of $CaCO_3$ by plankton (controlled by the availability of phosphorus) to the rate of supply of dissolved $CaCO_3$ from rivers becomes critical in controlling the depth of the CCD. If the ratio is high, most of the carbonate must undergo dissolution to maintain the balance between the input and removal of the calcium and carbonate species (otherwise the ocean would run out of Ca), and this will result in a shallow CCD. If the ratio is low, more of the carbonate will be preserved and the CCD will be found at greater depths (see also Drever, 1982). Thus, the CCD acts to balance the calcium (and carbonate) input–output mechanisms by responding to changes in the ratio between them.

The distribution of calcium carbonate in deep-sea sediments from the major oceans is known in some detail—see e.g. the Pacific Ocean (Berger *et al.*, 1976), the Indian Ocean (Kolla *et al.*, 1976) and the Atlantic Ocean (Biscaye *et al.*, 1976). For example, Biscaye *et al.* (1976) used over 1700 data points to construct a map showing the areal distribution of calcium carbonate in surface sediments from the Atlantic Ocean and adjacent waters, and this is illustrated in Fig. 15.4. The study confirmed the first-order relationship between the amount of calcium carbonate in deep-sea sediments and the water depth under which they were deposited, the highest concentrations being found on the Mid-Atlantic Ridge (MAR) and flanks, and on smaller topographic highs such as the Bermuda Rise, the Walvis Ridge and the Falkland Rise. This relationship was perturbed by several factors, however, which superimposed other controls on the concentrations of carbonate in the sediments. These included:

1 high surface productivity, where an enhanced supply of biogenic material can result in the accumulation of carbonate at depths greater than those under non-productive areas (e.g. in the Cape Verde Basin);

2 dilution by non-carbonate components, which can swamp the shell material;

3 enhanced dissolution, e.g. under the influence of an influx of corrosive, $CO_2$-rich, bottom waters in the basins on the western side of the ocean, which results in dissolution being so effective that carbonate concentrations are low at all water depths (e.g. in the Argentine Basin).

It may be concluded, therefore, that there is a general first-order relationship between calcium carbonate in the sediments and water depth in the Atlantic. However, the details in the carbonate distribution in the surface sediments are influenced by three rate processes associated with: (i) primary productivity; (ii) dilution of the carbonate components; and (iii) shell dissolution, all of which perturb the general carbonate–water-depth relationship in the Atlantic. Thus, both *thermodynamic* and *kinetic* factors affect the distribution of carbonates in Atlantic deep-sea sediments.

### 15.2.4.2 *Opaline shells*

The principal factors that control the distribution of opaline silica in deep-sea sediments can be summa-

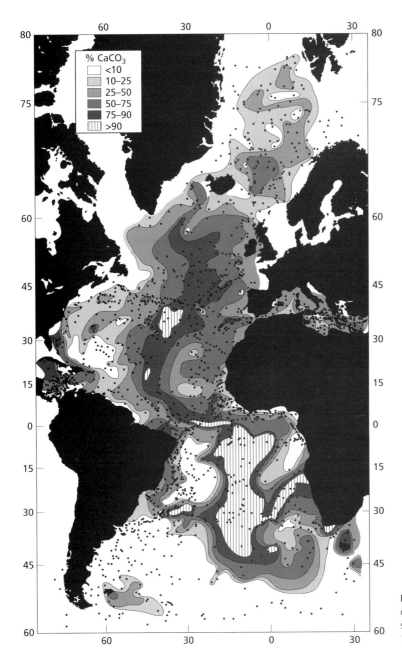

**Fig. 15.4** The present-day distribution of calcium carbonate in Atlantic Ocean surface sediments (from Biscaye *et al.*, 1976).

rized as follows. As sea water is universally undersaturated with respect to opaline silica, the extent to which this component is preserved in deep-sea sediments depends ultimately on the degree to which it has undergone, or escaped, dissolution. The preserva-

tion of opaline silica in deep-sea sediments therefore is influenced by the following factors.

1 The thermodynamic driving force, which is a function of the temperature, pressure and the $H_4SiO_4$ content of the bottom water.

**2** The rate of production of opaline organisms in the overlying waters, i.e. a steady-state situation develops in which the greater the amount of opal falling down the water column the greater the chance it will be preserved in the sediment. As a result, the highest concentrations of opal are found in the sediments under areas of coastal and equatorial upwelling.

**3** The extent to which the opaline remains are diluted with non-opaline material, which acts to decrease the rate at which opal dissolves because this depends, among other factors, on the amount of reactive surface area exposed by the skeletal particles (for a review of this topic—see van Cappellen & Qiu, 1997b).

Opal dissolution continues on the sea bed and in the sediment–interstitial-water complex, which can lead to the diffusion of dissolved silica out of bottom sediments.

The regeneration of silica from opaline shells transported to sediments depends on a complex interplay between sediment burial and mixing, dissolution of the shells, transport of silicic acid through the interstitial waters and the precipitation of authigenic silicate minerals. A number of recent investigations have provided data on the preservation and/or regeneration of biogenic silica in oceanic sediments. For example, Archer *et al.* (1993) reviewed the controls on opal preservation in tropical sediments. The ANTARES Programme (the French contribution to Southern JGOFS) has provided insights into the factors that control the distribution of biogenic silica in sediments from the Southern Ocean (see e.g. Rabouille *et al.*, 1997; van Cappellen & Qiu, 1997a,b). The thermodynamic driving force for the dissolution of opaline shells is the degree of undersaturation of the interstitial waters with respect to the dissolving material. In this context, van Cappellen & Qiu (1997a) concluded that in sediments that have a significant detrital input, the simultaneous reprecipitation of dissolved silica and dissolved aluminium prevents interstitial water silicic acid from reaching saturation with the dissolving biogenic silica. The authors therefore concluded that the principal oceanographic control on porewater silica build-up in sediments from the Southern Ocean is the ratio of the deposition fluxes of biogenic silica and detrital material.

The overall effect of the various fertility (production), dissolution and preservation processes is a very restricted, but highly distinctive, distribution of siliceous oozes in the oceans—see Fig. 15.2(b). This distribution correlates with the high primary productivity upwelling areas, and as a result the distribution of sediments rich in siliceous organisms is concentrated into three major bands.

**1** A Southern Hemisphere polar band, or ring, which nearly circles the globe. The sediments here are mainly diatomaceous oozes, and according to Calvert (1968) they account for ~80% of the total silica accumulation in the oceans.

**2** A Northern Hemisphere polar belt, which is found in the Pacific Ocean but is absent from the Atlantic and Indian Oceans.

**3** An equatorial belt in the Pacific Ocean, in which the sediments are relatively rich in radiolaria.

In addition to these three major bands, siliceous sediments are found in a number of coastal regions where upwelling occurs (e.g. in the Gulf of California and off the coast of Namibia). In general, diatoms are the dominant organisms in high-latitude and coastal siliceous sediments, whereas radiolarians predominate in the equatorial sediments.

### 15.2.5 The chemical significance of carbonates and opaline silica in marine processes

#### 15.2.5.1 *Carbonates*

The production, dissolution and sediment preservation patterns of plankton-produced calcium carbonate in the ocean system have important effects on the chemistry of sea water, and on the global $CO_2$ system (including climate); these have been discussed above and in Section 8.4.2. The formation of the skeletal carbonates, and their incorporation into deep-sea sediments, regulates the marine budget of calcium. Carbonate sedimentation also removes carbon from the oceans. The production of skeletal carbonates does not appear to play a major role in the marine budgets of elements other than calcium and carbon. However, Sr, which can be found in the carbonate lattice, is an exception to this and much of the Sr in deep-sea sediments is associated with carbonate material. Carbonate skeletal material undergoes transit down the water column, but is not a significant carrier for the nutrient-type trace metals (see Section 11.6.3.2). Overall, therefore, the carbonate skeletal material incorporated into deep-sea sediments tends to be relatively pure, and contains very low concentrations of trace metals other than Sr; however, trace-metal-rich oxides of Mn and Fe can be found in intimate association with carbonate shells (e.g. as infillings).

Because they contain information on the fertility, chemistry and climate of the oceans, carbonate deep-sea sediments have been of particular value in palaeo-oceanography. For example, data on water temperatures prevalent at the time of deposition have been obtained from ancient carbonates using temperature-dependent species distribution and $^{18}O$: $^{16}O$ ratios. Studies of this kind have played a large part in evaluating past oceanic environments in the CLIMAP programme; for a detailed treatment of the involvement of carbonates in palaeo-oceanography the reader is referred to the following key publications: Emiliani & Flint (1963), Broecker (1971), Luz & Shackleton (1975), Berger (1976), CLIMAP Project Members (1976), Schopf (1980), Vincent & Berger (1981), Shackleton (1982) and the volume edited by Hay (1974).

### 15.2.5.2 Opaline silica

The marine budget of silicon is more complicated than that of calcium, and the role played by biogenous sediments in the removal of silica has proved difficult to assess. In fact, the topic has given rise to considerable controversy and two schools of thought have emerged in the literature regarding the factors that control the marine silica budget.

**1 Biological removal.** Calvert (1968) proposed that biological processes involving the incorporation of opaline shell material into sediments are the most important pathway for the removal of silica from the oceans, with most of the removal (~80%) taking place around Antarctica. Calvert (1968) estimated that ~$3.6 \times 10^{14}\,g\,yr^{-1}$ of $SiO_2$ are taken out of sea water by biological processes, which amounts to ~83% of the generally accepted estimate for the river input of silica (i.e. $4.3 \times 10^{14}\,g\,yr^{-1}$).

**2 Non-biological removal.** Some authors have proposed that the removal of silica by organisms, and the subsequent burial in the sediments of that fraction which escapes dissolution, can account for only a relatively small fraction of the input of silica to the ocean system (see e.g. Wollast, 1974).

Between these two extremes, other authors have suggested that, although biological removal is certainly significant in the marine budget of silica, it is not the major control on it. For example, Burton & Liss (1968) estimated that biological processes remove $SiO_2$ at a rate of ~$1.9 \times 10^{14}\,g\,yr^{-1}$, which is ~44% of the amount added by rivers each year. Because of this type of imbalance, other (i.e. inorganic) mechanisms have had to be sought to account for the removal of silica from the oceans. These have included:

1 reactions with clay minerals, including reverse weathering reactions between dissolved silicon and degraded silicates (see Section 15.1.3);

2 incorporation into quartz overgrowths, or into authigenic aluminosilicates such as palygorskite and sepiolite;

3 reactions with particulate material during the early stages of estuarine mixing.

In an assessment of the marine silica budget, DeMaster (1981) critically examined the previous supply–removal estimates. He concluded that river run-off (~4.1 ($\pm 0.8$) $\times 10^{14}\,g\,yr^{-1}$) and hydrothermal emanations (~1.9 ($\pm 1.0$) $\times 10^{14}\,g\,yr^{-1}$) supply most of the silica brought to the marine environment. Although DeMaster (1981) was not able to construct an exact balance between supply and removal, he concluded that about two-thirds of the silica supplied to the oceans can be accounted for via the deposition of biogenic continental margin and deep-sea sediments, with >25% of the total input being taken up by sediments depositing beneath the Antarctic Polar Front. Details of the marine silica budget outlined by DeMaster (1981) are given in Table 15.2. An interesting feature of this budget is that although coastal upwelling areas, such as those found in the Gulf of California and Walvis Bay, have some of the highest silica accumulation rates in the World Ocean, their relatively small areal extent means that the siliceous sediments deposited there take up <5% of the total silica supplied to the marine environment.

From most of the budget estimates it is apparent, therefore, that biological processes involving the formation of opaline skeletal material, and its preservation in sediments, play an important, perhaps the major, role in the marine silica budget.

**Table 15.2** The marine budget of silica (units, $10^{14}$ g yr$^{-1}$).

| | | From DeMaster, 1981 | From DeMaster, personal communication |
|---|---|---|---|
| Supply | Rivers | $4.2 \pm 0.8$ | 4.3 |
| | Hydrothermal emanations | $1.9 \pm 1.0$ | 0.1 |
| | *Total* | $6.1 \pm 1.8$ | 4.4 |
| Removal | Deep sea | | |
| | Antarctic | 2.5 | 2.5 |
| | Bering Sea | 0.28 | 0.3 |
| | North Pacific | 0.15 | 0.3 |
| | Sea of Okhotsk | 0.14 | — |
| | Poorly siliceous Indian, Atlantic and Pacific sediments | <0.1 | <0.1 |
| | Peripheral Antarctic basins | <0.02 | — |
| | Equatorial Pacific | 0.01 | 0.01 |
| | *Subtotal* | 3.1–3.2 | 3.1–3.2 |
| | Continental margins | | |
| | Estuaries | <0.8 | |
| | Gulf of California | 0.10 | |
| | Walvis Bay | <0.11 | |
| | North America west coast basins | <0.10 | |
| | Peru and Chile coast | <0.06 | |
| | *Subtotal* | 0.1–0.2 | 0.5–1.7 |
| | Total deep-sea and continental margin removal | 3.2–4.4 | 3.6–4.9 |

## 15.2.6 Carbonate and opaline shell material in marine sediments: summary

The oceanic fertility dichotomy in which diatoms are favoured in highly productive regions and coccoliths in poorly productive regions, combined with differences in the oceanic chemistries of the carbonate and silica systems, has led to distinct geographical patterns in the distributions of calcareous and siliceous oozes. Thus, in general, the calcareous deposits are concentrated in relatively shallow open-ocean regions, whereas the siliceous deposits are found mainly at the edges of the oceans (with the exception of the radiolarian-rich band under the equatorial Pacific high-productivity zone). In addition to these global distribution patterns, Berger (1976) has made the important observation that there is a distinct tendency for carbonates to accumulate in the Atlantic (where they cover ~67.5% of the deep-sea floor, compared with ~54% in the Indian Ocean and ~36% in the Pacific), and for siliceous oozes to be deposited in the Pacific and Indian Oceans (where they cover ~15% and ~20% of the deep-sea floor, respectively, compared with only ~7% in the Atlantic). The author ascribes these differences to the inter-ocean basin fractionation of silica and lime as a consequence of the global deep-water circulation path. He points out that according to Redfield *et al.* (1963) the North Atlantic can be described as having an **anti-estuarine** type of circulation in which deep water flows outwards and is replaced at the surface sources, i.e. deep water is exchanged for shallow water. Vertical profiles of the nutrient elements exhibit higher concentrations at depth as a result of regeneration (see Section 9.1.2), so that an anti-estuarine system, which exchanges deep water for shallow water, becomes depleted in nutrients. Further, the bottom waters in such a system are young and oxygenated, and tend to be close to saturation

**Fig. 15.5** Water circulation patterns that influence the distributions of carbonate-rich and silica-rich sediments in the Atlantic and Pacific Oceans. (a) Schematic representation of the anti-estuarine and estuarine water circulation patterns in the Atlantic and Pacific Oceans (from Kennet 1982). In the Atlantic, young oxygenated bottom waters are formed at high latitudes, where deep water is exchanged for shallow water (anti-estuarine circulation type). The young oxygenated waters tend to be close to saturation with respect to calcite, which allows carbonate to accumulate over large parts of the ocean. In contrast, the Pacific receives older deep water at depth, i.e. it exchanges shallow water for deep water (estuarine circulation type). These older deep waters are rich in $CO_2$, i.e. they are relatively acidic, and as a result carbonate-poor sediments accumulate over much of the North Pacific. Silica-rich sediments accumulate in areas of upwelling. (b) Oxygen concentrations along longitudinal profiles in the central pacific and western Atlantic (from Berger, 1970). The oxygen concentration pattern shows the formation of young oxygenated bottom waters in the high-latitude North Atlantic (anti-estuarine circulation), and a progressive decrease towards the North Pacific where shallow water is exchanged for deep water (estuarine circulation).

with respect to calcite, thus allowing carbonate sediments to accumulate. In contrast, the North Pacific has an **estuarine** type of circulation, in which shallow water is exchanged for deep water, thus promoting upwelling. As a consequence, the waters of the North Pacific are enriched in nutrients and because they are older they are also enriched in $CO_2$. Thus, they tend to be undersaturated with respect to calcite even at shallow depths, which leads to enhanced calcite dissolution. In this type of system, therefore, both production and preservation processes favour the deposition of silica. The relationship between the North Atlantic anti-estuarine–carbonate-rich system and the North Pacific estuarine–silica-rich system is illustrated in Fig. 15.5(a,b).

## 15.3 'Hydrogenous' components: halmyrolysates and precipitates

### 15.3.1 Definition

There is some confusion in the literature over the terminology used to describe those components of

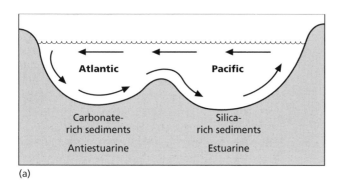

(a)

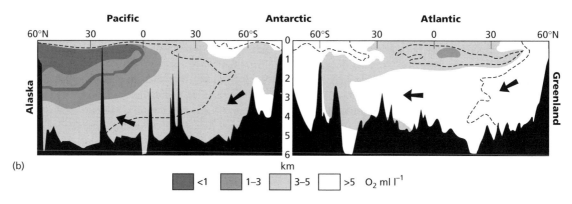

(b)

marine sediments which have been formed inorganically from constituents dissolved in sea water. In the initial 'geosphere of origin' classification outlined by Goldberg (1954) **hydrogenous** components were defined as those which result from the formation of solid material in the sea by inorganic reactions, i.e. by non-biological processes. However, these components have subsequently been reclassified into a number of separate types. For example, Chester & Hughes (1967) distinguished between (i) **primary hydrogenous material**, which is formed directly from material dissolved in sea water, and (ii) **secondary hydrogenous material**, which results from the submarine alteration of pre-existing minerals. Thus, both the Goldberg (1954) and the Chester & Hughes (1967) classifications would include hydrothermal components in the general 'hydrogenous' category because they are also formed from sea water (modified by water–rock interactions) by inorganic reactions. In a more rigorous classification, Elderfield (1976) subdivided hydrogenous components into two basic categories, termed precipitates and halmyrolysates. **Precipitates** are primary inorganic components formed directly from sea water. **Halmyrolysates** are secondary components formed as a result of reactions between sediment components (usually silicates) and sea water subsequent to *in situ* weathering but prior to diagenesis. Elderfield (1976) then identified four main groups of hydrogenous material in marine sediments.

1 Volcanic precipitates, which result from the introduction of elements into sea water from volcanic processes.

2 Supergene precipitates, which are precipitated from sea water, or interstitial water, but are non-volcanic in origin.

3 Lithogenous halmyrolysates, which are formed from the reaction of lithogenous components with sea water.

4 Volcanic halmyrolysates, which usually are formed as a result of seawater–basalt reactions.

In his classification, therefore, (1) and (2) are primary components, and (3) and (4) are secondary.

Further complications arise when a closer look is taken at the origins of the dissolved elements that are involved in the formation of the precipitates and halmyrolysates. It is clear from Section 14.1 that diagenesis involving the destruction of organic carbon can release dissolved elements either at, or close to, the sediment surface (oxic diagenesis), or at depth in

the interstitial waters from both the carbon carriers and from the components utilized as secondary oxidants (suboxic diagenesis). Oxic diagenesis therefore releases elements that have been removed from solution in sea water by their involvement in the oceanic biogeochemical cycles, and have been transported by the down-column vertical carbon-driven flux. In addition to this route for the transport of elements to the sediment surface, dissolved elements also can be removed directly from sea water, i.e. without undergoing a sediment surface-mediated carrier destruction process. In theory, therefore, these latter elements will be removed under conditions where there is no sediment substrate to promote either oxic or suboxic diagenesis, e.g. on exposed rock surfaces.

For the present purposes an attempt will be made to classify the dissolved elements that take part in inorganic component-forming reactions by distinguishing between them on the basis of their sources. Thus, the following categories of elements are identified.

1 **Direct seawater-derived** elements, which are further subdivided into (i) **hydrogenous** elements, i.e. those originating from the general background of elements dissolved in sea water, and (ii) **hydrothermal** elements, i.e. those originating from the debouching of hydrothermal solutions at the ridge-crest spreading centres (see Section 5.1).

2 **Oxic diagenetically derived** elements, i.e. those generated close to the sediment surface following their release on oxidative destruction of organic carbon.

3 **Suboxic diagenetically derived** elements, i.e. those originating from interstitial waters at some depth in the sediment following the destruction of organic carbon by secondary oxidants.

The sediment components formed from these various types of elements can then be classified into **precipitates** (primary) and **halmyrolysates** (secondary) as proposed by Elderfield (1976). This hydrogenous–hydrothermal–diagenetic trinity offers a convenient framework within which to describe those components of marine sediments that are formed in the oceanic environment by inorganic reactions involving dissolved elements, and it is this inorganic origin that distinguishes them from the biogenous components. As the individual components are described, an attempt will be made to identify any coupling between the *processes* that utilize these genetically different elements.

The *precipitates*, or primary inorganic components, in marine sediments include oxyhydroxides, carbonates, phosphates, sulphides, sulphates and evaporite minerals; and the *halmyrolysates*, or secondary inorganic components, include glauconite, chamosite, palagonite, montmorillonite (smectite) and the zeolites. Thus, there is a wide range of halmyrolysates and precipitates in the marine environment. However, the influence they have on marine sedimentation varies from one component to another, both *quantitatively*, in their role as sediment-forming material, and *geochemically*, in the extent to which they influence chemical processes in the oceans. In both of these senses, the most important inorganically produced components found in marine sediments are (i) ferromanganese nodules, (ii) ferromanganese oxyhydroxides, and (iii) ferromanganese ridge-crest iron and manganese oxides. Because of their geochemical importance, these ferromanganese deposits will therefore serve as prime examples of process-orientated components formed inorganically in the marine environment.

### 15.3.2 Ferromanganese deposits in the oceans

Without assuming, at least at this stage, any genetic association in the sequence, the geochemical importance of ferromanganese deposits will be considered in the order, encrustations → ferromanganese nodules → sediment ferromanganese oxyhydroxides → hydrothermal precipitates. The sequence is set up in this way in order to consider first a hydrogenous end-member ferromanganese component (encrustations), then mixed hydrogenous–hydrothermal–diagenetic ferromanganese components (nodules), and finally a hydrothermal end-member ferromanganese component (hydrothermal precipitates). In this way it should be possible to identify how the pure hydrogenous and hydrothermal signals operate, and also how they combine with each other, and with the diagenetic signal, to form a variety of ferromanganese components.

### 15.3.3 Ferromanganese encrustations

Ferromanganese encrustations, or crusts, are relatively thin deposits on submarine rock outcrops, or on objects such as boulders or volcanic slabs. For example, on the Blake Plateau there is a manganese oxide crust 'pavement' that covers an area of ~5000 km² (Pratt & McFarlin, 1966). Further, these crusts can be the typical ferromanganese deposit in seamount provinces. One particularly interesting feature of the crusts is that those which grow on exposed rock surfaces have acquired the elements necessary for their growth directly from the overlying sea water (*hydrogenous* supply) without receiving an input from sediment sources. In this sense, it will be shown below that they can be thought of as representing an end-member in the oceanic ferromanganese depositional sequence.

### 15.3.4 Ferromanganese nodules

#### 15.3.4.1 Occurrence

The first detailed description of the widespread occurrence of ferromanganese nodules in the oceans is usually attributed to Murray & Renard (1891) from collections made during the *Challenger Expedition* (1873–76). The nodules are present throughout the sediment column, but the highest concentrations are found on the surface. Here, they can cover vast areas of the sea floor in some regions; for example, ~50% of the sediment surface in the western Pacific has a blanket of nodules. They have been reported in association with many types of sediment, and Cronan (1977) identified two principal factors that control their abundance.

1 Rate of accumulation of the host sediment. The highest number of nodules are found on those deep-sea sediments that are deposited at relatively slow rates of around a few millimetres per 1000 yr (mm $10^3 \, yr^{-1}$); these include pelagic clays and siliceous oozes. High concentrations of nodules also can occur where sedimentation is inhibited as a result of current action, e.g. on the tops of seamounts.

2 The availability of suitable nuclei for oxide accretion. Various materials (e.g. a fragment of pumice, a shark's tooth, a foram skeleton, a piece of consolidated clay) can be utilized for this purpose, but volcanic nuclei are especially common, and this may explain the relatively high concentrations of nodules in areas of volcanism.

A generalized pattern of the distribution of ferromanganese nodules is illustrated in Fig. 15.6.

#### 15.3.4.2 Morphology

The nodules can have a wide variety of shapes and sizes. When attempting to assess the role they play in

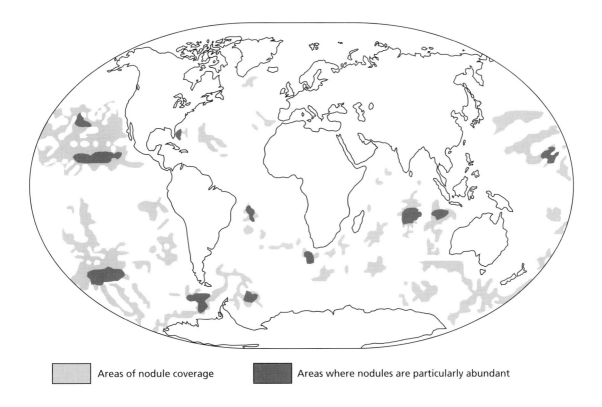

Areas of nodule coverage          Areas where nodules are particularly abundant

**Fig. 15.6** Generalized distribution of ferromanganese nodules in the World Ocean (from Cronan, 1980).

marine geochemistry it is useful to make a distinction between **macronodules** and **micronodules**. This is a purely arbitrary division in which macronodules have diameters in the centimetre (or greater) range, and micronodules have diameters in the millimetre (or less) range. However, the distinction has some meaning in the sense that, whereas the macronodules may be considered to be foreign bodies on or in the sediment host, the micronodules are scattered throughout the deposits and are much more intimately associated with the other sediment-forming components, i.e. they are part of the general sediment. Further, the micronodules often do not have a nucleus, and may in fact be composed of oxides that have sedimented directly from the water column.

Some typical macronodules are illustrated in Fig. 15.7. These macronodules have a wide variety of physical forms, and various schemes have been proposed for their morphological classification; for a detailed discussion of this topic see Raab & Meylan (1977). Perhaps the most widely used morphologi-

cal framework is that in which the nodules are referred to as being some variant of a basic discoidal–ellipsoidal–spherical shape pattern. It is now recognized that the shape adopted by a macronodule is related to the matter in which it has grown, and in particular to whether it has received its component elements (mainly Mn and Fe) from overlying sea water or from underlying interstitial water. Manheim (1965) has described the two nodule end-member morphologies that might arise from these different element supply mechanisms—see Fig. 15.8(a,b).

**1** In Fig. 15.8(a), the nodule is formed on an oxidizing substrate and has received only a minor supply of dissolved material from the underlying sediment. Most of the elements therefore are derived from the overlying sea water and the nodule grows slowly, with occasional overturning, so that it acquires a generally spherical shape.

**2** In Fig. 15.8(b), most of the elements are supplied from the interstitial waters below the sediment surface, following their suboxic release during the diagenetic sequence (see Section 14.1). They are then precipitated by oxidative processes at a suitable surface, in this case the nodule nucleus, which may be

**Fig. 15.7** Typical morphologies of ferromanganese macro-nodules from the Atlantic and Indian Oceans (photographs supplied by D.S. Cronan and S.A. Moorby; (c), (e) and (f) have also been reproduced in Cronan, 1980). (a–c) *Shapes* include (a) and (b), semi-spherical shapes with 'knobbly' surface; (c) flattened nodule. (d,e) *Surface textures*: (d) smooth; (e) granular. (f) *A typical internal structure*, showing concentric series of light and dark bands (see text).

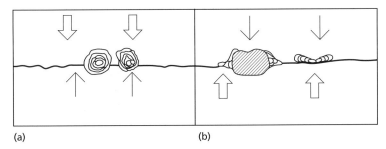

(a)                                                    (b)

**Fig. 15.8** Schematic representation of ferromanganese nodule end-member morphologies (from Manheim, 1965). The size of the arrows indicates the proportion and direction of metal supply. (a) This illustrates a typical situation in the open-ocean, with the nodules lying on an oxidized sediment substrate: here the dominant supply of metals is from the overlying sea water. (b) This illustrates a typical situation in nearshore and freshwater environments, with the nodules lying on a sediment substrate that is partly reducing in character: here the dominant supply of metals is via intersitial waters from below the substrate surface.

partly buried in the sediment. As the precipitation process continues, the sides nearest the source will take up the thickest layer of this bottom-derived metal input, whereas the upper sides will rely on a supply from the overlying sea water. As a result, nodules of this type will have a flattened discoidal shape.

These are, however, two extreme end-member morphologies, and a variety of intermediate nodule shapes can be found. In general, the extreme *seawater* end-member is really represented by encrustations on exposed rock surfaces, and the nodule (concretion) form closest to that of the seawater end-member is probably that found on the tops of current-washed seamounts. The extreme *diagenetic* end-member is best developed in nearshore and hemi-pelagic regions, where the diagenetic sequence can proceed to a late stage. Between these two extremes, individual nodules can have a supply of elements from both of the principal sources. Thus, the upper surfaces are supplied from the overlying sea water, whereas the lower surfaces are diagenetic in origin.

### 15.3.4.3 Growth rates

The rate at which a ferromanganese nodule grows is intimately related to the environment within which it forms, the slowest accretion rates of a few millimetres per million years ($mm\,10^6\,yr^{-1}$) being found for the hydrogenous end-member, intermediate rates of a few hundred millimetres per million years being characteristic of the diagenetic end-member, and the highest rates of a few thousand millimetres per million years being shown by the hydrothermal end-member crusts. If it is accepted that at least some pelagic nodules do grow at an average rate of a few millimetres per million years, then this is considerably slower than the rates at which the host sediments accumulate, i.e. a few millimetres per thousand years. The difference in the rates at which Mn is accreted in the nodules and host sediments has a number of important implications. For example, Broecker & Peng (1982) posed the question 'Why do the nodules appear to attract more than their fair share of manganese and iron?' The answer, as the authors pointed out, is that of course they do not, because the sediments with which they are associated assumulate manganese at a much faster rate than do the nodules themselves (see e.g. Bender *et al.*, 1970). Ku (1977) concluded, therefore, that the nodules do not grow because they have a special capability to attract manganese, but rather because some suitable nucleus escapes burial in the sediment. This suggests that, other factors being equal, the growth of ferromanganese oxides will depend largely on the length of time the nucleus has been in contact with the sources of the metals, and the rate at which the metals themselves are supplied.

### 15.3.4.4 Mineralogy

Much of our knowledge of the mineralogy of the manganese and iron phases in the nodules has come from the pioneering work of Buser & Grutter (1956). Subsequently, many other studies have been carried out on the mineralogy of the nodules, and for a detailed review of this topic the reader is referred to Burns & Burns (1977). The nodule mineralogy may be summarized as follows.

There are at least four mineral phases found in ferromanganese nodules and crusts: $\delta MnO_2$, 7 Å and 10 Å manganite, and iron hydroxide.

1 $\delta MnO_2$ forms aggregates of randomly orientated sheets.

2 7 Å manganite and 10 Å manganite have a double-layer structure consisting of ordered sheets of $MnO_2$ with disordered layers of metal ions, such as $Mn^{2+}$ and $Fe^{2+}$, coordinated with $O^{2-}$, $OH^-$ and $H_2O$, lying between them. In the 10 Å manganites the water and hydroxyls are probably present as discrete layers, whereas in the 7 Å manganites they form a single layer.

3 Amorphous iron hydroxide ($FeOOH \cdot nH_2O$), which may be converted to goethite ($\alpha$-$FeOOH$); other iron-rich phases, e.g. lepidocrocite, can also be present.

The manganese phases have been related to known minerals, and in the most commonly adopted terminology 10 Å manganite is referred to as toderokite and 7 Å manganite is termed birnessite. Thus, the principal mineral phases in the nodules are **$\delta MnO_2$, birnessite, toderokite** and some form of **hydrous iron oxide**.

There are various lines of evidence to suggest that, like the growth rate, the mineralogy of the manganese phases in the nodules (and crusts) is governed by the environment of formation, especially the prevailing oxidation conditions. For example, Arrhenius (1963) suggested that the degree of oxidation at formation decreases in the order $\delta MnO_2$ > birnessite > toderokite. This is reflected in the end-member nodule mineralogies; thus, nodules of the oxidized pelagic end-member type are relatively rich in $\delta MnO_2$, whereas the nearshore and hemi-pelagic end-members contain more toderokite. However, this is a very general distinction, and is by no means clear-cut. Nonetheless, it is a useful first assumption to assign $\delta MnO_2$ to nodules that receive much of their Mn from sea water, and toderokite to those that have a diagenetic origin.

### 15.3.4.5 Chemistry

Ferromanganese nodules are particularly rich in manganese. However, they also contain relatively large concentrations of a variety of trace metals. Cronan (1976) assessed the magnitudes of the enrichments of a series of elements in ferromanganese nodules relative to their average crustal abundances, and on this basis he was able to identify three groups of elements.

1 The **enriched** elements, which were divided into a number of subgroups depending on the extent of their enrichment:

(a) Mn, Co, Mo and Th, which are enriched in the nodules by a factor of more than 100;

(b) Ni, Ag, Ir and Pb, which are enriched by a factor of between 50 and 100;

(c) B, Cu, Zn, Cd, Yb, W and Bi, which have enrichments between 10 and 50;

(d) P, V, Fe, Sr, Y, Zr, Ba, La and Hg, which are enriched by less than a factor of 10.

2 The **non-enriched–non-depleted** elements; these include Na, Mg, Ca, Sn, Ti, Ga, Pd and Au.

3 The **depleted** elements, which include Al, Si, K, Sc, Cr; however, the depletions are only slight.

A compilation of the average abundances of elements in ferromanganese nodules from the major oceans is given in Table 15.3, together with average elemental enrichment factors. The data in this table give an indication of how the composition of the nodules varies on an inter-oceanic basis. In addition, there are considerable variations in the compositions of the nodules within specific oceanic regions. For example, Mero (1962) showed that the nodules in the Pacific can be categorized into a number of Mn-, Cu-, Ni-, Fe- and Co-rich 'ore provinces'. To some extent the variations in the chemical compositions of the nodules can be related to their mineralogy which, in turn, is a function of the depositional environment.

It may be concluded that ferromanganese nodules can be described in terms of two end-members, i.e. a **diagenetic** and a **seawater** end-member. It was shown in Section 14.2.3 that there is a lateral sequence of diagenetic environments on the deep-sea floor, from suboxic in marginal regions to oxic in the open ocean. This lateral sequence of environments exerts a control on the type of nodule formed. For example, Price & Calvert (1970) concluded that Pacific Ocean nodules have a continuous range of chemical and mineralogical compositions which extend between those of two end-members.

1 **Marginal end-member.** This has a disc-shaped morphology. Chemically, it is characterized by a high Mn : Fe ratio, and is relatively enriched in Mn, Cu, Ni

**Table 15.3** Average abundances of elements in ferromanganese oxide deposits from each of the major oceans (in wt.%), and enrichment factors for each element (from Cronan, 1976) (Superscript numbers denote powers of ten, e.g.$^{-6}$ = × $10^{-6}$).

| Element | Pacific Ocean | Atlantic Ocean | Indian Ocean | Southern Ocean | World Ocean average | Crustal abundance | Enrichment factor |
|---|---|---|---|---|---|---|---|
| B | 0.0277 | — | — | — | — | 0.0010 | 27.7 |
| Na | 2.054 | 1.88 | — | — | 1.9409 | 2.36 | 0.822 |
| Mg | 1.710 | 1.89 | — | — | 1.8234 | 2.33 | 0.782 |
| Al | 3.060 | 3.27 | 3.60 | — | 3.0981 | 8.23 | 0.376 |
| Si | 8.320 | 9.58 | 11.40 | — | 8.624 | 28.15 | 0.306 |
| P | 0.235 | 0.098 | — | — | 0.2244 | 0.105 | 2.13 |
| K | 0.753 | 0.567 | — | — | 0.6427 | 2.09 | 0.307 |
| Ca | 1.960 | 2.96 | 3.16 | — | 2.5348 | 4.15 | 0.610 |
| Sc | 0.00097 | — | — | — | — | 0.0022 | 0.441 |
| Ti | 0.674 | 0.421 | 0.629 | 0.640 | 0.6424 | 0.570 | 1.13 |
| V | 0.053 | 0.053 | 0.044 | 0.060 | 0.0558 | 0.0135 | 4.13 |
| Cr | 0.0013 | 0.007 | 0.0014 | — | 0.0014 | 0.01 | 0.14 |
| Mn | 19.78 | 15.78 | 15.12 | 11.69 | 16.174 | 0.095 | 170.25 |
| Fe | 11.96 | 20.78 | 13.30 | 15.78 | 15.608 | 5.63 | 2.77 |
| Co | 0.335 | 0.318 | 0.242 | 0.240 | 0.2987 | 0.0025 | 119.48 |
| Ni | 0.634 | 0.328 | 0.507 | 0.450 | 0.4888 | 0.0075 | 65.17 |
| Cu | 0.392 | 0.116 | 0.274 | 0.210 | 0.2561 | 0.0055 | 46.56 |
| Zn | 0.068 | 0.084 | 0.061 | 0.060 | 0.0710 | 0.007 | 10.14 |
| Ga | 0.001 | — | — | — | — | 0.0015 | 0.666 |
| Sr | 0.085 | 0.093 | 0.086 | 0.080 | 0.0825 | 0.0375 | 2.20 |
| Y | 0.031 | — | — | — | — | 0.0033 | 9.39 |
| Zr | 0.052 | — | — | 0.070 | 0.0648 | 0.0165 | 3.92 |
| Mo | 0.044 | 0.049 | 0.029 | 0.040 | 0.0412 | 0.00015 | 274.66 |
| Pd | $0.602^{-6}$ | $0.574^{-6}$ | $0.391^{-6}$ | — | $0.553^{-6}$ | $0.665^{-6}$ | 0.832 |
| Ag | 0.0006 | — | — | — | — | 0.000007 | 85.71 |
| Cd | 0.0007 | 0.0011 | — | — | 0.00079 | 0.00002 | 39.50 |
| Sn | 0.00027 | — | — | — | — | 0.00002 | 13.50 |
| Te | 0.0050 | — | — | — | — | — | — |
| Ba | 0.276 | 0.498 | 0.182 | 0.100 | 0.2012 | 0.0425 | 4.73 |
| La | 0.016 | — | — | — | — | 0.0030 | 5.33 |
| Yb | 0.0031 | — | — | — | — | 0.0003 | 10.33 |
| W | 0.006 | — | — | — | — | 0.00015 | 40.00 |
| Ir | $0.939^{-6}$ | $0.932^{-6}$ | — | — | $0.935^{-6}$ | $0.132^{-7}$ | 70.83 |
| Au | $0.266^{-6}$ | $0.302^{-6}$ | $0.811^{-7}$ | — | $0.248^{-6}$ | $0.400^{-6}$ | 0.62 |
| Hg | $0.82^{-4}$ | $0.16^{-4}$ | $0.15^{-6}$ | — | $0.50^{-4}$ | $0.80^{-5}$ | 6.25 |
| Tl | 0.017 | 0.0077 | 0.010 | — | 0.0129 | 0.000045 | 286.66 |
| Pb | 0.0846 | 0.127 | 0.070 | — | 0.0867 | 0.00125 | 69.36 |
| Bi | 0.0006 | 0.0005 | 0.0014 | — | 0.0008 | 0.000017 | 47.05 |

and Zn. Mineralogically, it is composed of both $\delta MnO_2$ and toderokite, and some varieties also contain birnessite. This end-member is typically found in marginal and hemi-pelagic sediments, and is equivalent to the **diagenetic** nodule identified above.

**2 Open-ocean end-member.** This has a generally spherical shape. Chemically it is typified by a low Mn : Fe ratio, and is relatively enriched in Fe, Co and Pb. Mineralogically, it consists mainly, or even exclusively, of $\delta MnO_2$. This end-member is best represented by ferromanganese crusts and nodules from seamounts, and is equivalent to the **seawater** nodule identified above.

### 15.3.4.6 *Formation*

The two end-member nodules are formed by two end-member *processes*, the marginal nodules acquiring most of their component metals from the interstitial waters of the host sediments (**diagenetic nodules**) and the open-ocean varieties taking most of their metals from the overlying water (**seawater nodules**). It must be stressed, however, that there is a continuous spectrum of oceanic ferromanganese nodules between these two extreme end-members. This is evidenced by the fact that many nodules from abyssal regions have both a seawater source (to their upper surfaces) and a diagenetic source (to their lower surfaces). Thus, many nodules are compositionally different on their seawater-facing and sediment-facing surfaces. None the less, the basic differences are still related to the two different prime sources of metals to nodules.

According to Dymond *et al.* (1984), 'the compositions of ferromanganese nodules respond in a consistent manner to the sea floor environment in which they form'. In this respect the two end-member (i.e. sea water and diagenetic) nodule source concept provides a useful framework for a discussion of how the nodules formed However, it was pointed out above that the components formed from both seawater and diagenetic sources can be subdivided into a number of genetic classes. For example, seawater sources can be either **hydrogenous** or **hydrothermal**, and diagenetic sources can be either **oxic** or **suboxic** in character. As a result, nodules having a seawater source can be subdivided into *hydrogenous* and *hydrothermal* types, and those having a diagenetic source can be subdivided into *oxic* and *suboxic* types. Hydrothermal processes will be considered separately in Section 15.3.6, and for the moment attention will be focused on the three nodule accretionary modes identified by Dymond *et al.* (1984). These are:

**1** that resulting from hydrogenous precipitation from sea water;
**2** that arising from oxic diagenetic processes;
**3** that associated with suboxic diagenetic processes.

Dymond *et al.* (1984) then tested their three-mode accretionary model with reference to the formation of ferromanganese nodules at three contrasting MANOP sites in the Pacific (sites H, S and R). This study will serve as a blueprint for a discussion of how the nodule-forming processes operate. To aid the discussion, model compositions of ferromanganese components arising from the three accretionary modes are given in Table 15.4.

*The seawater (or hydrogenous) nodule end-member.* This type of ferromanganese deposit is formed by the direct precipitation of colloidal metal oxides from sea water at the growth site, together with the accretion to the depositional surface of suspended Fe–Mn precipitates formed in the water column. The extreme example of a hydrogenous end-member ferromanganese deposit is provided by encrustations that are formed on exposed rock surfaces outside the direct influence of hydrothermal activity. Aplin & Cronan (1985a) described a series of such encrustations from the Line Islands (Central Pacific). The crusts have relatively low Mn : Fe ratios, and the principal minerals in them are $\delta MnO_2$ and amorphous $FeOOH \cdot nH_2O$; of these the $\delta MnO_2$ was the most important trace-metal-bearing phase, containing Co, Mo, Ni, Zn and Cd, with only Ba appearing to be associated specifically with the iron hydroxide phase. The authors concluded that the crusts had formed directly from the slow accumulation of trace-metal-enriched oxides, which had been deposited directly from the overlying water column, and that the trace-metal–mineral associations had resulted from the different scavenging behaviours of Mn and Fe oxides in the water.

The accretion of the hydrogenous ferromanganese deposits is constrained by the concentration of Mn and the kinetics of its oxidation, and Piper *et al.* (1984) have reviewed the processes involved in the formation of marine manganese minerals. According to these authors the Mn oxide that forms initially may be a metastable hausmannite, which undergoes ageing to $\delta MnO_2$ possibly by a disproportionation reaction as suggested by Hem (1978). Thus

$$3H_2O + 1/2O_2 + 3Mn^{2+} = Mn_3O_4 + 6H^+ \qquad (15.5)$$

and

$$Mn_3O_4 + 4H^+ = \delta MnO_2 + 2Mn^{2+} + 2H_2O \qquad (15.6)$$

Most Mn is present in sea water as dissolved $Mn^{2+}$, whereas the element is now thought to be found in nodules almost exclusively in the Mn(IV) state (Piper *et al.*, 1984). Goldberg & Arrhenius (1958) suggested that ferric oxides might provide a catalytic surface for the oxidation of $Mn^{2+}$ to $MnO_2$ and so initiate nodule formation. Once the process has started,

**Table 15.4** Chemical composition of 'end-member' ferromanganese nodules; units, $\mu g\,g^{-1}$.

| Element | Ferromanganese crusts | | Ferromanganese nodules | |
| | Hydrogenous | | Oxic | Suboxic |
| | Line Islands, Pacific* | MANOP Site H, Pacific† | MANOP Site H, Pacific† | MANOP Site H, Pacific† |
|---|---|---|---|---|
| Mn | 204 000 | 222 000 | 316 500 | 480 000 |
| Fe | 170 000 | 190 000 | 44 500 | 4900 |
| Co | 5500 | 1300 | 280 | 35 |
| Ni | 3900 | 5500 | 10 100 | 4400 |
| Zn | 590 | 750 | 2500 | 2200 |
| Cu | – | 1480 | 4400 | 2000 |
| Al | 16 000 | 11 800 | 27 100 | 7500 |
| Ti | 12 000 | 5300 | 1700 | 365 |
| Si | – | 52 000 | 59 000 | 16 300 |
| Na | – | 10 400 | 16 100 | 32 800 |
| K | – | 4900 | 8200 | 6200 |
| Mg | – | 10 400 | 23 000 | 13 800 |
| Ca | 26 000 | 26 000 | 15 200 | 12 500 |
| Mn : Fe | 1.2 | 1.2 | 7.1 | 98 |
| Co : Mn | 0.027 | 0.006 | 0.0009 | 0.00007 |
| Ni : Mn | 0.019 | 0.024 | 0.032 | 0.0092 |
| Cu : Mn | 0.0075 | 0.0022 | 0.020 | 0.0024 |
| Zn : Mn | 0.0029 | 0.0034 | 0.0079 | 0.0046 |

\* Data from Aplin & Cronan (1985a).
† Data from Dymond *et al.* (1984).

usually around a nucleus, the nodule will continue to grow, being constrained by the length of time the catalytic surface is exposed to sea water and the rate of supply of the component elements. According to Dymond *et al.* (1984), it is the flux of dissolved Mn to the site of oxidation that controls the **accretion rate** of ferromanganese nodules, and it is the relative flux of Mn compared with those of other cations that determines their **composition**. The hydrogenous precipitation of ferromanganese components is a relatively slow process, and Dymond *et al.* (1984) estimated an accretion rate of 1–2 mm $10^6$ yr$^{-1}$ for Pacific Ocean crusts, and the tops of nodules from MANOP site R.

Although the best end-member composition for the hydrogenous ferromanganese deposits is provided by crusts formed directly on to exposed rock surfaces, this type of accretion process also contributes to *all* nodule surfaces that are exposed to sea water, and it is evident that in addition to crusts, *nodules* having a large hydrogenous component can be found in the marine environment. Once a sediment substrate is introduced, however, the conditions for ferromanganese oxide formation change dramatically. In particular, the nodules can become partly buried and so can acquire a diagenetic input of dissolved metals. Although it is therefore convenient to treat diagenetic end-member nodules separately from the point of view of the principal processes involved in their generation, it must be stressed that in a strict sense most marine nodules will be mixed-source types.

*The diagenetic nodule end-member.* The sea-water–diagenetic metal-supply system can result in nodules exhibiting top–bottom compositional differences. The extent to which these differences are found depends on how far the diagenetic sequence has proceeded. This, in turn, depends on the magnitude of the down-column carbon flux, which controls the sediment substrate depositional environment (see Section 14.3). In this respect, Dymond *et al.* (1984) distinguished between nodules formed under conditions of **oxic** and **suboxic** diagenesis.

**1** *Oxic diagenesis.* In oxic diagenesis, degradable organic matter is destroyed by dissolved $O_2$, and at this stage particulate $MnO_2$ is not broken down for utilization as a secondary oxidant (see Section 14.1.2). There is no doubt that nodules formed on *oxic* substrates receive a proportion of their metals directly from sea water (hydrogenous supply). However, *direct* precipitation from sea water is not the sole, or even the dominant, route by which Mn and associated metals are supplied to nodules accreting on oxic sediment substrates, the reason being that the sediments themselves can act as metal sources. A number of mechanisms have been proposed to describe the manner in which the sediment source operates under conditions of oxic diagenesis.

(a) Release from labile carriers. During oxic diagenesis, the dissolution of labile organic carriers will release biologically bound trace metals into solution. Some of these trace metals will be lost by diffusion out of the sediments, but a fraction will be retained and so become available for incorporation into nodules that may form there. Callender & Bowser (1980) outlined a model for this; the general model is illustrated in Fig. 15.9(a) and is applied to Cu and Mn in Fig. 15.9(b).

(b) Adsorption–desorption reactions involving inorganic (non-labile) sediment phases. It is unlikely that labile carriers in (a) above will release sufficient Mn for nodule formation under all conditions of oxic diagenesis, and Lyle *et al.* (1984) proposed an alternative mechanism for the supply of Mn, and associated elements, to the lower surfaces of nodules under oxic substrate conditions. Using a solid-state speciation technique the authors identified a *loosely-held* trace-metal association (mainly consisting of metals in surface-sorbed binding fractions) in sediments. At MANOP sites M and H between ~2 and ~10% of the total Mn, ~10% of the total Ni and Cu, but <1% of the total Fe was present in the sediments in this sorbed association. The authors then outlined a two-stage adsorption–bioturbation model by which this sorbed 'pool' of transition metals is transferred to growing nodules. In stage 1, the solid particles containing the 'pool' are transported to the region around a growing nodule. In stage 2 there is desorption of metals from the 'pool' in the carrier particles and their uptake into the growing ferromanganese oxyhydroxide phases. The transfer is

thought to be brought about by equilibrium between the metals in solution in the interstitial waters and those in the 'pool', which can add metals to the waters. Growing nodule surfaces will attract the metals from the waters into their structures, however, thus taking them permanently out of solution. It is assumed that bioturbation brings fresh supplies of sorbed material to the reaction zones around growing nodules. The model is illustrated in Fig. 15.9(c).

(c) Reactions between amorphous Fe–Mn oxyhydroxides and authigenic silicates. Dymond *et al.* (1984) identified two mechanisms of this type that might have the potential to release nodule-forming elements in oxic sediments. The first of these involves a reaction between the oxyhydroxides and dissolving opal, to yield the smectite mineral **nontronite**. During this process $SiO_2$ and Fe are fixed in the silicate, but Mn (together with elements such as Ni, Cu and Co) is released. In the second mechanism, the alteration of volcanic glass to form the zeolite **phillipsite** raises the local pH and enhances the oxidation of $Mn^{2+}$, which has been adsorbed from solution on to oxide surfaces, thus promoting oxide growth (see also Bischoff *et al.*, 1981).

A number of mechanisms therefore can provide nodule-forming elements to sediments that are deposited under conditions of oxic diagenesis; in general, when the flux of biogenically associated labile elements is relatively high, mechanism (a) is operative, but when the flux is relatively low, mechanisms (b) and (c) become more important.

On the basis of data obtained during their investigation, Dymond *et al.* (1984) concluded that processes involving oxic diagenesis had operated during the formation of nodules at MANOP sites R and S. The nodules at these two sites provide an interesting example of how the influence of oxic diagenesis can vary according to the depositional environment. Nodules at both sites have a seawater (hydrogenous) component on their upper surfaces. At site R the flux of biogenic components was relatively low and the principal Mn mineral was $\delta MnO_2$. In contrast, there was a relatively strong biogenic flux to site S, and here oxic diagenesis becomes more important and a Cu- and Ni-rich toderokite was formed. The growth rates of the oxic nodules were estimated to be of the order of ~10–50 mm $10^6$ yr$^{-1}$.

**2** *Suboxic diagenesis.* In suboxic diagenesis, the dia-

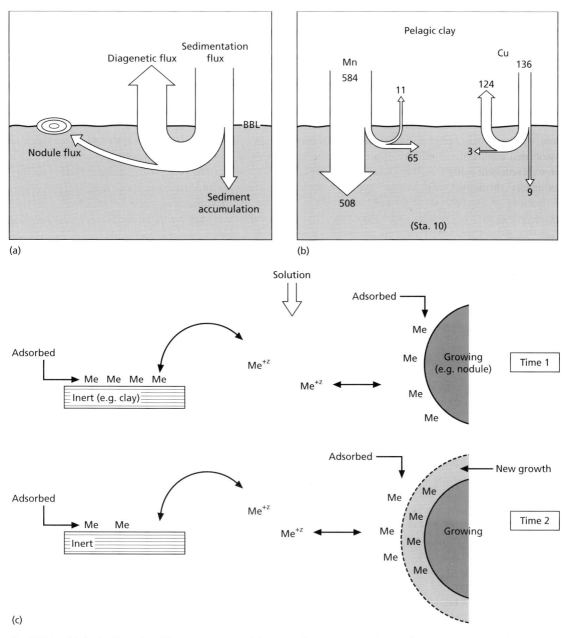

**Fig. 15.9** Models for the formation of ferromanganese nodules under conditions of oxic diagenesis. (a) General model for the early diagenetic release of trace metals (from Callender & Bowser, 1980). (b) Fluxes involved in the early diagenetic release of Mn and Cu in sediments from the equatorial Pacific (from Callender & Bowser, 1980). (c) Adsorption–bioturbation model for the transfer of loosely held elements to growing nodules (from Lyle et al., 1984); see text for details.

genetic sequence is driven by the utilization of secondary oxidants for the degradation of organic carbon, usually at a relatively shallow depth in the sediment (see Section 14.1.2). Manganese dioxide can act as one of these secondary oxidants and in the process $Mn^{2+}$ (and other trace elements associated with the oxides) is released into the interstitial waters where it can migrate upwards under concentration gradients to undergo reprecipitation to $MnO_2$ in the upper oxic zone (see Section 14.6.3), i.e. the elements reach the site of nodule formation from below. The depth at which this suboxic process takes place can be affected by the strength of the down-column carbon-mediated signal. For example, Dymond et al. (1984) reported the presence of suboxic diagenetic nodules at MANOP site H, but pointed out that Mn released from the utilization of $MnO_2$ was reprecipitated *below* the sediment surface on which the nodules were forming. To explain the formation of the suboxic diagenetic nodules, the authors therefore suggested that the site had received a supply of elements from pulses of biogenic debris, which had been sufficiently strong to decrease the depth at which suboxic diagenesis normally takes place.

In oxic diagenesis, the destruction of labile material from the organic carriers results in the supply of relatively large amounts of metals, such as Cu, Ni and Zn, giving rise to the formation of Cu-, Ni- and Zn-rich toderokite. The suboxic diagenesis of Mn oxides from the organic matter pulses, however, does not release such large amounts of Cu and Ni, and the Mn-rich toderokite formed is less stable than the Ni-, Cu- and Zn-rich variety, and can collapse on dehydration to yield birnessite. Dymond et al. (1984) estimated the overall growth rates of the suboxic nodules to be of the order of $\sim$100–200 mm $10^6$ yr$^{-1}$.

### 15.3.4.7 Ferromanganese nodules: summary

There are three general modes by which non-hydrothermal ferromanganese crusts and nodules accrete: these are the *seawater* mode, the *oxic diagenetic* mode and the *suboxic diagenetic* mode. The crusts represent an individual end-member formed by the seawater mode, but the nodules are usually top–bottom mixtures of the seawater mode and one, or both, of the diagenetic modes. The signatures associated with each of the three modes are summarized below.

1 Surface ferromanganese components formed by the seawater mode contain $\delta MnO_2$. They have low Mn:Fe ratios ($\sim$1), low Ni:Co ratios, and accrete at slow rates, which probably are in the order of $\sim$1–2 mm $10^6$ yr$^{-1}$. Ferromanganese crusts on exposed rock surfaces and nodules formed on the tops of seamounts, and other swept sediment surfaces, are formed by this mode. In addition, the upper surfaces of all nodules have a component derived from the seawater accretion mode.

2 Sediment surface ferromanganese components formed by the oxic diagenetic mode contain a Cu-, Ni- and Zn-rich toderokite. The growth-forming metals have been supplied by particulate matter via a variety of oxic diagenetic mechanisms, and the components have intermediate Mn:Fe rations (e.g. $\sim$5–10) and accrete at rates of a few tens of millimetres per million years. These components are found on the lower surfaces of nodules that grow on sediment substrates where the diagenetic sequence has not progressed beyond the oxic stage.

3 Sediment surface ferromanganese components formed by the suboxic diagenetic mode contain a Mn-rich toderokite, which may undergo transformation to birnessite. The metals necessary for the formation of these components are released from particulate matter during suboxic diagenesis. In some instances they may arise from the oxidative utilization of $MnO_2$ at depth in the sediment. If the $Mn^{2+}$ released in this manner is trapped below the sediment surface, however, a secondary source of suboxic $Mn^{2+}$ may be provided by pulses of biogenic material that are sufficiently strong to permit suboxic diagenesis to occur close to, or even at, the sediment surface. The components formed during suboxic diagenesis tend to have relatively high Mn:Fe ratios (often $\geqslant$20), and accrete episodically at rates that can be as high as several hundred millimetres per million years. Suboxic diagenetic components characterize the lower surfaces of nodules formed on marginal and hemi-pelagic sediments in which the diagenetic sequence has reached the $MnO_2$ oxidative utilization stage.

### 15.3.5 Sediment ferromanganese oxyhydroxides

In addition to macronodules, oceanic sediments contain micronodules and small-sized oxyhydroxides of a non-nodular origin. These are an intimate part of

**Table 15.5** Elemental ratios in sediment oxyhydroxides, ferromanganese crusts, $\delta MnO_2$-rich nodules and toderokite-rich nodules from the central Pacific (data from Aplin & Cronan, 1985b).

| Elemental ratio | Component | | | |
| | Sediment oxyhydroxide* | Fe–Mn crusts | $\delta MnO_2$-rich nodules | Toderokite-rich nodules |
|---|---|---|---|---|
| Mn:Fe | 0.99 | 1.15 | 0.97 | 2.6 |
| Cu:Mn | 0.020 | 0.009 | 0.013 | 0.032 |
| Ni:Mn | 0.019 | 0.020 | 0.020 | 0.040 |
| Co:Mn | 0.012 | 0.026 | 0.019 | 0.009 |
| Pb:Mn | 0.004 | 0.006 | 0.0048 | 0.002 |
| Zn:Mn | 0.007 | 0.003 | 0.003 | 0.004 |

* Leached fraction of sediment.

the sediment, and in the present section attention will be focused on the question 'Do they form part of a crust → nodule → sediment oxyhydroxide continuum, or is there a major decoupling between the nodule–sediment system?'

Aplin & Cronan (1985a,b) gave data on the chemical composition of a series of ferromanganese crusts, sediment oxyhydroxides, $\delta MnO_2$-rich macronodules and toderokite-rich macronodules from the central Pacific. The authors used a selective leaching technique to determine the compositions of the sediment oxyhydroxides, and then compared these with those of the crusts and nodules from the same sample suite. The data are summarized in Table 15.5, and two principal conclusions can be drawn from this investigation.

1 The chemical compositions of the crusts, sediment oxyhydroxides and $\delta MnO_2$-rich nodules are generally similar. There are, however, some differences between the three phases; for example, the Mn:Fe, Cu:Mn and Co:Mn ratios varied, with those of the $\delta MnO_2$-rich nodules being intermediate between those of the crusts and the sediment oxyhydroxides. According to Aplin & Cronan (1985a,b) this can be explained in the following manner. The crusts derive their elements largely from sea water (*hydrogenous* supply). In contrast, the oxyhydroxides acquire most of their elements from processes occurring in the sediments, and in the region studied much of this elemental input arises from the decay of labile biologically transported material during early diagenesis (*oxic diagenetic* supply). The two phases therefore represent a seawater and an oxic diagenetic end-member, respectively. The $\delta MnO_2$-rich nodules, which grow at the sediment surface, receive elements from both the

hydrogenous seawater and the oxic diagenetic sources; thus, they have a composition that is intermediate between those of the seawater and the oxic diagenetic end-members. In this crust → $\delta MnO_2$-rich nodule → sediment oxyhydroxide sequence it is therefore oxic diagenetic reactions that perturb the composition of the seawater end-member.

2 The chemical compositions of the toderokite-rich nodules were quite distinct from those of the members of the crust → $\delta MnO_2$-rich nodule → sediment oxyhydroxide sequence. For example, it can be seen from Table 15.5 that the Mn:Fe, Cu:Mn and Ni:Mn ratios are higher, and the Co:Mn ratio lower, in the toderokite-rich nodules than in the other components. These toderokite-rich nodules are indicative of more intense (i.e. suboxic) diagenesis, which involves the remobilization of $Mn^{2+}$ from the breakdown of Mn oxides, and Aplin & Cronan (1985b) suggested that in the area under investigation the toderokite, which is a Cu-, Ni- and Zn-poor variety, had precipitated directly from the sediment interstitial waters.

It would appear, therefore, that although $\delta MnO_2$-rich nodules can have both a seawater and a diagenetic supply of trace metals, there is a general continuum of compositions in the sequence crust → $\delta MnO_2$-rich nodule → sediment oxyhydroxides. A *major decoupling* is found, however, in the sediment–nodule system, at the seawater–diagenetic boundary, between the $\delta MnO_2$-rich and the toderokite-rich nodules. This results because suboxic diagenesis is a recycling process, which can release elements already incorporated into the sediments, and depends largely on the extent to which the diagenetic sequence has proceeded. Oxic diagenesis, e.g. when labile biologically transported elements are solubilized, can supply metals to both sediment oxyhydroxides and nodules. The decoupling break is strongest at the point where suboxic diagenesis releases $Mn^{2+}$, however, which can replace $Cu^{2+}$, $Ni^{2+}$ and $Zn^{2+}$ in toderokite and can result in the formation of a Cu-, Ni- and Zn-poor toderokite variety, which may subsequently form birnessite (see Section 15.3.4.6).

### 15.3.6 Hydrothermal ferromanganese deposits

The convection of sea water through freshly generated oceanic crust at the centres of sea-floor spreading is now recognized as being a process of gobal geochemical significance (see Sections 5.1 and 6.3). The elements that are solubilized during hydrothermal activity are subsequently removed from solution via solid phases, and the processes involved can be described in terms of the Red Sea system, a system in which the complete sequence of metal-rich precipitates can be found in sediments surrounding the discharge points of the mineralizing solutions.

In the Red Sea system hot saline brines become enriched in metals by reacting at high temperature with underlying evaporites, shales and volcanic rocks, and on discharge into sea water they gather in deeps in the central rift valley area. According to Cronin (1980) the Atlantic II Deep, an area of present-day metalliferous sediment deposition, can be used to illustrate the range of hydrothermal precipitates formed in the oceans. The general precipitation sequence from the hydrothermal solutions occurs in the order: **sulphides** (e.g. sphalerite, pyrite, chalcopyrite, marcasite, galena) → **iron silicates** (e.g. smectite, chamosite, amorphous silicates) → iron oxides → **manganese oxides**. This fractionation of the precipitates manifests itself in their spatial separation from the venting source; for example, the iron and manganese oxides sometimes form distinct haloes at a distance from the discharge point. Thus, marine hydrothermal deposits can be thought of as representing various stages in the evolution of the mineralizing solutions. It was pointed out in Section 5.1, however, that the fractionation sequence may commence before the hydrothermal mineralizing solutions actually reach the venting outlet. Thus, in the **white smokers** the mixing of sea water and hydrothermal fluids occurs at depth in the system, with the result that sulphides have been removed within the vents before the fluids are debouched into sea water at a relatively low temperature. Here, therefore, the sulphide stage in the sequence has been reached under the sea floor. In the **black smokers**, however, hydrothermal fluids are vented directly on to the sea floor at high temperature, and the sulphides are precipitated directly on to the sea bed. Some of these sulphides are used to build the venting chimneys and some, together with other still dissolved hydrothermally derived metals, are dispersed to carry on the sequence around, and away from, the vents.

Hydrothermal deposits have been found, either at the sediment surface or close to the sediment–basalt basement interface, at many marine locations; for example, Lalou (1983) identified 70 such locations. The metal compositions of a series of hydrothermal deposits from a number of the more important hydrothermal study regions are given in Table 15.6.

Hydrothermal sulphides are generally restricted either to within the vents themselves or to the immediate vicinity of the venting system. In contrast, Fe and Mn dispersion haloes can cover a much greater area, and ferromanganese oxyhydroxides, nodules and crusts are the characteristic sedimentary components of much open-ocean hydrothermal activity. It is of interest, therefore, to compare the compositions of these end-member *hydrothermal* ferromanganese components with those that have a *hydrogenous* origin. For this purpose, representative chemical and mineralogical analyses, together with growth rate characteristics, for a hydrogenous crust (column A), an oxic nodule (column B), a suboxic nodule (column C) and a ridge-centre hydrothermal crust (column D) are given in Table 15.7. When hydrothermal crusts are considered it must be remembered, however, that the parent hydrothermal activity is not confined to active centres of sea-floor spreading, but that it can also occur in other regions associated with volcanic activity. One such region can be found in island arcs and their associated marginal basins. For example, Moorby *et al.* (1984) described ferromanganese crusts recovered from several sites on the Tonga–Kermadec Ridge in the southwest Pacific island arc. The average chemical and mineralogical composition of these crusts is also included in Table 15.7 (column E).

Although most crusts are to some extent mixtures of end-member sources, it is apparent from the various data in Table 15.7 that there are a number of

**Table 15.6** Chemical compositions of hydrothermal deposits; units, $\mu g\,g^{-1}$.

|     | A*     | B†     | C‡    | D§     | E¶     | F‖     | G**    | H††    | I‡‡    | J§§    |
|-----|--------|--------|-------|--------|--------|--------|--------|--------|--------|--------|
| Mn  | 60 000 | 46 000 | 39 043| 430 000| 550 000| 470 000| 40 000 | 279 000| 410 000| 380 000|
| Fe  | 180 000| 141 000| 587   | 1600   | 2000   | 6600   | 232 900| 10 500 | 8000   | 27 000 |
| Al  | 5000   | 23 000 | —     | 1800   | —      | 2000   | 7900   | 12 700 | 9000   | 6900   |
| Ti  | 200    | —      | —     | —      | —      | 28     | —      | —      | 400    | 1060   |
| Co  | 105    | 64     | 19    | 24     | 39     | 13     | 10     | 82     | 33     | 30     |
| Ni  | 430    | 820    | 353   | 880    | 180    | 125    | 80     | 371    | 310    | 400    |
| Cu  | 730    | 910    | 43    | 450    | 50     | 80     | 76     | 206    | 120    | 80     |
| Zn  | 380    | 330    | —     | 540    | 2020   | 90     | 35     | 83     | 400    | 310    |
| Mo  | 30     | —      | —     | —      | —      | 540    | —      | —      | 900    | —      |
| V   | 450    | —      | —     | —      | —      | 110    | 152    | 214    | 110    | —      |

* Metal-rich sediment at crest of East Pacific Rise, data on a carbonate-free basis (Bostrom & Peterson, 1969).
† Metal-rich sediment, Bauer Deep, central Pacific, data on a carbonate-free basis (Dymond *et al.*, 1973).
‡ Hydrothermal deposits, TAG area, Mid-Atlantic Ridge (Scott *et al.*, 1974).
§ Hydrothermal deposits, TAG area, Mid-Atlantic Ridge (Toth, 1980).
¶ Hydrothermal deposits, Galapagos spreading centre (Moore & Vogt, 1976).
‖ Hydrothermal deposits, Galapagos mounds (Moorby & Cronan, 1983).
** Hydrothermal clay-rich deposit, FAMOUS area, Mid-Atlantic Ridge (Hoffert *et al.*, 1978).
†† Hydrothermal Fe-Mn concretions, FAMOUS area, Mid-Atlantic Ridge (Hoffert *et al.*, 1978).
‡‡ Hydrothermal deposits, southwest Pacific island arc system (Moorby *et al.*, 1984).
§§ Hydrothermal deposits, Gulf of Aden (Cann *et al.*, 1977).

**Table 15.7** Some general characteristics of oceanic Fe–Mn deposits; units, $\mu g\,g^{-1}$.

|     | Chemical composition | | | | |
|-----|--------|--------|--------|--------|--------|
|     | A* Hydrogenous crust | B* Oxic nodule | C* Suboxic nodule | D† Hydrothermal crust | E‡ Hydrothermal crust |
| Mn  | 222 000 | 316 500 | 480 000 | 410 000 | 550 000 |
| Fe  | 190 000 | 44 500  | 4900    | 8000    | 2000    |
| Co  | 1300    | 280     | 35      | 33      | 39      |
| Ni  | 5500    | 10 100  | 4400    | 310     | 180     |
| Cu  | 1480    | 4400    | 2000    | 120     | 50      |
| Zn  | 750     | 2500    | 2200    | 400     | 2020    |
| Mn : Fe | 1.2 | 7.1     | 98      | 51      | 275     |
| Principal mineralogy of Fe–Mn phases | $\delta MnO_2$ | $\delta MnO_2$, toderokite | Toderokite | Birnessite | Birnessite, toderokite |
| Approximate growth rates (mm $10^6\,yr^{-1}$) | 1–2 | 10–50 | 100–200 | 500 | 1000–2000 |

* Data from Dymond *et al.* (1984), see Table 15.4.
† Data from Moorby *et al.* (1984); non-spreading centre hydrothermal deposit, southwest Pacific island arc.
‡ Data from Moore & Vogt (1976); spreading centre hydrothermal deposit, Galapagos region.

important differences between the hydrothermal and hydrogenous crusts. These can be summarized as follows.

**1 Growth rates.** In general, the hydrogenous crusts appear to accrete at relatively slow rates, which are of the order of a few millilitres per million years. In contrast, the hydrothermal crusts grow at much faster rates, which are of the order of several hundred millimetres per million years. The principal reason for this difference is probably related to the fact that the hydrothermal crusts are formed in the vicinity of localized pulses of Mn-rich mineralizing solutions, whereas hydrogenous nodules have to rely on a smoothed-out background supply of Mn from normal sea water.

**2 Mineralogy.** Hydrogenous crusts are usually dominated by $\delta MnO_2$ phases. In contrast, the predominant minerals in the hydrothermal crusts are often toderokite or birnessite, i.e. minerals with relatively high Mn:Fe ratios; however, a wide range of manganese minerals has in fact been reported in hydrothermal crusts.

**3 Chemistry.** There are two principal differences between the chemical compositions of the hydrothermal and hydrogenous crusts. First, the hydrothermal crusts generally are depleted in Co, Ni and Cu compared with their hydrogenous counterparts; this probably is a result of the uptake of these elements in sulphides early in the hydrothermal sequence. Secondly, the hydrothermal crusts exhibit an extreme fractionation of Mn from Fe, evidenced in their relatively very high Mn:Fe ratios, which probably results from the removal of Fe in the sulphide, silicate and Fe oxide phases that precipitate before the Mn oxides in the hydrothermal pulses. Thus, the *crusts*, in which the constituent elements are derived directly from sea water, are enriched in Mn relative to Fe in hydrothermal regions; however, metal-rich *sediments*, which are found around active crestal regions, do not show this fractionation of Mn from Fe because they contain a variety of hydrothermal components, including Fe oxides.

One of the principal reasons for the differences between hydrogenous and hydrothermal ferromanganese components, especially with regard to the faster growth rates of the latter, is that the hydrothermal activity involves the local introduction of 'pulses' of Fe- and Mn-rich solutions into sea water. In contrast, hydrogenous components acquire their Fe and Mn from 'background' sea water. The two types of end-member ferromanganese deposits therefore have independent metal sources, e.g. ridge-crest metalliferous material having a local hydrothermal source and abyssal plain nodules having a 'background' hydrogenous source. Lalou (1983), however, presented the alternative view that there may be a coupling between hydrothermal activity and the genesis of ferromanganese nodules on a global scale. In this coupling it was proposed that a significant fraction of the hydrothermal flux, which occurs episodically in the form of non-steady-state 'pulses', can be transported for relatively large distances from the venting systems (see Section 6.3) and become the major source of Mn in many ferromanganese nodules. Mass balance calculations carried out by Lalou (1983) indicated that the number of active vents necessary to maintain the hydrothermal Mn source was unrealistic. However, the author suggested that the strength of the hydrothermal activity has varied over geological time, and that the present day is not a particularly active period.

Clearly, hydrothermal activity is an importance source for Fe and Mn (and some other metals) to sea water. However, there is no doubt that *end-member* ferromanganese deposits have a number of individual characteristics, which result largely from differences between the local 'pulsed' (hydrothermal) and smoothed out seawater 'background' (hydrogenous) metal sources. The question of the ocean-wide importance of hydrothermal sources to ferromanganese deposits therefore must remain unresolved at present.

### 15.3.7 Ferromanganese deposits in the oceans: summary

**1** Ferromanganese components in oceanic sediments consist of crusts, nodules and oxyhydroxides.

**2** The formation of these ferromanganese components can be related to a **hydrogenous–hydrothermal–diagenetic** Mn supply trinity.

**3** At the mid-ocean ridges the supply of Mn can be dominated by hydrothermal pulses. Away from the ridges, non-hydrothermal sources become more important, and there is an increasing contribution of elements to ferromanganese components from 'background' sea water, either directly (hydrogenous supply) or via diagenetic processes that involve a sediment substrate.

4 In open-ocean areas, oxic diagenesis releases elements close to the sediment surface, but in marginal regions suboxic diagenesis drives the processes involved in the liberation of elements following recycling at depth in the sediment column.

5 The ferromanganese components formed via this hydrogenous–hydrothermal–diagenetic trinity play an extremely important role in removing a variety of dissolved elements from sea water and also exert a major influence on the composition of deep-sea sediments.

## 15.4 Cosmogenous components

### 15.4.1 Definition

These are components that have been formed in outer space and have reached the surface of the Earth via the atmosphere. According to Chester & Aston (1976) the extraterrestrial material that has been identified in deep-sea sediments includes cosmic spherules and microtectites, together with cosmic-ray-produced radioactive and stable nuclides. However, these cosmic components form a very minor part of oceanic sediments, and for a detailed review of the topic the reader is referred to Brownlee (1981).

### 15.4.2 Cosmic spherules

Essentially, there are two kinds of spherules in deep-sea sediments, the iron and the stony types.

#### 15.4.2.1 The iron spherules

Small black magnetic spherules have been found in sediments from all the major oceans (see e.g. Brunn et al., 1955; Crozier, 1961; Millard & Finkelman, 1970). They were first described by Murray & Renard (1891), and it was these authors who suggested that they had a cosmic origin. This type of spherule has a density of $\sim 6\,g\,cm^{-3}$, and consists of a metallic (Fe–Ni) nucleus surrounded by a shell of magnetite and wüstite, the latter being an iron oxide that is virtually non-existent on Earth but is found in meteorite fusion crusts (see e.g. Murrell et al., 1980). According to Murray & Renard (1891), this type of cosmic spherule originated as molten droplets ablated from a meteorite. The droplets then underwent oxi-

dation, during which the outer magnetite shell was formed, as they entered the Earth's atmosphere. In general, this theory is still accepted, and these spherules are considered to have a cosmic origin (see e.g. Parkin et al., 1967; Blanchard & Davies, 1978; Blanchard et al., 1980). Estimates of the influx, or accretion, rates of interplanetary dust particles (IDPs) to the surface of the Earth vary widely, ranging from $>10^6$ to $<10^3\,ton\,yr^{-1}$; in one estimate, Esser & Turekian (1988) used osmium isotope systematics to derive a rate of $4.9 \times 10^4\,ton\,yr^{-1}$ for the accretion rate of C-1 carbonaceous chondrite material to the Earth's surface. Interplanetary dust of the zodical cloud is inhomogeneously distributed in space, being localized in bands related to asteroid families, and is confined in rings locked to the Earth's orbit (see e.g. Dermott et al., 1994). It has been suggested that variations in the Earth's orbit induce changes in the rate at which IDPs are accreted, higher accretion occurring during periods of high inclination and/or eccentricity, and Muller & MacDonald (1995) proposed that the accretion varies with a 100 kyr periodicity in the Earth's orbital inclination, which is linked into glacial cycles. Farley & Patterson (1995) used the $^3$He extraterrestrial flux to the sea floor to confirm the 100 kyr periodicity in the Atlantic; however, Marcantonio et al. (1996) did not find $^3$He evidence to support the 100 kyr periodicity in sediments from the equatorial Pacific. It is thought by some workers that the accretion rates of the iron spherules may actually approach zero at certain times. For example, Parkin et al. (1967) sampled clean air at Barbados and reported that iron spherules with diameters $>20\,\mu m$ were reaching the surface of the Earth at $\sim 350\,ton\,yr^{-1}$, but that most of this could in fact be accounted for by contamination effects. These authors suggested that variations in the accretion rates of the magnetic spherules, as recorded in deep-sea sediments and ice cores, result from the spasmodic availability of suitable parent meteorites in outer space.

#### 15.4.2.2 The stony spherules

These have a density of $\sim 3\,g\,cm^{-3}$ and diameters in the range $\sim 15$–$250\,\mu m$, although they often depart from a spherical shape. Essentially, they consist of fine-grained silicates (mainly olivine), magnetite and glass. It is thought generally that this type of spherule has originated from stony meteorites, although the actual mechanism involved in their formation is not clear. For example, Parkin *et al.* (1967) have suggested that they are in fact micrometeorites, which undergo relatively little physical alteration on passage through the atmosphere. Blanchard *et al.* (1980), however, showed that the stony spherules have similarities with chondrite fusion crusts, and concluded that they are products of the atmospheric heating of stony meteoroids, a theory that would account for the presence of the magnetite in them. On the basis of their deep-sea sediment storage, Murrell *et al.* (1980) have estimated the accretion rate of the stony spherules to be $\sim$90 ton yr$^{-1}$. Some of the stony spherules identified in deep-sea sediments, however, may have a terrestrial origin (see e.g. Fredricksson & Martin, 1963).

### 15.4.3 Microtectites

These are small bodies of green-black glass found in large numbers over a few restricted regions of the Earth's surface. They have been identified in bands in deep-sea sediments from the Indian Ocean and the equatorial Atlantic, and probably originate from locally associated strewn fields.

### 15.4.4 Cosmic-ray-produced nuclides

According to Chester & Aston (1976) cosmic-ray-produced stable and radioactive nuclides that have been identified in deep-sea sediments, and various ice deposits, include $^{36}Cl$, $^{26}Al$, $^{10}Be$, $^{64}Fe$, $^{3}He$, $^{4}He$, $^{40}Ar$ and $^{36}Ar$.

### 15.5 Summary

We have now described the compositions and origins of the various components that are found in oceanic sediments. In the following chapter, we will attempt to show how the chemical signals associated with the formation of these components combine together to impose a control on the overall chemical composition of the sediments found in the oceans.

### References

Aplin, A.C. & Cronan, D.S. (1985a) Ferromanganese oxide deposits from the central Pacific Ocean, I. Encrustations from the Line Islands Archipelago. *Geochim. Cosmochim. Acta*, **40**, 427–36.

Aplin, A.C. & Cronan, D.S. (1985b) Ferromanganese oxide deposits from the central Pacific Ocean, II. Nodules and associated sediments. *Geochim. Cosmochim. Acta*, **49**, 437–51.

Arrhenius, G. (1963) Pelagic sediments. In *The Sea*, M.N. Hill (ed.), Vol. 3, 655–727. New York: Interscience.

Bender, M.L., Ku, T.-L. & Broecker, W.S. (1970) Accumulation rates of manganese in pelagic sediments and nodules. *Earth Planet. Sci. Lett.*, **8**, 143–8.

Berger, W.H. (1968) Planktonic foraminifera: selective solution and paleoclimatic interpretation. *Deep Sea Res.*, **15**, 31–43.

Berger, W.H. (1970) Biogenous deep-sea sediments: fractionation by deep-sea circulation. *Geol. Soc. Am. Bull.*, **18**, 1385–402.

Berger, W.H. (1976) Biogenous deep sea sediments: production, preservation and interpretation. In *Chemical Oceanography*, J.P. Riley & R. Chester (eds), Vol. 5, 265–388. London: Academic Press.

Berger, W.H. & Winterer, E.L. (1974) Plate stratigraphy and the fluctuating carbonate line. In *Pelagic Sediments on Land and under the Sea*, K.J. Hsü & H.C. Jenkyns (eds), Oxford: Blackwell Scientific Publications, International Association of Sedimentologists, Special Publication 1, 11–48.

Berger, W.H., Adelseck, C.G. & Mayer, L.A. (1976) Distribution of carbonate in surface sediments of the Pacific Ocean. *J. Geophys. Res.*, **81**, 2617–27.

Biscaye, P.E. (1965) Mineralogy and sedimentation of Recent deep-sea clay in he Atlantic Ocean and adjacent seas and oceans. *Geol. Soc. Am. Bull.*, **76**, 830–32.

Biscaye, P.E., Kolla, V. & Turekian, K.K. (1976) Distribution of calcium carbonate in surface sediments of the Atlantic Ocean. *J. Geophys. Res.*, **81**, 2595–603.

Bischoff, J.L., Piper, D.Z. & Leong, K. (1981) The aluminosilicate fraction of North Pacific manganese nodules. *Geochim. Cosmochim. Acta*, **45**, 2047–63.

Blanchard, M.B. & Davies, A.S. (1978) Analysis of ablation debris from natural and artificial iron meteorites. *J. Geophys. Res.*, **83**, 1793–808.

Blanchard, M.B., Brownlee, D.E., Bunch, T.E., Hodge, P.W. & Kyte, F.T. (1980) Meteor ablation spheres from deep-sea sediments. *Earth Planet. Sci. Lett.*, **46**, 178–90.

Bostrom, K. & Peterson, M.N.A (1969) The origin of aluminium-poor ferromanganoan sediments in areas of high heat flow on the East Pacific rise. *Mar. Geol.*, **7**, 427–47.

Bostrom, K., Kraemer, T. & Gartner, S. (1973) Provenance and accumulation rates of opaline silica, Al, Ti, Fe, Mn, Cu, Ni and Co in Pacific pelagic sediments. *Chem. Geol.*, **11**, 123–48.

Broecker, W.S (1971) A kinetic model for the chemical composition of sea water. *Quat. Res.*, **1**, 188–207.

Broecker, W.S. & Peng, T.-H. (1982) *Tracers in the Sea*. Palisades, NY: Lamont-Doherty Geological Observatory.

Brownlee, D.E. (1981) Extraterrestrial components. In *The Sea*, C. Emiliani (ed.), Vol. 7, 733–62. New York: Interscience.

Brunn, A.F., Langer, E. & Pauly, H. (1955) Magnetic particles found by raking the deep sea bottom. *Deep Sea Res.*, 2, 230–46.

Burns, R.G. & Burns, V.M. (1977) Mineralogy. In *Marine Manganese Deposits*, G.P. Glasby (ed.), 185–248. Amsterdam: Elsevier.

Burton, J.D. & Liss, P.S. (1968) Processes of supply of dissolved silicon in the oceans. *Geochim. Cosmochim. Acta*, 37, 1761–73.

Buser, W. & Grutter, A. (1956) Uber die Natur der Mangankollen. *Schweiz. Mineral. Petrogr. Mitt.*, 36, 49–62.

Callender, E. & Bowser, C.J. (1980) Manganese and copper geochemistry of interstitial fluids from manganese-rich sediments of the northeastern equatorial Pacific Ocean. *Am. J. Sci.*, 280, 1063–96.

Calvert, S.E. (1968) Silica balance in the oceans and diagenesis. *Nature*, 219, 919–20.

Cann, J.R., Winter, C.K. & Pritchard, R.G. (1977) A hydrothermal deposit from the floor of the Gulf of Aden. *Mineral. Mag.*, 41, 193–99.

Chester, R. & Aston, S.R. (1976) The geochemistry of deep-sea sediments. In *Chemical Oceanography*, J.P. Riley & R. Chester (eds), Vol. 6, 281–390. London: Academic Press.

Chester, R. & Hughes, M.J. (1967) A chemical technique for the separation of ferromanganese minerals, carbonate minerals and adsorbed trace elements from pelagic sediments. *Chem. Geol.*, 3 199–212.

CLIMAP Project Members (1976) The surface of the ice age earth. *Science*, 191, 1131–7.

Cronan, D.S (1976) Manganese nodules and other ferromanganese oxide deposits. In *Chemical Oceanography*, J.P. Riley & R. Chester (eds), Vol. 5, 217–63. London: Academic Press.

Cronan, D.S. (1977) Deep-sea nodules: distribution and geochemistry. In *Marine Manganese Deposits*, G.P. Glasby (ed.), 11–44. Amsterdam: Elsevier.

Cronan, D.S. (1980) *Underwater Minerals*. London: Academic Press.

Crozier, D.S. (1961) Micrometeorite measurements — satellite and ground-level data compared. *J. Geophys. Res.*, 66, 2793–5.

Dermott, S.F., Jayaraman, S., Xu, Y.L., Gustafson, B.A.S. & Lou, J.C. (1994) A circumsolar ring of asteroidal dust in resonant lock with the Earth. *Nature*, 369, 719–23.

DeMaster, D.J. (1981) The supply and accumulation of silica in the marine environment. *Geochim. Cosmochim. Acta*, 45, 1715–32.

Drever, J.I. (1982) *The Geochemistry of Natural Waters*. Englewood Cliffs, NJ: Prentice-Hall.

Dymond, J.M., Corliss, J.B., Heath, G.R., Field, C.W., Dasch, E.J. & Veeh, H.H. (1973) Origin of metalliferous sediments from the Pacific Ocean. *Geol. Soc. Am. Bull.*, 84, 335–72.

Dymond, J.M., Lyle, M., Finney, B., *et al.* (1984) Ferromanganese nodules from MANOP Sites H, S and R — control of mineralogical and chemical composition by multiple accretionary processes. *Geochim. Cosmochim. Acta*, 48, 931–49.

Edmond, J.M. (1974) On the dissolution of carbonate and silicate in the deep ocean. *Deep Sea Res.*, 21, 455–80.

Elderfield, H. (1976) Hydrogenous material in marine sediments: excluding manganese nodules. In *Chemical Oceanography*, J.P. Riley & R. Chester (eds), Vol. 5, 137–215. London: Academic Press.

Emiliani, C. & Flint, R.F. (1963) The Pleistocene record. In *The Sea*, M.N. Hill (ed.), Vol. 3, 888–927. New York: Interscience.

Esser, B.K. & Turekian, K.K. (1988) Accretion rate of extraterrestrial particles determined from osmium isotope systematics of Pacific pelagic clay and manganese nodules. *Geochim. Cosmochim. Acta*, 52, 1383–8.

Farley, K.A. & Patterson, D.B. (1995) A 100-kyr periodicity in the flux of extraterrestrial $^3$He to the sea floor. *Nature*, 378, 600–3.

Fredericksson, K. & Martin, L.R. (1963) The origin of black spherules in Pacific islands, deep-sea sediments and Antarctic ice. *Geochim. Cosmochim. Acta*, 27, 241–51.

Goldberg, E.D. (1954) Marine geochemistry. Chemical scavengers of the sea. *J. Geol.*, 62, 249–55.

Goldberg, E.D. & Arrhenius, G.O.S. (1958) Chemistry of Pacific pelagic sediments. *Geochim. Cosmochim. Acta*, 13, 153–212.

Griffin, J.J. & Goldberg, E.D. (1963) Clay-mineral distribution in the Pacific Ocean. In *The Sea*, M.N. Hill (ed.), Vol. 3, 728–41. New York: Interscience.

Griffin, J.J., Windom, H. & Goldberg, E.D. (1968) The distribution of clay minerals in the World Ocean. *Deep Sea Res.*, 15, 433–59.

Hay, W.W. (ed.) (1974) *Studies in Paleo-oceanography*. Tulsa, OK: Society of Economic Paleontologists and Mineralogists, Special Publication 20.

Hem, J.D. (1978) Redox processes at surfaces of manganese oxide and their effects on aqueous metal ions. *Chem. Geol.*, 21, 199–218.

Hoffert, M., Perseil, A., Hekinian, R., *et al.* (1978) Hydrothermal deposits sampled by diving saucer in Transform Fault 'A' near 37°N on the Mid-Atlantic Ridge, FAMOUS area. *Oceanol. Acta*, 1, 73–86.

Holland, H.D. (1978) *The Chemistry of the Atmosphere and Oceans*. New York: Wiley.

Kennet, J.P. (1982) *Marine Geology*. Englewood Cliffs, NJ: Prentice-Hall.

Kolla, V.R., Be, A.W.H. & Biscaye, P.E. (1976) Calcium carbonate distribution in surface sediments of the Indian Ocean. *J. Geophys. Res.*, 81, 2605–16.

Ku, T.L. (1977) Rates of accretion. In *Marine Manganese*

*Deposits*, G.P. Glasby (ed.), 249–67. Amsterdam: Elsevier.

Lalou, C. (1983) Genesis of ferromanganese deposits: hydrothermal origin. In *Hydorthermal Processes at Seafloor Spreading Centres*, P.A. Rona, K. Bostrom, L. Laubier & K.L. Smith (eds), 503–34. New York: Plenum.

Li, Y.-H., Takahashi, T. & Broecker, W.S. (1969) Degree of saturation of $CaCO_3$ in the oceans. *J. Geophys. Res.*, **23**, 5507–25.

Luz, B. & Shackleton, N.J. (1975) $CaCO_3$ solution in the tropical East Pacific during the past 130 000 years. In *Dissolution of Deep-sea Carbonates*, W.V. Sliter, A.W.H. Be & W.H. Berger (eds), 142–50. Washington, DC: Cushman Foundation, Special Publication 13.

Lyle, M., Heath, G.R. & Robbins, J.M. (1984) Transport and release of transition elements during early diagenesis: sequential leaching of sediments from MANOP sites M and H, Part I. pH5 acetic acid leach. *Geochim. Cosmochim. Acta*, **48**, 1705–15.

Mackenzie, F.T. & Garrels, R.M. (1966) Chemical mass balance between rivers and oceans. *Am. J. Sci.* **264**, 507–25.

Manheim, F.T. (1965) Manganese–iron accumulations in the shallow marine environment. In *Symposium on Marine Geochemistry*, D.R. Shrink & J.T. Corless (eds), 217–76. Narragansett, RI: Narragansett Marine Laboratory, University of Rhode Island, Occasional Publication, no. 3-1965.

Marcantonio, F., Anderson, R.F., Stute, M., Kumar, N., Scholosser, P. & Mix, A. (1996) Extraterrestrial $^3$He as a tracer of marine sediment transport and accumulation. *Nature*, **383**, 705–7.

Mero, J.L. (1962) Ocean-floor manganese nodules. *Econ. Geol.*, **57**, 747–67.

Millard, H.T. & Finkelman, R.B. (1970) Chemical and mineralogical compositions of cosmic and terrestrial spherules from a marine sediment. *J. Geophys. Res.*, **75**, 2125–34.

Moorby, S.A. & Cronan, D.S. (1983) The geochemistry of hydrothermal and pelagic sediments from the Galapagos Hydrothermal Mounds DSDP Leg 70. *Mineral. Mag.*, **47**, 291–300.

Moorby, S.A., Cronan, D.S. & Clasby, G.P. (1984) Geochemistry of hydrothermal Mn-oxide deposits from the S.W. Pacific island arc. *Geochim. Cosmochim. Acta*, **48**, 433–41.

Moore, W.S. & Vogt, P.R. (1976) Hydrothermal manganese crust from two sites near the Galapagos spreading axis. *Earth Planet. Sci. Lett.*, **29**, 349–56.

Morse, J.W. & Berner, R.A. (1972) Dissolution kinetics of calcium carbonate in seawater: II. A kinetic origin for the lysocline. *Am. J. Sci.*, **272**, 840–51.

Muller, R.A. & MacDonald, G.J. (1995) Glacial cycles and orbital inclination. *Nature*, **377**, 107–8.

Murray, J. & Renard, A.F. (1891) *Deep-sea Deposits*. London: Scientific Report of the *Challenger* Expedition, no. 3 *HMSO*, London.

Murrell, M.T., Davies, P.A., Nishizumi, K. & Millard, H.T. (1980) Deep-sea spherules from Pacific clay: mass distribution and influx rate. *Geochim. Cosmochim. Acta*, **44**, 2067–74.

Parkin, D.W., Delany, A.C. & Delany, A.C. (1967) A search for airborne cosmic dust on Barbados. *Geochim. Cosmochim. Acta*, **31**, 1311–20.

Piper, D.Z., Basler, J.R. & Bischoff, J.L. (1984) Oxidation state of marine manganese nodules. *Geochim. Cosmochim. Acta*, **48**, 2347–55.

Pratt, R.M. & McFarlin, P.F. (1966) Manganese pavements on the Blake Plateau. *Science*, **151**, 1080–2.

Price, N.B. & Calvert, S.E. (1970) Compositional variation in Pacific Ocean ferromanganese nodules and its relationship to sediment accumulation rates. *Mar. Geol.*, **9**, 145–71.

Raab, W.J. & Meylan, M.A. (1977) Morphology. In *Marine Manganese Deposits*, G.P. Glasby (ed.), 109–46. Amsterdam: Elsevier.

Rabouille, C., Gaillard, J.-F., Treguer, P. & Vincendeau, M.-A. (1997) Biogenic silica recycling in surficial sediments across the Polar Front of the Southern Ocean (Indian Sector). *Deep Sea Res.*, **44**, 1151–76.

Redfield, A.C., Ketchum, B.H. & Richards, F.A. (1963) The influence of organisms on the composition of seawater. In *The Sea*, M.N. Hill (ed.), Vol. 2, 26–77. New York: Interscience.

Riley, J.P. & Chester, R. (1971) *Introduction to Marine Chemistry*. London: Academic Press.

Schopf, T.J.M. (1980) *Paleoceanography*. Cambridge, MA: Harvard University Press.

Scott, M.R., Scott, R.B., Rona, P.A., Butler, L.W. & Nalwak, A.J. (1974) Rapidly accumulating manganese deposit from the median valley of the Mid-Atlantic Ridge. *Geophys. Res. Lett.*, **1**, 355–8.

Shackleton, N.J. (1982) The deep-sea sediment record of climate variability. *Prog. Oceanogr.*, **11**, 199–218.

Sillen, L.G. (1961) The physical chemistry of sea water. In *Oceanography*, M. Sears (ed.), 549–81. American Association for the Advancement of Science, Publication 67.

Sillen, L.G. (1963) How has sea water got its present composition? *Sven. Kem. Tidskr.*, **75**, 161–77.

Sillen, L.G. (1965) Oxidation state of earth's ocean and atmosphere. I. A model calculation on earlier states. The myth of the 'probiotic soup'. *Ark. Kemi*, **24**, 431–44.

Sillen, L.G. (1967) Gibbs phase rule and marine sediments. In *Equilibrium Concepts in Natural Water Systems*, W. Stumm (ed.), 57–69. Washington, DC: American Chemical Society, Advances in Chemistry Series, No. 67.

Toth, J.R. (1980) Deposition of submarine crusts rich in manganese and iron. *Geol. Soc. Am. Bull.*, **91**, 44–54.

Van Cappellen, P. & Qiu, L. (1997a) Biogenic silica dissolution in sediments of the Southern Ocean. I. Solubility. *Deep Sea Res.*, **44**, 1109–28.

Van Cappellen, P. & Qiu, L. (1997b) Biogenic silica dissolution in sediments of the Southern Ocean. II. Kinetics. *Deep Sea Res.*, **44**, 1129–49.

Vincent, E. & Berger, W.H. (1981) Planktonic foraminifera and their use in paleoceanography. In *The Sea*, C. Emiliani (ed.), Vol. 7, 1025–119. New York: Interscience.

Wollast, R. (1974) The silica problem. In *The Sea*, E.D. Goldberg (ed.), Vol. 5, 359–92. New York: Wiley Interscience.

# 16 Unscrambling the sediment-forming chemical signals

We are now attempting to understand the factors that control the chemical composition of marine sediments, and from the point of view of the present volume attention will be focused largely on the upper few metres of the sediment column, because it is here that material reacts directly with sea water. In this context, therefore, the oceanic water column can be viewed as a medium through which chemical signals, or *fluxes*, are transmitted from above to the upper portions of the bottom sediments. In addition, signals are also transmitted to this part of the sediment column from below via interstitial waters.

A characteristic feature of deep-sea clays is that, relative to nearshore muds, they contain enhanced concentrations of some trace elements (e.g. Mn, Cu, Ni, Co, Pb). These 'excess' trace elements are further enhanced in pelagic sediments laid down at slow rates in areas remote from the land masses, and are even more enhanced in ferromanganese nodules and metalliferous active-ridge deposits. It was pointed out in Section 13.5 that some kind of sequential enhancement, or fractionation, is therefore occurring for the 'excess' trace elements within marine sediments. A central issue of marine geochemistry is to identify the processes by which such elemental enhancements and fractionations have arisen.

A number of authors have suggested that the overall chemical compositions of marine sediments can be considered to result from the contributions made by individual sediment *fractions* (see e.g. Krishnaswami, 1976; Bacon & Rosholt, 1982). However, each of the fractions contains a number of individual sediment-forming *components*. Because of this, an approach in which the formation of individual components is viewed in terms of *chemical signals* offers a potentially attractive insight into the factors that control the overall composition of marine sediments.

The major sediment-forming components have been described in Chapter 15, in which it became apparent that the processes involved in the generation of some components cannot be related to single chemical signals, but rather are the result of *signal coupling*. For example, ferromanganese nodules can acquire elements transmitted by signals associated with hydrogenous, hydrothermal and both oxic and suboxic diagenetic processes. To assess the chemical composition of marine sediments in terms of chemical signals it is therefore necessary to make a distinction between individual signals that are *potentially* able to give rise to individual components but that more often combine together to form multisource components. In the present approach, therefore, an attempt will be made to relate the chemistry of the sediments to chemical signals that are associated with the *processes* involved in the generation of the sediment-forming components. This is a purely artificial approach, but is adopted in order to provide a convenient framework within which to describe the factors that constrain the chemical composition of the sediments.

## 16.1 Definition of terminology

To simplify the often confusing terminology found in the literature, the chemical signals identified here will be defined on the basis of the processes that have been shown to be operative in the genesis of the sediment-forming components, i.e. the *process-oriented* approach will be adopted.

Initially, a distinction will be made between **detrital** (sometimes also termed lithogenous or refractory) and **non-detrital** (sometimes termed non-lithogenous, or non-refractory) components (see also Section 3.1.4). This two-component classification involves a fundamental geochemical division between two different types of elements.

1 Detrital elements are part of the crystalline mineral matrix, usually in lattice-held associations, and have

been carried through the mobilization–transportation cycle in a solid form. Thus, the detrital components, such as clay minerals, take part in particulate ↔ dissolved reactions, but the detrital matrix-associated elements themselves do not undergo exchange between the solid and dissolved states.

2 Non-detrital elements are not part of the mineral matrix, but have been removed from solution in association with either inorganic or organic hosts. Thus, these are the elements that take part in exchanges between the particulate and dissolved states.

It should be stressed that this classification takes no account of non-lattice → lattice transformations that can affect some elements, e.g. during diagenesis.

In the simple twofold classification, a single signal is involved in the formation of the non-detrital components. In reality, the non-detrital signal can be resolved into a number of individual signals associated with both the *supply* and *removal* of elements from solution. The nature of these individual non-detrital signals can be related to the processes involved in the generation of the sediment-forming components. In Chapter 15 it was shown that elements that were classified originally as 'hydrogenous' were related to a number of different processes, which were classified as being: (i) hydrogenous; (ii) hydrothermal; (iii) oxic diagenetic; and (iv) suboxic diagenetic in origin. It was also shown that there is a degree of **coupling** between some of the signals involved in the formation of the components; for example, that between hydrogenous ferromanganese crusts, sediment ferromanganese oxyhydroxides and $\delta MnO_2$-rich ferromanganese nodules, all of which receive their elements from overlying sea water. However, it was also shown that there was a major **decoupling** between these three components and toderokite-rich nodules, which were supplied mainly from interstitial waters following suboxic diagenesis at depth in the sediment column. Thus, a distinction could be made between elements supplied from the water column *above*, and those supplied from the interstitial water column *below*. In terms of the identification of chemical *signals* that result in the formation of sediment-forming components, it is necessary to take account of this **decoupling break** between seawater and interstitial-water sources. This is an extremely important distinction because hydrogenous, hydrothermal and oxic diagenetic processes, which release elements at or near the sediment surface, represent a **primary** source to the sediments, whereas suboxic mobilization at depth can involve a **recycled** supply of elements associated with the secondary oxidants that are utilized in suboxic diagenesis.

In the *process-orientated* approach used here for the classification of the chemical signals transmitted to marine sediments, the following signal types will therefore be identified.

1 The **detrital** signal. This is a background signal, which transmits elements carried in the crystalline matrix of lithogenous minerals.

2 The **non-detrital** signal. This transmits elements that have been removed from solution at some stage during their history. It is sub-divided into a number of categories.

(a) The **biogenous** signal. In the present context, this refers mainly to the signal associated with biological shell material, and excludes the organic carbon involved in both the down-column transport of trace metals to the sediment surface and subsequent diagenetic processes.

(b) The **authigenic** signal. The term authigenic will be used here to describe the primary 'background' signal transmitted through sea water. In an ocean-wide context this signal receives contributions from *all* sources that supply elements to sea water, but in the present definition elements from these sources are regarded as being smoothed out into a general *background* signal, i.e. one that excludes the effects of localized inputs, such as those arising from hydrothermal sources, which perturb steady-state conditions. The removal of elements from 'background' sea water involves processes associated with both the **hydrogenous** and **oxic diagenetic** signals. The hydrogenous signal can result in the inorganic formation of components via the removal of dissolved elements from sea water without the necessity of a sediment substrate, e.g. on to rock surfaces (see Section 15.3.4.6). In contrast, the oxic diagenetic signal requires the initial removal of a dissolved element from sea water on to a carrier phase and the down-column transport of the carrier, via the large organic aggregates (CPM + associated FPM) of the global carbon flux, to a **sediment** surface. Following this, some elements (e.g. 'nutrient-type') are released to become available for incorporation into the sediment as the labile fraction of the carriers is destroyed during oxic diagenesis. Other elements (e.g. 'scavenged-type') may be directly incorporated into the sediment in association with the FPM carriers; although the latter can be released and re-distributed within the sediment (see e.g., Fig. 15.9(c)). The principal difference between the hydrogenous and the oxic diagenetic signals, however, is that the latter requires diagenesis to be initiated on a sediment substrate. However, the hydrogenous and the oxic diagenetic signals are interlinked in that they both acquire their elements from the same general source, i.e. 'background' sea water, and reach the sediment surface directly from the overlying water column (a primary seawater source). In the present context, therefore, the hydrogenous and oxic diagenetic signals are combined into the authigenic signal, which may be defined here as 'that giving rise to components formed inorganically from elements originating in the overlying **background** sea water.

(c) The **diagenetic**, or sediment-recycled, signal. As used in the present context, this describes the signal transmitted through the interstitial water column following suboxic diagenetic mobilization; i.e. this recycled signal largely involves redox-sensitive elements (e.g. Mn, Fe, Cu).

(d) The **hydrothermal** signal. This applies, in the sense used here, to the seawater signal resulting from the pulsed debouching of high-temperature solutions at the ridge crests.

(e) The **contaminant** signal. This arises from the introduction of anthropogenic material into the oceans.

The detrital, authigenic and biogenous signals may be considered to operate on an ocean-wide basis and to transmit elements that have a primary 'background' seawater source. To establish a theoretical framework within which to unscramble the sediment-forming signals it therefore is convenient to envisage marine sediments as being formed by components originating from mixtures of these 'background' signals, upon which are superimposed perturbation **spikes** from more localized hydrothermal and contaminant seawater signals, and from the more widely occurring interstitial-water suboxic diagenetic signal. The elements associated with the suboxic diagenetic signal can have had a variety of primary sources, but the important point is that they have been recycled within the sediment complex itself. In the present context, therefore, any signal that is not associated

directly with background sea water is regarded as a spike. The question that must now be considered is 'how can these various signals be unscrambled?' A variety of techniques have been used for this purpose, and these include the following.

1 Interpretation of total sediment chemical analyses. These can be used, for example, to assess the relative amounts of major components such as clays, carbonates and opaline silica in oceanic sediments (see e.g. El Wakeel & Riley, 1961).

2 Spatial mapping of elemental concentrations. This can be used, for example, to establish elemental source–transport patterns (see e.g. Turekian & Imbrie, 1966).

3 Elemental accumulation rate comparisons. These can be used to determine the rates at which different components are formed (see e.g. Bender *et al.*, 1971).

4 Factor analysis (see e.g. Krishnaswami, 1976), isotope analysis (see e.g. Thompson *et al.*, 1984) and chemical leaching techniques (see e.g. Chester & Hughes, 1967). These can be utilized to assess the partitioning of elements between individual sediment phases.

Techniques such as those identified above will be introduced into the text as each of the individual chemical sediment-forming signals is described in the following sections.

## 16.2 The biogenous signal

The deposition of calcareous and siliceous shells in marine sediments, and the degree to which they are preserved, plays a dominant role in controlling the major element chemistry of the deposits. Siliceous sediments generally have a restricted distribution and, with a few important exceptions, are confined to the edges of the oceans. Carbonate sediments, however, are much more widespread and, in fact, the bulk composition of many deep-sea deposits may be considered to be made up of mixtures of carbonate and lithogenous, mainly clay, end-members.

The extent to which carbonate and siliceous shell materials affect the overall major element composition of deep-sea sediments can be estimated from the relative proportions of Al (lithogenous), Ca (carbonate) and Si (siliceous) present in them; see, for example, the major element analyses listed in Table 13.4. Thus, the chemical signals associated with the deposition of biogenous carbonate and siliceous shells can be unscrambled by relating them to the major element composition of the sediments. However, it was pointed out in Section 13.5 that, with the exception of Sr, which is found in the carbonate lattice, both carbonate and siliceous sediments are generally impoverished in those excess trace metals that are concentrated in deep-sea clays. Further, neither carbonate nor opaline shell phases act as significant carriers for the down-column transport of trace metals. It may be concluded, therefore, that the biogenous shells are not important trace-metal hosts and act rather to dilute metals associated with other sediment components. Indeed, it is common practice to express the concentrations of trace metals in sediments on a 'carbonate-free' basis in order to overcome the dilution effects of the shell matrix.

With respect to the transport of trace metals to the sediment surface from *background* sea water, and the incorporation of the excess trace-metal fractions into deep-sea clays, attention therefore can be focused largely on the detrital and authigenic signals. In a 'normal' deep-sea clay, i.e. one that has not been perturbed by major hydrothermal, contaminant or suboxic diagenetic spikes, the total concentration ($C_t$) of an element therefore may be expressed in terms of the sum of the contributions from detrital and authigenic components, thus:

$$C_t = C_d + C_a \qquad (16.1)$$

where $C$ is the concentration of an element and the suffixes t, d and a refer to the total, detrital and authigenic concentrations, respectively (see e.g. Krishnaswami, 1976). In terms of this two-component framework, an attempt will now be made to unscramble the detrital and authigenic signals.

## 16.3 The detrital signal

In the definition used here, the detrital signal transmits elements carried in the crystalline matrix of lithogenous minerals. Crust-derived weathering products, together with continental and submarine volcanic debris, make up most of the lithogenous material transported to the oceans over geological time. This lithogenous material comprises the bulk of deep-sea clays; for example, the oxides of aluminium and silicon alone account for ~80% of the total sediments. This crust-derived material is brought to the open ocean mainly along the fluvial and atmospheric routes, but it is important to remember that the material carried by these forms of transport consists of both detrital and non-detrital components. Clearly, therefore, the *bulk* composition of river and atmospheric particulates cannot be used to assess the composition of the detrital signal ($C_d$) transmitted to the oceans.

## 16.4 The authigenic signal

The geochemically important primary authigenic material in marine sediments consists mainly of ferromanganese crusts, nodules and sediment oxyhydroxides, which are formed by hydrogenous and oxic diagenetic processes, together with the population of elements associated with a wide variety of sediment components in intersheet and surface-adsorbed positions. The detailed chemistry of these primary authigenic components has been described in Section 15.3, and the average compositions of the crusts and nodules are summarized in Table 16.1. It can be seen from this table that these authigenic components are rich in those elements that are present in deep-sea clays in excess concentrations (see also Table 13.5). Some of the authigenic components, such as micronodules and oxyhydroxides, are an intimate part of the sediment complex, and in certain regions they can have a major influence on the composition of the total sediments with respect to the excess trace elements.

## 16.5 Unscrambling the detrital and authigenic signals

The bulk of the sediment-forming components in deep-sea clays are detrital in origin. Authigenic components are quantitatively far less important, but they

**Table 16.1** Typical compositions of primary authigenic components in deep-sea sediments; units, $\mu g\,g^{-1}$.

| Elements | Hydrogenous crusts* | Oxic nodules* | Suboxic nodules* | Pelagic clays† |
|---|---|---|---|---|
| Mn | 222 000 | 316 500 | 480 000 | 65 000 |
| Fe | 190 000 | 44 500 | 4900 | 12 500 |
| Co | 1300 | 280 | 35 | 116 |
| Ni | 5500 | 10 100 | 4400 | 293 |
| Cu | 1480 | 4400 | 2000 | 570 |
| Zn | 750 | 2500 | 2200 | — |
| Al | 11 800 | 27 100 | 7500 | 83 000 |

* Data from Table 15.7. † Data from Table 13.5.

are rich in the excess trace metals. However, the proportions in which the detrital and authigenic components are present in deep-sea sediments vary from one suite of deposits to another, and to evaluate their relative importance it is necessary to unscramble the signals associated with the two components. There are several ways in which this can be done, and some of these are discussed below.

*The direct approach using chemical leaching procedures to determine elemental partitioning.* In this approach, the detrital and authigenic fractions of a sediment are actually separated from each other

and analysed individually. For example, Chester & Hughes (1967) used a chemical leaching technique to establish the partitioning of elements in a North Pacific pelagic clay core, and Chester & Messiha-Hanna (1970) applied a similar technique to a series of North Atlantic deep-sea surface sediments. The data for the chemical compositions of the detrital and authigenic fractions of the sediments obtained in this way are given in Table 16.2.

*The indirect approach.* Several routes can be used in this approach.

**1 Background correction: detrital fraction.** When

**Table 16.2** Estimated chemical compositions of the detrital and authigenic fractions of deep-sea clays; units, $\mu g\,g^{-1}$.

(a)  Detrital fraction.

| Element | Nares Abyssal Plain red clays* | Pacific pelagic clays† | Nares Abyssal Plain red clays‡ | Bermuda Rise carbonates§ | Atlantic deep-sea sediments¶ | Pacific pelagic clay‖ | Average nearshore mud (Wedepohl, 1960) | Average shale (Wedepohl, 1968) |
|---|---|---|---|---|---|---|---|---|
| Mn | 578 | 2087 | 770 | 605 | 550 | 740 | 850 | 850 |
| Fe | 42 280 | 43 300 | 51 800 | 51 240 | — | 51 800 | 69 900 | 47 000 |
| Cu | 51 | 244 | 45 | 36 | 67 | 212 | 48 | 45 |
| Co | 24 | 48 | 24 | 23 | 12 | 16 | 13 | 19 |
| Ni | 65 | 92 | 56 | 65 | 63 | 46 | 55 | 68 |
| Zn | 111 | — | 117 | 124 | — | — | 95 | 95 |
| V | 158 | — | 153 | — | 120 | 92 | 130 | 130 |
| Cr | — | — | — | — | 72 | 91 | 100 | 90 |

* Estimated using background correction with respect to local grey clays (Thompson *et al.*, 1984).
† Estimated using background correction with respect to continental shales (Krishnaswami, 1976).
‡ Estimated using graphical procedure (Thompson *et al.*, 1984).
§ Estimated using graphical procedure (Bacon & Rosholt, 1982).
¶ Estimated using chemical leaching technique (Chester & Messiha-Hanna, 1970).
‖ Estimated using chemical leaching technique (Chester & Hughes, 1967).

(b)  Authigenic fraction.

| Element | Bermuda Rise carbonates* | Pacific pelagic clay† | Atlantic deep-sea sediments‡ | Atlantic deep-sea sediments§ | Pacific pelagic clays¶ |
|---|---|---|---|---|---|
| Mn | 4400 | 4380 | 3871 | 3112 | 7000 |
| Fe | — | — | — | — | 7500 |
| Cu | 110 | 101 | 110 | 128 | 225 |
| Co | 6 | 73 | 60 | — | 80 |
| Ni | 61 | 147 | 112 | 62 | 175 |
| Zn | 40 | — | — | — | — |

* Estimated using graphical procedure (Bacon & Rosholt, 1982).
† Estimated using chemical leaching technique (Chester & Hughes, 1967).
‡ Estimated using chemical leaching technique (Chester & Messiha-Hanna, 1970).
§ Estimated using chemical leaching technique (Thomas, 1987).
¶ Best estimate using a variety of techniques (Krishnaswami, 1976).

this route is adopted it is assumed that the chemical composition of the detrital fraction of deep-sea sediments can be estimated with respect to some reference material.

(a) Krishnaswami (1976) used the compositions of (i) continental shale; and (ii) nearshore sediments as being representative of the crust-derived (lithogenous) material transported to the marine environment; i.e. the baseline material upon which the 'excess' authigenic trace elements are superimposed. In terms of eqn (16.1), values of $C_t$ were obtained by direct analysis, and $C_d$ was assumed to be the same as that of either continental shales or nearshore muds. Thus, the composition of $C_a$ could be obtained from $C_t - C_d$. The compositions of the detrital fractions based on those of continental shales and nearshore muds are listed in Table 16.2(a) Col. 3, and the 'best estimate' for the derived authigenic fraction is given in Table 16.2(b) Col. 6.

(b) A different kind of background correction for the estimation of the detrital fraction was applied by Thompson *et al.* (1984) to a suite of deep-sea clays from the Nares Abyssal Plain in the northeast Atlantic. Two types of clay were found in this area; a **red clay** and a **grey clay**. On the basis of their total sediment analysis it was evident that the grey clays had only a negligible authigenic fraction, and it was assumed that for these sediments $C_t \approx C_d$. The data for the composition of the grey clays, which are therefore representative of the detrital fraction of deep-sea sediments, are given in Table 16.2(a) Col. 2.

**2 Graphical procedures: detrital and authigenic fractions.** Bacon & Rosholt (1982) used a graphical approach to estimate the compositions of the terrigenous (detrital) and pelagic (authigenic) components in sediments from the Bermuda Rise in the North Atlantic. These authors assumed that the total trace-metal content of the sediments could be expressed in terms of a two-component terrigenous (detrital) and pelagic (authigenic) system by the equation:

$$C_{tot} = f(C_{pel} - C_{ter}) + C_{ter} \tag{16.2}$$

where $C_{tot}$, $C_{pel}$ and $C_{ter}$ are the metal concentrations in the total sediment, the pelagic and the terrigenous components, respectively, and $f$ is the proportion of the pelagic component in the sediment. If $C_{ter}$ and $C_{pel}$ are constant, then a plot of $C_{tot}$ versus $f$ will have a linear relationship. Bacon & Rosholt (1982) extrapolated this relationship to give estimates of the compositions of a pure terrigenous ($f = 0$) and a pure pelagic

($f = 1$) component. The compositional data for the terrigenous (detrital) and pelagic (authigenic) end-member components obtained in this way are given in Table 16.2 (a Col. 5 and b Col. 2, respectively). In addition to their background correction method, Thompson *et al.* (1984) also used a graphical approach to estimate the detrital composition of their Nares abyssal plain sediments. The data obtained are included in Table 16.2(a) Col. 4, and the graphical approach used is discussed in more detail later in this section.

**3 Manganese nodule model: authigenic fraction.** Krishnaswami (1976) used this model, which is based on the assumption that the rates of precipitation for authigenic trace metals are the same in pelagic clays and ferromanganese nodules. Using compositional data obtained from nodules, Krishnaswami (1976) was then able to estimate the concentrations of a series of elements in the authigenic fractions of Pacific pelagic clays. These data are given in Table 16.2(b) Col. 6.

Several different estimates therefore are available for the compositions of the detrital and authigenic components present in deep-sea sediments (see Table 16.2). Perhaps the most striking feature to emerge from these data is that there is a generally good agreement between the estimates, especially for the composition of the detrital fraction in deep-sea sediments. Thus, it is possible, at least to a first-order approximation, to unscramble the detrital from the authigenic signal if the chemical composition of a total deep-sea sediment is known. The signal strengths will, of course, be dependent on the relative rates at which the components associated with the detrital and authigenic signals accumulate. The manner in which the proportions of the two components vary in oceanic sediments can be illustrated with respect to the Atlantic Ocean.

Chester & Messiha-Hanna (1970) and Thomas (1987) used chemical leaching procedures to investigate the partitioning of a series of elements in samples from the present-day Atlantic *sediment surface*. A compilation of their data, expressed in terms of the two-component detrital–authigenic classification, is given in Table 16.3, and the data are illustrated in Fig. 16.1. From Table 16.3 it can be seen that the average partitioning signatures of the elements vary considerably; for example, Al is ~80% detrital, whereas Mn is ~70% authigenic in character. These are ocean-wide averages, however, and the partitioning signatures of some elements vary with the environment of deposition of the host sediment. An example of this has been provided by Thomas (1987), who compared the partitioning signatures of a range of elements in a North Atlantic marginal (hemi-pelagic) sediment with those in a Mid-Atlantic Ridge sediment. The data are given in Table 16.4, and show that, whereas Al, Fe and Cr retain their detrital character (and Mn remains strongly authigenic) in sediments from both environments of deposition, Cu, Ni and Pb are considerably more authigenic in nature in the ridge than in the marginal sediments. Chester *et al.* (1988) used a six-stage sequential leaching technique, in which the non-detrital fraction was subdivided into a number of individual trace-metal hosts, to investigate the partitioning of Cu in North Atlantic deep-sea surface sediments. The results are illustrated in Fig. 16.2 in terms of an east–west, marginal → open-ocean → marginal, sediment transect. The principal partitioning trends can be summarized as follows.

**Table 16.3** Average detrital–authigenic partitioning of elements in North Atlantic deep-sea sediments (data from Chester & Messiha-Hanna (1970) and Thomas (1987)).

| Element | Percentage of total concentration | |
| --- | --- | --- |
| | Detrital | Authigenic |
| Al | 81 | 19 |
| Fe | 82 | 18 |
| V | 71 | 29 |
| Cr | 70 | 30 |
| Ni | 55 | 45 |
| Cu | 44 | 56 |
| Co | 42 | 48 |
| Mn | 32 | 68 |

**Table 16.4** Variations in the detrital–authigenic partitioning of elements in North Atlantic deep-sea sediments (data from Thomas, 1987).

| Element | Marginal sediment: percentage of total concentration | | Mid-Atlantic Ridge sediment: percentage of total concentration | |
| --- | --- | --- | --- | --- |
| | Detrital | Authigenic | Detrital | Authigenic |
| Al | 93 | 7 | 76 | 14 |
| Fe | 77 | 23 | 68 | 32 |
| Mn | 18 | 82 | 14 | 86 |
| Cu | 55 | 45 | 23 | 67 |
| Ni | 64 | 36 | 17 | 83 |
| Cr | 63 | 37 | 88 | 12 |
| Pb | 85 | 15 | 38 | 62 |

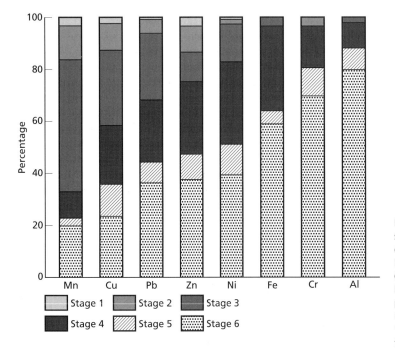

Fig. 16.1 The average partitioning of some elements in North Atlantic surface deep-sea sediments (data from Chester & Messiha-Hanna (1970) and Thomas (1987)). The diagram is intended to highlight the detrital–authigenic partitioning of the elements; stage 6 represents the detrital fraction and stages 1–5 represent individual authigenic host fractions.

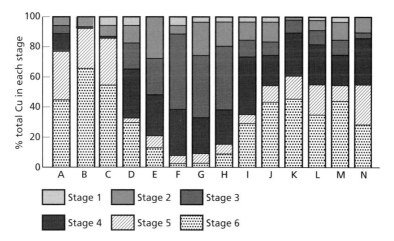

**Fig. 16.2** The partitioning signatures for total copper ($\Sigma Cu$) in Atlantic deep-sea sediments (from Chester *et al.*, 1988). The samples were collected on an east–west equatorial transect: A–D, eastern margins; E–H, Mid-Atlantic Ridge and ridge flanks; I–N, western margins. The Cu host associations are as follows: stage 1, loosely held or exchangeable associations; stage 2, carbonate and surface oxide associations; stage 3, easily reducible associations, mainly with (new) oxides and oxyhydroxides of manganese and amorphous iron oxides; stage 4, moderately reducible associations, mainly with (aged) manganese oxides and crystalline iron oxides; stage 5, organic association; stage 6, detrital, or residual, associations.

1 There is a general decrease in the proportion of detrital Cu (stage 6) away from the margins in the mid-ocean regions, as the influence of continentally derived material decreases.

2 The highest proportion of organically associated authigenic Cu (stage 5) is found in marginal sediments under regions of high surface-water productivity. In these regions there is a high sedimentation rate and an enhanced down-column flux. Under these conditions relatively high concentrations of organic matter, together with organically associated Cu, can be preserved in the sediments. This is an important finding because it suggests that in some marginal regions significant amounts of authigenic Cu can be stored in a relatively immobile form as the organic carriers escape oxic diagenesis at the sediment surface.

3 There is a clear trend in the distribution of Cu associated with the easily reducible oxides of stage 3 (mainly new manganese oxides), with the highest contributions being found in mid-ocean deposits around the Mid-Atlantic Ridge and ridge flanks and the lowest in sediments from the marginal areas. These mid-ocean sediments also contain enhanced concentrations of total Cu, and it is evident, therefore, that these result mainly from the element being incorporated into stage 3 oxide hosts following its release from the down-column carrier phases under oxic conditions. In addition, some of the released Cu will escape into sea water as an out-of-sediment flux (see Section 14.6.2.2).

This type of spatial variation in elemental partitioning signatures is extremely important in relation to understanding the processes that control the chemical compositions of deep-sea sediments, and can be explored further by considering Atlantic sediments in more detail and then using them to illustrate how the detrital and authigenic signals operate on an ocean-wide basis.

In a classic paper, Turekian & Imbrie (1996) provided data on the spatial distributions of Mn, Co, Cu, Ni and Cr in surface sediments from the North and South Atlantic. When the concentrations of Mn, Co, Cu and Ni were expressed on a carbonate-free basis there were clear patterns in their spatial distributions, the highest values being found in sediments from remote mid-ocean areas of low clay accumulation, and the lowest in deposits from the continentally adjacent abyssal plains—this is illustrated for Ni in Fig. 16.3(a). In contrast, there were no clear patterns in the spatial distribution of Cr, other than a number of high-concentration patches (e.g. around the

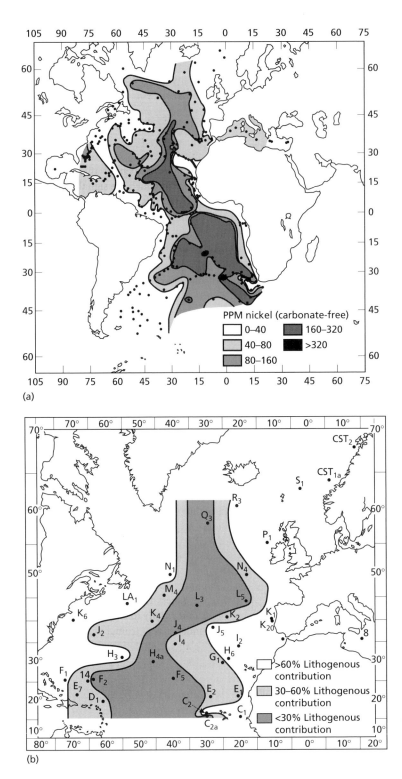

(a)

(b)

**Fig. 16.3** The distribution of Ni in Atlantic deep-sea sediments. (a) The distribution of total Ni (from Turekian & Imbrie, 1966). (b) The distribution of lithogenous (detrital) Ni (from Chester & Messiha-Hanna, 1970).

Antilles), which are related to the supply of Cr-rich detrital minerals. Thus, the investigation identified a **geographic fractionation** between those elements that have excess concentrations in pelagic clays (e.g. Mn, Co, Cu and Ni) and those that do not (e.g. Cr).

Chester & Messiha-Hanna (1970) used a two-stage sequential leaching technique to investigate the partitioning of a series of elements between the detrital and authigenic fractions of surface deep-sea sediments from the North Atlantic. A number of important conclusions can be drawn from this study, and these are summarized below.

1 Elements that have similar concentrations in pelagic clays and nearshore muds (see Table 13.5) are associated mainly with the detrital fraction of the sediments. For example, Fe, Cr, V and Al all have >70% of their total concentrations in the detrital fractions.

2 Elements such as Mn, Ni, Co and Cu, which are present in excess concentrations in pelagic clays, have a much higher proportion of their total concentrations in the authigenic associations. For example, >70% of the total Mn is present in the authigenic fractions (see above). It may be concluded, therefore, that the elements that are enhanced in pelagic clays are authigenic in origin and have been removed from solution at some time during their history.

3 There were clear trends in the spatial distributions of the authigenic, or excess, elements. This is illustrated for Ni in Fig. 16.3(b), from which it is evident that the proportion of detrital Ni decreases from c. 60% near to the continents to <30% in the mid-ocean areas. Thus, the highest proportions of authigenic (non-detrital) Ni are found in the remote regions. It may be concluded, therefore, that the enhanced total sediment concentrations of Ni reported by Turekian & Imbrie (1966) for the mid-ocean areas (Fig. 16.3a) are the result of an increased contribution from authigenic Ni.

4 The detrital fractions of the North Atlantic deep-sea sediments have a similar composition to those of the total samples of nearshore muds and continental shales—see Table 13.5. According to Chester & Aston (1976) these nearshore muds may be considered to represent an early stage in the adjustment of land-derived material to those processes that operate to produce trace-metal enhancements in the marine environment. This therefore supports their use as indicators of the composition of the detrital fraction

of deep-sea sediments, e.g. in the background correction method (see above).

A number of techniques have been described that can be used to unscramble the detrital and authigenic sediment-forming signals, and in terms of the twofold background-signal classification it has been demonstrated that the concentrations of the excess trace metals in deep-sea sediments result from the authigenic signal. It is now necessary to consider both the nature of the *driving force* behind this authigenic signal and the *manner* in which the signal itself operates in the oceans. Up to this point, spatial elemental distributions, together with a variety of unscrambling techniques, have been used to distinguish between the detrital and authigenic sediment components. In order to describe the various theories that have been proposed to explain how the authigenic signal itself operates, it is necessary to introduce another parameter into the system, and this is the rate at which the host sediments accumulate. The lithogenous, or clay, fraction in deep-sea sediments has a relatively low accumulation rate, generally in the range of a few millimetres per thousand years (see Section 13.1). Within this low accumulation pattern, however, there are considerable variations between sediments deposited in different oceanic areas. In the two-component detrital–authigenic classification, the detrital material is equivalent to the clay fraction, and the rate at which it accumulates obviously will affect the authigenic signal, as the latter can be swamped out at very fast detrital accumulation rates.

Originally, two general theories were proposed to explain how the 'detrital–authigenic' sediment-generating system operated on an ocean-wide basis.

*The trace-element-veil theory.* This theory was proposed by Wedepohl (1960), and the rationale underlying it is that the excess trace elements in deep-sea clays have a *constant supply* to all parts of the World Ocean, which is independent from, but superimposed upon, a *variable* clay accumulation rate. Thus, in the **constant-flux model** there is a veil of trace elements homogeneously distributed throughout the oceanic water column, from which the elements are removed by sediment components at *rates* that vary according to how fast the total deposits accumulate; i.e. the longer the components are in contact with sea water, the higher the authigenic trace-metal content of the sediment.

Turekian (1967) criticized the trace-element-veil theory on a number of grounds, the major criticism being directed at the fundamental underlying concept that the independent removal of trace metals from solution is essentially constant throughout the World Ocean. He pointed out that this implies that there is a homogeneity in the trace-metal composition of sea water, whereas in fact variations occur in the geographical distributions of the metals in both surface and deep waters (see Chapter 11).

Turekian (1967) developed an alternative theory to explain the incorporation of the excess trace elements in deep-sea sediments, and this is described below.

*The differential transport theory.* Turekian (1967) used spatial trace-metal-distribution data for Atlantic surface sediments to develop this theory. It was shown above that the essential features in the spatial distributions of the *excess* trace metals in the Atlantic deep-sea sediments are that they have their highest concentrations in open-ocean areas and their lowest concentrations on the continentally adjacent abyssal plains. Turekian (1967) related these distribution patterns to the processes that govern sediment transport on a global-ocean scale. To do this, he distinguished between two principal transport mechanisms, which operate on material that escapes the coastal zone. These were (i) lateral off-shelf movement along the sea bed by **bottom** transport, and (ii) movement down the water column by **vertical** transport; see also, the sediment transport vectors identified in Section 13.2, and the study of bottom-transport–vertical-transport flux reported by Grousset & Chesselet (1986), which is described in Section 13.2. In the differential transport theory it is proposed that each of the two transport mechanisms is associated with a specific trace-metal particulate population.

1 Relatively large-sized particulate material that escapes the estuarine environment is often dumped on the continental shelves, from which it can bleed out on to the deep-sea floor by processes such as turbidity current transport, which is largely responsible for the formation of the abyssal plains fringing the continental margins. These shelf-sediment particles contain relatively high concentrations of lithogenous components, such as quartz, and generally have low trace-metal concentrations, most of which are located in detrital material. Thus, Turekian (1967) identified a **trace-metal-poor** particulate fraction, which is initially deposited on the shelves and may be transported to hemi-pelagic areas by bottom transport.

2 The small-sized particles that escape the estuarine and coastal zones can be transported out into the open ocean by surface currents, where they are joined by particles transported directly via the atmospheric flux. Here, the particles enter the vertical flux and settle out down the water column to form pelagic deposits on remote topographic highs and other mid-ocean areas. These particles are composed of material such as manganese and iron oxyhydroxides, together with clay minerals that have oxide, and/or organic matter, coatings. As they have relatively large specific surface areas, the particles are actively involved in the scavenging of trace metals from solution as they are transported along the river → estuarine → open-ocean → water-column → sediment, or atmospheric → open-ocean → water-column → sediment, pathways. Thus, they form a **trace-metal-rich** particulate fraction.

Turekian (1967) therefore suggested that the deposition of the large-sized trace-metal-poor particles around the continents by **bottom** processes, and the deposition of the small-sized trace-metal-rich particles via **vertical** water-column settling in mid-ocean areas, was mainly responsible for the fractionation of the excess trace elements between hemi-pelagic and pelagic deep-sea sediments in the Atlantic.

One of the major differences between the trace-element-veil and the differential transport theories revolves around the *rates* at which the excess elements accumulate in deep-sea sediments.

Turekian (1967) wrote a general equation for the accumulation rate of a trace element in a deep-sea sediment. Elderfield (1976) modified this equation into a form that, using the terminology adopted in this volume, can be expressed in the following way:

$$\Sigma F_E = F_E + [E]_D F_D \qquad (16.3)$$

where $\Sigma F_E$ is the total accumulation rate of an element in a sediment, $F_E$ is the constant rate of addition of $E$ from solution in sea water to the sediment, and $[E]_D$ is the concentration of the element in the detrital component of the sediment, which accumulates at a rate $F_D$. Thus, a further step has been taken in defining the two-component detrital–authigenic concept for the distribution of trace elements in deep-sea sediments by linking the signals involved to sediment accumulation rates. Originally, in terms of the trace-element-veil theory, it was thought that the coupling between the two signals involved a *variable* detrital accumulation rate and a *constant* authigenic accumulation rate. This **constant-flux model** for the accumulation of authigenic trace metals is based on the assumption of a uniform authigenic deposition rate over the entire ocean floor. If the rate of authigenic deposition is in fact uniform, and the detrital accumulation variable, then there will be a negative correlation between the authigenic concentration ($C_a$) and the sediment accumulation rate ($S$), because in clay sediments the latter is dependent mainly on the deposition of detrital material. Krishnaswami (1976) has expressed this relationship as

$$C_a = \frac{K}{S_{\rho_s}} \qquad (16.4)$$

and from this the total sediment concentration would be

$$C_t = C_a + C_d = \frac{K}{S_{\rho_s}} + C_d \qquad (16.5)$$

where $K$ is the authigenic deposition rate (g cm$^{-2}$ yr$^{-1}$), $\rho_s$ is the *in situ* density of the sediment (g cm$^{-3}$) and $S$ is the sedimentation rate (cm yr$^{-1}$). This constant-flux model will therefore apply to elements that have (i) a homogeneous distribution in the oceans and (ii) residence times equal to or greater than the oceanic circulation times.

Krishnaswami (1976) used a variety of techniques to demonstrate that ~90% of the Mn, ~80% of the Ni and Co, and ~50% of the Cu in Pacific pelagic clays are authigenic in origin. In contrast, >90% of

the Sc, Ti and Th is detrital in character. The author then applied the constant-flux equations to his elemental concentration and sediment accumulation rate data and found, as predicted by the model, that there was a negative correlation between the detrital, or clay, sediment accumulation rates and the concentrations of Mn, Co, Ni, Fe and Cu in the Pacific pelagic sediments—see Fig. 16.4. These negative correlations applied to a series of different sediments, and Krishnaswami (1976) concluded that this was consistent with the constant-flux model of a *uniform* authigenic deposition superimposed on to a *variable* background detrital input. In contrast, he found that the concentrations of the detrital elements Sc, Ti and Th were independent of sediment accumulation rates—see Fig. 16.4. Krishnaswami (1976) also concluded that when the detrital sedimentation rate is high, i.e. ≥10 mm 10$^3$ yr$^{-1}$, the authigenic concentration will be small and $C_t \approx C_d$. When the detrital sedimentation rate is small, however, a pure authigenic fraction will be formed and $C_t \approx C_a$; for example, the extreme case for this would be found in hydrogenous ferromanganese crusts (see Section 15.3.3). From his data, Krishnaswami (1976) derived a best estimate of the authigenic deposition rate of a series of elements in the World Ocean, and these are given in Table 16.5.

Krishnaswami's data appeared to confirm the constant-flux model for the removal of authigenic trace metals from solution in sea water into pelagic clays, i.e. the **trace-element-veil theory** was approximated. However, the past two decades have seen a number of very important advances in our understanding of

**Table 16.5** Estimates of the accumulation rates of authigenic elements in sediments from the World Ocean; units, μg cm$^{-2}$ 10$^3$ yr$^{-1}$.

| Element | Estimated 'global' authigenic flux to deep sea (Krishnaswami, 1976) | Estimated authigenic flux to Bermuda Rise sediments (Bacon & Rosholt, 1982) | Estimated authigenic flux to Nares Abyssal Plain red clays* | Estimated authigenic flux to Nares Abyssal Plain red clays† |
|---|---|---|---|---|
| Mn | 500 | 4300 ± 1100 | 1220 | 1264 ± 110 |
| Fe | 800 | — | 1560 | 2737 ± 503 |
| Co | 5 | 7.2 ± 5.7 | 14 | 13 ± 2.9 |
| Ni | 10 | 46 ± 16 | 20 | 17 ± 3.5 |
| Cu | 8 | 76 ± 26 | 28 | 26 ± 2.5 |
| Zn | — | 17 ± 20 | 5.1 | 7.2 ± 1.3 |
| V | — | — | 6.2 | 6.1 ± 2.5 |

* Estimated using detrital background correction method (Thompson *et al.*, 1984).
† Estimated using constant-flux method (Thompson *et al.*, 1984).

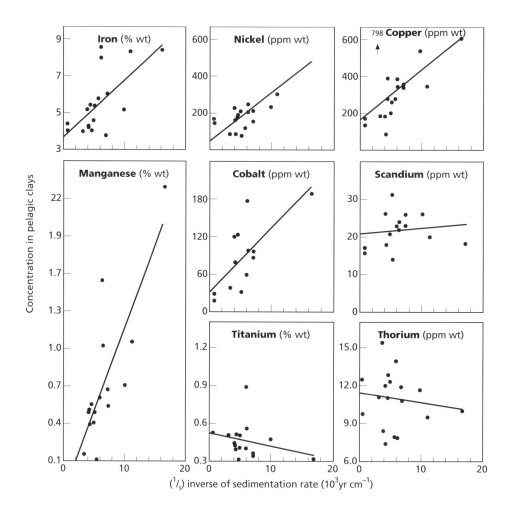

**Fig. 16.4** Scatter plots of elemental concentrations in Pacific pelagic clays against inverse sedimentation rates (from Krishnaswami, 1976).

how oceanic trace-metal cycles operate (see Chapters 11 & 12), and these can be combined with sediment studies to assess the validity of the constant-flux model. Two investigations will serve to illustrate this.

*Sediments from the Bermuda Rise.* Bacon & Rosholt (1982) carried out a study of the accumulation rates of a series of radionuclides and trace metals in carbonate-rich sediments from the Bermuda Rise (North Atlantic), an area in which the sediment regime is characterized by a rapid deposition of material transported by abyssal currents. The authors used

a two-component sediment matrix, consisting of a terrigenous clay (detrital) component and a pelagic (authigenic) component, for the interpretation of their radionuclide and trace-metal data. In terms of this matrix, it was assumed that the total metal content of the sediments was given by eqn (16.2), and the authors were able to show that the sediments in the region were formed by the deposition of (i) a terrigenous component transported to the North American Basin from continental sources (detrital), added to which was (ii) a rain of biogenic debris produced in the surface waters, which had an associated flux of trace metals scavenged from the water column (pelagic). The sediments therefore were a mixture of these two end-member components. The magnitudes of the pelagic, or authigenic, fluxes to the sediments were estimated by relating the trace-metal data to the *excess* $^{230}Th$ in the sediments. This $^{230}$Th excess is defined as 'unsupported' $^{230}$Th, and is obtained by subtracting the uranium-activity equivalent from the total $^{230}$Th activity. The rationale behind this approach is that $^{230}$Th is known to be produced in the oceans at a constant rate determined by the amount of $^{234}$U dissolved in sea water. If the decay of $^{234}$U in sea water is the sole source of the unsupported $^{230}$Th, or $^{230}$Th$_{ex}$, in the sediments it can be assumed that the $^{230}$Th has been scavenged from the water column, and so it can be used to estimate the rates at which trace metals associated with the same sediment fractions have also been scavenged from the water column. A summary of the trace-metal scavenging rates determined in this way by Bacon & Rosholt (1982) is given in Table 16.5, together with other authigenic flux estimates.

It is apparent from the data given in Table 16.5 that although the authigenic accumulation rate of Co on the Bermuda Rise is reasonably similar to that predicted as normal for the World Ocean by Krishnaswami (1976), those of Mn, Ni and Cu are considerably higher, thus throwing doubt on the constant-flux model. Bacon & Rosholt (1982) were unable to identify the reasons for the higher authigenic fluxes of Mn, Ni and Cu on the Bermuda Rise, but pointed out that they may have been affected by, among other factors, an enhanced trace-metal scavenging by resuspended TSM, which is at a high level in this region of the North Atlantic (see Section 10.2). The authors also considered the important relationship between the *residence time* of an element and the magnitude of its *authigenic flux*. They pointed out that $^{230}$Th has such a short oceanic residence time that it must be deposited almost entirely in the basin within which it is generated. Trace metals having longer residence times, however, might be expected to migrate greater distances from their source of supply, and perhaps accumulate preferentially in areas, such as the Bermuda Rise, that are particularly efficient **scavenging sinks**. Bacon & Rosholt (1982) suggested that this might explain the relatively high authigenic fluxes for Ni and Cu in their survey area, as both elements have residence times of several thousand years (see Section 11.2), and so could undergo large-scale oceanic migration. They concluded, however, that an inter-oceanic migratory hypothesis of this kind is unlikely to explain the differences in the magnitude of the Atlantic and Pacific authigenic Mn signals

because this element has a relatively short oceanic residence time, which is similar to that of $^{230}$Th. To understand the differences in the rates of authigenic Mn accumulation between Pacific and Atlantic deep-sea clays it is necessary to relate them to differences in the magnitudes of the Mn input fluxes to the two oceans. This is considered below, when it will be shown that the input of Mn to the Atlantic Ocean is considerably higher than that to the Pacific, which, if the metal is mainly removed in the ocean to which it is initially introduced, would result in higher authigenic fluxes to some Atlantic deep-sea clays.

On the basis of down-core data, Bacon & Rosholt (1982) concluded that the *concentrations* of trace metals in the sediments are controlled mainly by dilution of the authigenic fraction by varying inputs of the terrigenous (detrital) end-member; for example, the clay flux varied by a factor of around four. However, the authigenic *fluxes* of the metals and radionuclides scavenged from the *water column* are not sensitive to this variation in the clay flux, and in fact they have remained almost constant from glacial to interglacial periods; i.e. in the region itself the constant-flux model was approximated over time. Thus, the authigenic fluxes are not associated with the deposition of clay material, but are probably related to the deposition of biogenic carriers, i.e. in association with the down-column TSM fluxes—see Chapters 10, 11 and 12.

Three important overall conclusions therefore can be drawn from the study carried out by Bacon & Rosholt (1982).

1 Regions that are large accumulators of sediment can also act as accumulators of scavenged trace metals.

2 Although the authigenic fluxes of Mn, Ni and Cu on the Bermuda Rise differ from those in the Pacific, the constant-flux model appears to be approximated on a *regional* scale.

3 The trace-metal scavenging in the water column is controlled by biogenic rather than terrigenous phases.

*Sediments from the Nares Abyssal Plain.* Thompson *et al.* (1984) carried out an investigation into trace metal accumulation rates in a series of grey ($^{230}$Th$_{ex}$-poor) and red ($^{230}$Th$_{ex}$-rich) clays from the Nares Abyssal Plain (NAP) in the North Atlantic; the techniques they used to derive the composition of the detrital fractions of the clays have been described above. The sedimentation regime in the region was related to detrital material that had been deposited rapidly from distal turbidity currents to form the *grey* clays, and slowly deposited from nepheloid layers to give rise to the *red* clays. On the basis of the total sediment geochemistry the red clays were strongly enriched in Mn, Co, Cu, Ni, Zn and V relative to the grey clays. Despite their colour differences, however, Sr isotope evidence showed that the clay fractions of both sediments had the same terrigenous origin, and the trace-metal enhancement in the red clays was attributed to their scavenging from the water column; i.e. the slowly deposited red clays had received an additional trace-metal supply from the overlying sea water. Thus, the chemical composition of the clays is controlled by the mixing of the detrital and authigenic components, which, in turn, is controlled by their relative accumulation rates. As a result, in the grey clays $C_t \approx C_d$, which was the condition predicted by Krishnaswami (1976) when the detrital material accumulates at a sufficiently fast rate. For the red clays, however, the total metal flux ($F_t$) is the sum of the fluxes of the authigenic ($F_a$) and the detrital ($F_d$) components:

$$F_t = F_d + F_a \qquad (16.6)$$

$F_a$ is assumed to vary independently of $F_d$. Both $F_t$ and $F_d$ are the products of the total mean sediment accumulation rate ($S$) and the concentration of the metals in, respectively, the total sediment ($C_t$) and the detrital component ($C_d$). The net removal rate of the authigenic trace metals was then calculated by two routes:

**1**

$$F_a = (C_t - C_d)S \qquad (16.7)$$

where $C_d$ is assumed to be equal to the total metal content of the grey clays, i.e. $C_t = C_d$ (the background correction method—see above);

**2**

$$F_t = F_a + C_d S \qquad (16.8)$$

where a graphical procedure is used in which the total metal fluxes ($F_t = C_t S$) are plotted against sedimentation rates to give a regression line that allows both $F_a$ (the $S = 0$ intercept) and $C_d$ (the gradient) to be evaluated. This approach does not assume a knowledge of

$C_d$, but does assume that the authigenic flux is constant for each metal throughout the survey area, i.e. a *constant-flux regional model*.

The linearity of the plots confirmed the latter assumption, and the composition of the detrital fraction obtained from the gradients generally was similar to that obtained by assuming that it was equal to the total concentration of the grey clays. The form of the plots suggested that it is biogenic and not terrigenous (detrital) clays that control the authigenic fluxes, a conclusion similar to that reached by Bacon & Rosholt (1982) for the Bermuda Rise. These biogenic trace-metal carriers have been destroyed at the sediment surface by oxic diagenesis, thus highlighting the importance of including oxic diagenetic processes in the authigenic signal.

The authigenic fluxes for the NAP derived from the two equations showed good agreement for Mn, Cu, Co, Ni and Zn—see Table 16.5. Although these authigenic fluxes are reasonably constant over the NAP survey area, comparison with data for other regions shows that the magnitudes of the fluxes do vary, both *within* ocean basins (cf. those for sediments of the Bermuda Rise, Atlantic) and *between* ocean basins (cf. those for the Pacific clays).

It was therefore becoming increasingly apparent that although the *constant-flux* model for the accumulation of authigenic elements in sediments can be approximated *regionally*, it does not apply on an ocean-wide basis. Despite this, however, Thompson *et al.* (1984) pointed out that the *relative magnitudes* of the authigenic fluxes generally are the same in different oceanic areas, and that there is a tendency for them to decrease in the order observed for the NAP, i.e.

$$Fe > Mn > Cu \approx Ni > Co \approx V \approx Zn$$

Thompson *et al.* (1984) identified two factors that might be expected to control the relative magnitudes of the authigenic metal fluxes. These were (i) the relative values of their input fluxes to the oceans, i.e. their **geochemical abundances**, and (ii) the relative efficiency of the **scavenging processes** that remove them from sea water. On the assumption that the oceanic inputs of the metals are dominated by fluvial sources, Thompson *et al.* (1984) used the data provided by Martin & Whitfield (1983) to show that the relative dissolved concentrations of the metals in river water are

Fe > Mn > Cu > Ni > Co

This is the same order as that of their relative authigenic fluxes, although distortions by factors such as hydrothermal sources were not considered by the authors in their model. Thompson *et al.* (1984) concluded, therefore, that although there are large variations in the absolute *magnitudes* of the authigenic metal fluxes from one oceanic region to another, these do not *fractionate* the elements to an extent that removes the underlying pattern imposed by their geochemical abundances.

Transition metals that have relatively long residence times in sea water, e.g. Cu and Ni, will be expected to be transported further from their source areas before being removed from solution than would metals with shorter residence times, e.g. Fe and Mn. Despite the geochemical abundance control the authors were therefore able to identify differences in the magnitudes of the authigenic fluxes that do, in fact, reflect oceanic reactivities of the individual elements. To illustrate this, Thompson *et al.* (1984) made a direct comparison between the authigenic fluxes they derived for the NAP red clays and those estimated by Krishnaswami (1976) for a series of Pacific clays. To do this, the data for Mn, Cu and Ni from the two oceans were plotted in a graphical form—see Fig. 16.5. The plots reveal that there is a wide range in the authigenic accumulation rates for Mn despite the narrow range in sediment accumulation rates. In contrast, the ranges in the authigenic

accumulation rates for Cu and Ni are much smaller. Further, when derived from the data as displayed in Fig. 16.5, the authigenic fluxes for Cu and Ni in the Pacific clays are similar to those in the Atlantic clays at the NAP site, the ratios of the NAP:average Pacific fluxes being 0.75 and 1.2 for Cu and Ni, respectively. In contrast, the fluxes for Mn are lower for the Pacific clays, the ratio of the NAP:average Pacific fluxes being 2.5. The authors concluded that these features are consistent with the oceanic reactivities of the elements, Mn having a short oceanic residence time ($\sim$<100 yr) resulting from rapid surface-water scavenging, and Cu and Ni having longer residence times (several thousand years) owing to their involvement in the biogeochemical assimilation–regeneration cycle and release from bottom sediments. Thus, it would be expected that because of their longer residence times, inter-oceanic concentration variations for Cu and Ni will be smaller than those for Mn, i.e. source-strength effects will be smoothed out to a much greater degree for Cu and Ni than for Mn, so allowing the two former elements to be redistributed in both sea water and the authigenic fractions of deep-sea sediments.

It may be concluded, therefore, that the **geochemical abundances** of the metals are the principal factor governing their authigenic sediment fluxes, but that fractionation occurs in accordance with their **oceanic reactivities**. This results in the reactive elements (e.g. Mn) having higher fluxes in Atlantic relative to Pacific clays because the Atlantic has a greater fluvial input

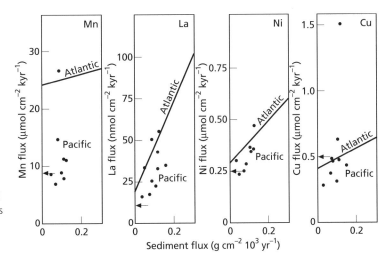

**Fig. 16.5** Comparison between the authigenic fluxes of elements in red clays from the Nares Abyssal Plain (Atlantic) and those of Pacific pelagic red clays. The ratios of the Atlantic to the average Pacific red clays are Mn = 2.5, La = 1.7, Ni = 1.2 and Cu = 0.75 (from Thompson *et al.*, 1984).

than the Pacific, and in the less reactive elements (e.g. Cu, Ni) having generally similar fluxes in the two oceans.

Bacon & Rosholt (1982) and Thompson *et al.* (1984) concluded that the settling of biogenic carriers in the form of organic aggregates is the main driving force behind the down-column transport of trace metals to the sediments on both the Bermuda Rise and the NAP. The carrier phases are destroyed by oxic diagenesis at the sediment surface and some of the associated metals are taken into authigenic components. The ocean-wide importance of these carriers has been demonstrated by other workers. For example, Aplin & Cronan (1985) compared the magnitudes of the biogenic carrier down-column fluxes of a series of metals with the rates at which they accumulate in sediments from the southwest equatorial Pacific. The relationship is illustrated graphically in Fig. 16.6(a), and demonstrates a close similarity between the down-column fluxes and sediment accumulation rates for Ni, Co, Ti, Mn and Fe; Cu is supplied in excess of its accumulation rate, which is in agreement with its release back into sea water across the sediment–water interface (see Section 14.6.3).

It is apparent, therefore, that down-column transport via biogenic carriers is the principal route by which authigenic trace metals reach the sediment surface. In view of this, it is not surprising that although the authigenic processes provide an ocean-wide background for the supply of non-detrital metals to sediments, the magnitudes of the authigenic fluxes of some elements vary both *between* and *within* ocean basins (see Section 12.1.1). These fluxes appear to be controlled mainly by geochemical abundances, but metal reactivities in the water column are superimposed on input source strengths. In terms of their oceanic reactivities, the scavenging-type elements have relatively short residence times and so undergo restricted transport between the major ocean basins. The nutrient-type elements have longer residence times, and so have a greater potential for interoceanic transfer. The down-column fluxes of these nutrient-type elements vary considerably throughout the oceans, however, depending largely on the magnitude of the down-column carbon-driven flux. For example, Collier & Edmond (1984) found a striking correlation between the scavenging rate of Cu in the water column and primary productivity in surface

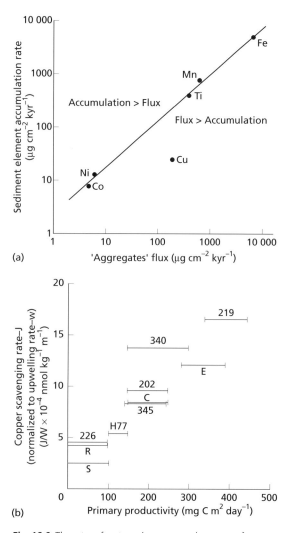

**Fig. 16.6** The rates of water-column removal processes for trace elements. (a) Comparison between biogenic carrier down-column fluxes and accumulation rates of elements in sediments from the southwest Pacific (from Aplin & Cronan, 1985). (b) Estimates of the primary production of organic carbon in surface waters ($^{14}$C uptake) versus the relative scavenging rate of Cu in the deep ocean; numbers and letters indicate original data sources (from Collier & Edmond, 1984).

waters—see Fig. 16.6(b). The authors concluded that this relationship emphasizes the importance of regions of high primary productivity in driving the vertical transport of elements in the oceans; i.e. these regions have relatively high down-column TSM fluxes—see Section 12.1.1. As a result, nutrient-type elements *can* be removed from the water column on time-scales that prevent their inter-oceanic transport, and this can lead to patches of high concentrations of these elements in sediments deposited under regions of intense primary production; for example, it will be shown in Section 16.7 that the authigenic fluxes of Cu in the Indo-Pacific Ocean are relatively high under the area of equatorial primary productivity, which leads to enhanced concentrations of the element in the underlying sediments in which the organic carriers are preserved (see e.g. Chester *et al.* (1988) and Section 14.3.2).

It may therefore be concluded that the authigenic signal does not operate in a manner that yields an ocean-wide *constant flux* for those authigenic elements which have relatively short residence times in sea water. Authigenic elements that have longer residence times, however, can have their dissolved concentrations smoothed out between the major oceans, and therefore may approach the *constant-flux model* as originally proposed by Wedepohl (1960) and supported by data from Krishnaswami (1976). Thus, we have come full circle, and the distributions of the elements in deep-sea sediments can be linked to the processes controlling their oceanic residence times, and their general seawater chemistries, discussed in Chapter 11.

## 16.6 Signal spikes

### 16.6.1 Introduction

In the previous sections, the sources of the excess trace metals found in deep-sea clays were assessed in terms of their background transport via the detrital and authigenic signals. This involved a two-component system in which (eqn 16.1)

$$C_t = C_d + C_a$$

Within the ocean system, however, a series of more localized signals can impose perturbation spikes on this background, such that

$$C_t = C_d + C_a + C_s \qquad (16.9)$$

where $C_s$ refers to the concentration of an element arising from the perturbation spikes. These spikes are transmitted through the water column mainly by the hydrothermal and contaminant signals, and through interstitial waters by the diagenetic signal.

### 16.6.2 The hydrothermal signal spike

The concepts underlying hydrothermal activity have been discussed in Chapter 5, and the processes involved in the generation of hydrothermal sediment-forming components have been described in Section 15.3.6. Attention now will be directed to considering the question 'How do these chemically specialized components affect the compositions of deep-sea sediments?' The hydrothemal emanations occur in the form of pulses, which are generated at the ridge-crest spreading centres. These metal-rich pulses give rise to a variety of hydrothermal components, but as far as deep-sea sediments in general are concerned it is the ferromanganese precipitates that are the principal manifestations of hydrothermal activity.

It has been known for many years that metal-rich sediments are found in association with the mid-ocean ridge system. For example, Murray & Renard (1891) and Revelle (1994) reported the presence of such deposits in the eastern Pacific, and El Wakeel & Riley (1961) gave details of Mn- and Fe-rich calcareous ooze from the vicinity of the East Pacific Rise (EPR). The major boost to the study of these metalliferous deposits came in the mid-1960s following the work of Bostrom and his co-workers. It was at this stage that the genesis of the deposits was linked to the formation of new crust at the ridge spreading centres, a theory that was given a firm foundation by the identification of hydrothermal venting systems on the ridge areas (see Section 5.1). Metal-rich deposits subsequently have been identified on, or in the vicinities of, the ridge system from various parts of the World Ocean. The locations where the deposits have been found include the Pacific Ocean (see e.g. Bender *et al.*, 1971; Dasch *et al.*, 1971; Dymond *et al.*, 1973; Piper, 1973; Sayles & Bischoff, 1973; Sayles *et al.*, 1975; Heath & Dymond, 1977; Marching & Grundlach, 1982), the Indian Ocean (see e.g. Bostrom *et al.*, 1969; Bostrom & Fisher, 1971) and the Atlantic Ocean (see e.g. Bostrom *et al.*, 1969; Cronan, 1972; Horowitz, 1973); compositional data for some of these metal-rich sediments are given in Table 16.6; see also Table 15.6.

Metal-rich deposits are formed on newly generated crust at the ridge spreading centres. During the process of sea-floor spreading this crust moves away from the ridges as the ocean basins are opened up, and the basement is covered by a blanket of normal

**Table 16.6** Chemical composition of some modern and ancient metalliferous sediments; units, $\mu g\,g^{-1}$, carbonate-free basis.

| Element | East Pacific Rise: crest* | East Pacific Rise: flanks* | Pacific: Northwest Nazca Plate† | | | East Pacific Rise: basal sediments‡ | Mid-Atlantic Ridge§ |
| | | | East Pacific | Bauer Deep | Central Basin | | |
|---|---|---|---|---|---|---|---|
| Fe | 180 000 | 105 000 | 302 000 | 158 300 | 121 100 | 200 700 | 76 000 |
| Mn | 60 000 | 30 000 | 99 200 | 57 500 | 39 600 | 60 600 | 4100 |
| Cu | 730 | 960 | 1450 | 1171 | 985 | 790 | — |
| Ni | 430 | 675 | 642 | 1066 | 1307 | 460 | — |
| Co | 105 | 230 | — | — | — | 82 | — |
| Pb | — | — | — | — | — | 100 | — |
| Zn | 380 | 290 | 594 | 413 | 311 | 470 | — |
| V | 450 | 240 | — | — | — | — | — |
| Hg | — | — | — | — | — | — | 414 |
| As | 145 | 65 | — | — | — | — | 174 |
| Mo | 30 | 113 | — | — | — | — | — |
| Cr | 55 | 32 | — | — | — | — | — |
| Al | 5000 | 46 300 | 5100 | 32 400 | 67 400 | 27 300 | 57 900 |

* Bostrom & Peterson (1969). † Heath & Dymond (1977). ‡ Cronan (1976). § Cronan (1972).

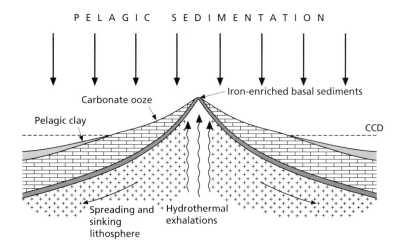

**Fig. 16.7** Schematic representation of a model for sediment accumulation at oceanic spreading ridge centres (from Davies & Gorsline 1976, after Broecker 1974 and Bostrom & Peterson 1969).

deep-sea sediments. As a result, although surface outcrops of metal-rich sediments are usually found mainly at the ridge crests, Bostrom & Peterson (1969) predicted that they should also form a layer on top of the oceanic basement at all locations. A general model illustrating this process is shown in Fig. 16.7. Confirmation that metalliferous sediments are the first material deposited on the basaltic basement was provided by cores obtained during the Deep Sea Drilling Project (DSDP), which showed that Mn- and Fe-rich EPR analogue deposits were indeed found at the base of the sediment column at many oceanic locations (see e.g. von der Borch & Rex, 1970; Cronan *et al.*, 1972; Cronan, 1973; Dymond *et al.*, 1973; Cronan & Garrett, 1973; Horder & Cronan, 1981).

It may be concluded, therefore, that Fe- and Mn-rich metalliferous deposits are usually the first type of sediment to be laid down on the oceanic crust at the spreading centres, thus forming a bottom layer on to which other types of deep-sea sediments accumulate. Although the hydrothermal precipitates have their source at the ridge-crest venting systems, the sediments formed there are rarely composed of pure hydrothermal material, but usually are mixtures of the metal-rich end-member precipitates and other sediment-forming components: mainly carbonate shells, which are preserved on the topographic highs. Marchig & Grundlach (1982) identified a *prototype* undifferentiated end-member formed at an initial stage in the evolution of hydrothermal material from the EPR. From the data given by these authors an estimate is therefore available of the composition of the Fe- and Mn-rich end-member that is precipitated from hydrothermal fluids on mixing with sea water (see Table 16.7). Some authors have suggested that pulses of these hydrothermal fluids can be dispersed on an ocean-wide basis (see e.g. Lalou, 1983), and it is clear that hydrothermal manganese signals *can* be transmitted for long distances via the global mid-depth water circulation patterns (see Section 15.3.6). The question that arises, therefore, is 'How widespread is the Fe- and Mn-rich hydrothermal end-member that is generated from these pulses?' To answer this question, it is necessary to unscramble the effects of the hydrothermal pulse signal from those of other sediment-forming signals.

Bostrom *et al.* (1969) addressed the problem of unscrambling the hydrothermal signal on an oceanic-wide basis. Various authors have identified a series of metal-rich deep-sea sediments associated with areas of high heat flow on the EPR. Relative to normal deep-sea sediments these deposits are enriched in elements such as Fe, Mn, Cu, Ni and Pb, which are associated with the colloidal-sized hydrothermal par-

ticles. Another important characteristic of the metal-liferous sediments is that they are depleted in lithogenous elements such as Al and Ti—see Table 16.6. This enrichment–depletion pattern was utilized by Bostrom *et al.* (1969) to characterize a **hydrothermal signature**. To quantify the signature, the authors used the ratio $Al:(Al + Fe + Mn)$ as an indicator of metalliferous sedimentation. In this way, the hydrothermal end-member was characterized as having $Al:(Al + Fe + Mn)$ ratios of $<10 \times 10^2$, and the normal deep-sea sediment end-member as having ratios $>60 \times 10^2$. Sediments that are mixtures of the two end-members have intermediate ratios. The authors then plotted values of the hydrothermal ratio for deep-sea sediments, and showed that the metalliferous deposits are indeed characteristic of the mid-ocean ridge system throughout the World Ocean—see Fig. 16.8. The lowest ratios, and the greatest coverage of hydrothermal sedimentation, were found in the Pacific Ocean around the EPR, and the influence of the hydrothermal material on surrounding sediments decreased in the order Pacific Ocean ridge system > Indian Ocean ridge system > Atlantic Ocean ridge system. This rank order correlates well with the relative rates of ocean-floor spreading, which can indicate the general degree to which a ridge is active; thus, the spreading rate in the Pacific is $\sim 2.0$–$6.0\,cm\,yr^{-1}$, in the South Atlantic and Indian Oceans it is $\sim 1.5$–$3.0\,cm\,yr^{-1}$, and in the North Atlantic it is $\sim 1.0$–$1.4\,cm\,yr^{-1}$.

Bostrom *et al.* (1969) therefore were able to identify the widespread occurrence of metalliferous deposits on the mid-ocean ridge system by unscrambling the hydrothermal signal in terms of the sediment $Al:(Al + Fe + Mn)$ ratio. Thus, hydrothermal deposits have relatively high *concentrations* of Fe and Mn. In addition, the hydrothermal deposits can be differentiated from 'normal' deep-sea sediments on the basis of the *rates* at which the Fe and Mn in them accumulate. This was considered in Section 15.3.6 with respect to ferromanganese nodules and crusts, and now will be discussed in terms of deep-sea sediments. It was shown in Section 16.5 that, although there is not a constant ocean-wide authigenic deposition rate for Mn in deep-sea sediments, the rates do appear to be in the range $\sim 1000$–$5000\,\mu g\,cm^{-2}\,10^3$ $yr^{-1}$ for the Atlantic, and somewhat lower for the Pacific. For example, Krishnaswami (1976) estimated an authigenic deposition rate for Mn of $\sim 500\,\mu g$ $cm^{-2}\,10^3\,yr^{-1}$ for Pacific pelagic clays, and Bender *et*

**Table 16.7** Chemical composition of a prototype undifferentiated hydrothermal precipitate*; units, $\mu g\,g^{-1}$, carbonate-free basis.

| Element | Concentration | Element | Concentration |
|---|---|---|---|
| Si | 36 000 | Mo | 134 |
| Al | 3600 | Zr | 73 |
| Mn | 91 800 | Nb | 2 |
| Fe | 281 000 | Rb | 9.3 |
| Ti | 300 | Ce | 13 |
| Cu | 1300 | Th | 61 |
| Zn | 393 | V | 1923 |
| Co | 111 | W | 21 |
| Cr | 30 | Y | 101 |
| Ni | 600 | La | 98 |

* Data for the $<63\,\mu m$ fractions of sediment cores from the East Pacific Rise (Marchig & Grundlach, 1982).

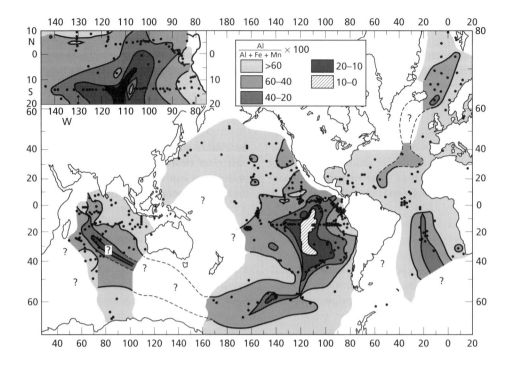

**Fig. 16.8** The distribution of the hydrothermal ratio, Al : (Al + Fe + Mn), in sediments from the World Ocean (from Bostrom *et al.*, 1969). Note that low ratios are associated with spreading ridges.

*al.* (1971) calculated an average total deposition rate for Mn of $\sim$1300 $\mu$g cm$^{-2}$ 10$^3$ yr$^{-1}$ for Pacific deep-sea sediments from the same oceanic region. In contrast, Bender *et al.* (1971) used radiochemical data to estimate that the accumulation rate of Mn in a core from the crest of the EPR was $\sim$35 000 $\mu$g cm$^{-2}$ 10$^3$ yr$^{-1}$, i.e. this Mn was accumulating almost 25 times faster than in normal deep-sea sediments. Dymond & Veeh (1975) gave data on the accumulation rates of Mn, Fe and Al in sediments on a transect extending west to east across the EPR, and clearly demonstrated enhanced accumulation rates for Mn and Fe in sediments deposited on the ridge crest. The Mn accumulation rates on the crest reached values of $\sim$28 000 $\mu$g cm$^{-2}$ 10$^3$ yr$^{-1}$, i.e. the same order of magnitude as those reported by Bender *et al.* (1971). The maximum rates for the accumulation of Fe on the EPR were $\sim$82 000 $\mu$g cm$^{-2}$ 10$^3$ yr$^{-1}$. According to Dymond & Veeh (1975), the authigenic accumulation rate for Fe in normal deep-sea sediments from this region is

$\sim$200 $\mu$g cm$^{-2}$ 10$^3$ yr$^{-1}$, and Krishnaswami (1976) estimated an average rate of $\sim$800 $\mu$g cm$^{-2}$ 10$^3$ yr$^{-1}$ for accumulation of Fe in Pacific pelagic clays. Clearly, therefore, the accumulation of Fe on the crest of the EPR is of the order of 100–400 times faster than it is in normal Pacific deep-sea sediments.

It may be concluded, therefore, that authigenic deposition cannot explain the enhanced accumulation rates of Mn and Fe on the crestal regions of the EPR. Rather, it would appear that the high accumulation rates are the result of an enhanced supply of the elements from a pulsed hydrothermal signal. However, as they form, hydrothermal precipitates can remove dissolved elements directly from sea water, so that not all the elements that accumulate at fast rates around the ridge crests necessarily have a hydrothermal origin.

Another way of characterizing the hydrothermal source of an element is by making a direct comparison between the compositions of hydrothermal solutions and those of metalliferous sediments deposited in the same area. Such an approach was adopted by Von Damm *et al.* (1985). These authors determined the concentrations of Mn, Fe, Ni, Cu, Zn, Co, Cd, Ag and Pb in hydrothermal fluids vented from the high-temperature system at 21°N on the EPR, and compared the elemental ratios in the fluids with those in associated metalliferous sediments. Iron was used as an indicator of hydrothermal sources and a comparison was then made between element : Fe ratios in the fluids and the sediments. The results may be summarized as follows.

1 For Co and Ni (which has an insignificant hydrothermal source), the ratios in the venting solutions were lower than in the sediments, implying that additional amounts of these elements are scavenged from sea water by the hydrothermal Fe–Mn precipitates.

2 For Mn, Zn, Cu and Ag, the ratios in the hydrothermal solutions were greater than, or equal to, those in the sediments, indicating that although the hydrothermal input is the major source of the elements a proportion of them may be lost to sea water, i.e. they may escape the immediate venting area.

3 The Pb : Fe ratios were very similar in both the venting solutions and the metalliferous sediments, implying a predominantly hydrothermal origin for the Pb.

The hydrothermal signal results from the emanation of mineralizing solutions at the ridge-crest spreading centres, which are delivered in the form of pulses. These metal-rich emanations impose *spikes* on the general background of biogenous, detrital and authigenic components that combine to form normal deep-sea sediments. The components derived from the hydrothermal pulses form an important class of metal-rich sediments. By unscrambling the hydrothermal signal, however, it has been demonstrated that the metalliferous deposits themselves have only a relatively small areal extent and, with respect to the sediment surface, they are largely confined to regions around the mid-ocean ridge system.

### 16.6.3 The contaminant signal spike

There are a number of texts that offer an extensive treatment of pollution in the marine environment, and for this reason it is not covered as an individual topic in the present volume; for a review of the subject, the reader is referred to Preston (1989). It is of interest at this stage, however, to demonstrate how contaminant spikes can influence the elemental chemistry of marine sediments.

#### 16.6.3.1 Nearshore sediments

Contaminant, or anthropogenically generated, elements are brought to the ocean reservoir by the same pathways that transport the material released naturally during crustal mobilization. Both spatial and temporal 'fingerprints' resulting from the inputs of contaminant elements have been recorded in nearshore sediments, which accumulate at relatively fast rates close to the sources of pollution. Various techniques can be used to identify elements transmitted by contaminant signals, and to unscramble them from natural background inputs. Two examples will serve to illustrate how such techniques can assess the effects that contaminant spikes can have on coastal deposits.

*The spatial distribution of contaminant elements in coastal sediments.* A large proportion of many of the elements associated with contaminant spikes have been introduced into sea water in a dissolved state and subsequently removed from solution; thus, they form part of the non-detrital sediment fraction. To separate this fraction from the crystalline mineral matrix, Chester & Voutsinou (1981) applied a chemical leaching technique to sediments from two Greek gulfs, one of which had received pollutant inputs and the other that was reasonably pristine in character. By using the sediments of the unpolluted gulf as a baseline, the authors were able to (i) identify contaminant trace metals; and (ii) map their spatial distributions in sediments from the polluted gulf (Thermaikos Gulf). To illustrate this, the spatial distribution of Zn in Thermaikos Gulf is shown in Fig. 16.9. There is a heavily industrialized region at the head of this gulf, and it is evident from Fig. 16.9 that the non-detrital concentrations of Zn are highest in sediments in this region and decrease out into the central gulf; thus, local inputs have imposed a fingerprint on the distribution of this trace metal in the surficial sediments of the polluted gulf.

*The temporal distribution of contaminant elements in coastal sediments.* Because of their relatively fast rates of deposition coastal sediments can sometimes provide a historical record of contaminant inputs. This was demonstrated by Chow *et al.* (1973) with respect to the deposition of Pb in coastal sediments off southern California. The authors carried out a

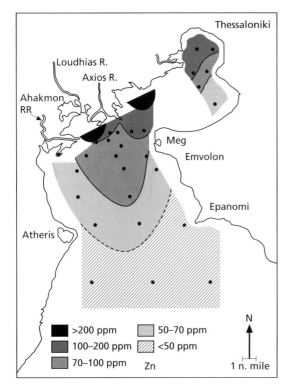

**Fig. 16.9** Distribution of non-detrital Zn in surface sediments from Thermaikos Gulf, Greece (from Chester & Voutsinou, 1981).

survey of the distribution of Pb in dated sediment columns from a number of basins in the area. To aid their interpretation of the data, Al (which has a mainly natural origin) was used as a normalizing element to establish the background levels of Pb in the sediments, and isotopic ratios were used to identify the sources of the Pb itself. Sediments were sampled from the following localities.

1 The Soledad Basin. This is remote from the prevailing winds blowing off southern California and is free from waste discharges, and therefore was used as a control area.

2 The San Pedro, Santa Monica and Santa Barbara Basins. These are inner basins in the Los Angeles area.

3 White Point. This is close to a waste outfall and provided sediments containing industrial and domestic wastes; however, the sediments here had accumulated so rapidly that only a recent age could be assigned to them over the interval studied.

The Pb concentrations, and Pb : Al ratios, in the sediments from the various basins are illustrated in Fig. 16.10, and the results of the study can be summarized as follows.

1 The Pb concentrations in the sediments of the Soledad Basin are generally invariant and are less than ~20 µg g$^{-1}$. Further, the surface deposits contain Pb which had an isotopic composition similar to that of weathered material from the Baja California province. The sediments in this basin therefore were used as a baseline against which to compare the Pb distributions in sediments from the other basins.

2 Relative to those of the Soledad Basin, the sediments of the inner basins contain higher concentrations of Pb in their upper portions, which decreased towards baseline levels at depths of ~8 cm. Below this depth, the isotopic ratios of the Pb were similar to those that represent the pre-pollution supply. The sediments in these basins are anoxic and had not suffered bioturbation, and it could be shown that the rates of contaminant Pb accumulation had started to increase in the 1940s. The increased inputs are clearly reflected in contaminant spikes in the down-column Pb profiles in the inner basin sediments.

3 The sediments from White Point have the highest Pb concentrations of several hundreds of micrograms per gram (µg g$^{-1}$), and Pb isotopic ratios that are similar to those in petrol (gasoline) sold in southern California.

### 16.6.3.2 Deep-sea sediments

At present, slowly accumulating deep-sea sediments have not yet recorded *major* inputs of most anthropogenically generated elements to the extent that concentration spikes are present. Evidence is now beginning to appear, however, which suggests that the situation is changing; for example, artificial radionuclides have been reported in the upper portions of deep-sea sediments (see e.g. Lapicque *et al.*, 1987).

It is elements that have a strong anthropogenic signal to the oceans, and a relatively short residence time in the water column, that have the greatest potential for being recorded in deep-sea sediments. Lead is a prime example of such an element:

1 it has a strong atmospheric anthropogenic signal to open-ocean surface waters (see Section 6.4);

2 it is a scavenging-type element that is rapidly

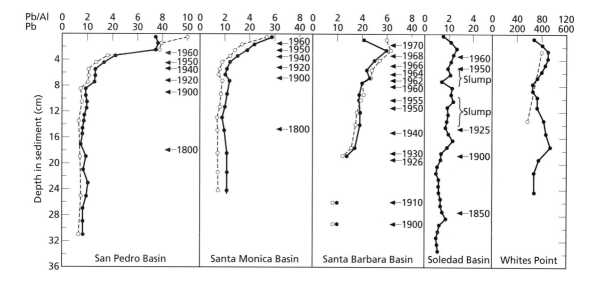

**Fig. 16.10** The down-column concentrations of Pb (solid circles), and Pb : Al ratios (open circles), in sediments from basins off the coast of southern California (from Chow *et al.*, 1973); concentrations in μg⁻¹.

removed from surface waters within a few years (see Section 11.6.3.1); and

3 it has a relatively short overall residence time in the oceans of around 100 yr (see Section 11.2).

The relationship between the atmospheric input of Pb to surface ocean waters and its effect on the water-column distribution of dissolved Pb is shown in Fig. 16.11(a). Most of the atmospheric Pb is anthropogenic in origin and has been delivered directly to open-ocean areas, thus by-passing the coastal sediment-trap barriers (see Section 11.4). The highest anthropogenic input of Pb is to the North Atlantic, where it is reflected in the large 'bulge' of dissolved Pb concentrations in the upper 1–2 km of the water column (Patterson, 1987). It is in the North Atlantic, therefore, that the effects of the input of anthropogenic Pb to surface waters are most likely to be recorded in surficial open-ocean sediments, and Veron *et al.* (1987) have reported data that do indeed offer evidence of recent Pb pollution in northeast Atlantic deep-sea sediments. These authors showed that Pb concentrations in the surficial sediments (21 and 15 μg g⁻¹ in the top 1 cm) of two short cores were higher than those in the underlying 10 cm (6.0 and

2.8 μg g⁻¹)—see Fig. 16.11(b). The amount of Pb stored in the surficial sediments was of the same order as the amount of Pb in the overlying water column, and the authors concluded that the surficial Pb spike was derived from anthropogenic sources.

It is apparent, therefore, that contaminant spikes are beginning to be identified in surficial deep-sea sediments, but it will be a considerable time before such spikes are recorded for elements that have relatively long oceanic residence times.

### 16.6.4 The diagenetic signal spike

Trace metals are transported down the water column from primary seawater sources in association with large-sized particulate organic carrier phases. Once they are deposited at the sediment surface diagenesis is initiated, and the carriers are either destroyed during oxic diagenesis, or buried and subsequently at least partially destroyed during suboxic diagenesis. Many of the reactions that occur in the upper sediment, and the benthic boundary layer, are driven by oxic diagenesis. Because this affects elements that have been removed from sea water and subsequently transported down the water column, oxic diagenesis has been classified as part of the authigenic signal. Some of the processes associated with oxic diagenesis can lead to the transport of elements *downwards* into the sediments through interstitial waters; this was illustrated in Section 14.6.3 with respect to Cu. The

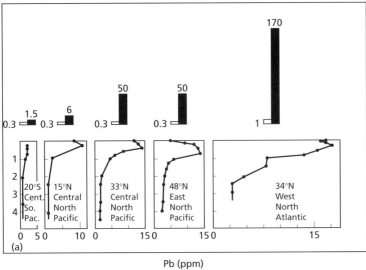

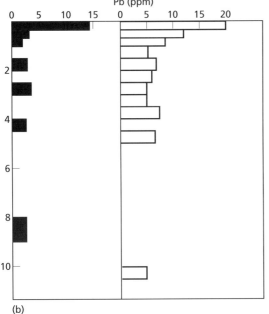

**Fig. 16.11** Anthropogenic effects on lead in the marine environment. (a) Comparisons of atmospheric lead inputs with lead concentration profiles in the open ocean (from Patterson, 1987). The histograms display net atmospheric lead input fluxes ($10^{-9}$ g cm$^{-2}$ yr$^{-1}$), after correction for recycled sea-spray; black columns, present-day fluxes; white columns, ancient fluxes. Graphs display the lead concentration in sea water ($10^{-9}$ g l$^{-1}$) against depth (km). The oceanic environments are ranked from the most pristine (central South Pacific) to the most anthropogenically contaminated (North Atlantic); this is reflected in the magnitudes of the atmospheric lead fluxes and the effect they have on the water-column profiles, both of which are greatest in the North Atlantic. (b) Depth distributions of lead in two deep-sea sediment cores from the northeast Atlantic (from Veron *et al.*, 1987); units in μg g$^{-1}$. The surficial samples in both cores contained excess Pb, which was attributed to an anthropogenic source.

point is, however, that these elements were originally transported down the seawater column and released during oxic diagenesis. The authigenic signal therefore is the basic background control on the *spatial* distribution of the chemical elements in surface deep-sea sediments, although the distribution can be perturbed by hydrothermal and contaminant spikes. The *vertical* distribution of elements added to the sediment column from seawater sources can, however, be

affected by remobilization and transport through interstitial waters associated with suboxic diagenesis. As defined in the present context, therefore, the diagenetic signal refers to that transmitted through the interstitial water column following suboxic diagenesis, and can affect both natural and contaminant elements.

During suboxic diagenesis dissolved trace metals are released from both organic carriers and secondary oxidants, and can reach concentrations in the interstitial waters in excess of those in the overlying sea water. As a result, the metals migrate upwards under the influence of the concentration gradients and either escape into overlying sea water or become trapped in the upper oxic sediment layer. When the metals are trapped in the oxic layer, suboxic diagenetic recycling through interstitial waters severely modifies both their vertical distributions in the sediment column and their concentrations in surface sediments; i.e. the trace metals from primary sources are redistributed within the sediment. This redistribution involves a **decoupling** break between sediment components formed from elements derived directly from sea water and those formed from elements derived from interstitial waters.

The post-depositional mobility and the migration of elements in suboxic diagenetic processes have been described in Section 14.6.2, and were shown to affect trace metals such as Mn, Fe, Co, Ni, Cu, Zn, V, U and Mo. The extent to which these metals are trapped in the upper sediments depends on the presence, and thickness, of the oxic layer, which in turn is dependent on the surface sediment redox environment. It was shown in Section 14.1.3 that there is a lateral, i.e. shelf → hemi-pelagic → pelagic, control on the depth of the sediment oxic layer, and as a result there is also a lateral constraint on the fate of trace metals released during suboxic diagenesis.

The behaviour of Mn during suboxic diagenesis was described in detail in Section 14.6.3, and this element will serve as an example to illustrate how the suboxic diagenetic signal can modify trace-metal concentration profiles in sediments. It has been known for many years that the profiles for manganese often display higher values close to the surface in *both* hemi-pelagic and pelagic deep-sea sediments (see e.g. Fig. 14.8). This has led to discrepancies in the oceanic manganese budget based on surface sediment data because the high surface concentrations could not be sustained by the magnitude of the primary seawater inputs (see below). It has been suggested that the key to understanding this lies in the fact that manganese responds to redox changes that involve its recycling from previously deposited Mn-containing components during suboxic diagenesis, and its subsequent enhancement in the upper oxic layers of the sediments; i.e. there is a major redistribution of manganese in the sediment column following suboxic diagenesis.

There is no doubt that manganese is recycled as a result of suboxic diagenesis (see Section 14.6.3), but the question that arises is 'Do these diagenetic recycling processes operate on a sufficiently global-ocean scale to account for the depth-depletion–surface-enrichment found in all types of oceanic sediments?' A number of authors have attempted to address this question. For example, Bender (1971) applied a simple diffusive model to interstitial water data and concluded that dissolved interstitial water $Mn^{2+}$ can only supply the Mn concentrated in the upper *pelagic* sediment column if it has to migrate through <1 m of sediment. Thus, for those pelagic deep-sea sediments in which the Mn-rich zone can extend over tens of metres it would appear that post-depositional migration cannot supply the excess Mn requirement, although it can contribute to the formation of ferromanganese nodules and individual Mn-rich bands (see also Section 14.6.3). Elderfield (1976) made a more detailed attempt to assess the influence of the diagenetic interstitial water transport of $Mn^{2+}$ on the supply of non-detrital Mn to the upper layers of marine sediments, and reached the same overall conclusions as Bender (1971). To do this, Elderfield (1976) used a number of approaches to estimate how much Mn in excess of that provided by fluvial transport was required to maintain the non-detrital concentration of the element in the upper layers of the sediments. This fluvial-excess non-detrital Mn was found to be accumulating at a rate in the range $\sim$300–2000 $\mu g\, cm^{-2}\, 10^3\, yr^{-1}$. The author then used a vertical advection–diffusion model to assess the interstitial water transport of Mn in both nearshore and deep-sea sediments. For nearshore sediments, he concluded that the estimated diagenetic flux was in agreement with the known accretion rate of Mn deposits, thus supporting diagenetic theories for the origin of shallow-water marine ferromanganese nodules. For deep-sea sediments, however, the maximum value of the surface diagenetic flux of Mn was calculated to be $\sim$70 $\mu g\, cm^{-2}\, 10^3\, yr^{-1}$ when the upper zone of the sediment was $\sim$20 cm thick, compared with the minimum estimate of 300 $\mu g\, cm^{-2}\, 10^3\, yr^{-1}$ for the fluvial-excess Mn accumulation rate. Much smaller diagenetic fluxes will be found for deep-sea sediments in which the upper oxic zone extends over several metres, and Elderfield (1976) concluded that suboxic diagenesis cannot supply all the excess Mn in such pelagic sediments and is, in fact, likely to be significant only for those deposits that have a thin (<25 cm) oxic layer. This type of thin layer is characteristic of a number of hemi-pelagic sediments, however, and for these deposits diagenetic Mn fluxes can be important in supplying the element to the upper layers; this general conclusion also has been supported by Sawlan & Murray (1983) on the basis of detailed interstitial-water Mn profiles (see Section 14.6.3).

It may be concluded, therefore, that the extent to which upward suboxic diagenetic fluxes of Mn contribute to the total Mn in surface sediment layers varies in the sequence shelf > hemi-pelagic > pelagic deep-sea sediments, in relation to the lateral diagenetic sequence. In general, therefore, suboxic diagenetic recycling can supply only sufficient manganese to satisfy the upper layer enhancement in those sediments that have an oxic layer ∼<20 cm in thickness. This generally is restricted to nearshore and hemi-pelagic sediments, and here when the oxic layer is present the remobilized Mn is trapped within the upper oxic layers of the sediment, and so imposes a diagenetic spike on the sediment column itself. In pelagic sediments, which have relatively thick surface oxic layers, however, diagenetic fluxes cannot supply the Mn required to generate the Mn-rich upper layer, and Elderfield (1976) concluded that in these deposits the fluvial-excess Mn probably has a hydrothermal origin; atmospheric sources also contribute to the fluvial-excess Mn. Nonetheless, hemi-pelagic sediments cover a significant area of the deep-sea floor, and here the top-loading of solid-phase Mn results from a major redistribution of the element in response to suboxic diagenetic recycling.

## 16.7 The ocean-wide operation of the sediment-forming signals

Most oceanic sediments are made up of a mixture of components of different origins, and so have received a variety of chemical signals. The nature of the individual chemical signals, and how they can be unscrambled from each other, has been described in the previous sections. In this way, it was possible, at least in a tentative manner, to understand the chemical processes that are active in the formation of oceanic sediments. From this it has become apparent that the chemical compositions of the sediments are governed by a complex of interrelating controls in which the end-member components derived from the various signals are mixed together. The controls that govern the signal strengths are related to the overall *pattern of sedimentation* in a particular oceanic area. Thus, to understand fully how these interrelating controls operate, account must be taken of the physical processes that move material around the ocean reservoir.

To conclude the treatment of the factors that control the chemical composition of marine sediments, an attempt therefore will be made to draw together the various lines of evidence discussed earlier in order to establish how the signals interact on a global scale and so fix the overall chemical compositions of the sediments. To do this, attention will be focused on the deposits at the sediment surface.

For most deep-sea sediments, the contaminant signal can be ignored at the present day, and the suboxic diagenetic signal recycles elements, most of which have already been deposited at depth in the sediment column. In order to obtain a first-order description of how the element-carrying signals interact on a global scale to control the composition of the surface sediments, attention will be concentrated on the *primary* signals, and for this purpose it will be assumed that

$$C_t = C_b + C_d + C_a + C_h \qquad (16.10)$$

where $C_t$ is the total content of an element in a sediment, and $C_b$, $C_d$, $C_a$ and $C_h$ are the biogenous, detrital, authigenic and hydrothermal contributions, respectively.

Relatively few studies on the distributions and accumulation rates of elements in deep-sea sediments have been made on an ocean-wide basis; however, those carried out by Bostrom and co-workers are ideal for the present purpose. Bostrom *et al.* (1973) combined concentration patterns with sediment accumulation rate data to identify the factors that control the distributions of a series of elements in surface deep-sea sediments from the Indo-Pacific Ocean. This study, which included elements having a number of different geochemical characteristics, therefore can be used as a basis for assessing how chemical sediment-forming signals interact on a quasi-global scale.

### 16.7.1 Biogenous elements

Silicon in the form of opaline silica will serve as an example of a biogenous element in deep-sea sediments. Bostrom *et al.* (1973) estimated the concentration of opaline silica from the total silica ($\Sigma SiO_2$) contents in the Indo-Pacific sediments by assuming that in crustal material $SiO_2 : Al_2O_3 = 3.0$, and then assigning any $SiO_2$ in excess of this to a biogenic source. The distribution of this biogenic silica in

**Fig. 16.12** Distributions and accumulation rates of elements in Indo-Pacific sediments (from Bostrom *et al.*, 1973); MB = minerogen basis. (a) (i) Distribution of $SiO_2$ in Indo-Pacific sediments (on a carbonate-, salt- and organic-matter-free basis (CSO)); (ii) accumulation rates of opaline silica in Pacific sediments. (b) (i) Distribution of Al in Indo-Pacific sediments (on a minerogen basis); (ii) accumulation rates of Al in Pacific pelagic sediments. (c) (i) Distribution of Fe in Indo-Pacific sediments (on a minerogen basis); (ii) accumulation rates of Fe in Pacific sediments. (d) Accumulation rates of Mn in Pacific sediments. (e) (i) Distribution of Cu in Indo-Pacific sediments (on a minerogen basis); (ii) accumulation rates of Cu in Pacific sediments.

Indo-Pacific sediments is shown in Fig. 16.12(a(i)), and clearly reflects the controls on the distribution of opaline silica that were discussed in Section 15.2.4.2. Thus, the greatest concentrations of biogenic silica are found in the high-latitude rings around Antarctica and in the northern Pacific, with enhanced concentrations in sediments under the equatorial high primary productivity zone also evident. The same general pattern also can be identified in the opaline silica accumulation rates (Fig. 16.12a(ii)), which demonstrates how the strength of the signal varies on an ocean-wide scale. It was shown in Section 15.2.3 that there is a fundamental oceanic dichotomy in which diatoms are favoured in highly productive regions and coccoliths in poorly productive regions, and that as a result siliceous deposits are found mainly at the edges of the oceans and calcareous deposits are concentrated in relatively shallow open-ocean regions. This dichotomy is evidenced when the distribution of opaline silica (Fig. 15.2b) is compared with that of calcium carbonate (Fig. 15.2a) in the Indo-Pacific deep-sea sediments. Thus, whereas opaline silica shows a pronounced 'ring' distribution around the

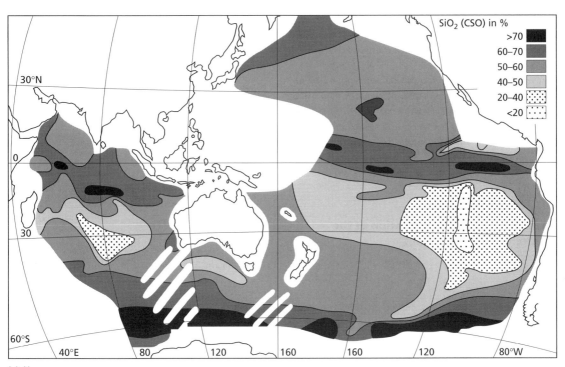

(a) (i)

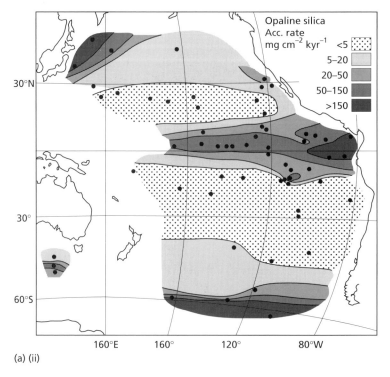

(a) (ii)

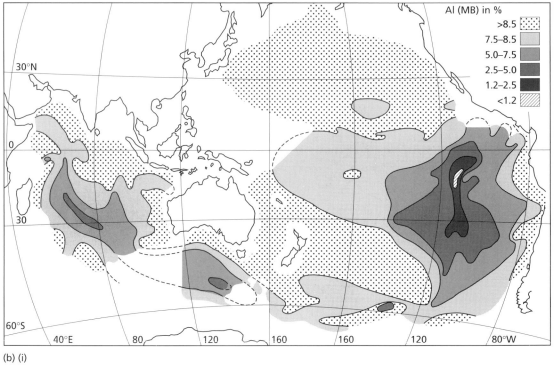

(b) (i)

**Fig. 16.12** *Continued*

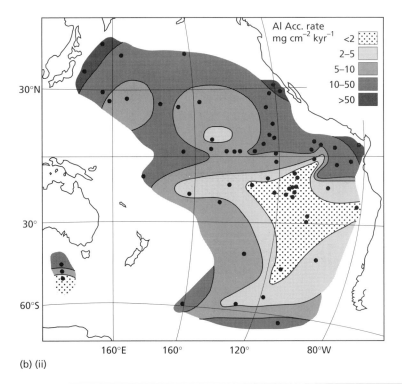

(b) (ii)

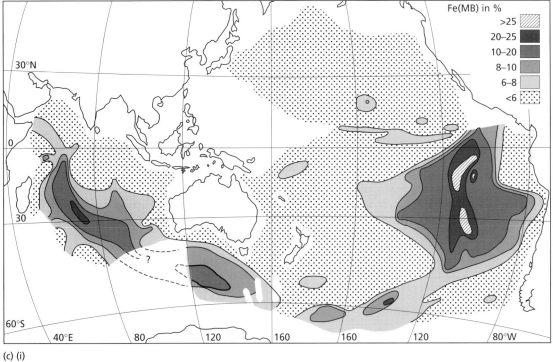

(c) (i)

**Fig. 16.12** *Continued*

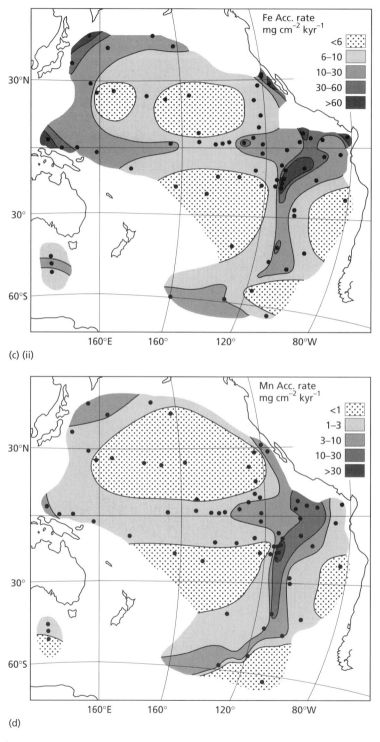

(c) (ii)

(d)

**Fig. 16.12** *Continued*

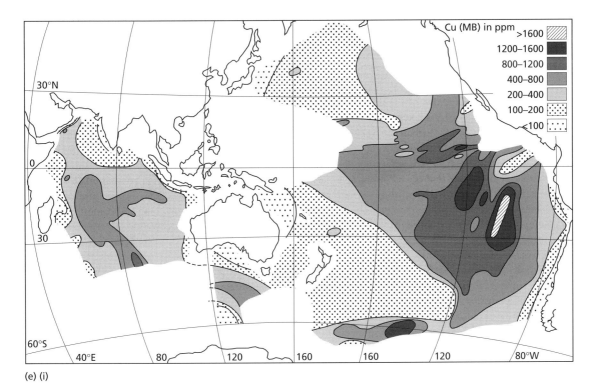

(e) (i)

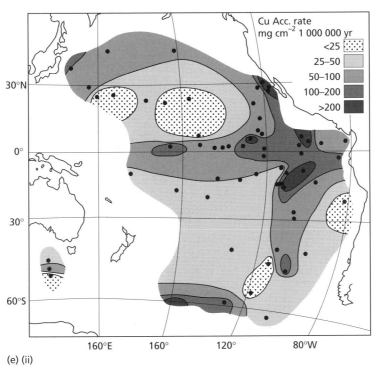

(e) (ii)

**Fig. 16.12** *Continued*

ocean margins, calcium carbonate reaches its highest concentrations on mid-ocean topographic highs.

### 16.7.2 Non-biogenous elements

Bostrom *et al.* (1973) expressed the concentrations of non-biogenous elements on a **minerogen** basis. This procedure, in which the non-biogenic component is expressed in terms of a series of oxides, is more rigorous than those that simply express the data on a carbonate-free basis, because it compensates for the diluting effects of both carbonate and opaline silica shells. In relation to the primary chemical signals that contribute to the total concentration of the non-biogenic elements, therefore

$$C_t = C_d + C_a + C_h \qquad (16.11)$$

where $C_t$ is the total content of an element in a sediment, and $C_d$, $C_a$ and $C_h$ are the detrital, authigenic and hydrothermal contributions, respectively.

The factors controlling the distributions of a number of non-biogenic elements in the Indo-Pacific deep-sea sediments are discussed below in terms of this general equation.

*Aluminium.* The Al in deep-sea sediments is almost, but not exclusively, associated with detrital material in which it is held in lattice positions within aluminosilicates; for example >80% of the total $Al_2O_3$ in deep-sea sediments is partitioned into the detrital phase (see Section 16.5). Thus, the distribution of the element is controlled mainly by the detrital signal, and the data provided by Bostrom *et al.* (1973) allow us to evaluate the manner in which this signal operates on a quasi-global scale. The distribution of total Al ($\Sigma$Al) in the Indo-Pacific is illustrated in Fig. 16.2(b(i)), from which it can be seen that the highest concentrations are found in sediments close to the continents and the lowest around the spreading ridge system. The strength of the detrital signal is perhaps better seen in the accumulation rate patterns for $\Sigma$Al in the Pacific sediments (Fig. 16.12b(ii)), and for this it is useful to treat the North and South Pacific separately. In the North Pacific, $\Sigma$Al accumulates at its highest rates around the continental margins, and decreases in a semi-regular manner towards the remote open-ocean areas. As a result, the strength of the detrital signal decreases in the order marginal sediments > hemi-pelagic sediments > normal open-

ocean pelagic sediments. Bostrom *et al.* (1973) showed that over most of the North Pacific the detrital material in the sediments has a continental source, and the $\Sigma$Al accumulation rate pattern has clearly been established in response to the distance of the depositional environments from the surrounding source areas. Oceanic crust and active ridge sources complicate the situation in the South Pacific, and an area of relatively low $\Sigma$Al accumulation is found around the EPR.

*Iron.* The distribution and accumulation rates of $\Sigma$Fe in the Indo-Pacific sediments are illustrated in Fig. 16.12(c(i) and (ii) respectively). Like $\Sigma$Al, much of the $\Sigma$Fe in normal deep-sea sediments is detrital in character (see Section 16.5), and this is apparent in the accumulation rates of the element in the North Pacific sediments, in which the patterns are very similar to those of $\Sigma$Al, i.e. high values adjacent to the continents, and a general decrease out towards open-ocean areas. It is apparent from Fig. 16.12(c(i), (ii)) that the open-ocean North Pacific is outside the major influence of the hydrothermal supply of Fe from the EPR, and here therefore

$$C_t(\text{Fe}) \cong C_d(\text{Fe}) + C_d(\text{Fe}). \qquad (16.12)$$

On the basis of the data given by Bostrom *et al.* (1973) (see Fig. 16.12c(ii)) it is apparent that in the open-ocean North Pacific, $\Sigma$Fe ($C_t$) is accumulating at $\sim$<6000–10 000 $\mu$g cm$^{-2}$ 10$^3$ yr$^{-1}$. According to Krishnaswami (1976), $\text{Fe}_a$ accumulates at $\sim$800 $\mu$g cm$^{-2}$ 10$^3$ yr$^{-1}$ in normal pelagic clays and so cannot account for the accumulation of the $\Sigma$Fe. Clearly, therefore, the detrital signal dominates the distribution of $\Sigma$Fe in the open-ocean North Pacific sediments. This is consistent with the conclusions reached by Chester & Hughes (1969) that $\cong$90% of the $\Sigma$Fe in a North Pacific deep-sea clay was detrital in origin. Further, in the North Pacific, both the distribution and accumulation rates of Fe are similar to those of the detritally dominated $\Sigma$Al. In the South Pacific, however, the situation is very different, and here the patterns for $\Sigma$Fe are almost a mirror image of those for the detrital element Al. Two features are apparent in the $\Sigma$Fe accumulation patterns in this region.

1 The dominant feature is the very high $\Sigma$Fe accumulation rate around the EPR, where it can reach values of >60 000 $\mu$g cm$^{-2}$ 10$^3$ yr$^{-1}$, which agree well with

those given in Section 16.6.2 for hydrothermal deposition.

2 A less well-developed feature is a band in the equatorial Pacific in which the $\Sigma$Fe accumulates at intermediate rates.

It would appear, therefore, that in the North Pacific the detrital signal is mainly responsible for the deposition of $\Sigma$Fe, whereas in the South Pacific there is a very pronounced hydrothermal spike that dominates the accumulation of $\Sigma$Fe.

*Manganese.* The accumulation rates of $\Sigma$Mn in the Indo-Pacific sediments are illustrated in Fig. 16.12(d). Manganese differs from both Al and Fe in that in deep-sea sediments an average of ~70–80% of the $\Sigma$Mn is non-detrital in origin (see Section 16.5). This non-detrital character is seen in the accumulation rate pattern for $\Sigma$Mn in the North Pacific, a region that is not strongly affected by hydrothermal inputs. Here, therefore

$$C_t(\text{Mn}) \cong C_d(\text{Mn}) + C_a(\text{Mn}). \qquad (16.13)$$

Although $\Sigma$Mn does accumulate faster closest to the continents, the effect is much less apparent than that for either $\Sigma$Al or $\Sigma$Fe, and over a very large area of the open-ocean North Pacific $\Sigma$Mn accumulates at rates $<1000\,\mu\text{g}\,\text{cm}^{-2}\,10^3\,\text{yr}^{-1}$. According to Krishnaswami (1976) $\text{Mn}_a$ accumulates in normal Pacific clays at a rate of ~$500\,\mu\text{g}\,\text{cm}^{-2}\,10^3\,\text{yr}^{-1}$, which begins to approach that of the $\Sigma$Mn accumulation rate in the sediments. In the North Pacific, therefore, the authigenic signal can supply a large fraction of the $\Sigma$Mn accumulating in the sediments, thus highlighting the difference between this element and Fe, for which the authigenic signal supplies a very much smaller fraction of the $\Sigma$Fe. In the South Pacific, however, the accumulation rate patterns of the two elements are generally similar, each showing a very high accumulation-rate patch around the EPR and a lesser one in the vicinity of the Equator. There is, however, a difference in the degree to which the hydrothermal spike influences the accumulation rates of $\Sigma$Mn and $\Sigma$Fe. Around the EPR, $\Sigma$Mn accumulates at rates $>30\,000\,\mu\text{g}\,\text{cm}^{-2}\,10^3\,\text{yr}^{-1}$, which is more than three times faster than over any other part of the entire Pacific; in contrast, the near-continental detrital accumulation rates for $\Sigma$Fe are of a similar magnitude to the hydrothermal accumulation rates. For $\Sigma$Mn, therefore, the detrital signal plays only a rela-

tively minor role in the accumulation of the element. Thus, over much of the North Pacific the authigenic signal dominates the accumulation of $\Sigma$Mn. In contrast, in the eastern South Pacific the hydrothermal signal is predominant, and results in accumulation rates that are at least three times greater than those found anywhere else in the ocean.

*Copper.* The distribution and accumulation rates of $\Sigma$Cu in the Indo-Pacific deep-sea sediments are illustrated in Fig. 16.12(e)(i) and (ii) respectively). Around 50% of the $\Sigma$Cu in Pacific pelagic clays appears to be non-detrital in origin (see e.g. Krishnaswami, 1976); i.e. the element has a partitioning signature that is intermediate between those of $\Sigma$Fe and $\Sigma$Mn. A number of features can be seen in the accumulation rate pattern of $\Sigma$Cu in Pacific deep-sea sediments.

1 There is a tendency for $\Sigma$Cu to accumulate faster around the continents, which presumably reflects the near-source strength of the detrital Cu signal.

2 In the North Pacific there is a general decrease in the accumulation rates of $\Sigma$Cu towards mid-ocean areas as the influence of the detrital signal decreases. Here, $\Sigma$Cu accumulates at rates of $<25\,\mu\text{g}\,\text{cm}^{-2}\,10^3\,\text{yr}^{-1}$. From the data provided by Krishnaswami (1976), authigenic Cu accumulates in normal Pacific pelagic clays at ~$8\,\mu\text{g}\,\text{cm}^{-2}\,10^3\,\text{yr}^{-1}$, so that on this basis around one-third of the $\Sigma$Cu arises from the authigenic signal; however, this is a minimum estimate, and the figure may well reach the 50% authigenic partitioning reported by Krishnaswami (1976).

3 In the South Pacific, the accumulation of $\Sigma$Cu, like those of $\Sigma$Mn and $\Sigma$Fe, is dominated by the hydrothermal signal, and around the EPR $\Sigma$Cu accumulates at rates of $>200\,\mu\text{g}\,\text{cm}^{-2}\,10^3\,\text{yr}^{-1}$, which is similar to the rate of accumulation associated with the detrital input from the USA mainland in the North Pacific.

4 There is a band of relatively high $\Sigma$Cu accumulation around the Equator, which is considerably more well developed than those for either $\Sigma$Mn or $\Sigma$Fe. This band corresponds to a zone of high primary productivity in the surface waters, and the enhanced accumulation of $\Sigma$Cu probably results from an accelerated down-column authigenic flux from the carbon-rich overlying waters (see Figs 9.6 & 16.6b). Bostrom *et al.* (1973) reported that the distribution and accumulation rate patterns of $\Sigma$Ni and $\Sigma$Co in the Indo-Pacific generally were similar to those for $\Sigma$Cu.

It may be concluded, therefore, that the distributions and accumulation rate patterns of elements in surface sediments from the Indo-Pacific can be used to illustrate how sediment-forming signals interact on a global scale. The main features in the distribution and accumulation rates of opaline silica in the sediments can be related to the biogenic signal, and those for $\Sigma Al$ can be explained in terms of the predominance of the detrital signal. Patterns in the distributions and accumulation rates of $\Sigma Fe$, $\Sigma Mn$ and $\Sigma Cu$ can be interpreted mainly in terms of the global interaction of the detrital, authigenic and hydrothermal signals. It must be remembered, however, that within the sediment column the suboxic diagenetic signal can redistribute redox-sensitive elements such as manganese, the overall effect decreasing in the order: shelf > hemi-pelagic > pelagic sediments (see Section 14.1.3).

## 16.8 Unscrambling the sediment-forming chemical signals: summary

**1** The chemical composition of marine sediments can be interpreted within a framework in which the elements are envisaged as being transmitted to the deposits via a series of signals associated with a variety of geochemical processes. The components making up the sediments are formed by elements that are transmitted either by individual signals or, more usually, by signal couplings.

**2** Detrital, biogenous and authigenic signals operate on an ocean-wide scale. These are the background signals, which carry the elements having a direct seawater source.

**3** A series of perturbation spikes from more localized hydrothermal and contaminant seawater signals, and from interstitial-water diagenetic signals, are superimposed on the background signals.

**4** Sediments from different oceanic environments record varying signal strengths. For example:

    (a) authigenic signals have their greatest influence on pelagic sediments deposited in regions away from the spreading ridges;

    (b) diagenetic signals are strongest in hemi-pelagic sediments, in which suboxic diagenesis takes place;

    (c) hydrothermal signals can dominate in areas in the vicinity of spreading ridges, where metal-rich sediments can be formed;

    (d) effects of contaminant signals are confined mainly to coastal deposits, but are starting to appear in deep-sea sediments.

## References

Aplin, A.C. & Cronan, D.S. (1985) Ferromanganese oxide deposits in the central Pacific Ocean, II. Nodules and associated elements. *Geochim. Cosmochim. Acta*, **49**, 437–51.

Bacon, M.P. & Rosholt, J.N. (1982) Accumulation rates of Th-230, Pa-231, and some transition metals on the Bermuda Rise. *Geochim. Cosmochim. Acta*, **46**, 651–66.

Bender, M.L. (1971) Does upward diffusion supply the excess in manganese in sediments? *J. Geophys. Res.*, **76**, 4212–15.

Bender, M.L., Broecker, W., Gornitz, V., *et al.* (1971) Geochemistry of three cores from the East Pacific Rise. *Earth Planet. Sci. Lett.*, **12**, 425–33.

Bostrom, K. & Fisher, D.E. (1971) Volcanogenic uranium, vanadium and iron in Indian Ocean sediments. *Earth Planet. Sci. Lett.*, **11**, 95–8.

Bostrom, K. & Peterson, M.N.A. (1969) The origin of aluminium-poor ferromanganoan sediments in areas of high heat flow on the East Pacific Rise. *Mar. Geol.*, 7, 427–47.

Bostrom, K., Peterson, M.N.A., Joensuu, O. & Fisher, D.E. (1969) Aluminium-poor ferromanganoan sediments on active oceanic ridges. *J. Geophys. Res.*, **74**, 3261–70.

Bostrom, K., Kraemer, T. & Gartner, S. (1973) Provenance and accumulation rates of opaline silica, Al, Ti, Fe, Mn, Cu, Ni and Co in Pacific pelagic sediments. *Chem. Geol.*, **11**, 123–48.

Broecker, W.S. (1974) *Chemical Oceanography*. New York: Harcourt Brace Jovanovich.

Chester, R. & Aston, S.R. (1976) The geochemistry of deep-sea sediments. In *Chemical Oceanography*, J.P. Riley & R. Chester (eds), Vol. 6, 281–390. London: Academic Press.

Chester, R. & Hughes, M.J. (1967) A chemical technique for the separation of ferrogmanganese minerals, carbonate minerals and adsorbed trace elements from pelagic sediments. *Chem. Geol.*, 3, 199–212.

Chester, R. & Hughes, M.J. (1969) The trace element geochemistry of a North Pacific pelagic clay core. *Deep Sea Res.*, **13**, 627–34.

Chester, R. & Messiha-Hanna, R.G. (1970) Trace element partition patterns in North Atlantic deep-sea sediments. *Geochim. Cosmochim. Acta*, **34**, 1121–8.

Chester, R. & Voutsinou, F.G. (1981) The initial assessment of trace metal pollution in coastal sediments. *Mar. Pollut. Bull.*, **12**, 84–91.

Chester, R., Thomas, A., Lin, F.J., Basaham, A.S. & Jacinto, G. (1988) The solid state speciation of copper in surface water particulates and oceanic sediments. *Mar. Chem.*, **24**, 261–92.

Chow, T.J., Bruland, K.W., Bertine, K., Soutar, A., Koide, M. & Goldberg, E.D. (1973) Lead pollution: records in Southern California coastal sediments. *Science*, **181**, 551–2.

Collier, R. & Edmond, J.M. (1984) The trace element geochemistry of marine biogenic particulate matter. *Prog. Oceanogr.*, **13**, 113–99.

Cronan, D.S. (1972) The Mid-Atlantic Ridge near 45°N, XVII: Al, As, Hg and Mn in ferruginous sediments from the median valley. *Can. J. Earth Sci.*, **9**, 319–23.

Cronan, D.S. (1973) Basal ferruginous sediments cored during Leg 16, Deep Sea Drilling Project. In *Initial Reports of the Deep-Sea Drilling Project*, Vol. XVI, 601–4. Washington, DC: US Government Printing Office.

Cronan, D.S. (1976) Basal metalliferous sediments from the eastern Pacific. *Geol. Soc. Am. Bull.*, **87**, 929–34.

Cronan, D.S. & Garrett, D.E. (1973) Distribution of elements in metalliferous Pacific sediments collected during the Deep Sea Drilling Project. *Nature*, **242**, 88–9.

Cronan, D.S., van Andel, Tj. H., Heath, G.R., *et al.* (1972) Iron-rich basal sediments from the eastern equatorial Pacific: Leg 16, Deep Sea Drilling Project. *Science*, **175**, 61–3.

Dasch, E.J., Dymond, J. & Heath, G.R. (1971) Isotopic analysis of metalliferous sediments from the East Pacific Rise. *Earth Planet. Sci. Lett.*, **13**, 175–80.

Davies, T.A. & Gorsline, D.S. (1976) Oceanic sediments and sedimentary processes. In *Chemical Oceanography*, J.P. Riley & R. Chester (eds), Vol. 5, 1–80. London: Academic Press.

Dymond, J. & Veeh, H.H. (1975) Metal accumulation rates in the southeast Pacific and the origin of metalliferous sediments. *Earth Planet. Sci. Lett.*, **28**, 13–22.

Dymond, J., Corless, J.B., Heath, G.R., Field, C.W., Dasch, E.J. & Veeh, H.H. (1973) Origin of metalliferous sediments from the Pacific Ocean. *Geol. Soc. Am. Bull.*, **84**, 3355–72.

Elderfield, H. (1976) Manganese fluxes to the oceans. *Mar. Chem.*, **4**, 103–32.

El Wakeel, S.K. & Riley, J.P. (1961) Chemical and mineralogical studies of deep-sea sediments. *Geochim. Cosmochim. Acta*, **25**, 110–46.

Grousset, F.E. & Chesselet, R. (1986) The Holocene sedimentary regime in the northern Mid-Atlantic Ridge region. *Earth Planet. Sci. Lett.*, **78**, 271–87.

Heath, G.R. & Dymond, J. (1977) Genesis and transformation of metalliferous sediments from the East Pacific Rise, Bauer Deep and Central Basin, northwest Nazca Plate. *Geol. Soc. Am. Bull.*, **88**, 723–33.

Horder, M.F. & Cronan, D.S. (1981) The geochemistry of some basal sediments from the western Indian Ocean. *Oceanol. Acta*, **4**, 213–21.

Horowitz, A. (1974) The geochemistry of sediments from the northern Reykjanes Ridge and the Iceland–Faroes Ridge. *Mar. Geol.*, **17**, 103–22.

Krishnaswami, S. (1976) Authigenic transition elements in Pacific pelagic clays. *Geochim. Cosmochim. Acta*, **40**, 425–34.

Lalou, C. (1983) Genesis of ferromanganese deposits: hydrothermal origin. In *Hydrothermal Processes at Seafloor Spreading Centres*, P.A. Rona, K. Bostrom, L. Laubier & K.L. Smith (eds), 503–34. New York: Plenum.

Lapicque, G., Livingston, H.D., Lambert, C.E., Bard, E. & Labeyrie, L.D. (1987) Interpretation of $^{239,240}$Pu in Atlantic sediments with a non-steady state input model. *Deep Sea Res.*, **34**, 1841–50.

Marchig, V. & Grundlach, H. (1982) Iron-rich metalliferous sediments on the East Pacific Rise: prototype of undifferentiated metalliferous sediments on divergent plate boundaries. *Earth Planet. Sci. Lett.*, **58**, 361–82.

Martin, J.M. & Whitfield, M. (1983) The significance of the river input of chemical elements to the oceans. In *Trace Metals in Sea Water*, C.S. Wong, E. Boyle, K.W. Bruland, J.D. Burton & E.D. Goldberg (eds), 256–96. New York: Plenum.

Murray, J. & Renard, A.F. (1891) *Deep-sea Deposits*. London: Scientific Report, *Challenger* Expedition, no. 3 HMSO, London.

Patterson, C. (1987) Global pollution measured by lead in mid-ocean sediments. *Nature*, **326**, 244.

Piper, D.Z. (1973) Origin of metalliferous sediments from the East Pacific Rise. *Earth Planet. Sci. Lett.*, **19**, 75–82.

Preston, M.R. (1989) Marine pollution. In *Chemical Oceanography*, J.P. Riley (ed.), Vol. 9, 53–196. London: Academic Press.

Revelle, R.R. (1944) Scientific results of the cruise VII of the 'Carnegie'. *Publ. Carnegie Inst.*, **556**, 1–180.

Sawlan, J.J. & Murray, J.W. (1983) Trace metal remobilization in the interstitial waters of red clay and hemipelagic marine sediments. *Earth Planet. Sci. Lett.*, **64**, 213–30.

Sayles, F.L. & Bischoff, J.L. (1973) Ferromanganoan sediments in the equatorial east Pacific. *Earth Planet. Sci. Lett.*, **19**, 330–6.

Sayles, F.L., Ku, T.-L. & Bowker, P.C. (1975) Chemistry of ferromanganoan sediments of the Bauer deep. *Geol. Soc. Am. Bull.*, **86**, 1423–31.

Thomas, A.R. (1987) *Glacial–interglacial variations in the geochemistry of North Atlantic deep-sea deposits*. PhD Thesis, University of Liverpool.

Thompson, J., Carpenter, M.S.N., Colley, S., Wilson, T.R.S., Elderfield, H. & Kennedy, H. (1984) Metal accumulation in northwest Atlantic pelagic sediments. *Geochim. Cosmochim. Acta*, **48**, 1935–48.

Turekian, K.K. (1967) Estimates of the average Pacific deep-sea clay accumulation rate from material balance considerations. *Prog. Oceanogr.*, **4**, 226–44.

Turekian, K.K. & Imbrie, J. (1966) The distribution of trace elements in deep-sea sediments of the Atlantic Ocean. *Earth Planet. Sci. Lett.*, **1**, 161–8.

Veron, A., Lambert, C.E., Isley, A., Linet, P. & Grousset, F.

(1987) Evidence of recent lead pollution in deep-north east Atlantic sediments. *Nature*, **326**, 278–81.

Von Damm, K.L., Edmond, J.M., Grant, B., Measures, C.J., Walden, B. & Weiss, R.F. (1985) Chemistry of submarine hydrothermal solutions at 21°N, East Pacific Rise. *Geochim. Cosmochim. Acta*, **49**, 2197–220.

Von der Borch, C.C. & Rex, R.W. (1970) Amorphous iron oxide precipitates in sediments cored during Leg 5, Deep Sea Drilling Project. In *Initial Reports of the Deep Sea Drilling Project*, Vol. 5, 541–4. Washington, DC: US Government Printing Office.

Wedepohl, K.H. (1960) Spureanalytische Untersuchungen an Tiefseetonen aus dem Atlantik. *Geochim. Cosmochim. Acta*, **18**, 200–31.

Wedepohl, K.H. (1968) Chemical fractionation in the sedimentary environment. In *Origin Distribution of the Elements*, L.H. Ahrens (ed.), 999–1016. Oxford: Pergamon.

# Part IV
# The Global Journey: Synthesis

# 17 Marine geochemistry: an overview

The 'global journey', which traced material from its sources, through the ocean reservoir, and to the sediment sink, has now been completed. In the present chapter the various strands will be brought together in an overview of the present state of the art in marine geochemistry.

## 17.1 How the system works

The question that was posed in the Introduction was 'How do the oceans work as a chemical system?' The route that was selected in an attempt to answer this question involved:

1 identifying the pathways followed by the material that entered the ocean reservoir from both external and internal sources, and quantifying the magnitudes of the fluxes associated with them;

2 describing the physical, biological and chemical processes that occur within the water column, and relating them to the fluxes that carry material to the sediment (and rock) sink; and

3 outlining the various processes that interact to control the composition of the sediments themselves. There are a wide variety of inorganic and organic components in sea water, but it was pointed out in the Introduction that in order to follow the route outlined above particular attention would be paid to a selected number of process-orientated trace elements, which could be used as tracers to establish how the ocean works as a chemical system. In order to summarize these processes, attention will again largely be focused on these process-orientated trace elements.

Essentially, the ocean system consists of two layers of water: a thin, warm, less dense surface layer, which caps a more dense, thick, cold, deep-water layer. The two layers are separated by the thermocline and pycnocline mixing barriers. The primary global-scale sources of material to the oceans are river run-off, atmospheric deposition and hydrothermal exhala-

tions, all of which supply both dissolved and particulate material to the ocean system. The system therefore is dominated by the large fluxes of material that enter it, and it has become apparent that the key to understanding the driving force behind many of the processes that operate in the water column lies in the particulate ↔ dissolved interactions which take place during the throughput of material from its sources to the sediment sink. During this source → sink journey the dissolved constituents encounter regions of relatively high particle concentrations, e.g. in estuaries (especially those having a turbidity maximum) and river plumes, in the sea-surface microlayer, under conditions of high primary production, in the regions of hydrothermal venting systems and in nepheloid layers generated by boundary currents at the western edges of the ocean basins. All these high-particle regions became zones of enhanced dissolved ↔ particulate reactivity. In addition, there is a background microcosm of particles dispersed throughout all the water column. The overall effect is that the ocean may be considered to be a particle-dominated system, and the composition of the seawater phase is controlled to a large extent by the 'great particle conspiracy'. It is this conspiracy that is the key to Forchhammer's 'facility with which the elements in sea water are made insoluble'.

Once they have reached the seawater system the dissolved and particulate elements are subjected to a complex series of transport–removal processes. Non-reactive elements will tend to behave in a generally conservative manner. Their distributions will be controlled by physical processes, such as water mass mixing, and their residence times in the ocean will be relatively long. As the degree of reactivity of an element increases it becomes progressively influenced by biogeochemical processes, and begins to behave in a non-conservative manner. The degree of reactivity exhibited by an element in sea water therefore exerts a basic control on how it moves through the ocean system. River run-off and atmospheric deposition both deliver dissolved and particulate material to the surface ocean via exchange across the estuarine–sea and air–sea interfaces. This surface ocean is a zone of relatively high particle concentration, the externally delivered particles being swamped in most areas by internally produced biological particles. Large-sized, biologically produced aggregates, which consist of a significant fraction of faecal material, sink from the surface layer. At relatively shallow depths a large fraction of the aggregates, usually >90%, undergoes destruction. However, it is the fraction that escapes destruction and falls to the sea bed that drives the principal transport of material to the sediment surface via the **global carbon flux**. The total oceanic particle population therefore consists of a wide particle-size spectrum. However, it is convenient to divide it into fine particulate matter (FPM), i.e. the suspended population, and coarse particulate matter (CPM), i.e. the sinking population. The production coupling between the populations involves a continuous series of aggregation–disaggregation processes, the end-point of which is to produce a fine particle population, and the down-column transport coupling between the two populations involves a piggy-back type of reversible FPM association with the large, fast-sinking CPM aggregates. Both the FPM and CPM particulate populations take part in a series of complex biologically and chemically mediated reactions, both in the surface waters and at depth in the water column. These particle-driven reactions are the principal control on the chemical composition of sea water, which is regulated by a balance between the rate of addition of a dissolved component and its rate of removal via sinking particulate material to the sediment sink; however, it must be remembered that uptake into the rock sink is important in the marine budgets of some elements.

Many trace elements in sea water have an oceanic residence time that is relatively short compared with both the rate at which they have been added to sea water over geological time and the holding time of the oceans. The controls on the short residence times of these dissolved elements are imposed by uptake reactions with the oceanic particle population, and a number of process-orientated trace elements were used in the text as examples of reactive oceanic components in order to illustrate how this 'great particle conspiracy' operates. Although it is difficult to distinguish between biologically dominated and inorganically dominated controls on dissolved trace elements in the water column, two general particle-association removal routes were distinguished. Both of these routes are ultimately driven by the global carbon flux, either directly, with carbon-associated carriers (**nutrient-type** trace metal removal reactions), or indirectly, via a coupling between the small-sized inorganic (FPM) and the large-sized carbon particles (CPM) (**scavenging-type** removal reactions). Although some trace metals do in fact exhibit mixed removal processes, the distinction does underline a fundamental dichotomy in the oceanic behaviour of a number of trace metals.

1 Scavenging-type trace metals, which show a surface-enrichment–depth-depletion profile, undergo reactions with the fine (FPM) fraction of oceanic TSM, and although these reactions often involve reversible equilibria, in which there is continuous exchange between the dissolved and particulate states, their residence times in the oceans tend to be relatively short, i.e. of the order of a few hundreds of years. These trace metals reach the sediment surface in association with their inorganic host particles.

2 Nutrient-type trace metals become involved in the major oceanic biogeochemical cycles and undergo a surface-depletion–subsurface-regeneration enrichment at depth in the water column. This recycling results in the nutrient-type elements having residence times that are relatively long compared with those of the scavenging-type elements and with the oceanic mixing time. For example, Bruland (1980) set a lower limit residence time for the nutrient-type elements of c. 5000 yr. The nutrient elements are carried to the sediment reservoir in association with the carbon fraction of the large-sized (CPM) oceanic TSM population. The fate of these elements depends on the diagenetic environment in the upper part of the sediment column.

(a) Under oxic conditions the carriers are rapidly destroyed to release their associated elements at the sediment surface.

(b) Under suboxic conditions, however, the carriers can be buried to release their associated elements at depth into the interstitial waters.

In addition to particle reactivity, water transport acts as a control on the distributions of dissolved elements in the ocean system. This has different effects on the nutrient-type and the scavenging-type elements. The trace-element particulate carriers sink to deep-water where the nutrient-type elements are released back into solution, and this imposes a fundamental control on their oceanic distributions. The oceanic deep-water 'grand tour' transports water from the main deep-water sources in the Atlantic through the Indian Ocean and finally to the North Pacific. As a result, components such as the nutrients and the nutrient-type trace elements, which have a deep-water recycling stage, build up in concentration in the deep-waters of the Pacific relative to those of the Atlantic. In contrast, the scavenging-type elements will become enriched in the deep Atlantic relative to the deep Pacific. Thus, the differences between the scavenging-type and nutrient-type elements are reflected not only in the manner in which they are taken out of solution but also in their water-column residence times. Elements are introduced into the oceans from both continental and oceanic crustal sources, but for most elements the continental source dominates. The **geochemical abundances** of the elements in the continental crust source material will therefore exert a fundamental control on their oceanic distributions. However, because they are removed relatively rapidly from the water column the scavenging-type elements tend to be fractionated from the nutrient-type elements. In this way, therefore, the concentration of a specific dissolved element in sea water will depend in part on its geochemical abundance in the crust, and in part on the efficiency with which it is removed to the solid phase, i.e. its **oceanic reactivity**.

The down-column transport of components from the surface ocean is dominated by the carbon flux, which varies in relation to the extent of primary production in the surface waters; thus, the strength of the flux is greatest under the regions of intense productivity, which are located mainly at the edges of the ocean basins. However, it must be remembered that a series of lateral mid-water and bottom fluxes can also transport material in the oceans. These are coupled to the vertical flux, so that the throughput of material to the sea bed involves a coupling between the vertical flux and a series of these lateral fluxes. It is the removal of dissolved components via the particulates transported by this flux combination that controls the elemental composition of sea water by delivering

material across the benthic boundary layer to the sediment sink and so taking it out of the water column.

The composition of oceanic sediments is therefore determined by the nature of the components transported by the various down-column and lateral fluxes, and by the relative strengths of the individual fluxes themselves. The overall result of these various factors is to set up a sediment regime in which the following general trends can be identified.

**1** During descent down the water column a large fraction of the organic matter from the surface layer is destroyed in the upper waters, but that which reaches the sediment surface is related to production so that the most organic-rich deposits are found fringing the continents under the areas of intense production.

**2** The major sediment-forming biogenic components are opaline silica and carbonate shell material. There is a fundamental biogeographical dichotomy in the distribution of planktonic populations that is evidenced in a coastal high-fertility regime, which is characterized by silica-secreting diatoms, and an oceanic low-fertility regime, which is characterized by carbonate-secreting coccoliths. The extent to which sea water is undersaturated with respect to calcium carbonate increases with depth so that away from the coastal regime its preservation is restricted mainly to mid-ocean topographic highs located above the carbonate compensation depth. In contrast, sea water is everywhere undersaturated with respect to silica and siliceous shells can accumulate only in those regions where the supply exceeds the rate of dissolution, i.e. under areas of high primary production at the ocean margins. The overall result of these dissolution processes is that siliceous deposits are generally found at the edges of the oceans, whereas carbonates are concentrated in mid-ocean areas. Thus, dissolution constraints on the preservation of shell material lead to an enhancement of the biogeographical open-ocean (carbonate)–coastal-ocean (silica) dichotomy.

**3** In open-ocean areas below the carbonate compensation depth the sediments are dominated by inorganic components and deep-sea clays are formed, which may be either lithogenous or hydrogenous in character.

A series of chemical signals is transmitted through the ocean system, and these combine together to control the composition of the bottom sediments. The background transport of material to sediments operates on a global-ocean scale, and the signals involved can be subdivided into biogenous, detrital and authigenic types. The authigenic signal is the primary background signal, which carries elements derived from solution in sea water to the sediment surface. For elements that have relatively short residence times in sea water, the authigenic signal does not operate on a constant-flux basis. Dissolved elements that have a longer residence time, however, can have their concentrations smoothed out between the major oceans, and their deposition may approach a constant-flux model. As a result, because they are removed relatively rapidly from the water column the scavenging-type elements tend to be fractionated from the nutrient-type elements in the sediments as well as in the water column (see above). Thus, Mn (a scavenging-type element) has an oceanic reactivity pattern imposed on its removal into sediments, with the result that the magnitude of its authigenic flux appears to be related directly to that of its input flux; for example, authigenic Mn is accumulating faster in the Atlantic, which has stronger fluvial and atmospheric fluxes, than in the Pacific. In contrast, less reactive elements, such as the nutrient-type Ni and the mixed-type Cu, generally have similar authigenic fluxes in the two oceans, i.e. the *geochemical abundance* (Mn) versus *oceanic reactivity* (Ni, Cu) control. In addition to the background signals, more localized signals, such as those associated with hydrothermal activity and anthropogenic effects, can transport components to the sediment surface from the overlying sea water.

Even when the components are actually incorporated into the sediments they have not simply been locked away in a static reservoir, but have in fact entered a diagenetically active and biogeochemically dynamic environment. The diagenetic reactions are controlled by the manner in which the sedimentary environment attempts to destroy organic carbon; i.e. even the small fraction of organic matter that survives the journey down the water column is subjected to further degradation in the sediment reservoir. The intensity of diagenesis is controlled by the amount of organic carbon that reaches the sediment surface, and there is a redox-driven diagenetic sequence in the sediments, in which a variety of oxidants are switched on as the previous one is exhausted, in the general order: oxygen > nitrate ≃ manganese oxides > iron oxides > sulphate. The extent to which this sequence progresses depends on the amount of organic matter reaching the sediment surface, and as this is related to the extent or primary production in the surface waters (supply rate) and the rate at which the sediments accumulate (burial rate), it results in a lateral diagenetic sequence, in which the diagenetic intensity decreases in the order: nearshore > hemi-pelagic > pelagic sediments. Components released in the diagenetic reactions are transported through the sediment interstitial waters. Some of these components are trapped in the solid phases and so can impose the diagenetic spike on the sediments. Others can escape back into sea water, however, thus providing a secondary, i.e. recycled, oceanic source. Thus, rather than acting as a static sink, sediments can recycle some elements, either (i) retaining but redistributing them in the sediment column; or (ii) losing them back to sea water. Despite these recycling processes, the sediment reservoir remains the ultimate sink for particulate material that flows through the ocean system, and so also acts as the major sink for particle-reactive elements that are removed from the dissolved phase.

It is apparent, therefore, that the major process that controls the dissolved-element composition of sea water is a balance between the rates at which the elements are added to the system and the rates at which they are removed by the throughput of particulate material that delivers them to the sediment sink. During their residence time in sea water, dissolved elements are transported by physical circulation, and undergo a series of biogeochemical reactions by which ultimately they are taken up by particulate matter, which also is transported by vertical and horizontal movements. Overall, therefore, it may be concluded that the oceanic chemical system is driven by a physical–chemical–biological **process trinity**. This process trinity operates on both the particulate and dissolved material introduced into the ocean reservoir and controls their passage through the system to the sediment sink. It also continues to influence the fate of the elements within the sediment sink itself; e.g. physical processes resuspend sediments into the water column, and biogeochemical processes are active in diagenesis.

## 17.2 Balancing the books

A **process-orientated** approach to marine geochemistry has been adopted in an attempt to elucidate the oceanic cycles of various constituents, and one possible way of answering the question 'To what extent have we understood how the oceans work as a chemical system?' is to establish accounting procedures that can be used to assess the quantitative aspect of the cycles, e.g. by attempting to construct mass balances for the system. A variety of physical and biogeochemical processes control both the removal of a dissolved element from sea water and its transfer to the deposition sink(s). Although these processes work differently for different elements, if the oceans are assumed to be in a steady state, then there should be a balance for any constituent between its *input* and its *output* rates, i.e. if sufficient data are available it should be possible to construct a mass balance for the constituent.

Various authors have attempted to produce overall mass balances for the dissolved constituents found in sea water. However, a few examples will serve to illustrate how the mass-balance approach has evolved in terms of recent advances in our understanding of how the ocean system works.

*Magnesium.* Magnesium is a major, salinity-contributing element in sea water, and it offers an excellent example of how modern research has led to the filling of gaps in the calculation of marine mass balances. Drever (1974) addressed what he called the 'magnesium problem' in sea water. This problem was identified as follows. Dissolved Mg is being brought to the oceans by rivers, and other sources, at a rate of

$\sim 1.3 \times 10^{14} \, g \, yr^{-1}$, and if it is assumed that the oceans are in a steady state then this should be balanced by the output. However, when Drever (1974) attempted to make a mass balance for Mg he found that the then known removal processes could account for only $\sim 50\%$ of the river input. Significantly, one of the reasons postulated to explain this Mg imbalance was that perhaps there was a $Mg^{2+}$ removal process that had not yet been studied intensely. We now know that basalt–seawater reactions can act as a $Mg^{2+}$ sink (see Chapter 5), and according to Thompson (1983) the total flux taken up by the basalt sink in the various kinds of rock–seawater reactions is $\sim 0.6 \times 10^{14} \, g \, yr^{-1}$; this is $\sim 45\%$ of the river flux, and approaches the imbalance of $\sim 0.72 \times 10^{14} \, g \, yr^{-1}$ found by Drever (1974). The old and new Mg marine mass balances are given in Table 17.1.

*Uranium.* A trace element that also had gaps in its marine balance until it was proposed that $\sim 50\%$ of its seawater input was taken up by low-temperature seawater–basalt reactions (see e.g. Bloch, 1980).

For examples of how the marine mass balances of other trace elements have been sharpened by advances in our understanding of oceanic biogeochemical processes, we can return to the combined experimental–modelling approach adopted by Collier & Edmond (1984) (see Section 11.6.3.2) and select examples of a nutrient-type element and a mixed nutrient-type–scavenging-type element.

**Table 17.1** The marine budget of magnesium (units, $10^{14} \, g \, yr^{-1}$).

| | | | | |
|---|---|---|---|---|
| Supply of $Mg^{2+}$ | Rivers | | 1.3 | |
| | Total | | 1.3 | |
| Removal of $Mg^{2+}$ | **1** Original budget* | | | Percentage river flux |
| | Carbonate formation | | 0.075 | |
| | Ion exchange | | 0.097 | |
| | Glauconite formation | | 0.039 | |
| | Mg–Fe exchange | | 0.29 | |
| | Burial of interstitial water | | 0.11 | |
| | Subtotal | | 0.61 | 47 |
| | **2** Modified budget | | | |
| | Hydrothermal activity | | 0.60 | |
| | Subtotal | | 0.60 | 46 |
| | Total removal | | 1.21 | 93 |

* After Drever (1974).

**Table 17.2** Maring geochemical cycle of cadmium (after Collier & Edmond, 1984).

| | | |
|---|---|---|
| Primary input fluxes | Rivers | 0.002 nmol cm$^{-2}$ yr$^{-1}$ |
| | Atmosphere | 0.004–0.017 nmol cm$^{-2}$ yr$^{-1}$ |
| Distribution of dissolved Cd | Surface ocean | 0.01 pmol cm$^{-3}$ |
| | Deep ocean | 0.9 pmol cm$^{-3}$ |
| Box model for surface particulate flux | | 0.34 nmol cm$^{-2}$ yr$^{-1}$ |
| Organic carrier model for surface particulate flux | | 0.5–1.0 nmol cm$^{-2}$ yr$^{-1}$ |
| Sediment trap fluxes | California Current | |
| | 35 m | 0.6 nmol cm$^{-2}$ yr$^{-1}$ |
| | 1500 m | 0.05 nmol cm$^{-2}$ yr$^{-1}$ |
| Advection—diffusion model for deep-water Cd distribution | Regeneration | 0.02–0.5 nmol cm$^{-2}$ yr$^{-1}$ |
| Sediment accumulation | Eastern tropical Pacific | 0.0006 nmol cm$^{-2}$ yr$^{-1}$ |

*Cadmium.* A nutrient-type trace element that has a shallow-water regeneration cycle. The marine biogeochemical cycle of Cd as outlined by Collier & Edmond (1984) is given in Table 17.2, and reveals that the cyclic component completely dominates the total particulate flux. Thus, the particulate flux leaving the surface water, as estimated from a box model, is ~38 ng (0.34 nmol) cm$^{-2}$ yr$^{-1}$, whereas that passing into sediment traps at 1500 m depth is only ~5.6 ng (0.05 nmol) cm$^{-2}$ yr$^{-1}$, i.e. ~15% of the out-of-surface flux. The sediment accumulation rate for Cd given in the model is ~0.07 ng cm$^{-2}$ yr$^{-1}$; this is ≪1% of the down-column particulate flux, but is reasonably close to that of the primary river flux (~0.23 ng (0.002 nmol) cm$^{-2}$ yr$^{-1}$), i.e. the fluxes in the cycle are moving towards a balance. It may be concluded, therefore, that recent advances in our understanding of the marine chemistry of Cd have allowed the main features in its cycle to be identified. In general terms, Cd is brought to the ocean mainly via river and atmospheric inputs to the surface ocean. Here, the element becomes involved in the major oceanic biogeochemical cycle in which it has a shallow regeneration phase similar to those of phosphate and nitrate. Cadmium enters this cycle via a labile carrier and the regeneration–recycling leads to the element having an oceanic residence time of the order of several thousand years.

*Copper.* An element with an oceanic distribution that shows both nutrient-type and scavenging-type characteristics, and because of the complexity of its distribution pattern Collier & Edmond (1984) pointed out that its oceanic cycle is difficult to describe in terms of a simple two-box model. However, Cu *is* involved in removal by a biogenic carrier in surface waters and the authors attempted to relate this particulate flux to the overall cycle of the element in the oceans. The details of their marine biogeochemical cycle for Cu are given in Table 17.3. The main difference between the cycles of Cu and Cd is that although Cu is removed in surface waters it also undergoes scavenging in deep water, and according to Collier & Edmond (1984) this leads to an additional down-column particulate flux (~64 ng (1 nmol) cm$^{-2}$ yr$^{-1}$—see Table 17.3), the actual amount being related to the magnitude of the flux of organic particles through the water column—see Section 16.5 and Fig. 16.6(b). In general terms, the particulate down-column flux of Cu (i.e. the surface flux + the deep-water scavenging flux) is ~150 ng cm$^{-2}$ yr$^{-1}$. As for Cd, this is considerably in excess of the surface water primary input of Cu, which is ~20 ng cm$^{-2}$ yr$^{-1}$ for the Pacific and ~29 ng cm$^{-2}$ yr$^{-1}$ for the Atlantic. However, a major difference between the cycles of Cu and Cd is that the main regeneration of Cu takes place at or close to the sediment surface. According to the data for the Pacific given in Table 17.3, the magnitude of this out-of-sediment Cu flux is in the range ~60–380 ng (1–6 nmol) cm$^{-2}$ yr$^{-1}$. Clearly, therefore, much of the down-column particulate Cu flux is regenerated in the upper sediment column and escapes back into sea water (see Section 14.6.3). Thus, the vertical profiles of Cu in the water column

**Table 17.3** Marine geochemical cycle of copper (after Collier & Edmond, 1984).

| | | |
|---|---|---|
| Primary input fluxes | Rivers | 0.3 nmol cm$^{-2}$ yr$^{-1}$ |
| | Atmosphere | 0.01–0.15 nmol cm$^{-2}$ yr$^{-1}$ |
| Distribution of dissolved Cu | Surface ocean | 1–1.5 pmol cm$^{-3}$ |
| | Deep ocean | 4.0 pmol cm$^{-3}$ |
| Box model for surface particulate flux | | 1.2–1.5 nmol cm$^{-2}$ yr$^{-1}$ |
| Organic carrier model for surface particulate flux | | 1.5–1.8 nmol cm$^{-2}$ yr$^{-1}$ |
| Sediment trap fluxes | Eastern Pacific through 500 m | 0.3–1.7 nmol cm$^{-2}$ yr$^{-1}$ |
| Scavenging fluxes from advection—diffusion models | | 0.3–1.7 nmol cm$^{-2}$ yr$^{-1}$ |
| Porewater fluxes | Eastern tropical Pacific | 1–6 nmol cm$^{-2}$ yr$^{-1}$ |

involve the removal of the dissolved element via biogenic carriers in the surface waters and by scavenging in deep waters, and the marine biogeochemical cycle of the element, like that of Cd, is dominated by a large recycling component; however, for Cu this also occurs in the sediment. The cycling involved in the shallow-water biogenic regeneration of Cd led to the element having a residence time in sea water that may be as high as several tens of thousands of years. However, the deep-water scavenging pulls down the residence time of Cu to around 4000–5000 yr (see Section 11.2). The primary input of Cu to the Atlantic is ~29 ng cm$^{-2}$ yr$^{-1}$, and the removal of dissolved Cu in an authigenic form to the sediment surface in this ocean ranges from ~25 to ~75 ng cm$^{-2}$ yr$^{-1}$ (see Table 16.5). The higher end of the range, however, was found for the Bermuda Rise under conditions of enhanced particle reactivity, and the value for the Nares Abyssal Plain is probably more representative of the Atlantic as a whole. Here the authigenic sediment flux of Cu was ~28 ng cm$^{-2}$ yr$^{-1}$; this is very close to the estimated primary input of Cu to the Atlantic (~29 ng cm$^{-2}$ yr$^{-1}$), so that again the magnitudes of the input and output fluxes in the Cu cycle begin to balance out.

The mass balances described above involved the surface- and deep-water reservoirs. In addition, it is necessary to distinguish between two subreservoirs, i.e. the ocean margins and the open-ocean (see Section 11.4). The ocean margins, which include estuaries and the continental shelf, make up only ~7% of the global ocean surface, but account for a significant fraction of oceanic primary production (probably around 20%) and are impacted by large terrigenous particulate and dissolved fluvial fluxes—for a discussion of the role of the ocean margins in global processes the reader is referred to the volume edited by Mantoura *et al.* (1991).

Martin & Thomas (1994) attempted to assess the fluxes and fates of the dissolved nutrient-type trace metals Cd, Cu, Ni and Zn in the ocean margins. The authors used a variety of approaches to estimate the input fluxes of the trace metals to the ocean margins from (i) the external 'telluric' source, i.e. fluvial and atmospheric; and (ii) the internal oceanic source, derived from upwelled water. It was not possible to quantify the potential additional source of trace-metal mobilization from shelf sediments, and as a result the overall mass balance calculations could not be closed. Despite this, two important conclusions could be drawn from the study.

1 A comparison of the data indicated that when the export and input fluxes are compared it appears that the standing crop of trace metals in the margin regions cannot be sustained by the telluric sources alone, but is highly dependent on dissolved inputs from the deep ocean.

2 Residence times of the trace metals in the margins were estimated by dividing the dissolved stock by the sum of the telluric inputs. The values derived in this way were strikingly low, the margin residence times ranging from 0.3–0.9 yr for Cd to 0.7–2.7 yr for Cu, Ni and Zn. On the basis of these data it may be concluded that the margin and open-ocean subsystems exchange trace metals more intensively than was at first thought.

## 17.3 Conclusions

The last few decades have seen a quantum leap in our understanding of how the ocean works as a chemical system. As a result, marine geochemists are now beginning to have at least a first-order understanding of many of the processes that drive oceanic chemistry. The present volume has described the manner in which these processes operate within a global ocean source–sink framework. The next step is to evaluate the chemistry of the ocean system in relation to planetary geochemistry. In this context it is of interest to summarize the concepts developed by Whitfield and his co-workers who have attempted to find a rationale for the composition of sea water that is based on fundamental chemical principles.

It was suggested in Section 16.5 that the concentration of an element in sea water is controlled by a combination of its input strengths (*geochemical abundance*) and its output strengths (*oceanic reactivity*). Many of the elements present in sea water have been released from crustal rocks, transported to the oceans, and then become incorporated into marine sediments. These sediments, however, are eventually recycled back to the continents during the processes of sea-floor spreading and mountain building; these processes are illustrated in Fig. 17.1, in which the

dynamic nature of the structural and spatial relationships between large-scale geological phenomena is shown. Because of these large-scale recycling processes Whitfield & Turner (1979) concluded that the same material has been cycled through the ocean system several times during the history of the Earth, and suggested that it therefore should be possible to use the partitioning of the elements between the ocean and the crustal rocks to gain an insight into the nature of sea water itself. The ocean system is assumed to be in a steady state, such that the rate of input of an element balances its rate of removal. The time an element resides in sea water can be expressed in terms of its mean oceanic residence time (MORT), $\bar{t}_Y$ (Section 11.2), which may be defined as

$$\bar{t}_Y = Y_S^0 / \bar{J}_Y^0 \qquad (17.1)$$

where $Y_S^0$ is the total mass of the element $Y$ dissolved in the ocean reservoir and $\bar{J}_Y^0$ is the mean flux of Y through the reservoir in unit time. The superscript zero emphasizes that the values refer to a system in steady state. The MORT is a measure of the reactivity of an element in the ocean system. This is because elements that are highly reactive will have low MORT values and a rapid throughput, whereas those that are unreactive will have high MORT values and will tend to accumulate in the system. The MORT values therefore are essential parameters that describe the steady-state composition of sea water, and Whitfield (1979) has suggested that Forchhammer's 'facility with which the elements in sea water are made

**Fig. 17.1** The recycling of oceanic sediments during the processes of sea-floor spreading and mountain building (from Degens & Mopper, 1976).

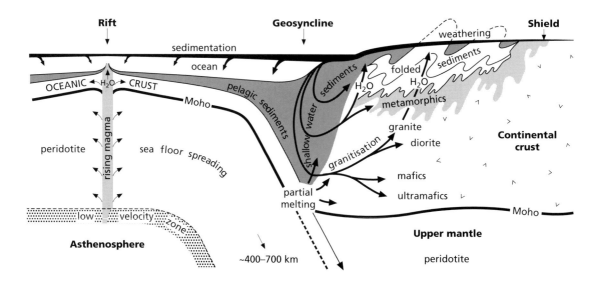

insoluble' has found a quantitative expression in the MORT concept. Thus, a MORT value is a direct measure of the ease with which an element is removed to the sediment sink by incorporation into the solid phase. This affinity of an element for the solid phase can be described by a coefficient that expresses its partitioning between the water and rock phases (par-

titioning coefficient, $K_Y$), which is calculated as the ratio between the mean concentration of the element in natural water to that in crustal rock. Whitfield & Turner (1979) found a linear relationship between the seawater–crustal-rock partition coefficient (log $K_Y$(SW)) and the MORT (log $t_Y$) values of elements (see Fig. 17.2a). It was therefore demonstrated

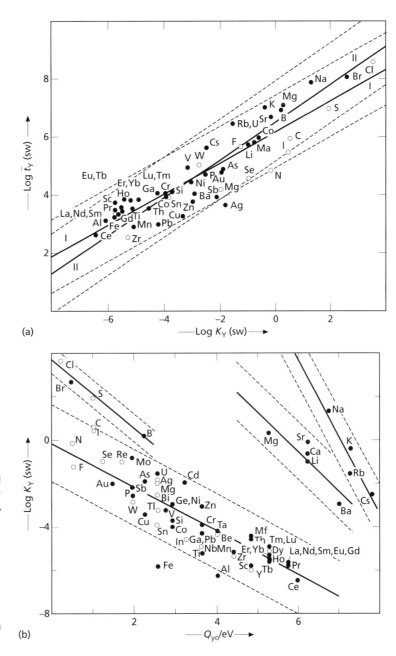

**Fig. 17.2** Relationships in the 'Whitfield ocean'. (a) The relationship between the mean oceanic residence time (MORT, $t_Y$), and the seawater–crustal-rock partition coefficient, $K_Y$ (SW) (from Turner et al., 1980). (b) The relationship between the seawater–crustal-rock partition coefficient, $K_Y$ (SW), and the electronegativity function, $Q_{YO}$ (from Turner et al., 1980). (c) Comparison between the observed, $Y_s$ (obs), and calculated, $Y_s$ (calc), compositions of sea water (from Whitfield, 1979).

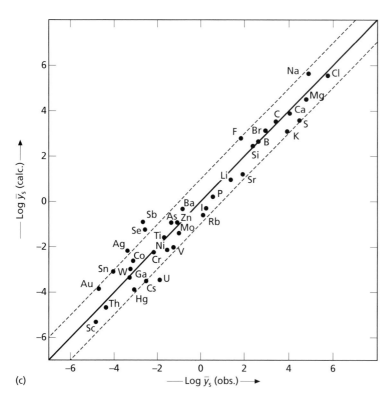

(c)

**Fig. 17.2** *Continued*

that the MORT value of an element is related directly to its partitioning between the oceans and crustal rocks. Thus, a relationship was established between the reactivity of an element in the oceans (MORT) and its long-term recycling through the global reservoir system ($K_Y$).

Whitfield & Turner (1979) suggested that the partitioning of elements between solid and liquid phases in sea water (and river water) could be rationalized using a simple electrostatic model, in which the fundamental chemical control on the solid–liquid partitioning coefficients is related to the electronegativity of an element. This can be quantified by an electronegativity function ($Q_{YO}$), which is a measure of the attraction that an oxide-based mineral lattice will exert on the element. The authors then showed that the electronegativity function ($Q_{YO}$) of an element can be related directly to its partition coefficient ($K_Y$). The relationship between crustal-rock–seawater partition coefficients and the electronegativity functions of the elements is illustrated in Fig. 17.2(b), from which it can be seen that the elements that are more strongly bound to the solid phase (high $Q_{YO}$ values)

have small partition coefficients (low $K_Y$ values). It is apparent, therefore, that the manner in which an element is partitioned between crustal rock and sea water is dependent on the extent to which it is attracted to the oxide-based mineral lattice. The correlations between $Q_{YO}$ and $K_Y$ thus offer a theoretical explanation for variations in the partition coefficients of the elements, which is based ultimately on differences in their electronic structures, which themselves are a function of their atomic number and so their chemical periodicity. Whitfield & Turner (1983) drew attention to the fact that this chemical periodicity, which involves a link between electronic structure and chemical behaviour, provides a rationalization of the inorganic chemistry of all the elements. There is therefore a fundamental regularity in the organization of the elements, which, as the authors pointed out, is sometimes forgotten when attempts are made to assess their behaviour in natural systems. For this reason, the correlation between the partition coefficients of the elements and their electronegativities represents an important step forward in our understanding of the chemistry of sea water. Further, the MORT–partition-coefficient–electronegativity-function relationships permit a number of the basic aspects of oceanic chemistry to be predicted on the basis of theoretical chemical concepts. For example, Whitfield (1979) derived a general equation relating to the MORT value and the electronegativity function of an element, and showed that MORT values derived from the electronegativity functions agreed with observed values within an order of magnitude; i.e. MORT values can be predicted reasonably well from a knowledge of the electrochemical properties of the elements. A second equation was proposed, which related the electronegativity function of an element to the global mean value of its river input, and this was used to estimate the composition of sea water. For most elements the estimated mean global composition of sea water again agreed with the observed values within an order of magnitude, even though the concentrations themselves range over 12 orders of magnitude; the predicted–observed sea-water composition comparison is illustrated in Fig. 17.2(c).

The MORT–partition-coefficient–electronegativity-function relationships developed by Whitfield and co-workers therefore provide a series of theoretical chemical concepts which suggest that the overall composition of sea water is controlled by geological processes that are governed by relatively simple geochemical rules. As a result, the concentration of an element in sea water is controlled by its abundance in the crust (*geochemical abundance*) and by the ease with which it can be taken into solid sedimentary phases (*oceanic reactivity*). Thus, we now have a wider theoretical framework within which to interpret the factors that control the chemical composition of sea water.

To come full circle, therefore, it may be concluded that the overall *composition* of sea water is controlled by relatively simple geochemical rules. The *distribution* of the elements within sea water, however, is dependent on the physical–biological–chemical process trinity that drives the ocean system. The present volume has been concerned with the manner in which the processes involved in this process trinity operate on the throughput of material in the ocean system. The ultimate aim of marine geochemistry must be to produce a rationale for oceanic chemistry based on fundamental chemical principles, which can therefore provide a set of general rules enabling the concentrations and behaviours of elements in sea water to be described within a coherent pattern. It is apparent that such an approach is already gaining ground. The models produced will be refined in the future as marine geochemists struggle towards an explanation of Forchhammer's 'facility with which the elements in sea water are made insoluble'.

This future promises to be exciting.

## References

Bloch, S. (1980) Some factors controlling the concentration of uranium in the World Ocean. *Geochim. Cosmochim. Acta*, **44**, 373–7.

Bruland, K.W. (1980) Oceanographic distributions of cadmium, zinc, nickel and copper in the North Pacific. *Earth Planet. Sci. Lett.*, **47**, 176–98.

Collier, R.W. & Edmond, J.M. (1984) The trace element geochemistry of marine biogenic particulate matter. *Prog. Oceanogr.*, **13**, 113–99.

Degens, E.T. & Mopper, K. (1976) Factors controlling the distribution and early diagenesis of organic matter in marine sediments. In *Chemical Oceanography*, J.P. Riley & R. Chester (eds), Vol. 5, 59–113. London: Academic Press.

Drever, J.I. (1974) The magnesium problem. In *The Sea*, E.D. Goldberg (ed.), Vol. 5, 337–57. New York: Wiley.

Mantoura, R.F.C., Martin, J.M. & Wollast, R. (eds) (1991) *Ocean Margin Processes in Global Change*. Chichester: Wiley.

Martin, J.M. & Thomas, A.J. (1994) The global insignificance of telluric input of dissolved trace metals (Cd, Cu, Ni and Zn) to ocean margins. *Mar. Chem.*, **46**, 165–78.

Thompson, G. (1983) Hydrothermal fluxes in the ocean. In *Chemical Oceanography*, J.P. Riley & R. Chester (eds), Vol. 8, 270–337. London: Academic Press.

Turner, D.R., Dickson, A.G. & Whitfield, M. (1980) Water–rock partition coefficients and the composition of natural waters: a reassessment. *Mar. Chem.*, **9**, 211–18.

Whitfield, M. (1979). The mean oceanic residence time (MORT) concept—a rationalization. *Mar. Chem.*, **8**, 101–23.

Whitfield, M. & Turner, D.R. (1979) Water–rock partition coefficients and the composition of river and seawater. *Nature*, **278**, 132–6.

Whitfield, M. & Turner, D.R. (1983) Chemical periodicity and the speciation and cycling of the elements. In *Trace Metals in Sea Water*, C.S. Wong, E. Boyle, K.W. Bruland, J.D. Burton & E.D. Goldberg (eds), 719–50. New York: Plenum.

# Index